Graduate Texts in Mathematics 88

Richard S. Pierce

Associative Algebras

Springer-Verlag
New York Heidelberg Berlin

Richard S. Pierce

Professor of Mathematics
University of Arizona
Tucson, Arizona 85721
USA

Editorial Board

P. R. Halmos

Managing Editor
Department of Mathematics
Indiana University
Bloomington, Indiana 47401
USA

F. W. Gehring

Department of Mathematics
University of Michigan
Ann Arbor, Michigan 48104
USA

C. C. Moore

Department of Mathematics
University of California
Berkeley, CA 94720
USA

AMS Subject Classification (1980): 16-01

Library of Congress Cataloging in Publication Data
Pierce, Richard S.
Associative algebras.
(Graduate texts in mathematics; 88)
Bibliography: p.
Includes index.
1. Associative algebras. I. Title. II. Series.
QA251.5.P5 512′.24 82-862 AACR2

© 1982 by Springer-Verlag New York Inc.
All rights reserved. No part of this book may be translated or reproduced in any form without permission from Springer-Verlag, 175 Fifth Avenue, New York, New York 10010, U.S.A.

Typeset in Hong Kong by Asco Trade Typesetting Ltd.
Printed and bound by R. R. Donnelley & Sons, Harrisonburg, VA.
Printed in the United States of America.

9 8 7 6 5 4 3 2 1

ISBN 0-387-90693-2 Springer-Verlag New York Heidelberg Berlin
ISBN 3-540-90693-2 Springer-Verlag Berlin Heidelberg New York

Preface

For many people there is life after 40; for some mathematicians there is algebra after Galois theory. The objective of this book is to prove the latter thesis. It is written primarily for students who have assimilated substantial portions of a standard first year graduate algebra textbook, and who have enjoyed the experience. The material that is presented here should not be fatal if it is swallowed by persons who are not members of that group.

The objects of our attention in this book are associative algebras, mostly the ones that are finite dimensional over a field. This subject is ideal for a textbook that will lead graduate students into a specialized field of research. The major theorems on associative algebras include some of the most splendid results of the great heros of algebra: Wedderburn, Artin, Noether, Hasse, Brauer, Albert, Jacobson, and many others. The process of refinement and clarification has brought the proof of the gems in this subject to a level that can be appreciated by students with only modest background. The subject is almost unique in the wide range of contacts that it makes with other parts of mathematics. The study of associative algebras contributes to and draws from such topics as group theory, commutative ring theory, field theory, algebraic number theory, algebraic geometry, homological algebra, and category theory. It even has some ties with parts of applied mathematics.

There is no intention to make this book an encyclopedia of associative algebra. Such a book would be a useful research tool, but it would not fit the needs of a novice mathematician. On the other hand, it is more than a rehash of existing expositions of the theory of associative algebras. The classical results of the subject are explored more deeply than in most student-oriented expositions of associative algebras, and the recent developments in the theory of algebras are liberally sampled. The serious student will find a substantial variety of challenges and rewards in the book.

Roughly speaking, the book is divided into two parts. Part one occupies chapters one through eleven. It could be called "the classical theory of associative algebras." This first part contains the basic structure and representation theorems for associative algebras: Wedderburn's Structure Theorem for Semisimple Algebras, Wedderburn's Principal Theorem, the structure of projective modules of Artinian algebras, and the recent work on representation types. Part two of the book concentrates on central simple algebras. It is organized around the concept of the Brauer group of a field. Chapter 12 builds the tools that are needed to construct the edifice of central simple algebras: the Jacobson Density Theorem, the Noether–Skolem Theorem, and the Double Centralizer Theorem. The topics that part two covers are fairly traditional: splitting fields, cohomological characterization of the Brauer group, cyclic algebras, the reduced norm and its applications, the Brauer groups of local and global fields, and finally an introduction to Amitsur's work on generic algebras.

The difficulty level of the book is a piecewise increasing graph. Each chapter begins with elementary material and escalates in complexity. The last few sections of each chapter contain the specialized and (usually) more difficult topics. At the same time, the median difficulty level of the chapters follows an increasing curve. Probably the best advice for readers of the book is to start at the beginning and plod through it to the end.

Every section of the book is equipped with at least one exercise. The exercises are included for the usual reasons: to keep the serious students awake; to ease the author's conscience pangs over omitted proofs; and to include results for which there is no room in the text. Most of the exercises are of the "follow your nose" variety. The non-trivial problems are accompanied by generous hints. In fact, some of the hints are so extensive that they might justifiably be called proofs.

Following an established tradition, we conclude this preface with acknowledgments and thanks to the friends who supported the preparation of the book. A list of these persons should include the names of a couple of dozen listeners who endured the author's lectures at the University of Connecticut, the University of Arizona, and the University of Hawaii. Most of these people will remain anonymous, but special mention is due to Javier Gomez, Oma Hamara, Eliot Jacobson, Bill Ullery, Bill Velez, and Kwang-Shang Wong whose eagle eyes found some of the numerous errors in the preliminary manuscript. Chuck Vinsonhaler deserves particular recognition for using several parts of the book as a basis for his own lectures. His suggestions and corrections have been extremely valuable.

The majority of credit for the completion of this book is owed to Marilyn Pierce. It was her patience and impatience that kept the project moving from its beginning to the end. She typed, corrected, and recorrected the whole manuscript. Her help and encouragement were always given generously, even though she has long held the author's solemn written promise never to write another book. It is to Marilyn that this book is dedicated.

Contents

CHAPTER 1

The Associative Algebra

Our objective in this chapter is to show off a few examples of algebras that occur naturally. After a brief orientation toward concepts and notation, the reader is introduced to group algebras, endomorphism algebras, matrix algebras, and quaternion algebras. Along the way, there is a brief digression, which contains a hint of the connection between algebraic geometry and the theory of finite dimensional algebras over a field.

1.1. Conventions

Throughout this book, the letter R will stand for a commutative ring with unity 1. The subjects of our study are R-algebras.

Definition. An *R-algebra* (or *algebra over R*) is a unital right R-module A on which is defined a bilinear mapping $A \times A \to A$ (denoted $(x,y) \mapsto xy$) that is associative ($x(yz) = (xy)z$ for all x, y, z in A), and there is a unity element 1_A in A that satisfies $1_A x = x 1_A = x$ for all $x \in A$.

The assumption that multiplication is bilinear is equivalent to the right and left distributive laws, plus

$$(xy)a = x(ya) = (xa)y \quad \text{for all } x, y \in A \text{ and } a \in R. \tag{1}$$

Any R-algebra is a ring with unity. Conversely, if A is a ring with unity and a right R-module that satisfies (1), then A is an R-algebra.

Any R-module A that is equipped with a bilinear mapping $A \times A \to A$ is called a *non-associative R-algebra*. Occasionally, this wider class of algebras will be considered.

The bilinearity of multiplication and the module identities for an R-algebra imply that the mapping $a \mapsto 1_A a$ is a ring homomorphism from R to the center of A. Conversely, if A is a ring with identity, then any homomorphism from R to the center of A imposes an R-module structure on A which turns A into an R-algebra. This observation provides an alternative definition of R-algebras that is sometimes convenient. If the mapping $a \mapsto 1_A a$ is injective, that is, A is a faithful R-module, then R can be identified with a subring of the center of A. Making this identification, we have $xa = ax$, and A becomes a left R-module. Even if A is not faithful as an R-module, a left R-module structure can be defined on A by decreeing that $ax = xa$, since R is commutative.

If no restriction is imposed on R, then the concept of an R-algebra is very general. Indeed, every associative ring with unity is a $\mathbb{Z}$-algebra. In the elementary part of the theory of algebras, the ring R of scalars plays almost no role. This will be the case in the first seven chapters of this book. When we deal with these elementary aspects of our subject, no harm will be done by using the term "algebra" rather than "R-algebra." Indeed, the word "ring" would usually be equally appropriate.

A major split between the theories of rings and algebras occurs when the ring of scalars is taken to be a field. In this case, we will speak of an F-algebra. The letter F will always designate a (commutative) field. Of course, other symbols may be used to denote a field.

If A is an F-algebra, then in particular A is a vector space over the field F. Thus, the module structure of A is determined by the dimension of A as an F-space. This dimension will be denoted by $\dim A$, or, if necessary, $\dim_F A$. Strictly speaking, $\dim A$ might be any cardinal number, but for our purposes it won't be necessary to distinguish orders of infinity. Thus, $\dim A$ is either a natural number or ∞.

Beside making the machinery of linear algebra accessible, the restriction to algebras over fields simplifies and enriches the theory of algebras in many ways. For example, the mapping $a \mapsto 1_A a$ imbeds F in A, provided only that A is *non-trivial*, that is, $1_A \neq 0$ (or equivalently, $|A| \neq 1$). Therefore, in cases of interest, F can be identified with a subring of the center of A. In particular, $1_A = 1$. (Even for algebras over a ring R, we will often use the symbol 1 for both the unity of R and the unity of A, provided this notation is not likely to cause confusion.)

The class of all R-algebras forms a category in which morphisms are simultaneously module and ring homomorphisms that preserve the unity element. Such mappings are called *algebra homomorphisms*. The concepts of *isomorphism*, *endomorphism*, and *automorphism* for R-algebras are defined in the expected way. As usual, we will write $A \cong B$ if there is an isomorphism from A to B. Of course, $\cong$ is an equivalence relation. Following an old algebraic tradition, the term *subalgebra* can have two meanings: (i) a subset of A that includes 0 and 1, and is closed under the addition, multiplication, and scalar operations of A; (ii) an R-algebra B that is a subset of A such that the inclusion mapping from B to A is a homomorphism. Of course, these two definitions of a subalgebra are just different perspectives of the same concept. *Products* in the category of R-algebras have an explicit description. They are obtained by endowing the cartesian product of a set of algebras with operations that are defined componentwise. The component projections are then algebra homomorphisms that satisfy the universal condition required of a product in any category. (See [45], p. 53.) Following a classical tradition, we will denote the product of a finite set $\{A_1, A_2, \ldots, A_n\}$ of algebras by $A_1 \dotplus A_2 \dotplus \cdots \dotplus A_n$.

If $\phi: A \to B$ is a homomorphism of algebras, then the *kernel* of ϕ, that is, $\operatorname{Ker}\phi = \{x \in A: \phi(x) = 0\}$, is a (two-sided) ring ideal and R-submodule of A. Conversely, if I is a ring ideal of A (notation: $I \lhd A$), then I is automatically an R-submodule, because $xa = x(1_A a) \in I$ when $x \in I$ and $a \in R$. It follows that the factor ring A/I is an R-algebra, and the *natural projection* mapping $\pi: A \to A/I$ is an algebra homomorphism with kernel I. The various homomorphism theorems of rings and modules are valid without changes for R-algebras. Perhaps the most important of these theorems is the (right) *factorization criterion*: if $\phi: A \to B$ and $\psi: A \to C$ are homomorphisms of R-algebras, and ϕ is surjective, then ψ factors through ϕ (that is, $\psi = \theta\phi$ for some homomorphism $\theta: B \to C$) if and only if $\operatorname{Ker}\phi \subseteq \operatorname{Ker}\psi$.

The term *ideal* in the context of algebras will mean two sided ideal. When right or left ideals are encountered, they will be called *right* (or *left*) *ideals*. Just as we noted in the case of ideals, every right (left) ring ideal of an R-algebra is also an R-submodule.

The definition of a module over an algebra is identical with the definition of a unital module over a ring. If A is an R-algebra and M is a right A-module, then M inherits an R-module structure: $ua = u(1_A a)$ for $u \in M$ and $a \in R$. Similarly, if M is a left A-module, then M is also a left R-module. In particular, every module over an F-algebra is a vector space, so that its additive structure is known.

If A and B are R-algebras, then an *A-B bimodule* is an algebraic system M that is simultaneously a left A-module and a right B-module, such that

$$(xu)y = x(uy) \quad \text{for all } x \in A, u \in M, \text{ and } y \in B; \tag{2}$$

$$au = ua \quad \text{for all } a \in R \text{ and } u \in M. \tag{3}$$

We will see later that the theory of A-B bimodules can be reduced to the study of modules over the tensor product of the opposite algebra of A with B. However, it is often convenient to deal directly with bimodules.

To simplify terminology, we will use the expression *A-bimodule* instead of A-A bimodule. Most of the bimodules that appear in the book are of this kind.

EXERCISE

Let A be a non-associative R-algebra. Denote $B = A \oplus R$, and define a mapping $\mu: B \times B \to B$ by $((x,a), (y,b)) \mapsto (xy + ay + xb, ab)$. Prove the following statements.

(a) μ is bilinear.

(b) If the multiplication on A is associative, then μ is an associative multiplication on B with $(0,1)$ as a unity element for B; thus B is an R-algebra.

(c) The multiplication on B is commutative if and only if A is commutative.

(d) $x \mapsto (x,0)$ is a bijective algebra homomorphism from A to an ideal of B.

(e) If A is an R-algebra, then $B \cong A \dotplus R$ as R-algebras.

1.2. Group Algebras

The study of associative algebras was partly motivated by the theory of group representations. The bridge between these subjects has its footings on the concept of a *group algebra.* A group algebra over R is an R-algebra that is constructed as a free R-module with a basis consisting of the elements of a group G, and with multiplication induced by the given multiplication in G. It is useful to generalize this construction to the case in which G is a monoid, that is, a set with an associative multiplication with respect to which there is a unity element.

Definition. Let G be a monoid, and suppose that R is a commutative ring with a unity element. Denote

$$RG = \{\xi \in R^G: \xi(x) = 0 \quad \text{for almost all } x \in G\}.$$

Define addition and scalar multiplication of elements in RG componentwise:

$$(\xi a + \eta b)(x) = \xi(x)a + \eta(x)b \quad \text{for } a, b \in R,\ \xi, \eta \in RG,\ x \in G.$$

Define multiplication in RG by convolution:

$$(\xi\eta)(x) = \sum \xi(y)\eta(z),$$

summed over the finite set of pairs $(y,z) \in G \times G$ such that $yz = x$ and $\xi(y)\eta(z) \neq 0$.

The main result of this section is that RG is an R-algebra, called the *convolution algebra* of G over R, or the *group algebra* of G over R when G is a group. The R-algebra identities for RG can be proved by straightforward computations, but a bit of cunning will shorten the work and provide extra information. First note that RG is closed under addition and scalar multiplication, and that these operations satisfy the module identities. It is also routine to verify that multiplication is bilinear. The more delicate properties of associativity and the existence of unity reflect the corresponding properties of the monoid G. To see that this is the case, define for each

$x \in G$ the function $\chi_x \in R^G$ by $\chi_x(z) = 0$ if $z \neq x$, and $\chi_x(x) = 1$. Clearly, if $\xi \in RG$, then $\xi = \sum \chi_x \xi(x)$, summed over all $x \in G$ such that $\xi(x) \neq 0$. It follows that RG is a free R-module, and $\{\chi_x : x \in G\}$ is a basis for RG. Moreover,

$$\chi_x \chi_y = \chi_{xy}. \tag{1}$$

Indeed, $\chi_x(u)\chi_y(v) = 0$ if $u \neq x$ or $v \neq y$, and $\chi_x(x)\chi_y(y) = 1$. Thus, $(\chi_x \chi_y)(z) = 0$ if $z \neq xy$, and $(\chi_x \chi_y)(xy) = 1$. It is clear from (1) and the next lemma that χ_1 is the unity element of RG, and RG is associative.

Lemma. *Let A be an R-module on which is defined a bilinear binary operation $(x,y) \mapsto xy$, that is, A is a non-associative R-algebra. Suppose that $X \subseteq A$ is such that X generates A as an R-module, and $x(yz) = (xy)z$ for all x, y, $z \in X$. Then A satisfies the associative law. If also there exists $x_1 \in X$ such that $x_1 y = y x_1 = y$ for all $y \in X$, then x_1 is a unity element of A.*

PROOF. Since $A = \sum_{x \in X} xR$, it is possible to write three typical elements of A in the forms $\sum_{i \in I} x_i a_i$, $\sum_{j \in J} x_j b_j$, $\sum_{k \in K} x_k c_k$, with a_i, b_j, $c_k \in R$; x_i, x_j, $x_k \in X$. The bilinearity of multiplication and the commutative-associative law of scalar multiplication give

$$\begin{aligned}((\sum_i x_i a_i)(\sum_j x_j b_j))(\sum_k x_k c_k) &= \sum_{i,j,k} (x_i x_j) x_k (a_i b_j c_k) = \sum_{i,j,k} x_i (x_j x_k)(a_i b_j c_k) \\ &= (\sum_i x_i a_i)((\sum_j x_j b_j)(\sum_k x_k c_k)).\end{aligned}$$

The proof that x_1 is a unity element follows the same pattern. □

By (1), the mapping $x \mapsto \chi_x$ is a monoid homomorphism; plainly, this mapping is injective. It is convenient and customary to identify x with the corresponding function χ_x. This convention allows us to simplify our notation: the elements of RG are the linear combinations $\sum x a_x$, where $x \in G$, $a_x \in R$, and the sum is over a finite subset of G. Ignoring the order of summands and the occurrence of zero summands, this representation is unique.

Proposition. *For a monoid G, RG is an R-algebra. As an R-module, RG is free with basis G. If A is an R-algebra and $\phi: G \to A$ is a homomorphism to the multiplicative monoid of A, then ϕ extends uniquely to an R-algebra homomorphism of RG to A.*

PROOF. Only the last assertion requires further proof. Any extension of ϕ would be a module homomorphism; hence it would satisfy

$$\phi(\sum x a_x) = \sum \phi(x) a_x. \tag{2}$$

Since G is a basis of RG, the formula (2) does define an extension of ϕ to a module homomorphism. Using distributivity and the hypotheses

$\phi(xy) = \phi(x)\phi(y)$, $\phi(1) = 1$, it is easy to check that (2) is a ring homomorphism. □

EXAMPLE. Let X be a set of symbols. The *free monoid* on X is the set G_X of all finite sequences $\mathbf{x}_0\mathbf{x}_1 \cdots \mathbf{x}_{m-1}$ of elements from X, including the empty sequence. Multiplication in G_X is defined by juxtaposing sequences: $(\mathbf{x}_0\mathbf{x}_1 \cdots \mathbf{x}_{m-1})(\mathbf{y}_0\mathbf{y}_1 \cdots \mathbf{y}_{n-1}) = \mathbf{x}_0\mathbf{x}_1 \cdots \mathbf{x}_{m-1}\mathbf{y}_0\mathbf{y}_1 \cdots \mathbf{y}_{n-1}$. Thus, the empty sequence is the unity element of G_X. It is clear that any mapping from X to a monoid H has a unique extension to a monoid homomorphism. Thus, every mapping from X to an R-algebra A extends uniquely to an R-algebra homomorphism of RG_X to A, that is, RG_X is the free R-algebra on X. We will denote RG_X by the usual notation $R\{X\}$ (or $R\{\mathbf{x}_1,\ldots,\mathbf{x}_n\}$ if X consists of the distinct symbols $\mathbf{x}_1, \ldots, \mathbf{x}_n$). The elements of $R\{X\}$ are non-commuting polynomials in the symbols of X with coefficients in R.

The familiar (commutative) polynomial ring $R[X]$ can be obtained by a similar construction: let H_X be the free commutative monoid on X; then $R[X] = RH_X$. As usual, if X consists of the distinct symbols $\mathbf{x}_1, \ldots, \mathbf{x}_n$, we will write $R[\mathbf{x}_1, \ldots, \mathbf{x}_n]$ for $R[X]$. In the case $n = 1$, $R[X]$ and $R\{X\}$ coincide with the algebra of polynomials in one variable.

EXERCISE

Let $\mathfrak{C}$ be a small category, that is, a category in which the objects form a set. Let $G = \bigcup \operatorname{Hom}_{\mathfrak{C}}(X, Y)$ be the set of all morphisms of G. Note that G is a partial semigroup. For an R-algebra A, denote

$$A\mathfrak{C} = \{\xi \in A^G: \xi(u) = 0 \quad \text{for almost all } u \in G\}.$$

Define $(\xi + \eta)(u) = \xi(u) + \eta(u)$, $(x\xi)(u) = x(\xi(u))$, $(\xi x)(u) = \xi(u)x$, and $(\xi\eta)(u) = \sum_{vw=u} \xi(v)\eta(w)$ for all $\xi, \eta \in A\mathfrak{C}$, $x \in A$, $u \in G$, with the convention that an empty sum is 0. Prove the following statements.

(a) $A\mathfrak{C}$ is an A-bimodule and a ring that satisfies the associativity conditions $x(\xi\eta) = (x\xi)\eta$, $(\xi x)\eta = \xi(x\eta)$, and $(\xi\eta)x = \xi(\eta x)$ for all $\xi, \eta \in A\mathfrak{C}$ and $x \in A$.

(b) If the set of objects in $\mathfrak{C}$ is finite, then A has a unity element.

(c) If $\mathfrak{C}$ is the category whose objects are the natural numbers $1, 2, \ldots, n$, and $\operatorname{Hom}_{\mathfrak{C}}(i,j) = \{\varepsilon_{ij}\}$ for $1 \le i,j \le n$, with $\varepsilon_{ij}\varepsilon_{jk} = \varepsilon_{ik}$ for all i, j, and k, then $A\mathfrak{C} \cong M_n(A)$.

1.3. Endomorphism Algebras

Let A be an R-algebra. If M and N are right or left A-modules, we will denote the set of all A-module homomorphisms from M to N by $\operatorname{Hom}_A(M,N)$. The set $\operatorname{Hom}_A(M,N)$ has the structure of an R-module if addition and scalar multiplication are defined pointwise: $(\phi + \psi)(u) = \phi(u) + \psi(u)$, $(\phi a)(u) = \phi(u)a$. If N coincides with M, then the composition of homomorphisms, $(\phi\psi)(u) = \phi(\psi(u))$, defines an associative bilinear product under which $\operatorname{Hom}_A(M,M)$ becomes an R-algebra with unity element id_M.

We will write $\mathbf{E}_A(M)$ for $\mathrm{Hom}_A(M,M)$, and call this algebra the *endomorphism algebra* of the module M.

If M and N are left A-modules, then it is sometimes advantageous to write $x\phi$ instead of $\phi(x)$ for $x \in M$, $\phi \in \mathrm{Hom}_A(M,N)$. In general, however, we will not follow this convention. Functions and mappings will be written on the left side of the object on which they operate. Exceptions to this rule will be signalized.

The operation of $\mathbf{E}_A(M)$ on M imposes a left $\mathbf{E}_A(M)$-module structure on M. In this way, a right A-module M is given an $\mathbf{E}_A(M)$-A bimodule structure: for $\phi \in \mathbf{E}_A(M)$, $u \in M$, $x \in A$, and $a \in R$, the equation $\phi(ux) = (\phi u)x$ is satisfied by virtue of the fact that ϕ is an A-module homomorphism, and $au = (id_M a)(u) = (id_M u)a = ua$ by the definition of scalar multiplication in $\mathbf{E}_A(M)$. If M is a left A-module, then by writing endomorphisms on the right, we can view M as an A-$\mathbf{E}_A(M)$ bimodule.

Suppose that A and B are R-algebras, and M is an A-B bimodule. For $x \in A$, define $\lambda_x: M \to M$ by $\lambda_x u = xu$. Then $\lambda_x \in \mathbf{E}_B(M)$: additivity is clear, and $\lambda_x(uy) = x(uy) = (xu)y = \lambda_x(u)y$ by 1.1(2). Similarly if $y \in B$, define $\rho_y: M \to M$ by $\rho_y u = uy$. Then $\rho_y \in \mathbf{E}_A(M)$.

The mapping $x \mapsto \lambda_x$ is easily seen to be a ring homomorphism, and the conditions 1.1(2) and (3) imply that it is in fact an R-algebra homomorphism: $\lambda_{xa}u = x(au) = x(ua) = (xu)a = (\lambda_x u)a = (\lambda_x a)u$ for $x \in A$, $a \in R$, $u \in M$.

The right scalar product mapping is not a ring homomorphism, but rather an antihomomorphism. In fact, $\rho_{xy}u = uxy = \rho_y(ux) = \rho_y\rho_x u$. It is often useful to think of ρ as a homomorphism of the opposite B^* of B to $\mathbf{E}_A(M)$. Recall that B^* is the R-algebra that is obtained from B by inverting the order of the factors in a product.

We will denote the homomorphism $x \mapsto \lambda_x (x \in A)$ and the antihomomorphism $y \mapsto \rho_y$ by λ and ρ respectively.

If M is a left A-module, then M can be viewed as an A-R bimodule, since R is commutative. In this case, λ is a *representation* of A, that is, an algebra homomorphism of A to $\mathbf{E}_R(M)$, where M is an R-module. Conversely, if $\phi: A \to \mathbf{E}_R(M)$ is a representation, then M is a left A-module, with $xu = \phi(x)u$, where $u \in M$, $x \in A$. The homomorphism $\lambda: A \to \mathbf{E}_R(M)$ associated with this A-module structure on M is of course the original representation ϕ. Consequently, there is a one-to-one correspondence between representations of an algebra A and left A-modules. There is a similar relation between right A-modules and representations of A^*.

A routine calculation shows that if ϕ and ψ are two homomorphisms of A to $\mathbf{E}_R(M)$, that is, representations on the same R-module, then ϕ and ψ induce isomorphic A-module structures on M if and only if there is a unit θ of $\mathbf{E}_R(M)$ such that $\psi(x) = \theta^{-1}\phi(x)\theta$ for all $x \in A$. More generally, there is a homomorphism from M with the module structure defined by ϕ to M with the module structure defined by ψ if and only if there is an R-module endomorphism $\theta \in \mathbf{E}_R(M)$ such that $\phi(x)\theta = \theta\psi(x)$ for all $x \in A$. The mapping θ is said to *intertwine* the representations ϕ and ψ.

Any R-algebra A can be considered as an A-bimodule by virtue of associativity and the identity 1.1(1). The corresponding homomorphisms λ and ρ of A to $\mathbf{E}_A(A)$ are respectively called the *left* and *right regular representations* of A.

Proposition. *The right and left regular representations of an R-algebra A are bijective. In particular, $A \cong \mathbf{E}_A(A)$ as R-algebras, where A is considered as a right A-module.*

PROOF. As we have observed, $\lambda: A \to \mathbf{E}_A(A)$ is an R-algebra homomorphism with kernel $\{x \in A: xA = 0\}$, and this ideal is 0 because A has a unity element. Now $\phi(x) = \phi(1 \cdot x) = \phi(1)x = \lambda_y x$, where $y = \phi(1)$. Thus, λ is surjective. A similar proof shows that ρ is bijective. □

EXERCISE

Let A be an R-algebra, and suppose that M is a right A-module. Define a right A-module structure on $\mathrm{Hom}_A(A,M)$ by $(\phi x)(y) = \phi(xy)$ for x and y in A and $\phi \in \mathrm{Hom}_A(A,M)$. Prove that $\mathrm{Hom}_A(A,M)$ is isomorphic to M as a right A-module.

1.4. Matrix Algebras

Given an R-algebra A and a natural number n, let $M_n(A)$ denote the set of all n by n matrices with entries in A. Then $M_n(A)$ is itself an R-algebra with the customary matrix addition, multiplication, and scalar operations by elements of R. We will not repeat the familiar definitions of these operations. However, it is perhaps interesting to observe that the algebra $M_n(A)$ can be constructed by a process that extends the definition in 1.2 of a convolution algebra. (See Exercise 1.2.) The algebra $M_n(A)$ is called the *n by n matrix algebra* over A.

In general, we will denote matrices (not necessarily square) by lower case Greek letters. In particular, ι or ι_n will designate the n by n identity matrix. Moreover, in discussing $M_n(A)$ for a fixed n and A, the matrix units will be denoted by ε_{ij}. That is, ε_{ij} stands for the n by n matrix with the unity of A at the row i, column j position, and the zero of A in all other entries. These matrix units are easily seen to satisfy the following rules of multiplication:

$$\text{for } 1 \le i,j,k,l \le n, \quad \varepsilon_{ij}\varepsilon_{kl} = 0 \quad \text{if } j \ne k; \quad \varepsilon_{ij}\varepsilon_{jl} = \varepsilon_{il}. \tag{1}$$

Occasionally it will be necessary to describe a matrix in terms of its entries. In this case we will use such notations as

$$\alpha = [x_{ij}], \quad \alpha = \begin{bmatrix} x_{11}x_{12} \\ x_{21}x_{22} \end{bmatrix}, \quad \alpha = \begin{bmatrix} x_{11}x_{12} & \cdots & x_{1n} \\ x_{21}x_{22} & \cdots & x_{2n} \\ \cdot \quad \cdot & \cdot & \cdot \\ x_{m1}x_{m2} & \cdots & x_{mn} \end{bmatrix}.$$

Usually the entries of a matrix are designated by doubly subscripted letters, in which case the first subscript indicates the row of the entry, and the second subscript denotes its column.

If A is not commutative, then $M_n(A)$ is not an A-algebra. Nevertheless, it is useful to introduce left and right scalar multiplication of matrices by the elements of A. If

$$\alpha = \begin{bmatrix} x_{11}x_{12} & \cdots & x_{1n} \\ x_{21}x_{22} & \cdots & x_{2n} \\ \cdot \quad \cdot & \cdot \quad \cdot & \cdot \\ x_{m1}x_{m2} & \cdots & x_{mn} \end{bmatrix}$$

is an m by n matrix with entries in the algebra A, and if $y \in A$, define

$$y\alpha = \begin{bmatrix} yx_{11}\, yx_{12} & \cdots & yx_{1n} \\ yx_{21}\, yx_{22} & \cdots & yx_{2n} \\ \cdot \quad \cdot & \cdot \quad \cdot & \cdot \\ yx_{m1}\, yx_{m2} & \cdots & yx_{mn} \end{bmatrix}, \quad \alpha y = \begin{bmatrix} x_{11}y\, x_{12}y & \cdots & x_{1n}y \\ x_{21}y\, x_{22}y & \cdots & x_{2n}y \\ \cdot \quad \cdot & \cdot \quad \cdot & \cdot \\ x_{m1}y\, x_{m2}y & \cdots & x_{mn}y \end{bmatrix}.$$

These scalar operations define an A-bimodule structure on the set of all m by n matrices with entries in A. In particular, $M_n(A)$ is a free A-module with a basis consisting of the matrix units:

$$[x_{ij}] = \sum_{ij} \varepsilon_{ij} x_{ij}, \tag{2}$$

and this representation is unique.

In addition to the associative laws that hold in a bimodule, the matrix algebras satisfy

$$(\alpha y)\beta = \phi(y\beta) \quad \text{for } \alpha, \beta \in M_n(A) \text{ and } y \in A. \tag{3}$$

From an abstract viewpoint, matrix algebras are special cases of endomorphism algebras. In fact, we will later prove (Corollary 3.4b) that $M_n(A) \cong \mathbf{E}_A(M)$, where M is the free right A-module on n generators. Therefore, matrix algebras can always be replaced by endomorphism algebras. However, in many situations, matrices can be used efficiently for calculations in which endomorphisms appear to be foreign.

Matrix algebras over division algebras play an important role in the general theory of algebras over fields. We conclude this section with a result that will later be incorporated into the fundamental structure theorem of Wedderburn.

Lemma. *If D is a division algebra, then $M_n(D)$ is simple for all $n \geq 1$.*

PROOF. Let I be a non-zero ideal of $M_n(D)$. We have to prove that if $\alpha = [x_{ij}] \in M_n(D)$, then $\alpha \in I$. Since $I \neq 0$, there exists a non-zero $\beta = [y_{ij}]$ in I, say $y_{rs} \neq 0$. By (1) and (2), $\alpha = \sum_{i,j} \varepsilon_{ij} x_{ij} = \sum_{i,j} (\varepsilon_{ir} \beta \varepsilon_{sj}) y_{rs}^{-1} x_{ij} \in I$, since I is a two sided ideal of $M_n(D)$. □

EXERCISE

Generalize Lemma 1.4 by proving that if A is a simple algebra, then $M_n(A)$ is simple for all $n \geq 1$. Hint. Show that if J is a non-zero ideal of $M_n(A)$, then $\{x \in A : \varepsilon_{11} x \in J\}$ is a non-zero ideal of A.

1.5. Finite Dimensional Algebras over a Field

If F is a field and A is an F-space with basis $x_1, \ldots, x_n$, then non-associative F-algebras can be defined by specifying the products

$$x_i x_j = \sum_{k=1}^{n} x_k a_{ij}^k, \quad a_{ij}^k \in F, \quad 1 \leq i,j \leq n. \tag{1}$$

Indeed, (1) extends uniquely to a bilinear product on A by the rule $(\sum_i x_i b_i)(\sum_j x_j c_j) = \sum_k x_k(\sum_{i,j} b_i c_j a_{ij}^k)$. The n^3 elements a_{ij}^k are called the *structure constants* of the multiplication that is defined by (1). (In this section, superscripts represent indices, not exponents.)

Every n-dimensional F-algebra A can be realized (up to isomorphism) by specifying suitable structure constants a_{ij}^k. On the other hand, not all choices of structure constants yield an associative multiplication with unity. Furthermore, different choices of the structure constants can give isomorphic algebras.

By Lemma 1.2, the multiplication in A is associative if and only if $x_i(x_j x_k) = (x_i x_j)x_k$ for all i, j, k in the range 1 to n. A straightforward calculation, using (1), shows that associativity is equivalent to the following relations among the structure constants:

$$\sum_{r=1}^{n} a_{ij}^r a_{rk}^s = \sum_{r=1}^{n} a_{jk}^r a_{ir}^s \quad \text{for } 1 \leq i,j,k,s \leq n. \tag{2}$$

A more sophisticated viewpoint sheds light on the identities (2). Corresponding to each linear transformation ϕ and F-basis $x_1, \ldots, x_n$ of A, associate the matrix $\alpha(\phi) = [a_{\phi j}^k]$ defined by $\phi(x_m) = \sum_{k=1}^n x_k a_{\phi j}^k$. It is well known (and will be proved) that the mapping $\phi \mapsto \alpha(\phi)$ is an isomorphism of $\mathbf{E}(A)$ to $M_n(F)$. Clearly, $[a_{ij}^k]$ is the matrix that is associated in this way with λ_{x_i} (with k as the row index and j as the column index), and $[a_{ij}^k]$ is also the matrix corresponding to ρ_{x_j} (with k as the row index and i as the column index). The equations (2) correspond to the conditions $\lambda_{x_i}\rho_{x_k}(x_j) = \rho_{x_k}\lambda_{x_i}(x_j)$ for $1 \leq i,j,k \leq n$. In other words, (2) is a coordinatized version of the condition that the left and right regular representations of A commute, which is clearly equivalent to associativity.

The easiest way to guarantee that equations (1) define an algebra with unity is to require that one of the basis elements, say x_1, acts as unity. Plainly this condition is equivalent to

$$a_{1j}^k = a_{j1}^k = \delta_{jk} \quad \text{for } 1 \leq j,k \leq n, \tag{3}$$

where δ_{jk} is the usual Kronecker delta. In other words, $\lambda_{x_1} = \rho_{x_1} = id_A$, the identity transformation on A. If (3) is satisfied, then so are the cases of (2) in which any of i, j, or k is 1. Finally, it is worth remarking that by requiring $x_1 = 1$, we have not restricted the class of F-algebras that can be constructed on A via the multiplication (1). In fact, if $n > 0$, then any n-dimensional F-algebra A is non-trivial, so that the unity element of A can serve as one element of a basis of A.

Next, consider the problem of non-uniqueness. When do two systems a_{ij}^k and b_{ij}^k of structure constants define isomorphic (non-associative) algebra structures on A? Clearly, this will be the case if and only if there is a non-singular linear transformation γ of A such that for $1 \leq i,j \leq n$, $\gamma(x_i * x_j) = \gamma(x_i) \circ \gamma(x_j)$, where $*$ denotes the product defined by a_{ij}^k and $\circ$ stands for the product given by b_{ij}^k. Let $\gamma(x_i) = \sum_{r=1}^n x_r c_i^r$ ($c_i^r \in F$) for $1 \leq i \leq n$ be the matrix representation of γ in terms of the distinguished coordinates. A straightforward calculation based on (1) yields

$$\sum_{r=1}^{n} c_r^k b_{ij}^r = \sum_{s,t=1}^{n} a_{st}^k c_i^s c_j^t, \quad 1 \leq i,j,k \leq n. \tag{4}$$

That is, the structure constants a_{ij}^k and b_{ij}^k give rise to isomorphic algebras if and only if there is a non-singular n by n matrix $[c_i^k]$ such that (4) is satisfied. Denoting $[c_i^k]^{-1} = [d_i^k]$, the equations (4) can be recast in the form

$$b_{ij}^k = \sum_{r,s,t=1}^{n} d_r^k a_{st}^r c_i^s c_j^t, \quad 1 \leq i,j,k \leq n. \tag{5}$$

We can put these equations in a coordinate-free form, using the left and right regular representations λ_{x_i}, ρ_{x_j} corresponding to the multiplications defined by a_{ij}^k and the representations λ'_{x_i}, ρ'_{x_j} corresponding to b_{ij}^k. The formulas (5) are equivalent to either of the systems:

$$\lambda'_{x_i} = \sum_{s=1}^{n} \gamma^{-1} \lambda_{x_s} \gamma c_i^s, \quad 1 \leq i \leq n;$$

$$\rho'_{x_j} = \sum_{t=1}^{n} \gamma^{-1} \rho_{x_t} \gamma c_j^t, \quad 1 \leq j \leq n.$$

If the multiplications defined by a_{ij}^k and b_{ij}^k are such that x_1 is a unity element, then any isomorphism between these structures must map x_1 to itself. In terms of the matrix $[c_i^k]$ of the isomorphism, this condition amounts to specifying the first column to be

$$\begin{bmatrix} 1 \\ 0 \\ \cdot \\ \cdot \\ \cdot \\ 0 \end{bmatrix}.$$

If x_1 is a unity for the multiplication defined by a_{ij}^k, and if b_{ij}^k is defined by (5) with $c_1^k = \delta_1^k$, then x_1 is also a unity for the multiplication defined by b_{ij}^k.

The discussion of this section can be formulated in the language of algebraic geometry. Fix an ordering of the triples (i,j,k), $1 \leq i,j,k \leq n$, and assign to each system of structure constants a_{ij}^k the point in n^3 dimensional affine F-space $A^{n^3}(F)$ whose coordinates are a_{ij}^k. This defines a one-to-one correspondence between $A^{n^3}(F)$ and the set of all non-associative F-algebras on A. The systems of equations (2) and (3) define a subvariety V (possibly reducible) of $A^{n^3}(F)$ consisting of those points that correspond to F-algebras for which x_1 is the unity element. The formulas (5) define a linear action of the general linear group $GL_n(F)$ on $A^{n^3}(F)$ such that the orbits of the action are in one-to-one correspondence with the isomorphism classes of non-associative algebras over F. The variety V is invariant under the subgroup $A_{n-1}(F)$ of $GL_n(F)$ that consists of those matrices whose first column is

$$\begin{bmatrix} 1 \\ 0 \\ \cdot \\ \cdot \\ \cdot \\ 0 \end{bmatrix},$$

that is, the $n - 1$ dimensional affine group. Consequently, the isomorphism classes of n dimensional F-algebras are in one-to-one correspondence with the orbits in V of the affine group $A_{n-1}(F)$.

Theoretically, the classification of the structure constants that we have described gives a complete survey of the isomorphism classes of n dimensional F-algebras. Practically speaking, however, this approach is useless for large values of n. Exercise 3 shows what happens if $n = 2$.

We conclude this section with one obvious consequence of the remarks that have been made here.

Proposition. *For any field F and natural number $n \geq 1$, the cardinal number of isomorphism types of n dimensional F-algebras is at most $|F|^{n^3}$.*

Exercises

1. Verify the formulas 1.5(2) and (4).
2. Prove that for every $n \geq 1$ there exist structure constants a_{ij}^k satisfying 1.5(2) and (3). (From the standpoint of algebraic geometry, this is not obvious, because (2) and (3) impose $n^4 + 2n^2 - 1$ conditions on n^3 indeterminates.)
3. Let F be a field with char $F \neq 2$. Use the methods of this section to classify all 2 dimensional F-algebras. In more detail:

   (a) show that every F-algebra with F-space structure $A = x_1F \oplus x_2F$ is isomorphic to an F-algebra such that $\lambda_{x_1} = id_A$, and λ_{x_2} has the matrix representation $\begin{bmatrix} 0 & a \\ 1 & 0 \end{bmatrix}$, $a \in F$;

(b) two such algebras with λ_{x_2} given by the matrices $\begin{bmatrix} 0 & a \\ 1 & 0 \end{bmatrix}, \begin{bmatrix} 0 & b \\ 1 & 0 \end{bmatrix}$ respectively are isomorphic if and only if $a = b = 0$ or $ab \in F^2 - \{0\}$;

(c) deduce that every 2 dimensional F-algebra is either a quadratic field extension of F, the ring product $F \dotplus F$, or an F-algebra with basis $1, x$, such that $x^2 = 0$.

4. The results of this section can be obtained without using coordinates, following the outline of this problem. Let A be a right R-module. Prove the following facts.

(a) There is a bijection between the set of all multiplications that define a non-associative algebra structure on A and the set of all R-module homomorphisms of $A \otimes A$ to A. (Tensor products are taken over R.)

(b) The multiplication corresponding to $\mu \in \mathrm{Hom}_R(A \otimes A, A)$ is associative if and only if the following diagram commutes

$$\begin{array}{ccc} (A \otimes A) \otimes A & \xrightarrow{\mu \otimes 1} & A \otimes A \\ & & \downarrow \mu \\ \alpha \downarrow & & A \\ & & \uparrow \mu \\ A \otimes (A \otimes A) & \xrightarrow{1 \otimes \mu} & A \otimes A, \end{array}$$

where α is the natural isomorphism $(A \otimes A) \otimes A \to A \otimes (A \otimes A)$ defined by $(x \otimes y) \otimes z \mapsto x \otimes (y \otimes z)$.

(c) The multiplication corresponding to $\mu \in \mathrm{Hom}_R(A \otimes A, A)$ admits a unity element if and only if there exists $\lambda \in \mathrm{Hom}_R(R, A)$ such that the two diagrams commute

$$\begin{array}{ccccc} A \otimes R & \xleftarrow{\beta_1} & A & \xrightarrow{\beta_2} & R \otimes A \\ id \otimes \lambda \downarrow & & id_A \downarrow & & \downarrow \lambda \otimes id \\ A \otimes A & \xrightarrow[\mu]{} & A & \xleftarrow[\mu]{} & A \otimes A, \end{array}$$

where β_1 and β_2 are the natural isomorphisms defined by $x \mapsto x \otimes 1$ and $x \mapsto 1 \otimes x$ respectively.

(d) μ_1 and μ_2 in $\mathrm{Hom}_R(A \otimes A, A)$ define isomorphic algebras if and only if there is an R-module automorphism γ of A such that $\mu_2 = \gamma^{-1}\mu_1(\gamma \otimes \gamma)$.

(e) Use the results of (a) to (d) to derive formulas (1), (2), and (5).

1.6. Quaternion Algebras

The history of associative algebras begins with Hamilton's discovery of the real quaternions in 1843. In this section we will define quaternion algebras over general fields. Some of the basic properties of these algebras will be derived, using straightforward computational arguments. Most of the results in this section will reappear as special cases of general theorems in later chapters. In Section 1.7, we will prove an important fact about quaternion algebras that does not generalize.

Throughout this section, F is to be a field whose characteristic is *not* 2. The analogues of the quaternion algebras over fields whose characteristic is 2 are defined differently. (See Exercise 2.)

Definition. Let a and b be non-zero elements of F. Let A be the four dimensional F-space with basis $1, \mathbf{i}, \mathbf{j}, \mathbf{k}$ and the bilinear multiplication defined by the conditions that 1 is a unity element, and

$$\mathbf{i}^2 = a, \quad \mathbf{j}^2 = b, \quad \mathbf{ij} = -\mathbf{ji} = \mathbf{k}. \tag{1}$$

As usual, the first two equations in (1) employ the conventional identification of F with the set of scalar multiples of the unity element in A. Assuming associativity, the remainder of the multiplication table for the basis of A follows directly from (1):

$$\mathbf{k}^2 = -ab, \quad \mathbf{ik} = -\mathbf{ki} = \mathbf{j}a, \quad \mathbf{jk} = -\mathbf{kj} = -\mathbf{i}b. \tag{2}$$

Conversely, with structure constants given by (1) and (2) (plus $1\mathbf{i} = \mathbf{i}1 = \mathbf{i}$, $1\mathbf{j} = \mathbf{j}1 = \mathbf{j}$, $1\mathbf{k} = \mathbf{k}1 = \mathbf{k}$), the condition 1.5(2) is satisfied, so that A is an associative F-algebra.

Notation and Terminology: $A = \left(\frac{a,b}{F}\right)$ is called a (generalized) *quaternion algebra* over F.

Hamilton's quaternions occur as the special case $\mathbb{H} = \left(\frac{-1,-1}{\mathbb{R}}\right)$.

Lemma. *For any non-zero a and b from F, $\left(\frac{a,b}{F}\right)$ is a simple algebra whose center is F.*

Proof. It is convenient to introduce the Lie bracket operation: $[x,y] = xy - yx$. If $x = c_0 + \mathbf{i}c_1 + \mathbf{j}c_2 + \mathbf{k}c_3 \in A = \left(\frac{a,b}{F}\right)$, then by (1) and (2), $[\mathbf{i},x] = \mathbf{j}(2ac_3) + \mathbf{k}(2c_2)$, $[\mathbf{j},x] = \mathbf{i}(-2bc_3) + \mathbf{k}(-2c_1)$, and $[\mathbf{k},x] = \mathbf{i}(2bc_2) + \mathbf{j}(-2ac_1)$. In particular, $x \in \mathbf{Z}(A)$ implies $[\mathbf{i},x] = [\mathbf{j},x] = [\mathbf{k},x] = 0$, so that $c_1 = c_2 = c_3 = 0$. Consequently, $\mathbf{Z}(A) = F$. Suppose that $0 \neq x \in I$, where I is an ideal of A. Since I is a two sided ideal containing x, it also includes the Lie triple products $[\mathbf{j},[\mathbf{i},x]] = \mathbf{i}(-4bc_2)$, $[\mathbf{k},[\mathbf{j},x]] = \mathbf{j}(4abc_3)$, and $[\mathbf{i},[\mathbf{k},x]] = \mathbf{k}(-4ac_1)$. If one of c_1, c_2, or c_3 is not 0, then I contains a unit of A; if $c_1 = c_2 = c_3 = 0$, then $0 \neq x = c_0$ is a unit belonging to I. In all cases, $I = A$. □

An F-algebra A is called *central* if $\mathbf{Z}(A) = F$. Thus, the quaternion algebras are central simple. It will be shown in Section 13.1 that every four dimensional central simple algebra over a field F with char $F \neq 2$ is necessarily a quaternion algebra. It will also follow from general theory that a quaternion algebra over F is either a division algebra or else is isomorphic to $M_2(F)$. This result is also outlined in Exercise 2 of Section 1.7. It is therefore natural to ask: for what choices of a and b in F is $\left(\frac{a,b}{F}\right)$ a division algebra? This turns out to be a difficult question for most fields, for example if $F = \mathbb{Q}$. However, the problem can be translated into a question concerning quadratic forms for which there is a substantial theory.

Write $\left(\frac{a,b}{F}\right) = A = F \bigoplus A_+$, where $A_+ = \mathbf{i}F \bigoplus \mathbf{j}F \bigoplus \mathbf{k}F$. The elements of A_+ are called *pure quaternions*. For $x = c_0 + z$, $c_0 \in F$, $z \in A_+$, define the conjugate of x to be $x^* = c_0 - z$. For $x,y \in A$, and $d \in F$, we have

$$(x + y)^* = x^* + y^*, \quad (xy)^* = y^*x^*, \quad x^{**} = x, \quad d^* = d. \tag{3}$$

With the exception of $(xy)^* = y^*x^*$, these equations are obvious. Linearity reduces the proof that $(xy)^* = y^*x^*$ to the finite set of cases in which $\{x,y\} \subseteq \{1,\mathbf{i},\mathbf{j},\mathbf{k}\}$. We relegate this chore to Exercise 1.

For $x \in A = \left(\frac{a,b}{F}\right)$, the *norm* of x is defined to be $\nu(x) = xx^*$. If $x = c_0 + \mathbf{i}c_1 + \mathbf{j}c_2 + \mathbf{k}c_3$, then $\nu(x) = c_0^2 - ac_1^2 - bc_2^2 + abc_3^2$ is obtained by direct computation. In particular, $\nu(x) \in F$, and $\nu(x) = \nu(x^*) = x^*x$.

$$\text{If } x,y \in A, \text{ and } d \in F, \text{ then } \nu(xy) = \nu(x)\nu(y),\ \nu(d) = d^2. \tag{4}$$

Indeed, by (3),

$$\nu(xy) = xy(xy)^* = xyy^*x^* = x\nu(y)x^* = xx^*\nu(y) = \nu(x)\nu(y);\ \nu(d) = d^2$$

is obvious.

Proposition. *The following conditions are equivalent for* $A = \left(\frac{a,b}{F}\right)$:

(i) *A is a division algebra;*
(ii) *$x \in A - \{0\}$ implies $\nu(x) \neq 0$;*
(iii) *if $(c_0,c_1,c_2) \in F^3$ satisfy $c_0^2 = ac_1^2 + bc_2^2$, then $c_0 = c_1 = c_2 = 0$.*

PROOF. (i) implies (ii), since $\nu(x)\nu(x^{-1}) = \nu(xx^{-1}) = \nu(1) = 1$ by (4); and (i) is a consequence of (ii), because $xx^*\nu(x)^{-1} = (x^*\nu(x)^{-1})x = 1$ whenever $\nu(x) \neq 0$. If $c_0^2 = ac_1^2 + bc_2^2$ with $(c_0,c_1,c_2) \neq (0,0,0)$, then $x = c_0 + \mathbf{i}c_1 + \mathbf{j}c_2 \neq 0$ and $\nu(x) = 0$. Therefore, (ii) implies (iii). Finally, (iii) implies (ii). In fact, suppose that $\nu(x) = 0$, where $x = d_0 + \mathbf{i}d_1 + \mathbf{j}d_2 + \mathbf{k}d_3$; that is, $d_0^2 - bd_2^2 = a(d_1^2 - bd_3^2)$. Then $a(d_1^2 - bd_3^2)^2 = (d_0^2 - bd_2^2)(d_1^2 - bd_3^2) = (d_0d_1 + bd_2d_3)^2 - b(d_0d_3 + d_1d_2)^2$. The hypothesis (iii) yields $d_1^2 - bd_3^2 = 0$, and therefore $d_1 = d_3 = 0$. Thus, $d_0^2 - bd_2^2 = 0$, so that $d_0 = d_2 = 0$. That is, $x = 0$. □

In the language of quadratic forms, the conditions (ii) and (iii) of the proposition state that $\mathbf{x}_0^2 - a\mathbf{x}_1^2 - b\mathbf{x}_2^2 + ab\mathbf{x}_3^2$ and $\mathbf{x}_0^2 - a\mathbf{x}_1^2 - b\mathbf{x}_2^2$ are *anisotropic*.

It is a consequence of the proposition that the Hamiltonian quaternions $\left(\frac{-1,-1}{\mathbb{R}}\right)$ form a division algebra: $c_0^2 = -c_0^2 - c_2^2$ implies $c_0 = c_1 = c_2 = 0$ for $c_0,c_1,c_2 \in \mathbb{R}$.

EXERCISES

1. Verify that $(xy)^* = y^*x^*$ for $x,y \in A = \left(\frac{a,b}{F}\right)$.

2. Let F be a field with char $F = 2$, and let a and b be non-zero elements of F. Define A to be the 4-dimensional F-algebra with basis $1,\mathbf{i},\mathbf{j},\mathbf{k}$, where multiplication is defined by the conditions (1) and (2). Prove the following facts.

   (a) A is a commutative F-algebra, and the mapping $x \mapsto x^2$ is a ring homomorphism of A to F.

   (b) If $a \notin F^2$, then A contains the field $K = F(\sqrt{a})$ as a subring, and $A = F(\sqrt{a},\sqrt{b})$, or A is isomorphic to the unique 2-dimensional K-algebra with non-zero nilpotent radical that was described in part c of Exercise 3, Section 1.5.

   (c) If $a \in F^2$ and $b \in F^2$, then A has a basis i, u, v, w such that $u^2 = v^2 = w^2 = 0$ and $uv = uw = vw = u + v + w$.

For the remaining problems of this section, assume that char $F \neq 2$. Let $a,b,c \in F^\circ$.

3. Prove that for every $a \in F^\circ$, the quaternion algebra $\left(\frac{a,1}{F}\right)$ is isomorphic to $M_2(F)$. Hint. Compute the multiplication table for the basis $e_{11} = (1/2)(1 - \mathbf{j})$, $e_{22} = (1/2)(1 + \mathbf{j})$, $e_{21} = (1/2a)(\mathbf{i} - \mathbf{k})$, $e_{12} = (1/2)(\mathbf{i} + \mathbf{k})$ of $\left(\frac{a,1}{F}\right)$.

4. Show that condition (ii) of Proposition 1.6 is equivalent to: $a \notin F^2$ and $b \notin N_{F(\sqrt{a})/F}(F(\sqrt{a}))$.

5. Let F be a finite field with q elements, and suppose that $a,b \in F^\circ$.

   (a) Prove that elements c_1 and c_2 exist in F satisfying $ac_1^2 + bc_2^2 = 1$. Hint. Show that the values taken by ac_1^2 and $1 - bc_2^2$ as c_1 and c_2 range independently over the q elements of F cannot all be different.

   (b) Deduce from (a) that $\left(\frac{a,b}{F}\right)$ is not a division algebra. This result is a special case of Wedderburn's theorem that every finite division algebra is a field.

1.7. Isomorphism of Quaternion Algebras

The fundamental problem of the theory of quaternion algebras is the question: when is $\left(\frac{a,b}{F}\right) \cong \left(\frac{a',b'}{F}\right)$? In this section, we will use the norm mapping to translate this problem into the language of quadratic forms. It is convenient to introduce the bilinear form that is obtained by polarizing the norm.

For x, y in $A = \left(\frac{a,b}{F}\right)$, define $\beta(x,y) = 1/2(\nu(x + y) - \nu(x) - \nu(y))$. If $x = c_0 + z$, $y = d_0 + w$, with $c_0, d_0 \in F$ and $z = \mathbf{i}c_1 + \mathbf{j}c_2 + \mathbf{k}c_3 \in A_+$, $w = \mathbf{i}d_1 + \mathbf{j}d_2 + \mathbf{k}d_3 \in A_+$, then $\beta(x,y) = 1/2((x + y)(x + y)^* - xx^* - yy^*) = 1/2(xy^* + yx^*) = 1/2((c_0 + z)(d_0 - w) + (d_0 + w)(c_0 - z)) =$

$c_0d_0 - 1/2(zw + wz) = c_0d_0 - ac_1d_1 - bc_2d_2 + abc_3d_3$. These equations show that β is a bilinear mapping of $A \times A$ to F that is symmetric ($\beta(x,y) = \beta(y,x)$) and non-singular ($\beta(x,y) = 0$ for all $y \in A$ implies $x = 0$). Moreover, $\nu(x) = \beta(x,x)$, and if F is identified with $1_A F = \mathbf{Z}(A)$, then

$$z,w \in A_+ \text{ implies } \beta(z,w) = -1/2(zw + wz) \text{ and } \nu(z) = -z^2. \tag{1}$$

Lemma. *Let* $A = \left(\frac{a,b}{F}\right)$ *and* $A' = \left(\frac{a',b'}{F}\right)$ *be quaternion algebras with the respective norms* ν *and* ν'. *As F-algebras, A is isomorphic to A′ if and only if there is a vector space isomorphism* ϕ *of* A_+ *to* A'_+ *such that* $\nu'(\phi(z)) = \nu(z)$ *for all* $z \in A_+$.

Proof. We start the proof with a characterization of A_+. If $x = c + z$ with $c \in F$ and $z \in A_+$, then $x^2 = c^2 + z^2 + z(2c) = c^2 - \nu(z) + z(2c)$. Thus, $x^2 \in F = \mathbf{Z}(A)$ if and only if $z = 0$ (hence $x \in \mathbf{Z}(A)$), or $c = 0$ (hence $x \in A_+$). This calculation shows that for $x \in A - \{0\}$,

$$x \in A_+ \text{ if and only if } x \notin \mathbf{Z}(A) \text{ and } x^2 \in \mathbf{Z}(A). \tag{2}$$

Naturally, A'_+ can be characterized in the same way. Therefore, since any algebra isomorphism $\phi: A \to A'$ satisfies $\phi(\mathbf{Z}(A)) = \mathbf{Z}(A')$, and $\phi(x^2) = \phi(x)^2$, it follows from (2) that $\phi(A_+) = A'_+$, and if $z \in A_+$, then by (1), $\nu'(\phi(z)) = -\phi(z)^2 = \phi(-z^2) = \phi(\nu(z)) = \nu(z)$. Thus, ϕ restricts to a norm preserving, vector space isomorphism of A_+ to A'_+. Conversely, suppose that $\phi: A_+ \to A'_+$ is a bijective linear mapping such that $\nu'(\phi(z)) = \nu(z)$ for all $z \in A_+$. To show that $A' \cong A$, we construct a basis of A' for which the structure constants are the same as the structure constants associated with the standard basis of A. By (1), $\phi(\mathbf{i})^2 = -\nu'(\phi(\mathbf{i})) = -\nu(\mathbf{i}) = a$. Similarly, $\phi(\mathbf{j})^2 = b$. Moreover, $\phi(\mathbf{i})\phi(\mathbf{j}) + \phi(\mathbf{j})\phi(\mathbf{i}) = -2\beta'(\phi(\mathbf{i}),\phi(\mathbf{j})) = -2\beta(\mathbf{i},\mathbf{j}) = \mathbf{ij} + \mathbf{ji} = 0$, using (1) and the fact that the bilinear form β' associated with ν' clearly satisfies $\beta'(\phi(z),\phi(w)) = \beta(z,w)$ for all z, w in A_+. Thus, $\phi(\mathbf{j})\phi(\mathbf{i})\phi(\mathbf{j}) = \phi(\mathbf{i})(-b)$, and $(\phi(\mathbf{i})\phi(\mathbf{j}))^2 = -ab$. It follows from (2) that $\phi(\mathbf{i})\phi(\mathbf{j}) \in A'_+$. In fact, $\phi(\mathbf{i})$, $\phi(\mathbf{j})$, $\phi(\mathbf{i})\phi(\mathbf{j})$ is a basis of A'_+: if $\phi(\mathbf{i})c_1 + \phi(\mathbf{j})c_2 + \phi(\mathbf{i})\phi(\mathbf{j})c_3 = 0$ with $c_1,c_2,c_3 \in F$, then $0 = \phi(\mathbf{i})(\phi(\mathbf{i})c_1 + \phi(\mathbf{j})c_2 + \phi(\mathbf{i})\phi(\mathbf{j})c_3) = ac_1 + \phi(\mathbf{i})\phi(\mathbf{j})c_2 + \phi(\mathbf{j})ac_3$ implies $c_1 = 0$; similarly, $c_2 = 0$, so that $c_3 = 0$ as well. Define the mapping $\psi: A \to A'$ by $\psi(1_A) = 1_{A'}$, $\psi(\mathbf{i}) = \phi(\mathbf{i})$, $\psi(\mathbf{j}) = \phi(\mathbf{j})$, and $\psi(\mathbf{k}) = \phi(\mathbf{i})\phi(\mathbf{j})$. The preceding discussion shows that ψ is an F-algebra isomorphism. □

In general, the mapping ψ that is defined in this proof will not coincide with ϕ. The class of isometries from A_+ to A'_+ is larger and more tractable than the class of algebra homomorphisms from A to A'.

The principal result of this section is obtained by translating the lemma to the language of quadratic forms. If $z = \mathbf{i}c_1 + \mathbf{j}c_2 + \mathbf{k}c_3 \in A_+$, then $\nu(z) = \Phi(c_1,c_2,c_3)$, where Φ is the ternary quadratic form $-a\mathbf{x}_1^2 - b\mathbf{x}_2^2 +$

$ab\mathbf{x}_3^2$. Similarly, $v'(z') = \Phi'(c_1',c_2',c_3')$, where $\Phi' = -a'\mathbf{x}_1^2 - b'\mathbf{x}_2^2 + a'b'\mathbf{x}_3^2$ and $z' = \mathbf{i}'c_1' + \mathbf{j}'c_2' + \mathbf{k}'c_3' \in A'_+$ (with 1, $\mathbf{i}'$, $\mathbf{j}'$, $\mathbf{k}'$ denoting the standard basis of A'). It is convenient to put these equations in matrix form. Denote

$$\alpha = \begin{bmatrix} -a & 0 & 0 \\ 0 & -b & 0 \\ 0 & 0 & ab \end{bmatrix}, \quad \alpha' = \begin{bmatrix} -a' & 0 & 0 \\ 0 & -b' & 0 \\ 0 & 0 & a'b' \end{bmatrix},$$

$$\xi = \begin{bmatrix} c_1 \\ c_2 \\ c_3 \end{bmatrix}, \quad \xi' = \begin{bmatrix} c_1' \\ c_2' \\ c_3' \end{bmatrix}.$$

Then, $v(z) = \xi^t\alpha\xi$ and $v'(z') = (\xi')^t\alpha'\xi'$, where the superscript t denotes matrix transposition. Moreover, if $w = \mathbf{i}d_1 + \mathbf{j}d_2 + \mathbf{k}d_3$ and $w' = \mathbf{i}'d_1' + \mathbf{j}'d_2' + \mathbf{k}'d_3'$, then $\beta(z,w) = \xi^t\alpha\eta$ and $\beta'(z',w') = (\xi')^t\alpha'\eta'$, where

$$\eta = \begin{bmatrix} d_1 \\ d_2 \\ d_3 \end{bmatrix} \quad \text{and} \quad \eta' = \begin{bmatrix} d_1' \\ d_2' \\ d_3' \end{bmatrix}.$$

Suppose that $\phi\colon A_+ \to A'_+$ is linear, say $[\phi(\mathbf{i}),\phi(\mathbf{j}),\phi(\mathbf{k})] = [\mathbf{i}',\mathbf{j}',\mathbf{k}']\delta$, where $\delta = [d_{ij}] \in M_3(F)$. The mapping ϕ is bijective if and only if δ is non-singular. If $z = \mathbf{i}c_1 + \mathbf{j}c_2 + \mathbf{k}c_3 = [\mathbf{i},\mathbf{j},\mathbf{k}]\xi$, then

$$\phi(z) = [\phi(\mathbf{i}),\phi(\mathbf{j}),\phi(\mathbf{k})]\xi = [\mathbf{i}',\mathbf{j}',\mathbf{k}']\delta\xi.$$

Similarly, $\phi(w) = [\mathbf{i}',\mathbf{j}',\mathbf{k}']\delta\eta$. Consequently,

$$\beta'(\phi(z),\phi(w)) = (\delta\xi)^t\alpha'(\delta\eta) = \xi^t(\delta^t\alpha'\delta)\eta.$$

Therefore, ϕ satisfies $v'(\phi(z)) = v(z)$ for all $z \in A_+$, or equivalently, $\beta'(\phi(z),\phi(w)) = \beta(z,w)$ for all z, $w \in A_+$ if and only if $\xi^t\alpha\eta = \xi^t(\delta^t\alpha'\delta)\eta$ for all ξ, η in F^3. Clearly, this last condition amounts to the equation $\alpha = \delta^t\alpha'\delta$. Our discussion is summarized by saying that there is an isometry of A_+ to A'_+ if and only if the matrices α and α' are congruent.

Proposition. *The quaternion algebras* $\left(\dfrac{a,b}{F}\right)$ *and* $\left(\dfrac{a',b'}{F}\right)$ *are isomorphic if and only if the quadratic forms* $a\mathbf{x}_1^2 + b\mathbf{x}_2^2 - ab\mathbf{x}_3^2$ *and* $a'\mathbf{x}_1^2 + b'\mathbf{x}_2^2 - a'b'\mathbf{x}_3^2$ *are equivalent.*

Two quadratic forms are called *equivalent* if it is possible to pass from one to the other by a non-singular linear change of variables. When the forms Φ and Φ' are represented as matrix products

$$\Phi(\mathbf{x}_1,\mathbf{x}_2,\mathbf{x}_3) = [\mathbf{x}_1,\mathbf{x}_2,\mathbf{x}_3]\alpha\begin{bmatrix} \mathbf{x}_1 \\ \mathbf{x}_2 \\ \mathbf{x}_3 \end{bmatrix}, \quad \Phi'(\mathbf{x}_1,\mathbf{x}_2,\mathbf{x}_3) = [\mathbf{x}_1,\mathbf{x}_2,\mathbf{x}_3]\alpha'\begin{bmatrix} \mathbf{x}_1 \\ \mathbf{x}_2 \\ \mathbf{x}_3 \end{bmatrix},$$

the condition that Φ and Φ' are equivalent amounts to the existence of a non-singular matrix $\delta \in M_3(F)$ such that $\alpha = \delta^t\alpha'\delta$. Hence, the proposition is just a reformulation of the lemma.

Corollary. *If a, b, and c are non-zero elements of F, then*

$$\left(\frac{ac^2,b}{F}\right) \cong \left(\frac{a,bc^2}{F}\right) \cong \left(\frac{a,b}{F}\right).$$

By taking $F = \mathbb{R}$ in the corollary, we conclude that the only real quaternion algebras are $\left(\frac{1,1}{\mathbb{R}}\right)$, $\left(\frac{1,-1}{\mathbb{R}}\right)$, and $\mathbb{H} = \left(\frac{-1,-1}{\mathbb{R}}\right)$. In fact, $\left(\frac{1,1}{\mathbb{R}}\right) \cong \left(\frac{1,-1}{\mathbb{R}}\right) \cong M_2(\mathbb{R})$ by Exercise 3, Section 1.6.

Exercises

Assume in all of these exercises that char $F \neq 2$.

1. (a) Let V be an n-dimensional F-space, and let $\beta\colon V \times V \to F$ be a non-degenerate, symmetric, bilinear mapping. Assume that x, y in V satisfy $\beta(x,x) = \beta(y,y) \neq 0$. Prove that there is a non-singular linear transformation ϕ of V such that $\phi(x) = y$ and $\beta(\phi(z),\phi(w)) = \beta(z,w)$ for all z and w in V (that is, ϕ is an isometry). Hint. Let $u = (1/2)(x + y)$, $v = (1/2)(x - y)$. Prove that $\beta(u,v) = 0$, and that either $\beta(u,u) \neq 0$ or $\beta(v,v) \neq 0$. In the former case, define $\phi(z) = 2(\beta(u,z)/\beta(u,u))u - z$, and if $\beta(u,u) = 0$, $\beta(v,v) \neq 0$, let $\phi(z) = z - 2(\beta(v,z)/\beta(v,v))v$.

   (b) Use the result of (a) to prove the Witt cancellation theorem: If Φ and Ψ are $(n-1)$-ary quadratic forms over F such that $a\mathbf{x}_1^2 + \Phi(\mathbf{x}_2, \ldots, \mathbf{x}_n)$ is equivalent to $a\mathbf{x}_1^2 + \Psi(\mathbf{x}_2, \ldots, \mathbf{x}_n)$, then Φ is equivalent to Ψ.

2. Let $a \notin F^2$, and denote $E = F(\sqrt{a})$.

   (a) Prove: $\left(\frac{a,b}{F}\right) \cong \left(\frac{a,c}{F}\right)$ if and only if $b/c \in N_{E/F}(E^\circ)$.

   Hint. Apply the Witt cancellation theorem to Proposition 1.7; then compute with 2 by 2 matrices.

   (b) Deduce from (a), Proposition 1.6, and Exercise 3, Section 1.6, that $\left(\frac{a,b}{F}\right)$ is either a division algebra or isomorphic to $M_2(F)$.

3. Let $a, b \in \mathbb{Z} - \{0\}$ be square free (i.e., not divisible by the square of a prime). Prove that $\left(\frac{a,b}{\mathbb{Q}}\right) \cong M_2(\mathbb{Q})$ if and only if

   (i) a and b are not both negative,
   (ii) a is congruent to a square modulo b, and
   (iii) b is congruent to a square modulo a.

   Hint. $\left(\frac{a,b}{\mathbb{Q}}\right) \cong M_2(\mathbb{Q})$ implies by Proposition 1.6 that there exist integers c_0,c_1,c_2 with no common prime divisors such that $c_0^2 = ac_1^2 + bc_2^2$. This clearly implies (i).

In order to prove (ii), it suffices by the Chinese Remainder Theorem to show that a is a quadratic residue mod p for all primes p that divide b. This conclusion is obtained by arguing that if p divides b, then either p divides a, or p does not divide c_1. The converse can be proved by induction on $|a| + |b|$. If $|a| + |b| = 2$, then a or b is 1 (by (ii)), hence the result comes from Exercise 3, Section 1.6. Assume therefore that $|a| \leq |b|$, $|b| \geq 2$. By (ii), there exist integers c, d, and e such that c is square free, $d > 0$, $|e| \leq (1/2)|b|$, and $e^2 - a = bcd^2$. Consequently, $|c| \leq |e|^2/|b| + |a|/|b| \leq (1/4)|b| + 1 < |b|$, $b/c = (e/cd)^2 - a(1/cd)^2 \in N_{\mathbb{Q}(\sqrt{a})/\mathbb{Q}}(\mathbb{Q}(\sqrt{a}))$. By Exercise 2, $\left(\frac{a,b}{\mathbb{Q}}\right) \cong \left(\frac{a,c}{\mathbb{Q}}\right)$, and the induction hypothesis applies to $\left(\frac{a,c}{\mathbb{Q}}\right)$.

4. (a) Prove: if p is a rational prime that is congruent to 3 (mod 4), then $\left(\frac{-1,p}{\mathbb{Q}}\right)$ is a division algebra.

(b) Prove: if p and q are distinct rational primes that are congruent to 3 (mod 4), then $\left(\frac{-1,p}{\mathbb{Q}}\right) \not\cong \left(\frac{-1,q}{\mathbb{Q}}\right)$.

Notes on Chapter 1

The study of group algebras has been, and still is, an area of active research. For a reasonably up-to-date survey of this topic, the book [61] of Passman is recommended. The discussion in Section 1.5 of finite dimensional algebras over fields is thoroughly classical. Benjamin Peirce's ground-breaking paper [62] defines an associative algebra as a finite dimensional vector space with an associative bilinear multiplication. The relization that algebraic geometry sheds light on the classification of algebras came later. Gerstenhaber's papers [36] and [37] pioneered this approach to the subject. Our discussion of quaternion algebras is modeled on the treatment in Lam's book [54]. Quaternion algebras over $\mathbb{Q}$ (or more generally any algebraic number field) admit a complete classification, based on the Hasse–Minkowski theorem. Serre's book [70] provides an elementary treatment of this theory for the rational field.

CHAPTER 2
Modules

The theory of modules over a ring or algebra grew out of the study of representations. However, as homological algebra developed, it became clear that module theory constitutes a good foundation on which to erect the structure of rings and algebras. On the basis of this dictum, we begin our formal development with this chapter on modules. The emphasis is on semisimple modules, since these structures lead to semisimple algebras, the fundamental building blocks for any theory of algebras. Highlights of the chapter are (1) a discussion of the lattice of submodules of a module, (2) Schur's Lemma, (3) a fundamental characterization of semisimple modules (Proposition 2.4), (4) a structure and uniqueness theorem for semisimple modules, and (5) an external characterization of finitely generated semisimple modules.

2.1. Change of Scalars

Throughout this chapter, A will stand for a non-trivial R-algebra. (A unital module over a trivial algebra consists of the zero element; such objects are uninteresting.) The ring R will seldom be mentioned, because its role in the elementary parts of module theory is negligible. In particular, the expression "R-algebra" will be shortened to "algebra."

Since A needn't be commutative, there is no natural way to identify left and right A-modules. On the other hand, the left and right sided theories are identical, so that it is enough to develop one of them. Generally, our choice is to consider right A-modules, and the term "A-module" should always be interpreted as "right A-module." Occasionally it is necessary to deal with left modules, particularly when bimodules are considered.

Any algebra A is itself a right and left A-module, with the scalar operation defined by the algebra product. The notation A_A and ${}_AA$ will often be used to indicate that A is being considered as a right (respectively left) A-module. The submodules of A_A are precisely the right ideals of A. Therefore, all the concepts and results on submodules of a module can be used for right and left ideals. We will later show (in Chapter 10) that the same comment applies to two-sided ideals: A-bimodules can be viewed as right modules over the "enveloping algebra" A^e of A, and the sub-bimodules of A are the two sided ideals.

There is a fairly elaborate theory of the class of all A-modules. We will not deal with this topic in a systematic way, but many of these categorical aspects of module theory will inevitably creep into our considerations. Readers who are familiar with category theory will recognize many old acquaintances. Uninitiated readers should not feel insecure, because categorical concepts will be introduced only in concrete forms.

One of the most useful techniques in the study of algebra involves comparing the modules over A with the modules over a related algebra B. In category theory, such a comparison can be dealt with abstractly via the notion of a functor. In later chapters, various special functors will be used for this purpose. However, one of the most useful devices for comparing modules over different algebras is completely elementary.

Let A and B be algebras, and suppose that $\theta\colon A \to B$ is an algebra homomorphism. If M is a right B-module, define scalar operations on M by the elements of A according to the rule $ux = u\theta(x)$ for $u \in M$, $x \in A$. A routine check shows that with these operations M becomes a right A-module. We will use the notation M_A (or if necessary M_θ) to designate M with the A-module operation defined in this way.

There are two important special cases of this change of scalar functor. The first occurs when A is a subalgebra of B, and θ is the inclusion homomorphism. The correspondence $M \to M_A$ is called a *forgetful functor*. The second case of importance is when $B = A/I$ for some ideal I of A, and θ is the projection homomorphism. The operation of A on the A/I-module M is then defined by

$$ux = u(x + I). \tag{1}$$

Lemma a. *Let $\theta\colon A \to B$ be an algebra homomorphism, and suppose that M and N are right B-modules:*

(i) $(M \oplus N)_A = M_A \oplus N_A$.
(ii) *If $\phi \in \mathrm{Hom}_B(M,N)$, then $\phi \in \mathrm{Hom}_A(M_A,N_A)$; if θ is surjective, then* $\mathrm{Hom}_B(M,N) = \mathrm{Hom}_A(M_A,N_A)$.
(iii) *If N is a submodule of M, then N_A is a submodule of M_A; if θ is surjective, then the sets* $\mathbf{S}(M)$ *and* $\mathbf{S}(M_A)$ *of the submodules of M and M_A are equal.*

We leave the proofs of these elementary facts as exercises.

If $\theta: A \to B$ is a surjective algebra homomorphism, then there is a convenient characterization of the A-modules that have the form M_A with M a B-module.

Definition. Let M be a right A-module, and suppose that X is a subset of M. The *annihilator* of X in A is

$$\operatorname{ann} X = \{x \in A: ux = 0 \quad \text{for all } u \in X\}.$$

There is an obvious analog of this definition for left A-modules.

The annihilator has many simple properties that we record for future reference.

Lemma b. *Let M, M', and $\{M_i: i \in J\}$ be right A-modules, $X \subseteq M$ and $Y \subseteq M$.*

(i) $\operatorname{ann} X$ *is a right ideal of A; if X is a submodule of M, then* $\operatorname{ann} X \lhd A$.
(ii) $X \subseteq Y$ *implies* $\operatorname{ann} X \supseteq \operatorname{ann} Y$.
(iii) *If $M \cong M'$, then* $\operatorname{ann} M = \operatorname{ann} M'$.
(iv) *If $M = \sum_{i \in J} M_i$, then* $\operatorname{ann} M = \bigcap_{i \in J} \operatorname{ann} M_i$.
(v) *If M is a right ideal of A, then* $\operatorname{ann}(A/M)$ *is the largest ideal K of A such that $K \subseteq M$.*

The properties (i) through (iv) follow directly from the definition of the annihilator. To obtain (v), note that by (i), $\operatorname{ann}(A/M)$ is an ideal of A that is clearly a subset of M. On the other hand, if $K \lhd A$ and $K \subseteq M$, then $(x + M)K \subseteq M$ for all $x \in A$, so that $K \subseteq \operatorname{ann}(A/M)$.

Proposition. *Let A and B be R-algebras, and $\theta: A \to B$ a surjective homomorphism. If N is a right A-module, then there is a right B-module M such that $N = M_A$ if and only if* $\operatorname{Ker} \theta \subseteq \operatorname{ann} N$.

PROOF. It is clear that $\operatorname{Ker} \theta \subseteq \operatorname{ann} M_A$. Conversely, if $\operatorname{Ker} \theta \subseteq \operatorname{ann} N$, then the equation $u\theta(x) = ux$ defines a valid scalar operation on N by the elements of $\operatorname{Im} \theta = B$. With this operation, N becomes a B-module M, and $N = M_A$ by definition. □

The proposition is most useful when $B = A/I$ for an ideal I of A. In this situation, an A-module N comes from an A/I-module if and only if $I \subseteq \operatorname{ann} N$. We will usually make no distinction between A/I-modules and A-modules N such that $I \subseteq \operatorname{ann} N$.

An A-module M is called *faithful* if $\operatorname{ann} M = 0$.

Corollary. *If I is an ideal of the algebra A, and N is a right A/I-module, then N is faithful as an A/I-module if and only if* $\operatorname{ann} N_A = I$.

PROOF. Clearly, $\operatorname{ann} N_{A/I} = (\operatorname{ann} N_A)/I$. □

EXERCISES

1. Prove Lemma a.

2. An A-module M is called *cyclic* if M is generated by a single element, that is, $M = uA$ for some $u \in M$. Prove the following statements.
   (a) If $M = uA$ is cyclic, then $M \cong A/N$, where $N = \operatorname{ann} u$.
   (b) If N is a right ideal of A, then A/N is a cyclic right A-module.
   (c) Let N be a right ideal of A. Denote $B = \{x \in A: xN \subseteq N\}$. Then B is a subalgebra of A, with $N \subseteq B$. Moreover $\mathbf{E}_A(A/N) \cong B/N$ by the mapping $\phi \mapsto \phi(1) + N$.

2.2. The Lattice of Submodules

For any A-module M, the collection $\mathbf{S}(M)$ of all submodules of M is partially ordered by the inclusion relation. Moreover, if $\{N_i: i \in J\}$ is any set of submodules of M, then $\bigcap_{i \in J} N_i$ is a submodule of M. (If $J = \varnothing$, then the empty intersection $\bigcap_{i \in J} N_i$ is defined to be M.) Plainly, this intersection is the largest submodule of M that is included in all of the N_i, that is, $\bigcap_{i \in J} N_i$ is the greatest lower bound of $\{N_i: i \in J\}$ with respect to the ordering relation of inclusion. The set $\{N_i: i \in J\}$ also has a least upper bound among the submodules of M. In general this least upper bound is not the set union, but rather the submodule that is generated by the union: $\sum_{i \in J} N_i = \{\sum_{k=1}^{m} u_k: u_k \in N_{i_k}\}$. In particular, the least upper bound of two submodules N and P of M is $N + P = \{u + v: u \in N, v \in P\}$. Any partially ordered set in which all subsets have a greatest lower bound and a least upper bound is called a *complete lattice*. Our discussion can be summarized by the statement that $\mathbf{S}(M)$ is a complete lattice.

Many fundamental properties of modules can be interpreted as facts about submodule lattices. Section 2.4 will provide a striking example of this phenomenon. There are a few lattice theoretic properties that hold in all lattices of the form $\mathbf{S}(M)$. The most important of these is the

Modular Law. *If N, P, and Q are submodules of M such that $N \subseteq Q$, then $N + (P \cap Q) = (N + P) \cap Q$.*

PROOF. Plainly, $N + (P \cap Q) \subseteq N + P$, and $N + (P \cap Q) \subseteq Q$ by the hypothesis that $N \subseteq Q$. Thus, $N + (P \cap Q) \subseteq (N + P) \cap Q$. On the other hand, if $u \in (N + P) \cap Q$, then $u = v + w$, where $v \in N$ and $w \in P$. Thus, $w = u - v \in P \cap (Q + N) = P \cap Q$, and $u = v + w \in N + (P \cap Q)$. □

The *modular law* (or *modularity*) is a fairly weak condition on a lattice. Some of the submodule lattices that interest us have the stronger property distributivity.

Lemma. *Let M be an A-module. The following identities (that is, valid equations for all choices of N, P, and Q) are equivalent for* $\mathbf{S}(M)$:

(i) $N \cap (P + Q) = (N \cap P) + (N \cap Q)$;
(ii) $N + (P \cap Q) = (N + P) \cap (N + Q)$;
(iii) $(N \cap P) + (P \cap Q) + (Q \cap N) = (N + P) \cap (P + Q) \cap (Q + N)$.

Proof. Two applications of (i) gives

$$\begin{aligned}(N + P) \cap (N + Q) &= ((N + P) \cap N) + ((N + P) \cap Q)\\ &= N + ((N \cap Q) + (P \cap Q))\\ &= N + (P \cap Q),\end{aligned}$$

which is (ii). Similarly, by several applications of (ii) we get

$$\begin{aligned}&(N \cap P) + (P \cap Q) + (Q \cap N)\\ &= (((N \cap P) + P) \cap ((N \cap P) + Q)) + (Q \cap N)\\ &= (P \cap (N + Q) \cap (P + Q)) + (Q \cap N)\\ &= (P + (Q \cap N)) \cap (N + Q) \cap (P + Q)\\ &= (P + Q) \cap (P + N) \cap (N + Q).\end{aligned}$$

Finally, it follows from (iii) and the modular law that

$$\begin{aligned}N \cap (P + Q) &= N \cap ((N + P) \cap (P + Q) \cap (Q + N))\\ &= N \cap ((N \cap P) + (P \cap Q) + (Q \cap N))\\ &= (N \cap P) + (Q \cap N) + (N \cap P \cap Q)\\ &= (N \cap P) + (N \cap Q).\end{aligned}$$

□

The submodule lattice $\mathbf{S}(M)$ is called *distributive* if it satisfies the identities (i), (ii), and (iii).

Remark. The inclusions

$$(N \cap P) + (N \cap Q) \subseteq N \cap (P + Q),\ N + (P \cap Q) \subseteq (N + P) \cap (N + Q),$$

and

$$(N \cap P) + (P \cap Q) + (Q \cap N) \subseteq (N + P) \cap (P + Q) \cap (Q + N)$$

are valid whether $\mathbf{S}(M)$ is distributive or not.

Proposition. *Let M be an A-module such that $\mathbf{S}(M)$ is not distributive. Then there exist distinct submodules P and Q of M such that $P/P \cap Q \cong Q/P \cap Q$ as A-modules.*

PROOF. Since $\mathbf{S}(M)$ is not distributive, there exist submodules M_0, M_1, M_2, M_3, and M_4 of M such that

$$\begin{aligned} M_0 &= (M_1 \cap M_2) + (M_2 \cap M_3) + (M_3 \cap M_1) \\ &\subset (M_1 + M_2) \cap (M_2 + M_3) \cap (M_3 + M_1) \\ &= M_4. \end{aligned}$$

Define

$$N = (M_1 \cap M_2) + (M_3 \cap (M_1 + M_2)),$$

$$P = (M_2 \cap M_3) + (M_1 \cap (M_2 + M_3)),$$

and

$$Q = (M_1 \cap M_3) + (M_2 \cap (M_1 + M_3)).$$

By the modular law,

$$\begin{aligned} P \cap Q &= ((M_2 \cap M_3) + (M_1 \cap (M_2 + M_3))) \cap ((M_1 \cap M_3) \\ &\quad + (M_2 \cap (M_1 + M_3))) \\ &= (M_1 \cap M_3) + (((M_2 \cap M_3) \\ &\quad + (M_1 \cap (M_2 + M_3))) \cap M_2 \cap (M_1 + M_3)) \\ &= (M_1 \cap M_3) + (M_2 \cap M_3) + (M_1 \cap (M_2 + M_3) \\ &\quad \cap M_2 \cap (M_1 + M_3)) \\ &= (M_1 \cap M_3) + (M_2 \cap M_3) + (M_1 \cap M_2) \\ &= M_0, \end{aligned}$$

and

$$\begin{aligned} P + Q &= ((M_2 \cap M_3) + (M_1 \cap (M_2 + M_3))) \\ &\quad + ((M_1 \cap M_3) + (M_2 \cap (M_1 + M_3))) \\ &= (M_1 \cap (M_2 + M_3)) + (M_2 \cap (M_1 + M_3)) \\ &= ((M_1 \cap (M_2 + M_3)) + M_2) \cap (M_1 + M_3) \\ &= (M_1 + M_2) \cap (M_2 + M_3) \cap (M_1 + M_3) \\ &= M_4 \end{aligned}$$

In particular, $P \neq Q$ because $M_0 \subset M_4$. Moreover, by symmetrical computations, $N \cap P = N \cap Q = M_0$, and $N + P = N + Q = M_4$. By the Noether Isomorphism $P/P \cap Q = P/P \cap N \cong P + N/N = M_4/N$. Similarly, $Q/P \cap Q \cong M_4/N$. Thus, $P/P \cap Q \cong Q/P \cap Q$. □

For an R-algebra A, we denote by $\mathbf{I}(A)$ the set of all ideals of A. It was noted in Section 2.1 that the ideals of A can be viewed as the submodules of A, considered as a right module over its enveloping algebra. (See Section

10.1.) It follows that $\mathbf{I}(A)$ is a complete, modular lattice. Moreover, if $\mathbf{I}(A)$ is not distributive, then there exist distinct ideals I and J in A such that $I/I \cap J \cong J/I \cap J$ as A-bimodules. These facts can also be established directly by making minor changes in the wording of our proofs in this section.

EXERCISES

1. Use the Noether Isomorphism Theorems and the Modular Law to prove *Zassenhaus's Lemma*: if N, N', M, and M' are submodules of an A-module such that $N' \subseteq N$ and $M' \subseteq M$, then $(N' + (M \cap N))/(N' + (M' \cap N)) \cong (M \cap N)/((M' \cap N) + (M \cap N')) \cong (M' + (M \cap N))/(M' + (M \cap N'))$.

2. Let $0 = M_0 \subset M_1 \subset \cdots \subset M_{n-1} \subset M_n = M$ and $0 = N_0 \subset N_1 \subset \cdots \subset N_{m-1} \subset N_m = M$ be chains of submodules of the A-module M. Use the Zassenhaus Lemma to prove that there exist refinements $0 = M_0' \subset M_1' \subset \cdots \subset M_{k-1}' \subset M_k' = M$ and $0 = N_0' \subset N_1' \subset \cdots \subset N_{k-1}' \subset N_k' = M$ of these chains (that is, every M_i is equal to some M_j' and every N_i is equal to some N_l'), and a permutation π such that $M_{j+1}'/M_j' \cong N_{\pi(j)+1}'/N_{\pi(j)}'$ for all $j < k$. This result is called the *Schreier Refinement Theorem*. It is a generalization of the Jordan–Hölder Theorem, which will be proved by a more elementary argument in Section 2.6.

3. Prove that $\mathbf{S}(M)$ is distributive if and only if: for $N, P, Q \in \mathbf{S}(M)$, if $N \cap P = N \cap Q$ and $N + P = N + Q$, then $P = Q$. In particular, complements are unique in $\mathbf{S}(M)$. (See Section 2.4.)

4. Assume that A is an F-algebra, where F is an infinite field. Prove that if M is a right A-module such that $\mathbf{S}(M)$ is not distributive, then $S(\mathbf{M})$ is infinite. Hint. Let P and Q be as in Proposition 2.2. It can be assumed without loss of generality that $P \cap Q = 0$. Let $\phi\colon P \to Q$ be an isomorphism. Show that for each $a \in F$, $N_a = \{u + \phi(u)a\colon u \in P\}$ is a submodule of M, and $N_a \cap N_b = 0$ for $a \neq b$ in F.

2.3. Simple Modules

Definition. A right or left module N is *simple* if N is not the zero module and the only submodules of N are 0 and N. A module M is *semisimple* if M is a direct sum of simple modules.

In the literature of ring theory, the term *irreducible module* is often used for simple module, and *completely reducible* has the same meaning as semisimple. The current trend in terminology seems to be moving in the direction of the adjectives "simple" and "semisimple." We will follow this fashion.

It is evident that a right ideal M of an algebra A is a simple A-module if and only if M is a *minimal right ideal*, that is, M is minimal in the set of non-zero right ideals of A. Of course, there is no guarantee that A has any minimal right ideals. There are none in the ring $\mathbb{Z}$ of integers, for example. On the other hand, the hypothesis that A has a non-zero unity element implies (by an argument using Zorn's Lemma) that A includes at least one *maximal right ideal*, that is, an ideal that is maximal in the set of proper right ideals of A. Moreover, if M is a maximal right ideal of A, then by the Correspondence Theorem, A_A/M is a simple module. Conversely, if A_A/M is simple, then M is a maximal right ideal.

Proposition. *For a non-zero right A-module N, the following conditions are equivalent:*

(i) *N is simple;*
(ii) *$uA = N$ for all non-zero $u \in N$;*
(iii) *$N \cong A_A/M$ for some maximal right ideal M of A.*

PROOF. (i) implies (ii), because $0 \neq u \in uA < N$ forces $uA = N$ by the simplicity of N. Conversely, since $N \neq 0$, (ii) implies that N is the unique non-zero submodule of N; that is, N is simple. As we noted before, the fact that (iii) implies (i) is a consequence of the Correspondence Theorem. To prove that (ii) implies (iii), let u be a non-zero element of N. By (ii), the mapping $x \mapsto ux$ is a surjective module homomorphism of A_A to N whose kernel M is a right ideal of A. Since (ii) implies (i), it follows that $A_A/M \cong N$ is simple. Therefore, M is a maximal right ideal of A. □

Schur's Lemma. *Let M and N be right A-modules. Suppose that $\phi: M \to N$ is a non-zero homomorphism.*

(i) *If M is simple, then ϕ is injective.*
(ii) *If N is simple, then ϕ is surjective.*

PROOF. Since $\phi \neq 0$, it follows that $\operatorname{Ker}\phi \neq M$ and $\operatorname{Im}\phi \neq 0$. Hence, M simple implies $\operatorname{Ker}\phi = 0$, and N simple implies $\operatorname{Im}\phi = N$. □

Corollary a. *If M and N are simple right A-modules, then either $M \cong N$ or* $\operatorname{Hom}_A(M,N) = 0$.

This corollary follows directly from Schur's Lemma because a bijective homomorphism is an isomorphism.

A right A-module N is *indecomposable* if $N \neq 0$, and N cannot be written as a direct sum of non-zero submodules: if $N = P \oplus Q$, then $P = 0$ or $Q = 0$. Indecomposable modules are very important in the theory of algebras. They will reappear often in later chapters.

Corollary b. *For a semisimple module N, the following conditions are equivalent.*

(i) *N is simple.*
(ii) $\mathbf{E}_A(N)$ *is a division algebra.*
(iii) *N is indecomposable.*

PROOF. If N is simple, then every non-zero endomorphism of N has an inverse by Schur's Lemma. That is, $\mathbf{E}_A(N)$ is a division algebra. If $\mathbf{E}_A(N)$ is a division algebra, then $id_N \neq 0$, so that $N \neq 0$. Moreover, for any direct sum decomposition $N = P \bigoplus Q$, there is an element $\pi \in \mathbf{E}_A(N)$ such that $\pi(N) = P$, $(1 - \pi)(N) = Q$, and $\pi^2 = \pi$; namely, π is the projection of N on P associated with the decomposition. Hence, $\pi(id_N - \pi) = 0$. The assumption that $\mathbf{E}_A(N)$ is a division algebra implies that $\pi = 0$ or $id_N - \pi = 0$. In these respective cases, $P = 0$ and $Q = 0$. Finally, (iii) implies (i), because the hypotheses that N is indecomposable and semisimple (that is, a direct sum of simple modules) are compatible only if N is simple. □

The term "Schur's Lemma" is often used to describe the implication in Corollary b that if N is simple, then $\mathbf{E}_A(N)$ is a division algebra. We will go beyond this custom and refer to the lemma and both of its corollaries as Schur's Lemma. This abuse of terminology shouldn't cause confusion.

EXERCISES

1. Determine all of the simple right A-modules for the following algebras:
   (a) $A = \mathbb{Z}$;
   (b) $A = \{a/n : a \in \mathbb{Z}, n \in \mathbb{N}, 2 \nmid n\}$;
   (c) $A = \mathbb{C}[\mathbf{x}]$;
   (d) $A = \mathbb{C}[\mathbf{x},\mathbf{y}]/(\mathbf{x}^2 + \mathbf{y}^2 - 1)$;
   (e) $A = \mathbb{R}$-algebra of real valued, continuous functions on $[0,1]$;
   (f) $A = \left\{\begin{bmatrix} a & 0 \\ b & c \end{bmatrix} : a, b, c \in F\right\}$ (F = any field).

2. Let A be an integral domain that is not a field. Denote the field of fractions of A by F. Consider F as an A-module. Prove that $\mathbf{E}_A(F_A) \cong F$, but F_A is not simple. Of course, F_A is not semisimple, so that this example does not contradict Corollary b.

2.4. Semisimple Modules

If N and P are submodules of an A-module M, then P is called a *complement* of N in $\mathbf{S}(M)$ if $N + P = M$ and $N \cap P = 0$. In other words, M is the inner direct sum of N and P. Plainly, the relation of being a complement is symmetric. In general, complements are not unique (but see Exercise 3, Section 2.2), and not all submodules have a complement in $\mathbf{S}(M)$. Our main result in this section is that the universal existence of complements characterizes semisimple modules.

Lemma. *Let M be a module such that $M = \sum_{i \in J} N_i$, where each N_i is a simple submodule of M. If $P \in \mathbf{S}(M)$, then there is a subset I of J such that $M = (\bigoplus_{i \in I} N_i) \bigoplus P$.*

PROOF. By Zorn's Lemma, there is a subset I of J such that the collection $\{N_i : i \in I\} \cup \{P\}$ is maximal with respect to independence: $(\sum_{i \in I} N_i) + P = (\bigoplus_{i \in I} N_i) \bigoplus P$. Let $M_1 = (\sum_{i \in I} N_i) + P$. The maximality of I implies that $M_1 \cap N_j \neq 0$ for all $j \in J$. Therefore, since each N_j is simple, $N_j \subseteq M_1$ for all $j \in J$. Hence, $M = \sum_{j \in J} N_j \subseteq M_1 \subseteq M$, and $M = (\bigoplus_{i \in I} N_i) \bigoplus P$. □

Proposition. *For a right A-module M, the following conditions are equivalent.*

(i) *M is semisimple.*
(ii) *$M = \sum \{N \in \mathbf{S}(M) : N$ is simple$\}$.*
(iii) *$\mathbf{S}(M)$ is a complemented lattice, that is, every submodule of M has a complement in $\mathbf{S}(M)$.*

PROOF. It is clear that (i) implies (ii) by the definition of semisimplicity. By the lemma, (ii) implies (iii). Also, the lemma shows that (ii) implies (i) (taking $P = 0$). Using the modular law, (iii) can be strengthened to (iv) if $P \in \mathbf{S}(M)$, then $\mathbf{S}(P)$ is complemented. Indeed, if $M_1 \in \mathbf{S}(P)$, then by (iii) there exists $M_2 \in \mathbf{S}(M)$ such that $M = M_1 \bigoplus M_2$. Hence, $P = P \cap (M_1 \bigoplus M_2) = M_1 \bigoplus (P \cap M_2)$ with $P \cap M_2 \in \mathbf{S}(P)$. To complete the proof, we deduce from (iv) that if Q is a proper submodule of M, then there is a simple submodule N of M such that $N \cap Q = 0$. This result implies (ii), and completes the figure eight of equivalences. Let $0 \neq u \in M - Q$. By Zorn's Lemma, it can be assumed that Q is maximal with the property $u \notin Q$. Apply (iv) with $P = M$ to obtain $N \in \mathbf{S}(M)$ such that $M = Q \bigoplus N$. Write $u = w + v$ with $w \in Q$, $v \in N$. Since $u \notin Q$, it follows that $v \neq 0$. In particular $N \neq 0$. If N_1 is a non-zero submodule of N, then the maximality of Q implies that $w + v = u \in Q + N_1 = Q \bigoplus N_1$. Thus, $v \in N_1$. In particular, two non-zero submodules of N have a non-zero intersection. On the other hand, $\mathbf{S}(N)$ is complemented by (iv). The only way to escape a contradiction is to conclude that $\mathbf{S}(N) = \{0, N\}$, that is, N is simple. □

Corollary a. *If M is semisimple and $P < M$, then P and M/P are semisimple.*

PROOF. By (iii), $M \cong P \bigoplus M/P$. Hence, P and M/P are semisimple by the equivalence of (iii), (iv), and (i). □

Corollary b. *A direct sum of semisimple modules is semisimple.*

This corollary is a direct consequence of the definition of semisimple modules.

It is occasionally useful to know when $\mathbf{S}(M)$ is a distributive lattice. If M is semisimple, the following result settles this question.

Corollary c. *Let M be a semisimple right A-module. Assume that $M = \bigoplus_{i \in J} N_i$, where each N_i is simple. Then $\mathbf{S}(M)$ is a distributive lattice if and only if $N_i \ncong N_j$ for all $i \neq j$ in J.*

PROOF. If N is a non-zero right A-module, and $Q = N \oplus N$, then $\mathbf{S}(Q)$ is not distributive. In fact, if $N_1 = \{(u,0) \in Q: u \in N\}$, $N_2 = \{(0,u) \in Q: u \in N\}$, and $N_3 = \{(u,u) \in Q: u \in N\}$, then $N_1 + N_2 = Q$, $N_3 \cap N_1 = N_3 \cap N_2 = 0$, and $N_3 \cong N \neq 0$; thus, $N_3 \cap (N_1 + N_2) = N_3 \supset 0 = (N_3 \cap N_1) + (N_3 \cap N_2)$. In the context of the corollary, this observation shows that if $\mathbf{S}(M)$ is distributive, then $N_i \ncong N_j$ for all $i \neq j$. To prove the converse implication, define $N(I) = \sum_{j \in I} N_j$ for each subset I of J. Since the sum $M = \bigoplus_{i \in J} N_i$ is direct, it is clear that $N(I_1 \cup I_2) = N(I_1) + N(I_2)$ and $N(I_1 \cap I_2) = N(I_1) \cap N(I_2)$ for any two subsets I_1 and I_2 of J. Thus, $\{N(I): I \subseteq J\}$ is a distributive sublattice of $\mathbf{S}(M)$. The proof is completed by showing that $\mathbf{S}(M) = \{N(I): I \subseteq J\}$. For $P < M$, define $I = \{i \in J: P \cap N_i \neq 0\}$. We will show that $P = N(I)$. Since N_i is simple, $P \cap N_i \neq 0$ implies that $N_i = P \cap N_i \subseteq P$. Hence, $N(I) \subseteq P$. It will be sufficient to prove that $P \cap N(J - I) = 0$, since the Modular Law then gives $P = P \cap (N(J - I) + N(I)) = P \cap N(J - I) + N(I) = N(I)$. Assume that $P \cap N(J - I) \neq 0$, and choose a set $K \subseteq J - I$ of smallest cardinality such that $P \cap N(K) \neq 0$. Clearly, K is finite, and $|K| \geq 2$ because $P \cap N_i = 0$ for all $i \in J - I$. For $i \in K$, let $\pi_i: N(K) \to N_i$ be the projection homomorphism that is associated with the direct decomposition $N(K) = \bigoplus_{j \in K} N_j$. Since $\mathrm{Ker}(\pi_i | P \cap N(K)) \subseteq P \cap N(K - \{i\})$, the minimality of $|K|$ implies that $\mathrm{Ker}(\pi_i | P \cap N(K)) = 0$. Therefore, since N_i is simple and $P \cap N(K) \neq 0$, it follows that π_i maps $P \cap N(K)$ isomorphically to N_i. The fact that $|K| > 1$ then gives a contradiction to the hypothesis that $N_i \ncong N_j$ for all $i \neq j$ in J. □

EXERCISE

Prove: if $M = \bigoplus_{i \in J} N_i$, where the N_i are right A-modules, and if $I_1 \subseteq J$, $I_2 \subseteq J$, then $(\sum_{i \in I_1} N_i) \cap (\sum_{i \in I_2} N_i) = \sum_{i \in I_1 \cap I_2} N_i$. This fact was used in the proof of Corollary c. It is most easily established using outer direct sums.

2.5. Structure of Semisimple Modules

By definition a semisimple right A-module has "nice" structure; it is a direct sum of simple modules. The purpose of this section is to establish the uniqueness of such direct sum representations. Some preliminary results are needed.

Lemma a. *Let $M = \bigoplus_{i \in J} N_i$ with each N_i a simple right A-module. Suppose that N is a simple right A-module for which there is a non-zero homomorphism*

$\phi: N \to M$. Then there exists $j \in J$ such that $M = \phi(N) \oplus (\bigoplus_{i \neq j} N_i)$ and $N \cong N_j$.

PROOF. By Lemma 2.4, there exists $J' \subseteq J$ such that $M = \phi(N) \oplus (\bigoplus_{i \in J'} N_i)$. Consequently, $\bigoplus_{i \in J - J'} N_i \cong M/(\bigoplus_{i \in J'} N_i) \cong \phi(N) \cong N$ (by Schur's Lemma and the assumption $\phi \neq 0$). It follows that $|J - J'| = 1$, which proves the lemma. □

For the rest of this section, let $\{N_i : i \in J\}$ be a set of representatives of the isomorphism classes of simple right A-modules. Thus, J is a non-empty indexing set, each N_i is a simple right A-module, and each simple right A-module is isomorphic to exactly one N_i.

Notation: For a right A-module M, denote by $M(i)$ the submodule $\sum\{N < M : N \cong N_i\}$.

Lemma b. *If M is a semisimple right A-module, then $M = \bigoplus_{i \in J} M(i)$.*

PROOF. Since M is semisimple, $M = \bigoplus_{i \in J} M_i$, where $M_i = \bigoplus_{j \in K_i} N_{ij}$, $N_{ij} \cong N_i$. Plainly, $M_i \subseteq M(i)$. It will be sufficient to prove that $M(i) \subseteq M_i$. Let $N_i \cong N < M$. Write $M = M_i \oplus M_i'$, $M_i' = \bigoplus_{j \neq i} M_j$, and let $\pi: M \to M_i'$ be the projection associated with this decomposition. It follows from Lemma a that $\pi(N) = 0$, that is, $N \subseteq M_i$. Since N was any submodule of M isomorphic to N_i, this proves the desired inclusion $M(i) \subseteq M_i$. □

Lemma c. *Let M and M' be semisimple right A-modules. If $\phi: M \to M'$ is a homomorphism, then $\phi(M(i)) \subseteq M'(i)$ for all $i \in J$.*

PROOF. If $N_i \cong N < M$, then by Schur's Lemma, either $\phi(N) = 0$ or $\phi(N) \cong N_i$. In both cases, $\phi(N) \subseteq M'(i)$. It follows that $\phi(M(i)) \subseteq M'(i)$. □

Proposition. *Let M and M' be semisimple right A-modules. Suppose that $M = \bigoplus_{i \in J} M(i)$ with $M(i) \cong \bigoplus \alpha_i N_i$ and $M' = \bigoplus_{i \in J} M'(i)$ with $M'(i) \cong \bigoplus \beta_i N_i$. Then M is isomorphic to M' if and only if the cardinal numbers α_i and β_i are equal for all $i \in J$.*

PROOF. Suppose that $\phi: M \to M'$ is an isomorphism. By Lemma c, $\phi(M(i)) = M'(i)$ for all i. Fix $i \in J$. For the proof that $\alpha_i = \beta_i$, we first consider the case in which α_i is finite, and use induction. If $\alpha_i = 0$, then $M'(i) = \phi(M(i)) = \phi(0) = 0$, so that $\beta_i = 0$. Suppose that $\alpha_i = m \geq 1$. Write $M(i) = N_{i1} \oplus \cdots \oplus N_{im-1} \oplus N_{im}$, $M'(i) = \bigoplus_{k \in I} N_{ik}'$, where $N_{ij} \cong N_{ik}' \cong N_i$ for all j and k, and $|I| = \beta_i$. By Lemma a, there exists $l \in I$ such that $\phi(N_{im}) \oplus (\bigoplus_{k \neq l} N_{ik}') = M'(i) = \phi(N_{im}) \oplus \phi(N_{i1} \oplus \cdots \oplus N_{im-1})$. Consequently, $N_{i1} \oplus \cdots \oplus N_{im-1} \cong \phi(N_{i1} \oplus \cdots \oplus N_{im-1}) \cong M'(i)/\phi(N_{im}) \cong \bigoplus_{k \neq l} N_{ik}'$. By the induction hypothesis, $m - 1 = |I - \{l\}|$ so that $\alpha_i =$

$m = |I| = \beta_i$. This completes the induction. If β_i is finite, the same proof applies using ϕ^{-1}. Therefore, assume that α_i and β_i are both infinite. By Proposition 2.3, there exist decompositions $M(i) = \bigoplus_{k \in K} u_k A$, $M'(i) = \bigoplus_{l \in L} u_l' A$, with $|K| = \alpha_i$ and $|L| = \beta_i$. The isomorphism ϕ induces a mapping λ from K to the set of finite subsets of L such that $\phi(u_k) \in \sum_{l \in \lambda(k)} u_l' A$, and $\bigcup_{k \in K} \lambda(k) = L$ since ϕ is surjective. Therefore, since L and K are infinite, $\beta_i = |L| \leq \sum_{k \in K} |\lambda(k)| \leq \aleph_0 \cdot |K| = |K| = \alpha_i$. By symmetry, $\alpha_i \leq \beta_i$. This completes the proof that $M \cong M'$ implies $\alpha_i = \beta_i$ for all $i \in J$. The converse is obvious. □

Exercises

1. Use Lemma a to give an inductive proof of the Proposition in the special case that M and M' are finitely generated.

2. Prove that if V is an infinite dimensional F-space, and $A = \mathbf{E}_F(V)$, then $A_A \cong A_A \oplus A_A$. Hint. $V \cong V \oplus V$ as F-spaces implies

$$\mathrm{Hom}_F(V, V) \cong \mathrm{Hom}_F(V, V \oplus V).$$

2.6. Chain Conditions

Most of the semisimple modules that we will enounter are *finite* direct sums of simple modules. The main proposition of this section establishes the equivalence of several finiteness conditions for semisimple modules. The proof of this result is based on standard results of module theory that have many applications outside the context of semisimple modules and algebras.

An A-module M is *Artinian* (*Noetherian*) if $\mathbf{S}(M)$ satisfies the *descending* (*ascending*) *chain condition*. That is, there are no infinite, strictly decreasing (increasing) sequences of submodules of M. Equivalently, M is Artinian (Noetherian) if every non-empty subset of $\mathbf{S}(M)$ includes a minimal (maximal) member.

In a general context, the Artinian and Noetherian properties are independent of each other. For example, $\mathbb{Z}_\mathbb{Z}$ is Noetherian but not Artinian, whereas, the $\mathbb{Z}$-module $\mathbb{Z}(p^\infty) = \{a/p^n : a \in \mathbb{Z}, n \in \mathbb{N}\}/\mathbb{Z}$ is Artinian, but not Noetherian. We will see that in the presence of semisimplicity these conditions are equivalent.

Lemma a. *Let M', M'', and N be submodules of the A-module M, with $M' \subseteq M''$. There is an exact sequence*

$$0 \to \frac{M'' \cap N}{M' \cap N} \to \frac{M''}{M'} \to \frac{M'' + N}{M' + N} \to 0.$$

PROOF. The Noether Isomorphism Theorems and the Modular Law give

$$\frac{M'' + N}{M' + N} = \frac{M'' + (M' + N)}{M' + N} \cong \frac{M''}{M'' \cap (M' + N)}$$

$$= \frac{M''}{M' + (M'' \cap N)} \cong \frac{(M''/M')}{(M' + (M'' \cap N))/M'},$$

and $(M' + (M'' \cap N))/M' \cong (M'' \cap N)/M' \cap (M'' \cap N) = (M'' \cap N)/(M' \cap N)$. □

Lemma b. *Let $0 \to N \to M \to P \to 0$ be an exact sequence of A-modules. The module M is Artinian (Noetherian) if and only if both N and P are Artinian (respectively, Noetherian).*

PROOF. There is no harm in assuming that $N \in \mathbf{S}(M)$ and $P = M/N$. In this case, $\mathbf{S}(N)$ is a sublattice of $\mathbf{S}(M)$, and by the Correspondence Theorem, $\mathbf{S}(P)$ is isomorphic to a sublattice of $\mathbf{S}(M)$. Therefore, if M is Artinian (Noetherian), so are N and P. Conversely, if $M_0 \supset M_1 \supset M_2 \supset \cdots$ is an infinite descending chain in $\mathbf{S}(M)$, then $M_0 \cap N \supseteq M_1 \cap N \supseteq M_2 \cap N \supseteq \cdots$ in $\mathbf{S}(N)$, and $(M_0 + N)/N \supseteq (M_1 + N)/N \supseteq (M_2 + N)/N \supseteq \cdots$ in $\mathbf{S}(P)$; and by Lemma a, at least one of these chains is infinite. Thus, if N and P are Artinian, so is M. The proof in the Noetherian case is similar. □

Lemma c. *Assume that the A-module M is Artinian and Noetherian. There is a sequence $0 = M_0 \subset M_1 \subset M_2 \subset \cdots \subset M_{n-1} \subset M_n = M$ such that all of the factor modules M_{i+1}/M_i, $i < n$, are simple.*

PROOF. If $M = 0$, then $0 = M_0 = M$. Assume that $M \neq 0$. Using the fact that M is Artinian, it is possible to construct (by induction) an increasing sequence $0 = M_0 \subset M_1 \subset M_2 \subset \cdots$ of submodules of M such that all of the factors M_{i+1}/M_i are simple. Indeed, if $M_0, M_1, \ldots, M_i$ have been obtained, and if $M_i \neq M$, then there exists a submodule M_{i+1} of M containing M_i such that M_{i+1}/M_i is a minimal, non-zero submodule of M/M_i, because M/M_i is Artinian by Lemma b. Since M is Noetherian, this inductive process must be blocked at some finite stage; that is, for some $n < \omega$, it must be the case that $M_n = M$. □

A chain $0 = M_0 \subset M_1 \subset M_2 \subset \cdots \subset M_{n-1} \subset M_n = M$ of submodules of M is called a *composition series* of M if M_{i+1}/M_i is simple for all $i < n$. The factor modules M_{i+1}/M_i are called the *composition factors* of the series. They are unique.

Jordan–Hölder Theorem. *If $0 = M_0 \subset M_1 \subset M_2 \subset \cdots \subset M_{n-1} \subset M_n = M$ and $0 = M'_0 \subset M'_1 \subset M'_2 \subset \cdots \subset M'_{k-1} \subset M'_k = M$ are composition series of the module M, then $n = k$, and there is a permutation π of $\{0, 1, 2, \ldots, n-1\}$ such that $M'_{j+1}/M'_j \cong M_{\pi(j)+1}/M_{\pi(j)}$ for all $j < n$.*

PROOF. Induce on n. If $n = 0$, then $M = 0$ and $k = 0$. Assume that $n > 0$. Consider the chain of submodules $0 = M'_0 \cap M_{n-1} \subseteq M'_1 \cap M_{n-1} \subseteq \cdots \subseteq M'_k \cap M_{n-1} = M_{n-1} = M'_0 + M_{n-1} \subseteq M'_1 + M_{n-1} \subseteq \cdots \subseteq M'_k + M_{n-1} = M_n$. For $j < k$, there is an exact sequence

$$0 \to \frac{M'_{j+1} \cap M_{n-1}}{M'_j \cap M_{n-1}} \to \frac{M'_{j+1}}{M'_j} \to \frac{M'_{j+1} + M_{n-1}}{M'_j + M_{n-1}} \to 0$$

by Lemma a. Since M'_{j+1}/M'_j is simple, exactly one of

$$\frac{M'_{j+1} \cap M_{n-1}}{M'_j \cap M_{n-1}}, \frac{M'_{j+1} + M_{n-1}}{M'_j + M_{n-1}}$$

is isomorphic to M'_{j+1}/M'_j and the other quotient is 0. Also, since M_n/M_{n-1} is simple, there is exactly one $i < k$ such that $M_{n-1} = M'_i + M_{n-1} \subset M'_{i+1} + M_{n-1} = M_n$. Thus, $M_n/M_{n-1} \cong M'_{i+1}/M'_i$, and $0 = M'_0 \cap M_{n-1} \subset M'_1 \cap M_{n-1} \subset \cdots \subset M'_{i-1} \cap M_{n-1} \subset M'_i \cap M_{n-1} = M'_{i+1} \cap M_{n-1} \subset \cdots \subset M'_k \cap M_{n-1} = M_{n-1}$ is a composition series of M_{n-1}. By the induction hypothesis, $k - 1 = n - 1$, and there is a bijective mapping $\pi\colon \{0, 1, \ldots, i-1, i+1, \ldots, k-1\}$ to $\{0, 1, \ldots, n-2\}$ such that $M'_{j+1}/M'_j \cong M_{\pi(j)+1}/M_{\pi(j)}$, for all $j \neq i$. The proof is completed by defining $\pi(i) = n - 1$. □

If a module M can be written as a finite direct sum of simple modules in two ways, say $M = N_1 \oplus N_2 \oplus \cdots \oplus N_n$ and $M = N'_1 \oplus N'_2 \oplus \cdots \oplus N'_k$, then the Jordan–Hölder Theorem applies to the composition series $0 \subset N_1 \subset N_1 + N_2 \subset \cdots \subset N_1 + N_2 + \cdots + N_n = M$ and $0 \subset N'_1 \subset N'_1 + N'_2 \subset \cdots \subset N'_1 + N'_2 + \cdots + N'_k = M$. It yields the conclusions $n = k$ and $N'_j \cong N_{\pi(j)}$, for some permutation π. In other words, the Jordan–Hölder Theorem leads to an elementary proof of the special case of Proposition 2.5 in which the direct sums are finite.

Terminology. Let M be a right A-module that is Artinian and Noetherian. By Lemma c, M has a composition series, and the length of this series is unique by the Jordan–Hölder Theorem. The number of composition factors in a composition series of M is called the *composition length* of M. This number will be denoted by $l(M)$. Two useful observations are clear consequences of this definition: $l(M) = 0$ if and only if $M = 0$; $l(M) = 1$ if and only if M is simple.

Corollary. *Let M, N, and P be Artinian and Noetherian modules. If $0 \to N \to M \to P \to 0$ is an exact sequence of module homomorphisms, then $l(M) = l(N) + l(P)$.*

PROOF. Without loss of generality, suppose that N is a submodule of M, and $P = M/N$. If $0 = N_0 \subset N_1 \subset \cdots \subset N_r = N$ is a composition series of N, and $0 = Q_0/N \subset Q_1/N \subset \cdots \subset Q_s/N = P$ is a composition series of P, then $0 = N_0 \subset N_1 \subset \cdots \subset N_r \subset Q_1 \subset \cdots \subset Q_s = M$ is a composition series of M. Thus, $l(M) = r + s = l(N) + l(P)$. □

Proposition. *For a semisimple right A-module M, the following conditions are equivalent.*

(i) *M is finitely generated as an A-module.*
(ii) $M = N_1 \oplus N_2 \oplus \cdots \oplus N_m$ *with each* N_j *simple, and* $0 \le m < \omega$.
(iii) *M is Artinian.*
(iv) *M is Noetherian.*
(v) *There exists* $m < \omega$ *such that if* $M_0 \subset M_1 \subset \cdots \subset M_k$ *is a finite, strictly increasing sequence of submodules of M, then* $k \le m$.

PROOF. The conditions (i) and (ii) are equivalent. Indeed, (ii) implies (i) by Proposition 2.3; and (ii) follows from each of the conditions (i), (iii), and (iv), because M is semisimple (that is, a direct sum of simple modules), and an infinite direct sum of non-zero modules cannot be finitely generated, Artinian, or Noetherian. Since simple modules are plainly Artinian and Noetherian, the conditions (iii) and (iv) both follow from (ii), using Lemma b and induction on m. Clearly, (v) implies that M is both Artinian and Noetherian. Conversely, if M is Artinian and Noetherian, then a strictly increasing sequence of submodules of M includes at most $l(M) + 1$ terms (by the Corollary, using induction). □

Some of the implications in the proposition are true for arbitrary modules. Our proof shows that (v) is equivalent to the conjunction of (iii) and (iv). Also, every Noetherian A-module M is finitely generated: otherwise, the axiom of choice would enable us to select an infinite sequence $u_1, u_2, u_3, \ldots$ of elements from M such that $u_1A \subset u_1A + u_2A \subset u_1A + u_2A + u_3A \subset \cdots$, thereby violating the ascending chain condition.

EXERCISES

1. Prove the converse of Lemma c: if the module M has a composition series, then M is Artinian and Noetherian.

2. (a) Assume that M is an Artinian module such that $\mathbf{S}(M)$ is not distributive. Prove that the modules P and Q in Proposition 2.2 can be chosen so that $P/P \cap Q$ and $Q/P \cap Q$ are simple. Hint. Let M_0, N, P, Q, and M_4 be chosen as in the proof of Proposition 2.2. Use the descending chain condition to obtain $N' \in \mathbf{S}(M)$ such that $M_0 \subset N' \subseteq N$ and N'/M_0 is simple. Show that $P' = P \cap (N' + Q)$ and $Q' = Q \cap (N' + P)$ have the required property.
   (b) Prove the same result under the assumption that M is Noetherian.

3. Assume that M is Artinian and Noetherian. Prove that if $\mathbf{S}(M)$ is distributive, then $\mathbf{S}(M)$ is finite. Hint. Otherwise, since M is Artinian, there is a minimal $N \in \mathbf{S}(M)$ such that $\mathbf{S}(N)$ is infinite. Use distributivity to show that if $P_1, P_2, P_3, \ldots$ are distinct maximal proper submodules of N, then $P_1 \supset P_1 \cap P_2 \supset P_1 \cap P_2 \cap P_3 \supset \cdots$, so that by the descending chain condition $\{P_i\}$ is finite. Use the fact that M is Noetherian to deduce the contradiction that $\mathbf{S}(N) = \bigcup_i \mathbf{S}(P_i) \cup \{N\}$ is finite.

4. It is possible to define a composition series for modules that are neither Artinian nor Noetherian. A *composition series* for the module M is a maximal chain $\{N_j$:

$j \in I\}$ in the family of all submodules of M. That is, for all $i, j \in I$, either $N_i \subseteq N_j$ or $N_j \subseteq N_i$; and if N is a submodule of M such that $N \notin \{N_j : j \in I\}$, then there exists $i \in I$ such that $N \nsubseteq N_i$ and $N_i \nsubseteq N$.

(a) Deduce from Zorn's Lemma that every module has a composition series. (This is essentially equivalent to deducing the Hausdorff Maximum Principle from Zorn's Lemma.)

(b) Prove that a composition series $\{N_j : j \in I\}$ for M is closed under unions and intersections: if $X \subseteq I$, then $\bigcup_{j \in X} N_j$ and $\bigcap_{j \in X} N_j$ are members of $\{N_j : j \in I\}$. In particular, 0 and M belong to the series. Deduce that for any $u \in M$, there is a largest N_i such that $u \notin N_i$ and a smallest N_j such that $u \in N_j$; show that N_j covers N_i (i.e. $N_i \subset N_j$; and $N_i \subseteq N_k \subseteq N_j$ implies $N_k = N_i$ or $N_k = N_j$), and N_j/N_i is simple.

(c) Let $\{N_j : j \in I\}$ be a chain of submodules of M. Prove that $\{N_j : j \in I\}$ is a composition series for M if and only if $\{N_j : j \in I\}$ is closed under unions and intersections, and for every i, j in I such that $N_i \subset N_j$, there exist k, l in I such that $N_i \subseteq N_k \subset N_l \subseteq N_j$ and N_l/N_k is simple.

(d) Let M be a countably infinite dimensional F-space. Let $\{u_n : n < \omega\}$ be a basis of M. Define $N_n = \sum_{m<n} u_m F$ for $n \leq \omega$. Prove that $\{N_n : n < \omega\}$ is a composition series for M. Let $\{v_r : r \in \mathbb{Q}, 0 \leq r \leq 1\}$ be a basis of M, indexed by the rational numbers between 0 and 1. For each real number x with $0 \leq x \leq 1$, define $P_x = \sum_{r \in \mathbb{Q}, r < x} v_r F$, and for $s \in \mathbb{Q}$ with $0 \leq s \leq 1$, define $\bar{P}_s = \sum_{r \in \mathbb{Q}, r \leq s} v_r F$. Prove that $\{P_x : x \in \mathbb{R}, 0 \leq x \leq 1\} \cup \{\bar{P}_s : s \in \mathbb{Q}, 0 \leq s \leq 1\}$ is also a composition series for M.

This example shows that the Jordan–Hölder Theorem does not carry over intact to the generalized form of a composition series.

2.7. The Radical

We conclude this chapter with a characterization of finitely generated semisimple modules.

Definition. Let M be an A-module. The *radical* of M is $\operatorname{rad} M = \bigcap\{N \in \mathbf{S}(M) : M/N \text{ is simple}\}$.

Of course, there might not be any submodule N of M such that M/N is simple. In this case, $\operatorname{rad} M$ is the intersection of the empty subset of $\mathbf{S}(M)$ which is M by convention.

Lemma a. *Let M be an A-module.*

(i) rad M *is a submodule of* M.
(ii) *If* $N \in \mathbf{S}(M)$, *then* $\operatorname{rad} M/N = 0$ *implies* $N \supseteq \operatorname{rad} M$.
(iii) $\operatorname{rad}(M/\operatorname{rad} M) = 0$.

These observations are routine consequences of the Correspondence Theorem.

Lemma b. *If M is a semisimple A-module, then* $\operatorname{rad} M = 0$.

PROOF. Let $M = \bigoplus_{i \in J} P_i$ with each P_i simple. Denote $N_j = \sum_{i \neq j} P_i \in \mathbf{S}(M)$. Then $M/N_j \cong P_j$ is simple, so that $\operatorname{rad} M \subseteq \bigcap_{j \in J} N_j = 0$. □

Proposition. *An A-module M is finitely generated and semisimple if and only if M is Artinian and* $\operatorname{rad} M = 0$.

PROOF. If M is finitely generated and semisimple, then M is Artinian by Proposition 2.6, and $\operatorname{rad} M = 0$ by Lemma b. Conversely, assume that M is Artinian and $\operatorname{rad} M = 0$. We can also suppose that $M \neq 0$. Since $\operatorname{rad} M = 0$, there is a non-empty set $\{N_i : i \in J\} \subseteq \mathbf{S}(M)$ such that M/N_i is simple for all $i \in J$, and $\bigcap_{i \in J} N_i = 0$. The Artinian property of M guarantees the existence of a module $N_1 \cap \cdots \cap N_m$ that is minimal in the family $\{N_{i_1} \cap \cdots \cap N_{i_k} : i_1, \ldots, i_k \in J\}$. Necessarily, $N_1 \cap \cdots \cap N_m = 0$; otherwise $N_1 \cap \cdots \cap N_m \nsubseteq N_i$ for some $i \in J$, which gives the contradiction $N_1 \cap \cdots \cap N_m \cap N_i \subset N_1 \cap \cdots \cap N_m$ to the minimality of $N_1 \cap \cdots \cap N_m$. Define $\phi: M \to (M/N_1) \bigoplus \cdots \bigoplus (M/N_m)$ by $\phi(u) = (u + N_1, \ldots, u + N_m)$. Plainly, ϕ is an A-module homomorphism with kernel $N_1 \cap \cdots \cap N_m = 0$. Consequently, M is isomorphic to a submodule of the semisimple module $M/N_1 \bigoplus \cdots \bigoplus M/N_n$, so that M is semisimple by Corollary 2.4a. By Proposition 2.6, M is also finitely generated. □

EXERCISES

1. Let M be a right A-module. The *socle* of M is

$$\operatorname{soc} M = \sum \{N \in \mathbf{S}(M) : N \text{ is simple}\}.$$

Prove the following facts.
   (a) $\operatorname{soc} M$ is a semisimple submodule of M.
   (b) If N is a semisimple submodule of M, then $N \subseteq \operatorname{soc} M$.
   (c) M is semisimple if and only if $\operatorname{soc} M = M$.
   (d) $\operatorname{soc}(\operatorname{soc} M) = \operatorname{soc} M$.
   (e) If $\operatorname{soc} M = M$, then $\operatorname{rad} M = 0$.
   (f) If $M \neq 0$ and M is Artinian, then $\operatorname{soc} M \neq 0$.
   (g) If M is Artinian, and $\phi: M \to N$ is a module homomorphism such that $\phi|\operatorname{soc} M$ is injective, then ϕ is injective.

2. Prove the following implications for $\mathbb{Z}$-modules.
   (a) $M = \mathbb{Z}$ implies $\operatorname{rad} M = \operatorname{soc} M = 0$.
   (b) $M = \mathbb{Q}$ implies $\operatorname{rad} M = M$ and $\operatorname{soc} M = 0$.
   (c) $M = \mathbb{Q}/\mathbb{Z}$ implies $\operatorname{rad} M = M$ and $0 \neq \operatorname{soc} M \neq M$.
   (d) $M = \{a/n : a \in \mathbb{Z}, n \in \mathbb{N}, 2 \nmid n\}$ implies $\operatorname{rad} M = 2M$ and $\operatorname{soc} M = 0$.

3. Let A be a finite dimensional F-algebra, and suppose that M is a finitely generated right (left) A-module. Define the *dual module* $M^\wedge$ of M to be $\operatorname{Hom}_F(M,F)$.
   (a) Prove that $M^\wedge$ is a finitely generated left (right) A-module with the scalar operations given by $(x\phi)(u) = \phi(ux)$ (respectively, $(\phi x)(u) = \phi(xu)$).

(b) Prove that $(M^{\wedge})^{\wedge} \cong M$.

(c) Prove that $P \mapsto P^{\sim} = \{\phi \in M^{\wedge} : \phi(P) = 0\}$ is an inclusion reversing bijective mapping from $\mathbf{S}(M)$ to $\mathbf{S}(M^{\wedge})$.

(d) Prove that $(\operatorname{soc} M)^{\sim} = \operatorname{rad}(M^{\wedge})$ and $(\operatorname{rad} M)^{\sim} = \operatorname{soc}(M^{\wedge})$, where $^{\sim}$ is defined in (c).

(e) Prove that if M and N are finitely generated right (left) A-modules, then $(M \oplus N)^{\wedge} \cong M^{\wedge} \oplus N^{\wedge}$.

Notes on Chapter 2

The material of this chapter is well described as "standard algebra." Our terminology and general viewpoint is borrowed from the book [21] of Cartan and Eilenberg. In particular, the fundamental Proposition 2.4 appears as I.4.1 in [21].

CHAPTER 3
The Structure of Semisimple Algebras

This chapter focuses on one of the early monuments of algebra: the Wedderburn structure theorem for semisimple algebras. Most of the other results in the chapter are preliminaries to the proof of Wedderburn's theorem. It should be added, however, that much of this preparation leads to basic tools of algebra that are needed for work in many areas of mathematics. One other "name theorem" is proved in Section 6. This is Maschke's theorem, which shows that semisimple algebras arise naturally in the theory of finite groups.

Throughout most of the chapter, A is an R-algebra. The most interesting case occurs when R is a field, but this extra restriction brings little simplification of proofs or sharpening of results.

3.1. Semisimple Algebras

We are now ready to introduce one of the central concepts in the theory of associative algebras.

Definition. An R-algebra A is *semisimple* if A is semisimple as a right A-module.

In more detail, A is semisimple if $A = \bigoplus_{i \in J} N_i$, where each N_i is a simple right A-module. Since N_i is a submodule of A_A, this means that N_i is a minimal right ideal. Moreover, the indexing set J must be finite by Proposition 2.6, since A_A is finitely generated by 1_A.

The definition of semisimple algebras that is given above is biased toward right modules. A similar definition can be made using left modules, and at this point there is no reason to suppose that the classes of left semisimple algebras and right semisimple algebras coincide. They do, however, and this fact will be clear from the structure theorem of Wedderburn. Even without knowing that semisimplicity is symmetric, it is evident that the right and left handed theories will run in parallel: just interchange "right" and "left," and reverse the order of factors that occur in any formula. As a matter of fact, the equivalence between right and left theories can be proved rigorously by switching from A to its opposite algebra. For this reason, we will restrict our attention to algebras that are "right semisimple" in the sense of the above definition. The adjectives "right" and "left" will modify the term "semisimple" only when they are later needed to prove their own dispensibility.

We begin the discussion of semisimple algebras with a reformulation of our characterization of finitely generated semisimple modules. Another definition is needed. An algebra A is called *right Artinian* (*Noetherian*) if A_A is Artinian (Noetherian). That is, the lattice of right ideals of A satisfies the descending (ascending) chain condition. In contrast to semisimplicity, these properties are not right-left symmetric. (See Exercise 2.)

Proposition a. *An algebra A is semisimple if and only if A is right Artinian and* $\operatorname{rad} A_A = 0$.

Every algebra is finitely generated as a right or left module by the unity element, so that this result is a corollary of Proposition 2.7.

A finite dimensional algebra A over a field F is automatically Artinian since $\mathbf{S}(A_A)$ is a sublattice of $\mathbf{S}(A_F)$, and the finite dimensionality of A_F implies that $\mathbf{S}(A_F)$ satisfies the descending chain condition of Proposition 2.6. In this important case, A is semisimple if and only if the radical $\operatorname{rad} A_A$ is zero. We will prove later that $\operatorname{rad} A_A = \operatorname{rad} {}_A A$, so that this subset of A is an ideal—the Jacobson radical.

The theory of modules over semisimple algebras is reduced to considering the structure of the algebras, as our next result shows.

Proposition b. *If A is a semisimple algebra, then every A-module is semisimple. Moreover, the simple A-modules are isomorphic to minimal right ideals of A, and all minimal right ideals of A are simple A-modules.*

Proof. Every free right A-module is isomorphic to a direct sum of copies of A_A. Therefore, free A-modules are semisimple by Corollary 2.4b. Since any A-module is a homomorphic image of a free A-module, the first part of the proposition is a consequence of Corollary 2.4a. By Proposition 2.3, every simple right A-module is isomorphic to A_A/M, where M is a maximal right ideal. The semisimplicity of A_A guarantees that M has a complement N in $\mathbf{S}(A_A)$, according to Proposition 2.4. Since M is a maximal right ideal, N is a minimal right ideal; and $N \cong A_A/M$. Thus, the minimal right ideals of A represent all isomorphism classes of simple modules. □

Corollary. *If A is a semisimple algebra, then every homomorphic image of A is semisimple.*

Proof. Let $\theta: A \to B$ be a surjective algebra homomorphism. The observations that were made in Section 2.1 show that B can be viewed as a right A-module. Plainly, $\operatorname{ann} B_A = \ker \theta$. By the proposition, B_A is semisimple; say $B_A = \bigoplus_{i \in J} N_i$, where each N_i is a simple A-module. Since $\operatorname{ann} N_i \supseteq \operatorname{ann} B_A = \ker \theta$, it follows from Proposition 2.1 and Lemma 2.1a that N_i is a simple B-module. Therefore, B is semisimple. □

The class of semisimple algebras is also closed under finite products, as we will see in the next section.

Exercises

1. Prove that a commutative R-algebra A is semisimple if and only if A is a finite product of fields.

2. Let A be the set of all 2 by 2 matrices in $\begin{bmatrix} \mathbb{Q} & \mathbb{R} \\ 0 & \mathbb{R} \end{bmatrix}$, that is, all matrices $\begin{bmatrix} a & x \\ 0 & y \end{bmatrix}$ with $a \in \mathbb{Q}$; $x, y \in \mathbb{R}$.
   (a) Prove that A is a $\mathbb{Q}$-subalgebra of $M_2(\mathbb{R})$.
   (b) Prove that if V is a $\mathbb{Q}$-subspace of $\mathbb{R}$, then $\begin{bmatrix} 0 & V \\ 0 & 0 \end{bmatrix}$ is a left ideal of A.
   (c) Deduce from (b) that A is neither left Artinian nor left Noetherian.
   (d) Prove that every right ideal of A has one of the following forms: 0, A, $\begin{bmatrix} \mathbb{Q} & \mathbb{R} \\ 0 & 0 \end{bmatrix}$, $\begin{bmatrix} 0 & \mathbb{R} \\ 0 & \mathbb{R} \end{bmatrix}$, or $\begin{bmatrix} 0 & x \\ 0 & y \end{bmatrix}\mathbb{R}$ for some fixed $(x, y) \in \mathbb{R}^2 - \{(0,0)\}$.
   (e) Deduce from (d) that A is both right Artinian and right Noetherian.

3. (a) Let A be an R-algebra, and $n \geq 1$. Prove that $M_n(A)$ is right Artinian (Noetherian) if and only if A is right Artinian (Noetherian). Hint. Show that $I \mapsto \varepsilon_{11} I$ and $N \mapsto \sum_{i=1}^{n} \varepsilon_{i1} N$ are inverse, order preserving correspondences between the right ideals of $M_n(A)$ and the submodules of $\varepsilon_{11} M_n(A)$.
   (b) Deduce that if R is an integral domain that is not a field, then $M_n(R)$ is not semisimple.

3.2. Minimal Right Ideals

Wedderburn's structure theorem for semisimple algebras is derived from Schur's Lemma, plus some information on minimal right ideals. This section collects some facts about these ideals.

Lemma a. *Let $A = A_1 \dotplus A_2$ be the inner product of algebras A_1 and A_2. Suppose that M is a right ideal of A_1.*

(i) *M is a right ideal of A.*
(ii) $\mathbf{E}_A(M) = \mathbf{E}_{A_1}(M)$.
(iii) $\mathrm{Hom}_A(M,A) = \mathrm{Hom}_A(M,A_1) = \mathrm{Hom}_{A_1}(M,A_1)$.
(iv) $\mathbf{S}(M_{A_1}) = \mathbf{S}(M_A)$.
(v) *M is a minimal right ideal of A if and only if M is a minimal right ideal of A_1.*
(vi) *Every minimal right ideal of A is a minimal right ideal of A_1 or a minimal right ideal of A_2.*

PROOF. The statements (i) to (v) are obtained easily from the observation that $MA_2 \subseteq A_1A_2 = 0$. The details of the argument are left for Exercise 1. Suppose that N is a minimal right ideal of A. For $i = 1, 2$, $NA_i \subseteq N \cap A_i < N$, so that either $N = N \cap A_i \subseteq A_i$, or $NA_i = N \cap A_i = 0$.It cannot be the case that $NA_1 = NA_2 = 0$, because otherwise $N = NA_1 + NA_2 = 0$. □

Corollary a. *If A_1 and A_2 are semisimple algebras, then $A_1 \dot{+} A_2$ is semisimple.*

Thus, the class of semisimple algebras is closed under finite products.

Lemma b. *If N is a minimal right ideal of the algebra A, and $x \in A$, then either $xN = 0$, or xN is a minimal right ideal of A such that $xN \cong N$ as A-modules.*

PROOF. The mapping $y \mapsto xy$ is a surjective A-module homomorphism from N to xN, so that the assertion is a consequence of Schur's Lemma. □

Lemma c. *Let N be a right ideal of the algebra A that satisfies $N^k = 0$. If P is a simple A-module, then $PN = 0$. Moreover, $N \subseteq \mathrm{rad}\, A_A$.*

PROOF. Since P is simple and $PN < P$, either $PN = 0$ or $PN = N$. The second option is impossible, because it leads to the contradiction $P = PN = PN^2 = \cdots = PN^k = 0$. In particular, if M is a maximal right ideal of A, then $(A_A/M)N = 0$, that is, $N \subseteq M$. Therefore, $N \subseteq \mathrm{rad}\, A_A$, the intersection of all maximal right ideals of A. □

Proposition. *The following conditions are equivalent for minimal right ideals N_1 and N_2 of the semisimple algebra A.*

(i) *$N_1 \cong N_2$ as A-modules.*
(ii) $N_1N_2 \neq 0$.
(iii) *There is an element $x \in A$ such that $N_1 = xN_2$.*

PROOF. If $\phi\colon N_1 \to N_2$ is an A-module isomorphism, then $\phi(N_1N_2) = \phi(N_1)N_2 = N_2^2 \neq 0$ by Lemma c. Thus, $N_1N_2 \neq 0$. Suppose that $x \in N_1$ is such that $xN_2 \neq 0$. Since N_1 is simple and xN_2 is a non-zero submodule of N_1, it follows that $N_1 = xN_2$. Finally, (i) follows from (iii) by Lemma b. □

Lemma d. *Suppose that A is a semisimple algebra, and $A_A = N_1 \oplus \cdots \oplus N_m$ with each N_i a minimal right ideal of A. If N is a minimal right ideal of A, then $N \cong N_i$ for some i with $1 \leq i \leq m$.*

Proof. Since A_A is semisimple, it follows from Proposition 2.4 that $A_A = N \oplus M$ for a suitable right ideal M. By Corollary 2.4a, M is also a semisimple A-module. The conclusion that $N \cong N_i$ for some i then follows from Proposition 2.5 (or from the Jordan–Hölder Theorem in this case). □

Corollary b. *If A is a semisimple algebra, then the number of isomorphism classes of simple A-modules is finite.*

This corollary is a consequence of Proposition 3.1b and Lemma d.

Exercises

1. Prove the statements (i) through (v) in Lemma a.
2. Let A be a semisimple algebra, and suppose that $A = A_1 \dotplus \cdots \dotplus A_r$, where each of the algebras A_i is simple. Prove that if N_i is a minimal right ideal of A_i and N_j is a minimal right ideal of A_j with $i \neq j$, then $\mathrm{Hom}_A(N_i,N_j) = 0$; in particular, $N_i \ncong N_j$.
3. An element e of an algebra A is *idempotent* if $e^2 = e$. Prove the following statements.
   (a) A is semisimple if and only if every right ideal of A has the form eA for some idempotent element $e \in A$.
   (b) If A is semisimple and $M < A_A$, then $M^2 = M$.
   (c) If A is semisimple and I is an ideal of A, then there is a unique *central idempotent* e (that is, $e \in \mathbf{Z}(A)$) such that $I = eA$. Conversely, if e is a central idempotent, then eA is an ideal of A.
   (d) If $1 = e_1 + \cdots + e_r$ with each e_i a central idempotent, and if $e_ie_j = 0$ for $i \neq j$, then $A = e_1A \dotplus \cdots \dotplus e_rA$.

3.3. Simple Algebras

The term "semisimple algebra" suggests a generalization of simple algebras, but in fact not all simple algebras are semisimple. (See Exercises 1 and 5.) In this section we will characterize the simple algebras that are semisimple and the semisimple algebras that are simple.

An algebra A is *simple* if $A \neq 0$ and $\mathbf{I}(A) = \{0,A\}$.

Proposition a. *For a simple algebra A, the following conditions are equivalent:*

(i) *A is semisimple;*
(ii) *A is right Artinian;*
(iii) *A has a minimal right ideal.*

Proof. (i) implies (ii) by Proposition 2.7, and it is evident that (ii) implies (iii). Assume that N is a minimal right ideal of A. Then $AN = \sum_{x \in A} xN$ is a non-zero ideal of A, so that $A = \sum_{x \in A} xN$ because A is simple. By Lemma 3.2b, each non-zero xN is a simple right A-module. Therefore, A is semisimple by Proposition 2.4. □

Proposition b. *For a semisimple algebra A, the following conditions are equivalent:*

(i) *A is simple;*
(ii) *all minimal right ideals of A are isomorphic;*
(iii) *all simple right A-modules are isomorphic.*

Proof. By Proposition 3.1b, every simple right A-module is isomorphic to a minimal right ideal of A, so that (ii) and (iii) are equivalent. Suppose that A is simple, and that N_1 and N_2 are minimal right ideals of A. Then $AN_1 = AN_2 = A$, as in the proof of Proposition a. Hence, $A(N_1 N_2) = (AN_1)N_2 = AN_2 = A$. In particular, $N_1 N_2 \neq 0$, so that $N_1 \cong N_2$ by Proposition 3.2. Conversely, assume that all of the minimal right ideals of A are isomorphic. Let J be a non-zero ideal of A. Since A is semisimple, there is a minimal right ideal N of A such that $N \subseteq J$. The assumption that all minimal right ideals of A are isomorphic, together with Proposition 3.2 and Proposition 2.4, yields $A = \sum\{xN: x \in A\} \subseteq J$. Therefore, A is simple. □

Corollary a. *Let A be a simple algebra, and suppose that N is a minimal right ideal of A. If M is a right A-module, then there is a unique cardinal number α such that $M \cong \bigoplus \alpha N$.*

This corollary follows directly from Propositions a and b, and Proposition 2.5.

Corollary b. *Let A be a finite dimensional, simple F-algebra, and suppose that M_1 and M_2 are right A-modules. Then $M_1 \cong M_2$ if and only if $\dim_F M_1 = \dim_F M_2$.*

Proof. Since A is finite dimensional, it is Artinian, and there is a minimal right ideal N of A with $\dim_F N$ finite. By Corollary a, $M_1 \cong \bigoplus \alpha N$ and $M_2 \cong \bigoplus \beta N$ for unique cardinal numbers α and β. Plainly, $M_1 \cong M_2$ if and only if $\alpha = \beta$; and since $\dim_F N < \infty$, $\alpha = \beta$ is equivalent to $\dim_F M_1 = \alpha \dim_F N = \beta \dim_F N = \dim_F M_2$. □

Example. Let $A = M_n(D)$ be the algebra of n by n matrices with entries in a division algebra D. For $1 \leq i \leq n$, define $N_i = \varepsilon_{ii} A$.
(i) N_i is a minimal right ideal of A.
(ii) $A_A = \bigoplus_{i=1}^n N_i$.

(iii) $N_i \cong N_j$ for all i and j.
(iv) A_A is simple and semisimple.
(v) $\mathbf{E}_A(N_i) \cong D$.

PROOF. Plainly, N_i is a right ideal of A. If $\alpha = \sum_{j,k} \varepsilon_{jk} z_{jk}$ with $z_{jk} \in D$, then $\varepsilon_{ii}\alpha = \sum_k \varepsilon_{ik} z_{ik}$. Hence, $N_i = \sum_k \varepsilon_{ik} D = \bigoplus_k \varepsilon_{ik} D$, and $A = \bigoplus_{i,k} \varepsilon_{ik} D = \bigoplus_i N_i$ as D-modules. If $\beta = \sum_k \varepsilon_{ik} z_k \in N_i$ with some $z_j \neq 0$, then $\beta(\sum_l \varepsilon_{jl} z_j^{-1} w_l) = \sum_l \varepsilon_{il} w_l$ for arbitrary $w_l \in D$. That is, if $0 \neq \beta \in N_i$, then $\beta A = N_i$, so that N_i is simple by Proposition 2.3. Moreover, $N_j = \varepsilon_{jj} A = \varepsilon_{ji}\varepsilon_{ij} A = \varepsilon_{ji} N_i$; hence, $N_j \cong N_i$ by Proposition 3.2. It follows from (i), (ii), (iii) and Lemma 3.2d that A_A is semisimple and its minimal right ideals are isomorphic. Therefore, A is simple by Proposition b. (This fact was proved more directly in Section 1.4.) For $z \in D$, the left multiplication mapping $\lambda_z \alpha = z\alpha$ is an A-module endomorphism of N_i, and $z \mapsto \lambda_z$ is an injective homomorphism from D to $\mathbf{E}_A(N_i)$. If $\phi \in \mathbf{E}_A(N_i)$, say $\phi(\varepsilon_{ii}) = \varepsilon_{ii}\beta$, then $\phi(\varepsilon_{ii}) = \phi(\varepsilon_{ii}^2) = \phi(\varepsilon_{ii})\varepsilon_{ii} = \varepsilon_{ii}\beta\varepsilon_{ii} = z\varepsilon_{ii}$ for some $z \in D$. Consequently, if $\alpha \in N_i$, then $\phi(\alpha) = \phi(\varepsilon_{ii}\alpha) = \phi(\varepsilon_{ii})\alpha = z\varepsilon_{ii}\alpha = z\alpha = \lambda_z\alpha$; that is, $\phi = \lambda_z$. Therefore, $D \cong \mathbf{E}_A(N_i)$. □

EXERCISES

1. Let V be an F-space that is countably infinite dimensional. Define $A = \mathbf{E}_F(V)$. Prove the following statements.

   (a) $I = \{\phi \in A \colon \dim \phi V < \infty\}$ is a maximal proper ideal of A, so that $B = A/I$ is a simple F-algebra.

   (b) B is neither right Artinian nor left Noetherian. Hint. Let $V = V_0 \supset V_1 \supset V_2 \supset \cdots$ be an infinite descending chain of subspaces such that $\dim_F V_i/V_{i+1} = \infty$. Denote $M_i = \{\phi \in A \colon \phi(V) \subseteq V_i\}$ and $N_i = \{\phi \in A \colon \dim \phi(V_i) < \infty\} \supseteq I$. Show that $(M_0 + I)/I \supset (M_1 + I)/I \supset (M_2 + I)/I \supset \cdots$ is an infinite descending chain of right ideals in B, and $N_0/I \subset N_1/I \subset N_2/I \subset \cdots$ is an infinite ascending chain of left ideals in B.

   (c) B is neither left Artinian nor right Noetherian.

2. Let A be an R-algebra, and $M < A_A$. Prove the following statements.

   (a) If $B = \{x \in A \colon xM \subseteq M\}$, then B is a subalgebra of A such that $M \lhd B$ and $\mathbf{E}_A(A_A/M\} \cong B/M$.

   (b) If $M = eA$, where e is idempotent, then $\mathbf{E}_A(M) \cong eAe$.

   (c) Use (b) to give a new proof of part (v) in the example.

3. A non-zero idempotent element e of the R-algebra A is called primitive if e cannot be written in the form $e = e_1 + e_2$ with e_1 and e_2 non-zero idempotents such that $e_1e_2 = e_2e_1 = 0$. Prove that if A is semisimple, then the following conditions on an idempotent element $e \in A$ are equivalent.

   (i) e is primitive.
   (ii) eA is a minimal right ideal of A.
   (iii) eAe is a division algebra.

4. Let F be a field. Denote $A = F \oplus F \oplus F$. Define a bilinear multiplication $A \times A \to A$ by $(a_1,b_1,c_1)\,(a_2,b_2,c_2) = (a_1a_2, b_1b_2, a_1c_2 + c_1b_2)$. Prove the following facts.
   (a) A is an F-algebra.
   (b) $e = (1,0,0)$ is an idempotent element of A.
   (c) The right ideal eA is not simple.
   (d) $\mathbf{E}_A(eA) \cong F$.
   Thus, the converse part of Schur's Lemma can fail even for finite dimensional A-modules, where A is a finite dimensional F-algebra.

5. The following example is called the *algebra of quantum mechanics*. It is simple and Noetherian, but not Artinian. In addition, it is a non-commutative domain. Let $V = \bigoplus_{n<\omega} u_n\mathbb{R}$. Define $\xi, \eta \in \mathbf{E}_{\mathbb{R}}(V)$ by $\xi(u_n) = u_{n+1}\sqrt{n+1}$ for $n < \omega$, $\eta(u_n) = 0$ if $n = 0$, and $\eta(u_n) = u_{n-1}\sqrt{n}$ if $n > 0$. Let A be the subalgebra of $\mathbf{E}_{\mathbb{R}}(V)$ that is generated by ξ and η. Prove the following facts.
   (a) $\eta\xi - \xi\eta = 1_A (= id_V)$.
   (b) $\eta\xi^n - \xi^n\eta = \xi^{n-1}n$, $\eta^n\xi - \xi\eta^n = \eta^{n-1}n$.
   (c) $\eta^n\xi^m = \sum_{j\geq 0}\xi^{m-j}\eta^{n-j}\binom{m}{j}\binom{n}{j}j!$ (where $\binom{m}{j} = 0$ for $j > m$, $\binom{n}{j} = 0$ for $j > n$).
   (d) As an $\mathbb{R}$-space, $A = \bigoplus_{m,n<\omega}\xi^m\eta^n\mathbb{R}$.
   For $0 \neq \phi = \sum \xi^m\eta^n a_{mn}$, define the (total) degree of ϕ to be $\operatorname{Deg}\phi = \max\{m + n : a_{mn} \neq 0\}$.
   (e) If ϕ, ψ, and $\phi + \psi \neq 0$, then $\operatorname{Deg}(\phi + \psi) \leq \max\{\operatorname{Deg}\phi, \operatorname{Deg}\psi\}$; if $\phi \neq 0$ and $\psi \neq 0$, then $\phi\psi \neq 0$, and $\operatorname{Deg}\phi\psi = \operatorname{Deg}\phi + \operatorname{Deg}\psi$.
   For $\phi \in A$, denote $\partial_\xi\phi = \eta\phi - \phi\eta$, $\partial_\eta\phi = \phi\xi - \xi\phi$.
   (f) ∂_ξ and ∂_η are linear transformations of A such that $\partial_\xi(\xi^m\eta^n) = \xi^{m-1}\eta^n m$ and $\partial_\eta(\xi^m\eta^n) = \xi^m\eta^{n-1}n$.
   (g) A is simple. Hint. If I is an ideal of A, then $\partial_\xi I \subseteq I$ and $\partial_\eta I \subseteq I$. If $\phi = \sum \xi^i\eta^j a_{ij}$ has degree $r + s$, then $\partial_\xi^r\partial_\eta^s\phi = r!s!a_{rs}$.
   Write $A = \bigoplus_{n<\omega}\mathbb{R}[\xi]\eta^n = \bigcup_{n<\omega}A_n$, where $A_n = \sum_{m\leq n}\mathbb{R}[\xi]\eta^m$. For $0 \neq \phi \in A$, define $\deg_\eta\phi = \min\{n : \phi \in A_n\}$. Thus, $\deg_\eta\phi = n$ if and only if $\phi = \pi\eta^n + \psi$, where $0 \neq \pi \in \mathbb{R}[\xi]$ and $\psi = 0$ or $\deg_\eta\psi < n$.
   (h) If $\pi \in \mathbb{R}[\xi]$, then $\eta^n\pi = \pi\eta^n + \psi$, where $\psi = 0$ or $\deg_\eta\psi < n$.
   (i) A is right Noetherian. Hint. Copy the proof that the ring of polynomials with coefficients in a Noetherian domain is Noetherian.

3.4. Matrices of Homomorphisms

In preparation for the proof of the Wedderburn theorem, we introduce a useful generalized matrix notation.

Let A be an R-algebra, and suppose that $(M_1, M_2, \ldots, M_n)$ is a sequence of right A-modules. Denote by

$$[\operatorname{Hom}_A(M_j,M_i)] = \begin{bmatrix} \operatorname{Hom}_A(M_1,M_1), \operatorname{Hom}_A(M_2,M_1), \ldots, \operatorname{Hom}_A(M_n,M_1) \\ \operatorname{Hom}_A(M_1,M_2), \operatorname{Hom}_A(M_2,M_2), \ldots, \operatorname{Hom}_A(M_n,M_2) \\ \cdots \\ \operatorname{Hom}_A(M_1,M_n), \operatorname{Hom}_A(M_2,M_n), \ldots, \operatorname{Hom}_A(M_n,M_n) \end{bmatrix}$$

the set of all n by n matrices

$$\begin{bmatrix} \phi_{11} & \phi_{12} & \cdots & \phi_{1n} \\ \phi_{21} & \phi_{22} & \cdots & \phi_{2n} \\ & & \cdots & \\ \phi_{n1} & \phi_{n2} & \cdots & \phi_{nn} \end{bmatrix},$$

in which $\phi_{ij} \in \mathrm{Hom}_A(M_j,M_i)$.

Define addition and scalar multiplication componentwise in

$$[\mathrm{Hom}_A(M_i,M_j)].$$

Define multiplication in $[\mathrm{Hom}_A(M_i,M_j)]$ by the usual rule of matrix multiplication: $[\phi_{ij}][\psi_{jk}] = [\chi_{ik}]$, where $\chi_{ik} = \sum_{j=1}^n \phi_{ij}\psi_{jk} \in \mathrm{Hom}_A(M_k,M_i)$.

Proposition. $[\mathrm{Hom}_A(M_j,M_i)]$ *is an R-algebra that is isomorphic to*

$$\mathbf{E}_A(M_1 \oplus M_2 \oplus \cdots \oplus M_n).$$

PROOF. Denote the direct sum $M_1 \oplus M_2 \oplus \cdots \oplus M_n$ by M. Let π_j: $M \to M_j$ and κ_j: $M_j \to M$ be the projection and injection homomorphisms associated with this direct sum. Then

$$\textstyle\sum_{j=1}^n \kappa_j\pi_j = id_M, \text{ and} \tag{1}$$

$$\pi_i\kappa_j = 0 \text{ when } i \neq j;\ \pi_j\kappa_j = id_{M_j}. \tag{2}$$

Define mappings α: $\mathbf{E}_A(M) \to [\mathrm{Hom}_A(M_j,M_i)]$ and β: $[\mathrm{Hom}_A(M_j,M_i)] \to \mathbf{E}_A(M)$ by $\alpha(\phi) = [\pi_i\phi\kappa_j]$ and $\beta([\phi_{ij}]) = \sum_{i,j=1}^n \kappa_i\phi_{ij}\pi_j$. A straightforward calculation using (1) and (2) shows that $\beta\alpha$ is the identity mapping on $\mathbf{E}_A(M)$ and $\alpha\beta$ is the identity on $[\mathrm{Hom}_A(M_j,M_i)]$. Moreover, α is an R-algebra homomorphism. For example, $\alpha(\phi\psi) = [\pi_i\phi\psi\kappa_k] = [\pi_i\phi(\sum_{j=1}^n \kappa_j\pi_j)\psi\kappa_k] = [\sum_{j=1}^n (\pi_i\phi\kappa_j)(\pi_j\psi\kappa_k)] = \alpha(\phi)\alpha(\psi)$ by (1). Thus, α is an isomorphism. □

Corollary a. *If A is an R-algebra, and M is a right A-module, then*

$$\mathbf{E}_A(\textstyle\bigoplus nM) \cong M_n(\mathbf{E}_A(M)).$$

Corollary b. *If A is an R-algebra, and M is the free right A-module on n generators, then* $\mathbf{E}_A(M) \cong M_n(A)$.

PROOF. Since $M \cong \bigoplus_n A_A$, Corollary a yields $\mathbf{E}_A(M) \cong M_n(\mathbf{E}_A(A_A))$; and $\mathbf{E}_A(A_A) \cong A$ by Proposition 1.3. □

Corollary c. *If A is an R-algebra, and if* $M_1, M_2, \ldots, M_n$ *are right A-modules such that* $\mathrm{Hom}_A(M_i,M_j) = 0$ *if* $i \neq j$, *then* $\mathbf{E}_A(\bigoplus_{i=1}^n M_i) \cong \mathbf{E}_A(M_1) \dotplus \mathbf{E}_A(M_2) \dotplus \cdots \dotplus \mathbf{E}_A(M_n)$.

PROOF. By the proposition, $\mathbf{E}_A(\bigoplus_{i=1}^n M_i)$ is isomorphic to

$$\begin{bmatrix} \mathrm{Hom}_A(M_1,M_1) & \mathrm{Hom}_A(M_2,M_1) & \cdots & \mathrm{Hom}_A(M_n,M_1) \\ \mathrm{Hom}_A(M_1,M_2) & \mathrm{Hom}_A(M_2,M_2) & \cdots & \mathrm{Hom}_A(M_n,M_2) \\ & \cdots & & \\ \mathrm{Hom}_A(M_1,M_n) & \mathrm{Hom}_A(M_2,M_n) & \cdots & \mathrm{Hom}_A(M_n,M_n) \end{bmatrix} =$$

$$\begin{bmatrix} \mathbf{E}_A(M_1) & & & \\ & \mathbf{E}_A(M_2) & & \\ & & \ddots & \\ & & & \mathbf{E}_A(M_n) \end{bmatrix} \cong \mathbf{E}_A(M_1) \dotplus \mathbf{E}_A(M_2) \dotplus \cdots \dotplus \mathbf{E}_A(M_n).$$

□

EXERCISE

Let $M = M_1 \oplus \cdots \oplus M_n$ and $N = N_1 \oplus \cdots \oplus N_m$ be A-modules. Define $[\mathrm{Hom}_A(M_j,N_i)]$ to be the set of all m by n matrices $[\phi_{ij}]$ with $\phi_{ij} \in \mathrm{Hom}_A(M_j,N_i)$. Define addition and scalar multiplication componentwise in $[\mathrm{Hom}_A(M_j,N_i)]$. Prove that $[\mathrm{Hom}_A(M_j,N_i)] \cong \mathrm{Hom}_A(M,N)$. Use this result to deduce that if $\mathrm{Hom}_A(P,Q) = 0$, then $\mathrm{Hom}_A(\bigoplus nP, \bigoplus mQ) = 0$ for all $m, n \in \mathbb{N}$.

3.5. Wedderburn's Structure Theorem

We are ready to assemble the parts of the main result of this chapter.

Theorem. *Let A be a right or left semisimple R-algebra.*

(i) *There exist natural numbers $n_1, \ldots, n_r$ and R-division algebras $D_1, \ldots, D_r$ such that*

$$A \cong M_{n_1}(D_1) \dotplus \cdots \dotplus M_{n_r}(D_r). \tag{1}$$

(ii) *The pairs $(n_1,D_1), \ldots, (n_r,D_r)$ for which* (1) *is satisfied are uniquely determined* (*to isomorphism*) *by A.*

(iii) *Conversely, if $n_1, \ldots, n_r \in \mathbb{N}$ and $D_1, \ldots, D_r$ are division algebras over R, then $M_{n_1}(D_1) \dotplus \cdots \dotplus M_{n_r}(D_r)$ is a right and left semisimple R-algebra.*

PROOF. (i) If A is right semisimple, then $A_A \cong M_1 \oplus \cdots \oplus M_r$, where M_i is a direct sum of n_i copies of a minimal right ideal N_i of A, chosen so that N_i is not isomorphic to N_j if $i \neq j$. By Lemma 2.5c, $\mathrm{Hom}_A(M_i,M_j) = 0$ if $i \neq j$. The isomorphism (1) follows from Proposition 1.3, Corollary 3.4c, and Corollary 3.4a: $A \cong \mathbf{E}_A(A_A) \cong \mathbf{E}_A(M_1) \dotplus \cdots \dotplus \mathbf{E}_A(M_r) \cong M_{n_1}(D_1) \dotplus \cdots \dotplus M_{n_r}(D_r)$, with $D_i = \mathbf{E}_A(N_i)$ a division algebra over R by Schur's

Lemma. The same result is obtained for left semisimple algebras by using minimal left ideals and right handed endomorphisms.

(ii) Assume that $A = A_1 \dotplus \cdots \dotplus A_s$, where $A_i \cong M_{k_i}(C_i)$ and C_i is a division algebra over R. By Example 3.3 and Lemma 3.2a, A_i is isomorphic as an A-module to a direct sum of k_i copies of a minimal right ideal P_i of A, with $P_i \subseteq A_i$, and $C_i \cong \mathbf{E}_{A_i}(P_i) = \mathbf{E}_A(P_i)$. Since each A_i is an ideal of A that contains P_i, Proposition 3.2 implies that $P_i \ncong P_j$ if $i \neq j$. The uniqueness of the decompositions of direct sums of simple modules that was proved in Proposition 2.5 gives the desired conclusions $s = r$ and (for a suitable ordering) $k_i = n_i$, $P_i \cong N_i$ as A-modules, and $C_i \cong \mathbf{E}_A(P_i) \cong \mathbf{E}_A(N_i) = D_i$.

(iii) By Example 3.3 and its left analog, each of the R-algebras $M_{n_i}(D_i)$ is right and left semisimple. Therefore, $M_{n_1}(D_1) \dotplus \cdots \dotplus M_{n_r}(D_r)$ is right and left semisimple by Corollary 3.2a. □

Parts (i) and (iii) of the theorem fulfill our promise to show that the classes of right semisimple and left semi-simple algebras coincide. Moreover, for simple algebras, the right and left descending chain conditions are equivalent.

Corollary a. *A right or left Artinian algebra A is simple if and only if $A \cong M_n(D)$ for a natural number n and a division algebra D. In this case, A determines n uniquely and D to within isomorphism.*

For some fields F, the only finite dimensional division algebras over F are commutative. In this case, the structure theorem gives somewhat sharper conclusions about finite dimensional semisimple algebras: the D_i are necessarily fields. The optimum result is obtained when F is algebraically closed.

Lemma. *Let F be an algebraically closed field. If D is a finite dimensional division algebra over F, then $D = F$.*

PROOF. Let $\dim_F D = m$. If $x \in D$, then the sequence $1, x, \ldots, x^m$ is linearly dependent, so that there is a monic polynomial $\Phi \in F[\mathbf{x}]$ of minimal degree such that $\Phi(x) = 0$. Since D is a division algebra, the fact that the degree of Φ is minimal implies that Φ is irreducible over F. Consequently, $\Phi = \mathbf{x} - a$ for some $a \in F$, because F is algebraically closed. That is, $x = a \in F$. □

Corollary b. *Let F be an algebraically closed field. A finite dimensional F-algebra A is semisimple if and only if $A \cong M_{n_1}(F) \dotplus \cdots \dotplus M_{n_r}(F)$, where $1 \leq n_1 \leq \cdots \leq n_r$ are uniquely determined by the isomorphism type of A. Moreover, A is simple if and only if $A \cong M_n(F)$, where $\dim_F A = n^2$.*

This corollary is a consequence of the lemma and the structure theorem.

EXERCISES

1. Let $A = A_1 \dotplus \cdots \dotplus A_r$ be a semisimple algebra with each A_i simple. Prove that the ideals of A are the sums $\sum_{i \in J} A_i$, where $J \subseteq \{1,2, \ldots ,r\}$.

2. Let $A = A_1 \dotplus \cdots \dotplus A_r$ be a semisimple algebra with each A_i simple. For each index i, let e_i be a central idempotent element of A such that $A_i = e_i A$. (See Exercise 3, Section 3.2.) For a right A-module M, denote $M_i = Me_i$. Prove the following statements.
   (a) If $i \neq j$, then $e_i e_j = 0$.
   (b) $M_i < M$, and $M = M_l \oplus \cdots \oplus M_r$.
   (c) ann $M_i \supseteq A_j$ for all $j \neq i$.
   (d) M is a faithful A-module if and only if $M_i \neq 0$ for $1 \leq i \leq r$.
   (e) M is a faithful A-module if and only if every minimal right ideal of A is isomorphic to a direct summand of M.

3. Prove that if A is a semisimple algebra and M is a finitely generated A-module, then $\mathbf{E}_A(M)$ is semisimple.

3.6. Maschke's Theorem

The Wedderburn Structure Theorem is an internal characterization of semisimple algebras. The theory of the radical that will be developed in Chapters 4 and 11 shows how important semisimple algebras are for the investigation of general algebras. In this section it will be shown that semisimple algebras are also encountered in a natural setting. They provide the foundation of the classical theory of group representations.

Maschke's Theorem. *Let G be a finite group, and suppose that F is a field. The group algebra FG is semisimple if and only if the characteristic of F does not divide the order of G.*

PROOF. Suppose that char F does not divide $n = |G|$. The crucial consequence of this hypothesis is that n (identified with $n \cdot 1 \in FG$) is a unit. By Proposition 2.4, it is sufficient to prove that if M is a right ideal of FG, then $FG = M \oplus N$ for some right ideal N of FG. This conclusion is obtained by proving that there is an FG-module homomorphism $\rho: FG \to M$ such that $\rho u = u$ for all $u \in M$. In fact, given such a ρ, $N = \operatorname{Ker} \rho = (1 - \rho)M$ is a right ideal such that $M \cap N = 0$ and $M + N = FG$. As a first approximation to ρ, we use the fact that M is a subspace of FG (with FG considered as a vector space over the field F) to get an F-space homomorphism $\pi: FG \to M$ satisfying $\pi u = u$ for all $u \in M$. Define ρ by "averaging" π over G. Explicitly, for $u \in FG$, let $\rho u = (\sum_{x \in G} \pi(ux)x^{-1})n^{-1}$. Clearly, ρ is an F-space homomorphism. However, more is true. For any $y \in G$, $\rho(uy) = (\sum_{x \in G} \pi(uyx)x^{-1})n^{-1} = ((\sum_{x \in G} \pi(u(yx))(yx)^{-1})y)n^{-1} = ((\sum_{yx \in yG = G} \pi(u(yx))(yx)^{-1})n^{-1})y = \rho(u)y$.

Therefore, ρ is an FG-module homomorphism. Finally, suppose that $u \in M$. Then $ux \in M$ for all $x \in G$ since M is a right ideal. Therefore, since $\pi|M = id_M$, we have $\rho u = (\sum_{x \in G} \pi(ux)x^{-1})n^{-1} = (\sum_{x \in G}(ux)x^{-1})n^{-1} = (\sum_{x \in G} u)n^{-1} = (un)n^{-1} = u$. Now consider the case in which the characteristic of F (which is necessarily a prime) divides the order n of G. This hypothesis has the consequence that the sum of n copies of an element of FG is zero. Let $e = \sum_{x \in G} x \in FG - \{0\}$. Then $ey = \sum_{x \in G} xy = \sum_{xy \in Gy = G} xy = e$ for all $y \in G$. Consequently, $e^2 = \sum_{y \in G} ey = \sum_{y \in G} e = en = 0$. Moreover, the right ideal N of FG that is generated by e coincides with eF. Hence, $N^2 = 0$. It follows from Lemma 3.2c that $0 \neq N \subseteq \operatorname{rad} FG$. Therefore, by Corollary 3.1a, FG is not semisimple. □

The implications of Maschke's Theorem will be explored superficially in the exercises of this section and various sections of later chapters. There are many excellent text-books and monographs that offer detailed expositions of relation between groups and algebras. The encyclopedic work [24] of Curtis and Reiner is recommended with special warmth.

Classical group representations of a finite group G make use of the complex group algebra $\mathbb{C}G$. However, many results of this theory can be generalized to representations that are defined in terms of a group algebra FG where F is any algebraically closed field whose characteristic does not divide the order of G. With this hypothesis, it follows from Maschke's Theorem and Corollary 3.5b that $FG \cong M_{n_1}(F) \dotplus \cdots \dotplus M_{n_r}(F)$ for suitable natural numbers $n_1, \ldots, n_r$. These numbers are the *degrees* of the irreducible representations of G, that is, the dimensions of the simple FG-modules. These degrees are determined by the structure of G, but there is no simple formula that produces the n_i from elementary invariants of G. On the other hand, the number r of simple factors of FG coincides with a standard numerical property of G.

Corollary. *Let G be a finite group whose order is not divisible by the characteristic of the algebraically closed field F. The group algebra FG is isomorphic to a product $M_{n_1}(F) \dotplus \cdots \dotplus M_{n_r}(F)$ of full matrix algebras over F, where r is the number of conjugate classes in G.*

By virtue of Maschke's Theorem and Corollary 3.5b, the only part of this corollary that requires attention is the assertion that r is the number of conjugate classes in G. This conclusion follows from two observations: as an F-space, the center $\mathbf{Z}(M_{n_1}(F) \dotplus \cdots \dotplus M_{n_r}(F))$ is r-dimensional; $\dim_F \mathbf{Z}(FG)$ is the number of conjugate classes in G. The first of these assertions is the content of Exercise 1. To prove the second claim, let $K_1, K_2, \ldots, K_m$ be the distinct conjugate classes in G. Thus, $\varnothing \neq K_i \subseteq G$; and if $x_i \in K_i$, then $K_i = \{y^{-1}x_i y : y \in G\}$. Denote $z_i = \sum_{x \in K_i} x \in FG$. An element $w = \sum_{x \in G} xa_x \in FG$ belongs to the center of FG if and only if $y^{-1}wy = w$ for all $y \in G$. An easy calculation shows that this condition imposes the requirement $a_{yxy^{-1}} = a_x$ on the coefficients of w. It follows that $w \in \mathbf{Z}(FG)$ exactly when w can be written as a sum $\sum_{i=1}^m z_i b_i$ with $b_i \in F$. In other words, $\mathbf{Z}(FG) = \bigoplus_{i=1}^m z_i F$, so that dimension of $\mathbf{Z}(FG)$ is m.

Exercises

1. Prove the following statements.

   (a) If $A_1, A_2, \ldots, A_r$ are algebras, then $\mathbf{Z}(A_1 \dotplus A_2 \dotplus \cdots \dotplus A_r) = \mathbf{Z}(A_1) \dotplus \mathbf{Z}(A_2) \dotplus \cdots \dotplus \mathbf{Z}(A_r)$.

   (b) $\mathbf{Z}(M_n(F)) = \iota_n F$.

   (c) $\dim_F \mathbf{Z}(M_{n_1}(F) \dotplus M_{n_2}(F) \dotplus \cdots \dotplus M_{n_r}(F)) = r$.

2. Prove that for every finite group G, $\mathbb{Z}G$ is not semisimple. Hint. Show that if p is a prime divisor of $|G|$, then $p\mathbb{Z}G$ is an ideal of $\mathbb{Z}G$ such that $\mathbb{Z}G/p\mathbb{Z}G$ is not semisimple.

3. Assume that char F does not divide the order of the finite group G. Denote $A = FG$. For a subgroup H of G, define $e_H = (\sum_{x \in H} x)n^{-1}$, where $n = |H|$. Prove the following statements.

   (a) e_H is an idempotent element of A such that $xe_H = e_H x = e_H$ for all $x \in H$ and $e_H x \neq e_H \neq xe_H$ for all $x \in G - H$.

   (b) If H is a normal subgroup of G, then $e_H \in \mathbf{Z}(A)$ and $e_H A$ is isomorphic to the group algebra over F of G/H.

   (c) If G is not the one element group, then A is not simple. Hint. Show $A = e_G A \dotplus (1 - e_G)A$, and $e_G A \cong F$.

4. Let G be the Klein 4-group $\{1,x,y,z\}$ where $x^2 = y^2 = z^2 = 1$, $xy = yx = z$, $xz = zx = y$, $yz = zy = x$. Thus, $G \cong \mathbb{Z}/2\mathbb{Z} \times \mathbb{Z}/2\mathbb{Z}$. Prove that if F is a field with char $F \neq 2$, then $FG \cong F \dotplus F \dotplus F \dotplus F$. Hint. Consider

$$e_0 = (1/4)(1 + x + y + z),\ e_1 = (1/4)(1 + x - y - z),$$
$$e_2 = (1/4)(1 - x + y - z),\ e_3 = (1/4)(1 - x - y + z).$$

5. Let $\mathbb{H} = \left(\frac{-1,-1}{\mathbb{R}}\right)$ be the quaternion algebra of Hamilton. The subgroup G of $\mathbb{H}^\circ$ consisting of $1, -1, \mathbf{i}, -\mathbf{i}, \mathbf{j}, -\mathbf{j}, \mathbf{k}, -\mathbf{k}$ is called the *quaternion group*. Prove that $\mathbb{R}G \cong \mathbb{H} \dotplus \mathbb{R} \dotplus \mathbb{R} \dotplus \mathbb{R} \dotplus \mathbb{R}$. Hint. Show that the center K of G is $\{1, -1\}$, G/K is isomorphic to the Klein 4-group, and $(1 - e_K)\mathbb{R}G \cong \mathbb{H}$.

Notes on Chapter 3

This chapter offers a fairly traditional presentation of a classical theorem. Wedderburn's Structure Theorem is such a beautiful result that it needs no slick embellishments. We have tried to give as clean a proof as possible. Putting aside the uniqueness portion of the theorem, the proof of the Wedderburn theorem is remarkably easy: it depends on (1) the definition of a semisimple algebra, (2) the matrix notation for endomorphisms of direct sums (Section 3.4), and (3) Schur's Lemma, a triviality by modern standards.

It is unfair to give Wedderburn all of the credit for the structure theorem.

In 1893, T. Molien published a result that is essentially the Wedderburn Theorem for finite dimensional complex algebras. Wedderburn's paper [76] on the structure of semisimple algebras appeared in 1907. It treated finite dimensional algebras over arbitrary fields. In 1927 Emil Artin extended Wedderburn's result to rings that satisfy the ascending and descending chain conditions. Finally, in 1939, C. Hopkins showed that the ascending chain condition is a consequence of the descending chain condition, which gives the present form of the structure theorem. Because of its historical background, the structure theorem is sometimes called Molien's Theorem or the Artin–Wedderburn Theorem. A colleague has facetiously suggested using the term "W.H.A.M. Theorem." However, the name "Wedderburn's Structure Theorem" is universally recognized, and it will be used in this book.

CHAPTER 4

The Radical

Wedderburn's Theorem shows that the class of semisimple algebras is very limited. On the other hand, Proposition 3.1a suggests that Artinian algebras are semisimple "up to a radical." In fact, this is the case. All that is missing from a proof is the result that rad A_A is an ideal. We will establish this fact in Section 4.1. The rest of the chapter is concerned with properties and characterizations of the radical, a theorem about nilpotent algebras, and the radicals of group algebras.

4.1. The Radical of an Algebra

A fundamental fact about the radical of an algebra is that it is an ideal. We begin with a result concerning the radical of a module. Recall that the radical rad M of an A-module M is the intersection of all submodules N of M such that M/N is simple.

Lemma. *Let M_1 and M_2 be right A-modules. If $\phi \in \mathrm{Hom}_A(M_1, M_2)$, then $\phi(\mathrm{rad}\, M_1) \subseteq \mathrm{rad}\, M_2$, and ϕ induces a homomorphism of $M_1/\mathrm{rad}\, M_1$ to $M_2/\mathrm{rad}\, M_2$.*

PROOF. If $N < M_2$, then ϕ induces an injection of $M_1/\phi^{-1}(N)$ to M_2/N. In particular, if M_2/N is simple, then either $\phi^{-1}(N) = M_1$ or $M_1/\phi^{-1}(N) \cong M_2/N$ is simple. In both cases, $\phi^{-1}(N) \supseteq \mathrm{rad}\, M_1$. Thus, $\phi^{-1}(\mathrm{rad}\, M_2) \supseteq \mathrm{rad}\, M_1$; that is, $\phi(\mathrm{rad}\, M_1) \subseteq \mathrm{rad}\, M_2$. The last statement is obvious. □

Proposition. *If A is a non-trivial R-algebra, then* rad A_A *is a proper (two sided) ideal of A.*

PROOF. The mapping $y \mapsto xy$ is an A-module endomorphism of A for each $x \in A$. By the lemma, $x(\text{rad}\, A_A) \subseteq \text{rad}\, A_A$, that is, $\text{rad}\, A_A \lhd A$. Since A has a unity element, it follows from Zorn's Lemma that there is a maximal right ideal M of A. Consequently, A_A/M is simple. Thus, $\text{rad}\, A_A \subseteq M \subset A$, so that $\text{rad}\, A_A$ is a proper ideal of A. □

Corollary a. *If A is a right Artinian R-algebra, then $A/\text{rad}\, A_A$ is a semisimple R-algebra.*

PROOF. It follows from Corollary 2.4a that $A/\text{rad}\, A_A$ is a semisimple A-module, and therefore it is also semisimple as an $(A/\text{rad}\, A_A)$-module (by Proposition 2.1). According to Lemma 2.7a, $\text{rad}(A/\text{rad}\, A_A) = 0$. Consequently, $A/\text{rad}\, A_A$ is a semisimple R-algebra by Proposition 3.1a. □

It is worth pointing out that this corollary makes sense by virtue of the fact that $\text{rad}\, A_A$ is an ideal, so that $A/\text{rad}\, A_A$ is an algebra.

There is another consequence of the lemma that will be needed in the next section.

Corollary b. *If M is a right A-module, then $M(\text{rad}\, A_A) \subseteq \text{rad}\, M$.*

PROOF. For each $u \in M$, the mapping $x \mapsto ux$ is an A-module homomorphism from A_A to M, so that by the lemma $u(\text{rad}\, A_A) \subseteq \text{rad}\, M$. □

EXERCISES

1. Let A be an algebra. Prove that a simple A-module is a simple $A/\text{rad}\, A_A$-module, and conversely every simple $A/\text{rad}\, A_A$-module is a simple A-module.
2. (a) Prove that if A is a right Artinian algebra, and M is a right A-module, then $\text{rad}\, M = M(\text{rad}\, A_A)$. Hint. Corollary 4.1b gives one inclusion; the reverse inclusion is obtained from the results in Sections 2.7 and 3.1.
   (b) Prove that $\text{soc}\, M = \{u \in M : u(\text{rad}\, A_A) = 0\}$.

4.2. Nakayama's Lemma

Several equivalent statements are called "Nakayama's Lemma." This section presents a couple of the most familiar versions of this keystone of ring theory.

The radical of a module is analogous to the Frattini subgroup of a group, and Nakayama's Lemma is a variant of the standard characterization of the Frattini subgroup. The connection between these topics is clear from the lemma of this section.

Lemma. *For an element u of an A-module M, the following conditions are equivalent.*

(i) $u \in \operatorname{rad} M$.
(ii) *If $N < M$ is such that $uA + N = M$, then $N = M$.*

PROOF. If $u \notin \operatorname{rad} M$, then there is a submodule N of M such that M/N is simple and $u \notin N$. In this case, $uA + N = M \neq N$. Therefore, (i) is a consequence of (ii). Conversely, suppose that $N < M$ exists with the property $uA + N = M \neq N$. Plainly, $u \notin N$. By Zorn's Lemma, there is a submodule P of M containing N that is maximal with the property $u \notin P$. If $P \subset Q < M$, then $u \in Q$, so that $M = uA + N \subseteq Q$. Hence, M/P is simple, $\operatorname{rad} M \subseteq P$, and therefore $u \notin \operatorname{rad} M$. □

Nakayama's Lemma for Modules. *Suppose that P is a submodule of the A-module M. Assume that P satisfies*

$$\text{for all submodules } N \text{ of } M, \text{ if } P + N = M, \text{ then } N = M; \tag{1}$$

then $P \subseteq \operatorname{rad} M$. Conversely, if $P < M$, $P \subseteq \operatorname{rad} M$, and either P or M is finitely generated as an A-module, then P satisfies (1).

PROOF. Suppose that there exists $u \in P - \operatorname{rad} M$. By the lemma, there is a submodule N of M such that $N \neq M = uA + N \subseteq P + N$. To prove the converse, suppose that $P < \operatorname{rad} M$, and $N < M$ is such that $P + N = M$. If M is finitely generated, then there is a finitely generated submodule Q of P such that $Q + N = M$. Thus, in all cases it can be assumed that P is finitely generated—say $P = u_1A + u_2A + \cdots + u_nA$. By using the lemma repeatedly, we obtain the desired conclusion $M = u_1A + u_2A + \cdots + u_nA + N = u_2A + \cdots + u_nA + N = \cdots = u_nA + N = N$. □

Nakayama's Lemma for Algebras. *For a right ideal P of the R-algebra A, the following conditions are equivalent.*

(i) $P \subseteq \operatorname{rad} A_A$.
(ii) *If M is a finitely generated right A-module, and $N < M$ satisfies $N + MP = M$, then $N = M$.*
(iii) $G = \{1 + x : x \in P\}$ *is a subset of A°.*

PROOF. The property (ii) follows from (i) by Nakayama's Lemma for Modules, and Corollary 4.1b. In order to deduce (iii) from (ii), let $x \in P$. Denote $y = 1 + x$. It follows that $1 = y - x \in yA + P$, so that $yA + P = A_A$. Since A_A is finitely generated by 1, it follows from (ii) that $yA = A$. In particular, $1 = yz = z + xz$ for some $z \in A$. Consequently, $z = 1 - xz \in G$ because $P < A_A$ and $x \in P$. This argument shows that every element of G has a right inverse in G. Therefore, G is a group and $G \subseteq A^\circ$. For the deduction of (i) from (iii), let $x \in P$. By the lemma, it suffices to show that if the right ideal N of A satisfies $xA + N = A$, then $N = A$. The hypothesis $xA + N = A$ implies that $1 = xz + y$ for some $z \in A$ and $y \in N$. Thus, $y = 1 + x(-z)$ with $x(-z) \in P$. By (iii), $y \in A^\circ$. Therefore, $N = A$, as required. □

By taking $N = 0$ and $P = \text{rad}\, A_A$, we obtain a corollary that is often called Nakayama's Lemma.

Corollary. *If M is a finitely generated right A-module such that $M(\text{rad}\, A_A) = M$, then $M = 0$.*

EXERCISES

1. Show that the statement in Corollary 4.2 implies Nakayama's Lemma for Algebras.
2. Let A be an algebra, and suppose that M and N are right A-modules with N finitely generated. Prove that if $\phi \in \text{Hom}_A(M,N)$ induces a surjective homomorphism of $M/M(\text{rad}\, A_A)$ to $N/N(\text{rad}\, A_A)$, then ϕ is surjective.
3. Prove that if A is an R-algebra that is finitely generated as an R-module, then $A(\text{rad}\, R_R) \subseteq \text{rad}\, A_A$. Hint. Show that for any finitely generated A-module M, if $M(A(\text{rad}\, R_R)) = M$, then $M = 0$, and apply Nakayama's Lemma.
4. Let A be the $\mathbb{Z}$-algebra $\{a/n : a \in \mathbb{Z}, n \in \mathbb{N}, p \nmid n\}$, where p is a fixed prime number. Let $\mathbb{Z}(p^\infty)$ be the group $G_p/\mathbb{Z}$, where $G_p = \{a/p^i : a \in \mathbb{Z}, i \in \mathbb{N}\}$. Prove that $\text{rad}\, A_A = pA$, and that $\mathbb{Z}(p^\infty)$ is a non-zero A-module such that $\mathbb{Z}(p^\infty)(\text{rad}\, A_A) = \mathbb{Z}(p^\infty)$.

4.3. The Jacobson Radical

We are now in a position to prove an assertion that was made in Section 3.1.

Lemma a. *If A is an R-algebra, then* $\text{rad}\, A_A = \text{rad}\, {}_AA$.

PROOF. By the left hand analogues of Propositions 4.1 and 4.2, $\text{rad}\, {}_AA < A_A$, and $\{1 + x : x \in \text{rad}\, {}_AA\} \subseteq A^\circ$. Nakayama's Lemma implies that $\text{rad}\, {}_AA \subseteq \text{rad}\, A_A$. By a symmetrical argument, $\text{rad}\, A_A \subseteq \text{rad}\, {}_AA$. □

The time has come to assign the radical its proper name and notation.

Definition. For an R-algebra A, the *Jacobson radical* of A is $\mathbf{J}(A) = \text{rad}\, A_A$.

Proposition. *The Jacobson radical of an algebra A is a two sided ideal $\mathbf{J}(A)$ of A that satisfies*

(i) $\mathbf{J}(A) = \bigcap\{M : M = \textit{maximal right ideal of } A\}$,

(ii) $\mathbf{J}(A) = \bigcap \{M: M = \text{maximal left ideal of } A\}$,
(iii) $\mathbf{J}(A) = \{x \in A: 1 + xy \in A^\circ \text{ for all } y \in A\}$,
(iv) $\mathbf{J}(A) = \{x \in A: 1 + yx \in A^\circ \text{ for all } y \in A\}$.

This proposition is a direct consequence of Lemma a and Nakayama's Lemma. Another characterization of the Jacobson radical is given in Exercise 4.

Throughout the rest of this book, the expression "radical of A" will refer to the Jacobson radical of the algebra A. Other radicals can be defined for rings and algebras, but they will not concern us.

There is a useful variation of the proposition.

Corollary a. *If M is a right or left ideal of the algebra A such that $1 + x \in A^\circ$ for all $x \in M$, then $M \subseteq \mathbf{J}(A)$. If also* $\operatorname{rad} A/M = 0$, *then* $M = \mathbf{J}(A)$.

Proof. The hypothesis that M is a right or left ideal and $1 + x \in A^\circ$ for all $x \in M$ implies that $M \subseteq \mathbf{J}(A)$ by the proposition. On the other hand, if $\operatorname{rad} A/M = 0$, then $\mathbf{J}(A) \subseteq M$ by Lemma 2.7a. □

An element x of an algebra A is called *nilpotent* if there is a natural number n such that $x^n = 0$. The next corollary generalizes Lemma 3.2c.

Corollary b. *If M is a right (or left) ideal of the algebra A such that every element of M is nilpotent, then $M \subseteq \mathbf{J}(A)$.*

Proof. If $x^n = 0$, then $(1 + x)(\sum_{0 \le i < n}(-x)^i) = (\sum_{0 \le i < n}(-x)^i)(1 + x) = 1$, so that Corollary b is a special case of Corollary a. □

There are a few general properties of the Jacobson radical that are often used. Here are two of them; another one is given in Exercise 2.

Lemma b. *Let A and B be R-algebras.*

(i) *If $\theta: A \to B$ is a surjective algebra homomorphism, then $\theta(\mathbf{J}(A)) \subseteq \mathbf{J}(B)$.*
(ii) $\mathbf{J}(A \dotplus B) = \mathbf{J}(A) \dotplus \mathbf{J}(B)$.

Proof. The inclusion (i) is a consequence of Lemmas 4.1 and 2.1a: $\theta(\mathbf{J}(A)) = \theta(\operatorname{rad} A_A) \subseteq \operatorname{rad} B_A = \operatorname{rad} B_B = \mathbf{J}(B)$. Applying this result to the projections of $A \dotplus B$ to A and B gives $\mathbf{J}(A \dotplus B) \subseteq \mathbf{J}(A) \dotplus \mathbf{J}(B)$. On the other hand, if $x \in \mathbf{J}(A)$ and $y \in \mathbf{J}(B)$, then $1_A + x \in A^\circ$ and $1_B + y \in B^\circ$ by the proposition. Therefore, $(1_A, 1_B) + (x,y)$ is a unit of $A \dotplus B$. Since $\mathbf{J}(A) \dotplus \mathbf{J}(B)$ is an ideal of $A \dotplus B$, it follows from Corollary a that $\mathbf{J}(A) \dotplus \mathbf{J}(B) \subseteq \mathbf{J}(A \dotplus B)$. □

We end this section with an example of a class of algebras that have zero radical.

EXAMPLE. If M is a semisimple A-module, then $\mathbf{J}(\mathbf{E}_A(M)) = 0$. When M is also finitely generated, $\mathbf{E}_A(M)$ is even semisimple by Exercise 3 of Section 3.5; and the radical is certainly 0 in this case. To prove the assertion in general, let $0 \neq \phi \in \mathbf{E}_A(M)$. If follows easily from Proposition 2.4 that there is a simple submodule N of M such that $\phi(N) \neq 0$. By Schur's Lemma, ϕ maps N isomorphically to $\phi(N)$. Another application of Proposition 2.4 gives the existence of $\pi \in \mathbf{E}_A(M)$ such that $\pi^2 = \pi$ and $\pi(M) = \phi(N)$. Let $\psi = (\phi|N)^{-1}\pi$. Plainly, $\psi \in \mathbf{E}_A(M)$ and $\phi\psi = \pi$. Since $\pi \neq 0$ and $\pi(1 - \phi\psi) = 0$, it follows that $1 - \phi\psi$ is not a unit of $\mathbf{E}_A(M)$. Therefore, $\phi \notin \mathbf{J}(\mathbf{E}_A(M))$ by the proposition.

EXERCISES

1. Determine $\mathbf{J}(A)$ for the following choices of the algebra A.
   (a) A is a (commutative) principal ideal domain with infinitely many prime ideals.
   (b) A is a commutative integral domain with a finite set $\{M_1, \ldots, M_n\}$ of maximal ideals.
   (c) $A = \{a/(2n - 1): a \in \mathbb{Z}, n \in \mathbb{N}\}$.
   (d) $A = \mathbb{Z}/n\mathbb{Z}$, where $n \in \mathbb{N}$.

2. Prove that if A is an algebra, then $\mathbf{J}(M_n(A)) = \{[x_{ij}] \in M_n(A): x_{ij} \in \mathbf{J}(A)$ for $1 \leq i, j \leq n\}$. Hint. Note that if $\alpha = [x_{ij}] = \sum_{i,j} \varepsilon_{ij} x_{ij}$, then $\sum_{k=1}^{n} \varepsilon_{ki} \alpha \varepsilon_{jk} = \iota_n x_{ij}$. Show that if $\iota_n x \in \mathbf{J}(M_n(A))$, then $x \in \mathbf{J}(A)$, and if $x \in \mathbf{J}(A)$, then $\varepsilon_{ij} x \in \mathbf{J}(M_n(A))$ for all i and j.

3. Prove that if A is an R-algebra such that $A^\circ \cup \{0\}$ is a division algebra, then $\mathbf{J}(A) = 0$. Deduce that if D is a division algebra, then $\mathbf{J}(D[\mathbf{x}_1, \ldots, \mathbf{x}_n]) = 0$ and $\mathbf{J}(D\{\mathbf{x}_1, \ldots, \mathbf{x}_n\}) = 0$.

4. An algebra A is *primitive* if there is a faithful, simple A-module. An ideal K of an algebra A is primitive if the factor algebra A/K is primitive. Prove the following statements.
   (a) If M is a maximal right ideal of A, then $\operatorname{ann}(A_A/M)$ is a primitive ideal of A.
   (b) $\mathbf{J}(A)$ is the intersection of all primitive ideals of A.

5. Prove the following statements for an algebra A.
   (a) If A is simple, then A is primitive.
   (b) If A is commutative and primitive, then A is a field.
   (c) If $A = \mathbf{E}_F(V)$, where V is an infinite dimensional F-space, then A is primitive, but not simple.

6. Let $\{A_i: i \in J\}$ be a set of non-trivial R-algebras. Denote by π_j the canonical projection of the product algebra $\prod_{i \in J} A_i$ onto A_j. A subalgebra B of $\prod_{i \in J} A_i$ is called a *subdirect product* of $\{A_i: i \in J\}$ if $\pi_j(B) = A_j$ for all $j \in J$. Prove: A is isomorphic to a subdirect product of $\{A_i: i \in J\}$ if and only if there is a set $\{K_i: i \in J\}$ of ideals of A such that $A/K_i \cong A_i$ for all $i \in J$, and $\bigcap_{i \in J} K_i = 0$. Deduce that $\mathbf{J}(A) = 0$ if and only if A is isomorphic to a subdirect product of primitive algebras.

4.4. The Radical of an Artinian Algebra

For most rings, the problem of finding the radical lies somewhere between difficult and impossible. However, the radical of a finite dimensional algebra over a field (or more generally of an Artinian algebra) is more accessible.

Proposition. *If A is a right or left Artinian algebra, then there is a natural number k such that $\mathbf{J}(A)^k = 0$.*

PROOF. $\mathbf{J}(A) \supseteq \mathbf{J}(A)^2 \supseteq \mathbf{J}(A)^3 \supseteq \cdots$ is a descending sequence of two sided ideals, so that by the right or left Artinian property, there is a natural number k such that $\mathbf{J}(A)^k = \mathbf{J}(A)^{k+1}$. If we could assume that $\mathbf{J}(A)^k$ is finitely generated as an A-module, it would then follow from Corollary 4.2 that $\mathbf{J}(A)^k = 0$. Thus, if A is Noetherian as well as Artinian, then the proposition is an easy application of Nakayama's Lemma. The fact is that A must be Noetherian, as we will show in the next section. However, the proof uses the result that $\mathbf{J}(A)^k = 0$ for some $k \in \mathbb{N}$, so that it is necessary to base our argument solely on the descending chain condition on (say) the right ideals of A. Assume that $\mathbf{J}(A)^k \neq 0$. In particular, the set of non-zero right ideals M of A such that $M\mathbf{J}(A) = M$ includes $\mathbf{J}(A)^k$. Therefore, there is a minimal M with these properties. Since $M = M\mathbf{J}(A) = M\mathbf{J}(A)^2 = \cdots = M\mathbf{J}(A)^k$, there is some $x \in M$ such that $x\mathbf{J}(A)^k \neq 0$. Plainly, $x\mathbf{J}(A)^k$ is a right ideal of A that is contained in M, and $(x\mathbf{J}(A)^k)\mathbf{J}(A) = x\mathbf{J}(A)^{k+1} = x\mathbf{J}(A)^k$. From the minimality of M, it follows that $M = x\mathbf{J}(A)^k \subseteq xA = M$. Therefore, M is finitely generated, which contradicts Nakayama's Lemma, because $0 \neq M = M\mathbf{J}(A)$. □

Corollary. *Let A be a right or left Artinian algebra. For a right or left ideal M of A, the following conditions are equivalent.*

(i) $M \subseteq \mathbf{J}(A)$.
(ii) *There is a natural number k such that $M^k = 0$.*
(iii) *All of the elements in M are nilpotent.*

PROOF. The fact that (i) implies (ii) is a consequence of the proposition, and it is evident that (ii) implies (iii). Finally, (i) follows from (iii) for any algebra by Corollary 4.3b. □

It follows from the corollary that every element in the Jacobson radical of a right or left Artinian algebra is nilpotent. The converse is not true: nilpotent elements need not belong to the radical. For example, if $n > 1$, then the matrix algebras $M_n(D)$ (D a division algebra) have many nilpotent elements (for example, all ε_{ij} with $i \neq j$ are nilpotent), but $\mathbf{J}(M_n(D)) = 0$.

Exercises

1. Let F be a field.

(a) Let A be the F-algebra of row-finite, infinite matrices

$$\begin{bmatrix} a_{11} & a_{12} & a_{13} & \cdots \\ a_{21} & a_{22} & a_{23} & \cdots \\ a_{31} & a_{32} & a_{33} & \cdots \\ \cdot & \cdot & \cdot & \cdots \\ \cdot & \cdot & \cdot & \cdots \end{bmatrix},$$

where $a_{ij} \in F$ for all i and j, and for all i, there exists m such that $a_{ij} = 0$ for all $j > m$. Prove that $\mathbf{J}(A) = 0$.

(b) Let B be the subalgebra of A that consists of the matrices with zeros below the main diagonal, that is, $a_{ij} = 0$ if $i > j$. Prove that $\mathbf{J}(B)$ is the set of matrices in B that have zeros on the main diagonal, that is, $a_{ii} = 0$ for all i. Show that every element in $\mathbf{J}(B)$ is nilpotent, but $\mathbf{J}(B)^k \neq 0$ for all $k \in \mathbb{N}$.

(c) Let C be the algebra of all infinite matrices

$$\begin{bmatrix} a_{11} & a_{12} & a_{13} & \cdots \\ 0 & a_{22} & a_{23} & \cdots \\ 0 & 0 & a_{33} & \cdots \\ \cdot & \cdot & \cdot & \cdots \\ \cdot & \cdot & \cdot & \cdots \end{bmatrix},$$

where $a_{ij} \in F$, and the elements below the main diagonal are zero. These matrices are not assumed to be row finite. Prove that $\mathbf{J}(C)$ consists of the matrices in C that have zeros on the main diagonal. Show that not all elements of $\mathbf{J}(C)$ are nilpotent.

2. Let A be a right Artinian algebra.

(a) Prove that $\operatorname{soc}({}_AA) = \operatorname{ann}(\mathbf{J}(A)_A)$ and $\operatorname{soc}(A_A) = \operatorname{ann}({}_A\mathbf{J}(A))$.

(b) Show that the algebra A that was defined in Exercise 4 of Section 3.3 is such that $\operatorname{soc}({}_AA) \neq \operatorname{soc}(A_A)$.

3. An ideal, right ideal, or left ideal I in an algebra A is *nilpotent* if $I^n = 0$ for some $n \geq 1$. This property of I will of course imply that the elements of I are nilpotent, but not conversely. (See Exercise 1(b).) If I has the property that all of its elements are nilpotent, then I is called *nil*. The purpose of this Exercise is to outline a proof that for Noetherian algebras all nil ideals are nilpotent.

(a) Show that if the R-algebra A contains a non-zero nilpotent right ideal I, then AI is a non-zero nilpotent two sided ideal.

(b) Deduce that if A contains a maximal nilpotent ideal I, then A/I contains no non-zero nilpotent right ideals.

(c) Prove that if xA is a nil right ideal, then Ax is a nil left ideal, and conversely.

(d) Let Ax be a nil left ideal. Let $y \in Ax$ be such that $\operatorname{ann} y$ is maximal in $\{\operatorname{ann} z : z \in Ax - \{0\}\}$, where $\operatorname{ann} y = \{w \in A : yw = 0\}$. Prove that $(Ay)^2 = 0$. Hint. Note that by maximality, either $wy = 0$ or $\operatorname{ann} wy = \operatorname{ann} y$ for $w \in A$. Using the fact that Ay is nil, show that this implies $ywy = 0$ for all $w \in A$.

(e) Deduce from (c) and (d) that if the Noetherian algebra A contains a non-zero nil right ideal, then A contains a non-zero nilpotent right ideal.

(f) Prove *Levitzki's Theorem.* If A is a right Noetherian algebra, then every nil right or left ideal of A is nilpotent.

(g) Let $A = F[\mathbf{x}_0,\mathbf{x}_1,\mathbf{x}_2, \ldots]$ be the F-algebra of polynomials in the commuting indeterminants $\mathbf{x}_0$, $\mathbf{x}_1$, $\mathbf{x}_2$, ... Let I be the ideal of A that is generated by $\mathbf{x}_0^2$, $\mathbf{x}_1^3$, $\mathbf{x}_2^4$, ... Prove that A/I contains nil ideals that are not nilpotent.

4.5. Artinian Algebras Are Noetherian

One of the nicest applications of Proposition 4.4 is the result given as the heading of this section. We will actually prove a more general statement concerning modules.

Proposition. *Let A be a right or left Artinian algebra. If M is an Artinian A-module, then M is Noetherian.*

PROOF. Denote $J = \mathbf{J}(A)$. Since A is Artinian, there is a natural number k such that $J^k = 0$ by Proposition 4.4. In particular, there is a smallest $n \in \mathbb{N}$ such that $MJ^n = 0$. (We consider the case of right A-modules; the proof for left modules is obtained by reflection from right to left.) Proceed by induction on n. If $n = 0$, then $0 = MJ^0 = MA = M$; and the zero module is plainly Noetherian. Let $n = 1$. The condition $MJ = 0$ means that M can be considered as a module over the algebra A/J. Since A/J is semisimple by Corollary 4.1a, every right (left) A/J-module is semisimple by Proposition 3.1b (using the fact that right and left semisimplicity coincide). Therefore, $M_{A/J}$ is Noetherian by Proposition 2.6. Since $\mathbf{S}(M_A) = \mathbf{S}(M_{A/J})$ by Lemma 2.1a, M_A is also Noetherian. Assume that $n > 1$. The induction step is based on Lemma 2.6b. Denote $N = MJ^{n-1} < M$. Then N is Artinian and $NJ = 0$, so that by the case $n = 1$, N is Noetherian. The factor module M/N is also Artinian and $(M/N)J^{n-1} = 0$. By the induction hypothesis, M/N is Noetherian. Consequently, M is Noetherian. □

Corollary a. *If the R-algebra A is right (left) Artinian, then A is right (respectively, left) Noetherian.*

Corollary b. *If A is a right Artinian R-algebra, then the following conditions on the right A-module M are equivalent:*

(i) *M is Artinian;*
(ii) *M is Noetherian;*
(iii) *M is finitely generated.*

PROOF. By the proposition, (i) implies (ii), and (iii) follows from (ii) by Proposition 2.6. (It was pointed out in Section 2.6 that finite generation of a module follows from the ascending chain condition with no additional hypothesis.) Finally, suppose that $N = u_1 A + \cdots + u_n A$. Then there is a surjective A-module homomorphism of $\bigoplus n\, A_A$ to M defined by

$$(x_1, \ldots, x_n) \mapsto \sum_{i=1}^{n} u_i x_i.$$

Since A_A is Artinian, it follows from Lemma 2.6b that $\bigoplus n\, A_A$ is Artinian, and consequently M is Artinian. □

EXERCISE

An ideal K in an R-algebra A is *prime* if $K \neq A$, and for any two ideals I and J of A, if $IJ \subseteq K$, then either $I \subseteq K$ or $J \subseteq K$. Prime ideals are less prominent in non-commutative ring theory than they are in the study of commutative rings, but this may be a reflection of the fact that the theory of commutative rings is more advanced than the theory of non-commutative rings. The purpose of this Exercise is to develop some elementary properties of prime ideals. In all parts of the Exercise, A is an R-algebra.

(a) Prove that an ideal K of A is prime if and only if $xAy \subseteq K$ implies $x \in K$ or $y \in K$ for all $x, y \in A$.

(b) Prove that if $X \subseteq A$ is closed under multiplication, $0 \notin X$, and $X \neq \varnothing$, and if K is maximal in the set of all ideals I of A such that $I \cap X = \varnothing$, then K is prime.

(c) Deduce from (b) that all maximal ideals are prime. The intersection of all prime ideals of A is called the *prime radical* of A, and it is denoted by $\mathbf{P}(A)$.

(d) Prove that if $x \in \mathbf{P}(A)$, then x is nilpotent. Consequently, $\mathbf{P}(A)$ is a nil ideal of A . Deduce that $\mathbf{P}(A) \subseteq \mathbf{J}(A)$.

(e) Use Levitzky's Theorem (Exercise 3(f), Section 4.4) to show that if $\mathbf{A}$ is right Noetherian, then $\mathbf{P}(A)$ is the maximum nilpotent ideal of A.

(f) Show that the intersection of any non-empty, totally ordered (by inclusion) collection of prime ideals is a prime ideal. Deduce that for any proper ideal I of A there is a prime ideal that is minimal in the set of all prime ideals containing I.

(g) Assume that A is Noetherian. Prove that if I is a proper ideal of A, then the set of minimal primes over I (that is, minimal in $\{K \lhd A : K \text{ prime}, I \subseteq K\}$) is finite. Hint. Consider a maximal counterexample I. Then I is not prime, so that $xAy \subseteq I$ for some $x \notin I$ and $y \notin I$. Show that the minimal primes over I are included among the primes that are minimal over $I + AxA$ or $I + AyA$.

(h) Prove that if A is Noetherian and every prime ideal of A is maximal, then $\mathbf{P}(A) = \mathbf{J}(A)$, and $A/\mathbf{J}(A)$ is a finite product of simple Noetherian algebras. Hint. The hypothesis that all prime ideals are maximal and (g) yield the conclusion that the set of all prime ideals of A is finite, say $\{K_1, K_2, \ldots, K_r\}$. Use the Chinese Remainder Theorem to conclude that $A/\mathbf{P}(A) \cong A/K_1 \dotplus A/K_2 \dotplus \cdots \dotplus A/K_r$, and deduce from Lemma 4.3b that $\mathbf{J}(A) \subseteq \mathbf{P}(A)$.

(i) Prove the following converse of (h): if $\mathbf{J}(A)$ is finitely generated as a right A-module, $\mathbf{J}(A)$ is nilpotent, and $A/\mathbf{J}(A)$ is a finite product of simple Noetherian algebras, then A is Noetherian and every prime ideal of A is maximal. Hint. Show that for all i, $\mathbf{J}(A)^i$ is finitely generated as a right A-module; then follow the idea in the proof of Proposition 4.5.

(j) Show that for a commutative algebra A, the following properties are equivalent:

(1) A is Noetherian and every prime ideal of A is maximal.
(2) $\mathbf{J}(A)$ is finitely generated and nilpotent, and $A/\mathbf{J}(A)$ is a finite product of fields.
(3) A is Artinian.

4.6. Nilpotent Algebras

Another application of Proposition 4.4 gives a characterization of nilpotent, finite dimensional algebras. The theorem comes from one of Wedderburn's late papers [77], but it is related to older results of Engel and Lie on nilpotent and solvable Lie algebras.

The proof of Wedderburn's theorem is based on Proposition 4.4, the structure theorem, and an elementary lemma that is obtained using the trace mapping for matrices. If $\alpha = [a_{ij}]$ is an n by n matrix with entries in a field F, the *trace* of α is

$$\operatorname{tr}\alpha = \sum_{i=1}^{n} a_{ii}.$$

Two properties of the trace are needed: the trace mapping is F-linear from $M_n(F)$ to F; if α is nilpotent, then $\operatorname{tr}\alpha = 0$. The first of these statements is an easy consequence of the definition. If $\alpha^m = 0$, then the minimum polynomial of α is $\mathbf{x}^k$ with $1 \le k \le m$ (because this polynomial divides $\mathbf{x}^m$), and the characteristic polynomial $\mathbf{x}^n - (\operatorname{tr}\alpha)\mathbf{x}^{n-1} + \cdots = \mathbf{x}^n$ (since the minimum polynomial and the characteristic polynomial have the same irreducible factors). Therefore, $\operatorname{tr}\alpha = 0$.

We use these properties of the trace to prove a fact that will be subsumed by the main result to this section.

Lemma. *There is no set of nilpotent matrices that spans $M_n(F)$ as an F-space.*

Proof. Otherwise, there exist nilpotent matrices $\alpha_1, \ldots, \alpha_r \in M_n(F)$, and $b_1, \ldots, b_r \in F$ such that $\varepsilon_{11} = \alpha_1 b_1 + \cdots + \alpha_r b_r$. By virture of the properties of the trace that were just mentioned, this equation leads to the contradiction $1 = \operatorname{tr}\varepsilon_{11} = (\operatorname{tr}\alpha_1)b_1 + \cdots + (\operatorname{tr}\alpha_r)b_r = 0$. □

Proposition. *Let A be a finite dimensional F-algebra. Suppose that B is a subspace of A that is closed under multiplication, and is spanned by a set of nilpotent elements. Then $B^k = 0$ for some $k \in \mathbb{N}$.*

Proof. Two reduction steps precede the main part of the proof. First, it can be assumed that

$$F \text{ is algebraically closed.} \tag{1}$$

To see this, let $x_1, \ldots, x_m$ be an F-basis of A, with $x_j x_k = \sum_{i=1}^m x_i c_{ijk}$, $c_{ijk} \in F$. Denote the algebraic closure of F by K. Form the K-algebra $A' = x_1 K \oplus \cdots \oplus x_m K$ with multiplication in A' defined by the structure constants $\{c_{ijk}\}$. Plainly, A is a subalgebra of $(A')_F$. Therefore, B is a subalgebra of $(B')_F$, where $B' = BK$. Thus, $(B')^k = 0$ implies that $B^k = 0$. It remains to note that B' satisfies the same hypotheses as B: B' is a K-subspace of A'; B' is closed under multiplication; B' is spanned by nilpotent elements. The next reduction adds the hypothesis

$$B \text{ is an ideal of } A. \tag{2}$$

To achieve this condition, just replace A by $B + 1_A F$. Since B is closed under multiplication, it is evidently an ideal of $B + 1_A F$. In order to complete the proof, it is sufficient to show

$$\text{if } A \text{ is semisimple, then } B = 0. \tag{3}$$

In fact A can be replaced by the semisimple algebra $A/\mathbf{J}(A)$, and the ideal B of A by the ideal $(B + \mathbf{J}(A))/\mathbf{J}(A)$ of $A/\mathbf{J}(A)$. Since $(B + \mathbf{J}(A))/\mathbf{J}(A)$ is a homomorphic image of B, it is spanned by nilpotent elements. Therefore, (3) leads to the conclusion that $B + \mathbf{J}(A) = \mathbf{J}(A)$, that is $B \subseteq \mathbf{J}(A)$. By Proposition 4.4, $B^k \subseteq \mathbf{J}(A)^k = 0$ for a suitable $k \in \mathbb{N}$. It remains to prove (3), using the added hypotheses (1) and (2). By Corollary 3.5b, $A = A_1 \dotplus \cdots \dotplus A_t$, where $A_i \cong M_{n_i}(F)$ is simple. Let $\pi_i: A \to A_i$ be the projection homomorphism. For each i, $\pi_i(B)$ is an ideal of A_i, so that either $\pi_i(B) = 0$ or $\pi_i(B) = A_i$, because A_i is simple. The second option is ruled out, since it implies that $A_i \cong M_{n_i}(F)$ is spanned by nilpotent elements, in contradiction with the lemma. Therefore, $B \subseteq \ker \pi_i$ for all i. Thus, $B \subseteq \bigcap_{i=1}^t \ker \pi_i = 0$, which proves (3). □

Exercises

1. Let $\alpha = [x_{ij}] \in M_2(\mathbb{H})$ be defined by $x_{11} = -1 + \mathbf{i}$, $x_{12} = \mathbf{j}$, $x_{21} = 2\mathbf{k}$, and $x_{22} = 1 + \mathbf{i}$. Show that $\alpha^2 = 0$, but $x_{11} + x_{22} \neq 0$. This example shows that the proof of the lemma cannot be generalized in a naive way to algebras of the form $M_n(D)$, where D is a division algebra, even though the lemma remains true (if D is finite dimensional over its center).

2. For an F-algebra A, denote by C_A the subspace of A that is spanned (as an F-space) by $\{xy - yx: x, y \in A\}$, and let N_A be the subspace of A that is spanned by the set of all nilpotent elements of A. Prove the following statements.
   (a) If $A = M_n(F)$, then $C_A = N_A = \{\alpha \in A: \operatorname{tr} \alpha = 0\}$. Hint. Show that $\{\varepsilon_{ij}: i \neq j\} \cup \{\varepsilon_{11} - \varepsilon_{jj}: j > 1\}$ spans $\{\alpha \in A: \operatorname{tr} \alpha = 0\}$, $\{\varepsilon_{ij}: i \neq j\} \cup \{\varepsilon_{11} - \varepsilon_{jj}: j > 1\} \subseteq C_A \cap N_A$, and $C_A \cup N_A \subseteq \{\alpha \in A: \operatorname{tr} \alpha = 0\}$.
   (b) If F is algebraically closed, and A is a finite dimensional, semisimple F-algebra, then $N_A = C_A$, and $\dim_F A/N_A = \dim_F \mathbf{Z}(A)$. Hint. See Exercise 1, Section 3.6.
   (c) If F is algebraically closed, and A is a finite dimensional F-algebra, then $N_A = C_A + \mathbf{J}(A)$, and $\dim_F A/N_A = \dim_F \mathbf{Z}(A/\mathbf{J}(A))$. Hint. Show that if $\pi: A \to A/\mathbf{J}(A)$

is the projection homomorphism, then $\pi(C_A) = C_{A/\mathbf{J}(A)}$ and $\pi(N_A) = N_{A/\mathbf{J}(A)}$. Use (b) and the inclusion $N_A \supseteq \mathbf{J}(A)$ (that follows from Proposition 4.4).

3. Assume that F is an algebraically closed field of prime characteristic p. For an F-algebra A, let C_A and N_A be defined as in Exercise 2. Prove the following facts.
 (a) If $x, y \in A$, then $(x + y)^p \equiv x^p + y^p \bmod C_A$. Hint. $(x + y)^p = \sum z_1 \cdots z_p$, where the sum is over all sequences $(z_1, \ldots, z_p)$ of x's and y's. Moreover, $z_1 z_2 \cdots z_p \equiv z_2 \cdots z_p z_1 \equiv \cdots \equiv z_p z_1 \cdots z_{p-1} \bmod C_A$. Use the fact that p is prime to show that the cyclic permutations of $(z_1, \ldots, z_p)$ are distinct unless $z_1 = z_2 = \cdots = z_p$.
 (b) If $z \in C_A$, then $z^p \in C_A$.
 (c) If A is finite dimensional, then $z^p \in N_A$ implies $z \in N_A$. Hint. Reduce the proof to the case in which A is $M_n(F)$. Write z in the form $c\varepsilon_{11} + \phi$, where $\operatorname{tr}\phi = 0$, and apply part (a) of Exercise 2.
 (d) If A is finite dimensional, then $z \in N_A$ if and only if $z^{p^k} \in C_A$ for some $k \in \mathbb{N}$.

4.7. The Radical of a Group Algebra

This section is concerned with the problem of describing the radical of a group algebra. What can be said about $\mathbf{J}(FG)$ when F is a field and G is a finite group? Maschke's Theorem is equivalent to the statement that if $\operatorname{char} F$ does not divide the order of G, then $\mathbf{J}(FG) = 0$. Therefore, assume that $\operatorname{char} F$ is a prime p that divides $n = |G|$. The best known result concerning $\mathbf{J}(FG)$ is a theorem that was discovered by Wallace [74]. The proof of this result is based on the theorem of Wedderburn that was established in the last section.

Proposition. *Let F be a field of prime characteristic p. Assume that G is a finite group that has a normal p-Sylow subgroup H. The Jacobson radical of the group algebra FG is $\mathbf{J}(FG) = \sum_{x \in H - \{1\}} (x - 1)FG$.*

PROOF. Denote $A = FG$. By Proposition 1.2, the projection homomorphism $\phi\colon G \to G/H$ extends linearly to a surjective F-algebra homomorphism $\phi\colon A = FG \to F(G/H)$. If $y_1, \ldots, y_m$ is a collection of left coset representatives of H, that is, $G = Hy_1 \,\dot\cup\, \cdots \,\dot\cup\, Hy_m$, and if $y \in G$, then $\phi(y) = \phi(y_i)$ if and only if $y \in Hy_i$. Therefore, for $z = \sum_{y \in G} ya_y \in A$, we have $\phi(z) = \sum_{y \in G} \phi(y)a_{\dot{y}} = \sum_{i=1}^m \phi(y_i)(\sum_{x \in H} a_{xy_i})$. In particular, if $\phi(z) = 0$, then $\sum_{x \in H} a_{xy_i} = 0$ for $1 \leq i \leq m$. This implies that $a_{y_i} = -\sum_{x \in H - \{1\}} a_{xy_i}$, so that $z = \sum_{i=1}^m \sum_{x \in H - \{1\}} (x - 1)y_i a_{xy_i} = \sum_{x \in H - \{1\}} (x - 1)(\sum_{i=1}^m y_i a_{xy_i}) \in \sum_{x \in H - \{1\}} (x - 1)A$. Conversely, if $z \in \sum_{x \in H - \{1\}}(x - 1)A$, then $\phi(z) \in \sum_{x \in H - \{1\}} (\phi(x) - \phi(1))\phi(A) = 0$. Therefore, $\operatorname{Ker}\phi = \sum_{x \in H - \{1\}} (x - 1)A = J$. This shows that J is an ideal of A such that $A/J \cong F(G/H)$. Since H is a p-Sylow subgroup of G, p does not divide $|G/H|$. Hence, $F(G/H)$ is semisimple by Maschke's Theorem. This fact implies that $J \supseteq \mathbf{J}(A)$. The goal of showing that $J = \mathbf{J}(A)$ will be reached by proving that $J^k = 0$ for some $k \in \mathbb{N}$. Denote

$B = \sum_{x \in H - \{1\}} (x - 1)F$. Plainly, B is a subspace of A. Also, B is closed under multiplication, because $(x - 1)(y - 1) = (xy - 1) - (x - 1) - (y - 1)$. If $|H| = p^l$, then, since char $F = p$, $(x - 1)^{p^l} = x^{p^l} - 1 = 0$ for all $x \in H$. Therefore, B is spanned by nilpotent elements. Proposition 4.6 implies that $B^k = 0$ for some $k \in \mathbb{N}$. The desired result $J^k = 0$ is a consequence of this fact, because $J = BA$ (obviously), and $BA = AB$ (since $(x - 1)y = y(y^{-1}xy - 1)$ and $y^{-1}xy \in H$ for $x \in H$, $y \in G$, by the normality of H). Indeed, $J^k = (BA)^k = B^k A^k = 0A = 0$. □

Corollary. *If H is a finite p-group, and F is a field of characteristic p, then* $\mathbf{J}(FH) = \sum_{x \in H - \{1\}} (x - 1)F$.

PROOF. If $x, y \in H$, then $(x - 1)y = (xy - 1) - (y - 1)$. Consequently, $\sum_{x \in H - \{1\}} (x - 1)FH = \sum_{x \in H - \{1\}} (x - 1)F$. The corollary therefore follows from the proposition. □

EXERCISES

1. Let G be a finite group and suppose that F is a field. Denote $A = FG$, and $I = \sum_{x \in G} (x - 1)F$.

   (a) Prove that $I \lhd A$, and $A/I \cong F$. Hint. Prove that I is the kernel of the *augmentation homomorphism* $A \to F$ that is defined by $\sum_{x \in G} x a_x \mapsto \sum_{x \in G} a_x$. For this reason, I is called the *augmentation ideal* of A.

   (b) Show that if the characteristic of F does not divide the order n of G, then $I = (1 - e)A$, where $e = (\sum_{x \in G} x)n^{-1}$.

2. Let F be an algebraically closed field of prime characteristic p, and suppose that G is a finite group. Denote the group algebra FG by A. Define the subspaces C_A and N_A of A by the prescription that was given in Exercise 2 of Section 4.6. Prove the following statements.

   (a) $C_A = \sum_{x, y \in G} (xy - yx)F$.

   (b) If $z = \sum_{x \in G} x a_x$, then $z \in C_A$ if and only if $\sum_{x \in K} a_x = 0$ for all conjugate classes K in G.

   An element x in a finite group G is called p-regular (p-singular) if the order of x is relatively prime to p (a power of p).

   (c) If $x \in G$, then $x = yz = zy$, where y is p-regular and z is p-singular.

   (d) If x and y are p-regular elements of G such that $x^p = y^p$ (x^p is conjugate to y^p), then $x = y$ (x is conjugate to y).

   (e) If $x = yz = zy$ with y a p-regular element and z a p-singular element, then $x \equiv y \bmod N_A$. Hint. Use Exercise 3, Section 4.6.

   (f) If $y_1, y_2, \ldots, y_m$ belong to distinct conjugate classes of p-regular elements in G, then $y_1 a_1 + y_2 a_2 + \cdots + y_m a_m \in N_A$, $a_1, a_2, \ldots, a_m \in F$, implies $a_1 = a_2 = \cdots = a_m = 0$. Hint. Use parts (a) and (d) of Exercise 3 in Section 4.6, together with (b) and (d) above.

   (g) If $y_1, y_2, \ldots, y_m$ are representatives of the conjugate classes of p-regular elements in G, then $A = N_A \oplus y_1 F \oplus \cdots \oplus y_m F$.

   (h) $\dim_F \mathbf{Z}(A/\mathbf{J}(A))$ is the number of conjugate classes of p-regular elements in G.

4.8. Ideals in Artinian Algebras

We conclude this chapter with some results on the ideal lattices of algebras. If A is semisimple, then $\mathbf{I}(A)$ is distributive. Lemma a proves this fact and slightly more. This result makes it possible to determine whether or not $\mathbf{I}(A)$ is distributive by studying the lattice of sub-bimodules of $\mathbf{J}(A)$, provided $A/\mathbf{J}(A)$ is semisimple. The criterion that is developed here will be used in Chapter 11.

Lemma a. *Let A be a semisimple algebra.*

(i) *If M is a right ideal of A, and N is a left ideal of A, then $MN = M \cap N$.*
(ii) *$\mathbf{I}(A)$ is a distributive lattice.*

Proof. (i) $MN \subseteq M$ since $M < A_A$, and $MN \subseteq N$ since $N < {}_AA$. Therefore, $MN \subseteq M \cap N$. Since A is semisimple, it follows from Proposition 2.4 that $A_A = M \oplus P$ for a suitable right ideal P of A. Consequently, $N = AN = MN + PN \subseteq MN + P$. By the modular law, $N \cap M \subseteq (MN + P) \cap M = MN + (P \cap M) = MN$.

(ii) If I, J, and K are ideals of A, then by (i), $I \cap (J + K) = I(J + K) = IJ + IK = (I \cap J) + (I \cap K)$. Thus, $\mathbf{I}(A)$ is distributive. □

The second part of this lemma can also be deduced from the result of Exercise 1 in Section 3.5.

Lemma b. *Let A be an algebra such that $A/\mathbf{J}(A)$ is semisimple. There exist surjective lattice homomorphisms $\rho : \mathbf{I}(A) \to \mathbf{I}(A/\mathbf{J}(A))$ and $\sigma : \mathbf{I}(A) \to \mathbf{S}({}_A\mathbf{J}(A)_A)$ (where $\mathbf{S}({}_A\mathbf{J}(A)_A)$ is the lattice of sub-bimodules of $\mathbf{J}(A)$), such that if I and J are ideals in A that satisfy $\rho(I) = \rho(J)$ and $\sigma(I) = \sigma(J)$, then $I = J$.*

Proof. Let $\rho : A \to A/\mathbf{J}(A)$ be the projection homomorphism. It follows from the fact that ρ is surjective that $\rho(I) \lhd A/\mathbf{J}(A)$ for all $I \in \mathbf{I}(A)$. If I and J are ideals of A, then $\rho(I + J) = \rho(I) + \rho(J)$, and $\rho(I \cap J) \subseteq \rho(I) \cap \rho(J) = \rho(I)\rho(J) = \rho(IJ) \subseteq \rho(I \cap J)$ by Lemma a. Thus, ρ is a lattice homomorphism. By the Correspondence Theorem, every ideal of $A/\mathbf{J}(A)$ has the form $I/\mathbf{J}(A) = \rho(I)$ for a suitable $I \in \mathbf{I}(A)$. That is, ρ is surjective. Define $\sigma : \mathbf{I}(A) \to \mathbf{S}(\mathbf{J}(A))$ by $\sigma(I) = I \cap \mathbf{J}(A)$. Plainly, $\sigma(I \cap J) = \sigma(I) \cap \sigma(J)$ and $\sigma(I + J) \supseteq \sigma(I) + \sigma(J)$ for $I, J \in \mathbf{I}(A)$. If $x + y \in \mathbf{J}(A)$, where $x \in I$ and $y \in J$, then $\rho(x) = -\rho(y) \in \rho(I) \cap \rho(J) = \rho(I \cap J)$. That is, $\rho(x) = -\rho(y) = \rho(z)$, where $z \in I \cap J$. Therefore, $x - z \in I \cap \mathbf{J}(A) = \sigma(I)$ and $y + z \in J \cap \mathbf{J}(A) = \sigma(J)$, so that $x + y = (x - z) + (y + z) \in \sigma(I) + \sigma(J)$. This calculation shows that $\sigma(I + J) \subseteq \sigma(I) + \sigma(J)$, which proves that σ is a lattice homomorphism. Every sub-bimodule of $\mathbf{J}(A)$ is an ideal of A, so that σ is surjective. Finally, assume that $\rho(I) = \rho(J)$ and $\sigma(I) = \sigma(J)$.

If $x \in I$, then there exists $y \in J$ such that $\rho(x) = \rho(y)$. Hence, $x - y \in (I + J) \cap \mathbf{J}(A) = \sigma(I + J) = \sigma(I) + \sigma(J) = \sigma(J) \subseteq J$, and $x = (x - y) + y \in J$. Similarly, if $x \in J$, then $x \in I$. Therefore, $I = J$. □

Proposition. *Let A be an Artinian algebra. The lattice of ideals of A is distributive if and only if the lattice $\mathbf{S}(\mathbf{J}(A))$ of sub-bimodules of $\mathbf{J}(A)$ is distributive.*

PROOF. Since $\mathbf{S}(\mathbf{J}(A))$ is a sublattice of $\mathbf{I}(A)$, it is evident that if $\mathbf{I}(A)$ is distributive, so is $\mathbf{S}(\mathbf{J}(A))$. Conversely, suppose that $\mathbf{S}(\mathbf{J}(A))$ is distributive. If I, J, and K are ideals of A, then $\sigma(I \cap (J + K)) = \sigma(I) \cap (\sigma(J) + \sigma(K)) = (\sigma(I) \cap \sigma(J)) + (\sigma(I) \cap \sigma(K)) = \sigma((I \cap J) + (I \cap K))$. Similarly, it follows from Lemma a that $\rho(I \cap (J + K)) = \rho((I \cap J) + (I \cap K))$. By Lemma b, $I \cap (J + K) = (I \cap J) + (I \cap K)$. □

Anyone who is familiar with the formalism of universal algebra will recognize Lemma b as the statement that $\mathbf{I}(A)$ is a subdirect product of $\mathbf{I}(A/\mathbf{J}(A))$ and $\mathbf{S}(\mathbf{J}(A))$. In particular, $\mathbf{I}(A)$ is isomorphic to a sublattice of the product of $\mathbf{I}(A/\mathbf{J}(A))$ and $\mathbf{S}(\mathbf{J}(A))$. This observation is the essential idea behind the proposition.

EXERCISES

1. (a) Prove that for an algebra A, the following conditions are equivalent.

   (i) if $M < A_A$ and $N < {}_AA$, then $M \cap N = MN$.
   (ii) For all $x \in A$, there exists $y \in A$ such that $xyx = x$.
   (iii) The principal right and left ideals of A are generated by idempotent elements.

   An algebra that satisfies these conditions is called (von Neumann) *regular*. Thus, by Lemma a, every semisimple algebra is regular. Prove the following statements.
   (b) If V is a vector space over the field F, then $\mathbf{E}_F(V)$ is regular.
   (c) If A is regular, then $\mathbf{S}(A_A)$, $\mathbf{S}({}_AA)$, and $\mathbf{I}(A)$ are distributive lattices.
   (d) If A is regular, then $\mathbf{J}(A) = 0$.
   (e) A regular algebra A is semisimple if and only if A is right (left) Artinian.
   (f) If A is regular, then $\mathbf{Z}(A)$ is regular. Hint. If $x \in \mathbf{Z}(A)$, and $y \in A$ is such that $xyx = x$, show that $z = yxy$ satisfies $xzx = x$, and $z \in \mathbf{Z}(A)$.

2. Let p be a prime, and suppose that $G = \langle x \rangle \times \langle y \rangle$ is the product of two cyclic groups of order p. Let F be a field of characteristic p, and denote the group algebra FG by A. Prove the following statements.
   (a) $\mathbf{J}(A) = (x - 1)A + (y - 1)A$.
   (b) $\mathbf{J}(A)^{p-1} = (x - 1)^{p-1}(y - 1)^{p-1}A$; $\mathbf{J}(A)^p = 0$.
   (c) $\mathbf{I}(A)$ is not distributive. Hint. Show that $I_1 = (x - 1)^{p-1}A$, $I_2 = (y - 1)^{p-1}A$, and $I_3 = ((x - 1)^{p-1} + (y - 1)^{p-1})A$ satisfy $I_1 + I_2 = I_1 + I_3 = I_2 + I_3$ and $I_1 \cap I_2 = I_1 \cap I_3 = I_2 \cap I_3 = \mathbf{J}(A)^{p-1}$.

Notes on Chapter 4

The first four sections of this chapter are adapted from Bass's treatment of the Jacobson radical in [17]. The results of Section 5 are due to Hopkins; they are now standard topics in every ring theory book. Proposition 4.7 is due to Wallace [74]. The various properties of the radical that are offered in the Exercises can be traced back to early research of Jacobson. A more complete discussion of the Jacobson radical can be found in Jacobson's book [46]. An exhaustive treatment of general radicals is given in Divinsky's monograph [28]. The results that appear as Exercises 2 and 3 of Section 4.6, and Exercise 2 of Section 4.7 are due to Brauer. They are important in the theory of modular representations of groups.

CHAPTER 5
Indecomposable Modules

This chapter is the beginning of an examination of algebras that are more general than the semisimple algebras. Our strategy is to generalize the process that led to the Wedderburn Structure Theorem. The appropriate substitutes for the basic building blocks of that theory—simple modules—are *indecomposable modules*. These modules are introduced in this chapter. The analogue of Schur's lemma provides a characterization of indecomposable modules in terms of their endomorphism algebras. The main result of the chapter is the Krull–Schmidt Theorem. It leads to the conclusion that finitely generated modules over an Artinian algebra decompose uniquely into direct sums of indecomposable modules. In short, the results of Chapter 2 have close analogues in the theory of modules over Artinian algebras.

5.1. Direct Decompositions

Throughout this section and the rest of Chapter 5, A is an R-algebra. The commutative ring R plays a minor role in the theory, and reference to it will generally be omitted.

An A-module N is *indecomposable* if $N \neq 0$, and the only direct summands of N are 0 and N, that is, if $N = P \oplus Q$, then either $P = 0$ or $Q = 0$. A module M is *decomposable* if $M = M_1 \oplus M_2$, where M_1 and M_2 are non-zero modules. Thus, the zero module is neither decomposable nor indecomposable.

Proposition. *If M is an A-module that is either Artinian or Noetherian, then M can be written as a finite direct sum of indecomposable A-modules.*

Proof. If $M = 0$, then the proposition is true by virtue of our convention that the empty sum is 0. Assume that $M \neq 0$. We first note (1) there is an indecomposable direct summand of M. In fact, if N is minimal among the non-zero summands of M, then N is evidently indecomposable. The existence of such a minimal N is obvious if M is Artinian; if M is Noetherian, then any complement of a maximal direct summand is minimal. By repeated application of (1), using the fact that the Artinian and Noetherian properties are inherited by submodules, we obtain

$$M = N_1 \oplus M_1 = N_1 \oplus N_2 \oplus M_2 = N_1 \oplus N_2 \oplus N_3 \oplus M_3 = \cdots,$$

with each N_i indecomposable and $M_1 \supset M_2 \supset M_3 \supset \cdots$. This sequence of decompositions terminates at step k only if $M_k = 0$, in which case $M = N_1 \oplus N_2 \oplus \cdots \oplus N_k$. Either the descending chain condition (applied to $M_1 \supset M_2 \supset M_3 \supset \cdots$) or the ascending chain condition (applied to $N_1 \subset N_1 + N_2 \subset N_1 + N_2 + N_3 \subset \cdots$) forces such a termination. □

Exercises

1. For each set X of rational primes, define N_X to be the localization of $\mathbb{Z}$ at the multiplicative set generated by X, that is, $N_X = \{a/n: a \in \mathbb{Z}, n \in \mathbb{N}, (n,p) = 1 \text{ for all primes } p \notin X\}$. Prove that N_X is an indecomposable $\mathbb{Z}$-module, and $N_X \cong N_Y$ implies $X = Y$. This example shows that there are uncountably many indecomposable $\mathbb{Z}$-modules. Of course, N_X is not finitely generated if $X \neq \varnothing$.

2. Let B be the Boolean ring of all subsets of an infinite set X, and define $A = B/I$, where I is the ideal whose elements are the finite subsets of X. Prove that A_A has no indecomposable direct summands. Thus, the proposition fails for A.

5.2. Local Algebras

It is obvious that simple modules are indecomposable. The converse is true for modules over semisimple algebras, but not in general. For the class of Artinian algebras there is a characterization of the finitely generated, indecomposable modules in terms of their endomorphism algebras that is analogous to the characterization of simple modules that is provided by Schur's lemma.

Definition. An algebra A is a *local algebra* if $A/\mathbf{J}(A)$ is a division algebra.

Note that if A is local, then $1_A \neq 0$, that is, A is non-trivial.

Proposition. *For a non-trivial algebra A, the following conditions are equivalent.*

(i) *A is a local algebra.*
(ii) $A - A^\circ \subseteq \mathbf{J}(A)$.
(iii) *$A - A^\circ$ is closed under addition.*

PROOF. (i) implies (ii). If $x \in A - \mathbf{J}(A)$, then (i) provides the existence of $y \in A$ such that $xy - 1 \in \mathbf{J}(A)$ and $yx - 1 \in \mathbf{J}(A)$. Therefore, by Proposition 4.3, $xy = 1 + (xy - 1) \in A^\circ$; similarly, $yx \in A^\circ$. It follows that $x \in A^\circ$. (ii) implies (iii). Since A is non-trivial, it is clear from Proposition 4.3 that no unit of A belongs to $\mathbf{J}(A)$. That is, $\mathbf{J}(A) \cap A^\circ = \varnothing$. Thus, (ii) is in fact equivalent to $A - A^\circ = \mathbf{J}(A)$, and (iii) is a consequence of the fact that $\mathbf{J}(A)$ is an ideal. (iii) implies (i). Suppose that $x \in A - \mathbf{J}(A)$. By Proposition 4.3, there exist elements y and z in A such that $1 + xy \in A - A^\circ$ and $1 + zx \in A - A^\circ$. Consequently $xy \in A^\circ$ and $zx \in A^\circ$, since otherwise $1 \in A - A^\circ$ by (iii). Therefore, x has both a right inverse and a left inverse in A, so that $x \in A^\circ$. This argument shows that $A - \mathbf{J}(A) \subseteq A^\circ$, from which it is clear that $A/\mathbf{J}(A)$ is a division algebra. □

Corollary a. *Let A be an algebra such that every non-unit of A is nilpotent. Then A is a local algebra.*

PROOF. Let $0 \neq x \in A - A^\circ$. By assumption, $x^k = 0$ for some smallest natural number $k > 1$. Then $xy \in A - A^\circ$ for all $y \in A$. Otherwise, $x^{k-1}(xy) = 0$ implies $x^{k-1} = 0$, contrary to the minimality of k. Thus, by hypothesis, every element of xA is nilpotent, so that $x \in xA \subseteq \mathbf{J}(A)$ by Corollary 4.3b. This proves that $A - A^\circ \subseteq \mathbf{J}(A)$. Therefore, A is local by the proposition. □

Corollary b. *If N is an A-module such that $\mathbf{E}_A(N)$ is a local algebra, then N is indecomposable.*

PROOF. The hypothesis that $\mathbf{E}_A(N)$ is local includes the condition that $id_N \neq 0$, so that $N \neq 0$. If $N = P \bigoplus Q$ with the associated projections $\pi : N \to P$ and $\rho : N \to Q$, then, since $\pi + \rho = id_N$ and $\mathbf{E}_A(N)$ is local, either π or ρ is a unit by the proposition. Since $\pi^2 = \pi$ and $\rho^2 = \rho$, it follows that $\pi = id_N$ or $\rho = id_N$; that is, $Q = 0$ or $P = 0$. □

EXERCISES

1. Prove that an algebra A is local if and only if A has a unique maximal right ideal.
2. Let A be a finite dimensional F-algebra. Suppose that B is a subalgebra of A and C is a non-trivial homomorphic image of A. Prove the following statements.
   (a) If A is a division algebra, then B and C are division algebras.
   (b) If A is a local algebra, then B and C are local algebras; moreover, $\mathbf{J}(B) = B \cap \mathbf{J}(A)$.

3. Prove that if A is a local algebra, then every cyclic A-module is indecomposable. Hint. See Exercise 2, Section 2.1.

4. Prove that if A is a finite dimensional F-algebra, and if M is a finitely generated A-module such that $M/M\mathbf{J}(A)$ is simple, then M is indecomposable. Hint. Show that there is a homomorphism θ of $\mathbf{E}_A(M)$ to $\mathbf{E}_A(M/M\mathbf{J}(A))$ such that every ϕ in $\operatorname{Ker}\theta$ is nilpotent. Then use Exercise 2.

5.3. Fitting's Lemma

Our aim in this section is to prove a converse of Corollary 5.2b. The desired result is a consequence of Fitting's Lemma, one of the basic tools of algebra. We begin with a lemma that has many applications of its own.

Lemma. *Let M be an A-module, and suppose that $\phi \in \mathbf{E}_A(M)$. Each of the following hypotheses implies that ϕ is an automorphism.*

(i) *M is Noetherian and ϕ is surjective.*
(ii) *M is Artinian and ϕ is injective.*

PROOF. Assume that M is Noetherian, and ϕ is surjective. Since $0 \subseteq \operatorname{Ker}\phi \subseteq \operatorname{Ker}\phi^2 \subseteq \cdots$, the ascending chain condition implies that $\operatorname{Ker}\phi^n = \operatorname{Ker}\phi^{n+1}$ for some $n \in \mathbb{N}$. That is $(\phi^n)^{-1}(\operatorname{Ker}\phi) = (\phi^{n+1})^{-1}(0) = \operatorname{Ker}\phi^{n+1} = \operatorname{Ker}\phi^n = (\phi^n)^{-1}(0)$. Since ϕ is surjective, so is ϕ^n. Therefore, $\operatorname{Ker}\phi = \phi^n(\phi^n)^{-1}(\operatorname{Ker}\phi) = \phi^n(\phi^n)^{-1}(0) = 0$. The proof that (ii) implies $\phi \in \mathbf{E}_A(M)^\circ$ is left as Exercise 1. □

Fitting's Lemma. *Let M be an A-module that is both Artinian and Noetherian. If $\phi \in \mathbf{E}_A(M)$, then there is a decomposition $M = P \oplus Q$ such that*

(i) *$\phi(P) \subseteq P$ and $\phi(Q) \subseteq Q$,*
(ii) *$\phi|P$ is an automorphism, and*
(iii) *$\phi|Q$ is nilpotent.*

PROOF. The assumption that M is Artinian and Noetherian applied to the chains $M \supseteq \phi(M) \supseteq \phi^2(M) \supseteq \cdots$ and $0 \subseteq \operatorname{Ker}\phi \subseteq \operatorname{Ker}\phi^2 \subseteq \cdots$ yields a natural number m such that $\phi^n(M) = \phi^m(M)$ and $\operatorname{Ker}\phi^n = \operatorname{Ker}\phi^m$ for all $n \geq m$. Define $P = \phi^m(M)$ and $Q = \operatorname{Ker}\phi^m$. Then $\phi(P) = \phi^{m+1}(M) = \phi^m(M) = P$, and $\phi(Q) = \phi(\operatorname{Ker}\phi^m) = \phi(\operatorname{Ker}\phi^{m+1}) \subseteq \operatorname{Ker}\phi^m = Q$. By the lemma, $\phi|P$ is an automorphism. Also, $\phi^m(Q) = \phi^m((\phi^m)^{-1}(0)) = 0$, so that $\phi|Q$ is nilpotent. Moreover, $P \cap Q = 0$, because $\phi|P \cap Q$ is both injective and nilpotent. Finally, $M = (\phi^m)^{-1}(\phi^m(M)) = (\phi^m)^{-1}(\phi^{2m}(M)) = (\phi^m)^{-1}(\phi^m(\phi^m(M))) = \phi^m(M) + \operatorname{Ker}\phi^m = P + Q$. □

Corollary. *If the A-module M is Artinian and Noetherian, then M is indecomposable if and only if* $\mathbf{E}_A(M)$ *is a local algebra.*

PROOF. If $\mathbf{E}_A(M)$ is local, then M is indecomposable by Corollary 5.2b. Assume that M is indecomposable. By Fitting's Lemma, every element of $\mathbf{E}_A(M)$ is nilpotent or a unit. Therefore, $\mathbf{E}_A(M)$ is a local algebra by Corollary 5.2a. □

For us, the most important case of this Corollary occurs when A is right Artinian, and M is a finitely generated right A-module. By Corollary 4.5b, these hypotheses guarantee that M is Artinian and Noetherian. When the corollary is used in this context, we will often omit a reference to Corollary 4.5b.

EXERCISES

1. Complete the proof of the lemma by showing that if M is Artinian and ϕ is injective, then ϕ is an automorphism.
2. Use the lemma to prove that if α and β are n by n matrices with elements in a field, and if $\alpha\beta = \iota_n$, then $\beta\alpha = \iota_n$.
3. Let V be an infinite dimensional F-space. Show that there exists an injective $\phi \in \mathbf{E}_F(V)$ that is not surjective, and there is a surjective ψ in $\mathbf{E}_F(V)$ that is not injective.
4. Use the lemma to prove that for a Noetherian algebra A, if $\bigoplus n A_A \cong \bigoplus m A_A$, then $n = m$.
5. Give an example of a finitely generated, indecomposable $\mathbb{Z}$-module M such that $\mathbf{E}_{\mathbb{Z}}(M)$ is not a local algebra.
6. Use Fitting's Lemma to prove that if $\alpha \in M_n(F)$, then there is a non-singular matrix $\gamma \in M_n(F)$ such that $\gamma^{-1}\alpha\gamma$ has the form

$$\begin{bmatrix} \beta & 0 \\ 0 & \delta \end{bmatrix},$$

where $\beta \in M_r(F)^\circ$, $\delta \in M_s(F)$ $(r + s = n)$, and $\delta^s = 0$.

5.4. The Krull–Schmidt Theorem

This section deals with the uniqueness of direct sum decompositions. The principal result is Azumaya's generalization of a classical theorem of Krull and Schmidt. The proof of this result is prefaced by two lemmas: a standard criterion for an exact sequence to split; a technical matrix calculation.

Lemma a. *For an exact sequence* $0 \to N \xrightarrow{\phi} M \xrightarrow{\psi} P \to 0$ *of A-modules, the following conditions are equivalent.*

(i) *There is a* $\chi \in \operatorname{Hom}_A(P,M)$ *such that* $\psi\chi = id_P$.
(ii) *There is a* $\theta \in \operatorname{Hom}_A(M,N)$ *such that* $\theta\phi = id_N$.
In these cases, $M = \operatorname{Im}\chi \bigoplus \operatorname{Ker}\psi = \operatorname{Im}\phi \bigoplus \operatorname{Ker}\theta$.

PROOF. If (i) holds, then $u = \chi\psi u + (u - \chi\psi u)$ and $\psi(u - \chi\psi u) = \psi u - \psi u = 0$ for all $u \in M$. Thus, $M = \operatorname{Im}\chi + \operatorname{Ker}\psi$. Moreover, $\operatorname{Ker}\psi \cap \operatorname{Im}\chi = \operatorname{Ker}(\psi|\operatorname{Im}\chi) = 0$, since $\psi\chi = id_P$. Also, because $\operatorname{Im}\phi = \operatorname{Ker}\psi$, and ϕ is injective, $\theta u = \phi^{-1}(u - \chi\psi u)$ defines a homomorphism from M to N such that $\theta\phi v = \phi^{-1}\phi v = v$ for all $v \in N$. The analogous proof that (ii) implies (i), and $M = \operatorname{Im}\phi \bigoplus \operatorname{Ker}\theta$ is consigned to Exercise 1. □

If the conditions (i) and (ii) of the lemma hold, then $0 \to N \to M \to P \to 0$ is called a split *exact sequence*. Moreover, if (i) is satisfied, then ψ is called a *split surjection*, and if (ii) prevails, then ϕ is called a *split injection*.

Lemma b. *Let* $M = M_1 \bigoplus M_2 = N_1 \bigoplus N_2$ *be direct sum decompositions of the A-module M. Assume that there is an automorphism* ϕ *of* M, *with*

$$\phi = \begin{bmatrix} \phi_{11} & \phi_{12} \\ \phi_{21} & \phi_{22} \end{bmatrix} \in \begin{bmatrix} \operatorname{Hom}_A(M_1,N_1) & \operatorname{Hom}_A(M_2,N_1) \\ \operatorname{Hom}_A(M_1,N_2) & \operatorname{Hom}_A(M_2,N_2) \end{bmatrix},$$

such that ϕ_{11} *is an isomorphism. Then* $M_2 \cong N_2$.

PROOF. Plainly,

$$\begin{bmatrix} id_{N_1} & 0 \\ -\phi_{21}\phi_{11}^{-1} & id_{N_2} \end{bmatrix} \text{ and } \begin{bmatrix} id_{M_1} & -\phi_{11}^{-1}\phi_{12} \\ 0 & id_{M_2} \end{bmatrix}$$

are automorphisms of M. Since ϕ is an automorphism, so is

$$\begin{bmatrix} id_{N_1} & 0 \\ -\phi_{21}\phi_{11}^{-1} & id_{N_2} \end{bmatrix} \begin{bmatrix} \phi_{11} & \phi_{12} \\ \phi_{21} & \phi_{22} \end{bmatrix} \begin{bmatrix} id_{M_1} & -\phi_{11}^{-1}\phi_{12} \\ 0 & id_{M_2} \end{bmatrix} = \begin{bmatrix} \phi_{11} & 0 \\ 0 & \psi \end{bmatrix},$$

where $\psi = \phi_{22} - \phi_{21}\phi_{11}^{-1}\phi_{12} \in \operatorname{Hom}_A(M_2,N_2)$. Thus, ψ is also an isomorphism. □

Proposition. *Let A be an R-algebra. Suppose that M and N are right A-modules with* $M = M_1 \bigoplus \cdots \bigoplus M_r$, $N = N_1 \bigoplus \cdots \bigoplus N_s$, *where* $\mathbf{E}_A(M_i)$ *and* $\mathbf{E}_A(N_j)$ *are local algebras for all i and j. If* $M \cong N$, *then* $r = s$ *and there is a permutation* σ *such that* $M_i \cong N_{\sigma(i)}$ *for* $1 \le i \le r$.

PROOF. Use induction on r, starting with $r = 0$, that is $M = 0$. For the base step of the induction $N \cong M = 0$, so that $s = 0$. (Note that by definition local algebras are non-trivial; hence $\mathbf{E}_A(N_j)$ local implies $N_j \neq 0$.) Assume that $r > 0$, and the proposition is valid for modules that can be written as a direct sum of fewer than r factors that have local endomorphism algebras. Without loss of generality, it can be supposed that $N = M$; just transfer the decomposition of N to the module M, using the isomorphism that is assumed to exist. Thus, we have

$$M = M_1 \oplus \cdots \oplus M_r = N_1 \oplus \cdots \oplus N_s.$$

Let $\pi_i\colon M \to M_i$, $\kappa_i\colon M_i \to M$; $\rho_j\colon M \to N_j$, $\lambda_j\colon N_j \to M$ be the canonical projections and injections that are associated with these decompositions of M. Then $id_M = \lambda_1\rho_1 + \cdots + \lambda_s\rho_s$ and $id_{M_1} = \pi_1\kappa_1 = \sum_{j=1}^{s} \pi_1\lambda_j\rho_j\kappa_1$. Since $\mathbf{E}_A(M_1)$ is a local algebra, it follows from Proposition 5.2 that $\phi = \pi_1\lambda_j\rho_j\kappa_1$ is a unit of $\mathbf{E}_A(M_1)$ for some index j. For notational convenience, order the decomposition $N_1 \oplus \cdots \oplus N_s$ so that $j = 1$. Let $\psi = \phi^{-1}\pi_1\lambda_1 \in \mathrm{Hom}_A(N_1, M_1)$ and $\chi = \rho_1\kappa_1 \in \mathrm{Hom}_A(M_1, N_1)$, so that $\psi\chi = id_{M_1}$. It follows from Lemma a that $N_1 = \mathrm{Ker}\,\psi \oplus \mathrm{Im}\,\chi$. However, since $\mathbf{E}_A(N_1)$ is local, N_1 is indecomposable by Corollary 5.2b. Thus, $N_1 = \mathrm{Im}\,\chi$, and $\chi = \rho_1\kappa_1$ is an isomorphism. Denote $M' = M_2 \oplus \cdots \oplus M_r$ and $N' = N_2 \oplus \cdots \oplus N_s$, so that $M = M_1 \oplus M' = N_1 \oplus N'$ with the corresponding canonical projections and injections $\pi_1\colon M \to M_1$, $\pi'\colon M \to M'$, $\kappa_1\colon M_1 \to M$, $\kappa'\colon M' \to M$; $\rho_1\colon M \to N_1$, $\rho'\colon M \to N'$, $\lambda_1\colon N_1 \to M$, $\lambda'\colon N' \to M$. The matrix

$$\begin{bmatrix} \rho_1\kappa_1 & \rho_1\kappa' \\ \rho'\kappa_2 & \rho'\kappa' \end{bmatrix}$$

corresponds to the composition of the isomorphisms $M_1 \oplus M' \to M$ and $M \to N_1 \oplus N'$ (defined by $(u_1,u') \mapsto u_1 + u'$ and $v \mapsto (\rho_1 v, \rho' v)$), so that it is an isomorphism. Since $\rho_1\kappa_1$ is also an isomorphism, it follows from Lemma b that $M_2 \oplus \cdots \oplus M_r = M' \cong N' = N_2 \oplus \cdots \oplus N_s$. The induction hypothesis applies to M' and N', and it completes the proof. □

Corollary a. *If M is a right A-module that is both Artinian and Noetherian, then $M = M_1 \oplus \cdots \oplus M_r$, where each M_i is an indecomposable A-module; this decomposition of M is unique up to isomorphism.*

The corollary follows directly from the proposition, Proposition 5.1, and Corollary 5.3. The classical Krull–Schmidt Theorem is a slight generalization of this corollary (to groups with operators).

Corollary b. *If A is a right Artinian R-algebra, then every finitely generated A-module is uniquely (to isomorphism) a finite direct sum of indecomposable A-modules.*

This result is obtained by using Corollary a with Corollary 4.5b.

EXERCISES

1. Complete the proof of Lemma a: show that (ii) implies (i) and $M = \operatorname{Im}\phi \oplus \operatorname{Ker}\theta$.

2. Let V be a four dimensional $\mathbb{Q}$-space with the basis w, x, y, z. Define M to be the $\mathbb{Z}$-submodule of V that is generated by $\{1/5^n w, 1/5^n x, 1/7^n y, 1/11^n z, 1/3(x + y), 1/2(x + z) : n \in \mathbb{N}\}$. Let M_1, M_2, M_3, and M_4 be the $\mathbb{Z}$-submodules of V defined by the respective generating sets: $\{1/5^n w : n \in \mathbb{N}\}$;
$\{1/5^n x, 1/7^n y, 1/11^n z, 1/3(x + y), 1/2(x + z) : n \in \mathbb{N}\}$;
$\{1/5^n(3w - x), 1/7^n y, 1/3(3w - x - y) : n \in \mathbb{N}\}$;
$\{1/5^n(2w - x), 1/11^n z, 1/2(2w - x - z) : n \in \mathbb{N}\}$.
Prove the following statements.
   (a) $M = M_1 \oplus M_2 = M_3 \oplus M_4$.
   (b) M_1, M_2, M_3, and M_4 are indecomposable.
   (c) $M_1 \ncong M_3$, $M_1 \ncong M_4$, $M_2 \ncong M_3$, $M_2 \ncong M_4$.
This example shows that the Krull–Schmidt Theorem fails for abelian groups that are not finitely generated.

3. In this problem, A is assumed to be a right Artinian algebra. All modules are finitely generated, right A-modules. The isomorphism class of a module M is denoted by (M). The collection of all classes (M) with M finitely generated is a set. Denote by $\mathbf{F}(A)$ the free $\mathbb{Z}$-module that has this set as a basis. The elements of $\mathbf{F}(A)$ are uniquely represented as finite sums $\sum_{i=k}^{k}(M_i)n_i$ with $n_i \in \mathbb{Z}$. Let $\mathbf{R}(A)$ be the subgroup of $\mathbf{F}(A)$ that is generated by all of the elements $(M_2) - (M_1) - (M_3)$ for which there is an exact sequence $0 \to M_1 \to M_2 \to M_3 \to 0$. Let $\mathbf{R}_0(A)$ be the subgroup of $\mathbf{F}(A)$ that is generated by all of the elements $(M_2) - (M_1) - (M_3)$ such that $M_2 \cong M_1 \oplus M_3$. Finally, define the quotient groups $\mathbf{K}(A) = \mathbf{F}(A)/\mathbf{R}(A)$ and $\mathbf{K}_0(A) = \mathbf{F}(A)/\mathbf{R}_0(A)$. The group $\mathbf{K}(A)$ is called the *Grothendieck group* of the category of finitely generated right A-modules; $\mathbf{K}_0(A)$ is called the *Krull–Schmidt–Grothendieck group* of the category. If M is a finitely generated right A-module, denote $[M] = (M) + \mathbf{R}(A)$, and $[M]_0 = (M) + \mathbf{R}_0(A)$. Prove the following facts.
   (a) There is a surjective homomorphism $\theta: \mathbf{K}_0(A) \to \mathbf{K}(A)$ such that $\theta([M]_0) = [M]$. If A is semisimple, then θ is an isomorphism.
   (b) $[M_1] + [M_2] = [M_1 \oplus M_2]$ and $[M_1]_0 + [M_2]_0 = [M_1 \oplus M_2]_0$.
   (c) Every element of $\mathbf{K}(A)$ (or $\mathbf{K}_0(A)$) can be represented in the form $[M] - [N]$ (respectively, $[M]_0 - [N]_0$).
   (d) $[M]_0 = [N]_0$ if and only if $M \cong N$. Hint. Use the definition of $\mathbf{R}_0(A)$ to prove that $[M]_0 = [N]_0$ if and only if $M \oplus P = N \oplus P$ for some finitely generated A-module P.
   (e) $\mathbf{K}_0(A)$ is isomorphic to the free $\mathbb{Z}$-module with the basis $\{(M) : M =$ finitely generated, indecomposable A-module$\}$.
   (f) If $0 = M_0 \subset M_1 \subset \cdots \subset M_{n-1} \subset M_n = M$ is a composition series, then $[M] = \sum_{i<n}[M_{i+1}/M_i]$ in $\mathbf{K}(A)$.
   (g) $\mathbf{K}(A)$ is generated as a $\mathbb{Z}$-module by $\{[N_1], \ldots, [N_r]\}$, where $N_1, \ldots, N_r$ are representatives of the distinct isomorphism classes of simple right A-modules.
   (h) $\mathbf{K}(A)$ is isomorphic to the free abelian group generated by $[N_1], \ldots, [N_r]$ (as in (g)). Hint. For $1 \le j \le r$, define $\pi_j: \mathbf{F}(A) \to \mathbb{Z}$ by $\pi_j((M)) =$ number of composition factors of M that are isomorphic to N_j. Show that $\operatorname{Ker}\pi_j \supseteq \mathbf{R}(A)$, so that π_j induces a homomorphism $\rho_j: \mathbf{K}(A) \to \mathbb{Z}$, and $\rho_j([N_i]) = \delta_{ij}$.

5.5. Representations of Algebras

One of the best reasons for studying modules is to get insight on group representations. The purpose of this section is to clarify the relation between the representations of an algebra A, and the modules over A. We will limit our considerations to algebras over a field F, since most applications of the theory fulfill this restriction.

Definition. A *matrix representation* of an F-algebra A is an algebra homomorphism θ of A to the F-algebra of all n by n matrices with entries in the field F.

The natural number n is called the *degree* of θ. It will be denoted by $\deg \theta$.

A representation θ of A is *faithful* if $\operatorname{Ker} \theta = 0$. In this case $\dim_F A \leq \dim_F M_n(F) = n^2$, where $n = \deg \theta$. In particular, A cannot have a faithful representation if it is infinite dimensional. (See Corollary b below for a converse result.)

The discussion of representations can be presented most efficiently by using some concepts of category theory. The first step in this program is to introduce an appropriate notion for the morphisms between representations. This turns out to be an idea that was used in the earliest work on group representations. Let θ and ψ be representations of the algebra A that have degrees n and m respectively. An n by m matrix α with entries in F *intertwines* θ and ψ if $\theta(x)\alpha = \alpha\psi(x)$ for all $x \in A$. The intertwining matrices play the roles of morphisms. For this reason we will use the notation $\alpha\colon \theta \to \psi$ to abbreviate the statement "α intertwines θ and ψ." If $\alpha\colon \theta \to \psi$ and $\beta\colon \psi \to \chi$ are matrices that intertwine representations of A, then $\theta(x)\alpha\beta = \alpha\psi(x)\beta = \alpha\beta\chi(x)$ for all $x \in A$. Thus, the matrix product $\alpha\beta$ intertwines θ and χ. This calculation shows that the composition of morphisms in the category of representations of A can be taken to be matrix multiplication in the reverse order, that is, $\alpha\beta = \beta \circ \alpha$. The required associativity of composition is then automatically satisfied. Finally, the identity matrix ι_n plainly intertwines a representation of degree n with itself, and it has the usual properties of an identity morphism. This discussion shows that the representations of an algebra A, together with the intertwining matrices, form a category. A point of caution should be made. It is not accurate to identity the morphisms of this category with matrices. Instead, the morphisms should be viewed as triples (θ,α,ψ), where $\alpha\colon \theta \to \psi$. A single matrix can intertwine many different pairs of representations, and fail to intertwine others. Nevertheless, we will use notation that suppresses the dependence of a morphism on its domain and target.

Representations θ and ψ of A are *equivalent* if they are isomorphic in the categorical sense. This means that there are morphisms $\alpha\colon \theta \to \psi$ and $\beta\colon \psi \to \theta$ whose compositions in both orders are identity morphisms. It is easy to show (using Lemma 5.3, for instance) that θ and ψ are equivalent if and only if $\deg\theta = \deg\psi$, and there is a non-singular (square) matrix α that intertwines θ and ψ. In other words, $\psi(x) = \alpha^{-1}\theta(x)\alpha$ for all $x \in A$. We will write $\theta \cong \psi$ if θ and ψ are equivalent. Clearly, $\cong$ is an equivalence relation.

Let θ be a representation of A, with $\deg\theta = n$. Use θ to define an A-module M_θ in the following way. As an F-space, $M_\theta = \bigoplus n F$. The scalar operation of A on M_θ is given by

$$[a_1, \ldots, a_n]x = [a_1, \ldots, a_n]\theta(x) \tag{1}$$

(matrix product on the right side) for all $x \in A$. A routine calculation shows that M_θ is indeed a right A-module. It is clear from (1) that

$$\dim_F M_\theta = \deg\theta \quad \text{and} \quad \operatorname{ann} M_\theta = \operatorname{Ker}\theta. \tag{2}$$

The correspondence $M\colon \theta \mapsto M_\theta$ is the object map of a functor into the category of right A-modules. In fact, suppose that $\alpha\colon \theta \to \psi$. Define $\mu_\alpha\colon M_\theta \to M_\psi$ by $\mu_\alpha([a_1, \ldots, a_n]) = [a_1, \ldots, a_n]\alpha$. Plainly, μ_α is an F-linear mapping from M_θ to M_ψ, and $\mu_\alpha([a_1, \ldots, a_n]x) = \mu_\alpha([a_1, \ldots, a_n]\theta(x)) = [a_1, \ldots, a_n]\theta(x)\alpha = [a_1, \ldots, a_n]\alpha\psi(x) = \mu_\alpha([a_1, \ldots, a_n])x$. That is, $\mu_\alpha \in \operatorname{Hom}_A(M_\theta, M_\psi)$. Clearly, $\mu_\alpha\mu_\beta = \mu_{\beta\alpha} = \mu_{\alpha\circ\beta}$. Thus, the mappings $\theta \mapsto M_\theta$, $\alpha \mapsto \mu_\alpha$ define a functor from the category of representations of A to the category of right A-modules. The basic properties of this functor are collected in the next result.

Proposition. *Let θ and ψ be representations of the F-algebra A.*

(i) *If $\alpha\colon \theta \to \psi$ and $\beta\colon \theta \to \psi$ satisfy $\mu_\alpha = \mu_\beta$, then $\alpha = \beta$.*
(ii) *If $\phi \in \operatorname{Hom}_A(M_\theta, M_\psi)$, then there exists $\alpha\colon \theta \to \psi$ such that $\phi = \mu_\alpha$.*
(iii) *If M is a right A-module such that $0 < \dim_F M = n \in \mathbb{N}$, then there is a representation χ of A such that $M \cong M_\chi$.*

Proof. If $\mu_\alpha = \mu_\beta$, then $[a_1, \ldots, a_n]\alpha = [a_1, \ldots, a_n]\beta$ for all $a_1, \ldots, a_n \in F$. Clearly, this can happen only if $\alpha = \beta$. If $\phi \in \operatorname{Hom}_A(M_\theta, M_\psi)$, then in particular, ϕ is a linear mapping between two row vector spaces. Thus, there is a matrix α such that $\phi([a_1, \ldots, a_n]) = [a_1, \ldots, a_n]\alpha$ for all $a_1, \ldots, a_n$ in F. Also, ϕ is an A-module homomorphism. Thus $[a_1, \ldots, a_n]\theta(x)\alpha = \phi([a_1, \ldots, a_n]x) = \phi([a_1, \ldots, a_n])x = [a_1, \ldots, a_n]\alpha\psi(x)$ for every $a_1, \ldots, a_n \in F$, so that α intertwines θ and ψ. By definition, $\mu_\alpha = \phi$. For the proof of (iii), choose a basis $u_1, \ldots, u_n$ of M. Define $\chi\colon A \to M_n(F)$ by the condition (written in matrix form)

$$[u_1, \ldots, u_n]x \;(=[u_1 x, \ldots, u_n x]) = [u_1, \ldots, u_n]\chi(x)^t, \tag{3}$$

where the superscript t denotes the matrix transpose. Easy calculations based on (3) show that $\chi(x + y) = \chi(x) + \chi(y)$, $\chi(xa) = \chi(x)a$, and $\chi(xy) = \chi(x)\chi(y)$, for all $x, y \in A$ and $a \in F$. Thus, χ is a representation of A. Define

$\phi\colon M \to \bigoplus n F = M_\chi$ by $\phi(v) = [a_1, \ldots, a_n]$, where $v = u_1 a_1 + \cdots + u_n a_n$. Clearly, ϕ is an F-space isomorphism that is characterized by $v = [u_1, \ldots, u_n]\phi(v)^t$. If $v \in M$ and $x \in A$, then $[u_1, \ldots, u_n]\phi(vx)^t = vx = [u_1, \ldots, u_n]\phi(v)^t x = [u_1, \ldots, u_n]x\phi(v)^t = [u_1, \ldots, u_n]\chi(x)^t\phi(v)^t$ by (3) and the fact that $F \subseteq \mathbf{Z}(A)$. Therefore, $\phi(vx) = \phi(v)\chi(x) = \phi(v)x$; that is, ϕ is an A-module isomorphism. □

Corollary a. *If θ and ψ are representations of A, then $\theta \cong \psi$ if and only if $M_\theta \cong M_\psi$.*

Proof. If $\alpha\colon \theta \to \psi$ is non-singular, then $\mu_\alpha\colon M_\theta \to M_\psi$ is a module homomorphism such that $\mu_{\alpha^{-1}} = (\mu_\alpha)^{-1}$. Conversely, an isomorphism from M_θ to M_ψ is given by μ_α, where $\alpha\colon \theta \to \psi$ is non-singular. □

Corollary b. *Let A be an F-algebra such that* $\dim_F A = n$. *There exists a faithful representation θ of A such that* $\deg \theta = n$.

Proof. By the proposition, there is a representation θ of A such that $A_A \cong M_\theta$. Thus, $\deg \theta = \dim_F M_\theta = \dim_F A = n$, and $\operatorname{Ker} \theta = \operatorname{ann} M_\theta = \operatorname{ann} A_A = 0$ by (2). □

The discussion of algebra representations in this section has a close parallel in the theory of group representations. If G is a group and F is a field, an *F-representation* of G is a group homomorphism θ of G to $GL_n(F)$, where $GL_n(F) = M_n(F)^\circ$ is the general linear group of non-singular n by n matrices with entries in F. As in the case of algebras, the F-representations of G form a category in which the morphisms are triples (θ,α,ψ), such that α is a matrix that intertwines θ and ψ: $\theta(x)\alpha = \alpha\psi(x)$ for all $x \in G$.

The basic observation to make about the category of F-representations of G is that it is isomorphic to the category of representations of the group algebra FG. Indeed, by Proposition 1.2, if θ is a group homomorphism of G to $GL_n(F)$, then θ has a unique extension to an algebra homomorphism of FG to $M_n(F)$. Conversely, any algebra homomorphism of FG to $M_n(F)$ restricts to a group homomorphism of G to $GL_n(F)$. Thus, there is a natural, one-to-one correspondence between F-representations of G and the representations of FG. This correspondence is a category isomorphism, since a matrix α intertwines the F-representations θ and ψ of G if and only if α intertwines the extensions of θ and ψ to FG. In short, the F-representations of G are completely interchangeable with the algebra representations of FG.

Exercises

1. Prove that if θ is a representation of degree n, ψ a representation of degree m, $\alpha\colon \theta \to \psi$ and $\beta\colon \psi \to \theta$ intertwine these representations, and $\alpha\beta = \iota_n$, $\beta\alpha = \iota_m$, then $m = n$ and $\beta = \alpha^{-1}$.

2. Prove that the scalar operations that are defined by (1) satisfy the module axioms.

3. Prove that the mapping χ that is defined by (3) is a module homomorphism.

4. Let $A = \left(\frac{a,b}{F}\right)$ be a quaternion algebra. Prove that there is a representation θ of A with $\deg \theta = 4$, such that $\theta(1) = \iota_4$,

$$\theta(\mathbf{i}) = \begin{bmatrix} 0 & 1 & 0 & 0 \\ a & 0 & 0 & 0 \\ 0 & 0 & 0 & -1 \\ 0 & 0 & -a & 0 \end{bmatrix}, \quad \theta(\mathbf{j}) = \begin{bmatrix} 0 & 0 & 1 & 0 \\ 0 & 0 & 0 & 1 \\ b & 0 & 0 & 0 \\ 0 & b & 0 & 0 \end{bmatrix}, \quad \theta(\mathbf{k}) = \begin{bmatrix} 0 & 0 & 0 & 1 \\ 0 & 0 & a & 0 \\ 0 & -b & 0 & 0 \\ -ab & 0 & 0 & 0 \end{bmatrix}.$$

Show that $M_\theta \cong A_A$.

5.6. Indecomposable and Irreducible Representations

If θ and ψ are representations of the algebra A with $\deg \theta = n$ and $\deg \psi = m$, then the *direct sum* of θ and ψ is the mapping $\theta \oplus \psi\colon A \to M_{n+m}(F)$ that is defined by

$$(\theta \oplus \psi)(x) = \begin{bmatrix} \theta(x) & 0 \\ 0 & \psi(x) \end{bmatrix}.$$

It is obvious that $\theta \oplus \psi$ is a representation of degree $n + m$.

A representation ψ of A is *indecomposable* if ψ cannot be written as a direct sum of two representations (of positive degree). The Krull–Schmidt Theorem can be translated to a fundamental property of representations. It is convenient to preface the result.

Lemma a. *If θ and ψ are representations of A, then $M_{\theta \oplus \psi} \cong M_\theta \oplus M_\psi$.*

PROOF. The mapping $([a_1, \ldots, a_n], [b_1, \ldots, b_m]) \mapsto [a_1, \ldots, a_n, b_1, \ldots, b_m]$ is obviously an F-space isomorphism of $M_\theta \oplus M_\psi$ to $M_{\theta \oplus \psi}$, and the scalar operations for these modules are defined in a way that makes this mapping a module isomorphism. □

Proposition a. (i) *θ is an indecomposable representation of A if and only if M_θ is an indecomposable A-module.*

(ii) *Every representation θ of A is equivalent to a finite direct sum of indecomposable representations.*

(iii) *If $\psi_1 \oplus \cdots \oplus \psi_r \cong \chi_1 \oplus \cdots \oplus \chi_s$ with all ψ_i and χ_j indecomposable, then $r = s$, and there is a permutation σ such that $\chi_i \cong \psi_{\sigma(i)}$ for all i.*

PROOF. If $\theta = \psi \oplus \chi$, then $M_\theta \cong M_\psi \oplus M_\chi$ by Lemma a. Thus, M_θ is not indecomposable. Conversely, if $M_\theta = N_1 \oplus N_2$ with $N_1 \neq 0$ and $N_2 \neq 0$, then by Proposition 5.5, $M_\theta \cong M_\psi \bigoplus M_\chi$ for suitable representations ψ and χ. By Lemma a and Corollary 5.5a $\theta \cong \psi \bigoplus \chi$. Thus, θ is not indecomposable. To prove (ii), note that M_θ is finite dimensional, hence Artinian and Noetherian. By Proposition 5.1, $M_\theta = N_1 \bigoplus \cdots \bigoplus N_r$, with the N_i finite dimensional, indecomposable A-modules. By Proposition 5.5, there are representations ψ_i of A such that $N_i \cong M_{\psi_i}$. Hence, $\theta \cong \psi_1 \bigoplus \cdots \bigoplus \psi_r$ as before. By (i), each ψ_i is indecomposable. The uniqueness statement (iii) is obtained from (i) and Corollary 5.4a: $\psi_1 \bigoplus \cdots \bigoplus \psi_r \cong \chi_1 \bigoplus \cdots \bigoplus \chi_s$ implies $M_{\psi_1} \bigoplus \cdots \bigoplus M_{\psi_r} \cong M_{\chi_1} \bigoplus \cdots \bigoplus M_{\chi_s}$ with all summands indecomposable; thus $r = s$, and $M_{\chi_i} \cong M_{\psi_{\sigma(i)}}$ (hence $\chi_i \cong \psi_{\sigma(i)}$ by Corollary 5.5a) for a suitable permutation σ. □

Any two representations of A can be intertwined by a zero matrix of the appropriate dimensions. In some cases, this is the only intertwining that is possible.

Proposition b. *For a representation θ of the F-algebra A, the following conditions are equivalent.*

(i) *θ is equivalent to a representation ψ of A such that for all $x \in A$, $\psi(x)$ has the form*

$$\begin{bmatrix} \psi_1(x) & * \\ 0 & \psi_2(x) \end{bmatrix},$$

where ψ_1 and ψ_2 are representations of A.

(ii) *There is a representation χ of A with $\deg\chi < \deg\theta$, and a non-zero intertwining $\alpha\colon \theta \to \chi$.*

(iii) *M_θ is not simple.*

Proof. If (i) is satisfied, then (ii) holds true with $\chi = \psi_1$. Indeed, if $r = \deg\psi_1$, $\deg\theta = n$, and 0 is the $n - r$ by r zero matrix, then $\begin{bmatrix} I_r \\ 0 \end{bmatrix}$ intertwines ψ and ψ_1. Since $\theta \cong \psi$, there is a non-zero intertwining $\alpha\colon \theta \to \psi_1$. Assume that (ii) is satisfied. By Proposition 5.5, $\mu_\alpha\colon M_\theta \to M_\chi$ is a non-zero homomorphism. If M_θ were simple, then μ_α would be injective by Schur's Lemma, so that by 5.5(2), $\deg\theta = \dim_F M_\theta \le \dim_F M_\chi = \deg\chi$, contrary to the hypothesis. Hence, M_θ is not simple. Thus, (ii) implies (iii). If M_θ is not simple, then there is a sub-module N of M_θ such that $0 \ne N \subset M_\theta$. Choose an F-space basis $u_1, \ldots, u_r, u_{r+1}, \ldots, u_n$ of M_θ in such a way that $u_{r+1}, \ldots, u_n$ is a basis of N. Thus, $1 \le r \le n - 1$. Define $\psi\colon A \to M_n(F)$ by $[u_1, \ldots, u_n]x = [u_1, \ldots, u_n]\psi(x)^t$. The proof of Proposition 5.5 shows that ψ is a representation of A such that $M_\psi \cong M_\theta$. Therefore, $\psi \cong \theta$ by Corollary 5.5a. Let $\psi(x) = [a_{ij}]$, so that by definition $u_i x = \sum_{j=1}^n u_j a_{ij}$. Since $N < M_\theta$ and $u_{r+1}, \ldots, u_n \in N$, it follows that $a_{ij} = 0$ if $1 \le j \le r < i \le n$. In other words, $\psi(x)$ has the form

$$\begin{bmatrix} \psi_1(x) & * \\ 0 & \psi_2(x) \end{bmatrix},$$

where $\psi_1 \colon A \to M_r(F)$ and $\psi_2 \colon A \to M_{n-r}(F)$ are suitable mappings. The fact that ψ is a representation of A implies by an easy calculation that ψ_1 and ψ_2 are also representations. □

A representation θ of A is *irreducible* if θ satisfies the negations of the conditions (i), (ii), and (iii) in Lemma b. In particular, θ is irreducible if and only if M_θ is simple.

We need another characterization of simple modules. The result is valid for R-algebras.

Lemma b. *Let N be a right A-module. If N is simple, then $\mathbf{J}(A) \subseteq \operatorname{ann} N$ and N is indecomposable. The converse is true if A is right or left Artinian: $\mathbf{J}(A) \subseteq \operatorname{ann} N$ and N is indecomposable implies N is simple.*

PROOF. If N is simple, then by Corollary 4.1b, $N\mathbf{J}(A) \subseteq \operatorname{rad} N = 0$. That is, $\mathbf{J}(A) \subseteq \operatorname{ann} N$. Clearly, N is indecomposable. For the converse, note that by Proposition 2.1, N can be viewed as an $A/\mathbf{J}(A)$-module. The assumption that A is Artinian guarantees that $A/\mathbf{J}(A)$ is semisimple; hence, N is semisimple by Proposition 3.1b. Since N is also indecomposable, it follows from Corollary 2.3b that N is simple. □

Corollary a. *If A is a semisimple F-algebra, then a representation θ of A is indecomposable if and only if θ is irreducible.*

Corollary b. *Let A be a right Artinian F-algebra. The number of equivalence classes of irreducible representations of A is the number of factors in a decomposition of $A/\mathbf{J}(A)$ as a product of simple algebras.*

PROOF. By Lemma b, there is a one-to-one correspondence between the isomorphism classes of simple A-modules, and the isomorphism classes of simple $A/\mathbf{J}(A)$-modules. If $A/\mathbf{J}(A) = A_1 \dotplus \cdots \dotplus A_r$ with each A_i simple, then by Proposition 3.1b and Lemma 3.2a, each simple $A/\mathbf{J}(A)$-module is isomorphic to a minimal right ideal of some A_i. All minimal right ideals in A_i are isomorphic by Proposition 3.3b, whereas, if $i \neq j$, then a minimal right ideal of A_i is not isomorphic to a minimal right ideal of A_j. The corollary therefore follows from Proposition b. □

Corollary c. *Let F be an algebraically closed field, and assume that A is a finite dimensional F-algebra. The number of irreducible representations of A is $\dim_F \mathbf{Z}(A/\mathbf{J}(A))$.*

PROOF. By Corollary 3.5b, $A \cong M_{n_1}(F) \dotplus \cdots \dotplus M_{n_r}(F)$. The number r is $\dim_F \mathbf{Z}(A/\mathbf{J}(A))$. (See Exercise 1, Section 3.6.) □

Exercises

In all of these exercises, assume that F is a field with char $F = 0$.

1. In this exercise, A is assumed to be a finite dimensional F-algebra. If θ is a representation of A, define the *character afforded by* θ to be the mapping $\chi_\theta: A \to F$ that is defined by $\chi_\theta(x) = \operatorname{tr}\theta(x)$. Prove the following statements.

   (a) χ_θ is an F-space homomorphism of A to F, and $\chi_\theta(1_A) = \deg\theta$.

   (b) $\chi_\theta(\mathbf{J}(A)) = 0$.

   (c) If $\theta \cong \psi$, then $\chi_\theta = \chi_\psi$.

   It follows from (c) and Proposition 5.5 that a character can be associated with each finite dimensional A-module M by defining $\chi_M(x) = \chi_\theta(x)$, where θ is a representation such that $M_\theta \cong M$. The character χ_M is said to be afforded by the module M.

   (d) If $0 \to N \to M \to P \to 0$ is a sequence of finite dimensional A-modules, then $\chi_M = \chi_N + \chi_P$. Hint. Copy the last part of the proof of Proposition b.

   (e) If M is a finite dimensional, right A-module, then $\chi_M = \chi_N$ for some semisimple module N.

   Let $\mathbf{X}(A)$ be the subgroup of $\operatorname{Hom}_F(A,F)$ that is generated by the set of all characters of representations of A.

   (f) $\mathbf{X}(A)$ is generated by $\{\chi_\theta: \theta \text{ is irreducible}\}$.

   (g) $\mathbf{X}(A/\mathbf{J}(A)) \cong \mathbf{X}(A)$ by the mapping $\bar{\chi} \mapsto \bar{\chi} \circ \pi$ ($\bar{\chi} \in \mathbf{X}(A/\mathbf{J}(A))$), where π: $A \to A/\mathbf{J}(A)$ is the projection homomorphism.

   (h) If $A = B \dotplus C$, then $\mathbf{X}(A) \cong \mathbf{X}(B) \bigoplus \mathbf{X}(C)$.

   (i) $\mathbf{X}(M_n(F)) \cong \mathbb{Z}$.

   (j) If F is algebraically closed, then $\mathbf{X}(A)$ is a free $\mathbb{Z}$-module with a basis that consists of the characters of irreducible representations.

2. Assume that F is an algebraically closed field. Let θ and ψ be irreducible representations of the F-algebra A. Suppose that $\alpha: \theta \to \psi$ intertwines. Prove that if $\alpha \neq 0$, then $\theta \cong \psi$ and $\alpha = \iota_n c$ for some $c \in F$, where $n = \deg\theta$.

3. Let G be a finite group, and suppose that F is algebraically closed. Assume that θ and ψ are F-representations of G (or equivalently, representations of FG) with $\deg\theta = n$ and $\deg\psi = m$. Prove the following statements.

   (a) If γ is an n by m matrix with entries in F, then $\alpha = \sum_{x\in G}\theta(x^{-1})\gamma\psi(x)$ intertwines θ and ψ.

   (b) Write $\theta(y) = [a_{ij}(y)]$, $\psi(y) = [b_{kl}(y)]$. Assume that θ and ψ are irreducible.

   (i) If $\theta \not\cong \psi$, then $\sum_{x\in G}a_{ir}(x^{-1})b_{sl}(x) = 0$ for all i, r, s, and l.

   (ii) $\sum_{x\in G}a_{ir}(x^{-1})a_{sj}(x) = \delta_{ij}\delta_{rs}c$, where $c \in F$ satisfies $nc = |G|$.

   Hint. Use Exercise 2 and (a) with $\gamma = \varepsilon_{rs}$ to obtain (i) and $\sum_{x\in G}a_{ir}(x^{-1})a_{sj}(x) = \delta_{ij}c_{rs}$ for some $c_{rs} \in F$. By the change of variable, $x \mapsto x^{-1}$, show that $\delta_{ij}c_{rs} = \delta_{rs}c_{ij}$, so that $c_{rs} = \delta_{rs}c$. Evaluate c by summing the equation $\sum_{x\in G}a_{ir}(x^{-1})a_{rj}(x) = \delta_{ij}c$ over $r = 1, \ldots, n$.

   (c) If θ and ψ are irreducible representations of G (that is, of FG), and if χ_θ and χ_ψ are the characters that are afforded by these representations, as in Exercise 1, then $\sum_{x\in G}\chi_\theta(x^{-1})\chi_\psi(x) = 0$ if $\theta \not\cong \psi$, and $\sum_{x\in G}\chi_\theta(x^{-1})\chi_\theta(x) = |G|$.

Notes on Chapter 5

Our treatment of the Krull–Schmidt Theorem proceeds by the classical ring theoretic route that was pioneered by Azumaya. The lattice theoretic approach due to Ore has not been considered. In recent years, there has been much research on general versions of the Krull–Schmidt Theorem, and it seems likely that the last word on this subject has yet to be uttered. The last two sections of the chapter tidy up some previous vague allusions to the close connection between the theories of group representations and associative algebras. The material in these sections is expository.

CHAPTER 6

Projective Modules over Artinian Algebras

The indecomposable modules that we encounter in studying algebra structure are very special: they are direct summands of A_A. In particular, these modules are projective.

The first four sections of this chapter present the structure and classification of projective modules over Artinian algebras. The last three sections are concerned with applications of projective modules to the theory of Artinian algebras. One of these is the promised structure theorem. Its proof follows the pattern developed in Chapter 3 for semisimple algebras, but the result obtained is far less satisfying than the Wedderburn Structure Theorem.

6.1. Projective Modules

There are two (or more) ways to define projective modules. We will use the least technical one.

Definition. An A-module P is *projective* if P is isomorphic to a direct summand of a free A-module.

Several basic properties of projective modules come as gifts with this definition.

Proposition a. *Let A be an algebra.*

(i) *Every free A-module is projective.*
(ii) *A direct sum of projective A-modules is projective.*
(iii) *A direct summand of a projective A-module is projective.*

This is obvious from the definition of a projective module. Not so obvious is a lifting property for homomorphisms from a projective module.

Proposition b. *Let A be an algebra, and suppose that M, N, and P are (right or left) A-modules such that P is projective. If $\theta: N \to M$ is a surjective homomorphism, then every $\phi \in \mathrm{Hom}_A(P,M)$ factors through N. That is, there exists $\psi \in \mathrm{Hom}_A(P,N)$ such that $\phi = \theta\psi$.*

The most easily remembered version of this proposition has the form of a commutative diagram

$$\begin{array}{ccccc} & & P & & \\ & {}^{\psi}\swarrow & \downarrow{}^{\phi} & & \\ N & \underset{\theta}{\to} & M & \to & 0\,. \end{array}$$

Still another statement of the conclusion of the proposition is: θ induces a surjective homomorphism $\mathrm{Hom}_A(P,N) \to \mathrm{Hom}_A(P,M)$ by $\psi \mapsto \theta\psi$.

PROOF. Since P is projective, there is a free A-module Q containing P as a submodule, and a homomorphism $\pi: Q \to P$ such that $\pi|P = id_P$. Let X be a basis of Q. Since θ is surjective, $\theta^{-1}\phi\pi(u) \neq \varnothing$ for all $u \in X$. By the axiom of choice, there is a mapping $\chi: X \to N$ such that $\theta\chi(u) = \phi\pi(u)$ for all $u \in X$. Since Q is freely generated by X, χ extends to a homomorphism of Q to N. Plainly, $\theta\chi = \phi\pi$. Hence, if $\psi = \chi|P$, then $\theta\psi = \phi$. □

Corollary a. *If N and P are A-modules with P projective, and if there is a surjective homomorphism $\theta: N \to P$, then $N = \mathrm{Ker}\,\theta \oplus Q$, where $Q \cong P$.*

PROOF. By Proposition b, there is a homomorphism $\psi: P \to N$ such that $\theta\psi = id_P$. By Lemma 5.4a, $N = \mathrm{Im}\,\psi \oplus \mathrm{Ker}\,\theta$, and $\mathrm{Im}\,\psi \cong P$. □

The proof of this corollary shows that any module P is projective if it enjoys the lifting property of Proposition b: take N to be a free module for which there is a surjective homomorphism $\theta: N \to P$; then P is isomorphic to a direct summand of N. The homomorphism lifting property of Proposition b therefore characterizes projective modules, and this property is often taken as the definition of a projective module.

Corollary b. *For an algebra A, the following conditions are equivalent.*

(i) *A is semisimple.*
(ii) *Every right A-module is projective.*

PROOF. If A is semisimple, then every right A-module is isomorphic to a direct sum of right ideals of A by Proposition 3.1b, and any right ideal of A is a direct summand of A_A by Proposition 2.4. Thus, every right A-module is projective by Proposition a. Conversely, assume that every right A-module is projective. If M is a right ideal of A, then the projectivity of A/M implies that M is a direct summand of A_A by Corollary a. By Proposition 2.4, A_A is semisimple. □

Since right and left semisimplicity are equivalent conditions, the same argument shows that A is semisimple if and only if every left A-module is projective.

Exercises

1. Prove that if A is a commutative, principal ideal domain, then every projective A-module is free. The same result is true if $A = F[\mathbf{x}_1, \ldots, \mathbf{x}_n]$, but this fact is a difficult theorem of commutative ring theory (the Quillen–Souslin proof of the Serre conjecture).

2. Let A be a commutative integral domain. Prove the following facts.
   (a) If P is a projective A-module, then P is torsion free; that is, $ua = 0$ for $u \in P$, $a \in A$ implies $u = 0$ or $a = 0$.
   (b) The following conditions are equivalent: A is a field; A is semisimple; every A-module is projective.

3. Show that if P is a projective A-module, then there is a free A-module Q such that $P \bigoplus Q \cong Q$. Hint. Let P be a direct summand of the free A-module N, and define $Q = \bigoplus \aleph_0 N$.

4. A right A-module Q is called *injective* if, for every diagram of right A-module homomorphisms

$$\begin{array}{ccccc} 0 & \to & M & \xrightarrow{\theta} & N \\ & & \phi\downarrow & & \\ & & Q & & \end{array}$$

   in which θ is injective, there is an A-module homomorphism $\psi: N \to Q$ such that $\phi = \psi\theta$. This definition is the categorical dual of the characterization of projective modules in Proposition 6.1b. Prove that Q is injective if for every A-module homomorphism ϕ of a right ideal of A to Q, there is an extension of ϕ to a homorphism of A_A to Q. Hint. Use the given hypothesis to show that if $M' < N$, $\phi' \in \operatorname{Hom}_A(M', Q)$, and $u \in N - M'$, then ϕ' can be extended to a homomorphism $\phi'': M' + uA \to Q$. Apply Zorn's Lemma.

5. Let A be an R-algebra, and suppose that M is a right A-module such that M is injective as an R-module. Denote $Q(M) = \operatorname{Hom}_R(A, M)$, with right A-module structure defined by $(\phi x)(y) = \phi(xy)$. Prove that $Q(M)$ is injective, and the left regular representation $u \mapsto \lambda_u$, with $\lambda_u(x) = ux$, is an injective A-module homomorphism of M to $Q(M)$. Hint. Apply the criterion of Exercise 4 to establish injectivity: if P is a right ideal of A and $\phi: P \to Q(M)$ is an A-module homomorphism, define $\psi: P \to M$ by $\psi(x) = \phi(x)(1_A)$. Use the hypothesis that M is R-injective to extend ψ to an R-module homomorphism χ of A to M. Let $\theta: A \to Q(M)$ be defined by $\theta(x)(y) = \chi(xy)$, and show that θ is an A-module homomorphism extending ϕ.

6. Let A be a finite dimensional F-algebra, and suppose that M is a finitely generated right A-module. Prove that there is a finitely generated, injective, right A-module that contains M as a submodule.

6.2. Homomorphisms of Projective Modules

In this section we will explore the relation between the homomorphisms from an A-module P to an A-module Q, and the homomorphisms from $P/P\mathbf{J}(A)$ to $Q/Q\mathbf{J}(A)$. When A is Artinian and P and Q are projective, the relation is very close indeed.

The results of Section 2.1 show that $P/P\mathbf{J}(A)$ and $Q/Q\mathbf{J}(A)$ can be considered either as A-modules or as $A/\mathbf{J}(A)$-modules. Since $\mathrm{Hom}_A(P/P\mathbf{J}(A), Q/Q\mathbf{J}(A)) = \mathrm{Hom}_{A/\mathbf{J}(A)}(P/P\mathbf{J}(A), Q/Q\mathbf{J}(A))$, we are free to choose whichever viewpoint offers the best perspective.

If A is an R-algebra and J is any ideal of A, then the mapping $P \mapsto P/PJ$ is the object mapping of a functor from the category of right A-modules to the category of right A/J-modules. In fact, if $\phi\colon P \to Q$ is an A-module homomorphism, then $\phi(PJ) \subseteq QJ$. Consequently, there is a unique A/J-module homomorphism $\bar{\phi}\colon P/PJ \to Q/QJ$ such that $\pi_Q\phi = \bar{\phi}\pi_P$, where π_P and π_Q are the projection homomorphisms $P \to P/PJ$ and $Q \to Q/QJ$. The mapping $\phi \mapsto \bar{\phi}$ is not only functorial, it is an R-module homomorphism.

Lemma a. *For all pairs (P,Q) of right A-modules, there is an R-module homomorphism $\theta = \theta(P,Q)\colon \mathrm{Hom}_A(P,Q) \to \mathrm{Hom}_{A/J}(P/PJ, Q/QJ)$ such that $\pi_Q\phi = \theta(\phi)\pi_P$. If $\phi \in \mathrm{Hom}_A(P,Q)$ and $\psi \in \mathrm{Hom}_A(N,P)$, then $\theta(N,Q)(\phi \circ \psi) = \theta(P,Q)(\phi)\theta(N,P)(\psi)$. In particular, $\theta(P,P)$ is an R-algebra homomorphism of $\mathbf{E}_A(P)$ to $\mathbf{E}_A(P/PJ)$.*

The proof of this lemma is Exercise 1.

Lemma b. *With the hypothesis and notation of Lemma a, if P is projective, then $\theta(P,Q)\colon \mathrm{Hom}_A(P,Q) \to \mathrm{Hom}_{A/J}(P/PJ, Q/QJ)$ is surjective.*

Proof. If $\bar{\phi} \in \mathrm{Hom}_{A/J}(P/PJ, Q/QJ) = \mathrm{Hom}_A(P/PJ, Q/QJ)$, then since P is projective and $\pi_Q\colon Q \to Q/QJ$ is surjective, it follows from Proposition 6.1b that there is an A-module homomorphism $\phi\colon P \to Q$ such that

$$\begin{array}{ccc} P & \xrightarrow{\pi_P} & P/PJ \\ \phi\downarrow & & \downarrow\bar{\phi} \\ Q & \xrightarrow[\pi_Q]{} & Q/QJ \end{array}$$

commutes. By definition, $\bar{\phi} = \theta(\phi)$. □

Proposition. *If A is a right Artinian algebra and P is a projective right A-module, then the functor θ induces an isomorphism*

$$\mathbf{E}_A(P)/\mathbf{J}(\mathbf{E}_A(P)) \cong \mathbf{E}_{A/\mathbf{J}(A)}(P/P\mathbf{J}(A)).$$

Moreover, if $\mathbf{J}(A)^k = 0$, then $\phi^k = 0$ for all $\phi \in \mathbf{J}(\mathbf{E}_A(P))$.

Proof. To simplify the notation, write J for $\mathbf{J}(A)$, $\bar{A}$ for A/J, and $\bar{P}$ for P/PJ. By Lemmas a and b, $\theta\colon \mathbf{E}_A(P) \to \mathbf{E}_A(\bar{P})$ is a surjective R-algebra homomorphism. If $\phi \in \operatorname{Ker}\theta$, then $\phi(P) \subseteq \operatorname{Ker}\pi_P = PJ$. Iterating this observation gives $\phi^k(P) \subseteq PJ^k$ for all $k \in \mathbb{N}$. Thus, every element of $\operatorname{Ker}\theta$ is nilpotent, so that $\operatorname{Ker}\theta \subseteq \mathbf{J}(\mathbf{E}_A(P))$ by Corollary 4.3b. It remains to show that $\mathbf{J}(\mathbf{E}_A(P)) \subseteq \operatorname{Ker}\theta$. Since A/J is semisimple, so is $\bar{P}$ by Proposition 3.1b. Thus, $\mathbf{J}(\mathbf{E}_A(\bar{P})) = 0$ by Example 4.3. That is $\mathbf{J}(\mathbf{E}_A(P)/\operatorname{Ker}\theta) = 0$ and $\mathbf{J}(\mathbf{E}_A(P)) \subseteq \operatorname{Ker}\theta$ by Lemma 4.3b. □

Corollary. *If A is a right Artinian algebra, and if P and Q are projective right A-modules, then P is isomorphic to Q if and only if $P/P\mathbf{J}(A)$ is isomorphic to $Q/Q\mathbf{J}(A)$.*

PROOF. If $\phi: P \to Q$ is an isomorphism, then the functorial property of θ implies that $\theta(\phi^{-1}) = \theta(\phi)^{-1}$. Hence, $P/P\mathbf{J}(A) \cong Q/Q\mathbf{J}(A)$. Conversely, if $P/P\mathbf{J}(A) \cong Q/Q\mathbf{J}(A)$, then there exist homomorphisms $\phi: P \to Q$ and $\psi: Q \to P$ such that $id_P - \psi\phi \in \operatorname{Ker}\theta = \mathbf{J}(\mathbf{E}_A(P))$, and

$$id_Q - \phi\psi \in \mathbf{J}(\mathbf{E}_A(Q)).$$

By Proposition 4.3, $\psi\phi = id_P - (id_P - \psi\phi) \in \mathbf{E}_A(P)^\circ$, so that ϕ has a left inverse. Similarly, ϕ has a right inverse. Thus, ϕ is an isomorphism. □

EXERCISES

1. Complete the proof of the lemma.
2. Show that the corollary remains true when the hypothesis that A is right Artinian is replaced by the weaker assumption that $\mathbf{J}(A)^k = 0$ for some $k \in \mathbb{N}$.
3. Let $A = \mathbb{Z}/4\mathbb{Z}$. Give an example of A-modules M and N such that $M/M\mathbf{J}(A) \cong N/N\mathbf{J}(A)$ and $M \not\cong N$.

6.3. Structure of Projective Modules

The results of the last section lead to classification and structure theorems for projective modules over Artinian algebras. Throughout this section, A is a right Artinian R-algebra, and P is a right A-module that is usually assumed to be projective.

The indecomposable direct summands of A_A are called *principal indecomposable* right A-modules. We emphasize that a principal indecomposable right A-module is a right ideal of A, and it is projective.

Lemma. *A direct summand P of A_A is indecomposable if and only if $P/P\mathbf{J}(A)$ is a simple $A/\mathbf{J}(A)$-module.*

PROOF. Since A is Artinian, $P < A_A$ implies that P is Artinian and Noetherian. By Corollary 5.3, P is indecomposable if and only if $\mathbf{E}_A(P)$ is a local algebra, that is, $\mathbf{E}_A(P)/\mathbf{J}(\mathbf{E}_A(P))$ is a division algebra. The lemma therefore follows from Proposition 6.2 and Corollary 2.3b (Schur's Lemma):

$$\mathbf{E}_A(P)/\mathbf{J}(\mathbf{E}_A(P)) \cong \mathbf{E}_{A/\mathbf{J}(A)}(P/P\mathbf{J}(A))$$

is a division algebra if and only if $P/P\mathbf{J}(A)$ is simple. (Note that $P/P\mathbf{J}(A)$ is semisimple because it is an $A/\mathbf{J}(A)$-module.) □

Proposition. *Let A be a right Artinian algebra. The mapping $P \mapsto P/P\mathbf{J}(A)$ defines a bijective correspondence between the isomorphism classes of principal indecomposable right A-modules and the isomorphism classes of simple right $A/\mathbf{J}(A)$-modules.*

PROOF. If P is a principal indecomposable right A-module, then $P/P\mathbf{J}(A)$ is a simple $A/\mathbf{J}(A)$-module by the lemma. It follows from Corollary 6.2 that $P \cong Q$ if and only if $P/P\mathbf{J}(A) \cong Q/Q\mathbf{J}(A)$. Therefore, $P \mapsto P/P\mathbf{J}(A)$ induces an injective mapping of isomorphism classes of principal indecomposable modules to simple modules. The mapping is also surjective. Indeed, if $A_A = P_1 \oplus \cdots \oplus P_n$ is a decomposition of A_A into a direct sum of principal indecomposable modules, then by the lemma, $A/\mathbf{J}(A) \cong P_1/P_1\mathbf{J}(A) \oplus \cdots \oplus P_n/P_n\mathbf{J}(A)$ is a decomposition of $A/\mathbf{J}(A)$ into a direct sum of simple right A-modules. Every isomorphism class of simple right A-modules is represented by some $P_i/P_i\mathbf{J}(A)$ because of Proposition 3.1b and Lemma 3.2d. □

Structure Theorem. *If A is a right Artinian algebra, then every projective right A-module is isomorphic to a direct sum of principal indecomposable right A-modules. This decomposition is unique to within isomorphism and the ordering of the factors.*

PROOF. Let P be a projective right A-module. Since A is right Artinian, $A/\mathbf{J}(A)$ is semisimple; so is $P/P\mathbf{J}(A)$, that is, $P/P\mathbf{J}(A) = \bigoplus_{i\in I} N_i$, where the N_i are simple. By the proposition, there exist principal indecomposable modules P_i such that $N_i \cong P_i/P_i\mathbf{J}(A)$ for each $i \in I$. Thus,

$$P/P\mathbf{J}(A) \cong \bigoplus_{i\in I} P_i/P_i\mathbf{J}(A) \cong (\bigoplus_{i\in I} P_i)/(\bigoplus_{i\in I} P_i)\mathbf{J}(A).$$

Therefore, the required result $P \cong \bigoplus_{i\in I} P_i$ follows from Proposition 6.1a and Corollary 6.2. In order to prove the uniqueness, suppose that

$$\bigoplus_{i\in I} P_i \cong \bigoplus_{j\in K} Q_j,$$

where all of the Q_j are principal indecomposable right A-modules. This assumption yields $\bigoplus_{i\in I}(P_i/P_i\mathbf{J}(A)) \cong \bigoplus_{j\in K}(Q_j/Q_j\mathbf{J}(A))$, where the $Q_j/Q_j\mathbf{J}(A)$ are simple. By Proposition 2.5, there is a bijective mapping $\sigma\colon K \to I$ such that $Q_j/Q_j\mathbf{J}(A) \cong P_{\sigma(j)}/P_{\sigma(j)}\mathbf{J}(A)$. Hence, $Q_j \cong P_{\sigma(j)}$. □

Corollary a. *If A is a right Artinian algebra, then every indecomposable, projective right A-module is isomorphic to a principal indecomposable right A-module.*

In particular, the indecomposable, projective A-modules are cyclic.

Corollary b. *Let A be a right Artinian algebra, and suppose that P is an indecomposable, projective right A-module. If N is a proper submodule of P, then $N \subseteq P\mathbf{J}(A)$.*

PROOF. If $N \nsubseteq P\mathbf{J}(A)$, then $N + P\mathbf{J}(A) = P$, because $P/P\mathbf{J}(A)$ is simple by the lemma. It was noted that P is finitely generated. Hence, $N = P$ by Nakayama's Lemma. □

EXERCISES

1. (a) Let A be an R-algebra. Prove that every finitely generated, projective right A-module is a direct summand of a free A-module of finite rank, that is, a module of the form $\bigoplus n A_A$, where $n \in \mathbb{N}$.
   (b) Use (a) and the Krull–Schmidt Theorem to give a short proof of the Structure Theorem for finitely generated projective modules.
2. Let A be a right Artinian algebra. Prove that every non-zero homomorphic image of an indecomposable, projective right A-module is indecomposable.
3. Let A be a right Artinian, local algebra. Prove that A_A is the only principal indecomposable right A-module. Deduce that every projective right A-module is free.

6.4. Idempotents

An element e of an algebra is called *idempotent* if $e^2 = e$. Up to now, we have avoided using idempotents, except in the Exercises. Standard arguments using idempotents have been replaced by proofs that are based on homomorphisms. However, there are many situations in which the use of idempotent elements is convenient and natural. Their usefulness in making concrete calculations is beyond question.

The purpose of this section is to supplement our treatment of projective modules over an Artinian algebra by showing where idempotents fit in this subject. A few applications of idempotents will be given in the last three sections of this chapter, and in several later parts of the book.

Proposition a. *A right ideal P of an algebra A is a direct summand of A_A if and only if there is an idempotent element $e \in A$ such that $P = eA$. In this case, if $P = Q_1 \oplus Q_2 \oplus \cdots \oplus Q_m$, then there exist idempotents e_i such that $e = e_1 + e_2 + \cdots + e_m$, $e_ie_j = 0$ for $i \neq j$, $e_ie = ee_i = e_i$, and $Q_i = e_iA$ for $1 \leq i \leq m$. In particular, P is decomposable if and only if there is an idempotent f with $0 \neq f \neq e$ and $ef = fe = f$.*

PROOF. Let $e \in A$ be an idempotent. If $eA = Q_1 \oplus Q_2 \oplus \cdots \oplus Q_m$, then there exist elements $e_i \in Q_i$ for $1 \le i \le m$ such that $e = e_1 + e_2 + \cdots + e_m$. If $x \in Q_i \subseteq eA$, then $x = ex = e_1x + e_2x + \cdots + e_mx$. Hence,

$$x - e_ix = \sum_{j \neq i} e_jx \in Q_i \cap \sum_{j \neq i} Q_j = 0.$$

In particular, $e_i = e_i^2$ and $e_je_i = 0$ for all $j \neq i$. It follows directly that $e_ie = ee_i = e_i$. Moreover, $Q_i = eQ_i = e_iQ_i \subseteq e_iA \subseteq Q_i$. If $f \in A$ is an idempotent such that $ef = fe = f$, then $eA = fA \oplus (e - f)A$. In fact, $fA = efA \subseteq eA$, $(e - f)A = e(e - f)A \subseteq eA$; $ex = fx + (e - f)x$ for all $x \in A$; and $fx + (e - f)y = 0$ implies $fx = f(fx + (e - f)y) = 0$, $(e - f)y = 0$. Applying these observations to the special case in which $e = 1_A$ proves the first assertion of the proposition, and completes the argument. □

An idempotent e of the algebra A is called *primitive* if eA is an indecomposable A-module. By the proposition, eA is projective. Thus, for a right Artinian algebra A, the idempotent e is primitive if and only if eA is a principal indecomposable module. A characterization of primitive idempotents follows directly from the proposition.

Corollary a. *An idempotent element e of the algebra A is primitive if and only if there is no idempotent f of A such that $0 \neq f \neq e$ and $ef = fe = f$.*

An equivalent formulation of this criterion for e to be primitive is the definition that was given in Exercise 3 of Section 3.3: if $e = f_1 + f_2$, where f_1 and f_2 are idempotents such that $f_1f_2 = f_2f_1 = 0$, then either $f_1 = 0$ or $f_2 = 0$.

There is a connection between idempotents and homomorphisms of ideals that is obtained from a generalization of Proposition 1.3.

Lemma. *Let P be a direct summand of A_A. If M is a right A-module, then* $\mathrm{Hom}_A(P,M) = \{\lambda_u|P\colon u \in M\}$.

The notation λ_u is used (as in Section 1.3) to denote the homomorphism from A_A to M that is left multiplication by u, that is $\lambda_u(x) = ux$. Plainly, $\{\lambda_u|P\colon u \in M\} \subseteq \mathrm{Hom}_A(P,M)$. To reverse the inclusion, write $P = eA$ with $e^2 = e$, in accordance with the proposition. If $\theta \in \mathrm{Hom}_A(P,M)$ and $x \in P$, then $\theta(x) = \theta(ex) = \theta(e)x = \lambda_{\theta(e)}(x)$. Thus, $\theta = \lambda_{\theta(e)}|P$.

Corollary b. *If e and f are idempotents of the R-algebra A, then* $\mathrm{Hom}_A(eA,fA) \cong fAe$ *as R-modules, and* $\mathbf{E}_A(eA) \cong eAe$ *as R-algebras. If A is right Artinian, so is eAe.*

PROOF. By the lemma, $\mathrm{Hom}_A(eA,fA) = \{\lambda_{fx}|eA\colon x \in A\}$. Clearly, $\lambda_{fx}|eA = \lambda_{fxe}|eA$. If $\lambda_{fxe}|eA = \lambda_{fye}|eA$, then $fxe = \lambda_{fxe}(e) = \lambda_{fye}(e) = fye$. Thus,

$z \mapsto \lambda_z|eA$ is a bijective map from fAe to $\mathrm{Hom}_A(eA,fA)$. Easy calculations show that this mapping is a module isomorphism and an algebra isomorphism if $e = f$. If N is a right ideal of eAe, then $NA < A_A$, and $NAe = NeAe = N$ since e is the unity element of eAe. Thus, the mapping $N \mapsto NA$ embeds the lattice of right ideals of eAe in $\mathbf{S}(A_A)$. In particular, if A is right Artinian, then eAe is right Artinian. □

Corollary c. *Let e be an idempotent element of the algebra A, and denote $P = eA$. The following conditions are equivalent for a right A-module M:*

(i) $\mathrm{Hom}_A(P,M) \neq 0$;
(ii) $MP \neq 0$;
(iii) $Me \neq 0$.

In the most important case of this corollary, the modules P and M are principal indecomposable right A-modules, where A is right Artinian.

Corollary d. *Let e and f be primitive idempotents in the right Artinian algebra A, such that $P = eA$ is not isomorphic to $Q = fA$. The following conditions are equivalent:*

(i) $\mathrm{Hom}_A(P,Q) \neq 0$;
(ii) $Q\mathbf{J}(A)P \neq 0$;
(iii) $f\mathbf{J}(A)e \neq 0$.

Proof. Since $Q\mathbf{J}(A)P = fA\mathbf{J}(A)eA = f\mathbf{J}(A)eA$, it is evident that (ii) and (iii) are equivalent. The assumption that $P \not\cong Q$ implies that $P/P\mathbf{J}(A) \not\cong Q/Q\mathbf{J}(A)$ by Proposition 6.3, and Schur's Lemma yields $\mathrm{Hom}_{A/\mathbf{J}(A)}(P/P\mathbf{J}(A), Q/Q\mathbf{J}(A)) = 0$. In particular, if $0 \neq \phi \in \mathrm{Hom}_A(P,Q)$, then $\phi(P) \subseteq Q\mathbf{J}(A)$, and $0 \neq \mathrm{Im}\,\phi = \phi(eP) = \phi(e)P \subseteq Q\mathbf{J}(A)P$. Conversely, $0 \neq f\mathbf{J}(A)e \subseteq fAe$ implies $\mathrm{Hom}_A(P,Q) \neq 0$ by Corollary b. □

The program of generalizing the Wedderburn Structure Theorem to Artinian algebras breaks down chiefly because of the existence of principal indecomposable modules P and Q that are not isomorphic, but $\mathrm{Hom}_A(P,Q) \neq 0$. It is useful to pinpoint this phenomenon; we do this by associating a particular graph to each Artinian algebra.

Definition. Let A be a right Artinian algebra. Suppose that $e_1, e_2, \ldots, e_r$ are primitive idempotents in A such that the right ideals $P_1 = e_1A$, $P_2 = e_2A, \ldots, P_r = e_rA$ represent the distinct isomorphism classes of principal indecomposable right A-modules. The *quiver* of A is the directed graph $\Gamma(A) = (V,E)$ with the vertex set $V = \{e_1, e_2, \ldots, e_r\}$ and the edge set $E = \{(e_i,e_j): e_i\mathbf{J}(A)e_j \neq 0\}$. (In general, we will follow Gabriel [34] in referring to finite directed graphs as quivers.)

It is convenient (and harmless) to call two quivers equal when they are only isomorphic, that is, there is a bijection between their vertex sets that maps the edge sets bijectively. The nature of the objects that are the vertices of $\Gamma(A)$ has no importance. Using primitive idempotents for vertices is convenient, but there is no canonical way to choose these idempotents, and it is usually not necessary to make a specific choice.

Proposition b. $\Gamma(A)$ *is independent of the choice of primitive idempotents for the vertex set. Moreover, if* $A \cong B$, *then* $\Gamma(A) = \Gamma(B)$.

PROOF. The content of the first assertion is: if e, e', f, and f' are primitive idempotents such that $eA \cong e'A$ and $fA \cong f'A$, then $e\mathbf{J}(A)f \neq 0$ if and only if $e'\mathbf{J}(A)f' \neq 0$. Clearly, $eA \cong e'A$ implies $e\mathbf{J}(A) \cong e'\mathbf{J}(A)$, so that

$$\mathrm{Hom}_A(fA, e\mathbf{J}(A)) \cong \mathrm{Hom}_A(f'A, e'\mathbf{J}(A)).$$

The required result is a consequence of Corollary c. The second statement of the proposition is a consequence of the first part, and our convention that isomorphic quivers are identified. □

As a rule of thumb, the quiver $\Gamma(A)$ measures the complexity of A. If A is semisimple, then clearly $\Gamma(A) = (V, \varnothing)$, the quiver with no edges. The converse is also true. (See Exercise 4.) If the vertex set of $\Gamma(A)$ is a singleton, then $A/\mathbf{J}(A)$ is simple by Proposition 3.3b. Such an algebra is called primary. The structure of primary Artinian algebras will be analyzed in the next section. The edge sets of commutative Artinian algebras have simple forms: the only edges are the loops (e_i, e_i) for which $\mathbf{J}(A)e_i \neq 0$. The structure of commutative Artinian algebras is correspondingly simple.

EXERCISES

1. Let A be an algebra. Suppose that K is an ideal of A such that every element of K is nilpotent. Let $\pi: A \to A/K$ be the natural projection homomorphism. Prove that if $\bar{e} \in A/K$ is idempotent, then there exists $e \in A$ such that e is idempotent and $\pi(e) = \bar{e}$. Hint. Choose $x \in A$ so that $\pi(x) = \bar{e}$. Denote $y = 1_A - x$, so that $xy = yx \in \operatorname{Ker} \pi = K$. Choose n so that $(xy)^n = 0$. The binomial expansion gives

$$\begin{aligned} 1_A &= (x+y)^{2n-1} \\ &= x^{2n-1} + \binom{2n-1}{1} x^{2n-2}y + \cdots + \binom{2n-1}{n-1} x^n y^{n-1} \\ &\quad + \binom{2n-1}{n} x^{n-1}y^n + \cdots + y^{2n-1}. \end{aligned}$$

Show that

$$e = x^n\left(x^{n-1} + \binom{2n-1}{1} x^{n-2}y + \cdots + \binom{2n-1}{n-1} y^{n-1}\right)$$

has the desired properties.

2. Let A be an algebra. Denote $B(A) = \{e \in \mathbf{Z}(A): e^2 = e\}$. Prove the following statements.

   (a) $B(A)$ is a Boolean algebra (i.e., an $\mathbb{F}_2$-algebra in which all elements are idempotent) under the multiplication inherited from A and addition given by $e \oplus f = e + f - 2ef$.

   (b) $B(A) = B(A/\mathbf{J}(A))$.

   (c) If A is Artinian, then $B(A)$ is finite.

3. Let A be a right Artinian algebra with the quiver $\Gamma(A) = (E, V)$, $E = \{e_1, e_2, \ldots, e_r\}$. For $1 \leq i \leq r$, denote $P_i = e_i A$. Prove the following statements.

   (a) If $i \neq j$, then $(e_j, e_i) \in E$ if and only if the top composition factor of P_i (that is, $P_i/P_i\mathbf{J}(A)$) is isomorphic to a composition factor of P_j.

   (b) $(e_i, e_i) \in E$ if and only if the top composition factor of P_i appears more than once in a composition series of P_i.

   (c) e_i and e_j are in the same connected component of $\Gamma(A)$ if and only if there is a sequence $i_1, i_2, \ldots, i_n$ with $n > 1$ such that $i_1 = i$, $i_n = j$, and for all $k < n$, P_{i_k} and $P_{i_{k+1}}$ have a composition factor in common.

4. Prove that if A is a right Artinian algebra such that the edge set of $\Gamma(A)$ is empty, then A is semisimple. Hint. Write $1_A = e_1 + \cdots + e_n$, where the e_i are primitive idempotents, and use the fact that $\mathbf{J}(A) = 1_A \mathbf{J}(A) 1_A$

6.5. Structure of Artinian Algebras

This section offers some applications of the results on projective modules to the structure theory of Artinian algebras. We begin with a theorem that gives a nice description of a special class of algebras. An R-algebra A is called *primary* if $A/\mathbf{J}(A)$ is simple.

Proposition a. *If A is a right Artinian, primary algebra, then all principal indecomposable right A-modules are isomorphic. Moreover, $A \cong M_n(B)$ for a unique natural number n and a right Artinian local algebra B that is unique to within isomorphism.*

PROOF. Since $A/\mathbf{J}(A)$ is semisimple and simple, all simple right $A/\mathbf{J}(A)$-modules are isomorphic by Proposition 3.3b. It follows from Proposition 6.3 that all principal indecomposable right A-modules are isomorphic. Consequently, $A_A \cong \bigoplus n\, P$ where n is a uniquely determined natural number and P is a principal, indecomposable right A-module that is unique to within isomorphism. By Proposition 1.3 and Corollary 3.4a,

$$A \cong \mathbf{E}_A(A_A) \cong \mathbf{E}_A(\bigoplus n\, P) \cong M_n(B),$$

where $B = \mathbf{E}_A(P)$. Since P is indecomposable, Artinian and Noetherian (by Corollary 4.5b), it follows from Corollary 5.3 that B is a local algebra. Moreover, $P = eA$ for some idempotent element e of A by Proposition 6.4a. Since A is right Artinian, it follows from Corollary 6.4b that $\mathbf{E}_A(eA)$ is right Artinian. That is, B is Artinian. If $A \cong A' = M_m(C)$ with C local, then $A' = \varepsilon_{11}A' \oplus \varepsilon_{22}A' \oplus \cdots \oplus \varepsilon_{mm}A'$, and (by Corollary 6.4b), $C \cong \varepsilon_{11}A'\varepsilon_{11} \cong \mathbf{E}_{A'}(\varepsilon_{11}A')$, so that $\varepsilon_{11}A'$ is a principal indecomposable right A'-module. The uniqueness of n and P imply $m = n$ and $C \cong B$. □

The converse of this proposition is outlined in Exercise 1.

Artinian algebras that are not primary can have principal indecomposable modules P and Q that are not isomorphic, but $\mathrm{Hom}_A(P,Q) \neq 0$. When such modules exist, the proof that led to the Wedderburn Structure Theorem breaks down. The quiver $\Gamma(A)$ of an Artinian algebra A is defined in such a way that it keeps track of the isomorphically distinct principal indecomposable modules P and Q such that $\mathrm{Hom}_A(P,Q) \neq 0$. It can be expected that the geometrical properties of $\Gamma(A)$ reflect the structure of A. The rest of this section develops one of the simplest connections between A and $\Gamma(A)$.

If $\Gamma_1 = (V_1,E_1)$ and $\Gamma_2 = (V_2,E_2)$ are quivers with disjoint vertex sets, then the disjoint union of Γ_1 and Γ_2 is $\Gamma_1 \mathbin{\dot\cup} \Gamma_2 = (V_1 \cup V_2, E_1 \cup E_2)$. If $\Gamma = (V,E)$ is a quiver, and $V = V_1 \mathbin{\dot\cup} V_2$, then Γ is the disjoint union $(V_1, E \cap (V_1 \times V_1)) \mathbin{\dot\cup} (V_2, E \cap (V_2 \times V_2))$ exactly in the case that there is no edge in E that joins a vertex of V_1 to a vertex of V_2. If Γ cannot be written as a disjoint union of two non-empty quivers, then Γ is *connected*. This means that all pairs of vertices in Γ can be joined by a path that consists of a sequence of edges (with their orientations ignored). It is geometrically plausible that every quiver has a unique representation as a disjoint union of connected quivers. This fact will be proved in Section 8.4.

Lemma a. *If the Artinian algebra A is a product $B \dotplus C$ of algebras, then* $\Gamma(A) = \Gamma(B) \mathbin{\dot\cup} \Gamma(C)$.

Proof. Let $\Gamma(B) = (V_1,E_1)$ and $\Gamma(C) = (V_2,E_2)$, with $V_1 = \{e_1, \ldots, e_r\}$, $V_2 = \{f_1, \ldots, f_s\}$. By definition, the elements e_i and f_j are primitive idempotents in B and C; they are also primitive idempotents in A: $e_iA = e_iB$ and $f_jA = f_jC$ are indecomposable. For suitable natural numbers m_i and n_j, there are isomorphisms

$$B_B \cong \bigoplus_{i=1}^{r} \bigoplus m_i e_i B,\ C_C \cong \bigoplus_{j=1}^{s} \bigoplus n_j f_j C.$$

Hence,

$$A_A = B_A \oplus C_A \cong \left(\bigoplus_{i=1}^{r} \bigoplus m_i e_i A\right) \oplus \left(\bigoplus_{j=1}^{s} \bigoplus n_j f_j A\right).$$

By the Krull–Schmidt Theorem, every principal indecomposable right A-module is isomorphic to a unique e_iA or f_jA. Since $\mathbf{J}(A) = \mathbf{J}(B) \dotplus \mathbf{J}(C)$ by Lemma 4.3b, it follows that $e_i\mathbf{J}(A)f_j = f_j\mathbf{J}(A)e_i = 0$ for all i and j; moreover, $e_i\mathbf{J}(A)e_k \neq 0$ if and only if $e_i\mathbf{J}(B)e_k \neq 0$, and $f_j\mathbf{J}(A)f_l \neq 0$ if and only if $f_j\mathbf{J}(C)f_l \neq 0$. This proves that $\Gamma(A) = (V_1 \cup V_2, E_1 \cup E_2) = \Gamma(B) \mathbin{\dot\cup} \Gamma(C)$. □

Lemma b. *If A is a right Artinian algebra such that $\Gamma(A) = \Gamma_1 \uplus \Gamma_2$, then there are right Artinian algebras B and C that satisfy $A = B \dotplus C$, $\Gamma(B) = \Gamma_1$, and $\Gamma(C) = \Gamma_2$.*

PROOF. Denote $\Gamma_1 = (V_1,E_1)$ and $\Gamma_2 = (V_2,E_2)$, so that $\Gamma(A) = (V_1 \cup V_2, E_1 \cup E_2)$. Let $A_A = B \oplus C$ with $B = \bigoplus_{i=1}^m P_i$, $C = \bigoplus_{j=1}^n Q_j$ be a decomposition of A_A as a direct sum of indecomposable modules that are grouped so that for $1 \leq i \leq m$, $P_i \cong eA$ for some $e \in V_1$, and for $1 \leq j \leq n$, $Q_j \cong fA$ where $f \in V_2$. Since $E_1 \subseteq V_1 \times V_1$ and $E_2 \subseteq V_2 \times V_2$, it follows from Corollary 6.4d that $\text{Hom}_A(P_i,Q_j) = \text{Hom}_A(Q_j,P_i) = 0$ for all i and j. Consequently, $\text{Hom}_A(B,C) = \text{Hom}_A(C,B) = 0$, and $BC = CB = 0$ by Corollary 6.4c. Thus, B and C are ideals of A, and $A = B \dotplus C$. If $e \in V_1$, then $eB = eA$ is an indecomposable B-module. If also $e' \in V_1$ with $e' \neq e$, then $eB \ncong e'B$. Similarly, the elements of V_2 are primitive idempotents that generate isomorphically distinct C-modules. It follows from Lemma a that $\Gamma(B) = \Gamma_1$ and $\Gamma(C) = \Gamma_2$. □

A non-trivial Artinian algebra B is called a *block* if $\Gamma(B)$ is a connected quiver.

Proposition b. *Let A be a right Artinian algebra.*

(i) *A is uniquely a product of blocks.*
(ii) *A block is indecomposable as an algebra.*

PROOF. The quiver $\Gamma(A)$ decomposes uniquely into a disjoint union $\Gamma_1 \uplus \cdots \uplus \Gamma_t$ of connected quivers. By Lemma b, $A = B_1 \dotplus \cdots \dotplus B_t$, where $\Gamma(B_i) = \Gamma_i$. Thus, each B_i is a block. The uniqueness of this decomposition is a special case of the result that is outlined in Exercise 2. The fact that blocks are indecomposable follows directly from Lemma a. □

EXERCISES

1. Prove that if B is a right Artinian local algebra, then $M_n(B)$ is a right Artinian primary algebra. Hint. Use Exercise 2, Section 4.3.

2. Let A be an algebra such that $A = B \dotplus C$. Prove the following statements.
   (a) If $M < A_A$, then $MB = M \cap B$, $MC = M \cap C$, and $M = M \cap B \oplus M \cap C$.
   (b) Every indecomposable right ideal of A is contained in either B or C.
   (c) If $A = B_1 \dotplus \cdots \dotplus B_r = C_1 \dotplus \cdots \dotplus C_s$ with the B_i and C_j indecomposable algebras, then $r = s$, and $C_j = B_{\sigma(j)}$, for some permutation σ.

3. Prove that a right Artinian algebra A is a finite product of primary algebras if and only if all of the edges in $\Gamma(A)$ are loops, that is, the edges have the form (e_i,e_i).

4. Let A be a commutative, Artinian algebra. Write $A_A = e_1A \oplus \cdots \oplus e_rA$, where each e_i is a primitive idempotent. Prove that $e_1A, \ldots, e_rA$ are all of the principal indecomposable A-modules, and $\text{Hom}_A(e_iA,e_jA) = 0$ if $i \neq j$. Deduce from Exercise 3 that A is a finite product of commutative local algebras.

6.6. Basic Algebras

An algebra B is called *reduced* if $B/\mathbf{J}(B)$ is a finite product of division algebras. In this section it is shown that for every right Artinian algebra A there is an associated reduced algebra B that shares many properties with A. The reduced algebra B is uniquely determined by A; it is called the *basic algebra* of A.

Throughout this section, it is assumed that A is a right Artinian R-algebra.

Lemma a. *Let $P_1, P_2, \ldots, P_r$ be principal indecomposable right A-modules; define $P = P_1 \oplus P_2 \oplus \cdots \oplus P_r$. The endomorphism algebra $\mathbf{E}_A(P)$ is reduced if and only if $P_i \ncong P_j$ for all $i \neq j$.*

Proof. By Proposition 6.2, $\mathbf{E}_A(P)/\mathbf{J}(\mathbf{E}_A(P)) \cong \mathbf{E}_{A/\mathbf{J}(A)}(P/P\mathbf{J}(A)) \cong \mathbf{E}_{A/\mathbf{J}(A)}(P_1/P_1\mathbf{J}(A) \oplus P_2/P_2\mathbf{J}(A) \oplus \cdots \oplus P_r/P_r\mathbf{J}(A))$. It follows from Lemma 6.3 that each $P_j/P_j\mathbf{J}(A)$ is a simple $A/\mathbf{J}(A)$-module. Hence, $\mathbf{E}_A(P)/\mathbf{J}(\mathbf{E}_A(P))$ is a product of division algebras if and only if $P_i/P_i\mathbf{J}(A) \ncong P_j/P_j\mathbf{J}(A)$ for all $i \neq j$ by Corollaries 3.4a and 3.4c, and Schur's Lemma. The lemma follows from Proposition 6.3. □

Lemma b. *Suppose that the right ideal P of A is a direct summand of A_A. Write $P = P_1 \oplus P_2 \oplus \cdots \oplus P_r$, where each P_i is indecomposable. The following conditions are equivalent.*

(i) *$P/P\mathbf{J}(A)$ is a faithful right $A/\mathbf{J}(A)$-module.*
(ii) *$AP = A$.*
(iii) *Every principal indecomposable right A-module is isomorphic to one (or more) of the modules P_j.*

Proof. Two observations reduce the proof to the case in which A is semisimple. First note that (ii) is equivalent to

(ii′) $(A/\mathbf{J}(A))(P + \mathbf{J}(A))/\mathbf{J}(A) = A/\mathbf{J}(A)$.

Clearly, (ii′) is equivalent to $AP + \mathbf{J}(A) = A$, which is the same condition as (ii) by Nakayama's Lemma. By Proposition 6.3, the condition (iii) can be replaced by

(iii′) every simple right $A/\mathbf{J}(A)$-module is isomorphic to one of the modules $P_j/P_j\mathbf{J}(A)$.

Since (i) is already a condition on $A/\mathbf{J}(A)$, it can be assumed that $\mathbf{J}(A) = 0$, A is semisimple, and principal indecomposable modules are simple. Write $A_A = AP \oplus Q$. By Lemma 3.2b, the simple submodules of AP are the minimal right ideals of A that are isomorphic to one of the P_j, and every simple submodule N of Q satisfies $PN = 0$. Thus, (i) implies (ii); and (ii) is equivalent to (iii) because every simple right A-module is isomorphic to a minimal right ideal of A by Proposition 3.1b. It is obvious that (ii) implies (i). □

Proposition a. *If A is a right Artinian R-algebra, then there is a right ideal P of A such that:*

(i) *P is a direct summand of A_A;*
(ii) *$AP = A$;*
(iii) *$\mathbf{E}_A(P)$ is a reduced R-algebra.*

An ideal P that satisfies (i), (ii), *and* (iii) *is unique up to isomorphism.*

PROOF. By Lemmas a and b, a right ideal P that is a direct summand of A_A will satisfy (ii) and (iii) if and only if $P = P_1 \oplus P_2 \oplus \cdots \oplus P_r$, where $\{P_1, P_2 \cdots P_r\}$ is a set of distinct representatives of the isomorphism classes of principal indecomposable right A-modules. Obviously, such a P exists, and it is unique to within isomorphism. □

If P is a right ideal of A that satisfies conditions (i), (ii), and (iii) of the proposition, then the algebra $B = \mathbf{E}_A(P)$ is called a *basic algebra* of A. Since P is unique to within isomorphism, so is B. Therefore, no harm is done by referring to B as *the* basic algebra of A.

EXAMPLE. Let A be a semisimple algebra. By the Wedderburn Structure Theorem, $A = A_1 \dotplus \cdots \dotplus A_r$, $A_i \cong M_{n_i}(D_i)$, where the D_i are division algebras. In fact, $D_i = \mathbf{E}_A(P_i)$, where P_i is a simple right A-module that is a direct summand of A_i. Let $P = P_1 \oplus \cdots \oplus P_r$. Then P is a direct summand of A, $AP = A$, and $\mathbf{E}_A(P) \cong D_1 \dotplus \cdots \dotplus D_r$ is reduced. Hence, $\mathbf{E}_A(P)$ is a basic algebra of A.

Proposition b. *If A is a right Artinian algebra, then the basic algebra B of A has the properties:*

(i) *B is Artinian;*
(ii) *there is a lattice isomorphism $\tau\colon \mathbf{I}(A) \to \mathbf{I}(B)$ such that $\tau(\mathbf{J}(A)) = \mathbf{J}(B)$ and τ is multiplicative, that is, $\tau(I_1 I_2) = \tau(I_1)\tau(I_2)$ for all ideals I_1 and I_2 of A;*
(iii) $\Gamma(B) = \Gamma(A)$.

PROOF. Let $B = \mathbf{E}_A(P)$, where P is a direct summand of A_A, $AP = A$, and $P = P_1 \oplus \cdots \oplus P_r$ for a set $\{P_1, \ldots, P_r\}$ of representatives of the isomorphism classes of principal indecomposable right A-modules. By Proposition 6.4a, there exist idempotents $e, e_1, e_2, \ldots, e_r$ in A such that $P = eA$, $P_i = e_i A$, $e = e_1 + e_2 + \cdots + e_r$, $e_i e_j = 0$ for $i \neq j$, and $ee_i = e_i e = e_i$ for $1 \leq i \leq m$. By Corollary 6.4b, B can be identified with eAe; for the rest of the proof, we make this identification. If N is a right ideal of B, then NA is a right ideal of A, and, as we noted in the proof of Corollary 6.4b, $N \mapsto NA$ is an inclusion preserving mapping from $\mathbf{S}(B_B)$ to $\mathbf{S}(A_A)$. In particular, since A is right Artinian, B is also right Artinian. Similarly, if $J \lhd B$, then $AJA \lhd A$, and $I \lhd A$ implies $eIe \lhd B$. Moreover, $eAJAe = BJB = J$; and $AeIeA = AeAIAeA = AIA = I$. Thus, $J \mapsto AJA$

and $I \mapsto eIe$ are inverse, inclusion preserving mappings between $\mathbf{I}(B)$ and $\mathbf{I}(A)$. It follows that the mapping $\tau(I) = eIe$ is a lattice isomorphism of $\mathbf{I}(A)$ to $\mathbf{I}(B)$. Moreover, $\tau(I_1 I_2) = eI_1 I_2 e = eI_1 AeAI_2 e = eI_1 eI_2 e = \tau(I_1)\tau(I_2)$. To prove that $\tau(\mathbf{J}(A)) = \mathbf{J}(B)$, it is sufficient (because τ is a lattice isomorphism) to note that $\mathbf{J}(A) = \bigcap\{M \in \mathbf{I}(A) : M \text{ is maximal in } \mathbf{I}(A)\}$ and $\mathbf{J}(B) = \bigcap\{N \in \mathbf{I}(B) : N \text{ is maximal in } \mathbf{I}(B)\}$. The representation of the radical of an Artinian algebra as an intersection of maximal (two-sided) ideals follows easily from Wedderburn's Structure Theorem. To show that $\Gamma(B) = \Gamma(A)$, observe that $B = Pe = P_1 e \bigoplus P_2 e \bigoplus \cdots \bigoplus P_r e$, where $P_i e = e_i Ae = e_i B$ is indecomposable because $\mathbf{E}_B(e_i B) \cong e_i Be_i = e_i Ae_i \cong \mathbf{E}_A(e_i A)$ is a local algebra. If $e_i B \cong e_j B$, then by Corollary 6.4b there exist elements x and y in B such that $e_i xe_j ye_i = e_i$. In this case, it also follows from Corollary 6.4b that $e_i A \cong e_j A$, so that $i = j$. Therefore, $e_1 B, e_2 B, \ldots, e_r B$ is a system of representatives of the isomorphism classes of principal indecomposable right B-modules. Thus, $\Gamma(B)$ has the same vertex set $\{e_1, e_2, \ldots, e_r\}$ as $\Gamma(A)$. Since $e_i \mathbf{J}(B)e_j = e_i e\mathbf{J}(A)ee_j = e_i \mathbf{J}(A)e_j$, $\Gamma(B)$ and $\Gamma(A)$ have the same set of edges; that is, $\Gamma(B) = \Gamma(A)$. □

EXERCISES

1. Let B be the basic algebra of a right Artinian algebra A. Prove the following statements.

   (a) If $A/\mathbf{J}(A) \cong M_{n_1}(D_1) \dot{+} \cdots \dot{+} M_{n_r}(D_r)$, where $D_1, \ldots, D_r$ are division algebras, then $B/\mathbf{J}(B) \cong D_1 \dot{+} \cdots \dot{+} D_r$.

   (b) If $A \cong A_1 \dot{+} A_2$, then $B \cong B_1 \dot{+} B_2$, where B_i is the basic algebra of A_i for $i = 1, 2$.

2. Let B be a reduced R-algebra. Denote by $\vec{e} = (e_1, e_2, \ldots, e_r)$ a sequence of primitive idempotent elements of B such that $e_i e_j = 0$ for $i \neq j$, and $1_B = e_1 + e_2 + \cdots + e_r$. Let $\vec{n} = (n_1, n_2, \ldots, n_r)$ be a sequence of positive integers. Let $M_{\vec{n}}(B,\vec{e})$ be the set of all block matrices $[\mu_{ij}]_{1 \le i,j \le r}$, where μ_{ij} is an n_i by n_j matrix with entries in $e_i Be_j$. Define componentwise addition and scalar multiplication by elements of R for the matrices in $M_{\vec{n}}(B,\vec{e})$, and define matrix multiplication as usual (noting $\mu_{ij}\mu_{jk}$ is an n_i by n_k matrix with entries in $e_i Be_j Be_k \subseteq e_i Be_k$). Then $M_{\vec{n}}(B,\vec{e})$ is an R-algebra that is called a *checkered matrix algebra*. Prove that every right Artinian R-algebra A is isomorphic to a checkered matrix algebra of the form $M_{\vec{n}}(B,\vec{e})$, where B is the basic algebra of A.

6.7. Representation Type

For Artinian algebras, the Krull–Schmidt Theorem shifts the problem of classifying finitely generated modules to the study of indecomposable modules. Unfortunately, the difficulties encountered with these modules are formidable. In this section, we will prove that "most" Artinian algebras have infinitely many isomorphism classes of finitely generated, indecomposable modules.

A right Artinian algebra A has *finite representation type* if there are finitely many isomorphism classes of finitely generated, indecomposable right A-modules. In the contrary case, A has *infinite representation type*. By Corollaries 2.3b and 3.2b, every semisimple algebra has finite representation type. Some examples of algebras that have finite representation type and are not semisimple will be given in the next chapter. However, the following result shows that such algebras are exceptional.

Theorem. *If A is a right Artinian algebra of finite representation type, then the lattice $\mathbf{I}(A)$ of all ideals in A is distributive.*

This theorem will be obtained from a lemma that gives sufficient conditions for a right Artinian algebra A to have finitely generated indecomposable modules of arbitrarily great length. The Proposition of Section 2.2 will be used to show that if $\mathbf{I}(A)$ is not distributive, then the conditions of the lemma are satisfied. A preliminary lemma isolates the non-computational aspects of the main construction.

Lemma a. *Let A and B be right Artinian R-algebras, with B a local algebra. Suppose that N is a non-zero B-A bimodule that is finitely generated and projective as an A-module, and assume that L is a proper sub-bimodule of N.*

(i) $C = \{\psi \in \mathbf{E}_A(N) : \psi(L) \subseteq L\}$ *is a B-bimodule and a subalgebra of* $\mathbf{E}_A(N)$.

(ii) *If C contains an ideal I such that the elements of I are nilpotent and $C = B \cdot id_N + I$, then C is a local algebra, and N/L is indecomposable as an A-module.*

Proof. The assertion (i) is obvious. To show that C is a local algebra under the hypotheses in (ii), it is sufficient by Proposition 5.2 to prove that if $x \in B, \psi \in I$, then $\phi = x \cdot id_N + \psi \in C - C^\circ$ if and only if $x \in B - B^\circ$. If x is a unit, then $x^{-1} \cdot \phi = id_N + (x^{-1} \cdot \psi)$, and $x^{-1} \cdot \psi \in I$ is nilpotent. Therefore, $x^{-1} \cdot \phi$ is a unit, and so is ϕ. Conversely, if $\phi \in C^\circ$, then $x \cdot id_N = \phi(1 - \phi^{-1}\psi)$ is a unit. In particular, x is not nilpotent, so that $x \in B^\circ$ because B is a local Artinian algebra. To prove that N/L is indecomposable, let $\pi: N \to N/L$ be the projection homomorphism. If $\phi \in C$, then $\pi\phi(L) = 0$; equivalently, $\operatorname{Ker} \pi \subseteq \operatorname{Ker} \pi\phi$. Thus, there is a unique $\bar{\phi} \in \mathbf{E}_A(N/L)$ such that $\pi\phi = \bar{\phi}\pi$. The mapping $\theta: \phi \mapsto \bar{\phi}$ is easily found to be an algebra homomorphism. Moreover, θ is surjective: for each $\bar{\phi} \in \mathbf{E}_A(N/L)$ there exists $\phi \in \mathbf{E}_A(N)$ such that $\pi\phi = \bar{\phi}\pi$ (because N is projective and π is surjective), and $\pi\phi(L) = \bar{\phi}\pi(L) = 0$ implies $\phi(L) \subseteq L$, that is, $\phi \in C$. By Lemma 4.3b, $\theta(\mathbf{J}(C)) \subseteq \mathbf{J}(\mathbf{E}_A(N/L))$. Therefore, $\mathbf{E}_A(N/L)/\mathbf{J}(\mathbf{E}_A(N/L))$ is a non-trivial (since $L \neq N$) homomorphic image of $C/\mathbf{J}(C)$, which is a division algebra because C is local. Thus $\mathbf{E}_A(N/L)$ is local, and N/L is indecomposable. □

We are ready to prove the existence lemma.

Lemma b. *Assume that the right Artinian algebra A contains a principal indecomposable right A-module P such that there are non-zero submodules M_1 and M_2 of P satisfying*

(i) *M_1 and M_2 are fully invariant in P, that is, $\phi(M_1) \subseteq M_1$ and $\phi(M_2) \subseteq M_2$ for all $\phi \in \mathbf{E}_A(P)$;*
(ii) *$M_1 \cong M_2$ as $\mathbf{E}_A(P)$-A bimodules;*
(iii) *$M_1 \cap M_2 = 0$.*

If $2 \leq n \in \mathbb{N}$, then there is an indecomposable A-module Q such that $l(Q) > n$.

PROOF. Denote $N = \bigoplus n P$ with corresponding coordinate projections π_i and injections κ_i for $1 \leq i \leq n$. Let $\tau: M_1 \to M_2$ be the bimodule isomorphism that is assumed to exist. Define $L_i = \operatorname{Im}(\kappa_i\tau - \kappa_{i+1})$ for $1 \leq i < n$, and $L = L_1 + \cdots + L_{n-1}$. Plainly, L is a submodule of N, and the length of L is at most $(n-1)l(M_1)$ by Corollary 2.6. Define $Q = N/L$. Since $l(N) = nl(P) \geq nl(M_1 + M_2) = nl(M_1 \oplus M_2) = 2nl(M_1)$ by (ii) and (iii), it follows from Corollary 2.6 again that $l(Q) \geq (n+1)l(M_1) > n$. The proof will be completed by using Lemma a to show that Q is indecomposable. For the application of the lemma, let $B = \mathbf{E}_A(P)$. Since P is indecomposable and finitely generated, B is a local algebra. The fact that B is right Artinian follows from the assumption that A is right Artinian and P is a summand of A_A, by the last part of Corollary 6.4b. Since M_1 and M_2 are fully invariant in P and τ is a bimodule homomorphism, L is a sub-bimodule of N. Thus $C = \{\psi \in \mathbf{E}_A(N): \psi(L) \subseteq L\}$ is a subalgebra of $\mathbf{E}_A(N)$, and a B-bimodule. It is convenient to represent the elements of $\mathbf{E}_A(N)$ as matrices, using the isomorphism $\mathbf{E}_A(N) = \mathbf{E}_A(\bigoplus n P) \cong M_n(B)$ that is defined by $\phi \mapsto [\phi_{ij}]$, where $\phi_{ij} = \pi_i\phi\kappa_j$. Define $I = \{\phi \in C: \phi_{ij} \in B - B^\circ \text{ for all } i \text{ and } j\}$. Since B is local, I is an ideal of C. Moreover, by Proposition 5.2, $B - B^\circ \subseteq \mathbf{J}(B)$, and $\mathbf{J}(B)$ is nilpotent, say $\mathbf{J}(B)^m = 0$. Therefore, if $\phi \in I$, then $[\phi_{ij}]^m = 0$, so that the elements of I are nilpotent. It remains to show that if $\phi \in C$, then there exists $\xi \in B$ and $\psi \in I$ such that $\phi = \xi \cdot id_N + \psi$. If $x \in M_1$ and $j < n$, then $\phi(\kappa_j\tau x - \kappa_{j+1}x) \in L$. Thus, there exist $y_{ij} \in M_1$ such that $\phi\kappa_j\tau x - \phi\kappa_{j+1}x = \sum_{1 \leq i < n}(\kappa_i\tau y_{ij} - \kappa_{i+1}y_{ij})$. When the projection mappings π_i are applied to this equation, we obtain $\phi_{ij}\tau x - \phi_{i\,j+1}x = \tau y_{ij} - y_{i-1\,j}$ if $1 < i < n$, $\phi_{1j}\tau x - \phi_{1\,j+1}x = \tau y_{1j}$, and $\phi_{nj}\tau x - \phi_{n\,j+1}x = -y_{n-1\,j}$. Since $\phi_{ij}\tau = \tau\phi_{ij}$ by (i) and $M_1 \cap M_2 = 0$ by (iii), these equations yield $\phi_{ij}x = \phi_{i+1\,j+1}x$ for $1 \leq i, j < n$, $\phi_{1j}x = 0$ for $1 < j \leq n$, and $\phi_{nk}x = 0$ for $1 \leq j < n$. It follows that $\phi_{ij}x = 0$ if $i \neq j$, and $\phi_{11}x = \cdots = \phi_{nn}x$. Let $\xi = \phi_{11} \in \mathbf{E}_A(P) = B$. Then $\phi_{ij}|M_1 = 0$ for $i \neq j$, and $(\phi_{ii} - \xi)|M_1 = 0$ for all i. Since $M_1 \neq 0$, it follows that $[\phi_{ij}] - \iota_n\xi = [\psi_{ij}]$, where $\psi_{ij} \in B - B^\circ$ for all i, j. This gives the desired conclusion that $\phi = \xi \cdot id_N + \psi$ with $\psi \in I$. □

We can now prove the theorem. Assume that $\mathbf{I}(A)$ is not distributive. By the remark that follows Proposition 2.2, there exist distinct ideals I, J, and K in A such that $I \cap J = K$ and $I/K \cong J/K$ as A-bimodules. If there are infinitely many isomorphism classes of indecomposable A/K-modules, then by Lemma 2.1a there are infinitely many isomorphism classes of indecomposable A-modules. For this reason, it can be assumed that $K = 0$. Let $A_A = P_1 \oplus \cdots \oplus P_m$ be a decomposition of A_A as a direct sum of indecomposable modules. Since $0 \neq I = AI = P_1 I + \cdots + P_m I$, there is a principal indecomposable right A-module P such that $M_1 = PI \neq 0$. Let $M_2 = PJ$. Plainly, M_1 and M_2 are submodules of P, and $M_1 \cap M_2 \subseteq I \cap J = 0$. If $\phi \in \mathbf{E}_A(P)$, then $\phi = \lambda_x|P$ for some $x \in P$ by Lemma 6.4. Thus, $\phi(M_1) = xPI \subseteq PI = M_1$. Similarly $\phi(M_2) \subseteq M_2$. Therefore, M_1 and M_2 are fully invariant in P. Since $I \cong J$ as A-bimodules, there exists a group isomorphism $\tau: I \to J$ such that $\tau(xy) = x\tau(y)$ if $x \in A$, $y \in I$ and $\tau(xy) = \tau(x)y$ if $x \in I$, $y \in A$. Hence, $\tau(M_1) = \tau(PI) = P\tau(I) = PJ = M_2$, and for $x \in A$, $y \in P$, $z \in M_1 \subseteq I$, $\tau(zx) = \tau(z)x$ and $\tau(\lambda_y z) = \tau(yz) = y\tau(z) = \lambda_y \tau(z)$. Therefore, τ is a bimodule isomorphism.

Exercises

1. Prove that the product $A \dotplus B$ of two right Artinian algebras has finite representation type if and only if A and B both have finite representation type. Hint. See Exercise 2, Section 3.5.

2. Prove that the following conditions are equivalent for a commutative Artinian algebra A.

   (i) A has finite representation type.
   (ii) $A \cong A_1 \dotplus \cdots \dotplus A_r$, where $F_i = A_i/\mathbf{J}(A_i)$ is a field and $\dim_{F_i} \mathbf{J}(A_i)/\mathbf{J}(A_i)^2 \leq 1$ for $1 \leq i \leq r$.
   (iii) $A \cong F_1[\mathbf{x}]/(\mathbf{x}^{k_1}) \dotplus \cdots \dotplus F_r[\mathbf{x}]/(\mathbf{x}^{k_r})$, where each F_i is a field, and $k_i \in \mathbb{N}$.

   Hint. Use Exercise 4, Section 6.5 along with Exercise 1 above to reduce the problem to the case in which A is a local algebra. Use Proposition 4.8 to conclude that (i) implies (ii), and pass from (ii) to (iii) by means of Nakayama's Lemma. Finally, (i) can be obtained from (iii) by using the fundamental theorem of modules over a principal ideal domain.

Notes on Chapter 6

Standard treatments of the projective modules over Artinian algebras use idempotents. We have avoided this procedure in the first three sections of the chapter, but the usefulness of idempotents in the theory of algebras should be clear from the last few sections.

Basic algebras were introduced by Nesbitt and Scott [60]; they attribute the concept to Brauer. The definition of a reduced algebra appears in Chapter 6 of the notes [20] by Brauer and Weiss. The basic algebra of an algebra A is also discussed in these notes, and the representation by checkered matrix algebras in Exercise 2, Section 6.6 is proved there. The theorem of Section 6.7 is due to J. P. Jans [50] (for finite dimensional algebras over a field) and R. Colby [23] (for Artinian algebras). Our proof is modeled on a result of S. E. Dickson in [27].

CHAPTER 7

Finite Representation Type

If A is right Artinian, then the finitely generated right A-modules can be constructed in an orderly way from the indecomposable modules, and the construction is unique by the Krull–Schmidt Theorem. The next step toward understanding A-modules is therefore in the direction of indecomposable modules, and this topic is currently the center of vigorous activity in ring theory. The aim of this chapter and the next chapter is to introduce the reader to the flavor of two lines that are being pursued by research mathematicians who are now working on the theory of modules.

Throughout this chapter, it is assumed without mention that A is a right Artinian algebra. (Example 7.1a is an exception to this convention.) All of the A-modules under consideration are finitely generated, hence Artinian and Noetherian.

7.1. The Brauer–Thrall Conjectures

It is convenient to introduce notation that will be used throughout the chapter. The class of all finitely generated right A-modules will be denoted by $\mathfrak{M}_A$, and the subclass of $\mathfrak{M}_A$ that consists of indecomposable modules will be represented by $\mathfrak{N}_A$. For each natural number k, let $\mathfrak{M}_A(k) = \{M \in \mathfrak{M}_A : l(M) = k\}$ and $\mathfrak{N}_A(k) = \{N \in \mathfrak{N}_A : l(N) = k\}$. For the discussion in this section it is useful to define n_A and $n_A(k)$ to be the cardinal numbers of isomorphism classes of modules in $\mathfrak{N}_A$ and $\mathfrak{N}_A(k)$ respectively. Thus, $n_A(1)$ is the number of isomorphism classes of simple A-modules, so that $1 \leq n_A(1) < \aleph_0$.

It is natural to ask what sequences $(n_A(1), n_A(2), n_A(3), \ldots)$ are obtained from various Artinian algebras. This question has not been answered, but some important properties of the sequences have been established in the last few years. Around 1950 two conjectures were made, apparently by Brauer and Thrall. The first published appearance of these conjectures was in the paper [50] by Jans.

First Brauer–Thrall Conjecture. *If n_A is infinite, then $n_A(k) \neq 0$ for infinitely many k.*

Second Brauer–Thrall Conjecture. *If n_A is infinite, then $n_A(k)$ is infinite for infinitely many k, provided $\mathbf{Z}(A)$ is infinite.*

The first of these conjectures was proved by A. V. Roiter for finite dimensional F-algebras in 1968. (See [68].) In 1974, M. Auslander gave a different proof that applies to Artinian algebras. The Second Brauer–Thrall conjecture was proved for finite dimensional algebras over an algebraically closed field in 1974 by L. A. Nazarova and A. V. Roiter in [59]. Their result was extended to finite dimensional algebras by C. M. Ringel. The work of Ringel has not yet been published. Most of this chapter is devoted to the proof of the first Brauer–Thrall conjecture. We will not discuss the second conjecture; the existing proofs of this result are long, and probably not in final form. The rest of this section provides examples of algebras that have finite and infinite representation types.

EXAMPLE A. If A is a commutative principal ideal domain, then every finitely generated A-module is uniquely a direct sum of cyclic A-modules of the form A/p^kA, where $p = 0$ or p is irreducible. Thus, $\mathfrak{N}_A(k)$ is the class of modules that are isomorphic to A/p^kA, where p is irreducible. In particular, if A is local and not a field, then $n_A(k) = 1$ for all k. Of course, A cannot be Artinian in this case.

EXAMPLE B. Let $G = \langle y \rangle$ be a cyclic group of order p^k, where p is a prime integer and $k \geq 1$. Suppose that $A = FG$, where F is a field of characteristic p. The mapping θ: $F[\mathbf{x}] \to A$ defined by $\mathbf{x} \mapsto y$ is a surjective algebra homomorphism whose kernel is the principal ideal generated by $\mathbf{x}^{p^k} - 1 = (\mathbf{x} - 1)^{p^k}$. Thus, a finitely generated, indecomposable A-module N is also a finitely generated, indecomposable $F[\mathbf{x}]$-module such that $(\mathbf{x} - 1)^{p^k} \in \operatorname{ann} N$ (by Proposition 2.1). It follows from Example a that the isomorphism classes of $\mathfrak{N}_A$ are represented by the cyclic modules $F[\mathbf{x}]/(\mathbf{x} - 1)$, $F[\mathbf{x}]/((\mathbf{x} - 1)^2)$, $\ldots$, $F[\mathbf{x}]/((\mathbf{x} - 1)^{p^k})$. Thus, $n_A(1) = n_A(2) = \cdots = n_A(p^k) = 1$ and $n_A(m) = 0$ for all $m > p^k$. In particular, FG has finite representation type.

EXAMPLE C. Let $G = \langle x \rangle \times \langle y \rangle$ be the product of two cyclic groups of prime order p: $|x| = |y| = p$. Suppose that $A = FG$, where F is a field of characteristic p. In Exercise 2 of Section 4.8, a proof that $\mathbf{I}(A)$ is not distributive was sketched. Thus, by Theorem 6.7 A has infinite representation type. See Exercise 2 for another proof.

Examples b and c are group algebras. The group algebras of finite representation type have been characterized in a satisfying way by D. G. Higman: if F is a field of prime characteristic p, and G is a finite group, then FG has finite representation type if and only if the Sylow p-subgroup of G is cyclic. The proof of Higman's Theorem uses ideas that will be introduced in chapters 9 and 10. The general case is based on the result for p-groups, which in its turn is a consequence of the examples b and c that we have just described.

It is a standard fact that if G is a finite p-group and H is a proper subgroup of G, then $H \subset N_G(H)$. From this result it follows that there is a normal subgroup M of index p in G such that $H \subseteq M$. Indeed, any maximal proper subgroup containing H must be normal and of index p, as is easily verified.

Lemma. *Let G be a finite p-group, and suppose that F is a field of characteristic p. If G is cyclic, then FG has finite representation type. If G is not cyclic, then FG has infinite representation type.*

PROOF. The first statement was proved in Example b. Assume that G is not cyclic. We will show that there is a normal subgroup N of G such that $G/N \cong H = \mathbb{Z}/p\mathbb{Z} \times \mathbb{Z}/p\mathbb{Z}$. By the remarks above, there is a normal subgroup M_1 of G such that $G/M_1 \cong \mathbb{Z}/p\mathbb{Z}$. Let $x \in G - M_1$. Since G is not cyclic, there is a normal subgroup M_2 of G such that $x \in M_2$ and $G/M_2 \cong \mathbb{Z}/p\mathbb{Z}$. Define $N = M_1 \cap M_2$. Since $x \in M_2 - M_1$, N is a proper subgroup of M_2, so that the index of N in G is at least p^2. On the other hand, the mapping $y \mapsto (yM_1, yM_2)$ is a homomorphism of G to $G/M_1 \times G/M_2 \cong H$, and the kernel of this homomorphism is N. Thus, $p^2 \le |G/N| \le |(G/M_1) \times (G/M_2)| \le p^2$, and $G/N \cong H$. In particular, there is a surjective homomorphism $\phi: G \to H$. By Proposition 1.2, ϕ extends to a surjective homomorphism of FG to FH. The discussion in Section 2.1 shows that every indecomposable FH-module can be viewed as an FG-module that is still indecomposable. Thus, by Example c, FG has infinite representation type. □

EXERCISES

1. Let A be a finite dimensional F-algebra. Use Proposition 5.5 to prove that if F is finite then $n_A(k)$ is finite for all $k \in \mathbb{N}$, and if F is infinite, then $n_A(k) \le |F|$ for all k.

2. This exercise outlines a direct proof of the result in Example c. We begin with an elaboration of Lemma 6.2a.

   (a) Let A be an algebra such that $\mathbf{J}(A)^k = 0$ for some $k \in \mathbb{N}$. Suppose that M is a right A-module. Let $\theta: \mathbf{E}_A(M) \to \mathbf{E}_A(M/M\mathbf{J}(A))$ be defined by $\theta(\phi) = \bar{\phi}$, where $\bar{\phi}(u + M\mathbf{J}(A)) = \phi(u) + M\mathbf{J}(A)$, as in Lemma 6.2a. Prove that if $\theta(\phi) = 0$, then $\phi^k = 0$. Show that if $\operatorname{Im}\theta$ is local, then $\mathbf{E}_A(M)$ is local and M is indecomposable.

For the rest of this problem, assume that $A = FG$, where $\operatorname{char} F = p > 0$, and $G = \langle x \rangle \times \langle y \rangle$ with $|x| = |y| = p$.

(b) Let P and Q be k dimensional F-spaces, and $M = P \bigoplus Q$. Let $\psi \in \operatorname{Hom}_F(P,Q)$ be an isomorphism, and suppose that $\chi \in \mathbf{E}_F(P)$. Prove that there is a unique A-module structure on M such that $u(x - 1) = \psi(u)$, $u(y - 1) = \psi\chi(u)$ for all $u \in P$, and $Q(x - 1) = Q(y - 1) = 0$. Show that $M\mathbf{J}(A) = Q$.

(c) Prove that if $\phi \in \mathbf{E}_A(M)$ and $\bar{\phi} = \theta(\phi)$, where $\theta: \mathbf{E}_A(M) \to \mathbf{E}_A(M/M\mathbf{J}(A))$ is defined as in (a), then $\bar{\phi}\chi = \chi\bar{\phi}$, where $M/M\mathbf{J}(A)$ is identified with P and $\bar{\phi}$ is viewed as a linear transformation of P. Show that every linear transformation of P that commutes with χ is of the form $\bar{\phi}$ for some $\phi \in \mathbf{E}_A(M)$.

(d) Prove that if the minimum polynomial $\Phi \in F[\mathbf{x}]$ of χ has degree k, then the image of θ is isomorphic to $F[\mathbf{x}]/(\Phi)$. Hint. View P as an $F[\mathbf{x}]$-module with $u\mathbf{x} = \chi(u)$ for all $u \in P$, and use Exercise 2, Section 2.1.

(e) Let $u_1, \ldots, u_k$ be an F-basis of P, and suppose that $\chi(u_i) = u_i a + u_{i+1}$ for $i < k$, $\chi(u_k) = u_k a$, where $a \in F$. Prove that the module M is indecomposable. Hint. Show that the minimum polynomial of χ is $(\mathbf{x} - a)^k$, and use the results of (a) and (d).

(f) Prove that if M' is an A-module that is constructed by the process described in (b) using ϕ and χ', where χ' is defined as in (e) with a replaced by $a' \in F$, and if $a' \neq a$, then $M' \ncong M$. Hint. Show that the existence of an isomorphism $\phi: M \to M'$ would imply that χ and χ' are similar, and would therefore have the same minimum polynomial.

(g) Prove that if F is infinite, then $n_A(k) = |F|$ for all even $k \in \mathbb{N}$.

7.2. Bounded Representation Type

The proposition of this section generalizes the characterization of indecomposable modules that was given in Corollary 5.3. It is the basis of the proof of the First Brauer–Thrall Conjecture. The proof of this result is an elementary induction that evolves from Corollary 2.6: if $0 \to N \to M \to P \to 0$ is an exact sequence of modules in $\mathfrak{M}_A$, then $l(M) = l(N) + l(P)$.

Lemma a. *Let $\phi: M \to N$ and $\psi: N \to P$ be homomorphisms of modules in $\mathfrak{M}_A$.*

(i) *$l(\phi(M)) \leq l(N)$, and equality holds if and only if ϕ is surjective.*
(ii) *$l(\phi(M)) \leq l(M)$, and equality holds if and only if ϕ is injective.*
(iii) *$l(\psi\phi(M)) \leq l(\psi(N))$, and equality holds if and only if*

$$\operatorname{Im}\phi + \operatorname{Ker}\psi = N.$$

(iv) *$l(\psi\phi(M)) \leq l(\phi(M))$, and equality holds if and only if*

$$\operatorname{Im}\phi \cap \operatorname{Ker}\psi = 0.$$

Proof. The statements (i) and (ii) are obtained by applying Corollary 2.6 to the exact sequences $0 \to \phi(M) \to N \to N/\phi(M) \to 0$ and $0 \to \operatorname{Ker}\phi \to$

$M \to \phi(M) \to 0$, using the fact that only the zero module has length 0. The result (iii) is a consequence of (i), applied to the homomorphism $\psi\phi$: $M \to \psi(N)$, since $\psi\phi(M) = \psi(N)$ is equivalent to $\operatorname{Im}\phi + \operatorname{Ker}\psi = N$. Similarly, (iv) is obtained by using (ii) with the homomorphism $\psi|\phi(M)$: $\phi(M) \to P$, because $\operatorname{Ker}\psi|\phi(M) = \operatorname{Im}\phi \cap \operatorname{Ker}\psi$. □

Lemma b. *Let*

$$Q_m \xrightarrow{\chi_m} Q_{m-1} \xrightarrow{\chi_{m-1}} Q_{m-2} \to \cdots \to Q_1 \xrightarrow{\chi_1} Q_0$$

be a sequence of homomorphisms of modules in $\mathfrak{N}_A$. *Assume that* $l(Q_i) \leq n$, *and either*

(i) χ_i *is not injective for* $1 \leq i \leq m$, *or*
(ii) χ_i *is not surjective for* $1 \leq i \leq m$.

If $m \geq 2^{n-1}$, *then* $\chi_1\chi_2 \cdots \chi_m = 0$.

Proof. It will be sufficient to show by induction on k that if $0 \leq k \leq n-1$, and $0 \leq i \leq m - 2^k$, then $l(\operatorname{Im}(\chi_{i+1}\chi_{i+2} \cdots \chi_{i+2^k})) \leq n - k - 1$. For the case $k = 0$, note that by Lemma a, $l(\operatorname{Im}\chi_{i+1}) < l(Q_{i+1}) \leq n$ if χ_{i+1} is not injective, and $l(\operatorname{Im}\chi_{i+1}) < l(Q_i) \leq n$ if χ_{i+1} is not surjective. In both cases, $l(\operatorname{Im}\chi_{i+1}) \leq n - 1$. Assume $k + 1 \leq n - 1$, $0 \leq i \leq m - 2^{k+1}$, and the induction hypothesis is valid for compositions of 2^k homomorphisms. Denote $\phi = \chi_{i+2^k+1} \cdots \chi_{i+2^{k+1}}$, $\psi = \chi_{i+1} \cdots \chi_{i+2^k}$, $M = Q_{i+2^{k+1}}$, $N = Q_{i+2^k}$, $P = Q_i$. Thus, we have a sequence of two homomorphisms $M \xrightarrow{\phi} N \xrightarrow{\psi} P$, with $l(\phi(M)) \leq n - k - 1$ and $l(\psi(N)) \leq n - k - 1$ by the induction hypothesis. To complee the induction step, it is required to show that $l(\operatorname{Im}\psi\phi) \leq n - k - 2$. Suppose on the contrary that $l(\operatorname{Im}\psi\phi) \geq n - k - 1$. It would then follow from Lemma a that $l(\operatorname{Im}\phi) = l(\operatorname{Im}\psi) = n - k - 1 = l(\operatorname{Im}\psi\phi)$, and $\operatorname{Im}\phi + \operatorname{Ker}\psi = N$, $\operatorname{Im}\phi \cap \operatorname{Ker}\psi = 0$, so that $N = \operatorname{Im}\phi \oplus \operatorname{Ker}\psi$. Since $N = Q_{i+2^k}$ is indecomposable and $l(\operatorname{Im}\phi) = n - k - 1 > 0$, it follows that $\operatorname{Im}\phi = N$ and $\operatorname{Ker}\psi = 0$. In particular, χ_{i+2^k+1} is surjective, and χ_{i+2^k} is injective. These two conclusions respectively contradict the alternative hypotheses that $\operatorname{Ker}\chi_j \neq 0$ for all j or $\operatorname{Im}\chi_j \neq Q_{j-1}$ for all j. The induction is therefore complete. □

Proposition. *Let*

$$Q_m \xrightarrow{\chi_m} Q_{m-1} \xrightarrow{\chi_{m-1}} Q_{m-2} \to \cdots \to Q_1 \xrightarrow{\chi_1} Q_0$$

be a sequence of homomorphisms of modules in $\mathfrak{N}_A$. *Assume that* $l(Q_i) \leq n$, *and none of the* χ_i *is an isomorphism. If* $m \geq 2^n$, *then* $\chi_1\chi_2 \cdots \chi_m = 0$.

Proof. Let $I = \{j\colon \chi_j \text{ is not injective}\}$ and $I' = \{j\colon \chi_j \text{ is not surjective}\}$. The assumption that none of the χ_j is an isomorphism translates to $I \cup I' = \{1, \ldots, m\}$. Thus, either $|I| \geq 2^{n-1}$ or $|I'| \geq 2^{n-1}$, so that there is a sequence $1 \leq j_1 < j_2 < \cdots < j_k \leq m$ with $k \geq 2^{n-1}$ such that either

$$\chi_{j_i} \text{ is not injective for } 1 \leq i \leq k, \text{ or} \tag{1}$$

$$\chi_{j_i} \text{ is not surjective for } 1 \leq i \leq k. \tag{2}$$

In the first case, let

$$\theta_i = (\chi_{j_{i-1}+1})(\chi_{j_{i-1}+2}) \cdots \chi_{j_i} \in \mathrm{Hom}_A(Q_{j_i}, Q_{j_{i-1}})$$

with the convention that $j_0 = 0$. If the second alternative holds, define

$$\theta_i = \chi_{j_i}\chi_{j_i+1} \cdots (\chi_{j_{i+1}-1}) \in \mathrm{Hom}_A(Q_{j_{i+1}-1}, Q_{j_i-1})$$

with the convention that $j_{k+1} = m + 1$. In these cases, none of the homomorphisms θ_i is injective (respectively, surjective). By Lemma b, $\theta_1\theta_2 \cdots \theta_k = 0$. Thus, $\chi_1\chi_2 \cdots \chi_m = 0$, because $\theta_1\theta_2 \cdots \theta_k$ divides $\chi_1\chi_2 \cdots \chi_m$. □

It is convenient to make a provisional definition: the representation type of the right Artinian algebra A is bounded by n if $n_A(k) = 0$ for all $k > n$; A has *bounded representation type* if the representation type of A is bounded by some n. Plainly, if A has finite representation type, then A has bounded representation type; the converse statement is the First Brauer–Thrall Conjecture.

Corollary. *If A has representation type that is bounded by n, and if*

$$Q_m \xrightarrow{\chi_m} Q_{m-1} \xrightarrow{\chi_{m-1}} Q_{m-2} \to \cdots \to Q_1 \xrightarrow{\chi_1} Q_0$$

is a sequence of homomorphisms of modules in $\mathfrak{N}_A$ such that χ_i is not an isomorphism for $1 \leq i \leq m$, then $m \geq 2^n$ implies $\chi_1\chi_2 \cdots \chi_m = 0$.

EXERCISES

1. Let A be a right Artinian algebra, and suppose that $P \in \mathfrak{N}_A$. Prove that $\mathbf{J}(\mathbf{E}_A(P)) = \{\phi \in \mathbf{E}_A(P) : \phi \text{ is not an isomorphism}\}$, and $\mathbf{J}(\mathbf{E}_A(P))$ is nilpotent.
2. Let A be right Artinian, and $M \in \mathfrak{M}_A$. Write $M = P_1 \oplus P_2 \oplus \cdots \oplus P_k$, where $P_i \in \mathfrak{N}_A$ for all i. Denote the projections and injections that correspond to this decomposition by π_i and κ_i respectively. Prove that $\mathbf{J}(\mathbf{E}_A(M)) = \{\phi \in \mathbf{E}_A(M)$: for $1 \leq i, j \leq k$, $\pi_i\phi\kappa_j$ is not an isomorphism$\}$. Show that $\mathbf{J}(\mathbf{E}_A(M))$ is nilpotent, and $\mathbf{E}_A(M)/\mathbf{J}(\mathbf{E}_A(M))$ is semisimple.

7.3. Sequence Categories

Let $P \in \mathfrak{M}_A$. A *P-sequence* is a short exact sequence $\sum: 0 \to N \to M \xrightarrow{\phi} P \to 0$ of module homomorphisms such that $M \in \mathfrak{M}_A$ and $N \to M$ is the inclusion mapping (hence, $N = \mathrm{Ker}\,\phi$). Strictly speaking, $\sum$ is completely determined by specifying the homomorphism $\phi: M \to P$, but the kernel N of ϕ plays an important part in the theory to be developed, and we prefer to keep it visible by using slightly redundant notation.

The P-sequence $\sum: 0 \to N \to M \overset{\phi}{\to} P \to 0$ is split if there is an A-module homomorphism $\psi: P \to M$ such that $\phi\psi = id_p$. By Lemma 5.4a, $\sum$ is split if and only if there is a homomorphism $\chi: M \to N$ such that $\chi|N = id_N$.

The notation $\mathfrak{E}(P)$ will designate the class of all P-sequences. There is a natural way to define morphisms of sequences that makes $\mathfrak{E}(P)$ a category. Let $\sum$: $0 \to N \to M \overset{\phi}{\to} P \to 0$ and $\sum'$: $0 \to N' \to M' \overset{\phi'}{\to} P \to 0$ be P-sequences. A morphism ψ of $\sum$ to $\sum'$ is an A-module homomorphism $\psi: M \to M'$ such that

$$\begin{array}{ccccccccc} 0 & \to & N & \to & M & \overset{\phi}{\to} & P & \to & 0 \\ & & \downarrow{\scriptstyle \psi|N} & & \downarrow{\scriptstyle \psi} & & \downarrow{\scriptstyle id_P} & & \\ 0 & \to & N' & \to & M' & \underset{\phi'}{\to} & P & \to & 0 \end{array}$$

commutes. The commutativity condition amounts to the requirement that $\phi'\psi = \phi$, since it will then follow that $\psi(N) = \psi(\phi^{-1}(0)) \subseteq (\phi')^{-1}(0) = N'$. Clearly, the composition of morphisms is a morphism, and id_M has the usual properties of an identity morphism of $\sum: 0 \to N \to M \to P \to 0$.

The terminology that goes with module homomorphisms will be used for the morphisms of $\mathfrak{E}(P)$. In particular, $\psi: \sum \to \sum'$ is a split injection or a split surjection if $\chi\psi = id_{\Sigma}$ (respectively, $\psi\chi = id_{\Sigma'}$) for some morphism $\chi: \sum' \to \sum$. In this case, ψ is evidently a split injection (surjection) when it is viewed as a module homomorphism. For split surjections the converse is true: if $\chi \in \mathrm{Hom}_A(M', M)$ satisfies $\psi\chi = id_{M'}$, then $\phi\chi = \phi'\psi\chi = \phi'$, so that χ is a morphism of P-sequences and $\psi\chi = id_{\Sigma'}$. It is a special case of this observation that ψ is an isomorphism in $\mathfrak{E}(P)$ if and only if it is a module isomorphism.

Lemma. *Let $\sum: 0 \to N \to M \overset{\phi}{\to} P \to 0$ be a P-sequence. Suppose that $N = \bigoplus_{i=1}^r Q_i$. Denote $M_i = M/\sum_{j\neq i} Q_j$, $N_i = N/\sum_{j\neq i} Q_j$, with $\pi_i: M \to M_i$ the natural projection homomorphism.*

(i) *$\pi_i|Q_i: Q_i \to N_i$ is an isomorphism.*
(ii) *There is a unique surjective homomorphism $\phi_i: M_i \to P$ such that $\phi_i\pi_i = \phi$, and $\mathrm{Ker}\,\phi_i = N_i$. Thus,*

$$\sum_i: 0 \to N_i \to M_i \overset{\phi_i}{\to} P \to 0$$

is a P-sequence and $\pi_i: \sum \to \sum_i$ is a morphism in $\mathfrak{E}(P)$.
(iii) *If $\sum_i$ is split for $1 \leq i \leq r$, then $\sum$ is split.*

Proof. The assertion (i) is obvious, and (ii) follows from definitions and the facts that π_i and ϕ are surjective, and $\mathrm{Ker}\,\pi_i = \sum_{j\neq i} Q_j \subseteq N = \mathrm{Ker}\,\phi$. To prove (iii), suppose that all $\sum_i$ are split, say $\chi_i: M_i \to N_i$ are homomorphisms such that $\chi_i|N_i = id_{N_i}$. Let $\kappa_i: Q_i \to N$ be the injection homomorphisms that go with the decomposition $N = \bigoplus_{i=1}^r Q_i$. Define $\chi =$

$\sum_{i=1}^{r} \kappa_i(\pi_i|Q_i)^{-1}\chi_i\pi_i \in \operatorname{Hom}_A(M,N)$. If $u \in Q_i$, then $\pi_i(u) \in N_i$ and $\pi_j(u) = 0$ for $j \neq i$. Thus, $\chi(u) = \kappa_i(\pi_i|Q_i)^{-1}\chi_i\pi_i u = \kappa_i(\pi_i|Q_i)^{-1}\pi_i u = \kappa_i u = u$. Since χ is a homomorphism and $N = \sum_{i=1}^{r} Q_i$, it follows that $\chi|N = id_N$. Thus, $\sum$ is split. □

Let $\sum$: $0 \to N \to M \to P \to 0$ be a P-sequence. We will say that $\sum$ is *simple* if $\sum$ is not split, and N is an indecomposable A-module. The motive for this choice of terminology is the analogy with Schur's Lemma that will emerge from Corollary 7.4a.

Proposition. *If $P \in \mathfrak{M}_A$ and P is not projective, then there is a simple P-sequence.*

PROOF. Since P is finitely generated, there is a finitely generated free A-module M and a surjective homomorphism ϕ: $M \to P$. Let $N = \operatorname{Ker}\phi$. Then $\sum$: $0 \to N \to M \xrightarrow{\phi} P \to 0$ is a P-sequence, and $\sum$ cannot be split: otherwise, P would be a summand of M, hence projective. It follows from the lemma that there is an indecomposable summand Q of N and a submodule M' of M such that $\sum'$: $0 \to Q \to M' \xrightarrow{\phi|M'} P \to 0$ is not split. Thus, $\sum'$ is a simple P-sequence. □

EXERCISES

1. Let $\sum: 0 \to N \to M \to P \to 0$ and $\sum': 0 \to N' \to M' \to P \to 0$ be P-sequences, and suppose that $\psi: \sum \to \sum'$ is a morphism. Prove the following statements.

   (a) If ψ is not injective (as a module homomorphism), then there exist distinct morphisms χ and χ' from the P-sequence $\sum'': 0 \to N \oplus \operatorname{Ker}\psi \to M \oplus \operatorname{Ker}\psi \xrightarrow{\phi\oplus 0} P \to 0$ to $\sum$ such that $\psi\chi = \psi\chi'$.

   (b) If ψ is not surjective, then there exist distinct morphisms θ and θ' from $\sum'$ to the P-sequence $\sum'''$: $0 \to M'/\operatorname{Im}\psi \to P \oplus (M'/\operatorname{Im}\psi) \xrightarrow{id_P\oplus 0} P \to 0$ such that $\theta\psi = \theta'\psi$.

   This exercise shows that the monomorphisms of $\mathfrak{E}(P)$ coincide with the injective homomorphisms, and the epimorphisms of $\mathfrak{E}(P)$ coincide with the surjective homomorphisms.

2. Let A be an algebra, and suppose that

$$\begin{array}{ccccccccc} 0 & \to & Q & \xrightarrow{\kappa} & M & \xrightarrow{\phi} & P & \to & 0 \\ & & \downarrow \chi & & \downarrow \psi & & \downarrow id_P & & \\ 0 & \to & Q' & \xrightarrow[\kappa']{} & M' & \xrightarrow[\phi']{} & P & \to & 0 \end{array} \tag{1}$$

   is a commutative diagram of right A-modules with exact rows. Map $\theta: Q' \oplus M \to M'$ by $\theta(v,w) = \kappa'(v) + \psi(w)$. Prove that θ is a surjective A-module homomorphism with $\operatorname{Ker}\theta = \{(\chi(u), -\kappa(u)): u \in Q\}$. Conversely, prove that if $0 \to Q \xrightarrow{\kappa} M \xrightarrow{\phi} P \to 0$ is a short exact sequence of right A-modules, and if χ: $Q \to Q'$ is a homomorphism, then there is a right A-module M' and module homomorphisms κ': $Q' \to M'$, ϕ': $M' \to P$, ψ: $M \to M'$ such that the resulting diagram (1) is commutative and has exact rows. Moreover, the sequence

$$\textstyle\sum': 0 \to Q' \underset{\kappa'}{\to} M' \underset{\phi'}{\to} P \to 0$$

is unique to within the equivalence relation defined by

$$\textstyle\sum' \cong \sum'': 0 \to Q' \underset{\kappa''}{\to} M'' \underset{\phi''}{\to} P \to 0$$

if there is an isomorphism $\theta\colon M' \to M''$ such that $\kappa'' = \theta\kappa'$ and $\phi' = \phi''\theta$, that is,

$$\begin{array}{ccccc} & \nearrow & M' & \searrow & \\ Q' & & \downarrow{\scriptstyle\theta} & & P \\ & \searrow & M'' & \nearrow & \end{array}$$

commutes. (However, the homomorphism ψ in (1) is not unique. It can be modified by adding $\kappa'\theta\phi$, $\theta \in \mathrm{Hom}_A(P,Q')$.)

7.4. Simple Sequences

In this section we develop properties of simple P-sequences. It is convenient to introduce the notation $\mathfrak{F}(P)$ for the class of all simple P-sequences, where $P \in \mathfrak{M}_A$. If P is not projective, then $\mathfrak{F}(P) \neq \varnothing$ by Proposition 7.3.

Short 5-Lemma. *Let*

$$\begin{array}{ccccccc} & & N_1 & \to & M_1 & \to & P_1 & \to 0 \\ & & {\scriptstyle\phi}\downarrow & & {\scriptstyle\theta}\downarrow & & {\scriptstyle\psi}\downarrow & \\ 0 & \to & N_2 & \to & M_2 & \to & P_2 & \end{array}$$

be a commutative diagram of module homomorphisms that has exact rows.

(i) *If ϕ and ψ are injective, then θ is injective.*
(ii) *If ϕ and ψ are surjective, then θ is surjective.*

This standard result can be proved easily by a diagram chase; this project is suggested as an Exercise. It is also a corollary of the "snake lemma" that will be proved in Section 11.3.

Lemma a. *Let $\sum\colon 0 \to N \to M \overset{\phi}{\to} P \to 0$ and $\sum'\colon 0 \to Q \to M' \overset{\phi'}{\to} P \to 0$ be P-sequences, with $\sum'$ simple. If $\psi\colon \sum \to \sum'$ is a morphism, then $\psi|N \neq 0$.*

Proof. If $\psi|N = 0$, then $\mathrm{Ker}\,\phi = N \subseteq \mathrm{Ker}\,\psi$, so that ψ factors through ϕ: $\psi = \theta\phi$ for some $\theta \in \mathrm{Hom}_A(P,M')$. Consequently, $\phi'\theta\phi = \phi'\psi = \phi$, and $\phi'\theta = id_P$ because ϕ is surjective. Since simple sequences are not split, this conclusion is a contradiction. □

Corollary a. *If $\sum \in \mathfrak{F}(P)$ and $\psi\colon \sum \to \sum$ is a morphism, then ψ is an isomorphism.*

PROOF. Let $\sum: 0 \to Q \to M \to P \to 0$, with Q indecomposable. If ψ is not an isomorphism, then $\psi|Q$ is not an isomorphism by the Short 5-Lemma. It would then follow from Proposition 7.2 that $\psi^m|Q = (\psi|Q)^m = 0$ if $m \geq 2^{l(Q)}$, which contradicts the lemma. □

It will clarify our discussion if we introduce a pre-ordering relation on the class $\mathfrak{F}(P)$. Write $\sum \geq \sum'$ if there is a morphism of $\sum$ to $\sum'$. The relation $\geq$ is transitive because the composition of morphisms is a morphism, and the identity morphism on $\sum$ secures reflexivity: $\sum \geq \sum$.

Corollary b. *If $\sum$ and $\sum'$ are in $\mathfrak{F}(P)$, then $\sum \geq \sum'$ and $\sum' \geq \sum$ if and only if $\sum \cong \sum'$ (that is, $\sum$ is isomorphic to $\sum'$).*

PROOF. If there exist morphisms $\psi: \sum \to \sum'$ and $\psi': \sum' \to \sum$, then by Corollary a, $\psi\psi'$ and $\psi'\psi$ are isomorphisms. Thus, ψ and ψ' are isomorphisms. The converse is obvious. □

The proof shows that if $\sum \cong \sum'$, then every morphism from $\sum$ to $\sum'$ is an isomorphism.

A simple P-sequence $\sum$ will be called minimal if $\sum$ is minimal in $\mathfrak{F}(P)$ with respect to $\geq$. In other words, $\sum \geq \sum'$ implies $\sum' \geq \sum$; or, by Corollary b, $\sum \geq \sum'$ implies $\sum \cong \sum'$.

Corollary c. *If $\sum \in \mathfrak{F}(P)$, then $\sum$ is minimal if and only if every morphism from $\sum$ to some $\sum' \in \mathfrak{F}(P)$ is an isomorphism.*

PROOF. If $\sum$ is minimal and $\psi: \sum \to \sum'$ is a morphism, then $\sum \cong \sum'$, and ψ is an isomorphism by the remark after the proof of Corollary b. The converse is obvious. □

There is a useful analogue of Corollary c for morphisms to P-sequences that are not simple.

Lemma b. *Let $\sum \in \mathfrak{F}(P)$ be minimal, and suppose that $\sum' \in \mathfrak{E}(P)$. If $\sum'$ is not split, then every morphism from $\sum$ to $\sum'$ is a split injection.*

PROOF. Since $\sum'$ is not split, it follows from Lemma 7.3 that there is a morphism $\psi': \sum' \to \sum''$, where $\sum'' \in \mathfrak{F}(P)$. If $\psi: \sum \to \sum'$ is a morphism, then $\psi'\psi: \sum \to \sum''$ is an isomorphism by Corollary c. If $\chi = (\psi'\psi)^{-1}\psi'$, then $\chi\psi = id_{\Sigma}$, so that ψ is a split injection. □

The fact that minimal, simple P-sequences exist is important. When A has bounded representation type, the existence question is easily settled.

Lemma c. *Assume that the representation type of A is bounded by n. Every totally ordered subset of $\mathfrak{F}(P)$ has at most 2^n isomorphically distinct members.*

Proof. For $0 \leq j \leq m$, let $\sum_j : 0 \to Q_j \to M_j \to P \to 0$ be a simple P-sequence. Assume that for each $j > 0$, $\psi_j : \sum_j \to \sum_{j-1}$ is a morphism that is not an isomorphism. By the Short 5-Lemma, $\psi_j | Q_j$ is not an isomorphism, and $(\psi_1|Q_1)(\psi_2|Q_2) \cdots (\psi_m|Q_m) = (\psi_1\psi_2 \cdots \psi_m)|Q_m \neq 0$ according to Lemma a. Since each Q_i is indecomposable and the representation type of A is bounded by n, it follows that $l(Q_j) \leq n$ for $0 \leq j \leq m$. Therefore, $m < 2^n$ by Lemma 7.2b. □

Proposition. *Assume that A has bounded representation type. If $\sum \in \mathfrak{F}(P)$, then there is a minimal $\sum' \in \mathfrak{F}(P)$ such that $\sum \geq \sum'$.*

This result is an obvious consequence of Lemma c.

Exercise

Prove the Short 5-Lemma.

7.5. Almost Split Sequences

An *almost split sequence* is a simple P-sequence $\sum : 0 \to Q \to M \overset{\phi}{\to} P \to 0$ such that P is indecomposable and if $\psi : N \to P$ is a module homomorphism that is not split surjective, then ψ factors through ϕ; that is, $\psi = \phi\chi$ for some $\chi \in \operatorname{Hom}_A(N, M)$. This last condition can be expressed as a diagram that is analogous to the characterization of projectivity:

$$\begin{array}{ccc} & & N \\ & \overset{\chi}{\swarrow} & \downarrow \psi \\ M & \underset{\phi}{\to} & P \to 0. \end{array}$$

Proposition. *Let $P \in \mathfrak{N}_A$, and suppose that $\sum : 0 \to Q \to M \overset{\phi}{\to} P \to 0$ is a simple P-sequence. The following conditions are equivalent.*

(i) *$\sum$ is an almost split sequence.*
(ii) *If $\sum' \in \mathfrak{F}(P)$, then $\sum' \geq \sum$.*
(iii) *$\sum$ is minimal in $\mathfrak{F}(P)$.*

Proof. Assume that $\sum$ is an almost split sequence. If

$$\textstyle\sum' : 0 \to Q' \to M' \overset{\phi'}{\to} P \to 0$$

is simple, then ϕ' is not a split surjection, and there is a homomorphism $\chi : M' \to M$ such that $\phi' = \phi\chi$. Thus, χ is a morphism from $\sum'$ to $\sum$, that is, $\sum' \geq \sum$. It is clear that (ii) is at least as strong as condition (iii). Assume that $\sum$ is minimal. Let $\psi: N \to P$ be a homomorphism that is not a split surjection. Define $\phi': M \bigoplus N \to P$ by $\phi'(u,v) = \phi(u) + \psi(v)$. Plainly, ϕ' is surjective. However, ϕ' is not split. Otherwise, there is a homomorphism $\chi': P \to M \bigoplus N$ such that $\phi'\chi' = id_p$, say $\chi'(w) = (\chi_1(w),\chi_2(w))$, where $\chi_1 \in \operatorname{Hom}_A(P,M)$ and $\chi_2 \in \operatorname{Hom}_A(P,N)$. It follows from the definition of ϕ' that $id_p = \phi\chi_1 + \psi\chi_2$. Since P is indecomposable, we conclude from Proposition 5.2 and Corollary 5.3 that one of $\phi\chi_1$ or $\psi\chi_2$ is a unit. In these cases, $\chi_1(\phi\chi_1)^{-1}$ splits ϕ, or ψ is surjective and is split by $\chi_2(\psi\chi_2)^{-1}$. Both options contradict the assumptions that ϕ and ψ are not split surjections. Denote $L = \operatorname{Ker}\phi'$, and

$$\sum{}': 0 \to L \to M \bigoplus N \overset{\phi'}{\to} P \to 0.$$

By the definition of ϕ', the injection $\kappa: M \to M \bigoplus N$ satisfies $\phi'\kappa = \phi$, so that $\kappa: \sum \to \sum'$ is a morphism. By Lemma 7.4b, κ is a split injection. That is, there is a morphism $\theta: \sum' \to \sum$ such that $\theta\kappa = id_\Sigma$. Define $\chi: N \to M$ by $\chi(v) = \theta(0,v)$ for $v \in N$. The definitions of χ and ϕ' give

$$\phi\chi(v) = \phi\theta(0,v) = \phi'(0,v) = \psi(v),$$

since θ is a morphism. Therefore $\sum$ is an almost split sequence. □

Corollary. *If the algebra A has bounded representation type, and $P \in \mathfrak{N}_A$ is not projective, then there is a P-sequence $\sum$ such that $\sum$ is an almost split sequence. Moreover, $\sum$ is unique to within an isomorphism in the category $\mathfrak{E}(P)$.*

Proof. Since P is not projective, $\mathfrak{F}(P)$ is not empty. Proposition 7.4 guarantees that $\mathfrak{F}(P)$ contains a minimal sequence $\sum$; and $\sum$ is an almost split sequence by the proposition, because P is indecomposable. Moreover, the proposition shows that a minimal member of $\mathfrak{F}(P)$ is a minimum, so that $\sum$ is unique to within isomorphism by Corollary 7.4b. □

Exercise

Let $0 \to Q \to M \overset{\phi}{\to} P \to 0$ be an almost split sequence. Suppose that $\theta \in \operatorname{Hom}_A(Q,N)$ is not a split injection. Prove that there exists $\chi \in \operatorname{Hom}_A(M,N)$ such that $\chi|Q = \theta$. Hint. Let $L = \{(\theta(w), -w): w \in Q\} \in \mathbf{S}(N \bigoplus M)$, $M' = (N \bigoplus M)/L$ with $\pi: N \bigoplus M \to M'$ the projection homomorphism. Define $\psi: M \to M'$ by $\psi(v) = \pi(0,v)$, and $\lambda: N \to M'$ by $\lambda(u) = \pi(u,0)$. Show that there is a surjective homomorphism $\phi': M' \to P$ such that $\phi'\pi(u,v) = \phi(v)$, and $\operatorname{Ker}\phi' = \lambda(N)$. Prove that $\psi: \sum \to \sum'$ is a morphism, where $\sum'$ is the P-sequence

$$0 \to \lambda(N) \to M' \overset{\phi'}{\to} P \to 0.$$

Show that λ is injective, and $\lambda\theta w = \psi w$ for $w \in Q$. Use the hypothesis that θ is not split to show that ψ is not split injective. Deduce from Lemma 7.4b that $\sum'$ is split, say $\tau: M' \to \lambda(N)$ satisfies $\tau|\lambda(N) = id_{\lambda(N)}$. Prove that $\chi = \lambda^{-1}\tau\psi: M \to N$ is the desired homomorphism.

The development that has been presented in Sections 7.3, 7.4, and 7.5 can be dualized in the categorical sense (that is, reverse the arrows). The exercise (together with its dual) shows that the concept of an almost split sequence is self dual.

Our discussion of almost split sequences has only scratched the surface of an area of active current research. Auslander and Reiten have proved the existence of almost split sequences under hypotheses that are far more general than those of the corollary in this section. Moreover, they have given alternative characterizations of these sequences, and made considerable progress toward an understanding of their structure. Basic references on these topics are the papers [14] and [15].

7.6. Almost Split Extensions

If P is projective, then there are no simple P-sequences. In particular, no almost split sequence can terminate with a projective module. This defect can be remedied by broadening our definition. It will avoid confusion if the terminology is also modified.

Let $P \in \mathfrak{N}_A$. An *almost split extension* of P is a homomorphism $\phi: M \to P$ such that $0 \to \operatorname{Ker} \phi \to M \overset{\phi}{\to} P \to 0$ is an almost split sequence if P is not projective, and ϕ is injective and $\operatorname{Im} \phi = P\mathbf{J}(A)$ if P is projective. An alternative definition of almost split extensions is given in Exercise 1. Two extensions of P, say $\phi: M \to P$ and $\phi': M' \to P$ are isomorphic if there is an isomorphism $\psi: M \to M'$ such that $\phi'\psi = \phi$. Isomorphism of extensions is obviously an equivalence relation.

Proposition a. *Assume that A is a right Artinian algebra with bounded representation type. If P is a finitely generated, indecomposable A-module, then there is an almost split extension $\phi: M \to P$ of P that is unique to within isomorphism. If $\psi \in \operatorname{Hom}_A(N,P)$ is a module homomorphism that is not split surjective, then there is a homomorphism $\chi: N \to M$ such that $\psi = \phi\chi$.*

PROOF. If P is not projective, then the existence and uniqueness statement reformulates Corollary 7.5. If P is projective, then the existence and uniqueness of $\phi: M \to P$ is obvious from the definition of an almost split extension. The second assertion of the proposition is part of the definition of almost split sequences in case P is not projective. If P is projective, then ψ cannot be surjective. Therefore, $\operatorname{Im} \psi \subseteq P\mathbf{J}(A)$ by Corollary 6.3b. Consequently, since ϕ is injective with image $P\mathbf{J}(A)$, it follows that $\chi = \phi^{-1}\psi \in \operatorname{Hom}_A(N,M)$, and $\phi\chi = \psi$. □

It is possible to construct indecomposable A-modules from almost split extensions. The process is not effective, using the tools of the theory that are now available. However, the construction is powerful enough to provide a proof of the First Brauer–Thrall Conjecture.

For each non-empty subclass $\mathfrak{p}$ of $\mathfrak{N}_A$, denote by $I(\mathfrak{p})$ the collection of all $Q \in \mathfrak{N}_A$ such that either $Q \cong P$ for some $P \in \mathfrak{p}$, or there is an almost split extension $M \to P$ with $P \in \mathfrak{p}$ and Q is isomorphic to a direct summand of M. Define classes $I_k(\mathfrak{p}) \subseteq \mathfrak{N}_A$ by induction on k according to the conditions.

$$I_0(\mathfrak{p}) = \mathfrak{p}, \; I_{k+1}(\mathfrak{p}) = I(I_k(\mathfrak{p})). \tag{1}$$

When $\mathfrak{p}$ is the class of all simple A-modules, we will write I_k for $I_k(\mathfrak{p})$.

Several obvious consequences of these definitions are worth recording:

$$\mathfrak{p} = I_0(\mathfrak{p}) \subseteq I_1(\mathfrak{p}) \subseteq I_2(\mathfrak{p}) \subseteq \dots; \tag{2}$$

$$I_l(I_k(\mathfrak{p})) = I_{l+k}(\mathfrak{p}) \quad \text{for all } l \text{ and } k. \tag{3}$$

Proposition b. *For all k, the number of isomorphism classes of modules in I_k is finite.*

Proof. It suffices to observe that if the number of isomorphism classes of modules in some $\mathfrak{p} \subseteq \mathfrak{N}_A$ is finite, then the number of isomorphism classes of modules in $I(\mathfrak{p})$ is also finite. This fact is a consequence of the Krull–Schmidt Theorem and the uniqueness of almost split extensions of each $P \in \mathfrak{N}_A$. (Note that this uniqueness holds without the assumption that A has bounded representation type; only Proposition 7.5 was used to show that almost split extensions are unique.) □

Exercises

1. Let $\phi: M \to P$ and $\phi': M' \to P$ be extensions of P, that is, module homomorphisms. If there exist homomorphisms $\psi: M \to M'$ and $\psi': M' \to M$ such that $\phi'\psi = \phi$, $\phi\psi' = \phi'$, and $\psi\psi' = id_{M'}$, then $M' \to P$ is called a summand of $M \to P$. The extension $M \to P$ is indecomposable if it has no summand other than $0 \to P$ and the extensions $M' \to P$ that are isomorphic to $M \to P$. Prove that the module homomorphism $\phi: M \to P$ is an almost split extension of P if and only if:
   (i) P is indecomposable;
   (ii) ϕ is not an isomorphism;
   (iii) $\phi: M \to P$ is indecomposable;
   (iv) if $\psi: N \to P$ is not a split surjection, then there exists $\chi \in \mathrm{Hom}_A(N, M)$ such that $\phi\chi = \psi$. Hint. First show that if $\phi: M \to \phi(M)$ splits, then ϕ is injective. Use this result, together with (i), (iii), and (iv), to show that if P is projective, then ϕ is injective and $\mathrm{Im}\,\phi = P\mathbf{J}(A)$. Show that if P is not projective, then (iv) implies that ϕ is surjective. Use Lemma 7.3 and (iv) to obtain a commutative diagram

$$\begin{array}{ccccccccc} 0 & \to & \mathrm{Ker}\,\phi & \to & M & \xrightarrow{\phi} & P & \to & 0 \\ & & \downarrow\uparrow & & {\scriptstyle\psi}\downarrow\uparrow{\scriptstyle\chi} & & \| & & \\ 0 & \to & Q & \to & M' & \underset{\phi'}{\to} & P & \to & 0 \end{array}$$

with exact rows, Q indecomposable, and the bottom sequence is not split. Note that $id_{M'} - \psi\chi$ maps M' to Q, and this homomorphism cannot map Q isomorphically to itself since $0 \to Q \to M' \to P \to 0$ does not split. Deduce that $id_{M'} - \psi\chi$ is nilpotent, from which it follows that $\psi\psi' = id_{M'}$ for some ψ': $M' \to M$ such that $\phi\psi' = \phi'$. The indecomposability of $M \to P$ yields the desired result that $0 \to \operatorname{Ker}\phi \to M \to P \to 0$ is an almost split sequence.

2. Assume that N is a direct summand of the A-module M. Prove that if $\mathbf{E}_A(M)$ is right Artinian (Noetherian), then $\mathbf{E}_A(N)$ is right Artinian (respectively, Noetherian). Hint. Let $\pi \in \operatorname{Hom}_A(M,N)$ and $\kappa \in \operatorname{Hom}_A(N,M)$ satisfy $\pi\kappa = id_N$. For each right ideal K of $\mathbf{E}_A(N)$, denote $K^* = \kappa K \operatorname{Hom}_A(M,N)$. Show that $K \mapsto K^*$ is an inclusion preserving, injective mapping from the lattice of right ideals of $\mathbf{E}_A(N)$ to the lattice of right ideals of $\mathbf{E}_A(M)$; in fact, $\pi K^* \kappa = K$.

3. Let A be a right Artinian algebra with finite representation type. Prove that if $M \in \mathfrak{M}_A$, then $\mathbf{E}_A(M)$ is right Artinian. Hint. Let $P_1, \ldots, P_k$ be a set of representatives of the isomorphism classes of finitely generated, indecomposable right A-modules. For $1 \leq i \leq k$, let ϕ_i: $M_i \to P_i$ be an almost split extension. Denote $N = M_1 \oplus \cdots \oplus M_k \oplus P_1 \oplus \cdots \oplus P_k$. Use Exercise 2 above, and Exercise 3 of Section 3.1 to prove that it suffices to show that $\mathbf{E}_A(N)$ is Artinian. Let $N = Q_1 \oplus \cdots \oplus Q_l$ be a decomposition of N into indecomposable modules, with corresponding projections π_r: $N \to Q_r$ and injections κ_r: $Q_r \to N$. It can be assumed that $Q_r \cong P_{\sigma(r)}$ for a suitable mapping σ: $\{1, \ldots, l\} \to \{1, \ldots, k\}$. Let $\psi \in \mathbf{J}(\mathbf{E}_A(N))$. Use Exercise 2 in Section 7.2 to prove that there exist homomorphisms $\chi_{rs} \in \operatorname{Hom}_A(Q_s, M_{\sigma(r)})$ such that $\psi = \sum_{1 \leq r,s \leq l} \kappa_r \phi_{\sigma(r)} \rho_{\sigma(r)} \lambda_{\sigma(r)} \chi_{rs} \pi_s$, where ρ_i: $N \to M_i$, λ_i: $M_i \to N$ are the projections and injections associated with the decomposition $N = M_1 \oplus \cdots \oplus M_k \oplus P_1 \oplus \cdots \oplus P_k$. Deduce that $\mathbf{J}(\mathbf{E}_A(N))$ is finitely generated as a right $\mathbf{E}_A(N)$-module, and complete the proof by using Exercise 2 of Section 7.2.

7.7. Roiter's Theorem

The tools that we need to prove the first Brauer–Thrall conjecture are now in our hands; they are the existence of almost split extensions and the bound on the lengths of compositions of morphisms (Corollary 7.2). Using these facts, it will be possible to prove that if the representation type of A is bounded by n, then every indecomposable A-module belongs to I_{2^n}, where I_k denotes the subclass of $\mathfrak{N}_A$ that was constructed in Section 7.6. Since the number of isomorphism classes of modules in each I_k is finite, this conclusion yields the Brauer–Thrall Conjecture.

Here is an outline of the proof of this result. The key lemma shows that if P and Q are indecomposable modules such that $Q \notin I_k(P)$, then there is a non-zero homomorphism from Q to P that is a composition $\chi_1 \cdots \chi_k$ with each χ_i a non-isomorphism between indecomposable modules. This result is obtained by induction on k, using the existence of almost split extensions to take the induction step. By Corollary 7.2, such a composition can exist only if $k < 2^n$. The theorem follows easily from this observation. As a technical device in the existence proof, we introduce a modified version of the *trace module* $\mathrm{Tr}(Q,P) = \sum\{\phi(Q)\colon \phi \in \mathrm{Hom}_A(Q,P)\}$ of Q in P: instead of using all homomorphisms to define this submodule of P, the sum is restricted to those $\phi(Q)$ such that ϕ is a composition $\chi_1 \cdots \chi_k$ described above. The existence of a non-zero composition is plainly equivalent to the non-vanishing of this modified trace. We now give the details of the proof.

Let $P, Q \in \mathfrak{N}_A$. For $k < \omega$, denote by $D_k(Q,P)$ the set of homomorphisms $\phi \in \mathrm{Hom}_A(Q,P)$ such that there is a sequence

$$Q = Q_{k+1} \xrightarrow{\chi_{k+1}} Q_k \to \cdots \to Q_1 \xrightarrow{\chi_1} Q_0 = P$$

in which all of the modules Q_i are indecomposable, the homomorphisms χ_i are not isomorphisms, and $\phi = \chi_1\chi_2 \cdots \chi_{k+1}$. This definition has two consequences.

$$\text{If } Q \not\cong P, \text{ then } D_0(Q,P) = \mathrm{Hom}_A(Q,P). \tag{1}$$

If the representation type of A is bounded by n, then $D_k(Q,P)$ consists of the zero homomorphism for all $k \geq 2^n - 1$. (2)

The statement (1) is obvious from the definition of $D_0(Q,P)$, and (2) is a reformulation of Corollary 7.2.

The modified trace module is defined for $k < \omega$ by $T_k(Q,P) = \sum\{\phi(Q)\colon \phi \in D_k(Q,P)\}$. The properties (1) and (2) translate to:

$$T_k(Q,P) < \mathrm{Tr}(Q,P); \text{ if } Q \not\cong P, \text{ then } T_0(Q,P) = \mathrm{Tr}(Q,P); \tag{3}$$

If the representation type of A is bounded by n, then $T_k(Q,P) = 0$ for all $k \geq 2^n - 1$. (4)

Another useful fact comes easily from the definition of $T_k(Q,P)$.

If $N, P, Q \in \mathfrak{N}_A$, and $\psi \in \mathrm{Hom}_A(N,P)$ is not an isomorphism, then $\psi(T_k(Q,N)) \subseteq T_{k+1}(Q,P)$. (5)

Indeed, it is clear that $\{\psi\phi\colon \phi \in D_k(Q,N)\} \subseteq D_{k+1}(Q,P)$. Hence, $\psi(\mathrm{Tr}(Q,N)) = \psi(\sum\{\phi(Q)\colon \phi \in D_k(Q,N)\}) = \sum\{\psi\phi(Q)\colon \phi \in D_k(Q,N)\} \subseteq \sum\{\chi(Q)\colon \chi \in D_{k+1}(Q,P)\} = T_{k+1}(Q,P)$.

Lemma. *Assume that A has bounded representation type. If $P, Q \in \mathfrak{N}_A$ are such that $Q \notin I_{k+1}(P)$, then $T_k(Q,P) = \mathrm{Tr}(Q,P)$.*

PROOF. Induce on k. If $k = 0$, then the hypothesis $Q \notin I_{k+1}(P)$ includes the condition $Q \not\cong P$. Thus, $T_0(Q,P) = \mathrm{Tr}(Q,P)$ by (3). Assume that $k \geq 1$ and the lemma holds for $k - 1$. Our objective is to show that if $\psi \in \mathrm{Hom}_A(Q,P)$, then $\psi(Q) \subseteq T_k(Q,P)$. It will then follow that $\mathrm{Tr}(Q,P) = \sum\{\psi(Q)\colon \psi \in \mathrm{Hom}_A(Q,P)\} \subseteq T_k(Q,P)$, which implies that $T_k(Q,P) = \mathrm{Tr}(Q,P)$ by (3). Since $Q \not\cong P$, the homomorphism ψ is not an isomorphism. Therefore, ψ is not a split surjection, because Q is indecomposable. This observation enables us to use the machinery of almost split extensions. Let $\phi\colon M \to P$ be an almost split extension of P. The existence of ϕ is guaranteed by Proposition 7.6a. Write $M = P_1 \oplus \cdots \oplus P_r$ with each P_i indecomposable. Denote the projection and injection homomorphisms that are associated with this decomposition by $\pi_i: M \to P_i$ and $\kappa_i: P_i \to M$. We will use the identity $\sum_{i=1}^r \kappa_i\pi_i = id_M$ and the fact that $\phi\kappa_i$ is not an isomorphism. (Otherwise, $\kappa_i(\phi\kappa_i)^{-1} \in \mathrm{Hom}_A(P,M)$ splits ϕ, which is contrary to the definition of an almost split extension.) Note also that $Q \notin I_k(P_i)$. In fact, $P_i \in I_1(P)$ by construction; therefore, $Q \in I_k(P_i)$ would imply that $Q \in I_{k+1}(P)$ by (3) of Section 7.6, which is contrary to the assumption of the lemma. By the induction hypothesis, $\mathrm{Tr}(Q,P_i) = T_{k-1}(Q,P_i)$. Since $\psi \in \mathrm{Hom}_A(Q,P)$ is not a split surjection, there is a homomorphism $\chi: Q \to M$ such that $\psi = \phi\chi$. Therefore,

$$\psi(Q) = \phi\chi(Q) = \sum_{i=1}^r \phi\kappa_i\pi_i\chi(Q) \subseteq \sum_{i=1}^r \phi\kappa_i(\mathrm{Tr}(Q,P_i))$$

$$= \sum_{i=1}^r \phi\kappa_i(T_{k-1}(Q,P_i)) \subseteq T_k(Q,P)$$

by (5). As we observed earlier, this inclusion proves the lemma. □

Proposition. *If the representation type of the right Artinian algebra A is bounded by n, then $I_m = \mathfrak{N}_A$ for all $m \geq 2^n$.*

Proof. Let $Q \in \mathfrak{N}_A$. Since Q is right Noetherian, there is a maximal submodule N of Q. Let P be the simple module Q/N. Plainly, $\mathrm{Tr}(Q,P) = P \neq 0$. Thus, by (4), $T_k(Q,P) \neq \mathrm{Tr}(Q,P)$ if $k + 1 \geq 2^n$. The lemma gives the desired conclusion that $Q \in I_m(P) \subseteq I_m$ for $m = k + 1 \geq 2^n$. □

Theorem (Roiter, Auslander, Brauer, Thrall). *A right Artinian algebra of bounded representation type has finite representation type.*

The theorem is a consequence of the proposition and Proposition 7.6b.

Exercise

Let A be a finite dimensional F-algebra. Prove that the following conditions are equivalent.

(a) A has finite representation type for right A-modules.
(b) A has finite representation type for left A-modules.
(c) There exists $n \in \mathbb{N}$ such that $l(\mathrm{soc}\, P) \leq n$ for all $P \in \mathfrak{N}_A$.
(d) There exists $n \in \mathbb{N}$ such that $l(P/\mathrm{rad}\, P) \leq n$ for all $P \in \mathfrak{N}_A$.

Hint. The equivalence of (a) and (b), and of (c) and (d) follow easily from Exercise 3, Section 2.7. Clearly, (a) implies (c). Assume (c). Show that there is a positive integer m such that $l(\mathrm{Hom}_F(A,\mathrm{soc}\, P)) \leq m$ for all $P \in \mathfrak{N}_A$. Deduce from Exercise 5 of Section 6.1 and Exercise 1 of Section 2.7 that $l(P) \leq m$ for all $P \in \mathfrak{N}_A$. Apply the theorem.

Notes on Chapter 7

The history of the Brauer–Thrall Conjectures was outlined in Section 7.1. The exposition in this chapter follows the paper [80] of K. Yamagata, which in turn was based on Auslander's work in [11]. One of the essential parts of the proof, Proposition 7.2, comes from the paper [39] of M. Harada and Y. Sai. The result that is outlined in the Exercise of Section 7.7 is a theorem of Curtis and Jans.

Other proofs of the First Brauer–Thrall Conjecture have been given by M. M. Kleiner and A. V. Roiter in [53], and by S. O. Smalø. The paper [72] by Smalø also gives a proof for finite dimensional F-algebras of the Sesqui–Brauer–Thrall Conjecture: if $n_A(k) \geq \aleph_0$, then $n_A(m) \geq n_A(k)$ for infinitely many m.

CHAPTER 8
Representation of Quivers

This chapter introduces another aspect of the current research on representation of algebras. This line of work began with the papers [34] and [35] of P. Gabriel. He gave an explicit construction of the indecomposable modules for certain finite dimensional F-algebras. The most surprising part of Gabriel's result is a link between the representation theory of algebras and the Dynkin diagrams that occur in the study of semisimple Lie algebras. This relation between associative and Lie algebras was clarified by Bernstein, Gel'fand and Ponomarev in [18]; they showed that many algebraic problems can be formulated as questions about the representations of quivers. The characterization of finite representation type for certain associative algebras and the structure theory of semisimple Lie algebras are such problems.

Our aim in this chapter is to convey the flavor of this exciting new development in representation theory. It can serve as an introduction to a growing literature on the representation of quivers and related matrix problems.

8.1. Constructing Modules

We begin this chapter by introducing a class of algebras that have very simple structure and representation theories. In fact, it will be possible to give an explicit construction of the finite dimensional indecomposable modules for the algebras of this class. The difficulty arises in trying to characterize the isomorphism classes of indecomposable modules. That problem will occupy our attention throughout the rest of the chapter.

The results that are obtained in this chapter fall short of characterizing finite dimensional algebras that have finite representation types. That problem has not yet been solved. However, by combining the theorem that will be proved here with Theorem 6.7, a result from Morita Theory (Proposition 9.6), and the Wedderburn Principal Theorem, we will arrive in Section 11.8 at a characterization of the algebras of finite representation type in a fairly natural class of algebras.

It is useful to introduce notation that will be kept throughout the chapter. Let $\Gamma = (V,E)$ be a quiver, that is, a finite directed graph, with vertex set $V = \{1, 2, \ldots, r\}$. As usual, F denotes a field. Later it will be assumed that F is infinite.

To define the class of algebras that provide the motif for the chapter, let A be a commutative, semisimple F-algebra with a basis consisting of the orthogonal idempotent elements $e_1, e_2, \ldots, e_r$:

$$A = e_1F \oplus e_2F \oplus \cdots \oplus e_rF,\ e_ie_j = \delta_{ij}e_i,\ \textstyle\sum_{i=1}^r e_i = 1_A. \tag{1}$$

We wish to attach a radical to A. Let N denote the F-space with the basis $\{w_{ij}: (i,j) \in E\}$, where E is the edge set of the quiver Γ. As an algebra (without unity) in its own right, N is given the zero multiplication. Thus, if N is to be the radical of an algebra for which A is the semisimple quotient, then N will be a left and right A-module. Explicitly, the module and ring structure on N is defined by

$$e_kw_{ij} = w_{ij}e_l = 0 \quad \text{if} \quad k \neq i \quad \text{and} \quad j \neq l, \tag{2}$$
$$e_iw_{ij} = w_{ij}e_j = w_{ij};\ w_{ij}w_{kl} = 0.$$

Lemma a. *Let $B_\Gamma = B = N \oplus A$ be endowed with the multiplication defined on the basis $\{e_i: i \in v\} \cup \{w_{ij}: (i,j) \in E\}$ by* (1) *and* (2). *Then B is an F-algebra with unity element $1_B = 1_A$, $\mathbf{J}(B) = N$, and $B/\mathbf{J}(B) \cong A$. The elements $e_1, e_2, \ldots, e_r$ form a complete set of primitive idempotents with corresponding principal indecomposable modules $P_i = e_iB = e_iF \oplus \bigoplus_j \{w_{ij}F: (i,j) \in E\}$. The mapping $i \mapsto e_i$ defines a graph isomorphism of Γ to $\Gamma(B)$. The lattice $\mathbf{I}(B)$ of ideals of B is distributive.*

Proof. The only non-zero products of three basis elements have the form $e_i(e_ie_i) = (e_ie_i)e_i$, $e_i(w_{ij}e_j) = (e_iw_{ij})e_j$, $(e_ie_i)w_{ij} = e_i(e_iw_{ij})$, or $w_{ij}(e_je_j) = (w_{ij}e_j)e_j$. Thus, B is associative. It is clear from (1) and (2) that 1_A is the unity element of B. By (2), N is an ideal of B such that $N^2 = 0$, and $B/N \cong A$ is semisimple. Thus, $\mathbf{J}(B) = N$. Using (1) and $1_B = e_1 + e_2 + \cdots + e_r$, we obtain $B = e_1B \oplus e_2B \oplus \cdots \oplus e_rB$. Since $e_iB/e_i\mathbf{J}(B) = e_iF$ is simple in A, it follows that e_iB is a principal indecomposable module. Thus, $e_1, e_2, \ldots, e_r$ is a complete set of primitive idempotents in B. Plainly, $e_i\mathbf{J}(B)e_j = 0$ if $(i,j) \notin E$, and $e_i\mathbf{J}(B)e_j = w_{ij}F$ if $(i,j) \in E$. Therefore, $i \mapsto e_i$ is an isomorphism of Γ to $\Gamma(B)$. To prove that $\mathbf{I}(B)$ is distributive, it is sufficient by Proposition 4.6 to show that the lattice of sub-bimodules of $\mathbf{J}(B)$ is distributive; in other words, $\mathbf{S}(N)$ is distributive, when N is viewed as an A-bimodule. By (2), $N = \bigoplus_{(i,j)\in E} w_{ij}F$ is a decomposition of N into a direct sum of simple A-bimodules; and $w_{ij}F \cong w_{kl}F$ as bimodules only if $i = k$ and $j = l$: $e_i(w_{ij}F)e_j \neq 0$ implies $e_i(w_{kl}F)e_j \neq 0$, so that $i = k$ and $j = l$. Thus, by Corollary 2.4c (generalized to bimodules by means of Proposition 10.1), $\mathbf{S}(N)$ is distributive. □

The algebra $B = B_\Gamma$ that is described in the lemma depends only on the quiver Γ. Conversely, Γ is recovered from B in the form of $\Gamma(B)$. The lemma therefore shows that every quiver can be realized as $\Gamma(B)$ for a suitable finite dimensional F-algebra B.

Since $\mathbf{I}(B)$ is distributive, the possibility that B has finite representation type is not excluded by Theorem 6.7. It will turn out, however, that the representation type of B is finite only when Γ satisfies certain restrictions.

If M is a finitely generated right B-module, then M is a finite dimensional F-space, because B is finite dimensional. For $1 \leq i \leq r$, denote $M_i = Me_i$. Clearly, M_i is a subspace of M, and an elementary calculation based on (1) proves that $M = M_1 \oplus M_2 \oplus \cdots \oplus M_r$. Moreover, e_i acts as the zero mapping on M_k for each $k \neq i$, and as the identity on M_i. If $(i, j) \in E$, then $M_k w_{ij} = 0$ for $k \neq i$, and $M_i w_{ij} = M_i w_{ij} e_j \subseteq M_j$ by (2). Thus, the scalar multiplication of w_{ij} on M is uniquely determined by a linear mapping ψ_{ij} of M_i to M_j: $\psi_{ij}(u) = uw_{ij}$ for $u \in M_i$.

For $1 \leq i \leq r$, denote $M_{i1} = \{u \in M_i: u\mathbf{J}(B) = 0\}$. Plainly, M_{i1} is a subspace of M_i. Since $\mathbf{J}(B)^2 = 0$, $M_i \cap M\mathbf{J}(B)$ is a subspace of M_{i1}. In particular, $\psi_{ij}(M_i) \subseteq M_{j1}$ for all i and j. Denote $M_{i0} = M_i/M_{i1}$. Since the kernel of ψ_{ij} includes M_{i1}, it follows that ψ_{ij} induces a linear mapping ϕ_{ij} of M_{i0} to M_{j1} by the rule $\phi_{ij}(u + M_{i1}) = \psi_{ij}(u) = uw_{ij}$. If $u \in M_i$, then $u\mathbf{J}(B) = 0$ if and only if $uw_{ij} = 0$ for $1 \leq j \leq r$. Thus, for $1 \leq i \leq r$,

$$\bigcap_j \operatorname{Ker} \phi_{ij} = 0, \tag{3}$$

where the intersection is over $\{j: (i, j) \in E\}$.

This process can be reversed to show that the data $\{M_{i0}, M_{i1} : 1 \leq i \leq r\}$, $\{\phi_{ij}: (i, j) \in E\}$ can be used to construct a finite dimensional B-module.

Lemma b. *Let $\{M_{i0}: 1 \leq i \leq r\}$ and $\{M_{i1} : 1 \leq i \leq r\}$ be sets of finite dimensional F-spaces; assume that $\{\phi_{ij}: (i,j) \in E\}$ is a set of linear mappings, $\phi_{ij}: M_{i0} \to M_{j1}$, such that the condition* (3) *is satisfied. Define $M_i = M_{i0} \oplus M_{i1}$ and $M = \bigoplus_{i=1}^r M_i$. Then M is a right B-module with scalar operations:*

$$\begin{aligned} &ue_i = u \text{ for } u \in M_i,\ ue_j = 0 \text{ for } u \in M_i \text{ and } j \neq i;\\ &uw_{kj} = 0 \text{ for } u \in M_{i1},\ k \text{ and } j \text{ arbitrary};\\ &uw_{kj} = 0 \text{ for } u \in M_{i0},\ k \neq i, \text{ and } uw_{ij} = \phi_{ij}(u)\\ &\text{for } u \in M_{i0}. \end{aligned} \tag{4}$$

Moreover, $M_i = Me_i$, $M_{i1} = \{u \in M_i: u\mathbf{J}(B) = 0\}$, $M_{i0} \cong M_i/M_{i1}$, and the linear mapping of M_i/M_{i1} to M_{j1} induced by the scalar multiplication by w_{ij} is $u + M_{i1} \mapsto \phi_{ij}(u)$.

Proof. The conditions (4) extend by linearity to scalar operations on M by the elements of B. The two distributive laws for module operations follow from the nature of this extension process and linearity of ϕ_{ij}. The associative law of scalar multiplication is a consequence of (1) and (2), together with the fact that $\mathbf{J}(B)$ maps $\sum_{i=1}^{r} M_{i0}$ to $\sum_{i=1}^{r} M_{i1}$, and annihilates $\sum_{i=1}^{r} M_{i1}$. Finally, $u1_B = u$ for all $u \in M$ by (1) and (4). The only part of the last statement of the lemma that is not a direct consequence of (4) is the assertion that $M_{i1} = \{u \in M_i : u\mathbf{J}(B) = 0\}$. This equality follows from the hypothesis that $\{\phi_{ij} : (i,j) \in E\}$ satisfies (3). □

Lemma b gives a recipe for constructing all B-modules from data consisting of a set of vector spaces and linear mappings between some of these spaces. It is natural to ask how the data reflect homomorphisms of modules.

Let $\theta: M \to M'$ be a homomorphism of finitely generated B-modules. Then $\theta(Me_i) = \theta(M)e_i \subseteq M'e_i$, so that $\theta_i = \theta | M_i$ is a linear mapping from M_i to M_i'. Moreover, $\theta_{i1} = \theta_i | M_{i1}$ maps to M_{i1}', and induces a linear mapping $\theta_{i0}: M_{i0} \to M_{i0}'$. For $(i,j) \in E$, let $\phi_{ij}: M_{i0} \to M_{j1}$, and $\phi_{ij}': M_{i0}' \to M_{j1}'$, denote the mappings induced by w_{ij}. If $u + M_{i1} \in M_{i0}$, then $\theta_{j1}\phi_{ij}(u + M_{i1}) = \theta(uw_{ij}) = \theta(u)w_{ij} = \phi_{ij}'\theta_{i0}(u + M_{i1})$; hence, $\theta_{j1}\phi_{ij} = \phi_{ij}'\theta_{i0}$. Conversely, every system of linear mappings satisfying this commutativity property comes from a module homomorphism.

Lemma c. *Let $\{\theta_{i0} : 1 \le i \le r\}$ and $\{\theta_{i1} : 1 \le i \le r\}$ be sets of linear mappings $\theta_{i0} : M_{i0} \to M_{i0}'$, $\theta_{i1} : M_{i1} \to M_{i1}'$ such that if $(i,j) \in E$, then*

$$\theta_{j1}\phi_{ij} = \phi_{ij}'\theta_{i0}. \tag{5}$$

Then there is a B-module homomorphism $\theta: M \to M'$ such that $\theta_{i1} = \theta | M_{i1}$, and $\theta_{i0}(u + M_i) = \theta(u) + M_{i1}$ for all $u \in M_i$. The mappings θ_{i0} and θ_{i1} are all isomorphisms if and only if θ is an isomorphism.

Proof. Define linear mappings $\theta_i: M_i \to M_i'$ so that

$$\begin{array}{ccccccccc} 0 & \to & M_{i1} & \to & M_i & \to & M_{i0} & \to & 0 \\ & & {\scriptstyle\theta_{i1}}\downarrow & & {\scriptstyle\theta_i}\downarrow & & {\scriptstyle\theta_{i0}}\downarrow & & \\ 0 & \to & M_{i1}' & \to & M_i' & \to & M_{i0}' & \to & 0 \end{array}$$

commutes. This is possible because short exact sequences of vector spaces split. If $u \in M_i$, then (5) yields $\theta_j(uw_{ij}) = \theta_j(\phi_{ij}(u + M_{i1})) = \phi_{ij}'(\theta_{i0}(u + M_{i1})) = \phi_{ij}'(\theta_i(u) + M_{i1}') = \theta_i(u)w_{ij}$. Thus, $\theta = \theta_1 \oplus \theta_2 \oplus \cdots \oplus \theta_r$ is a B-module homomorphism of M to M'. If θ_{i0} and θ_{i1} are isomorphisms, so is θ_i by the Short Five-Lemma. It follows easily that θ is an isomorphism if and only if all of the θ_{i0} and θ_{i1} are isomorphisms. □

A minor precaution is in order. Lemma c is an existence theorem; the homomorphism θ is generally not unique, and there is no standard construction of this mapping.

EXERCISES

1. Prove that B_Γ is local if and only if $|V| = 1$, and B_Γ is simple if and only if $|V| = 1$ and $|E| = 0$.

2. Prove that B_Γ is commutative if and only if every edge of Γ is a loop, that is, $(i,j) \in E$ implies $i = j$.

3. Let Γ be such that $|V| = |E| = 1$. Prove that the isomorphism classes of B_Γ-modules are in one-to-one correspondence with the pairs (m,n) of non-negative integers such that $m \leq n$. Hint. If M is the module corresponding to the data $(\{M_{10}, M_{11}\}, \{\phi_{11}\})$ in Lemma b, assign to M the pair $(\dim_F M_{10}, \dim_F M_{11})$.

8.2. Representation of Quivers

The results of Section 8.1 suggest that it may be useful to isolate and study certain structures that are associated with quivers. We now begin that project.

Throughout this chapter, $\Gamma = (V,E)$ denotes a quiver whose vertex set is $V = \{1, 2, \ldots, r\}$ for notational convenience. As usual, F is a field.

Definition. An F-representation of Γ is a pair (M,ϕ) consisting of a set $M = \{M_i : i \in V\}$ of finite dimensional F-spaces, and a set $\phi = \{\phi_{ij} : (i,j) \in E\}$ of linear mappings $\phi_{ij} : M_i \to M_j$.

The class of all F-representations of Γ will be denoted by $\mathfrak{R}(\Gamma)$ (or $\mathfrak{R}(\Gamma,F)$ when it is necessary to identify F).

The class $\mathfrak{R}(\Gamma)$ is made into a category by defining a morphism $\theta : (M,\phi) \to (M',\phi')$ to be a set $\{\theta_i : i \in V\}$ of linear mappings $\theta_i : M_i \to M_i'$ such that

$$\begin{array}{ccc} M_i & \xrightarrow{\phi_{ij}} & M_j \\ {\scriptstyle\theta_i}\downarrow & & \downarrow{\scriptstyle\theta_j} \\ M_i' & \xrightarrow[\phi_{ij}']{} & M_j' \end{array}$$

commutes for all pairs $(i,j) \in E$. If $\theta : (M,\phi) \to (M',\phi')$ and $\theta' : (M',\phi') \to (M'',\phi'')$ are morphisms, then so is $\theta'\theta = \{\theta_i'\theta_i : i \in V\}$; and $id_{(M,\phi)} = \{id_{M_i} : i \in V\}$ is an automorphism of (M,ϕ). Thus, the requisities for a category are fulfilled.

The sets $\mathrm{Hom}((M,\phi), (M',\phi'))$ of morphisms from (M,ϕ) to (M',ϕ') have an F-space structure that is defined by $\theta_1 + \theta_2 = \{\theta_{1i} + \theta_{2i} : i \in V\}$ and $\theta a = \{\theta_i a : i \in V\}$, where $\theta_1, \theta_2, \theta \in \mathrm{Hom}((M,\phi), (M',\phi'))$ and $a \in F$. A routine calculation establishes the commutativity conditions for $\theta_1 + \theta_2$ and θa, and it is clear that these operations make $\mathrm{Hom}((M,\phi), (M',\phi'))$ an F-space. Also, it is easy to verify that composition of morphisms is bilinear. In particular, $\mathbf{E}(M,\phi) = \mathrm{Hom}((M,\phi), (M,\phi))$ is an F-algebra. The mapping $\theta \mapsto \theta_1 \oplus \theta_2 \oplus \cdots \oplus \theta_r$ is plainly an embedding of $\mathbf{E}(M,\phi)$ in $\mathbf{E}(M_1 \oplus M_2 \oplus \cdots \oplus M_r)$, so that $\mathbf{E}(M,\phi)$ is finite dimensional.

Direct sums are defined in $\mathfrak{R}(\Gamma)$ by $(M,\phi) \bigoplus (M',\phi') = (M \bigoplus M', \phi \bigoplus \phi')$, where $M \bigoplus M' = \{M_i \bigoplus M_i' : i \in V\}$ and $\phi \bigoplus \phi' = \{\phi_{ij} \bigoplus \phi_{ij}' : (i,j) \in E\}$. It is easy to check that this definition of a direct sum is satisfactory from the standpoint of category theory. Since Γ is finite, and the spaces M_i are finite dimensional, it is obvious that every $(M,\phi) \in \mathfrak{R}(\Gamma)$ can be written as a finite direct sum of indecomposable objects:

$$(M,\phi) = (M_1,\phi_1) \bigoplus (M_2,\phi_2) \bigoplus \cdots \bigoplus (M_s,\phi_s),$$

where $(M_k,\phi_k) \neq (0,0)$, and $(M_k,\phi_k) = (M_k',\phi_k') \bigoplus (M_k'',\phi_k'')$ implies $(M_k',\phi_k') = (0,0)$ or $(M_k'',\phi_k'') = (0,0)$ for $1 \leq k \leq s$. The notation $(0,0)$ in this context abbreviates the sets of zero dimensional spaces and zero mappings.

The arguments of Section 5.3 can be modified to prove that (M,ϕ) is indecomposable if and only if $\mathbf{E}(M,\phi)$ is a local algebra. The proof in Section 5.4 of the Krull–Schmidt Theorem can be used to show that decomposition into indecomposable objects is unique to within isomorphism. A fuller outline of this argument is given in the hint to Exercise 2.

Validity of the Krull–Schmidt Theorem for the category $\mathfrak{R}(\Gamma)$ makes it meaningful to discuss *quivers of finite representation type*. To be precise, a quiver Γ has finite F-representation type if there are finitely many isomorphism classes of indecomposable objects in $\mathfrak{R}(\Gamma,F)$; otherwise, the F-representation type of Γ is infinite.

It turns out that the finiteness of the representation type of a quiver Γ doesn't depend on F. However, the proofs of this fact use different techniques for finite and infinite fields. To avoid technical complications, we will consider only the case in which F in infinite. Throughout the rest of this chapter, the standing hypothesis that F is infinite will be in effect. Once this restriction is made, all references to F can be safely omitted. Our results are independent of F. In particular, it is permissible to use expressions such as "representation of Γ" and "finite representation type" without mention of F.

EXAMPLE. For i and k in V, define $P_i^{(k)} = 0$ if $i \neq k$, and $P_k^{(k)} = F$. For all $(i,j) \in E$, let ϕ_{ij} be the zero mapping from $P_i^{(k)}$ to $P_j^{(k)}$. Then $(P^{(k)},0)$ is (obviously) an indecomposable object in $\mathfrak{R}(\Gamma)$.

In fact, $(P^{(k)},0)$ is simple in the sense that non-zero morphisms with domain $(P^{(k)},0)$ are injective. We will call $(P^{(k)},0)$ the *simple representation* of Γ at k. Usually, $\mathfrak{R}(\Gamma)$ includes (categorically) simple objects that are not of this form.

The notation $(P^{(k)},0)$ is slightly ambiguous, since it does not hint that the set of zero morphisms denoted by 0 really depends on the orientation of Γ. If $\Gamma' = (V',E')$ is a quiver such that $V' = V$, but $E' \neq E$, then the simple representation $(P^{(k)},0)$ associated with Γ' is different from the $(P^{(k)},0)$ for Γ: in the first case, $0 = \{0_{ij} : (i,j) \in E'\}$; for Γ, $0 = \{0_{ij} : (i,j) \in E\}$.

In order to state the main theorem on the representation of quivers, a couple of items of notation are needed. The graph associated with a quiver $\Gamma = (V,E)$ will be denoted by Γ^u. Thus, Γ^u is the undirected graph with vertex set V, and edge set E, but with the orientation of edges ignored; the ordered pairs (i,j) and (j,i) are identified. If (i,j) and (j,i) both occur as edges of Γ, then these ordered pairs give rise to a double edge between i and j in Γ^u.

Every quiver and graph can be represented pictorially (in an unlimited number of ways) by a plane geometric figure consisting of directed or undirected lines between points. We will refer to such a figure as "the" diagram of Γ or Γ^u. For example, if $\Gamma = (\{1,2,3,4\}, \{(12),(23),(34),(43),(13),(14),(24),(33)\})$, then the diagram of Γ is

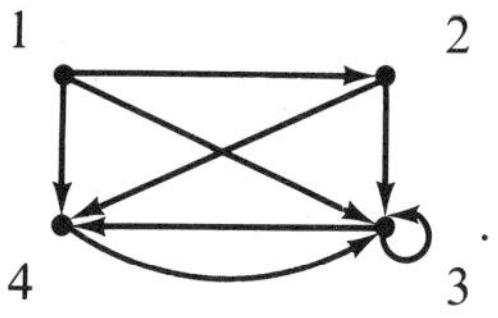

The diagram of Γ^u is obtained from the diagram of Γ by omitting arrowheads. Thus, for the above example, Γ^u has the diagram

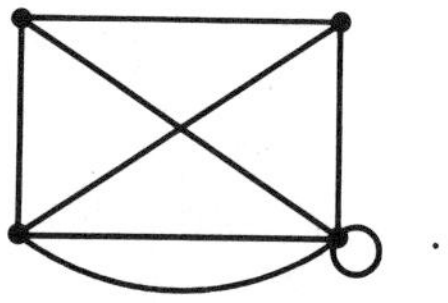

Theorem. *The quiver Γ has finite representation type if and only if the diagram of the graph Γ^u is a disjoint union of the following diagrams.*

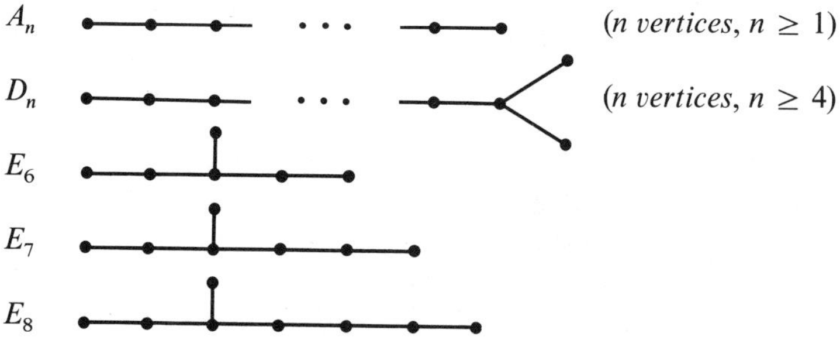

The diagrams in this list are familiar from the classification theory of Lie algebras. They are the Dynkin diagrams of the simple Lie algebras of types A_n, D_n, and E_k for $6 \leq k \leq 8$. We will see that they arise in studying the representations of quivers for the same reason that they occur in Lie algebra theory; they classify certain lattices of points in Euclidean spaces.

EXERCISES

1. Prove that $\mathfrak{R}(\Gamma)$ is an abelian category for every quiver Γ. In more detail, show that $\mathrm{Hom}((M,\phi), (N,\psi))$ is an abelian group such that composition distributes across sums; the direct sum $(M,\phi) \bigoplus (N,\psi)$ is both a product and a coproduct; a morphism $\theta: (M,\phi) \to (N,\psi)$ is a monomorphism (epimorphism) if and only if each θ_i is injective (surjective); and every morphism can be written as a composition $\theta'\theta''$, where θ' is a monomorphism and θ'' is an epimorphism.

2. Let (M,ϕ) be an F-representation of the quiver Γ.
   (a) Prove that $\mathbf{E}(M,\phi)$ is a finite dimensional F-algebra.
   (b) Prove that (M,ϕ) is indecomposable if and only if $\mathbf{E}(M,\phi)$ is a local algebra. Hint. Generalize the proof of Fitting's Lemma to the category $\mathfrak{R}(\Gamma)$, and use (a).
   (c) Prove that the Krull–Schmidt Theorem is valid for representations of quivers. Hint. Use Exercise 1 together with (b) to translate the proof of Proposition 5.4 to the context of quivers.

8.3. Application to Algebras

The purpose of this section is to show how Theorem 8.2 can be used to determine the representation types of the algebras that were described in Section 8.1.

A quiver $\Gamma = (V,E)$ is called *separated* (or *bipartite*) if V is a disjoint union $V_0 \cup V_1$, and all edges begin at a vertex of V_0 and end on V_1. That is, if $(i,j) \in E$, then $i \in V_0$ and $j \in V_1$. We will call V_0 the set of *sources* and V_1 the set of *sinks* of Γ. If a vertex of does not lie on any edge, then it can be either in V_0 or V_1 as convenience dictates.

Associated with any quiver $\Gamma = (V,E)$ is the separated quiver $\Gamma^s = (V^s,E^s)$, where $V^s = V \times \{0,1\}$ and $E^s = \{((i,0), (j,1)): (i,j) \in E\}$. In this case, $V_0 = \{(i,0): i \in V\}$ and $V_1 = \{(i,1): i \in V\}$. If B is an Artinian algebra with $\Gamma = \Gamma(B)$, then the separated quiver Γ^s corresponding to Γ is called the *separated quiver* of B. Denote this separated quiver by $\Gamma^s(B)$.

If Γ is a separated quiver with the set V_0 of sources, then a representation (M,ϕ) of Γ is called *reduced* if for all $i \in V_0$, the intersection over all j such that $(i,j) \in E$ of $\mathrm{Ker}\,\phi_{ij}$ is zero. For example, the simple representation $(P^{(k)},0)$ of Γ defined in Example 11.2 is reduced if and only if $k \in V_1$.

Lemma a. *Let Γ be a separated graph with the set V_0 of sources. Suppose that (M,ϕ) and (M',ϕ') are representations of Γ.*

(i) *$(M,\phi) \cong (\bigoplus_{k \in V_0} \bigoplus n_k (P^{(k)},0)) \bigoplus (N,\psi)$, where (N,ψ) is reduced, and $n_k = \dim_F(\bigcap_l \mathrm{Ker}\,\phi_{kl})$ for all $k \in V_0$.*
(ii) *$(M,\phi) \bigoplus (M',\phi')$ is reduced if and only if (M,ϕ) and (M',ϕ') are both reduced.*
(iii) *An indecomposable representation is either reduced or it is isomorphic to a simple representation $(P^{(k)},0)$ for some $k \in V_0$.*

Proof. (i) For $k \in V_0$, write $M_k = Q_k \bigoplus N_k$, where $Q_k = \bigcap_{(k,l)\in E} \operatorname{Ker} \phi_{kl}$, and let $\psi_{kl} = \phi_{kl}|N_k$. Since $\operatorname{Ker} \phi_{kl} \supseteq Q_k$ and $\operatorname{Im} \phi_{kl} \subseteq M_l$ with $l \in V_1 = V - V_0$, it is clear that $\phi_{kl} = 0_{kl} \bigoplus \psi_{kl}\colon Q_k \bigoplus N_k \to 0 \bigoplus M_l$. Hence, $(M,\phi) = (Q,0) \bigoplus (N,\psi)$, and $(Q,0) = \bigoplus_{k\in V_0} \bigoplus n_k P^{(k)}$. The statement (ii) follows from the definition of a reduced representation because $\bigcap_{(k,l)\in E} \operatorname{Ker}((\phi \bigoplus \phi')_{kl}) = (\bigcap_{(k,l)\in E} \operatorname{Ker} \phi_{kl}) \bigoplus (\bigcap_{(k,l)\in E} \operatorname{Ker} \phi'_{kl})$, and (iii) is a consequence of (i) and (ii). □

Lemma b. *Let $B = B_\Gamma$ be the F-algebra that was defined in Section 8.1. There is a bijection between the isomorphism classes of finitely generated, indecomposable right B-modules, and the isomorphism classes of indecomposable representations (M,ϕ) of the separated quiver $\Gamma^s(B)$ such that (M,ϕ) is not isomorphic to a simple representation of the form $(P^{(i,0)},0)$. In particular, B has finite representation type if and only if $\Gamma^s(B)$ has finite representation type.*

Proof. For each finitely generated right B-module M, the construction described in Section 8.1 yields a representation $\mathbf{R}(M) = (\{M_{i0}, M_{i1}: 1 \le i \le r\}, \{\phi_{ij}: ((i,0),(j,1)) \in E^s\})$ of the separated quiver $\Gamma^s(B)$. By 8.1 (3), $\mathbf{R}(M)$ is reduced. By Lemma 8.1c, every module homomorphism $\theta\colon M \to N$ gives rise to a morphism $\mathbf{R}(\theta) = \{\theta_{i0}, \theta_{i1} : 1 \le i \le r\}$ of $\mathbf{R}(M)$ to $\mathbf{R}(N)$. An easy calculation shows that $\mathbf{R}$ is a functor from the category of finitely generated right B-modules to $\mathfrak{R}(\Gamma^s(B))$. Moreover, by Lemma 8.1c, $\mathbf{R}$ is full: it maps $\operatorname{Hom}_B(M,N)$ onto $\operatorname{Hom}(\mathbf{R}(M), \mathbf{R}(N))$. Lemmas 8.1b and c also show that the mapping $M \mapsto \mathbf{R}(M)$ carries isomorphism classes of B-modules to isomorphism classes of objects in $\mathfrak{R}(\Gamma^s(B))$, and every isomorphism class of reduced objects in $\mathfrak{R}(\Gamma^s(B))$ includes some $\mathbf{R}(M)$. These observations, together with two simple consequences of the construction of $\mathbf{R}(M)$, namely

$$\mathbf{R}(M \bigoplus M') = \mathbf{R}(M) \bigoplus \mathbf{R}(M'), \tag{1}$$

and

$$M \neq 0 \text{ if and only if } \mathbf{R}(M) \neq (0,0), \tag{2}$$

lead easily to the assertion of the lemma. In fact, if M is a finitely generated, indecomposable B-module, then $\mathbf{R}(M)$ is indecomposable: $\mathbf{R}(M) \neq (0,0)$ by (2), and $\mathbf{R}(M) = (N',\phi') \bigoplus (N'',\phi'')$ implies that there are finitely generated B-modules M' and M'' satisfying $\mathbf{R}(M') \cong (N',\phi')$ and $\mathbf{R}(M'') \cong (N'',\phi'')$; by (1), $\mathbf{R}(M' \bigoplus M'') \cong \mathbf{R}(M)$, so that $M \cong M' \bigoplus M''$, and $M' = 0$ or $M'' = 0$ (hence $(N',\phi') = (0,0)$ or $(N'',\phi'') = (0,0)$) by the indecomposability of M. Conversely, if M is decomposable, then so is $\mathbf{R}(M)$ by (1) and (2). □

This lemma and Theorem 8.2 determine the representation types of the algebras that were discussed in Section 8.1.

Proposition. *Let Γ be a quiver, and Γ^s the separated quiver that is associated with Γ. If $B = B_\Gamma$ is the F-algebra that was defined in Lemma 8.1a, then B has finite representation type if and only if Γ^s has a diagram that is the disjoint union of Dynkin diagrams of types A_n, D_n, E_6, E_7, or E_8.*

In Chapter 11 we will reformulate this proposition in a more interesting way.

One aspect of the proposition is worth a comment. The proof of Theorem 8.2 that will be given in this chapter is constructive. It provides a recipe for making the indecomposable representations out of simple representations of quivers that are related to Γ. The process translates to an algorithm that constructs the indecomposable B_Γ-modules, provided B_Γ has finite representation type.

EXERCISES

1. Let B be the algebra that was constructed in Section 8.1. Prove that the isomorphism classes of simple B-modules correspond bijectively to the isomorphism classes of the simple representations $(P^{(i,1)},0)$ of the separated quiver $\Gamma^s(B)$.
2. Let $\Gamma = (V,E)$ be a quiver. For each vertex $i \in V$, define $d_\Gamma^-(i) = |\{j \in V: (i,j) \in E\}|$ and $d_\Gamma^+(i) = |\{j \in V: (j,i) \in E\}|$. Let $B = B_\Gamma$ be the F-algebra that was defined in Section 8.1. Prove that if B has finite representation type, then $d_\Gamma^-(i) \le 3$, $d_\Gamma^+(i) \le 3$, and $d_\Gamma^-(i) + d_\Gamma^+(i) \le 5$ for all vertices i of Γ. Give an example of an algebra B for which $d_\Gamma^-(i) + d_\Gamma^+(i) = 5$, and B has finite representation type.
3. Find all quivers $\Gamma = (V,E)$ with $V = \{1, 2, 3\}$ such that the separated quiver Γ^s corresponding to Γ has finite representation type.

8.4. Subquivers

In this section, we take a short step toward the proof of Theorem 8.2. Our first objective is to show that if the diagram of the quiver Γ is not a disjoint union of the Dynkin diagrams A_n, D_n and E_k $(6 \le k \le 8)$, then Γ has infinite representation type; the second half of the proof establishes the converse of this conclusion.

We start with an observation that will be used several times.

Lemma a. *Let $\Gamma = (V,E)$ be a quiver such that for some natural number n there are infinitely many isomorphism classes of representations (M,ϕ) satisfying $\sum_{i \in V} \dim_F M_i \le n$. Then Γ has infinite representation type.*

PROOF. If $(M,\phi) = (M^{(1)},\phi^{(1)}) \oplus \cdots \oplus (M^{(s)},\phi^{(s)})$, then $\sum_{i \in V} \dim_F M_i = \sum_{k=1}^s \sum_{i \in V} \dim_F M_i^{(k)}$. In particular, if the $M^{(k)}$ are not 0, then $s \le \sum_{i \in V} \dim_F M_i$. Thus, if the number of isomorphism classes of indecomposable representations of Γ is finite, then so is the total number of isomorphism classes of (M,ϕ) such that $\sum_{i\in V}\dim_F M_i \le n$. □

Let $\Gamma = (V,E)$ be a quiver. A *subquiver* of Γ is a quiver $\Gamma' = (V',E')$ such that $V' \subseteq V$ and $E' \subseteq E$. For any subset V' of V, there is a largest subquiver Γ' of Γ that has vertex set V': $\Gamma' = (V', E \cap (V' \times V'))$. This maximum subquiver is called the *full subquiver* on V', and it is denoted by $\Gamma | V'$.

If $\Gamma' = (V',E')$ and $\Gamma'' = (V'',E'')$ are quivers with disjoint vertex sets, then $\Gamma' \dot\cup \Gamma'' = (V' \cup V'', E' \cup E'')$ is a quiver that is called the disjoint union of Γ' and Γ''. Clearly, $\Gamma' = (\Gamma' \dot\cup \Gamma'')|V'$ and $\Gamma'' = (\Gamma' \dot\cup \Gamma'')|V''$. A quiver $\Gamma = (V,E)$ is connected if it is not a disjoint union of non-empty quivers. Thus, Γ is connected if and only if $V = V' \dot\cup V''$ and $V' \neq \varnothing \neq V''$ implies the existence of $i \in V'$ and $j \in V''$ such that either $(i,j) \in E$ or $(j,i) \in E$.

For vertices k and l of the quiver $\Gamma = (V,E)$, define $k \equiv l$ if $k = l$ or there is a path in the diagram of Γ^u (consisting of edges from E) that joins k and l. It is easy to see that $\equiv$ is the smallest equivalence relation on V that includes E. If $V_1, \ldots, V_r$ are the distinct equivalence classes of $\equiv$, then $\Gamma = (\Gamma|V_1) \dot\cup \cdots \dot\cup (\Gamma|V_r)$ is the unique decomposition of Γ as a disjoint union of connected subquivers. In particular, Γ is connected if and only if each pair of vertices in Γ can be joined by a path in the diagram of Γ^u.

Let $\Gamma' = (V'E')$ be a subquiver of $\Gamma = (V,E)$. If (M,ϕ) is a representation of Γ, then the *restriction* $\delta(M,\phi) = (M',\phi')$ of (M,ϕ) to Γ' is defined by $M_i' = M_i$ for $i \in V'$ and $\phi_{ij}' = \phi_{ij}$ for $(i,j) \in E'$. On the other hand, for each $(M',\phi') \in \mathfrak{R}(\Gamma')$, define the *extension* $\varepsilon(M',\phi') = (M,\phi)$ of (M',ϕ') to Γ by $M_i = M_i'$ for $i \in V'$, $M_i = 0$ for $i \in V - V'$; $\phi_{ij} = \phi_{ij}'$ for $(i,j) \in E'$, and $\phi_{ij} = 0$ for $(i,j) \in E - E'$. Plainly, $\delta\varepsilon(M',\phi') = (M',\phi')$. The object maps δ and ε extend to functors between $\mathfrak{R}(\Gamma)$ and $\mathfrak{R}(\Gamma')$ by putting $(\delta\theta)_i = \theta_i$ for $i \in V'$; $(\varepsilon\theta)_i = \theta_i$ for $i \in V'$, $(\varepsilon\theta)_i = 0$ for $i \in V - V'$. It is easy to see that two objects in $\mathfrak{R}(\Gamma')$ are isomorphic if and only if their extensions to Γ are isomorphic. Moreover, $(M',\phi') \in \mathfrak{R}(\Gamma')$ is indecomposable if and only if $\varepsilon(M',\phi')$ is indecomposable.

Lemma b. *Let Γ be a quiver.*

(i) *If the subquiver Γ' of Γ has infinite representation type, then Γ has infinite representation type.*
(ii) *If $\Gamma = \Gamma' \dot\cup \Gamma''$, then Γ has finite representation type if and only if Γ' and Γ'' have finite representation type.*

Proof. Let δ' and δ'' be the restrictions of Γ to Γ' and Γ'' respectively. Similarly, denote by ε' and ε'' the extensions from Γ' and Γ'' to Γ. If $(M,\phi) \in \mathfrak{R}(\Gamma)$, then $(M,\phi) = \varepsilon'\delta'(M,\phi) \oplus \varepsilon''\delta''(M,\phi)$. Therefore, the indecomposable objects of $\mathfrak{R}(\Gamma)$ are extensions of indecomposable objects in $\mathfrak{R}(\Gamma')$ and $\mathfrak{R}(\Gamma'')$. It follows that if Γ' and Γ'' have finite representation types, then so does Γ. The rest of the lemma follows from our previous observations. □

By the second part of Lemma b, we can limit our study of representation types to connected quivers. The first part of the lemma leads to a more interesting result. A quiver $\Gamma = (V,E)$ is called a *loop* if $|V| = |E| = 1$; and Γ is a *cycle* if Γ is a loop, or Γ is connected, and each vertex is an endpoint of exactly two edges of Γ. With suitable labeling, Γ is a loop if and only if $V = \{1, 2, \ldots, r\}$ $(r \geq 1)$, and $E = \{(i_1,j_1), (i_2,j_2), \ldots, (i_r,j_r)\}$, where $(i_k,j_k) = (k,k+1)$ or $(k+1,k)$ for $k < r$, and $(i_r,j_r) = (r,1)$ or $(1,r)$. The unoriented diagram of a loop has one of the forms

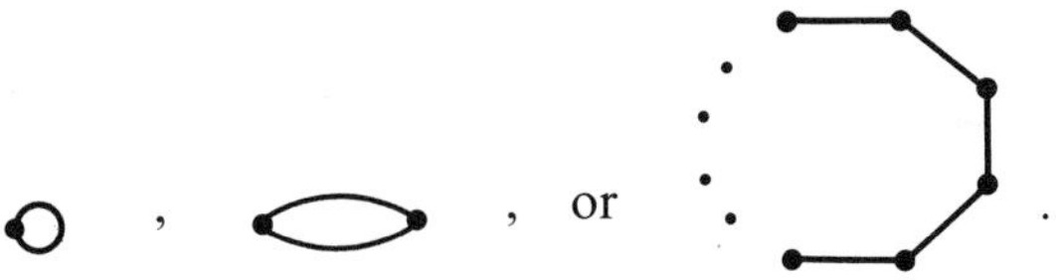

Lemma c. *Every cycle has infinite representation type.*

PROOF. Let $\Gamma = (V,E)$ be a cycle with $V = \{1, 2, \ldots, r\}$ and $E = \{(i_1,j_1), (i_2,j_2), \ldots, (i_r,j_r)\}$, as described above. Define $M_i = F$ for $1 \leq i \leq r$. For each $a \in F$, define $\phi^{(a)} = \{\phi^{(a)}_{i_kj_k} : 1 \leq k \leq r\}$ by $\phi^{(a)}_{i_kj_k} = id_F$ for $k < r$, and $\phi^{(a)}_{i_rj_r}(c) = ac$. If θ: $(M,\phi^{(a)}) \to (M,\phi^{(b)})$ is an isomorphism then the commutativity conditions $\phi^{(b)}_{i_kj_k}\theta_{i_k} = \theta_{j_k}\phi^{(a)}_{i_kj_k}$ for $k < r$ imply $\theta_1 = \theta_2 = \cdots = \theta_r$, and

$$b\theta_{i_r}(1) = \phi^{(b)}_{i_rj_r}\theta_{i_r}(1) = \theta_{j_r}\phi^{(a)}_{i_rj_r}(1) = \theta_{j_r}(a) = a\theta_{j_r}(1).$$

Hence, $b = a$. This proves that $(M,\phi^{(a)}) \cong (M,\phi^{(b)})$ only if $a = b$. Since F is infinite, it follows from Lemma a that Γ has infinite representation type. □

A quiver Γ is called *acyclic* if Γ has no cyclic subquivers. Lemmas b and c yield the main result of this section.

Proposition. *Every quiver that has finite representation type is acyclic.*

EXERCISE

Let $\Gamma = (\{1\}, \{(1,1)\})$ be a loop. Show that if (M,ϕ) and (M,ϕ') are representations of Γ in the same space M, then $(M,\phi) \cong (M,\phi')$ if and only if the linear transformations $\phi = \phi_{11}$ and $\phi' = \phi'_{11}$ of $M = M_1$ are similar. Thus, the problem of classifying the representations of a loop amounts to the classification of matrices with respect to similarity.

8.5. Rigid Representations

A representation (M,ϕ) of the quiver $\Gamma = (V,E)$ is *rigid at the vertex* k if every automorphism $\theta = \{\theta_i: i \in V\}$ of (M,ϕ) is such that θ_k is a scalar multiple of the identity mapping. For example, if $\dim_F M_k = 1$, then (M,ϕ) is rigid at k. The next result shows why the notion of rigidity is relevant to the study of representation type.

Lemma. *Let (M,ϕ) be a representation of the quiver $\Gamma = (V,E)$. Assume that (M,ϕ) is rigid at the vertex k, and that $\dim_F M_k > 1$. Define $\Gamma' = (V',E')$, where $V' = V \cup \{l\}$, $l \notin V$, and $E' = E \cup \{(l,k)\}$. The quiver Γ' has infinite representation type.*

Proof. For each non-zero F-space homomorphism $\psi\colon F \to M_k$, define the representation (M',ϕ^ψ) of Γ' by $M_i' = M_i$ for $i \in V$, $M_l' = F$; $\phi_{ij}' = \phi_{ij}$ for $(i,j) \in E$, and $\phi_{lk}' = \psi$. We will prove:

$$(M',\phi^\psi) \cong (M',\phi^\chi) \quad \text{implies} \quad \chi = \psi \cdot a \quad \text{for some} \quad a \in F. \tag{1}$$

Since F is infinite and $\dim_F M_k > 1$, it will follow from Lemma 8.4a that Γ' has infinite representation type. If $\theta'\colon (M',\phi^\psi) \to (M',\phi^\chi)$ is an isomorphism, then $\theta = \{\theta_i' : i \in V\}$ is an automorphism of (M,ϕ). Since (M,ϕ) is rigid at k, there exists $b \in F$ such that $\theta_k'(u) = u \cdot b$ for all $u \in M_k = M_k'$, Also, θ_l' is an F-space automorphism of F, so that for all $c \in F$, $\theta_l'(c) = dc$, where $d = \theta_l'(1) \neq 0$. The commutativity condition $\theta_k'\psi = \theta_k'\phi_{lk}^\psi = \phi_{lk}^\chi\theta_l' = \chi\theta_l'$ yields $\chi(c) = \chi\theta_l'(cd^{-1}) = \theta_k'\psi(cd^{-1}) = \psi(cd^{-1})b = \psi(c)d^{-1}b$. Hence, $\chi = \psi \cdot a$, where $a = d^{-1}b$. □

In many cases, the rigidity of a representation at one or more vertices can be proved by elementary computations. We illustrate this method with four examples of increasing complexity. In all of these examples, the representation spaces are subspaces of one space M, the homomorphisms ϕ_{ij} are inclusion mappings, and the rigidity (at all vertices) is established by proving that the only linear transformations θ of M such that $\theta(M_i) \subseteq M_i$ for all $i \in V$ are the scalar multiplications by elements of F. A representation of this kind is called a *poset* representation of Γ.

Example a. Let Γ be the quiver with the diagram

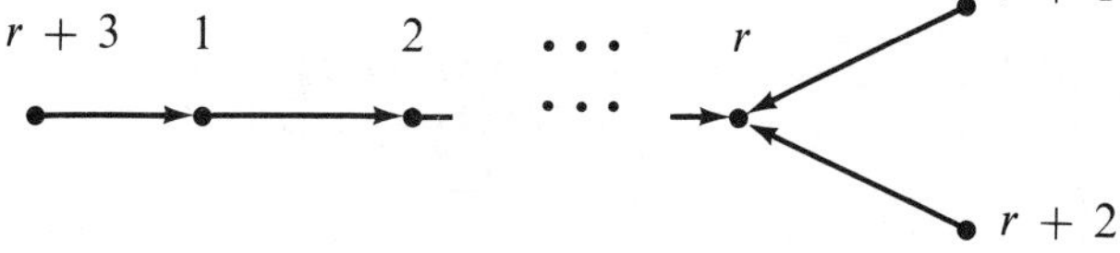

$, r \geq 1.$

Define the representation (M,ϕ) of Γ by $M_1 = M_2 = \cdots = M_r = u_1F \oplus u_2F$, $M_{r+1} = u_1F$, $M_{r+2} = u_2F$, $M_{r+3} = (u_1 + u_2)F$; and all of the ϕ_{ij} are inclusion homomorphisms. If $\theta \in \mathbf{E}(M,\phi)$, then the commutativity conditions $\phi_{ij}\theta_i = \theta_j\phi_{ij}$ imply that $\theta_1 = \theta_2 = \cdots = \theta_r = \theta \in \mathbf{E}(M)$, and $\theta_i = \theta \mid M_i$ for $i = r + 1$, $r + 2$, and $r + 3$. Hence, $\theta(u_1) \in u_1F$, $\theta(u_2) \in u_2F$, and $\theta(u_1 + u_2) \in (u_1 + u_2)F$. If $\theta(u_1) = u_1a_1$ and $\theta(u_2) = u_2a_2$, then $u_1a_1 + u_2a_2 = \theta(u_1 + u_2) \in (u_1 + u_2)F$ yields $a_1 = a_2$; that is, $\theta = id \cdot a_1$. This conclusion implies that (M,ϕ) is rigid at all vertices.

Corollary a. *For every $r \geq 1$, the quiver*

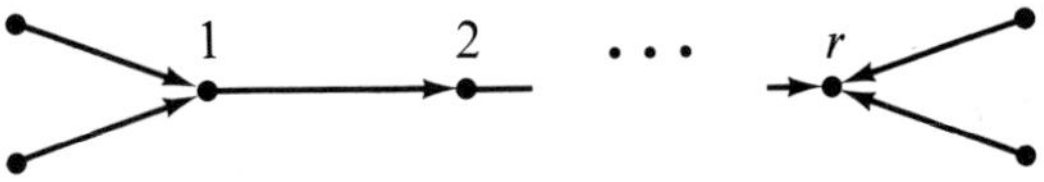

has infinite representation type.

The corollary is a consequence of the lemma, because the representation (M,ϕ) that was constructed in Example a satisfies $\dim_F M_1 = 2$.

EXAMPLE B. Let the diagram of the quiver Γ be

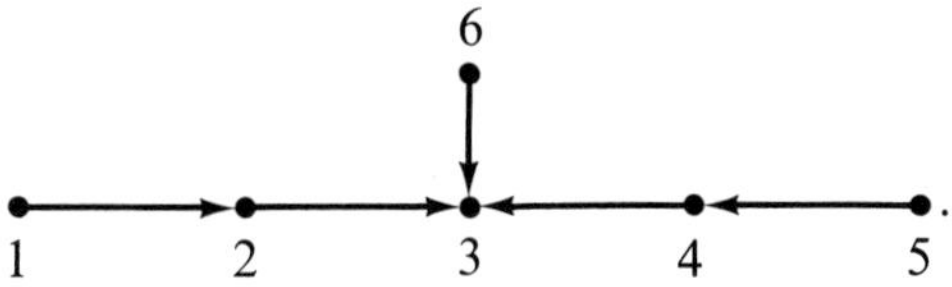

Define the poset representation (M,ϕ) of Γ by $M_1 = u_1F$, $M_2 = u_1F \bigoplus u_2F$, $M_3 = u_1F \bigoplus u_2F \bigoplus u_3F$, $M_4 = u_2F \bigoplus u_3F$, $M_5 = u_3F$, and $M_6 = (u_1+u_2)F \bigoplus (u_2+u_3)F$. Note that $M_2 \cap M_4 = u_2F$, $M_6 \cap M_2 = (u_1+u_2)F$, and $M_6 \cap M_4 = (u_2 + u_3)F$. If $\theta \in \mathbf{E}(M,\phi)$, then $\theta_i = \psi \mid M_i$, where $\psi = \theta_3 \in \mathbf{E}(M_3)$. Thus, $\psi(M_i) \subseteq M_i$ for $1 \leq i \leq 6$. Since $\{N \in \mathbf{S}(M_3) : \psi(N) \subseteq N\}$ is a sublattice of $\mathbf{S}(M_3)$, any subspace N of M_3 that is in the sublattice of $\mathbf{S}(M_3)$ generated by $\{M_1, M_2, \ldots, M_6\}$ has the property that $\psi(N) \subseteq N$. In particular, $\psi(M_2 \cap M_4) \subseteq M_2 \cap M_4$, $\psi(M_6 \cap M_2) \subseteq M_6 \cap M_2$, and $\psi(M_6 \cap M_4) \subseteq M_6 \cap M_4$. Hence, there exist a_1, a_2, a_3, b, and c in F such that $\psi(u_1) = u_1a_1$, $\psi(u_2) = u_2a_2$, $\psi(u_3) = u_3a_3$, $\psi(u_1 + u_2) = (u_1 + u_2)b$, and $\psi(u_2 + u_3) = (u_2 + u_3)c$. These conditions imply that ψ is a scalar multiple of id_{M_3}. Hence, (M,ϕ) is rigid.

Corollary b. *The quiver whose diagram is*

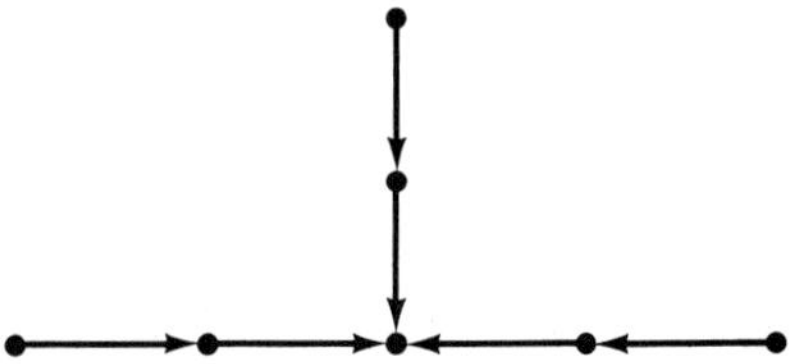

has infinite representation type.

EXAMPLE C. Let Γ be the quiver with the diagram

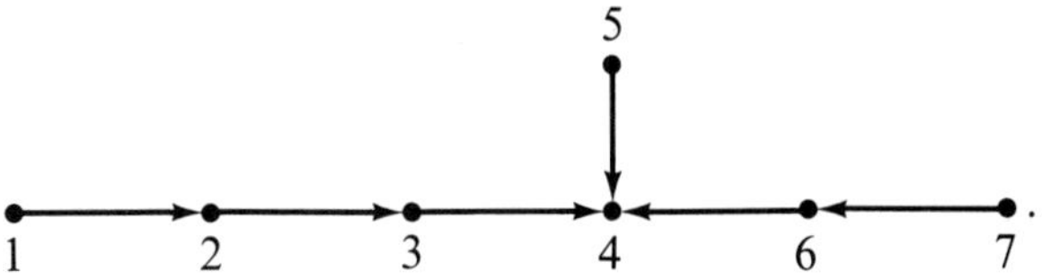

Define a poset representation of Γ by $M_1 = u_1F$, $M_2 = u_1F \oplus u_2F$, $M_3 = u_1F \oplus u_2F \oplus u_3F$, $M_4 = u_1F \oplus u_2F \oplus u_3F \oplus u_4F$, $M_5 = u_3F \oplus u_4F$, $M_6 = (u_1 + u_2)F \oplus (u_1 + u_3)F \oplus (u_1 + u_4)F$, and $M_7 = (u_2 - u_3)F \oplus (u_1 + u_4)F$. Note that $M_7 \subseteq M_6$, as required, and the other mandated inclusions obviously hold. Moreover, $M_3 \cap M_5 = u_3F$, $M_5 \cap M_6 = (u_3 - u_4)F$, $M_2 \cap M_6 = (u_1 + u_2)F$, $M_3 \cap M_7 = (u_2 - u_3)F$, $M_7 \cap (M_1 + M_5) = (u_1 + u_4)F$, and $M_6 \cap (M_1 + (M_3 \cap M_5)) = (u_1 + u_3)F$. If $\theta \in \mathbf{E}(M,\phi)$, then $\theta_i = \psi | M_i$, where $\psi = \theta_4$ satisfies $\psi(u_1) = u_1 a_1$, $\psi(u_3) = u_3 a_3$, $\psi(u_3 - u_4) = (u_3 - u_4)b_1$, $\psi(u_1 + u_2) = (u_1 + u_2)b_2$, $\psi(u_2 - u_3) = (u_2 - u_3)b_3$, $\psi(u_1 + u_4) = (u_1 + u_4)b_4$, and $\psi(u_1 + u_3) = (u_1 + u_3)b_5$ for suitable a_i and b_j in F. Routine computation leads to the conclusion that $\psi = id_{M_4} \cdot a_1$, so that (M,ϕ) is rigid.

Corollary c. *The quiver whose diagram is*

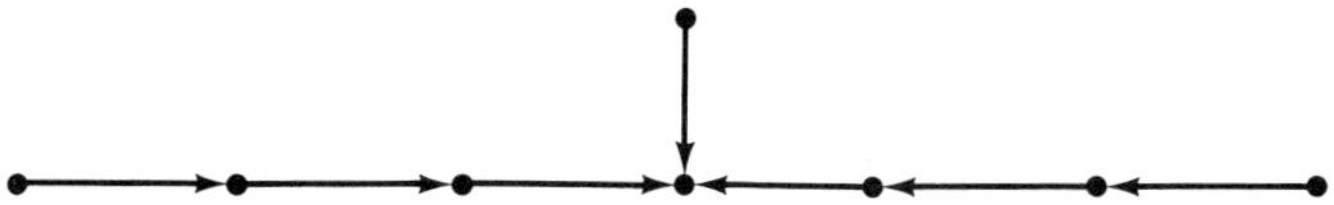

has infinite representation type.

Example d. Let Γ be the quiver with the diagram

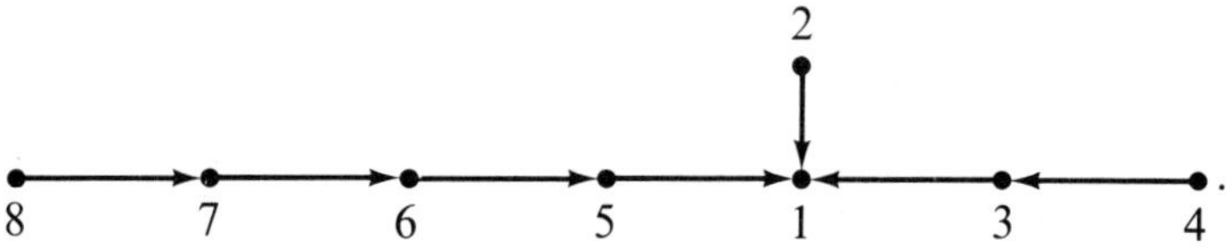

Define a poset representation of Γ by

$$M_1 = u_1F \oplus u_2F \oplus u_3F \oplus u_4F \oplus u_5F \oplus u_6F,$$

$$M_2 = u_3F \oplus u_5F \oplus u_6F,$$

$$M_3 = u_1F \oplus u_2F \oplus u_3F \oplus u_4F,$$

$$M_4 = u_1F \oplus u_2F,$$

$$M_5 = (u_1 + u_3)F \oplus (u_2 + u_3)F \oplus (u_3 + u_4)F \oplus (u_3 + u_5)F \oplus (u_3 + u_6)F,$$

$$M_6 = (u_2 + u_3)F \oplus (u_1 - u_4)F \oplus (u_1 - u_5)F \oplus (u_5 - u_6)F,$$

$$M_7 = (u_1 + u_2 + u_3 - u_4)F \oplus (u_4 - u_5)F \oplus (u_2 + u_3 + u_5 - u_6)F,$$

and

$$M_8 = (u_1 - u_4 - u_5 + u_6)F \oplus (u_1 - 2u_5 + u_6)F.$$

An easy check shows that the required inclusions $M_8 \subset M_7 \subset M_6 \subset M_5 \subset M_1$, $M_4 \subset M_3 \subset M_1$, and $M_2 \subset M_1$ occur. The proof that if $\theta \in \mathbf{E}(M,\phi)$, then $\theta_i = id_{M_i} \cdot a$ for some $a \in F$ follows the pattern that was established in Examples a, b, and c. The argument is based on the fact that θ_1 maps the following one dimensional subspaces of M_1 into themselves:

$$M_2 \cap M_3 = u_3F,$$

$$M_4 \cap M_5 = (u_1 - u_2)F,$$

$$M_2 \cap M_6 = (u_5 - u_6)F,$$

$$M_3 \cap M_7 = (u_1 + u_2 + u_3 - u_4)F,$$

$$M_6 \cap (M_4 + (M_2 \cap M_3)) = (u_2 + u_3)F,$$

$$M_8 \cap ((M_6 \cap M_3) + (M_6 \cap M_2)) = (u_1 - u_4 - u_5 + u_6)F,$$

and

$$M_8 \cap (M_2 + M_4) = (u_1 - 2u_5 + u_6)F.$$

It is also necessary to use the invariance under θ_1 of the two dimensional spaces M_4 and $M_2 \cap M_5 = (u_5 - u_6)F \oplus (u_3 + u_6)F$. We leave the details of the proof as an Exercise.

A more interesting question is where does this strange example come from? The choices of the dimensions of the spaces M_i are crucial for success. From a different point of view that will be explained later, these choices are natural. The rest of the construction is largely based on trial and error, carried out within certain guidelines.

Corollary d. *The quiver whose diagram is*

has infinite representation type.

Our inclusion in this section of these four examples was not done for perversity. We will show that these four examples, together with Proposition 8.4 and a result that will be established in Section 8.7, yield a proof of the first half of Theorem 8.2.

Proposition. *If the quiver Γ has finite representation, then Γ is a disjoint union of quivers whose unoriented diagrams are among the Dynkin diagrams A_n, D_n, and $E_k (6 \le k \le 8)$.*

Proof. The result from Section 8.7 that we need is the fact that the representation type of a quiver does not depend on its orientation. (See Corollary 8.7.) In particular, if Γ' is a quiver whose unoriented diagram coincides with the unoriented diagram of one of the quivers in Corollaries a, b, c, or d, then Γ' has infinite representation type. Thus, the hypothesis that Γ has finite representation type implies by Lemma 8.4b that no subquiver of Γ has an unoriented diagram like the diagrams of the quivers in Corollaries a, b, c, and d. Moreover, Γ is acyclic by Proposition 8.4. By limiting our attention to the connected components, it can be assumed that Γ is connected. Corollary 8.5a implies that no vertex of Γ is an endpoint of more than three edges, and at most one vertex is an endpoint of three edges. The diagram of Γ must therefore have the form

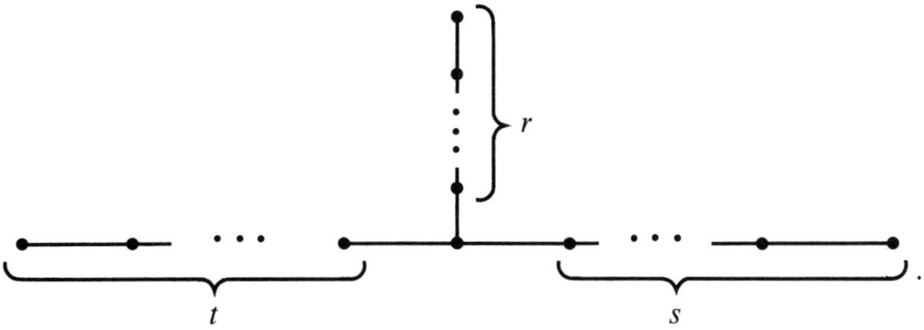

We can assume that $r \le s \le t$. By Corollary b, $r \le 1$. If $r = 0$, then the diagram of Γ is A_n with $n = s + t + 1$. Assume that $r = 1$. By Corollary c, $1 \le s \le 2$. If $s = 1$, then the diagram of Γ is D_n with $n = t + 3$. Assume that $s = 2$. By Corollary d, $2 \le t \le 4$. In these three cases, the diagram of Γ is E_k, where $6 \le k \le 8$. □

EXERCISE

Complete the proof that the quiver in Example d is rigid.

8.6. Change of Orientation

Throughout this section, $\Gamma = (V,E)$ is an acyclic quiver. The vertex i is called a *source* (*sink*) in Γ if $(j,i) \notin E$ (respectively, $(i,j) \notin E$) for all $j \in V$.

Lemma a. *Every non-empty acyclic quiver has sources and sinks.*

PROOF. Assume that Γ has no sources. Since $V \neq \emptyset$, there exists $i_0 \in V$. By assumption, i_0 is not a source. Thus, there exists $i_1 \in V$ such that $(i_1,i_0) \in E$. Since Γ is acyclic, $i_1 \neq i_0$. The fact that i_1 is not a source implies the existence of $i_2 \in V$ such that $(i_2,i_1) \in E$. If $i_2 = i_1$ or $i_2 = i_0$, then Γ would contain a cycle. Hence, $i_2 \neq i_0, i_1$. Repeating this selection process $|V| + 1$ times plainly gives a contradiction. Thus, there is a source in Γ. Similarly, Γ includes a sink. □

Corollary. *If Γ is an acyclic quiver, then the vertices of Γ can be labeled $V = \{1, 2, \ldots, r\}$ in such a way that $(i, j) \in E$ implies $i < j$.*

Proof. Induce on $r = |V|$. If $r = 1$, then $E = \varnothing$, and there is nothing to prove. Assume that $r > 1$, and the corollary is true for quivers with s vertices whenever $s < r$. Let V' be the set of sinks in Γ, and $V'' = V - V'$. By Lemma a, $V' \neq \varnothing$, so that $|V''| = s < r = |V|$. By the induction hypothesis, there is a labeling $V'' = \{1, 2, \ldots, s\}$ of the vertices of $\Gamma|V''$ such that if $(i, j) \in E \cap (V'' \times V'')$, then $i < j$. Label $V' = \{s + 1, \ldots, r\}$ arbitrarily. If $(i,j) \in E$, then i is not a sink. Hence, $i \in V''$. If $j \in V''$, then $i < j$; if $j \in V'$, then $i \leq s < s + 1 \leq j$. □

The quiver Γ will be called *standardized* if $V = \{1,2, \ldots,r\}$, and $i < j$ for all pairs $(i,j) \in E$. By the corollary, every acyclic quiver can be standardized, generally in many ways. Note that if Γ is standardized, then 1 is a source in Γ and r is a sink in Γ.

For each vertex k of the quiver Γ, define $\rho_k\Gamma = (V,E_k)$, where $E_k = \{(i,j) \in E: i \neq k \neq j\} \cup \{(j,k): (k,j) \in E\} \cup \{(k,i): (i,k) \in E\}$. Geometrically, $\rho_k\Gamma$ is the quiver that is obtained from Γ by reversing the orientation of all edges that have k as an endpoint. Plainly, the unoriented graphs associated with Γ and $\rho_k\Gamma$ are identical.

The transformation $\Gamma \mapsto \rho_k\Gamma$ can be iterated, using various vertices. The resulting quivers have the same unoriented graph as Γ, but different orientations. The principal result of this section is that it is possible to obtain every orientation in this way.

Lemma b. *Let Γ be a standardized quiver with r vertices.*

(i) *If $1 \leq k < r$, then k is a sink and $k + 1$ is a source in $\rho_k\rho_{k-1} \cdots \rho_1\Gamma$.*
(ii) *If $1 < k \leq r$, then k is a source and $k - 1$ is a sink in $\rho_k\rho_{k+1} \cdots \rho_r\Gamma$.*
(iii) $\rho_r\rho_{r-1} \cdots \rho_1\Gamma = \rho_1\rho_2 \cdots \rho_r\Gamma = \Gamma$.

Proof. These statements follow from the assumption that Γ is standardized together with the observation that if $k_1, k_2, \ldots, k_s$ are distinct vertices of Γ, and $\rho_{k_1}\rho_{k_2} \cdots \rho_{k_s}\Gamma = (V,E')$, then $(i,j) \in E'$ if and only if: $(i,j) \in E$ and either none or both of i, j occur among $k_1, k_2, \ldots, k_s$; or $(j,i) \in E$ and exactly one of i, j occurs among $k_1, k_2, \ldots, k_s$. □

Proposition. *Let Γ and Γ' be acyclic quivers that satisfy $\Gamma^u = (\Gamma')^u$. There is a sequence $k_1, k_2, \ldots, k_n$ of vertices of Γ such that for $1 \leq l \leq n$, k_l is a source in $\rho_{k_{l+1}}\rho_{k_{l+2}} \cdots \rho_{k_n}\Gamma$, and $\rho_{k_1}\rho_{k_2} \cdots \rho_{k_n}\Gamma = \Gamma'$.*

Proof. The hypothesis $\Gamma^u = (\Gamma')^u$ implies that $\Gamma = (V,E)$ and $\Gamma' = (V,E')$, where $(i, j) \in E$ if and only if $(i, j) \in E'$ or $(j,i) \in E'$. It suffices to prove the special case of the proposition in which there is exactly one edge $(i, j) \in E$

such that $(j,i) \in E'$; that is, $E' = (E - \{(i,j)\}) \cup \{(j,i)\}$. Iterating the result for this special case will yield the full proposition. Let $\Gamma_0 = (V, E - \{(i,j)\})$. If Γ_0 is written as the disjoint union of connected components, then i and j must belong to distinct components: if i and j could be joined by a path in Γ_0, then this path, together with (i,j) would form a cyclic subquiver of Γ, which is contrary to the assumption that Γ is acyclic. Thus, $\Gamma_0 = \Gamma_1 \cup \Gamma_2$, where $\Gamma_1 = (V_1,E_1)$, $\Gamma_2 = (V_2,E_2)$, $i \in V_1$, and $j \in V_2$. Then $V = V_1 \cup V_2$, and $E = E_1 \cup E_2 \cup \{(i,j)\}$. By the corollary, it can be assumed that $V_1 = \{1, 2, \ldots, s\}$ is listed so that Γ_1 is standardized. Let $V_2 = \{s + 1, s + 2, \ldots, r\}$. If $1 \leq k \leq s < l \leq r$, then $(k,l) \in E$ if and only if $k = i$ and $l = j$. By Lemma b, k is a source in $\rho_{k-1}\rho_{k-2} \cdots \rho_1\Gamma_1$ for $1 \leq k \leq s$, and $\rho_s \rho_{s-1} \cdots \rho_1 \Gamma_1 = \Gamma_1$. Since $\Gamma_0 = \Gamma_1 \cup \Gamma_2$, it follows that k is a source in $\rho_{k-1}\rho_{k-2} \cdots \rho_1\Gamma$, and $\rho_s\rho_{s-1} \cdots \rho_1\Gamma = (V, ((E - \{(i,j)\}) \cup \{(j,i)\}) = \Gamma'$. □

EXERCISES

1. Let $\Gamma = (V,E)$ be an unoriented graph with at most double edges, that is, for $i \neq j$ in Γ, there are at most two edges that join i and j. Assume that for each $i \in V$ there is at most one loop at i. Prove that the edge set E can be oriented in such a way that Γ becomes a quiver. Determine the number of ways in which Γ can be oriented to form a quiver.

2. Let $\Gamma = (V,E)$ be a quiver. An oriented cycle in Γ is a sequence $i_1, i_2, \ldots, i_n$ $(n \geq 1)$ of vertices such that (i_1,i_2), (i_2,i_3), $\ldots$, (i_{n-1},i_n), (i_n,i_1) are all edges. Prove that a non-empty quiver has a standard orientation if and only if there are no oriented cycles in Γ.

8.7. Change of Representation

In this section, we will show that the orientation reversals defined in Section 8.6 are accompanied by a change of representations. This fact is the key step in the proof of both parts of Theorem 8.2.

Notation. (i) Let i be a source in the quiver $\Gamma = (V,E)$. For a representation (M,ϕ) of Γ, define $S_i^-(M,\phi) = (N,\psi) \in \mathfrak{R}(\rho_i\Gamma)$ by $N_k = M_k$ for $k \neq i$, $N_i = \bigoplus_{(i,j)\in E} M_j/\operatorname{Im}\prod_j \phi_{ij} = \operatorname{Coker}(\prod_j \phi_{ij})$, $\psi_{kj} = \phi_{kj}$ for $k \neq i \neq j$, and (for $(i,k) \in E$) let ψ_{ki} be the composition $N_k = M_k \to \bigoplus_{(i,j)\in E} M_j \to N_i$ of the inclusion of M_k into $\bigoplus_{(i,j)\in E} M_j$, and the natural projection of $\bigoplus_{(i,j)\in E} M_j$ onto N_i. If $\theta\colon (M,\phi) \to (M',\phi')$ is a morphism in $\mathfrak{R}(\Gamma)$, define $S_i^-(\theta) = \chi\colon S_i^-(M,\phi) \to S_i^-(M',\phi')$ by $\chi_k = \theta_k$ for $k \neq i$, and $\chi_i\colon N_i = \operatorname{Coker}(\prod_j \phi_{ij}) \to \operatorname{Coker}(\prod_j \phi'_{ij}) = N'_i$ so that

$$\begin{array}{ccc} \bigoplus_{(i,j)\in E} M_j & \xrightarrow{\oplus_j\theta_j} & \bigoplus_{(i,j)\in E} M'_j \\ \downarrow & & \downarrow \\ N_i & \xrightarrow[\chi_i]{} & N'_i \end{array} \tag{1}$$

commutes.

(ii) Let i be a sink in Γ. For $(M,\phi) \in \mathfrak{R}(\Gamma)$, define $S_i^+(M,\phi) = (N,\psi) \in \mathfrak{R}(\rho_i\phi)$ by $N_k = M_k$ for $k \neq i$, $N_i = \operatorname{Ker}(\coprod_j \phi_{ji}) < \bigoplus_{(j,i)\in E} M_j$, $\psi_{kj} = \phi_{kj}$ for $k \neq i \neq j$, and (for $(k,i) \in E$) let ψ_{ik} be the composition

$$N_i \to \bigoplus_{(j,i)\in E} M_j \to M_k = N_k.$$

If $\theta\colon (M,\phi) \to (M',\phi')$ is a morphism in $\mathfrak{R}(\Gamma)$, define

$$S_i^+(\theta) = \chi\colon S_i^+(M,\phi) \to S_i^+(M',\phi')$$

by $\chi_k = \theta_k$ for $k \neq i$, and $\chi_i\colon N_i = \operatorname{Ker}(\coprod_j \phi_{ji}) \to \operatorname{Ker}(\coprod_j \phi'_{ji}) = N'_i$ is such that

$$\begin{array}{ccc} N_i & \xrightarrow{\chi_i} & N'_i \\ \downarrow & & \downarrow \\ \bigoplus_{(i,j)\in E} M_j & \xrightarrow[\bigoplus_j \theta_j]{} & \bigoplus_{(i,j)\in E} M'_j \end{array} \tag{2}$$

commutes.

The fact that $S_i^-(M,\phi)$ and $S_i^+(M,\phi)$ are representations of $\rho_i\Gamma$ is evident: the definition of quiver representations imposes practically no restrictions. However, to see that $S_i^-(\theta)$ and $S_i^+(\theta)$ are morphisms requires a bit of checking. Assume that i is a source in Γ. Denote $S_i^-(\theta) = \{\chi_j\colon j \in V\}$. If $j \neq i \neq k$, then $\chi_k\psi_{jk} = \psi'_{jk}\psi_j$ is a restatement of the corresponding commutativity condition on θ. The fact that $\chi_i\psi_{ki} = \psi'_{ki}\chi_k$ for $k \neq i$ is seen by chasing arrows in the diagram

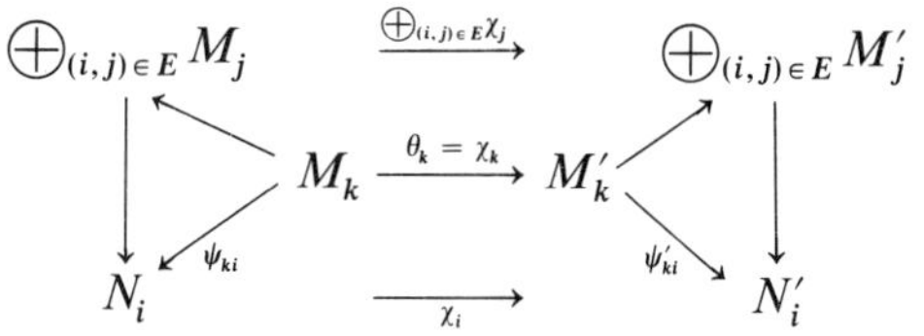

where the unlabeled arrows are natural injections or projections. Similar considerations show that $S_i^+(\theta)$ is a morphism when i is a sink in Γ.

Lemma. *Let Γ be an acyclic quiver, and suppose that (M,ϕ) and (M',ϕ') are representations of Γ.*

(i) *S_i^- and S_i^+ are functors from $\mathfrak{R}(\Gamma)$ to $\mathfrak{R}(\rho_i\Gamma)$ in the respective cases that i is a source or sink in Γ.*
(ii) $S_i^\pm((M,\phi) \bigoplus (M',\phi')) \cong S_i^\pm(M,\phi) \bigoplus S_i^\pm(M',\phi')$.
(iii) *If i is a source in Γ, then $(M,\phi) \cong S_i^+S_i^-(M,\phi) \bigoplus (P,0)$, where $P_j = 0$ for $j \neq i$ and $P_i = \operatorname{Ker}(\coprod_j \phi_{ij})$.*
(iv) *If i is a sink in Γ, then $(M,\phi) \cong S_i^-S_i^+(M,\phi) \bigoplus (Q,0)$, where $Q_j = 0$ for $j \neq i$ and $Q_i = \operatorname{Coker}(\coprod_j \phi_{ji})$.*

Proof. The equations $S_i^\pm(\theta'\theta) = S_i^\pm(\theta')S_i^\pm(\theta)$ are easily seen by putting together diagrams of the kind that are shown in (1) and (2). Therefore, (i) is verified, since clearly

$$S_i^\pm(id_{(M,\phi)}) = id_{S_i^\pm(M,\phi)}.$$

A routine calculation shows that $S_i^{\pm}(\theta + \theta') = S_i^{\pm}(\theta) + S_i^{\pm}(\theta')$, from which (ii) is an easy consequence. For the proof of (iii), denote $S_i^+ S_i^-(M,\phi)$ by (N,χ). A check of the definitions given above shows that $N_k = M_k$ for $k \neq i$, $N_i = \mathrm{Im}(\prod_j \phi_{ij}) < \bigoplus_{(i,j)\in E} M_j$, $\chi_{jk} = \phi_{jk}$ for $j \neq i$, $(j,k) \in E$, and for $(i,k) \in E$, χ_{ik} is the restriction to N_i of the projection from $\bigoplus_{(i,j)\in E} M_j$ to M_k (that is, if $x = (\ldots x_k \ldots) \in N_i$, then $\chi_{ik}(x) = x_k$). Since $P_i = \mathrm{Ker}(\prod_j \phi_{ij})$, the sequence of F-space homomorphisms $0 \to P_i \to M_i \xrightarrow{\prod_j \phi_{ij}} N_i \to 0$ is exact. Thus, there is an F-space homomorphism $\psi\colon M_i \to N_i$ such that $\theta_i(u) = ((\prod_j \phi_{ij})(u), \psi(u))$ is an isomorphism of M_i to $N_i \oplus P_i$. For $j \neq i$, define $\theta_j(u) = (u,0) \in N_j \oplus 0 = N_j \oplus P_j$. Then $\theta\colon (M,\phi) \to (N,\chi) \oplus (P,0)$ is an isomorphism: if $(i,k) \in E$, then $(\chi_{ik} \oplus 0)\theta_i(u) = (\chi_{ik}(\ldots \phi_{ik}(u) \ldots), 0) = (\phi_{ik}(u),0) = \theta_k \phi_{ik}(u)$; clearly, $(\chi_{jk} \oplus 0)\theta_j = \theta_k \phi_{jk}$ if $j \neq i$. The proof of (iv) is the categorical dual of the proof of (iii). □

Proposition. *Let Γ be an acyclic quiver. Assume that (M,ϕ) is an indecomposable representation of Γ. Let i be a source (sink) in Γ.*

(i) *If $(M,\phi) \cong (P^{(i)}, 0)$, where $(P^{(i)},0)$ is the simple representation of Γ at i, then $S_i^-(M,\phi) = (0,0)$ $(S_i^+(M,\phi) = (0,0))$.*
(ii) *If $(M,\phi) \ncong (P^{(i)},0)$, then $S_i^-(M,\phi) = (N,\psi)$ $(S_i^+(M,\phi) = (N,\psi))$ is indecomposable, $S_i^+ S_i^-(M,\phi) \cong (M,\phi)$ $(S_i^- S_i^+(M,\phi) \cong (M,\phi))$, and $\dim_F N_j = \dim_F M_j$ for $j \neq i$, $\dim_F N_i = \sum_{(i,j)\in E} \dim_F M_j - \dim_F M_i$.*
(iii) *Γ has finite representation type if and only if $\rho_i\Gamma$ has finite representation type.*

Proof. We will prove the statements (i), (ii), and (iii) for the case in which i is a source. The parenthetical assertions, that apply when i is a sink, are obtained by the standard trick of reversing the arrows that represent morphisms. The statement (i) is elementary: $N_j = M_j \cong P_j^{(i)} = 0$ for $j \neq i$, and N_i is a factor space of $\bigoplus_{(i,j)\in E} M_j$, which is 0 since Γ has no loops. Assume that $(M,\phi) \ncong (P^{(i)},0)$. It follows from part (iii) of the lemma that $(M,\phi) \cong S_i^+ S_i^-(M,\phi) \oplus \bigoplus m(P^{(i)},0)$. Since (M,ϕ) is indecomposable and not isomorphic to $(P^{(i)},0)$, it follows that $(M,\phi) \cong S_i^+ S_i^-(M,\phi)$. In particular, $S_i^-(M,\phi) \neq (0,0)$. Thus, we can write $S_i^-(M,\phi) = (N,\psi) \oplus (M',\phi')$, where (N,ψ) is indecomposable and $S_i^+(N,\psi) \neq (0,0)$. Then $(M,\phi) \cong S_i^+ S_i^-(M,\phi) \cong S_i^+(N,\psi) \oplus S_i^+(M',\phi')$. The indecomposability of (M,ϕ) implies $S_i^+(M',\phi') = (0,0)$. Consequently, $(N,\psi) \oplus (M',\phi') = S_i^-(M,\phi) \cong S_i^- S_i^+(N,\psi)$, so that $(M',\phi') = (0,0)$: if $S_i^- S_i^+(N,\psi) = (N',\psi')$, then by part (iv) of the lemma, $\dim_F M_j' = \dim_F N_j' - \dim_F N_j \leq 0$ for all $j \in V$. To prove the last part of (ii), use the fact that (M,ϕ) is reduced by Lemma 8.3a. In particular, $\prod_j \phi_{ij}$ is injective, and $\bigoplus_{(i,j)\in E} M_j \cong M_i \oplus N_i$ as F-spaces. Hence, $\dim_F N_i = \sum_{(i,j)\in E} \dim_F M_j - \dim_F M_i$. By the lemma and (ii), there is a bijection between the non-simple, indecomposable representations of Γ and the non-simple indecomposable representations of $\rho_i\Gamma$. Since $\mathfrak{R}(\Gamma)$ and $\mathfrak{R}(\rho_i\Gamma)$ have finitely many isomorphism classes of simple objects, the statement (iii) is clear. □

We can now pay the debt that was incurred in the proof of Proposition 8.5.

Corollary. *Let Γ and Γ' be quivers such that $\Gamma^u = (\Gamma')^u$. The representation type of Γ' is finite if and only if Γ has finite representation type.*

PROOF. Since $\Gamma^u = (\Gamma')^u$, either Γ or Γ' are both acyclic, or neither of them is acyclic. In the latter case, Γ and Γ' have infinite representation types by Proposition 8.4. If Γ and Γ' are acyclic, then Proposition 8.6 assures the existence of a sequence $k_1, k_2, \ldots, k_n$ of vertices such that $\Gamma' = \rho_{k_1}\rho_{k_2}\cdots\rho_{k_n}\Gamma$, and k_l is a source in $\rho_{k_{l+1}}\rho_{k_{l+2}}\cdots\rho_{k_n}\Gamma$ for $1 \le l \le n$. It follows by repeated applications of part (iii) of the Proposition that the representation types of Γ and Γ' are the same. □

EXERCISES

1. Prove that if i is a sink in Γ, and θ: $(M,\phi) \to (M',\phi')$ is a morphism, then $S_i^+(\theta)$ is a morphism.
2. Prove statement (iv) of the lemma.
3. Prove the parenthetical statements in the Proposition.

8.8. The Quadratic Space of a Quiver

Proposition 8.7 provides a way to construct indecomposable representations from simple representations: if i is a source in the acyclic quiver Γ, and $k \neq i$, then $S_i^-(P^{(k)},0)$ is an indecomposable representation of $\rho_i\Gamma$, and if (i,k) is an edge of Γ, then $S_i^-(P^{(k)},0)$ is not simple. This process can be iterated to construct a substantial supply of indecomposable representations.

In this section and the next one we will set up some geometric machinery that keeps track of the representations that are produced in this way. As usual, $\Gamma = (V,E)$ denotes an acyclic quiver with the vertex set labeled $V = \{1, 2, \ldots, r\}$. Let $\{E\} = \{\{i,j\}: (i,j) \in E\}$; in other words, $\{E\}$ is the edge set of the unoriented graph Γ^u. Associate with Γ the r-dimensional rational vector space $U_\Gamma = \bigoplus_{i\in V} v_i\mathbb{Q}$, and a quadratic mapping ϕ_Γ of U_Γ to $\mathbb{Q}$ defined by

$$\phi_\Gamma\Big(\sum_{i\in V} v_i a_i\Big) = \sum_{i\in V} a_i^2 - \sum_{\{i,j\}\in\{E\}} a_i a_j. \tag{1}$$

The quadratic form that defines ϕ_Γ is

$$\Phi_\Gamma(\mathbf{x}_1, \mathbf{x}_2, \ldots, \mathbf{x}_r) = \sum_{i=1}^{r} \mathbf{x}_i^2 - \sum_{\{i,j\}\in\{E\}} \mathbf{x}_i\mathbf{x}_j. \tag{1'}$$

The bilinear mapping obtained from ϕ_Γ by polarization will be needed. It is defined by

$$\beta_\Gamma\Big(\sum_{i\in V} v_i a_i, \sum_{i\in V} v_i b_i\Big) = \sum_{i\in V} a_i b_i - (1/2) \sum_{\{i,j\}\in\{E\}} (a_i b_j + a_j b_i). \tag{2}$$

Lemma a. *If the (unoriented) diagram of* Γ *is one of the Dynkin diagrams* A_n, D_n, *or* E_k $(6 \le k \le 8)$, *then the quadratic form* Φ_Γ *is positive definite.*

PROOF. For $m \ge 1$, denote

$$\begin{aligned}\Psi_m(\mathbf{x}_1, \ldots, \mathbf{x}_m) &= -\mathbf{x}_1\mathbf{x}_2 - \mathbf{x}_2\mathbf{x}_3 - \cdots - \mathbf{x}_{m-1}\mathbf{x}_m \\ &\quad + \mathbf{x}_1^2 + \cdots + \mathbf{x}_{m-1}^2 + ((m-1)/2m)\mathbf{x}_m^2 \\ &= \sum_{k=1}^{m-1} (1/2)k(k+1)((1/(k+1))\mathbf{x}_{k+1} - (1/k)\mathbf{x}_k)^2.\end{aligned}$$

Evidently, Ψ_m is positive semidefinite, and $\Psi_m(a_1, \ldots, a_m) = 0$ only if $a_1 = (1/2)a_2 = \cdots = (1/m)a_m$. If Γ has the diagram

A_n: 

, $n \ge 1$,

then $\Phi_\Gamma(\mathbf{x}_1, \ldots, \mathbf{x}_n) = ((n+2)/2n)\mathbf{x}_n^2 + \Psi_n(\mathbf{x}_1, \ldots, \mathbf{x}_n)$ is positive definite. If Γ has the diagram

D_n: 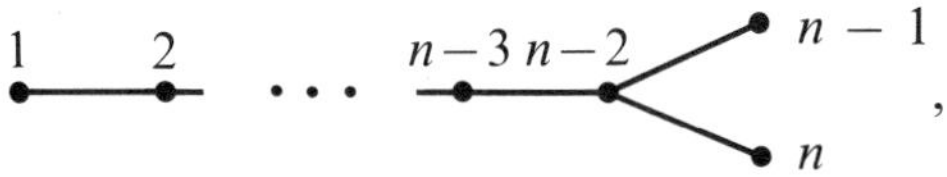

, $n \ge 4$,

then $\Phi_\Gamma(\mathbf{x}_1, \ldots, \mathbf{x}_n) = \Psi_{n-2}(\mathbf{x}_1, \ldots, \mathbf{x}_{n-2}) + \Psi_2(\mathbf{x}_{n-1}, \mathbf{x}_{n-2}) + \Psi_2(\mathbf{x}_n, \mathbf{x}_{n-2}) + (1/2(n-2))\mathbf{x}_{n-2}^2$ is positive definite. Finally, if the diagram of Γ is

E_k: 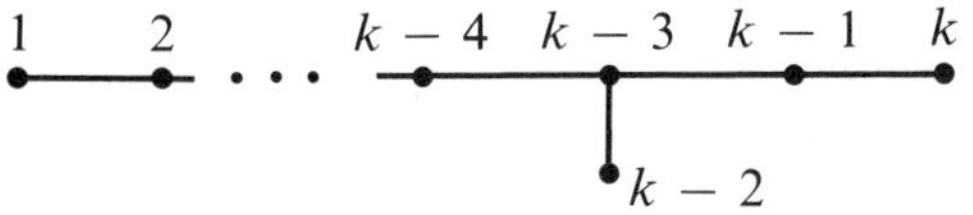

, $6 \le k \le 8$,

then

$$\begin{aligned}\Phi_\Gamma(\mathbf{x}_1, \ldots, \mathbf{x}_k) &= \Psi_{k-3}(\mathbf{x}_1, \ldots, \mathbf{x}_{k-3}) + \Psi_2(\mathbf{x}_{k-2}, \mathbf{x}_{k-3}) \\ &\quad + \Psi_3(\mathbf{x}_k, \mathbf{x}_{k-1}, \mathbf{x}_{k-3}) + ((9-k)/12(k-3))\mathbf{x}_{k-3}^2\end{aligned}$$

is positive definite. □

Proposition. *If* Γ *has finite representation type, then* Φ_Γ *is positive definite.*

PROOF. By Proposition 8.5, $\Gamma = \Gamma_1 \cup \Gamma_2 \cup \cdots \cup \Gamma_k$ where each quiver Γ_j has a diagram that is one of the Dynkin diagrams A_n, D_n, or E_k $(6 \le k \le 8)$. It follows from the definitions that have been given that $U_\Gamma = U_{\Gamma_1} \oplus U_{\Gamma_2} \oplus \cdots \oplus U_{\Gamma_k}$ and $\phi_\Gamma = \phi_{\Gamma_1} \oplus \phi_{\Gamma_2} \oplus \cdots \oplus \phi_{\Gamma_k}$. That is, if $w = w_1 + w_2 + \cdots + w_k$ with $w_j \in U_{\Gamma_j}$, then $\phi_\Gamma(w) = \phi_{\Gamma_1}(w_1) + \phi_{\Gamma_2}(w_2) + \cdots + \phi_{\Gamma_k}(w_k)$. By the lemma, each ϕ_{Γ_j} is positive definite. Thus, ϕ_Γ is positive definite. □

The converse of this proposition will follow from Proposition 8.5 and Theorem 8.9. However, it is not difficult to prove this converse directly (see Exercise 1), and a less roundabout proof of Proposition 8.5 can be based on this fact. The lemma that we need is this: if Γ has finite representation type (hence, Γ is acyclic), then Φ_Γ is positive definite. A geometrical argument due to Tits can be used to establish this lemma. If Φ_Γ is not positive definite, then there exist non-negative integers k_i, not all zero, such that $\phi_\Gamma(\sum_{i \in V} v_i k_i) \leq 0$. For $1 \leq i \leq r$, let M_i be an F-space of dimension k_i. The representations (M,ϕ) of Γ such that $M = \{M_i: 1 \leq i \leq r\}$ are in one-to-one correspondence with the elements of the F-space $N = \bigoplus_{(i,j) \in E} \mathrm{Hom}_F(M_i, M_j)$. By our construction, $\dim_F N = \sum_{(i,j) \in E} k_i k_j$. The algebraic group $G = \prod_{i \in V} GL_{k_i}(F)$ acts on N by $(\ldots \phi_{ij} \ldots) \mapsto (\ldots \theta_j \phi_{ij} \theta_i^{-1} \ldots)$, and every nonzero scalar multiple of 1_G acts trivially. Thus, N can be viewed as a G/F°-module. Plainly, the orbits of G/F° in N correspond bijectively to the isomorphism classes of representations of Γ on the spaces $\{M_i: 1 \leq i \leq r\}$. By Lemma 8.4a, the representation type of Γ is infinite if there are infinitely many orbits. Now, geometrical intuition comes into play. It is reasonable to suppose that the number of G/F° orbits in N will be infinite if the "dimension" of G/F° is less than the "dimension" of N. To make this reasoning sound requires an interpretation of the word "dimension." The appropriate meaning comes from algebraic geometry: G/F° and N are algebraic sets whose dimensions are respectively $(\sum_{i=1}^r k_i^2) - 1$ and $\sum_{(i,j) \in E} k_i k_j$. By the choice of $k_1, k_2, \ldots, k_r$, $\dim N - \dim G/F^\circ = 1 - \phi_\Gamma(\sum_{i=1}^r v_i k_i) \geq 1$. The desired conclusion that G/F° has infinitely many orbits in N can then be proved. Some hints for filling in the details of this sketch are provided with Exercise 2.

Definition. Let (M,ϕ) be a representation of the acyclic quiver $\Gamma = (V,E)$. The *dimension vector* of (M,ϕ) is the vector in $U_\Gamma = \bigoplus_{i \in V} v_i \mathbb{Q}$ given by

$$\mathrm{Dim}(M,\phi) = \sum_{i \in V} v_i(\dim_F M_i).$$

To formulate the essential properties of Dim in a convenient way, we introduce some notation that will be explored more systematically in the next section. We will denote by W_Γ the subgroup of U_Γ consisting of vectors with integral components: $W_\Gamma = \bigoplus_{i \in V} v_i \mathbb{Z}$. A vector $w = \sum_{i \in V} v_i a_i$ is positive if $w \neq 0$, and $a_i \geq 0$ for all $i \in V$. Denote the set of positive vectors in W_Γ by W_Γ^+. For each $i \in V$, define a linear transformation σ_i of U_Γ by $\sigma_i(w) = w - v_i(2\beta_\Gamma(w,v_i))$. When ϕ_Γ is positive definite, σ_i is the usual reflection in the plane perpendicular to v_i. In fact, $\phi_\Gamma(v_i) = 1$, so that $\sigma_i(w) = w - v_i(2\beta_\Gamma(w,v_i)/\phi_\Gamma(v_i))$. If $w = \sum_{i \in V} v_i a_i$, then an easy calculation yields

$$\sigma_i(w) = \sum_{i \neq j \in V} v_j a_j + v_i((\sum_{\{i,j\} \in \{E\}} a_j) - a_i). \tag{3}$$

Lemma b. *Let (M,ϕ) be a representation of the acyclic quiver $\Gamma = (V,E)$.*

(i) $(M,\phi) = (0,0)$ *if and only if* $\mathrm{Dim}\,(M,\phi)$ *is the zero vector. If* $(M,\phi) \neq (0,0)$, *then* $\mathrm{Dim}\,(M,\phi) \in W_\Gamma^+$.
(ii) $\mathrm{Dim}(M,\phi) = v_k$ *if and only if* $(M,\phi) \cong (P^{(k)},0)$.
(iii) *If i is a source in Γ, (M,ϕ) is indecomposable, and* $(M,\phi) \not\cong (P^{(i)},0)$, *then* $\mathrm{Dim}(S_i^-(M,\phi)) = \sigma_i \mathrm{Dim}\,(M,\phi)$.
(iv) *If i is a sink in Γ, (M,ϕ) is indecomposable, and* $(M,\phi) \not\cong (P^{(i)},0)$, *then* $\mathrm{Dim}(S_i^+(M,\phi)) = \sigma_i \mathrm{Dim}(M,\phi)$.

The statements (i) and (ii) come directly from the definition of the dimension vector. The properties (iii) and (iv) are restatements of a portion of Proposition 8.7 by virtue of (3).

Exercises

1. Give a direct proof that if Γ is an acyclic quiver such that Φ_Γ is positive definite, then the diagram of Γ is a disjoint union of the Dynkin diagrams A_n, D_n, and E_k $(6 \le k \le 8)$. Hint. If Γ' is a subquiver of Γ, then $\Phi_{\Gamma'}$ is positive definite. Use this observation to prove that Γ cannot contain a subquiver whose unoriented diagram is the same as the unoriented diagram of the quiver in Corollary 8.5a. Hence, the diagram of Γ must be like the figure in the proof of Proposition 8.5. Thus,

$$\Phi_\Gamma(\mathbf{x}_1, \ldots, \mathbf{x}_{1r}, \mathbf{x}_{21}, \ldots, \mathbf{x}_{2s}, \mathbf{x}_{31}, \ldots, \mathbf{x}_{3t}, \mathbf{y}) =$$
$$\Psi_{r+1}(\mathbf{x}_{11}, \ldots, \mathbf{x}_{1r}, \mathbf{y}) + \Psi_{s+1}(\mathbf{x}_{21}, \ldots, \mathbf{x}_{2s}, \mathbf{y}) + \Psi_{t+1}(\mathbf{x}_{31}, \ldots, \mathbf{x}_{3t}, \mathbf{y}) +$$
$$\mathbf{y}^2(1 - (r/2(r+1)) - (s/2(s+1)) - (t/2(t+1))),$$

where Ψ_m is defined as in the proof of Lemma 8.8a. Assume without loss of generality that $r \le s \le t$. Deduce that if Φ_Γ is positive definite, then either $r = 0$, or $r = s = 1$, or $r = 1$, $s = 2$, and $2 \le t \le 4$.

2. Let $\Gamma = (V,E)$ be an acyclic quiver. Assume that there is a set $\{M_i : i \in V\}$ of finite dimensional F-spaces such that $n = \sum_{i \in V} d_i^2 \le \sum_{(i,j)\in E} d_i d_j = m$, where $d_i = \dim_F M_i$. As usual, assume that the field F is infinite. Fill in the details of the following sketch that Γ has infinite representation type. Denote $G_0 = \prod_{i\in V} GL_{d_i}(F)$, and $G = \{(\alpha,a): \alpha = (\ldots \alpha_i \ldots) \in G_0, a = (\prod_{i \in V} \det \alpha_i)^{-1}\} \subseteq F^{n+1}$. The *affine ring* of G is

$$A(G) = \{\Theta | G: \Theta \in F[\ldots \mathbf{x}_{rs}^{(i)} \ldots, \mathbf{z}]\} \cong F[\ldots \mathbf{x}_{rs}^{(i)} \ldots, \mathbf{z}]/((\textstyle\prod_{i\in V} \det [\mathbf{x}_{rs}^{(i)}])\mathbf{z} - 1).$$

If $\phi = (\ldots \phi_{ij} \ldots) \in F^m$, then G operates on F^m by

$$\Theta(\alpha,a)\phi = (\ldots \alpha_j \phi_{ij} \alpha_i^{ad}(\textstyle\prod_{k\neq i} \det \alpha_k) a \ldots).$$

Note that $\alpha_i^{ad}(\prod_{k \ne i} \det \alpha_k) a = \alpha_i^{-1}$. For a fixed $\phi \in F^m$, define $\psi_\phi(\alpha,a) = \Theta(\alpha,a)\phi$. The mapping ψ_ϕ induces an algebra homomorphism $\psi_\phi^*: F[\ldots \mathbf{y}_{rs}^{(i,j)} \ldots] \to A(G)$ by $(\psi_\phi^* \Phi)\,(\alpha,a) = \Phi(\psi_\phi(\alpha,a))$. Prove the following statements.

(a) If $\mathrm{Ker}\,\psi_\phi^* \neq 0$, then there are infinitely many orbits of the action of G on F^m, in which case Γ has infinite representation type.

Denote $\vec{\mathbf{y}} = (\mathbf{y}_1, \ldots, \mathbf{y}_m)$, $\vec{\mathbf{x}} = (\mathbf{x}_1, \ldots, \mathbf{x}_n)$, and $\vec{\mathbf{x}}^p = (\mathbf{x}_1^p, \ldots, \mathbf{x}_n^p)$, where

$p = \sum_{i \in V} d_i$. Let $R = \{\sum_{j=0}^{l} \Phi_j(\vec{\mathbf{x}})\, \mathbf{z}^j : \Phi_j(\vec{\mathbf{x}})$ homogeneous of degree $pj\} \subseteq F[\vec{\mathbf{x}},\mathbf{z}]$, and $T = \{\sum_{j=0}^{l} \Phi_j(\vec{\mathbf{x}}^p)\mathbf{z}^j : \Phi_j \in F[\vec{\mathbf{x}}]$ homogeneous of degree $j\}$, a subring of R.

(b) $F[\vec{\mathbf{x}}] \cong T$.

Let K be the fraction field of R, and L the fraction field of T.

(c) $\operatorname{Tr.deg} K/F = \operatorname{Tr.deg} L/F = n$.

(d) If there is a subring S of R and $0 \neq \Phi(\vec{\mathbf{x}},\mathbf{z}) \in S$ such that $S/(\Phi(\vec{\mathbf{x}},\mathbf{z})) \cong F[\vec{\mathbf{y}}]$, then $m < n$. Hint. Show that $m + 1 \leq \operatorname{Tr.deg} E/F \leq \operatorname{Tr.deg} K/F = n$, where E is the fraction field of S.

Denote $\Theta(\vec{\mathbf{x}},\mathbf{z}) = (\prod_{i \in V} \det[\mathbf{x}_{rs}^{(i)}])\mathbf{z} - 1$.

(e) (Θ) is a prime ideal in $F[\vec{\mathbf{x}},\mathbf{z}]$. Moreover, if $\Phi \in F[\vec{\mathbf{x}},\mathbf{z}]$ satisfies $\Phi(\alpha,a) = 0$ for all $(\alpha,a) \in G$, then $\Phi \in (\Theta)$.

(f) If $\Phi[\vec{\mathbf{x}},\mathbf{z}] \in F[\vec{\mathbf{x}},\mathbf{z}]$ satisfies $\Phi(b\alpha,b^{-p}a) = \Phi(\alpha,a)$ for all $(\alpha,a) \in G$ and $b \in F$, then $\Phi(\vec{\mathbf{x}},\mathbf{z}) \equiv \Phi'(\vec{\mathbf{x}},\mathbf{z}) \bmod (\Theta)$, where $\Phi'(\vec{\mathbf{x}},\mathbf{z}) \in R$.

Deduce from (d), (e), and (f) that if $n \leq m$, then $\operatorname{Ker} \psi_\phi^* \neq 0$.

8.9. Roots and Representations

We are near the end of the proof of Theorem 8.2. Some facts about positive definite inner product spaces are needed. For convenience, the proofs of these results are given, even though they are standard fare in Lie algebra theory.

In this section, $U = \bigoplus_{i=1}^{r} v_i\mathbb{Q}$ denotes an r dimensional $\mathbb{Q}$-space with a distinguished basis $v_1, v_2, \ldots, v_r$. The space V is partially ordered by $\sum v_i a_i \leq \sum v_i b_i$ if $a_i \leq b_i$ for $1 \leq i \leq r$. In particular, $w = \sum v_i a_i$ is positive if $w \neq 0$ and $a_i \geq 0$ for all i; and w is negative if $-w$ is positive. The sets of positive and negative vectors are denoted by U^+ and U^- respectively.

Let $W = \bigoplus_{i=1}^{r} v_i\mathbb{Z}$ be the set of vectors in U that have integral components with respect to the basis $v_1, v_2, \ldots, v_r$. Plainly W is a finitely generated subgroup of U. Denote $W^+ = W \cap U^+$, and $W^- = W \cap U^-$.

Assume that ϕ is a positive definite quadratic mapping of U; that is, $\phi: U \to \mathbb{Q}$ satisfies $\phi(wa) = \phi(w)a^2$ for $w \in U$, $a \in \mathbb{Q}$, $\phi(w) > 0$ for $0 \neq w \in U$, and $\beta(w,w') = (1/2)(\phi(w + w') - \phi(w) - \phi(w'))$ is a symmetric, bilinear mapping. We also make the assumption that $\phi(v_1) = \phi(v_2) = \cdots = \phi(v_r) = 1$, and $\phi(u) \in \mathbb{Z}$ for all $u \in W$. By Proposition 8.8, these hypotheses are satisfied if $U = U_\Gamma$ and $\phi = \phi_\Gamma$, where Γ is a quiver that has finite representation type. The bilinear mapping β associated with ϕ may not map $W \times W$ to $\mathbb{Z}$, but clearly, $2\beta(u,u') \in \mathbb{Z}$ if u, $u' \in W$.

If w and w' are in U and $w' \neq 0$, then

$$\phi(w) - \beta(w,w')^2/\phi(w') = \phi(w - w'(\beta(w,w')/\phi(w'))) > 0$$

unless $w = w'(\beta(w,w')/\phi(w'))$. This observation proves Schwartz's inequality:

$$\beta(w,w')^2 \leq \phi(w)\phi(w') \quad \text{for all } w, w' \in U, \tag{1}$$

and the inequality is strict unless w and w' are linearly dependent.

For each $w \in U$, denote $w^\perp = \{u \in U: \beta(u,w) = 0\}$. Then $w^\perp$ is a subspace of U. Note that

$$\bigcap_{i=1}^{r} v_i^\perp = 0. \tag{2}$$

In fact, if $u \in \bigcap_{i=1}^r v_i^\perp$, then $\beta(u,w) = 0$ for all $w \in U$, since $\{v_1, \ldots, v_r\}$ is a basis of U. In particular, $\phi(u) = \beta(u,u) = 0$, so that $u = 0$.

If $w \in U - \{0\}$, then $\beta(w,w) = \phi(w) > 0$, so that $w \notin w^\perp$. On the other hand, $u - w(\beta(w,u)/\phi(w)) \in w^\perp$, so that $U = w\mathbb{Q} \bigoplus w^\perp$. In particular, $\dim_{\mathbb{Q}} w^\perp = r - 1$. A routine induction using the dimension formula $\dim(U_1 \cap U_2) = \dim U_1 + \dim U_2 - \dim(U_1 + U_2)$ yields the estimate $\dim_{\mathbb{Q}}(v_{i_1}^\perp \cap \cdots \cap v_{i_n}^\perp) \geq r - n$. In particular, there is a non-zero vector $u_i \in \bigcap_{j \neq i} v_j^\perp$. By (2), $u_i \notin v_i^\perp$. Therefore, u_i can be normalized so that $\beta(u_i,v_i) = 1$. This discussion establishes the existence of a "dual basis" $\{u_1, u_2, \ldots, u_r\} \subseteq U$ satisfying

$$\beta(u_i,v_j) = \delta_{ij} \quad \text{for} \quad 1 \leq i,j \leq r. \tag{3}$$

For $1 \leq i \leq r$, define linear transformations $\sigma_i: U \to U$ by

$$\sigma_i(w) = w - v_i(2\beta(w,v_i)),$$

as in Section 8.8. Since $\phi(v_i) = 1$, it follows that

$$\sigma(v_i) = -v_i, \quad \text{and} \quad \sigma_i(w) = w \quad \text{for all} \quad w \in v_i^\perp. \tag{4}$$

That is, σ_i is the reflection in the plane $v_i^\perp$. The equation (4) implies

$$\sigma_i^2 = id_U, \tag{5}$$

and

$$\beta(\sigma_i w, \sigma_i w') = \beta(w,w') \quad \text{for} \quad w, w' \in U. \tag{6}$$

Since 2β maps $W \times W$ to $\mathbb{Z}$, each σ_i is an automorphism of W.

Let G be the subgroup of $GL(U)$ that is generated by $\{\sigma_i: 1 \leq i \leq r\}$. By (5), every element of G can be written as a product $\sigma_{i_1}\sigma_{i_2} \cdots \sigma_{i_n}$. The group G is called the *Weyl group* of $(U, \phi; v_1, \ldots, v_r)$; in the case that $U = U_\Gamma$ and $\phi = \phi_\Gamma$, we will refer to G as the Weyl group of the quiver Γ. It follows from (6) that the elements of G are orthogonal transformations:

$$\beta(\tau w, \tau w') = \beta(w,w') \quad \text{for all } \tau \in G \text{ and } w, w' \in U. \tag{7}$$

Since the reflections σ_i map W to itself, so do the transformations of G:

$$\tau(W) = W \quad \text{for all } \tau \in G. \tag{8}$$

Define $Y = \{\tau(v_i): \tau \in G, 1 \leq i \leq r\}$. If $U = U_\Gamma$ and $\phi = \phi_\Gamma$, then the elements of Y are called *roots* of Γ. In particular, the basis elements v_i are called *simple roots* of Γ. Note that if $w = \tau(v_i) \in Y$, then $-w = \tau\sigma_i v_i \in Y$.

Lemma a. *G is a finite group, and Y is a finite subset of $\{w \in W: \phi(w) = 1\}$.*

PROOF. Let $X = \{w \in W: \phi(w) = 1\}$. By (7) and (8), G permutes the elements of X. The action of G on X is faithful because $G \subseteq GL(U)$ and $\{v_1, \ldots, v_r\} \subseteq X$. Moreover, $Y \subseteq X$. Thus, to prove the lemma it suffices to show that X is finite. If $w = \sum v_i a_i \in X$, then $a_i \in \mathbb{Z}$ for $1 \leq i \leq r$, and $a_i^2 = \beta(u_i,w)^2 \leq \phi(u_i)\phi(w) = \phi(u_i)$ by (1) and (3). Therefore $|X| \leq \prod_{i=1}^r (2\phi(u_i) + 1)$. □

Lemma b. *Let* $\tau = \sigma_r\sigma_{r-1} \cdots \sigma_1 \in G$.

(i) *If* $u \in U$ *satisfies* $\tau u = u$, *then* $u = 0$.
(ii) *If* $w \in U$, *then there is an integer* k *such that* $0 \leq k < |G|$, *and* $\tau^k w$ *is not positive.*

PROOF. (i) By (2), it is sufficient to prove: if $\sigma_r \cdots \sigma_{i+1}\sigma_i u = u$, then $u \in v_i^\perp$ (hence $\sigma_i u = u$ and $\sigma_r \cdots \sigma_{i+1} u = u$). Denote $\rho = \sigma_r \cdots \sigma_{i+1}$, so that $\rho\sigma_i u = u$ by hypothesis. It follows from (3) that $\sigma_j u_i = u_i$ for $j \neq i$, and $\sigma_i u_i = u_i - 2v_i$. Thus, $\beta(u_i,u) = \beta(u_i,\rho\sigma_i u) = \beta(\rho\sigma_i\sigma_i\rho^{-1}u_i,\rho\sigma_i u) = \beta(\sigma_i\rho^{-1}u_i,u) = \beta(\sigma_i u_i,u) = \beta(u_i,u) - 2\beta(v_i,u)$. Consequently, $u \in v_i^\perp$, as required.

(ii) By Lemma a, G is finite, so that $\tau^h = id_G$ for a positive integer $h \leq |G|$. By (i), $h > 1$. It follows that $u = w + \tau w + \tau^2 w + \cdots + \tau^{h-1} w$ satisfies $\tau u = u$. By (i), $u = 0$. In particular, not all of the vectors $\tau^k w$, $0 \leq k < h$, are positive. □

Henceforth, assume that $U = U_\Gamma$ and $\phi = \phi_\Gamma$, where Γ is a quiver such that ϕ_Γ is positive definite. This assumption has the consequence that

$$\beta(v_i,v_j) \leq 0 \quad \text{for} \quad i \neq j. \tag{9}$$

It is easy to see that the previous hypotheses that were imposed on ϕ, together with the inequalities (9) imply that $\phi = \phi_\Gamma$ for a suitable quiver Γ. (See Exercise 1.)

Lemma c. *Assume that* $\phi = \phi_\Gamma$ *is positive definite. Let* $w \in W^+$ *satisfy* $\phi(w) = 1$. *For* $1 \leq i \leq r$:

(i) *if* $w = v_i$, *then* $\sigma_i w = -w$;
(ii) *if* $w \neq v_i$, *then* $\sigma_i w > 0$.

PROOF. The assertion (i) restates part of (4). To prove (ii), note that by (1), $|\beta(w,v_i)| \leq 1$. Thus, $2\beta(w,v_i) = 0, \pm 1$, or ± 2. If $\beta(w,v_i) \leq 0$, then $\sigma_i w \geq w > 0$. If $2\beta(w,v_i) = 2$, then $\beta(w,v_i)^2 = \phi(w)\phi(v_i)$. In this case, it follows from (1) and the hypotheses $w > 0$, $\phi(w) = 1$ that $w = v_i$. Finally, if $2\beta(w,v_i) = 1$, then $\sigma_i(w) = w - v_i > 0$. Otherwise, $w = \sum_{j \neq i} v_j a_j$ with all $a_j \geq 0$, and $1/2 = \beta(w,v_i) = \sum_{j \neq i} \beta(v_j,v_i)a_j \leq 0$ by (9). □

Corollary. *For each non-negative integer* m, *write* $m = jr + k$ *with* $0 \leq k < r$; *define* $\tau_m = \sigma_k \cdots \sigma_1(\sigma_r \cdots \sigma_1)^j$.

(i) *If* $w \in U^+$ *then* $\tau_m w \in U^+$ *and* $\tau_{m+1} w \notin U^+$ *for some* $m \geq 0$.
(ii) *Let* $w \in W^+$, $\phi(w) = 1$, *and suppose that* $\tau_m w \in U^+$, $\tau_{m+1} w \notin U^+$. *If* $m = jr + k$ *with* $0 \leq k < r$, *then* $w = \tau_m(v_{k+1})$.

These results follow directly from Lemmas b and c together with (7) and (8).

Theorem. *Let* Γ *be an acyclic quiver such that* ϕ_Γ *is positive definite. The dimension mapping* $(M,\phi) \to \text{Dim}(M,\phi)$ *defines a bijection between the isomorphism classes of indecomposable representations of* Γ *and the positive roots of* Γ.

PROOF. By Corollary 8.6 it can be assumed that $\Gamma = (V,E)$ is standardized, where $V = \{1, 2, \ldots, r\}$. Let $(M,\phi) \in \mathfrak{R}(\Gamma)$ be indecomposable. Denote $w = \text{Dim}(M,\phi)$. By Lemma 8.8b, $w \in W^+$. Let m be the integer satisfying the conditions of part (i) of the corollary, say $m = jr + k$ with $0 \leq k < r$. The minimality of m implies that $\sigma_{k'} \cdots \sigma_1(\sigma_r \cdots \sigma_1)^{j'} w$ is a positive root, but not a simple root whenever $j' < j$ or $j' = j$ and $k' < k$. Since Γ is standardized, it follows from Lemmas 8.6b, 8.8b, and Proposition 8.7 that $(M',\phi') = S_k^- \cdots S_1^-(S_r^- \cdots S_1^-)^j(M,\phi)$ is a well defined, indecomposable representation of $\Gamma' = \rho_k^- \cdots \rho_1^-(\rho_r^- \cdots \rho_1^-)^j\Gamma$, and $k + 1$ is a source in Γ'. Since $\sigma_{k+1}\text{Dim}(M',\phi') = \tau_{m+1}w$ is not positive, we infer from Lemma 8.8b that $(M',\phi') = (P^{(k+1)},0)$. Repeated use of Proposition 8.7 and Lemma 8.8b gets us back to (M,ϕ) and w: $(M,\phi) \cong (S_1^+ \cdots S_r^+)^j S_1^+ \cdots S_k^+(P^{(k+1)},0)$ and $w = \text{Dim}(M,\phi) = \tau_m^{-1}v_{k+1}$ is a positive root of Γ. Since m is uniquely determined by w, it follows that if (N,ψ) is another indecomposable representation of Γ such that $\text{Dim}(N,\psi) = \text{Dim}(M,\phi)$, then $(N,\psi) \cong (M,\phi)$. It remains to prove that every positive root w of Γ has the form $\text{Dim}(M,\phi)$ for some indecomposable representation (M,ϕ) of Γ. By Lemma a, $w \in W^+$ and $\phi(w) = 1$. Hence, according to part (ii) of the corollary, $w = \tau_m^{-1}(v_{k+1})$ for a minimal non-negative integer $m = jr + k, 0 \leq k < r$. By Lemma 8.8b, $w = \text{Dim}(M,\phi)$, where $(M,\phi) = (S_1^+ \cdots S_r^+)^j S_1^+ \cdots S_k^+(P^{(k+1)},0)$. The fact that (M,ϕ) is indecomposable follows from Proposition 8.7, using the minimality of m (as in the first part of the proof). □

EXERCISES

1. Let $U = \bigoplus_{i=1}^r v_i\mathbb{Q}$, $W = \bigoplus_{i=1}^r v_i\mathbb{Z}$, and suppose that $\phi\colon U \to \mathbb{Q}$ is a positive definite quadratic mapping such that $\phi(v_i) = 1$ for $1 \leq i \leq r$, and $\phi(W) \subseteq \mathbb{Z}$, as in Section 8.9. Assume that the bilinear mapping β corresponding to ϕ satisfies $\beta(v_i,v_j) \leq 0$ for all $i \neq j$. Prove that there is a quiver $\Gamma = (V,E)$ with $V = \{1,2, \ldots, r\}$ such that $\phi = \phi_\Gamma$. Hint. Prove that $\beta(v_i,v_j) > -1$ for all $i \neq j$.

2. Let Γ be a quiver such that Φ_Γ is positive definite. Prove that the set of all roots of Γ is $\{u \in W\colon \phi_\Gamma(u) = 1\}$. Thus, finding the roots of Γ is the same as determining the integral solutions of the Diophantine equation $\Phi_\Gamma(\mathbf{x}_1, \ldots, \mathbf{x}_r) = 1$. Hint. Use

Lemmas 8.9b and c, and Corollary 8.9 to prove that the set Y of all roots of Γ is $\{u \in W^+ \cup W^- : \phi_\Gamma(u) = 1\}$. Then use the inequality (9) of Section 8.9 to show that if $u \in W$ and $\phi_\Gamma(u) = 1$, then $u \in W^+ \cup W^-$.

3. (a) Let $\Gamma = (V,E)$, $V = \{1, 2, \ldots, n\}$, be the quiver with the diagram

A_n:

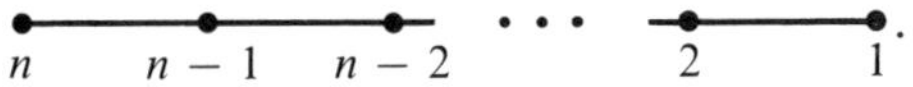

Prove that the positive roots of Γ in $U_\Gamma = \bigoplus_{i=1}^n v_i\mathbb{Q}$ are the vectors $v_i + v_{i+1} + \cdots + v_j$, where $1 \leq i \leq j \leq n$.

(b) Let Γ have the diagram

D_n:

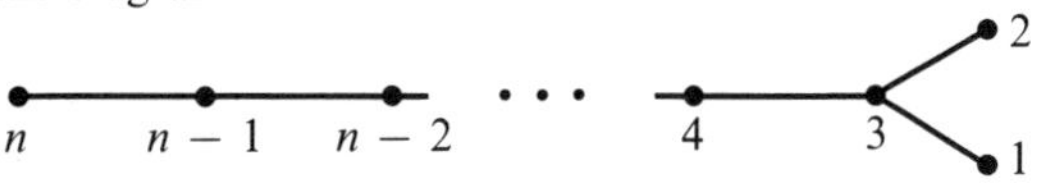

$(n \geq 4)$.

Prove that the positive roots of Γ in U_Γ are the vectors $v_1, v_2, \ldots, v_n$; $v_i + v_{i+1} + \cdots + v_j$, $1 \leq i < j \leq n$, $j \geq 3$; $v_1 + v_3 + \cdots + v_j$, $j \geq 3$; and $v_1 + v_2 + 2(v_3 + \cdots + v_i) + v_{i+1} + \cdots + v_j$, $3 \leq i < j \leq n$.

Hint. It suffices to verify that the vectors w in these lists satisfy $\phi_\Gamma(w) = 1$, and that the lists are closed under the reflections σ_k, $1 \leq k \leq n$.

4. Let Γ be a quiver whose unoriented diagram is the Dynkin diagram E_6. Prove that Γ has 36 positive roots, and find these vectors. Do the same thing for quivers whose diagrams are E_7 and E_8: there are 63 positive roots for E_7, and 120 for E_8.

Hint. The obvious embeddings of A_4, A_5, and D_5 in E_6 can be used to determine most of the positive roots. The rest can be found by evaluating $\sigma_i(w)$, where w is a known root such that $\beta_\Gamma(v_i,w) < 0$.

5. Let $B = B_\Gamma$ be the F-algebra that was defined in Section 8.1, corresponding to the quiver $\Gamma = (\{1, 2, 3\}, \{(12), (21), (23), (32), (31)\})$. Prove that B has finite representation type, and use the method that led to the proof of Theorem 8.9 to determine representatives of the 18 isomorphism classes of indecomposable B-modules.

Notes on Chapter 8

Our exposition of the representation theory of quivers is based on the paper [18] of Bernstein, Gel'fand, and Ponomarev. A minor innovation is the elementary proof of the fact that indecomposable quivers of finite representation type have the diagrams A_n, D_n, or E_k, based on the Examples 8.5a, b, c, and d. The usual proof due to Tits has conceptual advantages, but the rigorous presentation of Tits's argument (outlined in Exercise 2, Section 8.8) is difficult for readers with meager backgrounds in commutative ring theory.

There are several lines of research that start with the papers [34] and [35] of Gabriel and the work of Bernstein, Gel'fand, and Ponomarev cited above. Mention should be made of the work of Dlav and Ringel in [30] and [31] and of Ringel in [67]. These papers extend Gabriel's results to quivers with weighted edges (called Species) and treat representations that are defined by vector spaces over a division algebra. With this extended generality, it is possible to characterize certain finite dimensional algebras over an arbitrary field that have finite representation types. It turns out that all of the Dynkin diagrams arise in this context. Another problem that has been pursued by Nazarova is the construction of indecomposable representations for some quivers of infinite type. This turns out to be possible for indecomposable quivers whose diagrams are cycles, or have one of the forms shown in Corollaries 11.5a, b, c, or d. Such quivers are said to have tame representation type. The rest are called wild.

A more complete discussion of the current work on representation of quivers and related topics is given in Roiter's address at the 1978 International Congress of Mathematicians (see [69]).

CHAPTER 9
Tensor Products

Tensor products add a new dimension to the study of associative algebras. The part of the theory of algebras that does not involve tensor products is a purely additive subject; the tensor product introduces a multiplication.

The purpose of this chapter is to describe this powerful tool, and to show how it shapes the study of algebras. The results that are presented here, especially Proposition 9.2c, Corollary 9.3b, and Proposition 9.4b, will be used often in the later chapters. These theorems play an important part in the theory of central simple algebras, even though their proofs are easy. Indeed, none of the theorems in this chapter are deep. The machinery of tensor products is primarily a convenient formalism. Its usefulness illustrates the value of robust definitions and notation.

The last two sections of the chapter go beyond the standard materials in Sections 9.1 through 9.4. The induced modules defined in Section 9.5 are important in the theory of group representations. However, we only use this concept to prove half of Higman's Theorem on group algebras of finite representation type. The last section of this chapter provides a brief introduction to Morita equivalence of algebras. The simple properties of Morita equivalence that we prove here will be needed in the last part of Chapter 11.

9.1. Tensor Products of R-modules

Most applications of tensor products in the theory of algebras involve products over the commutative scalar ring R. In this section we outline the basic results on the tensor products of R-modules. The next section deals with R-algebras.

Definition. Let M and N be R-modules. A *tensor product* of M and N is an R-module $M \otimes N$, together with a bilinear mapping $M \times N \to M \otimes N$, denoted by $(u,v) \mapsto u \otimes v$ such that:

(i) $M \otimes N$ is generated as an R-module by $\{u \otimes v: u \in M, v \in N\}$;
(ii) (Universality) if $\Phi: M \times N \to P$ is a bilinear mapping of R-modules (that is, $\Phi(u,*): N \to P$ and $\Phi(*,v): M \to P$ are module homomorphisms for all $u \in M$ and $v \in N$), then there is a homomorphism $\phi: M \otimes N \to P$ such that $\phi(u \otimes v) = \Phi(u,v)$ for all $u \in M$ and $v \in N$.

This deceptively simple definition has many facets that are worth advertising. The hypothesis that $(u,v) \mapsto u \otimes v$ is bilinear implies four identities that are used repeatedly in dealing with tensor products:

$$u \otimes (v_1 a + v_2 b) = (u \otimes v_1)a + (u \otimes v_2)b; \tag{1}$$

$$(u_1 a + u_2 b) \otimes v = (u_1 \otimes v)a + (u_2 \otimes v)b; \tag{2}$$

$$u \otimes 0 = 0 \otimes v = 0; \tag{3}$$

$$ua \otimes v = (u \otimes v)a = u \otimes (va). \tag{4}$$

The assumption that $M \otimes N$ is generated by the elements $u \otimes v$ (called *rank one tensors*) leads via (4) to the conclusion that every element of $M \otimes N$ has a representation in the form $u_1 \otimes v_1 + \cdots + u_n \otimes v_n$. However, in most cases there is no natural canonical expression for the elements in a tensor product. This circumstance is sometimes a source of difficulty, but it is usually possible to avoid arguments involving arbitrary elements of $M \otimes N$ by using a simple observation.

Lemma a. *If ϕ and ψ are module homomorphisms of $M \otimes N$ to P such that $\phi(u \otimes v) = \psi(u \otimes v)$ for all $u \in M$ and $v \in N$, then $\phi = \psi$.*

In fact, $\operatorname{Ker}(\phi - \psi)$ is a submodule of $M \otimes N$ that includes all rank one tensors; hence $\operatorname{Ker}(\phi - \psi) = M \otimes N$ and $\phi = \psi$.

Corollary a. *The homomorphism ϕ in clause* (ii) *of the definition of $M \otimes N$ is unique.*

Corollary b. *If $M \otimes N$ and $M \otimes' N$ are tensor products of M and N, then there is a unique isomorphism $\phi: M \otimes N \to M \otimes' N$ such that $\phi(u \otimes v) = u \otimes' v$ for all $u \in M$ and $v \in N$.*

PROOF. The existence of a unique homomorphism $\phi: M \otimes N \to M \otimes' N$ such that $\phi(u \otimes v) = u \otimes' v$ is a consequence of the bilinearity of $\otimes'$ and the universality of $\otimes$. Similarly, there is a homomorphism $\psi: M \otimes' N \to M \otimes N$ such that $\psi(u \otimes' v) = u \otimes v$. It follows from Lemma a that $\phi\psi$ and $\psi\phi$ are identity mappings. □

The uniqueness result in Corollary b allows us to speak about "the" tensor product of M and N. This is a standard practice that we will follow.

Property (i) in the definition of tensor products can be used to show that a homomorphism is surjective: if $\phi: P \to M \otimes N$ is such that $\{u \otimes v: u \in M, v \in N\} \subseteq \operatorname{Im} \phi$, then ϕ is surjective.

Theorem. *The tensor product of two R-modules M and N exists.*

Here is an outline of the construction of $M \otimes N$. Let $F(M \times N)$ be the free R-module with the basis $M \times N$. Define $M \otimes N$ to be $F(M \times N)/G(M,N)$, where $G(M,N)$ is a suitably defined submodule of $F(M \times N)$. The rank one tensors are the cosets $u \otimes v = (u,v) + G(M,N)$, $u \in M$, $v \in N$. In order that $(u,v) \mapsto u \otimes v$ is bilinear, it is necessary and sufficient that $G(M,N)$ include all elements of the form

$$\begin{gathered}(u,v_1a + v_2b) - (u,v_1)a - (u,v_2)b\\ (u_1a + u_2b,v) - (u_1,v)a - (u_2,v)b.\end{gathered} \tag{5}$$

This motivates the definition: $G(M,N)$ is the submodule of $F(M \times N)$ that is generated by all elements of the form (5). Condition (i) of the definition is satisfied since the pairs (u,v) generate $F(M \times N)$, hence their images $u \otimes v$ generate $M \otimes N$. Let $\Phi: M \times N \to P$ be bilinear. Since $F(M \times N)$ is free on $M \times N$, there is an extension ψ of Φ to a homomorphism of $F(M \times N)$ to P. The bilinearity of Φ guarantees that all elements of the form (5) are in $\operatorname{Ker} \psi$. Therefore, ψ can be factored through the projection from $F(M \times N)$ to $M \otimes N$. This yields a homomorphism $\phi: M \otimes N \to P$ such that $\phi(u \otimes v) = \phi((u,v) + G(M,N)) = \psi((u,v)) = \Phi(u,v)$.

Lemma b. *Let M_1, M_2, N_1, and N_2 be R-modules. If $\phi: M_1 \to M_2$ and $\psi: N_1 \to N_2$ are module homomorphisms, then there is unique module homomorphism $\phi \otimes \psi: M_1 \otimes N_1 \to M_2 \otimes N_2$ such that for $u \in M_1$, $v \in N_1$,*

$$(\phi \otimes \psi)(u \otimes v) = \phi(u) \otimes \psi(v). \tag{6}$$

The following identities hold for these homomorphisms (with suitable domains and ranges).

(i) $(\phi' \otimes \psi')(\phi \otimes \psi) = \phi'\phi \otimes \psi'\psi$.
(ii) $id_M \otimes id_N = id_{M \otimes N}$.
(iii) $\phi \otimes (\psi_1 a + \psi_2 b) = (\phi \otimes \psi_1)a + (\phi \otimes \psi_2)b$,
$(\phi_1 a + \phi_2 b) \otimes \psi = (\phi_1 \otimes \psi)a + (\phi_2 \otimes \psi)b$.
(iv) $\phi \otimes 0 = 0 \otimes \psi = 0$.
(v) $(\phi a) \otimes \psi = \phi \otimes (\psi a) = (\phi \otimes \psi)a$.

Since ϕ and ψ are module homomorphisms, it follows from (1) and (2) that $(u,v) \mapsto \phi(u) \otimes \psi(v)$ is bilinear. The existence and uniqueness of $\phi \otimes \psi$ therefore follows from the definition of $M_1 \otimes N_1$. The identities (i)–(v) are obtained from Lemma a, using calculations that are based on (6), (1), (2), (3), and (4).

Proposition a. *Let M, M_1, M_2, N, N_1, N_2, and P be R-modules.*

(i) $M \otimes (N_1 \oplus N_2) \cong (M \otimes N_1) \oplus (M \otimes N_2)$ *by an isomorphism that maps* $u \otimes (v_1, v_2)$ *to* $(u \otimes v_1, u \otimes v_2)$.

(ii) $M \otimes (N \otimes P) \cong (M \otimes N) \otimes P$ *by an isomorphism that maps* $u \otimes (v \otimes w)$ *to* $(u \otimes v) \otimes w$.

(iii) $M \otimes N \cong N \otimes M$ *by an isomorphism that maps* $u \otimes v$ *to* $v \otimes u$.

(iv) $M \otimes R \cong M$ *and* $R \otimes M \cong M$ *by isomorphisms that map* $u \otimes a$ *and* $a \otimes u$ *to* ua.

The statement (i) can be proved using the identities of Lemma b and the characterization of the direct sum in terms of the projection and injection homomorphisms, or just by noting that the bilinear mapping $(u,(v_1,v_2)) \mapsto (u \otimes v_1, u \otimes v_2)$ fulfills the conditions that are required to make $(M \otimes N_1) \oplus (M \otimes N_2)$ a tensor product of M and $N_1 \oplus N_2$. Two applications of the universality condition produce a homomorphism $\phi: M \otimes (N \otimes P) \to (M \otimes N) \otimes P$ that satisfies $\phi(u \otimes (v \otimes w)) = (u \otimes v) \otimes w$. By symmetry, there is a homomorphism from $(M \otimes N) \otimes P$ to $M \otimes (N \otimes P)$ that sends $(u \otimes v) \otimes w$ to $u \otimes (v \otimes w)$. Thus, ϕ is an isomorphism, as in the proof of Corollary b. The proof of (iii) is similar, in fact easier. Finally, the bilinearity of $(u,a) \mapsto ua$ leads to a homomorphism ϕ of $M \otimes R$ to M that satisfies $\phi(u \otimes a) = ua$; and $\psi(u) = u \otimes 1$ is a homomorphism such that $\phi\psi(u) = u$ and $\psi\phi(u \otimes a) = ua \otimes 1 = u \otimes a$ by (4). Thus, ϕ is an isomorphism.

We now prove the fundamental exactness property of the tensor product.

Proposition b. *If* $M_1 \xrightarrow{\phi} M_2 \xrightarrow{\psi} M_3 \to 0$ *is an exact sequence of R-modules, then for any R-module N the sequence* $M_1 \otimes N \xrightarrow{\chi} M_2 \otimes N \xrightarrow{\theta} M_3 \otimes N \to 0$ *is exact, where* $\chi = \phi \otimes id_N$ *and* $\theta = \psi \otimes id_N$.

Proof. Plainly, $\operatorname{Im}\theta$ includes all rank one tensors, so that θ is surjective. Moreover, $\psi\phi = 0$ implies $\theta\chi = 0$; hence $\operatorname{Im}\chi \subseteq \operatorname{Ker}\theta$. Let π be the natural projection of $M_2 \otimes N$ to $M_2 \otimes N/\operatorname{Im}\chi$. Since $\psi^{-1}(0) \otimes N = \operatorname{Im}\phi \otimes N \subseteq \operatorname{Im}\chi = \operatorname{Ker}\pi$, the formula $\Phi(u_3,v) = \pi(\psi^{-1}u_3 \otimes v)$ describes a well defined bilinear mapping of $M_3 \times N$ to P. Hence, there is a homomorphism $\lambda: M_3 \otimes N \to P$ such that $\lambda(u_3 \otimes v) = \pi(\psi^{-1}u_3 \otimes v)$. In particular, $\lambda(\psi(u_2) \otimes v) = \pi(u_2 \otimes v)$, so that $\lambda\theta = \pi$ by Lemma a. Therefore, $\operatorname{Ker}\theta \subseteq \operatorname{Ker}\pi = \operatorname{Im}\chi$. □

If $0 \to M_1 \to M_2 \to M_3 \to 0$ is a short exact sequence, then in general $0 \to M_1 \otimes N \to M_2 \otimes N$ is not exact. (See Exercise 4.) In one important case, full exactness is preserved by the tensor product.

Corollary c. *If* $0 \to M_1 \xrightarrow{\phi} M_2 \xrightarrow{\psi} M_3 \to 0$ *is split exact, then* $0 \to M_1 \otimes N \xrightarrow{\chi} M_2 \otimes N \xrightarrow{\theta} M_3 \otimes N \to 0$ *is exact.*

In fact, if $\tau: M_2 \to M_1$ is such that $\tau\phi = id_{M_1}$, then $(\tau \otimes id_N)(\phi \otimes id_N) = id_{M_1 \otimes N}$, so that χ is injective.

An important special case of Corollary c is the result that the tensor product of vector spaces (that is, F-modules) preserves exactness. Indeed, every short exact sequence of vector spaces splits.

Proposition c. *If M and N are F-spaces with bases $\{u_i: i \in I\}$ and $\{v_j: j \in J\}$, then $\{u_i \otimes v_j: (i,j) \in I \times J\}$ is a basis of $M \otimes N$. In particular,* $\dim M \otimes N = (\dim M)(\dim N)$.

Proof. If $M_1 < M$ and $N_1 < N$, then by Corollary c the inclusion mappings induce an injective homomorphism $M_1 \otimes N_1 \to M \otimes N$. Therefore, since the rank one tensors span $M \otimes N$ and linear independence is defined in terms of finite sets, we can assume that $I = \{1,\ldots,m\}$ and $J = \{1,\ldots,n\}$ are finite. Consequently,

$$M \otimes N = (\textstyle\bigoplus_{i=1}^m u_iF) \otimes (\bigoplus_{j=1}^n v_jF) \cong \bigoplus_{i,j}(u_iF \otimes v_jF) \cong \bigoplus_{i,j}(u_i \otimes v_j)F$$

by Proposition a. □

The proof of this result illustrates a defect in the standard notation for tensor products. The expression $u_i \otimes v_j$ depends not only on the elements u_i and v_j, but also on the ambient modules N and M. For R-modules, the fact that the elements $u_i \otimes v_j$ are distinct, non-zero, and linearly independent in $M_1 \otimes N_1$ would not guarantee that the corresponding elements (also denoted by $u_i \otimes v_j$) in $M \otimes N$ retain these properties. In the case under consideration, we are saved in the proposition by the exactness of the tensor product over a field, that is, the fact that $M_1 \otimes N_1 \to M \otimes N$ is injective.

In a few sections of this chapter and Chapter 10 it will be necessary to deal with tensor products of A-modules, where A is an algebra that may not be commutative. The definition of such products is slightly more complicated than the definition of the tensor product of two R-modules.

Let A be an R-algebra. Suppose that M is a right A-module, and N is a left A-module. Then M and N can be viewed as R-modules by restricting scalar operations to R. A bilinear mapping Φ of $M \times N$ to an R-module P is called *balanced* if $\Phi(ux,v) = \Phi(u,xv)$ for all $u \in M$, $v \in N$, and $x \in A$. The *tensor product over A* of M and N is an R-module $M \otimes_A N$, together with a balanced, bilinear mapping $M \times N \to M \otimes_A N$ (denoted $(u,v) \mapsto u \otimes v$) such that $M \otimes_A N$ is generated as an R-module by $\{u \otimes v: u \in M, v \in N\}$; and if $\Phi: M \times N \to P$ is a balanced, bilinear mapping, then there is an R-module homomorphism $\phi: M \otimes_A N \to P$ satisfying $\phi(u \otimes v) = \Phi(u,v)$ for all $u \in M$ and $v \in N$.

In general, the tensor product $M \bigotimes_A N$ cannot be considered as an A-module. An exception to this assertion occurs when M or N is a bimodule, as we will see in Section 9.5. In particular, this is the case when A is commutative, since the right scalar operation can then be used to define a left scalar operation and vice versa. However, if A is a commutative R-algebra that properly contains R, then $M \bigotimes_A N$ and $M \bigotimes_R N$ are generally different. (When M and N are modules over two or more commutative rings, we will often write $M \bigotimes_R N$ instead of $M \bigotimes N$ to avoid confusion.)

Most of the results given in this section generalize very easily to $M \bigotimes_A N$. Exceptions to this rule are the associative and commutative laws in Proposition a: these isomorphisms make sense (and are valid) only for bimodules. The identities (1)–(4) are satisfied in $M \bigotimes_A N$, provided $a \in R$ and $b \in R$. The fact that $(u,v) \mapsto u \bigotimes v$ is balanced gives a stronger version of (4):

$$(ux) \bigotimes v = u \bigotimes (xv) \qquad \text{for } u \in M, v \in N, x \in A. \tag{4'}$$

Exercises

1. Establish the following isomorphisms of $\mathbb{Z}$-modules:
   (a) $\mathbb{Q} \bigotimes \mathbb{Q} \cong \mathbb{Q}$;
   (b) $\mathbb{Q} \bigotimes (\mathbb{Q}/\mathbb{Z}) \cong \mathbb{Q}/\mathbb{Z} \bigotimes \mathbb{Q}/\mathbb{Z}$;
   (c) for any abelian group M and $n \geq 1$, $M \bigotimes (\mathbb{Z}/n\mathbb{Z}) \cong M/nM$;
   (d) $(\mathbb{Z}/m\mathbb{Z}) \bigotimes (\mathbb{Z}/n\mathbb{Z}) \cong \mathbb{Z}/k\mathbb{Z}$, where k is the greatest common divisor of m and n.

2. Establish the following isomorphisms.
   (a) For any field F, $F[\mathbf{x}] \bigotimes_F F[\mathbf{x}] \cong F[\mathbf{x},\mathbf{y}]$.
   (b) $\mathbb{Q} \bigotimes_{\mathbb{Z}} M_n(\mathbb{Z}) \cong M_n(\mathbb{Q})$.

3. Prove that not all elements of the tensor product $F[\mathbf{x}] \bigotimes_F F[\mathbf{x}]$ have rank one.

4. (a) Show that when the exact sequence $0 \to \mathbb{Z} \to \mathbb{Q} \to \mathbb{Q}/\mathbb{Z} \to 0$ is tensored over $\mathbb{Z}$ with $\mathbb{Q}/\mathbb{Z}$, the resulting sequence is no longer exact.
   (b) Define $\phi\colon \mathbb{Z}/2\mathbb{Z} \to \mathbb{Z}/4\mathbb{Z}$ by $\phi(1 + 2\mathbb{Z}) = 2 + 4\mathbb{Z}$. Prove that ϕ is an injective homomorphism, but $id_{\mathbb{Z}/2\mathbb{Z}} \bigotimes \phi\colon \mathbb{Z}/2\mathbb{Z} \bigotimes \mathbb{Z}/2\mathbb{Z} \to \mathbb{Z}/2\mathbb{Z} \bigotimes \mathbb{Z}/4\mathbb{Z}$ is the zero map, whereas $\mathbb{Z}/2\mathbb{Z} \bigotimes \mathbb{Z}/2\mathbb{Z} \cong \mathbb{Z}/2\mathbb{Z} \cong \mathbb{Z}/2\mathbb{Z} \bigotimes \mathbb{Z}/4\mathbb{Z}$.

5. Prove that if M is a right A-module and N is a left A-module, then the tensor product $M \bigotimes_A N$ exists and is unique (to isomorphism). Hint. Modify the proof of the theorem by letting $G(M,N)$ be generated by all elements that have one of the forms $(u, v_1a + v_2b) - (u,v_1)a - (u,v_2)b$, $(u_1a + u_2b, v) - (u_1,v)a - (u_2,v)b$, $(ux,v) - (u,xv)$, where $u, u_1, u_2 \in M$, $v, v_1, v_2 \in N$, $a, b \in R$, and $x \in A$.

6. Let the R-module M be a union of a directed family of submodules, say $M = \bigcup_{i \in I} M_i$, where $i, j \in I$ implies the existence of $k \in I$ such that $M_i \cup M_j \subseteq M_k$. For each $i \in I$, let $\kappa_i\colon M_i \to M$ be the inclusion homomorphism. Use the proof of the existence of $M \bigotimes N$ to show that $M \bigotimes N$ is the union of the submodules $\operatorname{Im}(\kappa_i \bigotimes id_N)$. Deduce that if $\{M_j : j \in J\}$ is a set of R-modules and N is an R-module, then $(\bigoplus_{j \in J} M_j) \bigotimes N \cong \bigoplus_{j \in J} M_j \bigotimes N$.

9.2. Tensor Products of Algebras

If A and B are R-algebras, then they are also R-modules. Therefore, we can form the tensor product $A \otimes B$. The results of this section show that $A \otimes B$ is an R-algebra with a suitably defined multiplication, and that this tensor product algebra has an internal characterization in terms of subalgebras.

Proposition a. *If A and B are R-algebras, then there is a multiplication operation on $A \otimes B$ that satisfies*

$$(x_1 \otimes y_1)(x_2 \otimes y_2) = x_1 x_2 \otimes y_1 y_2. \tag{1}$$

The multiplication is associative, and $1_A \otimes 1_B = 1_{A \otimes B}$.

PROOF. For $x_1 \in A$ and $y_1 \in B$, let λ_{x_1} and λ_{y_1} be the left multiplication endomorphisms of A and B corresponding to x_1 and y_1. By Lemma 9.1b, $\lambda_{x_1} \otimes \lambda_{y_1} \in \mathbf{E}_R(A \otimes B)$ satisfies $(\lambda_{x_1} \otimes \lambda_{y_1})(x_2 \otimes y_2) = x_1 x_2 \otimes y_1 y_2$. Moreover, $(x_1, y_1) \mapsto \lambda_{x_1} \otimes \lambda_{y_1}$ is a bilinear mapping of $A \times B$ to $\mathbf{E}_R(A \otimes B)$. Thus, there is an R-module homomorphism $\phi: A \otimes B \to \mathbf{E}_R(A \otimes B)$ such that $\phi(x_1 \otimes y_1) = \lambda_{x_1} \otimes \lambda_{y_1}$. Define $(A \otimes B) \times (A \otimes B) \to A \otimes B$ by $(z,w) \mapsto zw = \phi(z)(w)$. Since ϕ is a homomorphism of R-modules, and $\phi(z) \in \mathbf{E}_R(A \otimes B)$, the mapping is bilinear, that is, a multiplication operation on $A \otimes B$. By construction, $(x_1 \otimes y_1)(x_2 \otimes y_2) = \phi(x_1 \otimes y_1)(x_2 \otimes y_2) = (\lambda_{x_1} \otimes \lambda_{y_1})(x_2 \otimes y_2) = x_1 x_2 \otimes y_1 y_2$; that is, (1) is satisfied. It follows easily from (1) and Lemma 1.2 that the multiplication is associative. Moreover, $1_A \otimes 1_B$ is the unity element of $A \otimes B$ by (1) and Lemma 9.1a. □

Corollary a. *Let A, B, and C be R-algebras.*

(i) $(A \dotplus B) \otimes C \cong (A \otimes C) \dotplus (B \otimes C)$.
(ii) $(A \otimes B) \otimes C \cong A \otimes (B \otimes C)$.
(iii) $A \otimes B \cong B \otimes A$.
(iv) $A \otimes R \cong R \otimes A \cong A$.

In this corollary, $\cong$ denotes the isomorphism relation in the category of R-algebras. The corresponding isomorphisms were obtained for R-modules in Proposition 9.1b. The proofs that the mappings described in Proposition 9.1b involve applications of the formula (1), and appeals to Lemma 9.1a. This task is the content of Exercise 1.

Lemma a. *The mappings $\kappa_A: A \to A \otimes B$ and $\kappa_B: B \to A \otimes B$ defined by $\kappa_A(x) = x \otimes 1_B$ and $\kappa_B(y) = 1_A \otimes y$ are algebra homomorphisms such that:*

(i) *$\kappa_A(A) \cup \kappa_B(B)$ generates $A \otimes B$ as an R-algebra;*
(ii) *$\kappa_A(x)\kappa_B(y) = \kappa_B(y)\kappa_A(x)$ for all $x \in A$ and $y \in B$.*

If A and B are F-algebras, then κ_A and κ_B are injective; moreover, if $\{x_i : i \in I\}$ is a basis of A and $\{y_j : j \in J\}$ is a basis of B, then $\{\kappa_A(x_i)\kappa_B(y_j) : (i,j) \in I \times J\}$ is a basis of $A \otimes B$.

PROOF. The bilinearity of $\otimes$ together with (1) imply that κ_A and κ_B are algebra homomorphisms. By (1), $\kappa_A(x)\kappa_B(y) = x \otimes y = \kappa_B(y)\kappa_A(x)$, which yields (i) and (ii). The last statement of the lemma is a consequence of Proposition 9.1c. □

If X is a subset of the algebra A, then the *centralizer of X in A* is defined to be

$$\mathbf{C}_A(X) = \{y \in A : xy = yx \quad \text{for all} \quad x \in X\}.$$

This familiar concept will be used frequently in the following chapters. It is convenient to record some obvious consequences of the definition.

Lemma b. *Let X and Y be subsets of the algebra A, and suppose that B is a subalgebra of A.*

(i) *$\mathbf{C}_A(X)$ is a subalgebra of A with $\mathbf{Z}(A) \subseteq \mathbf{C}_A(X)$.*
(ii) *If $X \subseteq Y$, then $\mathbf{C}_A(Y) \subseteq \mathbf{C}_A(X)$.*
(iii) *$X \subseteq \mathbf{C}_A(Y)$ if and only if $Y \subseteq \mathbf{C}_A(X)$; in particular, $X \subseteq \mathbf{C}_A(\mathbf{C}_A(X))$.*
(iv) *$B \cap \mathbf{C}_A(B) = \mathbf{Z}(B)$.*
(v) *$\mathbf{C}_A(X) = A$ if and only if $X \subseteq \mathbf{Z}(A)$.*

We now prove a universality property of tensor products of algebras. This result leads to the characterization theorem.

Proposition b. *Let A, B, and C be R-algebras. If $\phi: B \to A$ and $\psi: C \to A$ are algebra homomorphisms such that $\psi(C) \subseteq \mathbf{C}_A(\phi(B))$, then there is a unique algebra homomorphism $\theta: B \otimes C \to A$ that satisfies*

$$\theta(x \otimes y) = \phi(x)\psi(y) \tag{2}$$

for $x \in B$ and $y \in C$. In particular, $\phi = \theta\kappa_B$ and $\psi = \theta\kappa_C$.

PROOF. Since ϕ and ψ are R-module homomorphisms, the mapping $(x,y) \mapsto \phi(x)\psi(y)$ is bilinear. Thus, there is an R-module homomorphism that satisfies (2). By (1) and (2), $\theta((x_1 \otimes y_1)(x_2 \otimes y_2)) = \phi(x_1x_2)\psi(y_1y_2) = \phi(x_1)\phi(x_2)\psi(y_1)\psi(y_2) = \phi(x_1)\psi(y_1)\phi(x_2)\psi(y_2) = \theta(x_1 \otimes y_1)\theta(x_2 \otimes y_2)$, since $\psi(C) \subseteq \mathbf{C}_A(\phi(B))$. Therefore, θ is an algebra homomorphism. □

Corollary b. *If $\phi: B \to B_1$ and $\psi: C \to C_1$ are algebra homomorphisms, then $\phi \otimes \psi: B \otimes C \to B_1 \otimes C_1$ is an algebra homomorphism. If B, B_1, C, and C_1 are F-algebras, and if ϕ and ψ are injective, then $\phi \otimes \psi$ is injective.*

The corollary follows from Lemma a and Proposition b.

Proposition c. *If A, B, and C are F-algebras, then $B \otimes C \cong A$ if and only if A contains subalgebras B' and C' such that*

(i) *$B' \cong B$ and $C' \cong C$ as F-algebras,*
(ii) *$C' \subseteq \mathbf{C}_A(B')$, and*
(iii) *there exist bases $\{x_i : i \in I\}$ of B' and $\{y_j : j \in J\}$ of C' such that $\{x_i y_j : (i,j) \in I \times J\}$ is a basis of A.*

If A is finite dimensional, then (iii) *can be replaced by*

(iv) *A is generated as an F-algebra by $B' \cup C'$ and* $\dim A = (\dim B)(\dim C)$.

PROOF. By Lemma a, the conditions (i), (ii), (iii), and (iv) are necessary. Assume that there are subalgebras B' and C' of A that satisfy (i) and (ii). Let $\phi: B \to B'$ and $\psi: C \to C'$ be the isomorphisms that are promised by (i). By (ii) and Proposition b, there is an algebra homomorphism $\theta: B \otimes C \to A$ such that $\theta(x \otimes y) = \phi(x)\psi(y)$ for $x \in B$ and $y \in C$. By Lemma a and (iii), θ maps a basis of $B \otimes C$ bijectively to a basis of A; so that θ is an isomorphism in this case. If A is finite dimensional, and (iv) is satisfied, then θ is surjective because $\theta(B \otimes C)$ is a subalgebra of A that includes the generating set $B' \cup C'$; and θ is injective because $\dim A = (\dim B)(\dim C) = \dim B \otimes C$ by Lemma a. □

EXAMPLE. Let G_1 and G_2 be finite groups, and suppose that $G = G_1 \times G_2$ is the product of G_1 and G_2. If F is any field, then $FG \cong FG_1 \otimes FG_2$.

PROOF. Write $A = FG$. Consider G_1 and G_2 as subgroups of G, so that each element of G has a unique representation in the form xy with $x \in G_1$, $y \in G_2$. Let A_1 be the subspace that is spanned by G_1, and let A_2 be the subspace that is spanned by G_2. It is clear that A_1 and A_2 are subalgebras of A such that $A_1 \cong FG_1$ and $A_2 \cong FG_2$. Since $xy = yx$ for $x \in G_1$ and $y \in G_2$, we see that $A_1 \subseteq \mathbf{C}_A(A_2)$. Plainly, $A_1 \cup A_2$ generates A as an F-algebra. Finally, $\dim A = |G| = |G_1||G_2| = (\dim A_1)(\dim A_2)$. Thus, by Proposition c, $FG = A \cong A_1 \otimes A_2 \cong FG_1 \otimes FG_2$. □

EXERCISES

1. Complete the proof of Corollary a.

2. Let $\mathbb{H} = \left(\frac{-1,-1}{\mathbb{R}}\right)$ be Hamilton's quaternion algebra. Prove that $\mathbb{H} \otimes \mathbb{C} \cong M_2(\mathbb{C})$, considered as $\mathbb{R}$-algebras. Hint. Consider the $\mathbb{R}$-subalgebra of $M_2(\mathbb{C})$ that consists of all matrices that have the form

$$\begin{bmatrix} a & b \\ -\bar{b} & \bar{a} \end{bmatrix}$$

with $a, b \in \mathbb{C}$, and $\bar{a}$ is the complex conjugate of a. The general results of Chapter 13 will make this exercise a triviality.

3. If $\alpha = [a_{ij}] \in M_n(F)$ and $\beta = [b_{kl}] \in M_m(F)$, then the *tensor product of the matrices* α *and* β is the matrix $\alpha \otimes \beta$, whose rows and columns are indexed by the pairs (i,k) with $1 \le i \le n$ and $1 \le k \le m$, such that the element in row (i,k) and column (j,l) is $a_{ij}b_{kl}$. Prove that the mapping $(\alpha,\beta) \mapsto \alpha \otimes \beta$ induces an F-algebra isomorphism of $M_n(F) \otimes M_m(F)$ to $M_{nm}(F)$.

4. Let $\phi: A \to M_n(F)$ and $\psi: A \to M_m(F)$ be representations of the F-algebra A. Define $\theta: A \to M_{nm}(F)$ by $\theta(x) = \phi(x) \otimes \psi(x)$ (where $\otimes$ denotes the matrix tensor product that is defined in Exercise 3). Prove that θ is a representation of A whose character χ_θ is the point-wise product $\chi_\phi\chi_\psi$. (See Exercise 1 in Section 5.6.) Deduce that $\mathbf{X}(A)$ is a commutative ring.

5. Let the characteristic of the field F be different from 2. Prove that if $a, b, c \in F^\circ$, then $\left(\frac{a,b}{F}\right) \otimes \left(\frac{a,c}{F}\right) \cong \left(\frac{a,bc}{F}\right) \otimes M_2(F)$. Hint. Let $\{1, \mathbf{i}, \mathbf{j}, \mathbf{k}\}$ and $\{1, \mathbf{i}', \mathbf{j}', \mathbf{k}'\}$ be the standard bases of $\left(\frac{a,b}{F}\right)$ and $\left(\frac{a,c}{F}\right)$ respectively. Denote $1 = 1 \otimes 1$, $\mathbf{i}'' = \mathbf{i} \otimes 1$, $\mathbf{j}'' = \mathbf{j} \otimes \mathbf{j}'$, $\mathbf{k}'' = \mathbf{k} \otimes \mathbf{j}'$, $\mathbf{i}''' = 1 \otimes \mathbf{j}'$, $\mathbf{j}''' = \mathbf{i} \otimes \mathbf{k}'$, and $\mathbf{k}''' = (\mathbf{i} \otimes \mathbf{i}')(-c)$. Show that $1, \mathbf{i}'', \mathbf{j}'', \mathbf{k}''$ is a basis of a subalgebra B of $A = \left(\frac{a,b}{F}\right) \otimes \left(\frac{a,c}{F}\right)$ such that $B \cong \left(\frac{a,bc}{F}\right)$, and $1, \mathbf{i}''', \mathbf{j}''', \mathbf{k}'''$ is a basis of a subalgebra C of A such that $C \cong \left(\frac{c,-a^2c}{F}\right)$. Show that $A \cong B \otimes C$. Deduce from Exercise 2 of Section 1.7 that $C \cong M_2(F)$.

9.3. Tensor Products of Modules over Algebras

Let A and B be R-algebras. If M is a right A-module and N is a right B-module, then M and N are also R-modules, and the tensor product $M \otimes N$ can be endowed with the structure of an $A \otimes B$-module. This section explores the homological aspects of this construction.

Lemma. *If M is a right A-module and N is a right B-module, then $M \otimes N$ is a right $A \otimes B$-module with scalar operations that satisfy*

$$(u \otimes v)(x \otimes y) = ux \otimes vy \tag{1}$$

for all $u \in M$, $v \in N$, $x \in A$, and $y \in B$.

The proof that there is a right module operation of $A \otimes B$ on $M \otimes N$ satisfying (1) can be copied almost verbatim from the proof of Proposition 9.2a. It is a useful exercise to give the details of this argument.

Proposition. *Let M_1 and M_2 be right A-modules, and suppose that N_1 and N_2 are right B-modules.*

(i) *If* $\phi \in \mathrm{Hom}_A(M_1,M_2)$ *and* $\psi \in \mathrm{Hom}_B(N_1,N_2)$, *then* $\phi \otimes \psi \in \mathrm{Hom}_{A\otimes B}(M_1 \otimes N_1, M_2 \otimes N_2)$.

(ii) *The mapping* $(\phi,\psi) \mapsto \phi \otimes \psi$ *induces an R-module homomorphism* θ: $\mathrm{Hom}_A(M_1,M_2) \otimes \mathrm{Hom}_B(N_1,N_2) \to \mathrm{Hom}_{A\otimes B}(M_1 \otimes N_1, M_2 \otimes N_2)$.

(iii) θ: $\mathbf{E}_A(M_1) \otimes \mathbf{E}_B(N_1) \to \mathbf{E}_{A\otimes B}(M_1 \otimes N_1)$ *is an algebra homomorphism.*

PROOF. Since ϕ and ψ are module homomorphisms, $(\phi \otimes \psi)((u \otimes v)(x \otimes y)) = \phi(ux) \otimes \psi(vy) = \phi(u)x \otimes \psi(v)y = ((\phi \otimes \psi)(u \otimes v))(x \otimes y)$. The assertion (i) therefore follows from Lemma 9.1a. Since $(\phi,\psi) \mapsto \phi \otimes \psi$ is bilinear, the existence of θ is a consequence of the universality property of tensor products. To avoid confusion in the proof of (iii) it is helpful to denote a rank one tensor in $\mathbf{E}_A(M_1) \otimes \mathbf{E}_B(N_1)$ by $\phi \otimes' \psi$. Thus, by definition, $\theta(\phi \otimes' \psi) = \phi \otimes \psi$. Consequently, by 9.2(1), $\theta((\phi_1 \otimes' \psi_1)(\phi_2 \otimes' \psi_2)) = \theta(\phi_1\phi_2 \otimes' \psi_1\psi_2) = \phi_1\phi_2 \otimes \psi_1\psi_2 = (\phi_1 \otimes \psi_1)(\phi_2 \otimes \psi_2) = \theta(\phi_1 \otimes' \psi_1)\theta(\phi_2 \otimes' \psi_2)$. It follows that θ is an algebra homomorphism. □

In general, the homomorphism θ is neither injective nor surjective (Exercise 3). However, in the cases that interest us most, θ is an isomorphism.

EXAMPLE. θ maps $\mathrm{Hom}_A(A,M) \otimes \mathrm{Hom}_B(B,N)$ isomorphically to $\mathrm{Hom}_{A\otimes B}(A \otimes B, M \otimes N)$. This is to be expected because $\mathrm{Hom}_A(A,M) \cong M$, $\mathrm{Hom}_B(B,N) \cong N$, and $\mathrm{Hom}_{A\otimes B}(A \otimes B, M \otimes N) \cong M \otimes N$ by the Exercise of Section 1.3. However, a bit of care is needed. Define σ_A: $\mathrm{Hom}_A(A,M) \to M$ by $\sigma_A(\phi) = \phi(1_A)$. It follows from Lemma 6.4 that σ_A is an isomorphism. A straightforward computation (using Lemma 9.1a) shows that the diagram

$$\begin{array}{ccc} \mathrm{Hom}_A(A,M) \otimes \mathrm{Hom}_B(B,N) & \xrightarrow{\theta} & \mathrm{Hom}_{A\otimes B}(A \otimes B, M \otimes N) \\ & {\scriptstyle \sigma_A \otimes \sigma_B} \searrow \quad M \otimes N \quad \swarrow {\scriptstyle \sigma_{A\otimes B}} & \end{array}$$

is commutative. Therefore, θ is an isomorphism.

Corollary a. *Suppose that* M_1 *and* M_2 *are right A-modules,* N_1 *and* N_2 *are right B-modules, and* M_1 *and* N_1 *are free with finite bases. The homorphism* θ: $\mathrm{Hom}_A(M_1,M_2) \otimes \mathrm{Hom}_B(N_1,N_2) \to \mathrm{Hom}_{A\otimes B}(M_1 \otimes N_1, M_2 \otimes N_2)$ is *an isomorphism. In particular,* $\mathbf{E}_A(M_1) \otimes \mathbf{E}_B(N_1) \cong \mathbf{E}_{A\otimes B}(M_1 \otimes N_1)$.

PROOF. Since M_1 and N_1 are free, there exist natural numbers m and n such that $M_1 \cong \bigoplus m\, A$ and $N_1 = \bigoplus n\, B$. By the example, Proposition 9.1a, and the additivity of the Hom functor, $\mathrm{Hom}_A(M_1,M_2) \otimes \mathrm{Hom}_B(N_1,N_2) \cong \mathrm{Hom}_A(\bigoplus m\ A,M_2) \otimes \mathrm{Hom}_B(\bigoplus n\ B,N_2) \cong (\bigoplus m\ \mathrm{Hom}_A(A,M_2)) \otimes (\bigoplus n\ \mathrm{Hom}_B(B,N_2)) \cong \bigoplus mn\, (\mathrm{Hom}_A(A,M_2) \otimes \mathrm{Hom}_B(B,N_2)) \cong \bigoplus mn\, \mathrm{Hom}_{A\otimes B}(A \otimes B, M_2 \otimes N_2) \cong \mathrm{Hom}_{A\otimes B}(\bigoplus mn\ A \otimes B, M_2 \otimes N_2) \cong \mathrm{Hom}_{A\otimes B}$

$(M_1 \otimes N_1, M_2 \otimes N_2)$. The ingredient that is missing from this proof is a routine verification that θ defines the composite isomorphism. An alternative inductive proof is outlined in Exercise 4. □

Corollary b. $M_m(A) \otimes M_n(B) \cong M_{mn}(A \otimes B)$.

This result is a restatement of the last part of Corollary a, using the isomorphism $\mathbf{E}_A(\bigoplus m\, A) \cong M_m(A)$ that was proved in Corollary 3.4b.

EXERCISES

1. Prove the lemma.

2. Let P be a free right A-module with the basis $\{u_i : i \in I\}$, and let Q be a free right B-module with the basis $\{v_j : j \in J\}$. Assume that $A \otimes B$ is a non-trivial algebra. Prove that $P \otimes Q$ is a free $A \otimes B$-module with the basis $\{u_i \otimes v_j : (i,j) \in I \times J\}$.

3. Let $A = B = \mathbb{Z}$. Prove the following statements concerning the homomorphism θ that was defined in the lemma.
   (a) $\theta: \mathrm{Hom}_{\mathbb{Z}}(\mathbb{Q},\mathbb{Z}) \otimes \mathrm{Hom}_{\mathbb{Z}}(\mathbb{Q},\mathbb{Q}) \to \mathrm{Hom}_{\mathbb{Z}}(\mathbb{Q} \otimes \mathbb{Q}, \mathbb{Z} \otimes \mathbb{Q})$ is not surjective.
   (b) $\theta: \mathbf{E}_{\mathbb{Z}}(\mathbb{Q}/\mathbb{Z}) \otimes \mathbf{E}_{\mathbb{Z}}(\mathbb{Q}) \to \mathbf{E}_{\mathbb{Z}}((\mathbb{Q}/\mathbb{Z}) \otimes \mathbb{Q})$ is not injective.

4. Let M_1', M_1'', and M_2 be right A-modules, and suppose that N_1 and N_2 are right B-modules. Prove that if the homomorphisms

$$\theta': \mathrm{Hom}_A(M_1',M_2) \otimes \mathrm{Hom}_B(N_1,N_2) \to \mathrm{Hom}_{A\otimes B}(M_1' \otimes N_1, M_2 \otimes N_2)$$

$$\theta'': \mathrm{Hom}_A(M_1'',M_2) \otimes \mathrm{Hom}_B(N_1,N_2) \to \mathrm{Hom}_{A\otimes B}(M_1'' \otimes N_1, M_2 \otimes N_2)$$

are isomorphisms, then so is

$$\theta: \mathrm{Hom}_A(M_1' \oplus M_1'',M_2) \otimes \mathrm{Hom}_B(N_1,N_2)$$
$$\to \mathrm{Hom}_{A\otimes B}((M_1' \oplus M_1'') \otimes N_1, M_2 \otimes N_2).$$

Use this result (and its right analog) to give an inductive proof of the corollary.

5. Let S and T be contravariant functors from R-modules to abelian groups such that if $N \to M \to P \to 0$ is exact, then so are $0 \to S(P) \to S(M) \to S(N)$ and $0 \to T(P) \to T(M) \to T(N)$. Let $\theta: S \to T$ be a natural transformation of functors such that $\theta_M: S(M) \to T(M)$ is an isomorphism whenever M is a finitely generated free module. Prove that θ_M is an isomorphism for all finitely presented modules M. Use this result to show that the homomorphism θ of the proposition is an isomorphism under the hypothesis that M_1 and N_1 are finitely presented.

9.4. Scalar Extensions

Tensor products play another role in the study of algebras. They are used to extend the domain of scalars from R to a commutative ring that contains

R as a subring. More generally, it is possible to pass from an R-algebra to an S-algebra, where S is any commutative R-algebra.

Proposition a. *Let A be an R-algebra. If S is a commutative R-algebra, then $A \otimes S$ is an S-algebra whose product satisfies*

$$(x \otimes c)(y \otimes d) = xy \otimes cd \tag{1}$$

for all $x, y \in A$ and $c, d \in S$. The scalar operations by elements of S on $A \otimes S$ is defined by

$$zc = z(1_A \otimes c) \tag{2}$$

for all $z \in A \otimes S$ and $c \in S$.

PROOF. By Proposition 9.2a, $A \otimes S$ is an R-algebra, and $\kappa_S\colon S \to A \otimes S$ (defined by $\kappa_S(c) = 1_A \otimes c$) is an R-algebra homomorphism such that $\kappa_A(A) \subseteq \mathbf{C}_{A\otimes S}(\kappa_S(S))$. Also, $\kappa_S(S) \subseteq \mathbf{C}_{A\otimes S}(\kappa_S(S))$ because S is commutative. Therefore, $A \otimes S = \mathbf{C}_{A\otimes S}(\kappa_S(S))$ by Lemmas 9.2a and b. That is, $\kappa_S(S) \subseteq \mathbf{Z}(A \otimes S)$. This inclusion guarantees that the S-module operation given by (2) imposes an S-algebra structure on $A \otimes S$. □

If A is an R-algebra and S is a commutative R-algebra, then we will write A^S for $A \otimes S$ when this tensor product is to be viewed as an S-algebra. It should be mentioned that the notation A_S is used for $A \otimes S$ in many papers on associative algebras, especially in the early literature.

The distributive and associative laws (Corollary 9.2a) have important consequences for scalar extensions.

Corollary a. *Let A and B be R-algebras. If S is a commutative R-algebra and T is a commutative S-algebra, then*

(i) $(A \dotplus B)^S \cong A^S \dotplus B^S$,
(ii) $(A \otimes_R B)^S \cong A^S \otimes_S B^S$,
(iii) $(A^S)^T \cong A^T$.

PROOF. The first isomorphism is a direct consequence of Corollary 9.2a(i). The proofs of (ii) and (iii) use a minor extension of the associative law: if M is an R-module, and if N and P are S-modules, then N and $N \otimes_S P$ can be viewed as R-modules, and $M \otimes_R (N \otimes_S P) \cong (M \otimes_R N) \otimes_S P$. Using this result, we get $(A \otimes_R B)^S = (A \otimes_R B) \otimes_R S \cong (A \otimes_R S) \otimes_R B \cong (A \otimes_R (S \otimes_S S)) \otimes_R B \cong ((A \otimes_R S) \otimes_S S) \otimes_R B \cong A^S \otimes_S B^S$, and $(A^S)^T = (A \otimes_R S) \otimes_S T \cong A \otimes_R (S \otimes_S T) \cong A \otimes_R T = A^T$. □

Lemma. *Let A be an F-algebra with the basis $\{x_i : i \in I\}$. If E is a field extension of F, then $\{x_i \otimes 1_E : i \in I\}$ is a basis of A^E. In particular, $\dim_E A^E = \dim_F A$.*

PROOF. Let $\{c_j: j \in J\}$ be an F-space basis of E. By Proposition 9.1c, $\{x_i \otimes c_j: (i,j) \in I \times J\}$ is an F-space basis of $A \otimes E$. By (2), $x_i \otimes c_j = (x_i \otimes 1)c_j$, so that $\{x_i \otimes 1: i \in I\}$ spans A^E. Suppose that $\sum_i (x_i \otimes 1)d_i = 0$, with $d_i \in E$. Write $d_i = \sum_j c_j a_{ji}$, with suitable $a_{ji} \in F$. It follows that

$$\sum_{i,j} (x_i \otimes c_j)a_{ji} = \sum_i (x_i \otimes 1)d_i = 0,$$

so that $a_{ji} = 0$ for all j and i. Hence, $d_i = 0$ for all $i \in I$. □

This lemma can be formulated more simply for finite dimensional algebras.

Corollary b. *Let A be an n-dimensional F-algebra with the basis $x_1, x_2, \ldots, x_n$ and the corresponding structure constants a_{ij}^k. If E/F is a field extension, then A^E is isomorphic to the n-dimensional E-algebra with the basis $x_1, x_2, \ldots, x_n$ and the corresponding structure constants a_{ij}^k.*

This result is an easy consequence of the lemma. The isomorphism is the obvious one that maps $x_i \otimes 1$ to x_i.

A useful special case of Corollary b occurs when A is a quaternion algebra: if E/F is a field extension and $a, b \in F^\circ$, then $\left(\frac{a,b}{F}\right)^E = \left(\frac{a,b}{E}\right)$.

Proposition b. *Let A be an F-algebra, and suppose that E/F is a field extension. An E-algebra B is isomorphic to A^E if and only if there is an F-subalgebra A' of B such that*

(i) *$A' \cong A$ as F-algebras, and*
(ii) *there is an F-space basis of A' that is also an E-space basis of B.*

If $\dim_F A$ *is finite, then* (ii) *can be replaced by*

(iii) *$A'E = B$ and* $\dim_E B = \dim_F A$.

PROOF. If $B \cong A^E$, then (i) and (ii) are satisfied by Proposition 9.2b and the lemma. It is evident that conditions (ii) and (iii) are equivalent if $\dim_F A < \infty$. Assume that (i) and (ii) are satisfied. Let $E' = \{1_B c: c \in E\}$. Plainly, E' is an F-subalgebra of B that is isomorphic to E, and $E' \subseteq \mathbf{Z}(B) \subseteq \mathbf{C}_B(A')$. If $\{x_i: i \in I\}$ is an F-space basis of A', and $\{c_j: j \in J\}$ is an F-space basis of E', then $\{x_i c_j: (i,j) \in I \times J\}$ is an F-space basis of B. In fact, by (ii), $\sum_{i,j} x_i c_j F = \sum_i x_i E = B$; and $\sum_{i,j} x_i c_j a_{ij} = 0$ with $a_{ij} \in F$ implies $\sum_j c_j a_{ij} = 0$ for all i, so that $a_{ij} = 0$ for all $i \in I$, $j \in J$. By Proposition 9.2b, there is an F-algebra isomorphism $\theta: A \otimes E \to B$ such that $\theta(1_A \otimes c) = 1_B c$ for all $c \in E$. Hence, if $z \in A \otimes E$ and $c \in E$, then $\theta(zc) = \theta(z(1 \otimes c)) = \theta(z)(1_B c) = \theta(z)c$, that is, θ is an E-algebra isomorphism. □

EXAMPLE. Let A be a simple field extension of F, say $A = F(d)$. Denote the minimum polynomial of d over F by $\Phi(\mathbf{x})$. We will show that $A^E \cong$

$E[\mathbf{x}]/K$, where $K = \Phi(\mathbf{x})E[\mathbf{x}]$. Let $A' = (F[\mathbf{x}] + K)/K \cong F[\mathbf{x}]/K \cap F[\mathbf{x}]$. Plainly, $K \cap F[\mathbf{x}]$ is a proper ideal of $F[\mathbf{x}]$ that contains $\Phi(\mathbf{x})F[\mathbf{x}]$. However, $\Phi(\mathbf{x})F[\mathbf{x}]$ is a maximal ideal of $F[\mathbf{x}]$ because $F[\mathbf{x}]/\Phi(\mathbf{x})F[\mathbf{x}] \cong F(d)$ is a field. Thus, $K \cap F[\mathbf{x}] = \Phi(\mathbf{x})F[\mathbf{x}]$, so that $A' \cong A$. The isomorphism $A^E \cong E[\mathbf{x}]/K$ therefore follows from Proposition b, since $A'E = E[\mathbf{x}]/K$ and $\dim_E E[\mathbf{x}]/K = \deg \Phi(\mathbf{x}) = \dim_F A$.

This example presages a connection between separability and the behavior of fields under scalar extension that we will study in the next chapter. If the polynomial $\Phi(\mathbf{x})$ is separable, then it factors into distinct irreducible components in $E[\mathbf{x}]$. The Chinese Remainder Theorem yields the conclusion that A^E is isomorphic to a product of fields. However, the situation is different if $\Phi(\mathbf{x})$ is inseparable. For example, suppose that $\operatorname{Char} F = p$ and $d = a^{1/p}$, where $a \in F - F^p$. In this case, $\Phi(\mathbf{x}) = \mathbf{x}^p - a$. If the extension field E also contains d, then $\Phi(\mathbf{x}) = (\mathbf{x} - d)^p$ in $E[\mathbf{x}]$. Consequently, $E[\mathbf{x}]/K$ has non-zero radical: $\mathbf{J}(E[\mathbf{x}]/K) = (\mathbf{x} - d)E[\mathbf{x}]/K$.

EXERCISES

1. Prove the generalized associative law that was invoked in the proof of Corollary a: if M is an R-module, and if N and P are S-modules, where S is a commutative R-algebra, then $M \otimes_R (N \otimes_S P) \cong (M \otimes_R N) \otimes_S P$.

2. Prove that if A is a simple field extension of F, then A/F is separable if and only if A^E is semisimple for all field extensions E/F.

3. Use Proposition a to simplify the first reduction step in the proof of Proposition 4.6.

4. Let A be a simple field extension of F. Assume that E/F is a finite Galois extension. Use the result of the example to prove that A^E is a finite product of field extensions of E that are isomorphic as F-algebras.

9.5. Induced Modules

In this section, tensor products are used to convert A-modules to B-modules, where A is a subalgebra of B. This construction has important applications in the theory of group representations. For the first time, we must use tensor products over non-commutative algebras.

Lemma a. *Let A and B be R-algebras. If M is a right A-module and N is an A-B bimodule, then $M \otimes_A N$ is a right B-module with scalar operations that satisfy*

$$(u \otimes v)y = u \otimes (vy) \tag{1}$$

for all $u \in M$, $v \in N$, and $y \in B$.

Proof. If $y \in B$, define $\Phi_y\colon M \times N \to M \otimes_A N$ by $\Phi_y(u,v) = u \otimes vy$. Plainly, Φ_y is R-bilinear, and $\Phi_y(ux,v) = ux \otimes vy = u \otimes x(vy) = u \otimes (xv)y = \Phi_y(u,xv)$. Thus, there is an R-module endomorphism ϕ_y of $M \otimes_A N$ that satisfies $\phi_y(u \otimes v) = u \otimes vy$. Routine calculations with rank one tensors give $\phi_{ya+zb} = \phi_y a + \phi_z b$ and $\phi_z \phi_y = \phi_{yz}$. Thus, $M \otimes_A N$ is a right B-module with the scalar operations $wy = \phi_y(w)$ for $w \in M \otimes_A N$ and $y \in B$. Moreover, $(u \otimes v)y = \phi_y(u \otimes v) = u \otimes (vy)$. □

The mirror image of this argument shows that if M is a B-A bimodule and N is a left A-module, then $M \otimes_A N$ is a left B-module. In case M and N are both bimodules, then $M \otimes_A N$ is also a bimodule. Indeed, suppose that M is a B-A bimodule and N is an A-C bimodule. Then $M \otimes_A N$ is a left B-module and a right C-module. If $y \in B$, $z \in C$, $u \in M$, and $v \in N$, then $y((u \otimes v)z) = yu \otimes vz = (y(u \otimes v))z$, which implies that the associativity condition for a bimodule is satisfied. Finally, if $a \in R$, $u \in M$, and $v \in N$, then $a(u \otimes v) = (au) \otimes v = (ua) \otimes v = u \otimes (av) = (u \otimes v)a$.

It is now possible to make sense of a *generalized associativity law* for tensor products over algebras.

Corollary a. *Let A and B be R-algebras. If M is a right A-module, N is an A-B bimodule, and P is a left B-module, then $M \otimes_A (N \otimes_B P) \cong (M \otimes_A N) \otimes_B P$ as R-modules. If also M or P is a bimodule, then the isomorphism is a module isomorphism; and it is a bimodule isomorphism if M and P are both bimodules.*

The proof of this corollary is just an elaboration of the proof of Proposition 9.1a.

For us the most important case of the lemma occurs when N is an R-algebra B that contains A as a subalgebra. Plainly, B can be considered as an A-B bimodule. Therefore, if M is a right A-module, then $M \otimes_A B$ is a right B-module that is *induced* by M. It is customary to denote $M \otimes_A B$ by M^B.

Lemma b. *Let A, B, and C be R-algebras, with A a subalgebra of B and B a subalgebra of C. Assume that M and N are right A-modules.*

(i) $(M \oplus N)^B \cong M^B \oplus N^B$.
(ii) $(M^B)^C \cong M^C$.
(iii) $M^A \cong M$.

Moreover if M and N are bimodules, then the isomorphisms (i), (ii), *and* (iii) *are bimodule isomorphisms.*

The proof of this lemma is essentially the same as the proof of Corollary 9.4a. The formula (1) is used to show that the isomorphisms preserve the scalar operations.

If A is a subalgebra of B, then the forgetful functor $N \mapsto N_A$ that was described in Section 2.1 maps B-modules to A-modules. By composing this restriction of the scalar operations with the induction mapping, we get correspondences $M \mapsto (M^B)_A$ and $N \mapsto (N_A)^B$.

Lemma c. *Let A, B, and C be R-algebras, with A a subalgebra of B. Suppose that M is a right A-module and N is a right B-module. There exist homomorphisms $\nu_M\colon M \to (M^B)_A$ (of right A-modules), and $\mu_N\colon (N_A)^B \to N$ (of right B-modules) such that*

$$\nu_M(u) = u \otimes 1_B \qquad \text{for all } u \in M, \text{ and} \tag{2}$$

$$\mu_N(v \otimes 1_B) = v \qquad \text{for all } v \in N. \tag{3}$$

If M is a C-A bimodule, then ν_M is a C-module homorphism; if N is a C-B bimodule, then μ_N is a C-module homorphism.

Proof. It is clear that (2) defines an R-module homomorphism. If $u \in M$ and $x \in A$, then $\nu_M(ux) = ux \otimes 1_B = u \otimes x = (u \otimes 1_B)x = \nu_M(u)x$ by 9.1(4′) and (1). Thus, ν is an A-module homomorphism. If M is a C-A bimodule, then it is obvious from (2) that ν_M is a C-module homomorphism. The mapping $N_A \times B \to N$ that is defined by $(v,y) \mapsto vy$ is plainly R-bilinear and balanced (relative to the scalar operations of elements in A). Hence, there is an R-module homomorphism $\mu_N\colon N_A \otimes_A B \to N$ such that $\mu_N(v \otimes y) = vy$. By (1), μ_N is a B-module homomorphism, and if N is a C-B bimodule then μ_N is evidently a C-module homomorphism. □

We will usually write ν for ν_M and μ for μ_N. These mappings can be used to relate the representation types of A and B.

Proposition a. *Let A and B be Artinian algebras such that A is a subalgebra of B, and B is finitely generated as a right A-module.*

(i) *Assume that for every right A-module M, the homomorphism $\nu_M\colon M \to (M^B)_A$ is split injective. If B has finite representation type, then so does A.*
(ii) *Assume that for every right B-module N, the homomorphism $\mu_N\colon (N_A)^B \to N$ is split surjective. If A has finite representation type, then so does B.*

Proof. We will prove (i); the proof of (ii) is similar. (See Exercise 2.) Let $N_1, N_2, \ldots, N_k$ be representatives of the isomorphism classes of finitely generated indecomposable B-modules. Since B_A is a finitely generated A-module, so is each $(N_i)_A$. By the Krull–Schmidt Theorem, each $(N_i)_A$ is uniquely a finite direct sum of indecomposable A-modules. It will be sufficient to prove that every finitely generated indecomposable A-module M is isomorphic to a direct summand of some $(N_i)_A$. Write $M^B \cong \bigoplus_{i=1}^k \bigoplus m_i N_i$, where $m_i \geq 0$. Since $\nu_M\colon M \to (M^B)_A$ is split injective, M is isomorphic to a direct summand of $(M^B)_A \cong \bigoplus_{i=1}^k \bigoplus m_i(N_i)_A$. The Krull–Schmidt

Theorem yields the required conclusion that M is a direct summand of some $(N_i)_A$, because M is indecomposable. □

To use Proposition a, we have to know when μ and ν are split homomorphisms. The splitting of ν will be discussed here; μ will be handled in Section 10.8.

Proposition b. *Let A be a subalgebra of the R-algebra B. The following conditions are equivalent.*

(i) *$B = A \oplus N$, where N is a right and left A-submodule of B.*
(ii) *For every R-algebra C and C-A bimodule M, $\nu: M \to (M^B)_A$ is a split injective C-A bimodule homomorphism.*

Proof. (i) implies (ii). By virtue of (i), there is an A-bimodule homomorphism $\pi: B \to A$ such that $\pi|A = id_A$. Since π is a left A-module homomorphism, the mapping of $M \times B$ to M that is defined by $(u,y) \mapsto u\pi(y)$ is bilinear and balanced. Thus, there is a homomorphism ρ: $M^B \to M$ such that $\rho(u \otimes y) = u\pi(y)$. Clearly, ρ is a left C-module homomorphism; it is a right A-module homomorphism since π is a right A-module homomorphism. Finally, $1_B = 1_A \in A$, so that $\rho\nu(u) = \rho(u \otimes 1_B) = u1_A = u$. Therefore, ν is split injective. The property (i) is the special case of (ii) in which $C = A$, and $M = A$ is considered as an A-bimodule. □

The essential part of Proposition b can be stated succinctly: if ν_A is split injective, then ν_M is split injective for all A-modules M. It will be convenient to call B a *split extension* of A if A is a subalgebra of B such that B_A is a finitely generated A-module, and $B = A \oplus N$, where N is a right and left A-submodule of B. The term cleft extension is sometimes used for a similar concept.

Example. If H is a subgroup of the finite group G, then FG is a split extension of FH for every field F. In fact, $FG = FH \oplus N$, where $N = \sum_{y \in G-H} yF$; and N is a sub-A-bimodule of FG because $x \in H$ and $y \in G - H$ implies $xy \in G - H$ and $yx \in G - H$.

Corollary b. *Let A and B be Artinian algebras such that B is a split extension of A. If B has finite representation type, then so does A.*

This corollary comes directly from Propositions a and b. By combining Corollary a, the example, and Lemma 7.1, we obtain half of Higman's characterization of the group algebras of finite representation type.

Corollary c. *Let p be a prime, F a field of characteristic p, and G a finite group such that FG has finite representation type. Then the Sylow p-subgroups of G are cyclic.*

EXERCISES

1. Prove Corollary a.

2. Prove the second statement of Proposition a. Hint. Let $M_1, M_2, \ldots, M_k$ be representatives of the isomorphism classes of finitely generated indecomposable A-modules. Show that if N is a finitely generated, indecomposable B-module, then $N_A = \bigoplus_{j=1}^{k} \bigoplus n_j M_j$ for suitable $n_j \geq 0$. Use the assumption that μ_N is split surjective to conclude that N is isomorphic to a direct summand of some $(M_j)^B$.

3. Let A and B be finite dimensional F-algebras. Assume that A is a subalgebra of B, and that B is a free left A-module. Fix an A-module basis $y_1, \ldots, y_r$ of B, that is, $B = Ay_1 \oplus \cdots \oplus Ay_r$. Suppose that M is a finitely generated right A-module, so that M is finite dimensional as an F-space. In (c) and (d), assume that char $F = 0$.
   (a) Prove that if $\{u_i : 1 \leq i \leq n\}$ is an F-basis of M, then $\{u_i \otimes y_j : 1 \leq i \leq n, 1 \leq j \leq r\}$ is an F-basis of $M \otimes_A B = M^B$.
   (b) For $z \in B$, define $\xi(z) \in M_r(A)$ by the matrix equation

$$\begin{bmatrix} y_1 z \\ \cdot \\ \cdot \\ \cdot \\ y_r z \end{bmatrix} = \xi(z) \begin{bmatrix} y_1 \\ \cdot \\ \cdot \\ \cdot \\ y_r \end{bmatrix}.$$

Prove that for $z \in B$, $\chi_{M^B}(z) = \chi_M(\text{tr}\, \xi(z))$, where χ_M and χ_{M^B} are the characters of the representations of A and B that are afforded by M and M^B respectively. Show that tr $\xi(z)$ does not depend on the choice of the basis $y_1, \ldots, y_r$.
   (c) Let G be a finite group, and suppose that H is a subgroup of G. Let M be a right FH-module, that affords the character χ. Extend χ to a mapping $\chi_0 : G \to F$ by $\chi_0(z) = \chi(z)$ for $z \in H$, and $\chi_0(z) = 0$ for $z \in G - H$. Let ψ be the character of G that is afforded by M^{FG}. Prove that $\psi(z) = |H|^{-1} \sum_{y \in G} \chi_0(yzy^{-1})$ for all $z \in G$.
   Terminology. The character ψ that is afforded by M^{FG} is said to be *induced* from the character that is afforded by the FH-module M. It is customary to denote ψ by χ^G, when M affords χ.
   (d) Let G be a finite group, and suppose that H is a subgroup of G. Let N be a right FH-module, that affords the character ψ. Prove that the character χ that is afforded by N_{FH} satisfies $\chi(x) = \psi(x)$ for all $x \in H$, that is, $\chi = \psi | FH$.
   For obvious reasons, χ is called the *restriction* of ψ to H, and it is denoted by ψ_H.

9.6. Morita Equivalence

Two algebras A and B are called *Morita equivalent* if they have equivalent module categories. In this case, the isomorphism classes of indecomposable A-modules stand in one-to-one correspondence with the isomorphism classes of indecomposable B-modules. In particular, A has finite representation type if and only if B has finite representation type. Thus, the problem of determining the representation type of an algebra A can sometimes be simplified by passing to a more tractable algebra B that is Morita equivalent to A.

The appropriate setting for the study of Morita equivalence is category theory. In order to minimize the use of category theory, we will only prove a special case of the main theorem on Morita equivalence. The result that is obtained here will be sufficient for our needs. A systematic exposition of Morita's theory can be found in Bass's book [17].

Lemma. *Let P be a right ideal of the R-algebra A such that P is a direct summand of A_A, and $AP = A$. Define $B = \mathbf{E}_A(P)$. Consider P as a B-A bimodule. For each right A-module M, denote the right B-module* $\mathrm{Hom}_A(P,M)$ *by $S(M)$; for each right B-module N, denote the right A-module $N \otimes_B P$ by $T(N)$.*

(i) *$TS(M) \cong M$ as A-modules by a mapping such that $\theta \otimes u \mapsto \theta(u)$.*
(ii) *$N \cong ST(N)$ as B-modules by the mapping $u \mapsto \phi_u$, where $\phi_u(x) = u \otimes x$.*

PROOF. (i) The bilinear mapping $(\theta,y) \mapsto \theta(y)$ of $\mathrm{Hom}_A(P,M) \times P$ to M is balanced, since $(\theta\alpha)(y) = \theta(\alpha y)$ by the definition of the right action of B on $\mathrm{Hom}_A(P,M)$. Thus, there is a homomorphism ψ of the tensor product $TS(M) = \mathrm{Hom}_A(P,M) \otimes_B P$ to M, such that $\psi(\theta \otimes y) = \theta(y)$. Plainly, ψ is an A-module homomorphism. To prove that ψ is an isomorphism, we will construct ψ^{-1}. Since $A = AP$, there are finite sets of elements $x_i \in A$ and $y_i \in P$ such that $1_A = \sum_{i=1}^m x_i y_i$. For $u \in M$, define

$$\psi'(u) = \sum_{i=1}^m (\lambda_{ux_i}|P) \otimes y_i.$$

The definition yields $\psi\psi'(u) = \sum_{i=1}^m ux_iy_i = u$; and

$$\lambda_{\theta(yx_i)}(z) = \theta(yx_i)z = \theta(yx_iz) = (\theta\lambda_{yx_i})(z)$$

implies

$$\begin{aligned}\psi'\psi(\theta \otimes y) &= \psi'(\theta(y)) = \sum_{i=1}^m (\lambda_{\theta(y)x_i}|P) \otimes y_i = \sum_{i=1}^m (\lambda_{\theta(yx_i)}|P) \otimes y_i \\ &= \sum_{i=1}^m \theta(\lambda_{yx_i}|P) \otimes y_i = \sum_{i=1}^m \theta \otimes (\lambda_{yx_i}|P)y_i \\ &= \sum_{i=1}^m \theta \otimes yx_iy_i = \theta \otimes y.\end{aligned}$$

Hence, $\psi^{-1} = \psi'$.

(ii) Let $\mu: N \otimes_B P \to N$ be the R-module homomorphism that is defined by $\mu(v \otimes y) = v(\lambda_y|P)$. The bilinear mapping $(v,y) \mapsto v(\lambda_y|P)$ is balanced, since $(v(\lambda_x|P), y)$ and $(v, (\lambda_x|P)y)$ both map to $v(\lambda_{xy}|P)$, and $B = \{\lambda_x|P: x \in P\}$ by Lemma 6.4. Therefore, the homomorphism μ exists. Denote by χ the mapping $v \mapsto \phi_v$ of N to $\mathrm{Hom}_A(P, N \otimes_B P) = ST(N)$. Clearly, χ is an R-module homomorphism; it is a B-module homomorphism, because $\chi(v\alpha)(y) = v\alpha \otimes y = v \otimes \alpha y = \chi(v)(\alpha y) = (\chi(v)\alpha)(y)$ for all $\alpha \in B$. If $\chi(v) = 0$, then $v(\lambda_y|P) = \mu(v \otimes y) = 0$ for all $y \in P$. Therefore, by Lemma 6.4, $vB = 0$, and $v = 0$. Hence, χ is injective. If $\theta \in \mathrm{Hom}_A(P, N \otimes_B P)$, then by Lemma 6.4, there exists $w = \sum_{j=1}^n v_j \otimes y_j \in N \otimes_B P$ such that

$\theta = \lambda_w|P$. (In fact, it can be assumed that $n = 1$, but to see this requires some thought.) Let $v = \mu(w) = \sum_{j=1}^{n} v_j(\lambda_{y_j}|P)$. If $z \in P$, then $\chi(v)(z) = v \otimes z = \sum_{j=1}^{n} v_j(\lambda_{y_j}|P) \otimes z = \sum_{j=1}^{n} v_j \otimes (\lambda_{y_j}|P)z = \sum_{j=1}^{n} v_j \otimes y_j z = wz = \theta(z)$. Consequently, $\chi(v) = \theta$. This argument shows that χ is surjective. □

The correspondences $M \to S(M)$ and $N \to T(N)$ are object maps for functors between the categories of right A-modules and right B-modules. That is, if $\phi\colon M \to M'$ is an A-module homomorphism, then there is an associated B-module homomorphism $S(\phi)\colon S(M) \to S(M')$ defined by $S(\phi)(\theta) = \phi\theta$. Similarly, if $\psi\colon N \to N'$ is a B-module homomorphism, then $T(\psi) = \psi \otimes id_P\colon T(N) \to T(N')$. It is evident that S and T preserve composition of homomorphisms, and they send the identity mappings to identity mappings. It follows in particular that $M \cong M'$ implies $S(M) \cong S(M')$, and $N \cong N'$ implies $T(N) \cong T(N')$. It is also clear that S and T are additive: $S(\phi_1 + \phi_2) = S(\phi_1) + S(\phi_2)$ and $T(\psi_1 + \psi_2) = T(\psi_1) + T(\psi_2)$. From this observation, an easy argument shows that $S(M_1 \oplus M_2) \cong S(M_1) \oplus S(M_2)$ and $T(M_1 \oplus M_2) \cong T(M_1) \oplus T(M_2)$. It can be shown (see Exercise 1) that S and T define a categorical equivalence between the categories of right A-modules and right B-modules. By definition, this is the assertion that A and B are *Morita equivalent*.

Proposition. *Let A be a right Artinian algebra, and suppose that B is a basic algebra of A. There is a one-to-one correspondence between the isomorphism classes of right A-modules and the isomorphism classes of right B-modules, such that indecomposable modules correspond to indecomposable modules and finitely generated A-modules correspond to finitely generated B-modules.*

PROOF. By the definition following Proposition 6.6a, there is a right ideal P of A such that P is a direct summand of A_A, $AP = A$, and $B = \mathbf{E}_A(P)$. The lemma gives mappings S and T of A-modules to B-modules and back, such that $TS(M) \cong M$ and $ST(N) \cong N$ for each right A-module M and every right B-module N. Since S and T are functorial, they induce inverse bijections between isomorphism classes. The additivity of S and T imply that $S(M)$ is indecomposable if and only if M is indecomposable. Suppose that M is a finitely generated right A-module, say $M = \sum_{k=1}^{s} u_k A$. As in the proof of the lemma, write $1_A = \sum_{i=1}^{m} x_i y_i$, where $x_i \in A$, $y_i \in P$. Then $S(M) = \text{Hom}_A(P,M)$ is generated as a right B-module by $\{\lambda_{u_k x_i}|P\colon 1 \le k \le s, 1 \le i \le m\}$. In fact if $\theta \in \text{Hom}_A(P,M)$, then by Lemma 6.4b, there exists $w = \sum_{k=1}^{s} u_k z_k (z_k \in A)$ in M, such that $\theta = \lambda_w|P$. Since $w = \sum_{k=1}^{s}\sum_{i=1}^{m} u_k x_i y_i z_k$ with $y_i z_k \in P$, it follows that $\theta = \sum_{k=1}^{s}\sum_{i=1}^{m} (\lambda_{u_k x_i}|P)(\lambda_{y_i z_k}|P) \in \sum_{k=1}^{s}\sum_{i=1}^{m} (\lambda_{u_k x_i}|P)B$. Finally, assume that N is a finitely generated right B-module, say $N = \sum_{j=1}^{t} v_j B$. By Proposition 6.4a, there is an idempotent $e \in P$ such that $P = eA$. Consequently, $T(N) = N \otimes_B P = \sum_{j=1}^{t} v_j B \otimes_B P = \sum_{j=1}^{t} v_j \otimes_B BP = \sum_{j=1}^{t} v_j \otimes P = \sum_{j=1}^{t} (v_j \otimes e)A$. That is, $\{v_j \otimes e\colon 1 \le j \le t\}$ generates the right A-module $T(N)$. □

Corollary. *If A is a right Artinian algebra, and B is a basic subalgebra of A, then A has finite representation type if and only if B has finite representation type.*

EXERCISES

1. With the hypotheses and notation as they were in the lemma, prove that the isomorphism $TS(M) \cong M$ and $N \cong ST(N)$ are natural. That is, if $\phi: M_1 \to M_2$ and $\psi: N_1 \to N_2$ are module homomorphisms, then the following diagrams commute.

$$\begin{array}{ccc} M_1 & \cong & TS(M_1) \\ \phi \downarrow & & \downarrow TS(\phi) \\ M_2 & \cong & TS(M_2) \end{array} \qquad \begin{array}{ccc} ST(N_1) & \cong & N_1 \\ ST(\psi) \downarrow & & \downarrow \psi \\ ST(N_2) & \cong & N_2 \end{array}$$

2. (a) Let A and B be R-algebras, and suppose that M is a right A-module, N is an A-B bimodule, and P is a right B-module. Prove that $\mathrm{Hom}_B(M \bigotimes_A N, P) \cong \mathrm{Hom}_A(M, \mathrm{Hom}_B(N,P))$ as R-modules. Hint. For $\phi \in \mathrm{Hom}_B(M \bigotimes_A N, P)$, $u \in M$, and $v \in N$, define $(\theta(\phi)(u))(v) = \phi(u \bigotimes v)$. Prove that $(\theta\phi)(u) \in \mathrm{Hom}_B(N,P)$, $\theta(\phi) \in \mathrm{Hom}_A(M, \mathrm{Hom}_B(N,P))$, and

$$\theta: \mathrm{Hom}_B(M \bigotimes\nolimits_A N, P) \to \mathrm{Hom}_A(M, \mathrm{Hom}_B(N, P))$$

is an isomorphism of R-modules.

(b) Deduce that if A is a subalgebra of B, M is a right A-module, and P is a right B-module, then $\mathrm{Hom}_B(M^B,P) \cong \mathrm{Hom}_A(M,P_A)$.

(c) Assume that F is an algebraically closed field, A and B are finite dimensional, semisimple F-algebras such that A is a subalgebra of B, M is a simple right A-module, and P is a simple right B-module. Let $P_A = \bigoplus m\, M \bigoplus Q$, where M is not a direct summand of Q, and $M^B = \bigoplus n\, P \bigoplus N$, where P is not a direct summand of N. Prove that $m = n$. Hint. Use (b), Schur's Lemma, and the exercise of Section 3.4 to obtain $\bigoplus n\, F \cong \bigoplus m\, F$.

The result (c) is the analogue for algebras of the *Frobenius Reciprocity Theorem* in the classical theory of group representations. It is easy to deduce the Frobenius Theorem from (c).

Notes on Chapter 9

The first four sections of this chapter provide an orderly development of standard tensor product theory. The same material can be found in most first year graudate algebra textbooks. The applications of tensor products in Sections 9.5 and 9.6 are perhaps less familiar. The material in Section 9.5 is slanted toward the theory of group representations, particularly the proof of Higman's Theorem. Our discussion of Morita equivalence in Section 9.6 is skimpy, but perhaps it gives a hint of the usefulness of a categorical approach to classical algebra.

CHAPTER 10
Separable Algebras

This chapter introduces a class of algebras that enjoys some of the attractive properties of semisimple algebras. These are the separable algebras. For F-algebras, separability is more restrictive than semisimplicity. One purpose of this chapter is to give an effective characterization of separable algebras over fields. In the course of obtaining this characterization, we will establish some properties of separable algebras that are important even when they are applied only to semisimple algebras.

The definition of separable algebras uses concepts that were introduced by topologists for the study of manifolds. It is remarkable that the ideas coming from homological algebra are so fruitful when they are applied to ring theory. A principal objective of this chapter is to give some feeling for the power of these abstract methods. They enable us to give elegant proofs of some very deep results.

10.1. Bimodules

Section 9.5 provided a hint of the importance of bimodules in the study of algebras. This chapter and the next one will confirm the central position of the bimodule concept in algebra theory. In a sense, bimodules are no more general than modules. The aim of this section is to explain how it is possible to treat bimodules as modules.

If A is an R-algebra, the *opposite algebra of* A is the R-algebra A^* that coincides with A in its R-module structure, and has a multiplication operation $\circ$ that is defined by $x \circ y = yx$. A routine calculation shows that A^* is an R-algebra with $1_{A^*} = 1_A$.

Definition. The *enveloping algebra* of the R-algebra A is

$$A^e = A^* \bigotimes A.$$

It is important to note that the definition of A^e depends in an essential way on the scalar ring R. If A is viewed as an algebra over another commutative ring S (for example, a subring of R or $\mathbf{Z}(A)$), then the associated enveloping algebra $A^* \bigotimes_S A$ is different from A^e. This is the first chapter of the book in which the role of R is more than incidental. Therefore, we will take the care to speak of R-algebras rather than simply algebras.

The multiplication operation in A^e satisfies

$$(x_1 \bigotimes y_1)(x_2 \bigotimes y_2) = x_2 x_1 \bigotimes y_1 y_2. \tag{1}$$

Indeed, by 9.2(1), $(x_1 \bigotimes y_1)(x_2 \bigotimes y_2) = (x_1 \circ x_2) \bigotimes y_1 y_2 = x_2 x_1 \bigotimes y_1 y_2$. In general the opposite algebra of A will occur only as a factor of the enveloping algebra. Therefore, (1) will make it possible to avoid the use of the product symbol $\circ$.

Let A be an R-algebra. Recall that M is an A-bimodule if M is a right A-module and a left A-module such that

$$(xu)y = x(uy), \quad \text{and} \tag{2}$$

$$au = ua \tag{3}$$

for all $u \in M$, $x, y \in A$, and $a \in R$. Of course, au and ua are abbreviations of $(1_A a)u$ and $u(1_A a)$ respectively. Equation (3) shows that the R-module structure of A is important for the discussion of bimodules. When A can be viewed as an algebra over different commutative rings R and S, it is necessary to distinguish between A_R-bimodules and A_S-bimodules.

Proposition. *Let A be an R-algebra. If M is an A-bimodule, then M is a right A^e-module with scalar multiplication that satisfies*

$$u(x \bigotimes y) = (xu)y = x(uy) \tag{4}$$

for $x, y \in A$, $u \in M$. Conversely, every right A^e-module is an A-bimodule with $xu = u(x \bigotimes 1_A)$, $ux = u(1_A \bigotimes x)$. If M and N are A-bimodules, then $\mathrm{Hom}_{A-A}(M,N) = \mathrm{Hom}_{A^e}(M_{A^e},N_{A^e})$.

In short, the categories of A-bimodules and right A^e-modules are isomorphic. We will freely switch back and forth between A-bimodules and A^e-modules, choosing the most convenient formulation of an argument or statement in various situations.

The proof of the proposition is similar to arguments that were used in Section 9.2. If M is an A-bimodule, then by (3) the mapping $(x,y) \mapsto (u \mapsto xuy)$ is bilinear from $A^* \times A$ to $\mathbf{E}_R(M)$. This mapping induces a unique R-module homomorphism $\phi\colon A^e \to \mathbf{E}_R(M)$. For $z \in A^e$ and $u \in M$, denote $uz = \phi(z)(u)$. By construction, (4) is satisfied, and it follows from (1) and (4) that $u((x_1 \otimes y_1)(x_2 \otimes y_2)) = x_2 x_1 u y_1 y_2 = (u(x_1 \otimes y_1))(x_2 \otimes y_2)$. The other module identities are easy consequences of our definitions. The proof of the converse is similar, using $(x \otimes 1_A)(1_A \otimes y) = (1_A \otimes y)(x \otimes 1_A)$ and $1_{A^*}a \otimes 1_A = 1_{A^*} \otimes 1_A a$ to obtain (2) and (3). Note that the functors ${}_AM_A \mapsto M_{A^e}$ and $M_{A^e} \mapsto {}_AM_A$ are mutually inverse. The last assertion of the proposition follows routinely from (4).

Corollary. *If A is an R-algebra, then A is a right A^e-module, and A^e is an A-bimodule.*

Obviously, A is an A-bimodule and A^e is a right A^e-module, so that the corollary follows from the proposition.

EXERCISES

1. Let A and B be R-algebras. Prove that there is a one-to-one correspondence between the class of B-A bimodules and the class of right $B^* \otimes A$-modules. Show that this correspondence is given by a category isomorphism, that is, there is a functor S from the category of B-A modules such that S^{-1} exists and S^{-1} is a functor.

2. Prove that if A and B are R-algebras, then $(A \dotplus B)^* = A^* \dotplus B^*$, $(A \otimes B)^* = A^* \otimes B^*$, and $(A \otimes B)^e \cong A^e \otimes B^e$.

3. Let G be a group, and suppose that R is a commutative ring. Prove that the group ring $A = RG$ satisfies $A^* \cong A$. Show that if G is finite, and R is a field, then $A^e \cong R(G \times G)$.

4. An *involution* of an R-algebra A is an R-module automorphism $x \mapsto x^\tau$ of A such that $(xy)^\tau = y^\tau x^\tau$ and $x^{\tau\tau} = x$ for all $x, y \in A$. Prove that if there is an involution of A, then $A^* \cong A$. Deduce that $A^* \cong A$ for the following algebras: $A = M_n(F)$, $A = \left(\frac{a,b}{F}\right)$.

10.2. Separability

We are ready to meet the object of our affections in this chapter.

Definition. An R-algebra A is *separable* if A is a projective right A^e-module.

The scalar ring R enters this definition via the dependence of A^e on R. An algebra may be separable as an R-algebra, but not separable as an S-algebra for certain subrings S of R.

It will be useful to have several characterizations of separable algebras. One of these involves a special case of the mapping μ that was defined in Section 9.5. For $z \in A^e$, define $\mu(z) = 1_A z$. This definition takes advantage of the right A^e-module structure of A. Plainly, μ is an A^e-module homomorphism, therefore also an A-bimodule homomorphism. If A is not commutative, then μ is not a ring homomorphism. The description of μ takes a more familiar form in terms of rank one tensors:

$$\mu(x \otimes y) = xy. \tag{1}$$

It is clear from (1) that μ is surjective. Thus, we have a short exact sequence

$$0 \to \operatorname{Ker} \mu \to A^e \xrightarrow{\mu} A \to 0. \tag{2}$$

The homomorphism μ is called the *augmentation* (or *augmentation mapping*) of A. The right A^e-module $\operatorname{Ker} \mu$ is sometimes called the *augmentation module* of A.

Proposition. *For an R-algebra A, the following properties are equivalent.*

(i) *A is separable.*
(ii) *The sequence* (2) *is split exact.*
(iii) *There exists $e \in A^e$ such that $\mu(e) = 1_A$ and $xe = ex$ for all $x \in A$ (considering A^e as an A-bimodule).*

PROOF. Condition (i) states that A is A^e-projective, so that (i) implies (ii) by Corollary 6.1a. Conversely, if (2) is split exact, then $A^e \cong A \oplus \operatorname{Ker} \mu$ as A^e-modules, so that A is separable. The sequence (2) is split exact if and only if there exists $\phi \in \operatorname{Hom}_{A^e}(A, A^e)$ such that $\mu\phi = id_A$. Given such a ϕ, let $e = \phi(1_A)$. Then $\mu(e) = \mu\phi(1_A) = 1_A$, and $xe = x\phi(1_A) = \phi(x1_A) = \phi(1_A x) = \phi(1_A)x = ex$, since ϕ is a bimodule homomorphism. Therefore, (iii) holds. Conversely, if $e \in A^e$ satisfies (iii), and $\phi: A \to A^e$ is defined by $\phi(x) = ex$, then $\phi(xy) = e(xy) = (ex)y = \phi(x)y$ and $\phi(yx) = e(yx) = (ey)x = (ye)x = y(ex) = y\phi(x)$, that is, ϕ is a bimodule homomorphism. Moreover, $\mu(\phi(x)) = \mu(ex) = \mu(e)x = 1_A x = x$. Thus, (ii) is split exact. □

An element $e \in A^e$ that satisfies condition (iii) of the proposition is called a *separating* (or *separability*) *idempotent* for A. Use of the term "idempotent" is justified, as the next result shows.

Lemma. *Let A be an R-algebra that is generated as an R-module by $\{z_i : i \in I\}$.*

(i) $\operatorname{Ker} \mu = \sum_{i \in I} (z_i \otimes 1 - 1 \otimes z_i) A^e$.
(ii) *if $e \in A^e$ satisfies $\mu(e) = 1_A$, then e is a separating idempotent for A if and only if $e \operatorname{Ker} \mu = 0$.*
(iii) *If $e \in A^e$ is a separating idempotent for A, then $e^2 = e$.*
(iv) *$e = \sum_{j=1}^n x_j \otimes y_j$ is a separating idempotent for A if and only if $\sum_{j=1}^n x_j y_j = 1_A$ and $\sum_{j=1}^n z_i x_j \otimes y_j = \sum_{j=1}^n x_j \otimes y_j z_i$ for all $i \in I$.*

PROOF. By linearity, it can be assumed that $A = \{z_i : i \in I\}$. For the proof of (i), suppose that $\mu(w) = 0$, where $w = \sum_{j=1}^m z_j \otimes y_j$, that is, $\sum_{j=1}^m z_j y_j = 0$

by (1). Consequently, $w = \sum_{j=1}^{m} (z_j \otimes 1 - 1 \otimes z_j)(1 \otimes y_j) \in \sum_{i \in I} (z_i \otimes 1 - 1 \otimes z_i)A^e$. The reverse inclusion is clear because $\mu(z_i \otimes 1 - 1 \otimes z_i) = z_i - z_i = 0$, and μ is an A^e-module homomorphism. (ii) follows from (i), since $z_i e - e z_i = e(z_i \otimes 1 - 1 \otimes z_i)$. Since $\mu(1_{A^e} - e) = \mu(1 \otimes 1 - e) = 1 - 1 = 0$, it is a consequence of (ii) that $e - e^2 = e(1 - e) = 0$. The statement (iv) is a reformulation of the definition of a separating idempotent. □

Example A. If R is a commutative ring and n is a positive integer, then $M_n(R)$ is separable.

Proof. Fix j between 1 and n. Let $e = \sum_{i=1}^{n} \varepsilon_{ij} \otimes \varepsilon_{ji} \in M_n(R)^e$. Then $\mu(e) = \sum_{i=1}^{n} \varepsilon_{ij}\varepsilon_{ji} = \sum_{i=1}^{n} \varepsilon_{ii} = \imath_n$, and $\sum_{i=1}^{n} \varepsilon_{ij} \otimes \varepsilon_{ji}\varepsilon_{kl} = \varepsilon_{kj} \otimes \varepsilon_{jl} = \sum_{i=1}^{n} \varepsilon_{kl}\varepsilon_{ij} \otimes \varepsilon_{ji}$. By the lemma, e is a separating idempotent for $M_n(R)$. □

Example B. Let G be a finite group whose order n is a unit of the commutative ring R. The group algebra RG is a separable R-algebra.

Proof. Define $e = (\sum_{x \in G} x^{-1} \otimes x)n^{-1}$. Then $\mu(e) = (\sum_{x \in G} 1)n^{-1} = 1$, and $(\sum_{x \in G} x^{-1} \otimes xy)n^{-1} = (\sum_{x \in G} y(xy)^{-1} \otimes xy)n^{-1} = (\sum_{x \in G} yx^{-1} \otimes x)n^{-1}$, so that e is a separating idempotent for RG. □

Exercises

1. Determine which of the following algebras are separable over the given ring.
   (a) $\mathbb{Q}$ as a $\mathbb{Z}$-algebra.
   (b) $F[\mathbf{x}]/(\mathbf{x}^p)$ as an F-algebra, where F is a field of characteristic p.
   (c) $\mathbb{Z}[\mathbf{x}]$ as a $\mathbb{Z}$-algebra.
   (d) $\mathbb{Z}[\mathbf{x}]/(\mathbf{x}^2 + 1)$ as a $\mathbb{Z}$-algebra.
   (e) $\mathbb{Z}G$ as a $\mathbb{Z}$-algebra, where G is a finite (non-trivial) group.
   (f) $\mathbb{C}$ as an $\mathbb{R}$-algebra.
   (g) $\mathbb{C}$ as a $\mathbb{Q}$-algebra.
   (h) D is a division algebra with $\mathbf{Z}(D) = F$, where F is a field, and $\dim_F D = \infty$. Consider D as an F-algebra.
   (i) $M_n(\mathbb{Q})$ as a $\mathbb{Z}$-algebra.
   (j) $E_{\mathbb{Z}}(Q)$, where Q is a free abelian group of infinite rank, considered as a $\mathbb{Z}$-algebra.

2. Let e be a separating idempotent for A. Prove that $A^e = (1 - e)A^e \oplus eA^e$, with $(1 - e)A^e = \operatorname{Ker} \mu$, and $eA^e \cong A$ as right A^e-modules.

10.3. Separable Algebras Are Finitely Generated

This assertion is only approximately true. The correct statement is our main result of this section.

Proposition. *Let A be a separable R-algebra that is projective as an R-module. Then A is a finitely generated R-module.*

PROOF. The first step of the proof uses the R-projectivity of A to construct three collections of objects, all of which are indexed by the same set I, $\{\chi_i \in \operatorname{Hom}_R(A^*,R): i \in I\}$, $\{\phi_i \in \operatorname{Hom}_R(A^e,A): i \in I\}$, $\{z_i \in A^e: i \in I\}$, such that for all $i \in I$, x, $y \in A$, and $z \in A^e$,

$$\phi_i(x \otimes y) = y\chi_i(x); \tag{1}$$

$$\phi_i(zy) = \phi_i(z)y; \tag{2}$$

$$I(z) = \{i \in I: \phi_i(z) \neq 0\} \quad \text{is finite, and } z = \textstyle\sum_{i \in I(z)} z_i\phi_i(z). \tag{3}$$

This apparatus, together with the existence of a separating idempotent e for A, leads easily to the conclusion that A_R is finitely generated. In fact, let $e = \sum_{k=1}^n x_k \otimes y_k$. By definition, $\mu(e) = 1_A$, and $ey = ye$ for all $y \in A$. Hence, by (2), $I(ye) \subseteq I(e)$, and it follows from (3) and (1) that $y = y1_A = y\mu(e) = \mu(ye) = \mu(\sum_{i \in I(e)} z_i\phi_i(ye)) = \mu(\sum_{i \in I(e)} z_i\phi_i(\sum_{k=1}^n yx_k \otimes y_k)) = \mu(\sum_{i \in I(e)} z_i \sum_{k=1}^n y_k\chi_i(yx_k)) = \sum_{i \in I(e)} \sum_{k=1}^n \mu(z_i)y_k\chi_i(yx_k)$, because the augmentation μ is a bimodule homomorphism. Thus, A is generated as an R-module by the finite set $\{\mu(z_i)y_k: i \in I(e), 1 \leq k \leq n\}$. It remains to construct the homomorphisms χ_i, ϕ_i, and the elements z_i. The method used is a slight elaboration of the construction of dual bases of projective modules. Since A is R-projective, so is A^*. Thus, there is a free R-module P with a basis $\{u_i: i \in I\}$ such that A^* is a direct summand of P; that is, $A^* < P$ and there is a projection homomorphism $\pi: P \to A^*$ with $\pi|A^* = id_{A^*}$. Since P is free on $\{u_i: i \in I\}$, there exist unique R-module homomorphisms $\chi_i: A^* \to R$ such that $I_0(x) = \{i \in I: \chi_i(x) \neq 0\}$ is finite for every $x \in A^*$, and $x = \sum_{i \in I_0(x)} u_i\chi_i(x)$. The mappings χ_i are just the coordinate projections restricted to A^*. We then have

$$x = \textstyle\sum_{i \in I_0(x)} \pi(u_i)\chi_i(x) \tag{4}$$

for all $x \in A^*$. Since $(x,y) \mapsto y\chi_i(x)$ is clearly bilinear from $A^* \times A$ to A, the existence of $\phi_i \in \operatorname{Hom}_R(A^e,A)$ satisfying (1) is guaranteed by the universal property of tensor products. Moreover, $\phi_i((x \otimes y)w) = \phi_i(x \otimes yw) = yw\chi_i(x) = y\chi_i(x)w = \phi_i(x \otimes y)w$, so that (2) is also satisfied. In other words, ϕ_i is a right A-module homomorphism. Finally, define $z_i = \pi(u_i) \otimes 1_A \in A^e$. It suffices to prove (3) in the case $z = x \otimes y$. By (1), $I(x \otimes y) \subseteq I_0(x)$ is finite, and by (4), $x \otimes y = \sum_{i \in I_0(x)} \pi(u_i)\chi_i(x) \otimes y = \sum_{i \in I_0(x)} \pi(u_i) \otimes y\chi_i(x) = \sum_{i \in I(x \otimes y)} (\pi(u_i) \otimes 1_A)\phi_i(x \otimes y) = \sum_{i \in I(x \otimes y)} z_i\phi_i(x \otimes y)$. □

Corollary. *If A is a separable F-algebra, then A is finite dimensional.*

This result is a consequence of the proposition, because every F-space is free, hence projective.

EXERCISE

Prove that if $\mathbb{Q}$ is viewed as a $\mathbb{Z}$-algebra, then $\mathbb{Q}$ is separable but not finitely generated as a $\mathbb{Z}$-module.

10.4. Categorical Properties

The characterization of projective modules that was given in Proposition 6.1b leads to another description of separable algebras. The connection between separable and semisimple algebras is an easy consequence of this result.

Let M be an A-bimodule. Define

$$M^{(A)} = \{u \in M: xu = ux \quad \text{for all} \quad x \in A\}.$$

If M is viewed as a right A^e-module, then by Lemma 10.2, $M^{(A)} = \{u \in M: u \operatorname{Ker} \mu = 0\}$, where μ is the augmentation of A. Plainly, $M^{(A)}$ is an R-submodule of M; and if $\chi \in \operatorname{Hom}_{A^e}(M,N)$, then $\chi(M^{(A)}) \subseteq N^{(A)}$. In other words, $M \mapsto M^{(A)}$ is a functor from the category of A-bimodules to the category of R-modules. The functor acts on morphisms by sending χ to $\chi|M^{(A)}$. A routine check shows that this functor is left exact: if $0 \to M \to N \to P$ is exact, then so is $0 \to M^{(A)} \to N^{(A)} \to P^{(A)}$ exact. We will show that the functor is exact precisely in the case that A is separable.

Lemma. *Let A be an R-algebra. If M is a right A^e-module, then* $\operatorname{Hom}_{A^e}(A,M) \cong M^{(A)}$ *as R-modules by the mapping $\theta_M(\phi) = \phi(1_A)$. For each $\chi \in \operatorname{Hom}_{A^e}(M,N)$, the diagram*

$$\begin{array}{ccc} \operatorname{Hom}_{A^e}(A,M) & \xrightarrow{\phi \mapsto \chi\phi} & \operatorname{Hom}_{A^e}(A,N) \\ {\scriptstyle \theta_M}\downarrow & & \downarrow{\scriptstyle \theta_N} \\ M^{(A)} & \xrightarrow[\chi|M^{(A)}]{} & N^{(A)} \end{array}$$

commutes.

PROOF. If $\phi \in \operatorname{Hom}_{A^e}(A,M)$, then $\phi(1_A)z = \phi(1_A z) = \phi(\mu(z)) = 0$ for all $z \in \operatorname{Ker} \mu$. Hence, $\phi(1_A)\operatorname{Ker} \mu = 0$; that is, $\phi(1_A) \in M^{(A)}$. Clearly, θ_M is an R-module homomorphism. If $\phi(1_A) = 0$, then for all $x \in A$, $\phi(x) = \phi(1_A)x = 0$ because ϕ is an A-bimodule homorphism. Thus, θ_M is injective. If $u \in M^{(A)}$, then λ_u is a bimodule homomorphism from A to M and $u = \lambda_u(1_A) = \theta_M(\lambda_u)$. Thus, θ_M is an isomorphism. The commutativity of the diagram is equivalent to the observation that $\theta_N(\chi\phi) = (\chi\phi)(1_A) = \chi(\phi(1_A)) = \chi(\theta_M(\phi))$. □

Proposition. *The R-algebra A is separable if and only if the functor $M \to M^{(A)}$ is exact.*

The essence of this proposition is that the separability of A is equivalent to the validity of the statement that if $\chi \in \operatorname{Hom}_{A^e}(M,N)$ is surjective, then $\chi(M^{(A)}) = N^{(A)}$. By the lemma, this implication coincides with the requirement that every surjective homomorphism $\chi: M \to N$ of A^e-modules induces a surjective homomorphism $\operatorname{Hom}_{A^e}(A,M) \to \operatorname{Hom}_{A^e}(A,N)$ that is defined by $\phi \mapsto \chi\phi$. According to Proposition 6.1b, the last condition is equivalent to the statement that A is a projective A^e-module, that is, a separable R-algebra.

Our main application of this proposition is based on the description of $P^{(A)}$ for a particular A-bimodule P.

EXAMPLE. Let M and N be right A-modules. Define right and left scalar operations of the elements of A on $\operatorname{Hom}_R(M,N)$ by

$$(x\phi)(u) = \phi(ux), \quad (\phi x)(u) = \phi(u)x.$$

Routine calculations show that $\operatorname{Hom}_R(M,N)$ becomes an A-bimodule with these scalar operations, and it is evident from the definitions that $\operatorname{Hom}_R(M,N)^{(A)} = \operatorname{Hom}_A(M,N)$.

Corollary a. *Assume that A is a separable R-algebra. If P is a right A-module that is projective as an R-module, then P is projective as an A-module.*

PROOF. Let $\chi: M \to N$ be a surjective homomorphism of right A-modules. Since P is projective as an R-module, the mapping $\psi: \phi \mapsto \chi\phi$ is a surjective homomorphism of $\operatorname{Hom}_R(P,M)$ to $\operatorname{Hom}_R(P,N)$ by Proposition 6.1b. The assumption that χ is an A-module homomorphism implies that ψ is a bimodule homomorphism:

$$\psi(\phi x)(v) = \chi((\phi x)(v)) = \chi(\phi(v)x) = \chi(\phi(v))x = ((\psi\phi)x)(v),$$

and

$$\psi(x\phi)(v) = \chi((x\phi)(v)) = \chi(\phi(vx)) = (x(\chi\phi))(v) = (x(\psi\phi))(v).$$

By the example and the proposition, $\operatorname{Hom}_A(P,M) = \operatorname{Hom}_R(P,M)^{(A)} \to \operatorname{Hom}_R(P,N)^{(A)} = \operatorname{Hom}_A(P,N)$ is surjective. Thus, P is projective as an A-module. □

Corollary b. *Every separable F-algebra is semisimple.*

Since all vector spaces are projective, it follows from Corollary a that every right module over a separable F-algebra A is projective. Therefore, A is semisimple by Corollary 6.1b.

EXERCISES

1. Prove that the functor $M \mapsto M^{(A)}$ from the category of A-bimodules to the category of R-modules is left exact.

2. Verify the assertion in the example that $\mathrm{Hom}_R(M,N)$ is an A-bimodule.

3. Prove that if A is an R-algebra, then $A^{(A)} = \mathbf{Z}(A)$.

10.5. The Class of Separable Algebras

In this section we will use Proposition 10.4 to prove three closure properties of separable R-algebras.

Proposition a. *If K is an ideal of the separable R-algebra A, then A/K is a separable R-algebra.*

PROOF. By Proposition 2.1, the A/K-bimodules M coincide with the A-bimodules M such that $KM = MK = 0$. Moreover, $(x + K)u = xu$ and $u(x + K) = ux$ for all $u \in M$ and $x \in A$. Therefore, $M^{(A/K)} = M^{(A)}$ for every A/K-bimodule. If $\chi\colon M \to N$ is a surjective A/K-bimodule homomorphism, then χ is also an A-bimodule homomorphism. Therefore, $\chi(M^{(A/K)}) = \chi(M^{(A)}) = N^{(A)} = N^{(A/K)}$ and A/K is separable by Proposition 10.4. □

Proposition b. *Let A and B be R-algebras. The product $A \dotplus B$ is separable if and only if A and B are separable.*

PROOF. If $A \dotplus B$ is separable, then A and B are separable by Proposition a. For the proof of the converse, it can be assumed that $C = A \dotplus B$ is the outer product of A and B. Denote $e = (1_A,0)$, $f = (0,1_B)$. If M is a C-bimodule, then $M = M_{11} \oplus M_{12} \oplus M_{21} \oplus M_{22}$ (as C-bimodules), where $M_{11} = eMe$, $M_{12} = eMf$, $M_{21} = fMe$, and $M_{22} = fMf$. Plainly, M_{11} is an A-bimodule and M_{22} is a B-bimodule. Moreover, $M_{12}^{(C)} = 0$: if $u \in M_{12}^{(C)}$, then $u = (1,0)u = u(1,0) = u(0,1)(1,0) = 0$. Similarly, $M_{21}^{(C)} = 0$. Hence, $M^{(C)} = M_{11}^{(C)} \oplus M_{22}^{(C)} = M_{11}^{(A)} \oplus M_{22}^{(B)}$. If $\chi\colon M \to N$ is a surjective C-bimodule homomorphism, then $\chi(M_{11}) = N_{11}$ and $\chi(M_{22}) = N_{22}$. The hypothesis that A and B are separable implies that $\chi(M_{11}^{(A)}) = N_{11}^{(A)}$ and $\chi(M_{22}^{(B)}) = N_{22}^{(B)}$. Therefore, $\chi(M^{(C)}) = N^{(C)}$. It follows that C is separable. □

The final result in this section concerns tensor products of separable algebras: the class of separable R-algebras is closed under tensor products. We will prove a more general technical result.

If B is a subalgebra of A and M is an A-bimodule, then M is also a B-bimodule. Moreover, it is clear that $M^{(A)} \subseteq M^{(B)}$.

Proposition c. *Let B and C be subalgebras of the R-algebra A such that $C \subseteq \mathbf{C}_A(B)$ and $B \cup C$ generates A as an R-algebra.*

(i) *If M is an A-bimodule, then $M^{(C)}$ is a B-bimodule and $(M^{(C)})^{(B)} = M^{(A)}$.*
(ii) *If B and C are separable, then A is separable.*

PROOF. It is evident that the statement (ii) follows from (i) by using Proposition 10.4 twice. If $u \in M^{(C)}$ and $x \in B$, then

$$y(xu) = (yx)u = (xy)u = x(yu) = x(uy) = (xu)y$$

for all $y \in C$. Therefore, $xu \in M^{(C)}$. Similarly, $ux \in M^{(C)}$, so that $M^{(C)}$ is a B-bimodule. Moreover, $u \in (M^{(C)})^{(B)}$ if and only if $xu = ux$ for all $x \in B \cup C$. Since $\{x \in A: xu = ux\}$ is a subalgebra of A, and $B \cup C$ generates A as an R-algebra, it follows that $(M^{(C)})^{(B)} = M^{(A)}$. □

Corollary. *If B and C are separable R-algebras, then $B \otimes C$ is a separable R-algebra.*

EXERCISES

1. Prove that if K is an ideal of the separable R-algebra A, then $\mathbf{Z}(A/K) = (\mathbf{Z}(A) + K)/K$. Hint. Note that the projection $A \to A/K$ induces a surjective homomorphism $A^{(A)} \to (A/K)^{(A/K)}$, and deduce the result from Exercise 3 of Section 10.4.
2. Give an alternative proof of Proposition b by showing that if e is a separating idempotent for A and f is a separating idempotent for B, then $e + f$ is a separating idempotent for $A \dotplus B$.

10.6. Extensions of Separable Algebras

In this section S will denote a commutative R-algebra. We will study the relation between the conditions that A is a separable R-algebra and A^S is a separable S-algebra, where $A^S = A \otimes S$ is obtained from A by scalar extension.

Proposition a. *If A is a separable R-algebra, then A^S is a separable S-algebra.*

PROOF. The homomorphism $x \mapsto x \otimes 1_S$ of A to A^S endows each A^S-bimodule M with an A-bimodule structure that is defined by $xu = (x \otimes 1)u$, $ux = u(x \otimes 1)$ for all $u \in M$, $x \in A$. Moreover, $M^{(A^s)} = M^{(A)}$. Indeed, if $u \in M$ and $c \in S$, then $(1 \otimes c)u = (1_{A^s}c)u = u(1_{A^s}c) = u(1 \otimes c)$ because M is a bimodule over an S-algebra. Hence, $u \in M^{(A^s)}$ if and only if $(x \otimes 1)u = u(x \otimes 1)$ for all $x \in A$, that is, $u \in M^{(A)}$. Let $\chi: M \to N$ be a surjective A^S-bimodule homomorphism. Therefore, $\chi(M^{(A^s)}) = \chi(M^{(A)}) = N^{(A)} = N^{(A^s)}$, by Proposition 10.4 and the hypothesis that A is separable. Since χ is an arbitrary surjective A^S-bimodule homomorphism, Proposition 10.4 gives the desired conclusion that A^S is a separable S-algebra. □

There is a converse to this proposition, but it requires a mild restriction on the algebra S.

Proposition b. *Assume that there is an R-module homomorphism ϕ of S to R such that $\phi(1_S) = 1_R$. If A is an R-algebra for which A^S is a separable S-algebra, then A is a separable R-algebra.*

Proof. We need some preliminary work in order to pass between A-bimodules and A^S-bimodules. According to Lemma 9.3, if M is an A-bimodule, then $M^S = M \otimes S$ is an A^S-bimodule such that for all $u \in M$, $x \in A$, and $c, d \in S$,

$$(x \otimes c)(u \otimes d) = xu \otimes cd, \quad (u \otimes d)(x \otimes c) = ux \otimes dc. \tag{1}$$

As in Proposition a, M^S is also an A-bimodule with

$$xw = (x \otimes 1)w, \quad wx = w(x \otimes 1) \tag{2}$$

for all $w \in M^S$ and $x \in A$. In particular, if $u \in M$, $x \in A$, and $c \in S$, then

$$x(u \otimes c) = xu \otimes c \quad \text{and} \quad (u \otimes c)x = ux \otimes c. \tag{3}$$

Define $\psi_M: M^S \to M$ by $\psi_M(u \otimes c) = u\phi(c)$, where $\phi: S \to R$ is the given R-module homomorphism; that is, ψ_M is the composition $M \otimes S \xrightarrow{id \otimes \phi} M \otimes R \to M$. It follows by calculation from (3) that ψ_M is an A-bimodule homomorphism. Moreover,

$$\psi_M((M^S)^{(A^S)}) = M^{(A)}. \tag{4}$$

Indeed, it is clear from (2) that $(M^S)^{(A^S)} \subseteq (M^S)^{(A)}$, so that $\psi_M((M^S)^{(A^S)}) \subseteq M^{(A)}$ (because ψ_M is an A-bimodule homomorphism). To obtain the reversed inclusion, note that if $u \in M^{(A)}$, then $u \otimes 1_S \in (M^S)^{(A^S)}$ by (1); and $\psi_M(u \otimes 1_S) = u\phi(1_S) = u1_R = u$. We can now show that the criterion of Proposition 10.4 is satisfied. Let $\chi: M \to N$ be a surjective homomorphism of A-bimodules. We wish to prove that $\chi(M^{(A)}) = N^{(A)}$. By Propositions 9.1b and 9.3, $\chi \otimes id_S: M^S \to N^S$ is a surjective homomorphism of A^S-bimodules. The separability of A^S gives

$$(\chi \otimes id_S)((M^S)^{(A^S)}) = (N^S)^{(A^S)}. \tag{5}$$

If $u \in M$ and $c \in S$, then $\psi_N(\chi \otimes id_S)(u \otimes c) = \psi_N(\chi(u) \otimes c) = \chi(u)\phi(c) = \chi(u\phi(c)) = \chi\psi_M(u \otimes c)$; thus $\psi_N(\chi \otimes id_c) = \chi\psi_M$. This commutativity relation, together with (4) and (5), gives the desired conclusion $N^{(A)} = \psi_N((N^S)^{(A^S)}) = \psi_N((\chi \otimes id_S)((M^S)^{(A^S)})) = \chi(\psi_M((M^S)^{(A^S)})) = \chi(M^{(A)})$. □

Corollary. *An F-algebra A is separable if and only if A is finite dimensional, and for every field E that contains F as a subfield, A^E is semisimple.*

Proof. If A is separable, then $\dim_F A$ is finite by Corollary 10.3, and A^E is semisimple for every field extension E/F by Proposition a and Corollary 10.4b. To prove the converse, let E be an algebraic closure of F. By assumption A^E is semisimple, and $\dim_E A^E = \dim_F A < \infty$, using Lemma 9.4. The structure of A^E is therefore determined by Corollary 3.5b: $A^E \cong M_{n_1}(E)$

$\dotplus \cdots \dotplus M_{n_r}(E)$ for suitable $n_i \in \mathbb{N}$. It is a consequence of Example 10.2a and Proposition 10.5b that A^E is a separable E-algebra. Since $E = 1_E F \bigoplus N$ for some F-subspace N of E, there is an F-space homomorphism $\phi: E \to F$ that satisfies $\phi(1_E) = 1_F$. By Proposition b, A is a separable F-algebra. □

EXERCISES

1. Let $\phi: T \to S$ be a homomorphism of commutative R-algebras, and suppose that A is an R-algebra. Prove that if A^T is a separable T-algebra, then A^S is a separable S-algebra. Hint. ϕ induces an R-algebra homomorphism of A^T to A^S. Show that if M is an A^S-bimodule, then M is an A^T-bimodule such that $M^{(A^T)} = M^{(A^S)}$.
2. Prove that if A is a separable R-algebra and M is a maximal ideal of R, then A/AM is a separable R/M-algebra. Hint. Show that $A/AM \cong A \bigotimes R/M$. (If A is finitely generated as an R-module, the converse is true. However, the known proofs of this fact use results of commutative ring theory that are fairly sophisticated.)

10.7. Separable Algebras over Fields

The criterion in Corollary 10.6 for separability of an F-algebra is not usually easy to apply. In this section we give an alternative condition that reduces the problem of recognizing separable algebras to the consideration of finite field extensions.

Lemma a. *Let A be an R-algebra. Suppose that S is a subring of R.*

(i) *If A is separable as an S-algebra, then A is separable as an R-algebra.*
(ii) *If A is separable as an R-algebra and R is separable as an S-algebra, then A is separable as an S-algebra.*

PROOF. Both parts of the lemma are proved by suitable applications of Proposition 10.4. It is clear that every A_R-bimodule is also an A_S-bimodule. Thus, if A is separable as an S-algebra, and $\chi: M \to N$ is a surjective homomorphism of A_R-bimodules, then $\chi(M^{(A)}) = N^{(A)}$. It follows that A is separable as an R-algebra. To prove (ii), let M be an A_S-bimodule. Then M is an R_S-bimodule, and by Proposition 10.5c, $M^{(R)}$ is an A_R-bimodule such that $(M^{(R)})^{(A)} = M^{(A)}$. If $\chi: M \to N$ is a surjective homomorphism of A_S-bimodules, then χ is also an R_S-bimodule homomorphism. Since R is a separable S-algebra, it follows from Proposition 10.4 that $\chi(M^{(R)}) = N^{(R)}$. That is, $\chi|M^{(R)}$ is a surjective homomorphism of A_R-bimodules. Another application of Proposition 10.4 gives the desired result

$$\chi(M^{(A)}) = \chi((M^{(R)})^{(A)}) = (N^{(R)})^{(A)} = N^{(A)}.$$

Thus, A is a separable S-algebra. □

Lemma b. *A finite field extension A/F is separable if and only if A is a separable F-algebra.*

PROOF. If A/F is a finite separable field extension, then A is a simple extension of F, say $A = F(c)$. Also, the minimum polynomial Φ of c over F is separable. To prove that A is a separable F-algebra we use the criterion of Corollary 10.6. Let E/F be an arbitrary field extension. By Example 9.4, A^E is isomorphic as an E-algebra to $E[\mathbf{x}]/\Phi(\mathbf{x})E[\mathbf{x}]$. Since Φ is separable, it factors in $E[\mathbf{x}]$ as a product $\Phi_1 \cdots \Phi_r$ of distinct irreducible polynomials. The Chinese Remainder Theorem gives an isomorphism

$$E[\mathbf{x}]/\Phi(\mathbf{x})E[\mathbf{x}] \cong E[\mathbf{x}]/\Phi_1(\mathbf{x})E[\mathbf{x}] \dotplus \cdots \dotplus E[\mathbf{x}]/\Phi_r(\mathbf{x})E[\mathbf{x}].$$

Thus, A^E is a finite product of fields, that is, A^E is semisimple. It follows from Corollary 10.6 that A is separable. Assume that A/F is not separable. The characteristic of F must be a prime p. Moreover, the set L of elements in A that are separable over F is a proper subfield of A such that A/L is purely inseparable. Since A is finite dimensional, L is contained in a maximal proper subfield K of A. If $d \in A - K$, then $A = K(d)$ by the maximality of K. The minimum polynomial of d over L has d as its only root because A/L is purely inseparable; thus, the minimum polynomial of d over K is $(\mathbf{x} - d)^{p^n}$ for some $n \geq 1$. (It can be shown that $n = 1$.) By Example 9.4, $A \bigotimes_K A \cong A[\mathbf{x}]/((\mathbf{x} - d)A[\mathbf{x}])^{p^n}$ has non-zero radical. It is a consequence of the Corollaries 10.4b and 10.5 that A is not a separable K-algebra. Lemma a implies that A is not separable as an F-algebra. □

Corollary a. *Let A be a separable F-algebra. If K is a subfield of $\mathbf{Z}(A)$, then K/F is a separable field extension.*

PROOF. Let E/F be a field extension. By Corollary 9.2b K^E is isomorphic to an E-subalgebra of the center of A^E. Since A^E is semisimple by Corollary 10.4b, there are no non-zero nilpotent elements in $\mathbf{Z}(A^E)$. Thus, K^E is a finite dimensional commutative E-algebra in which there are no non-zero nilpotent elements. Hence, K^E is semisimple. By Corollary 10.6, K is a separable F-algebra. The corollary follows from Lemma b. □

Proposition. *An F-algebra A is separable if and only if $A \cong A_1 \dotplus \cdots \dotplus A_r$ where each A_i is a finite dimensional, simple F-algebra and $\mathbf{Z}(A_i)/F$ is a separable field extension.*

PROOF. If A is separable, then A is semisimple. The Wedderburn Structure Theorem provides the decomposition $A \cong A_1 \dotplus \cdots \dotplus A_r$ with A_i simple for all i. By Proposition 10.5b, each A_i is a separable F-algebra. Thus, $\dim_F A_i < \infty$. It is well known (and it will be proved in Chapter 12) that the center of a simple algebra is a field. Hence, $\mathbf{Z}(A_i)/F$ is a separable field extension by Corollary a. To prove the converse, we borrow another result from Chapter 12: every finite dimensional simple F-algebra A is a separable $\mathbf{Z}(A)$-algebra. This fact, together with Lemmas a and b imply that the algebras A_i are separable F-algebras. Thus, A is a separable F-algebra by Proposition 10.5b. □

Corollary b. *Let F be a perfect field. An F-algebra A is separable if and only if A is finite dimensional and semisimple.*

EXERCISES

1. Prove that an F-algebra A is separable if there is a finite, separable field extension E/F such that $A^E = M_{n_1}(E) \dotplus \cdots \dotplus M_{n_r}(E)$ for suitable natural numbers $n_1, \ldots, n_r$. (Later it will be possible to prove the converse of this exercise.)

2. Let F be a field of prime characteristic p. Assume that A/F is a finite field extension that is not separable.
   (a) Prove that there exist elements $x_1, \ldots, x_k$ in A that are linearly independent over F, but $x_1^p, \ldots, x_k^p$ is linearly dependent over F.
   (b) Use the result (a) to give an alternative proof of the converse implication in Lemma b: if A/F is not separable, then A is not a separable F-algebra. Hint. Let $E = F(a_1^{1/p}, \ldots, a_k^{1/p})$, and show that $z = \sum_{i=1}^k x_i \otimes a_i^{1/p}$ is a non-zero nilpotent element in A^E for suitable $a_i \in F$.

3. Let A be a non-trivial R-algebra that is free as an R-module. Assume that S is a subring of R such that A is separable as an S-algebra.
   (a) Prove that $A^* \otimes_S A$ is a free $R \otimes_S R$-module.
   (b) Deduce from the separability of A_S that A is projective as an $R \otimes_S R$-module.
   (c) Use the assumption that A is a non-zero free R-module and the definition of separability to conclude that R is a separable S-algebra.

10.8. Separable Extensions of Algebras

This section provides a brief introduction to a generalization of the concept of separable field extension. If B is an R-algebra and A is a subalgebra of B, then B can be viewed as a B-A bimodule and as an A-B bimodule. Therefore, $B \otimes_A B$ is a B-bimodule by Lemma 9.5a. Equivalently, $B \otimes_A B$ is a right B^e-module, where $B^e = B^* \otimes_R B$ as usual. By Lemma 9.5c, there is a B-bimodule homomorphism $\mu = \mu_B \colon B \otimes_A B \to B$ such that $\mu(x \otimes y) = xy$. Obviously, μ is surjective.

Definition. The R-algebra B is a *separable extension of A* (or more simply, B is *A-separable*) if μ_B is a split surjection of B^e-modules.

Explicitly, B is A-separable if there is a B-bimodule homomorphism ϕ: $B \to B \otimes_A B$ such that $\mu\phi = id_B$.

It is a consequence of Lemma 10.7b that if A is a subfield of B and B is a field, then B is a separable extension of A in the sense of the definition if and only if B/A is finite and separable according to the familiar definition of field theory.

If $A = R$, then B is A-separable if and only if B is a separable R-algebra. Indeed, the B-bimodule structure of $B \otimes_A B = B \otimes B$ that was defined in Section 9.5 coincides with the B-bimodule structure of $B^e = B^* \otimes B$ that comes from the right B^e-module structure that is induced by multiplication. In other words, as B-bimodules $B \otimes B$ and B^e are identical.

It is not difficult to show that if B is a separable R-algebra, then B is a separable extension of all of its subalgebras. More generally, if B is a separable extension of A and C is a subalgebra of B that contains A, then B is a separable extension of C. Moreover, separable extensions are transitive: if C is a separable extension of A and B is a separable extension of C, then B is a separable extension of A. The proofs of these facts are sketched in Exercise 1. It is clear from Lemma 9.5b that every algebra is a separable extension of itself.

To obtain an example of a separable extension of R-algebras in which the larger algebra is not a separable R-algebra we will use a generalization of Proposition 10.2.

Lemma. *The R-algebra B is a separable extension of its subalgebra A if and only if there is an element $e \in B \otimes_A B$ such that $\mu_B(e) = 1_B$ and $xe = ex$ for all $x \in B$.*

The proof of the equivalence of conditions (ii) and (iii) in Proposition 10.2 carries over to the context of the lemma without substantive changes.

EXAMPLE. Let H be a Sylow p-subgroup of the finite group G. Suppose that F is a field of characteristic p. If $B = FG$ and $A = FH$, then B is A-separable.

PROOF. Let $G = Hx_1 \cup \cdots \cup Hx_r$ be a decomposition of G into a disjoint union of right cosets of H. Since H is a Sylow p-subgroup of G, the index r of H in G is not divisible by p. Thus, we can define $e = (\sum_{i=1}^r x_i^{-1} \otimes_A x_i)r^{-1} \in B \otimes_A B$. The definition of μ gives $\mu(e) = (\sum_{i=1}^r 1_B)r^{-1} = 1_B$. Moreover, if $y \in G$, then there is a permutation π of $\{1, \ldots, r\}$ and elements $u_i \in H$ such that $x_i y = u_i x_{\pi(i)}$ (therefore $x_i^{-1}u_i = yx_{\pi(i)}^{-1}$) for $1 \le i \le r$. Thus, $ey = (\sum_{i=1}^r x_i^{-1} \otimes x_i y)r^{-1} = (\sum_{i=1}^r x_i^{-1} \otimes u_i x_{\pi(i)})r^{-1} = (\sum_{i=1}^r x_i^{-1}u_i \otimes x_{\pi(i)})r^{-1} = (\sum_{i=1}^r yx_{\pi(i)}^{-1} \otimes x_{\pi(i)})r^{-1} = ye$. By the lemma, B is A-separable. By Maschke's theorem, B is not semisimple (therefore, not separable) unless H is the one element group. □

The concept of a separable extension of algebras enables us to answer a question that was posed in Section 9.5.

Proposition. *Let A be a subalgebra of the R-algebra B. The following conditions are equivalent.*

(i) *B is a separable extension of A.*
(ii) *For every R-algebra C and C-B bimodule N, $\mu_N: (N_A)^B \to N$ is a split surjective homomorphism of C-B bimodules.*

PROOF. Plainly, (i) is the special case of (ii) in which N and C coincide with B. Assume that B is A-separable. We have the diagram of bimodule homomorphisms

$$\begin{array}{ccc} N \otimes_B (B \otimes_A B) & \xrightarrow{id_N \otimes \mu_B} & N \otimes_B B \\ \theta \downarrow & & \downarrow \chi \\ N \otimes_A B & \xrightarrow[\mu_N]{} & N \end{array}$$

in which θ is the composition

$$N \otimes_B (B \otimes_A B) \to (N \otimes_B B) \otimes_A B \xrightarrow{\chi \otimes id_B} N \otimes_A B$$

and $\chi: N \otimes_B B \to N$ is defined as in Proposition 9.1b; thus, $\theta(u \otimes (x \otimes y)) = ux \otimes y$, and $\chi(u \otimes x) = ux$. By Lemma 9.5b, χ is an isomorphism. An easy calculation with rank one tensors shows that the diagram commutes. Since B is A-separable, there is a B-bimodule homomorphism $\phi: B \to B \otimes_A B$ such that $\mu_B \phi = id_B$. If we define $\psi = \theta(id_N \otimes \phi)\chi^{-1}$, then ψ is a C-B bimodule homomorphism from N to $N \otimes_A B$ such that $\mu_N \psi = \mu_N \theta(id_N \otimes \phi)\chi^{-1} = \chi(id_N \otimes \mu_N)(id_N \otimes \phi)\chi^{-1} = \chi(id_N \otimes id_B)\chi^{-1} = id_N$. Therefore, μ_N is a split surjection. □

Corollary. *Let A be a right Artinian R-algebra and suppose that B is an R-algebra that contains A as a subalgebra and is finitely generated as a right A-module. Assume that B is a separable extension of A. If A has finite representation type, then B has finite representation type.*

This corollary follows from the proposition and Proposition 9.5a.

Higman's Theorem. *Let F be a field of prime characteristic p. If G is a finite group, then the group algebra FG has finite representation type if and only if the Sylow p-subgroups of G are cyclic.*

PROOF. Let H be a Sylow p-subgroup of G. If FG has finite representation type, then H is cyclic by Corollary 9.5c. If H is cyclic, then FH has finite representation type by Lemma 7.1. It follows from the corollary and the example that FG also has finite representation type. □

EXERCISES

1. Let A and C be subalgebras of the R-algebra B with $A \subseteq C \subseteq B$.

(a) Prove that if N and M are B-bimodules, then there is a surjective B-bimodule homomorphism $\theta: M \otimes_A N \to M \otimes_C N$ such that $\theta(u \otimes_A v) = u \otimes_C v$.

(b) Use (a) to show that if B is a separable extension of A, then B is a separable extension of C. Hint. Show that the augmentation $B \otimes_A B \to B$ is the composition $B \otimes_A B \xrightarrow{\theta} B \otimes_C B \xrightarrow{\mu} B$.

(c) Prove that if B is a separable extension of C and C is a separable extension of A, then B is a separable extension of A. Hint. Prove that the composition $B \otimes_A B \cong B \otimes_C C \otimes_A C \otimes_C B \xrightarrow{id \otimes \mu_c \otimes id} B \otimes_C C \otimes_C B \cong B \otimes_C B \xrightarrow{\mu_B} B$ is the augmentation $B \otimes_A B \to B$, and deduce that if μ_C and μ_B are split surjections, then so is $B \otimes_A B \to B$.

2. Prove the lemma.

Notes on Chapter 10

The importance of separable algebras over fields has been recognized for more years than most living mathematicians can include in their productive lives. However, the broadening of separability to algebras over commutative rings is a "fairly" recent development. The credit for this development is usually assigned to the paper [13] of M. Auslander and O. Goldman, but as usual there were several forerunners of this work. The fundamental paper [16] of G. Azumaya must be mentioned; it is a classic of modern algebra.

Our discussion in this chapter gives just a hint of the theory of separable algebras. We have followed a part of the monograph [25] of F. DeMeyer and E. Ingraham. A reading of the complete work [25] will provide substantially deeper understanding of separability. The notion of relative separability is a natural idea that is implicit in the papers [43] of Higman and [51] of Jans.

CHAPTER 11

The Cohomology of Algebras

We have reached another stage of machinery building in our development of the theory of associative algebras. This time the formalism of the cohomology of algebras is introduced. The reader is warned that the ratio of definitions to theorems in the first five sections of the chapter is very high. However, the cohomology of associative algebras plays an important part in the study of central simple algebras, as we will see in Chapter 14. In this chapter the cohomology theory is used to give a streamlined proof of the Wedderburn–Malcev Principal Theorem, one of the landmarks in the theory of associative algebras. The chapter ends with a discussion of the Principal Theorem in the general theory of associative algebras. The results on extensions enables us to formulate the work of Chapter 8 in a more natural way.

Anyone who has a low tolerance for diagram chasing and abstract nonsense is advised to skip lightly through the first part of this chapter. One of the virtues of cohomology is that its usefulness rests on a small number of properties. The four statements in Section 11.2 and the interpretation of the first cohomology group in Section 11.5 are sufficient tools for most applications of the theory. Familiarity with these results is certainly sufficient to understand the uses of cohomology that are made in this book.

11.1. Hochschild Cohomology

This section gives the definitions that make a cohomology theory of associative algebras. As usual, A denotes an R-algebra. The letters M, N, and P will designate A-bimodules.

A mapping $\Phi: A^n \to M$ is multilinear if it is R-linear in each component, that is, $\Phi(x_1, \ldots, x_i a + y_i b, \ldots, x_n) = \Phi(x_1, \ldots, x_i, \ldots, x_n)a + \Phi(x_1, \ldots,$

$y_i, \ldots, x_n)b$ for all $x_1, \ldots, x_i, y_i, \ldots, x_n$ in A and $a, b \in R$. If Φ and Ψ are multilinear mappings from A^n to M and $c \in R$, then the mappings $(\Phi + \Psi)(x_1, \ldots, x_n) = \Phi(x_1, \ldots, x_n) + \Psi(x_1, \ldots, x_n)$ and $(\Phi c)(x_1, \ldots, x_n) = \Phi(x_1, \ldots, x_n)c$ are multilinear. Thus, the set of all multilinear mappings from A^n to M is an R-module under the addition and scalar multiplication defined in this way. We will denote this module by $C_R^n(A,M)$, $C^n(M)$, or C^n, depending on how much information can be inferred from the context. It is convenient to identify $C_R^0(A,M)$ with M. (This is a natural extension of our notation if A^0 is interpreted as the one element set $\{\varnothing\}$.) Thus, C^n is defined for all non-negative integers n. The elements of $C_R^n(A,M)$ are called *n-cochains on A with values in M*.

If $\phi: M \to N$ is a homomorphism of A-bimodules, then ϕ induces R-module homomorphisms $\phi^{(n)}: C_R^n(A,M) \to C_R^n(A,N)$ by $(\phi^{(n)}\Phi)(x_1, \ldots, x_n) = \phi(\Phi(x_1, \ldots, x_n))$ and $\phi^{(0)} = \phi$. This definition obviously satisfies the composition property $(\psi\phi)^{(n)} = \psi^{(n)}\phi^{(n)}$ that is required to make $\phi \mapsto \phi^{(n)}$ a functor.

For $n > 0$, an alternative definition is available. Let $A^{\otimes n}$ denote the tensor product of n copies of A, considered as an R-module. Map $\tau: \mathrm{Hom}_R(A^{\otimes n},M) \to C_R^n(A,M)$ by $\tau\psi(x_1, \ldots, x_n) = \psi(x_1 \otimes \cdots \otimes x_n)$. An easy induction on n yields the results that $(x_1, \ldots, x_n) \mapsto x_1 \otimes \cdots \otimes x_n$ is multilinear, $\{x_1 \otimes \cdots \otimes x_n: x_i \in A\}$ generates $A^{\otimes n}$ as an R-module, and $A^{\otimes n}$ has the universality property for multilinear mappings of A^n; that is, if $\Phi: A^n \to M$ is multilinear, then there is a homomorphism $\phi: A^{\otimes n} \to M$ such that $\Phi(x_1, \ldots, x_n) = \phi(x_1 \otimes \cdots \otimes x_n)$ for all $(x_1, \ldots, x_n) \in A^n$. In other words, the mapping τ is an R-module isomorphism. We can therefore translate properties of $\mathrm{Hom}_R(A^{\otimes n},M)$ to $C_R^n(A,M)$ by using the isomorphism τ. The validity of this strategem is based on the fact that τ is a functor isomorphism: if $\phi: M \to N$ is a homomorphism (of R-modules), then

$$\begin{array}{ccc} \mathrm{Hom}_R(A^{\otimes n},M) & \xrightarrow{\phi_*} & \mathrm{Hom}_R(A^{\otimes n},N) \\ \tau\downarrow & & \downarrow\tau \\ C_R^n(A,M) & \xrightarrow[\phi^{(n)}]{} & C_R^n(A,N) \end{array}$$

commutes. Indeed, $(\tau(\phi_*\psi))(x_1, \ldots, x_n) = \phi\psi(x_1 \otimes \cdots \otimes x_n) = \phi^{(n)}\tau\psi(x_1, \ldots, x_n)$.

The construction of the cochain modules $C_R^n(A,M)$ does not use the algebra structure of A or the bimodule structure of M. However, to get a cohomology theory, a coboundary operator is needed. The multiplicative structure takes the principal role in the definition of the coboundary.

Definition a. The n'th *coboundary homomorphism* is the mapping $\delta^{(n)}$ (or $\delta_M^{(n)}$) from $C_R^n(A,M)$ to $C_R^{n+1}(A,M)$ that is defined by

(i) $(\delta^{(0)}u)(x) = xu - ux$, and
(ii) $(\delta^{(n)}\Phi)(x_1, x_2, \ldots, x_n, x_{n+1}) = x_1\Phi(x_2, \ldots, x_{n+1}) + \sum_{i=1}^{n}(-1)^i \Phi(x_1, \ldots, x_ix_{i+1}, \ldots, x_{n+1}) + (-1)^{n+1}\Phi(x_1, \ldots, x_n)x_{n+1}$ for $n \geq 1$.

The special cases of this definition that will be most important in our study of algebras occur for $n = 0$, 1, and 2. For convenience we record the explicit forms of the coboundary homomorphisms $\delta^{(1)}$ and $\delta^{(2)}$.

$$(\delta^{(1)}\Phi)(x_1,x_2) = x_1\Phi(x_2) - \Phi(x_1x_2) + \Phi(x_1)x_2. \tag{1}$$

$$\begin{aligned}(\delta^{(2)}\Phi)(x_1,x_2,x_3) = x_1\Phi(x_2,x_3) - \Phi(x_1x_2,x_3) + \\ \Phi(x_1,x_2x_3) - \Phi(x_1,x_2)x_3.\end{aligned} \tag{2}$$

Lemma. *For all $n \geq 0$, $\delta^{(n)}$ is an R-module homomorphism from $C_R^n(A,M)$ to $C_R^{n+1}(A,M)$ such that $\delta^{(n+1)}\delta^{(n)} = 0$. Moreover, if $\phi\colon M \to N$ is a homomorphism of A-bimodules, then the following diagram commutes.*

$$\begin{array}{ccc} C_R^n(A,M) & \xrightarrow{\phi^{(n)}} & C_R^n(A,N) \\ \delta_M^{(n)}\downarrow & & \downarrow\delta_N^{(n)} \\ C_R^{n+1}(A,M) & \xrightarrow[\phi^{(n+1)}]{} & C_R^{n+1}(A,N) \end{array}$$

PROOF. It is clear that $\delta^{(n)}\Phi\colon A^{n+1} \to M$ is multilinear. Thus, $\delta^{(n)}$ maps C^n to C^{n+1}. Obviously, $\delta^{(n)}$ is an R-module homomorphism. The commutativity of the diagram results from a routine computation, using the assumption that ϕ is a bimodule homomorphism. The proof that $\delta^{(n+1)}\delta^{(n)} = 0$ is a tedious computation. If $n = 0$, then $(\delta^{(1)}\delta^{(0)}u)(x_1,x_2) = x_1(\delta^{(0)}u)(x_2) - (\delta^{(0)}u)(x_1x_2) + (\delta^{(0)}u)(x_1)x_2 = x_1(x_2u - ux_2) - (x_1x_2u - ux_1x_2) + (x_1u - ux_1)x_2 = 0$. When $n \geq 1$, the computation of $\delta^{(n+1)}\delta^{(n)}\Phi(x_1, \ldots, x_{n+2})$ leads to a sum in which each of the following terms occurs twice, once with the coefficient 1, and once with -1: $x_1x_2\Phi(x_3, \ldots, x_{n+2})$; $x_1\Phi(x_2, \ldots, x_ix_{i+1}, \ldots, x_{n+2})$, $2 \leq i \leq n+1$; $\Phi(x_1, \ldots, x_ix_{i+1}, \ldots, x_jx_{j+1}, \ldots, x_{n+2})$, $1 \leq i < j-1 \leq n$; $\Phi(x_1, \ldots, x_ix_{i+1}x_{i+2}, \ldots, x_{n+2})$, $1 \leq i \leq n$; $\Phi(x_1, \ldots, x_ix_{i+1}, \ldots, x_{n+1})x_{n+2}$, $1 \leq i \leq n$; $\Phi(x_1, \ldots, x_n)x_{n+1}x_{n+2}$. Thus, the sum is 0. When it is done with pencil and paper, the calculation is more convincing than when it is printed; so we will consign that chore to the exercises. □

Definition b. Denote $Z_R^n(A,M) = \operatorname{Ker}\delta^{(n)}$ and $B_R^n(A,M) = \operatorname{Im}\delta^{(n-1)}$ for $n \geq 1$, $B_R^0(A,M) = 0$. The n'th *Hochschild cohomology module* of A with coefficients in M is the factor module

$$H_R^n(A,M) = Z_R^n(A,M)/B_R^n(A,M).$$

This definition makes sense, because $\operatorname{Im}\delta^{(n-1)} \subseteq \operatorname{Ker}\delta^{(n)}$ (for $n \geq 1$) by the lemma.

As in the case of C^n, we will use the notation $Z^n(M)$ or Z^n, $B^n(M)$ or B^n, and $H^n(M)$ or H^n for $Z_R^n(A,M)$, $B_R^n(A,M)$, and $H_R^n(A,M)$ when it is safe to do so. The elements of Z^n, B^n, and H^n are respectively called *n-cocycles*, *n-coboundaries*, and *n-cohomology classes*.

If $\phi\colon M \to N$ is a bimodule homomorphism, then it follows from the lemma that $\phi^{(n)}$ maps cocycles to cocycles and coboundaries to coboundaries.

This observation proves the first part of the next proposition; the second part is clear from the remark preceding Definition a; and the third part is obvious.

Proposition. *Let ϕ: $M \to N$, and ψ: $N \to P$ be A-bimodule homomorphisms. Define $\phi_*^{(n)}$: $H_R^n(A,M) \to H_R^n(A,N)$ by*

$$\phi_*^{(n)}(\Phi + B_R^n(A,M)) = \phi^{(n)}(\Phi) + B_R^n(A,N).$$

(i) $\phi_*^{(n)}$ *is an R-module homomorphism.*
(ii) $(\psi\phi)_*^{(n)} = \psi_*^{(n)}\phi_*^{(n)}$.
(iii) $(id_M)_*^{(n)} = id_{H^n}$.

In the language of category theory, this proposition asserts that $H_R^n(A,*)$ is a covariant functor from the category of A-bimodules to the category of R-modules. The next few sections are devoted to establishing the basic properties of the sequence of functors $\{H^n\}$.

If A is a group algebra, or more generally, if A is a free R-module on a basis X that is closed under multiplication, then the cohomology modules of A can be obtained more economically than by the construction in Definition b. Map $C_R^n(A,M)$ to M^{X^n} by $\Phi \mapsto \Phi|X^n$. If M^{X^n} is made into an R-module using pointwise operations, this mapping is a homomorphism. Since Φ is multilinear, $\Phi|X^n = 0$ implies $\Phi = 0$. Finally, the assumption that X is an R-basis of A implies that any mapping of X^n to M extends uniquely to a multilinear mapping of A^n to M. Therefore, $C_R^n(A,M)$ is isomorphic to M^{X^n} under the restriction mapping. Since X is closed under multiplication, Definition a defines a coboundary homomorphism from M_{X^n} to $M_{X^{n+1}}$. Plainly, the restriction mapping carries cocycles to cocycles and coboundaries to coboundaries. Thus, we can consider n-cohomology classes as cosets of mappings Φ: $X^n \to M$ that satisfy $\delta^{(n)}\Phi = 0$. In particular, if X is finite, R is Noetherian, and M is finitely generated as an R-module, then it is easy to see that the cohomology modules $H_R^n(A,M)$ are finitely generated.

EXERCISES

1. Prove by calculation that $\delta^{(n+1)}\delta^{(n)} = 0$.

2. A multilinear mapping Ψ: $A^n \to M$ is *normalized* if $\Psi(x_1, \ldots, x_n) = 0$ whenever $x_i = 1_A$ for at least one index i. Prove the following statements.
   (a) If $\Psi \in C^n$ is normalized, then $\delta^{(n)}\Psi$ is normalized.
   (b) If $\Phi \in C^n$, then there exists $\Psi \in C^n$ such that Ψ is normalized and $\Phi - \Psi \in B^n$. Hint. Define inductively $\Phi_0 = \Phi$, $\Phi_i = \Phi_{i-1} - \delta^{(n-1)}\Psi_i$, where $\Psi_i(x_1, \ldots, x_{n-1}) = (-1)^{i-1}\Phi_{i-1}(x_1, \ldots, x_{i-1}, 1_A, x_i, \ldots, x_{n-1})$. Show by induction on $i \geq 0$ that if one or more of $x_1, \ldots, x_i$ is 1_A, then $\Phi_i(x_1, \ldots, x_i, x_{i+1}, \ldots, x_n) = 0$. Conclude that $\Psi = \Phi_n$ has the required properties.
   (c) $H^n \cong Z_0^n/B_0^n$, where Z_0^n and B_0^n are the submodules of Z^n and B^n that consist of normalized cocycles and coboundaries.

11.2. Properties of Cohomology

It is rather astonishing that the cohomology modules of an algebra play a fundamental role in the structure theory of the algebra. The intuitive content of Definition 11.1b is almost nil. However, as we progress through this chapter, our appreciation of this new tool will grow.

In this section, the most important properties of the cohomology modules will be presented. It is surprising that most facts about cohomology can be derived from four basic results. Indeed, when certain "naturality conditions" are assumed, these four basic properties uniquely determine the cohomology modules.

Theorem. Zero Dimensional Cohomology. *If M is an A-bimodule, then*

$$H_R^0(A,M) \cong M^{(A)}.$$

Additivity. *If ϕ and ψ are A-bimodule homomorphisms of M to N, and $a, b \in R$, then for all $n < \omega$, $(\phi a + \psi b)_*^{(n)} = \phi_*^{(n)} a + \psi_*^{(n)} b$.*

The Long Exact Sequence. *Assume that A_R is projective. Let*

$$\sum : 0 \to N \xrightarrow{\phi} M \xrightarrow{\psi} P \to 0$$

be a short exact sequence of A-bimodules. For each $n < \omega$, there is an R-module homomorphism $\partial^{(n)}$: $H_R^n(A,P) \to H_R^{n+1}(A,N)$ (that depends on $\sum$) such that the following sequence is exact:

$$0 \to H_R^0(A,N) \xrightarrow{\phi_*^{(0)}} H_R^0(A,M) \xrightarrow{\psi_*^{(0)}} H_R^0(A,P) \xrightarrow{\partial^{(0)}} H_R^1(A,N) \to \cdots$$
$$\to H_R^n(A,N) \xrightarrow{\phi_*^{(n)}} H_R^n(A,M) \xrightarrow{\psi_*^{(n)}} H_R^n(A,P) \xrightarrow{\partial^{(n)}} H_R^{n+1}(A,N) \to \cdots.$$

Coinduced Bimodules. *For a left A-module M, let $P = \mathrm{Hom}_R(A,M)$. Then P is an A-bimodule with the scalar operations $(x\theta)(y) = x\theta(y)$ and $(\theta x)(y) = \theta(xy)$. For all $n > 0$, $H_R^n(A,P) = 0$. If M is an A-bimodule, then there is an injective bimodule homomorphism ϕ: $M \to P$ defined by $\phi(u)(x) = ux$.*

The first two parts of this result are easy consequences of definitions. By definition, $B_R^0(A,M) = 0$, and $Z_R^0(A,M) = \{u \in M\colon \delta^{(0)}(u) = 0\} = \{u \in M\colon xu - ux = 0 \text{ for all } x \in A\} = M^{(A)}$. To obtain the additivity, note that if $\Phi \in C_R^n(A,M)$, then $(\phi a + \psi b)^{(n)}(\Phi) = \phi^{(n)}(\Phi)a + \psi^{(n)}(\Phi)b$. Hence, $\Phi \in Z_R^n(A,M)$ implies

$$\begin{aligned}
(\phi a + \psi b)_*^{(n)}(\Phi + B_R^n(A,M)) &= (\phi a + \psi b)^{(n)}\Phi + B_R^n(A,N) \\
&= \phi^{(n)}\Phi a + \psi^{(n)}\Phi b + B_R^n(A,N) \\
&= \phi_*^{(n)}(\Phi + B_R^n(A,M))a + \psi_*^{(n)}(\Phi + B_R^n(A,M))b \\
&= (\phi_*^{(n)}a + \psi_*^{(n)}b)(\Phi + B_R^n(A,M)).
\end{aligned}$$

The long exact sequence will be constructed in the next section by means of the Snake Lemma. We conclude this section with a proof that coinduced bimodules have trivial cohomology.

Using the fact that θ is a homomorphism of R-modules, a routine check establishes the assertion that P is an A-bimodule. It is also easy to see that if M is an A-bimodule, then ϕ is a bimodule homomorphism that is injective since $\phi(u)(1) = u$. It remains to show that $H_R^{n+1}(A,P) = 0$ for all $n < \omega$. Define $\sigma^{(n)}: C_R^{n+1}(A,P) \to C_R^n(A,P)$ by

$$(\sigma^{(n)}\Phi)(x_1, \ldots, x_n)(y) = \Phi(x_1, \ldots, x_n, y)(1).$$

If $n > 0$, then the compositions $\sigma^{(n)}\delta^{(n)}$ and $\delta^{(n-1)}\sigma^{(n-1)}$ are R-module homomorphisms of $C_R^n(A,P)$ to itself, such that $(-1)^{n+1}(\sigma^{(n)}\delta^{(n)} - \delta^{(n-1)}\sigma^{(n-1)})$ is the identity mapping. In fact,

$$\begin{aligned}(\sigma^{(n)}\delta^{(n)}\Phi)(x_1, \ldots, x_n)(y) &= (\delta^{(n)}\Phi)(x_1, \ldots, x_n, y)(1)\\ &= x_1\Phi(x_2, \ldots, x_n, y)(1)\\ &\quad + \sum_{i=1}^{n-1}(-1)^i\Phi(x_1, \ldots, x_ix_{i+1}, \ldots, x_n, y)(1)\\ &\quad + (-1)^n\Phi(x_1, \ldots, x_{n-1}, x_ny)(1)\\ &\quad + (-1)^{n+1}\Phi(x_1, \ldots, x_n)(y)\end{aligned}$$

(using the definition of the right scalar product in P), and

$$\begin{aligned}(\delta^{(n-1)}\sigma^{(n-1)}\Phi)(x_1, \ldots, x_n)(y) &= x_1\sigma^{(n-1)}\Phi(x_2, \ldots, x_n)(y)\\ &\quad + \sum_{i=1}^{n-1}(-1)^i(\sigma^{(n-1)}\Phi)(x_1, \ldots, x_ix_{i+1}, \ldots, x_n)(y)\\ &\quad + (-1)^n((\sigma^{(n-1)}\Phi)(x_1, \ldots, x_{n-1})x_n)(y) = x_1\Phi(x_2, \ldots, x_n, y)(1)\\ &\quad + \sum_{i=1}^{n-1}(-1)^i\Phi(x_1, \ldots, x_ix_{i+1}, \ldots, x_n, y)(1)\\ &\quad + (-1)^n\Phi(x_1, \ldots, x_{n-1}, x_ny)(1).\end{aligned}$$

Subtracting the respective sides of these equations gives

$$\sigma^{(n)}\delta^{(n)} - \delta^{(n-1)}\sigma^{(n-1)} = (-1)^{n+1}id.$$

Thus, if $n \geq 1$ and $\Phi \in Z_R^n(A,P)$, then

$$\begin{aligned}\Phi &= (-1)^{n+1}(\sigma^{(n)}\delta^{(n)}\Phi - \delta^{(n-1)}\sigma^{(n-1)}\Phi)\\ &= \delta^{(n-1)}((-1)^n\sigma^{(n-1)}\Phi) \in B_R^n(A,P);\end{aligned}$$

that is, $H_R^n(A,P) = 0$.

EXERCISES

1. Show that the isomorphism $H_R^0(A,M) \cong M^{(A)}$ is natural. That is, if $\phi: M \to N$ is a bimodule homomorphism, then the following diagram commutes.

$$\begin{array}{ccc} H^0(A,M) & \xrightarrow{\phi_*^{(0)}} & H^0(A,N) \\ \downarrow & & \downarrow \\ M^{(A)} & \xrightarrow[\phi]{} & N^{(A)} \end{array}$$

2. Prove that if $\{M_i : i \in I\}$ is a non-empty set of A-bimodules, then $H_R^n(A, \bigoplus_{i \in I} M_i) \cong \bigoplus_{i \in I} H_R^n(A,M_i)$.

3. Let $\theta: A \to B$ be a homomorphism of R-algebras. For a B-bimodule M, define left and right scalar operations on M by the elements of A as in Section 2.1. Show that M becomes an A-bimodule under these operations, still denoted by M. Define $\theta^{(n)}: C_R^n(B,M) \to C_R^n(A,M)$ by $(\theta^{(n)}\Phi)(x_1, \ldots, x_n) = \Phi(\theta x_1, \ldots, \theta x_n)$. Show that $\theta^{(n)}\Phi$ is multilinear for $\Phi \in C_R^n(B,M)$, and that the following diagram commutes:

$$\begin{array}{ccc} C_R^n(B,M) & \xrightarrow{\delta^{(n)}} & C_R^{n+1}(B,M) \\ {\scriptstyle \theta^{(n)}}\downarrow & & \downarrow{\scriptstyle \theta^{(n+1)}} \\ C_R^n(A,M) & \xrightarrow{\delta^{(n)}} & C_R^{n+1}(A,M). \end{array}$$

Deduce that $\theta^{(n)}$ induces an R-module homomorphism $\theta^{*(n)}: H_R^n(B,M) \to H_R^n(A,M)$. Show that if $\phi: M \to N$ is a homomorphism of B-bimodules, then ϕ is also a homomorphism when M and N are viewed as A-bimodules, and

$$\begin{array}{ccc} H_R^n(B,M) & \xrightarrow{\phi_*^{(n)}} & H_R^n(B,N) \\ {\scriptstyle \theta^{*(n)}}\downarrow & & \downarrow{\scriptstyle \theta^{*(n)}} \\ H_R^n(A,M) & \xrightarrow[\phi_*^{(n)}]{} & H_R^n(A,N) \end{array}$$

commutes.

4. Let B and C be R-algebras. Denote $A = B \dotplus C$. Let M be a B-C bimodule. Consider M as an A-bimodule by defining $CM = MB = 0$. Prove that $H_R^n(A,M) = 0$ for all n. Use this result to show that for every A-bimodule N, $H_R^n(A,N) \cong H_R^n(B, 1_B N 1_B) \bigoplus H_R^n(C, 1_C N 1_C)$ for all $n < \omega$.

11.3. The Snake Lemma

Let

$$\begin{array}{ccccccc} & & N_1 & \xrightarrow{\psi_1} & N_2 & \xrightarrow{\psi_2} & N_3 & \to 0 \\ & & {\scriptstyle \phi_1}\downarrow & & {\scriptstyle \phi_2}\downarrow & & {\scriptstyle \phi_3}\downarrow & \\ 0 \to & & M_1 & \xrightarrow[\chi_1]{} & M_2 & \xrightarrow[\chi_2]{} & M_3 & \end{array} \tag{1}$$

be a commutative diagram of module homomorphisms with exact rows. There exist module homomorphisms $\psi_{1*}, \psi_{2*}, \chi_{1*}, \chi_{2*}$, and ∂ such that

$$\begin{array}{c} \operatorname{Ker} \phi_1 \xrightarrow{\psi_{1*}} \operatorname{Ker} \phi_2 \xrightarrow{\psi_{2*}} \operatorname{Ker} \phi_3 \xrightarrow{\partial} \\ \operatorname{Coker} \phi_1 \xrightarrow{\chi_{1*}} \operatorname{Coker} \phi_2 \xrightarrow{\chi_{2*}} \operatorname{Coker} \phi_3 \end{array} \tag{2}$$

is exact. If ψ_1 is injective, then so is ψ_{1*}; if χ_2 is surjective, then so is χ_{2*}.

The term "Snake Lemma" comes from the diagram that illustrates this result.

$$\begin{array}{ccccc} \operatorname{Ker}\phi_1 & \overset{\psi_{1*}}{\to} & \operatorname{Ker}\phi_2 & \overset{\psi_{2*}}{\to} & \operatorname{Ker}\phi_3 \\ {\scriptstyle\kappa_1}\downarrow & & {\scriptstyle\kappa_2}\downarrow & & {\scriptstyle\kappa_3}\downarrow \\ N_1 & \overset{\psi_1}{\to} & N_2 & \overset{\psi_2}{\to} & N_3 \\ {\scriptstyle\phi_1}\downarrow & & {\scriptstyle\phi_2}\downarrow & & {\scriptstyle\phi_3}\downarrow \\ M_1 & \underset{\chi_1}{\to} & M_2 & \underset{\chi_2}{\to} & M_3 \\ {\scriptstyle\pi_1}\downarrow & & {\scriptstyle\pi_2}\downarrow & & {\scriptstyle\pi_3}\downarrow \\ \operatorname{Coker}\phi_1 & \underset{\chi_{1*}}{\to} & \operatorname{Coker}\phi_2 & \underset{\chi_{2*}}{\to} & \operatorname{Coker}\phi_3 \end{array} \qquad \partial \tag{3}$$

The mappings κ_1, κ_2, and κ_3 are inclusion homomorphisms, and π_1, π_2, and π_3 are projection homomorphisms. Our aim is to prove the existence of homomorphisms ψ_{1*}, ψ_{2*}, χ_{1*}, χ_{2*}, and ∂ that make the sequence (2) exact, and the diagram (3) commutative. The commutativity requirement is satisfied if and only if $\psi_{i*} = \psi_i|\operatorname{Ker}\phi_i$, $\chi_{i*}(u + \operatorname{Im}\phi_i) = \chi_i(u) + \operatorname{Im}\phi_{i+1}$ for $i = 1,2$. These definitions of ψ_{i*} and χ_{i*} are legitimate because $\psi_i(\operatorname{Ker}\phi_i) \subseteq \operatorname{Ker}\phi_{i+1}$, $\chi_i(\operatorname{Im}\phi_i) \subseteq \operatorname{Im}\phi_{i+1}$ by the commutativity of (1). The exactness of (2) at $\operatorname{Ker}\phi_2$ and $\operatorname{Coker}\phi_2$ follows from the exactness and commutativity of (1) by traditional diagram chasing. For example,

$$\begin{aligned} \operatorname{Ker}\chi_{2*} &= \pi_2\pi_2^{-1}\chi_{2*}^{-1}(0) = \pi_2\chi_2^{-1}\pi_3^{-1}(0) = \pi_2\chi_2^{-1}\phi_3(N_3) \\ &= \pi_2\chi_2^{-1}\phi_3\psi_2(N_2) = \pi_2\chi_2^{-1}\chi_2\phi_2(N_2) = \pi_2(\phi_2(N_2) + \chi_2^{-1}(0)) \\ &= \pi_2\chi_1(M_1) = \chi_{1*}\pi_1(M_1) = \operatorname{Im}\chi_{1*}. \end{aligned}$$

Similarly, $\operatorname{Ker}\psi_{2*} = \operatorname{Im}\psi_{1*}$. The definitions of κ_1, π_3, and the commutativity of (3) make it clear that if ψ_1 is injective, then ψ_1^* is injective; and if χ_2 is surjective, then χ_{2*} is surjective. The rest of the proof consists of constructing ∂ in such a way that (2) is exact. To make this chore as painless as possible, we digress. For a commuting square of module homomorphisms

$$\sum: \begin{array}{ccc} N_1 & \overset{\psi_1}{\to} & N_2 \\ {\scriptstyle\phi_1}\downarrow & & \downarrow{\scriptstyle\phi_2} \\ M_1 & \underset{\chi_1}{\to} & M_2 \end{array},$$

define $\operatorname{Ker}\sum = (\operatorname{Ker}\chi_1\phi_1)/(\operatorname{Ker}\phi_1 + \operatorname{Ker}\psi_1)$ and

$$\operatorname{Im}\sum = (\operatorname{Im}\phi_2 \cap \operatorname{Im}\chi_1)/(\operatorname{Im}\phi_2\psi_1).$$

Lemma. *If*

$$\begin{array}{ccccc} N_1 & \overset{\psi_1}{\to} & N_2 & \overset{\psi_2}{\to} & N_3 \\ {\scriptstyle\phi_1}\downarrow & {\scriptstyle\Sigma_1} & {\scriptstyle\phi_2}\downarrow & {\scriptstyle\Sigma_2} & {\scriptstyle\phi_3}\downarrow \\ M_1 & \underset{\chi_1}{\to} & M_2 & \underset{\chi_2}{\to} & M_3 \end{array}$$

is commutative with exact rows, then $\operatorname{Im}\sum_1 \cong \operatorname{Ker}\sum_2$.

PROOF. For $x \in N_2$: $\phi_2(x) \in \operatorname{Im}\phi_2 \cap \operatorname{Im}\chi_1$ if and only if $x \in \operatorname{Ker}\chi_2\phi_2$; and $\phi_2(x) \in \operatorname{Im}\phi_2\psi_1$ if and only if $x \in \operatorname{Im}\psi_1 + \operatorname{Ker}\phi_2 = \operatorname{Ker}\psi_2 + \operatorname{Ker}\phi_2$. Thus, ϕ_2 induces an isomorphism of $\operatorname{Ker}\sum_2$ to $\operatorname{Im}\sum_1$. □

To complete the proof of the Snake Lemma, we augment the diagram (3) with $P = \operatorname{Coker}\psi_{2*}$ and $Q = \operatorname{Ker}\chi_{1*}$.

$$\begin{array}{ccccccccccc}
 & & & & & & & & 0 & \to P & \\
 & & & & & & & & \downarrow & \Sigma_1 \; \| & \\
 & & & & \operatorname{Ker}\phi_1 & \to & \operatorname{Ker}\phi_2 & \to & \operatorname{Ker}\phi_3 & \to P & \to 0 \\
 & & & & \downarrow & & \downarrow & \Sigma_3 & \downarrow & \Sigma_2 \; \downarrow & \\
 & & & & N_1 & \to & N_2 & \to & N_3 & \to 0 & \\
 & & & & \downarrow & \Sigma_5 & \downarrow & \Sigma_4 & \downarrow & & \\
 & & 0 & \to & M_1 & \to & M_2 & \to & M_3 & & \\
 & & \downarrow \; \Sigma_7 & & \downarrow & \Sigma_6 & \downarrow & & \downarrow & & \\
0 & \to & Q & \to & \operatorname{Coker}\phi_1 & \to & \operatorname{Coker}\phi_2 & \to & \operatorname{Coker}\phi_3 & & \\
 & & \| \; \Sigma_8 & & \downarrow & & & & & & \\
 & & Q & \to & 0 & & & & & &
\end{array}$$

The lemma gives $P = \operatorname{Im}\sum_1 \cong \operatorname{Ker}\sum_2 \cong \operatorname{Im}\sum_3 \cong \operatorname{Ker}\sum_4 \cong \operatorname{Im}\sum_5 \cong \operatorname{Ker}\sum_6 \cong \operatorname{Im}\sum_7 \cong \operatorname{Ker}\sum_8 = Q$. The composition $\operatorname{Ker}\phi_3 \to P \to Q \to \operatorname{Coker}\phi_1$ is the desired homomorphism ∂.

We can now prove the existence of the Long Exact Sequence that was described in Section 11.2. By assumption, A is projective as an R-module. It follows easily from Proposition 9.1a that $A^{\otimes n}$ is also projective. Therefore, since $0 \to N \xrightarrow{\phi} M \xrightarrow{\psi} P \to 0$ is exact, so is

$$0 \to \operatorname{Hom}_R(A^{\otimes n}, N) \to \operatorname{Hom}_R(A^{\otimes n}, M) \to \operatorname{Hom}_R(A^{\otimes n}, P) \to 0.$$

As we observed in Section 11.1, this means that $0 \to C^n(N) \xrightarrow{\phi^{(n)}} C^n(M) \xrightarrow{\psi^{(n)}} C^n(P) \to 0$ is exact. It follows from Lemma 11.1 that the diagram

$$\begin{array}{ccccccc}
0 \to & C^n(N) & \to & C^n(M) & \to & C^n(P) & \to 0 \\
 & \downarrow \delta^{(n)} & & \downarrow \delta^{(n)} & & \downarrow \delta^{(n)} & \\
0 \to & C^{n+1}(N) & \to & C^{n+1}(M) & \to & C^{n+1}(P) & \to 0
\end{array}$$

commutes and has exact rows. By applying the Snake Lemma to this diagram (and using $B^0 = 0$ in the case $n = 0$), we conclude that for all $n < \omega$, the sequences $0 \to Z^n(N) \to Z^n(M) \to Z^n(P)$ and

$$C^n(N)/B^n(N) \to C^n(M)/B^n(M) \to C^n(P)/B^n(P) \to 0$$

are exact, where the homomorphisms are induced by $\phi^{(n)}$ and $\psi^{(n)}$. Since $B^n \subseteq \operatorname{Ker}\delta^{(n)}$ and $\operatorname{Im}\delta^{(n)} \subseteq Z^{n+1}$, the coboundary homomorphism induces $\delta_*^{(n)}\colon C^n/B^n \to Z^{n+1}$ by $\delta_*^{(n)}(\Phi + B^n) = \delta^{(n)}\Phi$. Clearly, $\operatorname{Ker}\delta_*^{(n)} = H^n$ and $\operatorname{Coker}\delta_*^{(n)} = H^{n+1}$. The homomorphisms $\delta_*^{(n)}$ give rise to the following diagram with exact rows that is easily seen to be commutative.

$$\begin{array}{ccccc} C^n(N)/B^n(N) & \overset{\phi_*^{(n)}}{\to} & C^n(M)/B^n(M) & \overset{\psi_*^{(n)}}{\to} & C^n(P)/B^n(P) \to 0 \\ \downarrow \delta_*^{(n)} & & \downarrow \delta_*^{(n)} & & \downarrow \delta_*^{(n)} \\ 0 \to Z^{n+1}(N) & \underset{\phi^{(n+1)}}{\longrightarrow} & Z^{n+1}(M) & \underset{\psi^{(n+1)}}{\longrightarrow} & Z^{n+1}(P) \end{array}$$

By applying the Snake Lemma to this diagram we obtain a connecting homomorphism $\partial^{(n)}$ such that $H^n(N) \overset{\phi_*^{(n)}}{\to} H^n(M) \overset{\psi_*^{(n)}}{\to} H^n(P) \overset{\partial^{(n)}}{\to} H^{n+1}(N) \overset{\phi_*^{(n+1)}}{\longrightarrow} H^{n+1}(M) \overset{\psi_*^{(n+1)}}{\longrightarrow} H^{n+1}(P)$ is exact. A routine check shows that the homomorphisms $\phi_*^{(n)}$, $\psi_*^{(n)}$, $\phi_*^{(n+1)}$, and $\psi_*^{(n+1)}$ given by the Snake Lemma coincide with the homomorphisms of cohomology modules that are constructed from ϕ and ψ in Proposition 11.1. Thus, these finite sequences can be fused to produce all except the initial segment $0 \to H^0(N) \to H^0(M)$ of the Long Exact Sequence. Since $H^0 = Z^0$, it follows that $\phi_*^0 = \phi|Z^0(N)$ is injective. The proof of the Long Exact Sequence Theorem is complete.

Exercises

1. Let
$$\begin{array}{l} \sum_1 : 0 \to N_1 \overset{\phi_1}{\to} M_1 \overset{\psi_1}{\to} P_1 \to 0 \\ \qquad\quad \chi_1 \downarrow \quad \chi_2 \downarrow \quad \chi_3 \downarrow \\ \sum_2 : 0 \to N_2 \underset{\phi_2}{\to} M_2 \underset{\psi_2}{\to} P_2 \to 0 \end{array}$$
be a commutative diagram of A-bimodules with exact rows. Prove that for all $n \geq 0$, the following diagram commutes.
$$\begin{array}{ccc} H^n(P_1) & \overset{\partial^{(n)}}{\to} & H^{n+1}(N_1) \\ \chi_{3*}^{(n)} \downarrow & & \downarrow \chi_{1*}^{(n)} \\ H^n(P_2) & \underset{\partial^{(n)}}{\to} & H^{n+1}(N_2) \end{array}$$
Hint. Stack the diagram (3) that goes with $\sum_1$ on top of the diagram that goes with $\sum_2$, and connect the corresponding modules by vertical homomorphisms induced by χ_1, χ_2, and χ_3. Check commutativity around the various loops.
2. Use the Snake Lemma to prove the Short 5-Lemma:
   (a) If ϕ_1 and ϕ_3 in the diagram (1) are injective, then ϕ_2 is injective.
   (b) If ϕ_1 and ϕ_3 in the diagram (1) are surjective, then ϕ_2 is surjective.
3. Prove that the tensor product of two projective R-modules is projective.

11.4. Dimension

The geometrical notion of dimension can be translated to an algebraic setting by using the Hochschild cohomology groups. It is a standard result of topology that the sequence of cohomology groups of an n-dimensional manifold becomes zero beyond n, while $H^n \neq 0$ for suitably chosen coefficient domains. The similarity between the definitions and the properties of Hochschild cohomology and the topological cohomology theories suggests that an analogous concept of dimension might be a fruitful invariant for algebras. This idea provides the theme of this section and the two that follow it. In particular, we will show that zero dimensional algebras are the separable algebras that were studied in Chapter 10.

We begin with a useful lemma that also motivates the definition of dimension.

Lemma. *Let A be an R-algebra that is projective as an R-module, and suppose that n is a natural number such that $H_R^n(A,N) = 0$ for all A-bimodules N. If $m \geq n$, then $H_R^m(A,M) = 0$ for all A-bimodules M.*

PROOF. Induction makes it sufficient to prove that $H_R^{n+1}(A,M) = 0$ for all M. By Theorem 11.2, there is a short exact sequence $0 \to M \to P \to N \to 0$ of A-bimodules in which $H_R^k(A,P) = 0$ for all $k > 0$. This property of P, together with the hypothesis of the lemma gives the following segment of the Long Exact Sequence:

$$0 = H_R^n(A,N) \to H_R^{n+1}(A,M) \to H_R^{n+1}(A,P) = 0.$$

Thus, $H_R^{n+1}(A,M) = 0$. □

Definition. Let A be a non-trivial R-algebra. The *dimension* of A is

$$\operatorname{Dim} A = \sup\{n: H_R^n(A,M) \neq 0 \quad \text{for some } A\text{-bimodule } M\}.$$

If, for every $n < \omega$, there is an A-bimodule M such that $H_R^n(A,M) \neq 0$, then the dimension of A is ∞. It must be pointed out that the supremum in this definition is not applied to the empty set. In fact. $H_R^0(A,M) = M^{(A)}$; in particular, $H_R^0(A,A) = A^{(A)} = \mathbf{Z}(A) \neq 0$, since A is non-trivial.

For R-algebras A such that A_R is projective, it follows from the lemma that if $H_R^n(A,M) = 0$ for all bimodules M, then $\operatorname{Dim} A < n$. We conclude this section by translating the condition for an algebra to have dimension less than two into more familiar ideas.

The elements of $Z_R^2(A,M)$ are the bilinear mappings $\Phi: A \times A \to M$ such that $\delta^{(2)}\Phi = 0$, that is,

$$x\Phi(y,z) - \Phi(xy,z) + \Phi(x,yz) - \Phi(x,y)z = 0 \tag{1}$$

for all x, y, $z \in A$. A bilinear mapping that satisfies (1) is called a *factor set of A* with values in M. Such mappings occur in the study of group and algebra extensions. The factor sets of the form $\delta^{(1)}\phi$, where $\phi: A \to M$ is linear, are called *split factor sets*. By Definition 11.1a, $\delta^{(1)}\phi$ is a factor set Φ, defined by

$$\Phi(x,y) = x\phi(y) - \phi(xy) + \phi(x)y. \tag{2}$$

The lemma yields the following result.

Corollary. *If the R-algebra A is a projective R-module, then* $\operatorname{Dim} A \leq 1$ *if and only if every factor set of A with values in any bimodule M is split.*

EXERCISE

Prove the uniqueness theorem for cohomology of algebras: Assume that $\{T^n : n < \omega\}$ is a sequence of covariant functors from A-bimodules to R-modules satisfying the basic conditions of Section 11.2 and the naturality conditions of the Exercises 1 of Sections 11.2 and 11.3; then there exists a class of isomorphisms $\{\theta_M^{(n)} : n < \omega, M \text{ an } A\text{-bimodule}\}$, $\theta_M^{(n)}: H_R^n(A,M) \to T^n(M)$ such that:

(a) if $\phi: M \to N$ is a bimodule homomorphism, then

$$\begin{array}{ccc} H_R^n(A,M) & \xrightarrow{\phi_*^{(n)}} & H_R^n(A,N) \\ \theta_M^{(n)}\downarrow & & \downarrow\theta_N^{(n)} \\ T^n(M) & \xrightarrow{T^{(n)}(\phi)} & T^n(N) \end{array}$$

commutes;

(b) if $\sum: 0 \to N \to M \to P \to 0$ is exact, then

$$\begin{array}{ccc} H_R^n(A,P) & \xrightarrow{\partial^{(n)}} & H_R^{n+1}(A,N) \\ \theta_P^{(n)}\downarrow & & \downarrow\theta_N^{(n+1)} \\ T^n(P) & \xrightarrow[\partial^{(n)}]{} & T^n(N) \end{array}$$

commutes. Hint. Define $\theta_M^{(n)}$ for all M by induction on n, using the shifting technique that was introduced in the lemma. The proofs of the naturality conditions (a) and (b) require three dimensional diagrams.

11.5. Zero Dimensional Algebras

Our aim in this section is to prove that if an R-algebra A is a projective R-module, then A is separable if and only if $\operatorname{Dim} A = 0$.

Definition. Let M be an A-bimodule. A *derivation* of A to M is an R-module homomorphism $\phi: A \to M$ that satisfies Leibniz's rule:

$$\phi(xy) = x\phi(y) + \phi(x)y \quad \text{for all} \quad x, y \in A. \tag{1}$$

A derivation ϕ of A to M is an *inner derivation* if there exists $u \in M$ such that $\phi(x) = xu - ux$ for all $x \in A$.

Plainly, the derivations of A to M are exactly the elements of $Z_R^1(A,M)$, while the inner derivations are the elements of $B_R^1(A,M)$. Thus, the derivations of A to M form an R-module with operations defined pointwise, and the set of inner derivations is a submodule of derivations. The quotient of these modules is $H_R^1(A,M)$. Thus, $H_R^1(A,M) = 0$ if and only if every derivation of A is inner.

Let $\mu: A^e \to A$ be the augmentation homomorphism of right A^e-modules that was defined in Section 10.2. Define $\kappa: A \to A^e$ by $\kappa(x) = x \otimes 1 - 1 \otimes x$. Plainly, κ is an R-module homomorphism such that $\mu\kappa = 0$. Hence, $\operatorname{Im} \kappa \subseteq \operatorname{Ker} \mu$.

Lemma. *Let M be an A-bimodule, or equivalently a right A^e-module.*

(i) *If $\phi \in \mathrm{Hom}_{A^e}(\mathrm{Ker}\,\mu, M)$, then $\phi\kappa$ is a derivation of A to M.*
(ii) *The mapping $\theta: \phi \mapsto \phi\kappa$ is an isomorphism of $\mathrm{Hom}_{A^e}(\mathrm{Ker}\,\mu, M)$ to $Z_R^1(A,M)$.*
(iii) $\theta^{-1}(B_R^1(A,M)) = \{\psi | \mathrm{Ker}\,\mu : \psi \in \mathrm{Hom}_{A^e}(A^e, M)\}$.

PROOF. (i) Plainly, $\phi\kappa$ is an R-module homomorphism. Moreover, $\phi\kappa(xy) = \phi(xy \otimes 1 - 1 \otimes xy) = \phi(xy \otimes 1 - x \otimes y) + \phi(x \otimes y - 1 \otimes xy) = \phi(x(y \otimes 1 - 1 \otimes y)) + \phi((x \otimes 1 - 1 \otimes x)y) = x\phi\kappa(y) + \phi\kappa(x)y$.

(ii) By (i), θ maps $\mathrm{Hom}_{A^e}(\mathrm{Ker}\,\mu, M)$ to $Z_R^1(A,M)$, and θ is obviously an R-module homomorphism. If $\phi\kappa = 0$, then $\phi(x \otimes 1 - 1 \otimes x) = 0$ for all x. By Lemma 10.2, $\phi = 0$. Let $\chi: A \to M$ be a derivation. Define $\psi: A^e \to M$ to be the R-module homomorphism that is determined by the condition $\psi(x \otimes y) = \chi(x)y$. If $\phi = \psi | \mathrm{Ker}\,\mu$, then ϕ is an R-module homomorphism from $\mathrm{Ker}\,\mu$ to M such that for $\sum_{j=1}^n x_j \otimes y_j \in \mathrm{Ker}\,\mu$ and $z, w \in A$, we have $\phi((\sum x_j \otimes y_j)(z \otimes w)) = \phi(\sum zx_j \otimes y_j w) = \sum \chi(zx_j)y_j w = \sum z\chi(x_j)y_j w + \sum \chi(z)x_j y_j w = z(\sum \chi(x_j)y_j)w + \chi(z)\mu(\sum x_j \otimes y_j)w = z(\phi(\sum x_j \otimes y_j))w = \phi(\sum x_j \otimes y_j)(z \otimes w)$. That is, $\phi \in \mathrm{Hom}_{A^e}(\mathrm{Ker}\,\mu, M)$. Finally, if $x \in A$, then $\theta(\phi)(x) = \phi\kappa(x) = \phi(x \otimes 1 - 1 \otimes x) = \chi(x) - \chi(1)x = \chi(x)$, since $\chi(1) = \chi(1^2) = \chi(1) + \chi(1)$.

(iii) If $\phi \in \mathrm{Hom}_{A^e}(\mathrm{Ker}\,\mu, M)$ and $u \in M$, then $\phi\kappa = \delta^{(0)}u$ if and only if $\phi(x \otimes 1 - 1 \otimes x) = xu - ux = \lambda_u(x \otimes 1 - 1 \otimes x)$, where $\lambda_u \in \mathrm{Hom}_{A^e}(A^e, M)$ is the left multiplication homomorphism $\lambda_u(z) = uz$ for all $z \in A^e$. Thus, since $\mathrm{Ker}\,\mu$ is generated as an A^e-module by $\{x \otimes 1 - 1 \otimes x : x \in A\}$, it follows that the condition $\theta(\phi) \in B^1(M)$ is equivalent to $\phi = \lambda_u | \mathrm{Ker}\,\mu$ for some $u \in M$, which proves (iii) because $\mathrm{Hom}_{A^e}(A^e, M) = \{\lambda_u : u \in M\}$ by Lemma 6.4. □

Proposition. *For a non-trivial R-algebra A, $H_R^1(A,M) = 0$ for all A-bimodules M if and only if A is a separable R-algebra.*

PROOF. By Proposition 10.2, A is separable if and only if the sequence of A^e-modules $0 \to \mathrm{Ker}\,\mu \to A^e \to A \to 0$ is split exact, that is, there exists $\psi \in \mathrm{Hom}_{A^e}(A^e, \mathrm{Ker}\,\mu)$ such that $\psi | \mathrm{Ker}\,\mu$ is the identity homomorphism on $\mathrm{Ker}\,\mu$. Thus, A is separable if $H_R^1(A, \mathrm{Ker}\,\mu) = 0$ by the lemma. Conversely, if there is an extension $\psi \in \mathrm{Hom}_{A^e}(A^e, \mathrm{Ker}\,\mu)$ of the identity mapping of $\mathrm{Ker}\,\mu$, then every $\phi \in \mathrm{Hom}_{A^e}(\mathrm{Ker}\,\mu, M)$ has the form $\chi | \mathrm{Ker}\,\mu$, where $\chi = \phi\psi \in \mathrm{Hom}_{A^e}(A^e, M)$. By the lemma, $Z_R^1(A,M) = B_R^1(A,M)$ for every A-bimodule M. □

Corollary a. *If A is a non-trivial R-algebra that is projective as an R-module, then* $\mathrm{Dim}\,A = 0$ *if and only if A is a separable R-algebra.*

Corollary b. *If A is a separable R-algebra that is projective as an R-module, then every factor set of A is split.*

EXERCISES

1. For a field F, let B be the three dimensional F-algebra with the basis 1_B, e, and x such that $e^2 = e$, $ex = x$, $xe = 0$, and $x^2 = 0$. Prove that B is an associative F-algebra, and $\operatorname{Dim} B = 1$. Hint. Prove that if Φ is a normalized 2-cocycle, then $\Phi = \delta^{(1)}\phi$, where $\phi(e) = 2e\Phi(e,e) - \Phi(e,e)$, and $\phi(x) = e\Phi(x,e) - \Phi(e,x)$.

2. Let A be an R-algebra such that every derivation of A into a bimodule is inner. Without using Corollary 11.4, prove that every factor set of A is split. In more detail, let $\Phi: A \times A \to M$ be a factor set. Let $0 \to M \to \operatorname{Hom}_R(A,M) \xrightarrow{\pi} N \to 0$ be the short exact sequence of A-bimodules that was constructed in Section 11.2. Define $\psi\Phi: A \to \operatorname{Hom}_R(A,M)$ by $(\psi\Phi)(x)(y) = \Phi(x,y)$ and $\chi\Phi = \pi\psi\Phi: A \to N$. Show that $\chi\Phi$ is a derivation, therefore inner by assumption. Use this fact to obtain $\phi \in \operatorname{Hom}_R(A,M)$ such that $\Phi(x,y) = x\phi(y) - \phi(xy) + \phi(x)y$ for all $x, y \in A$.

11.6. The Principal Theorem

As an application of the cohomology theory, we prove one of the most important results in the theory of associative algebras.

Theorem (Wedderburn, Malcev). *Let B be an R-algebra that satisfies:*
(a) $\operatorname{Dim} B/\mathbf{J}(B) \leq 1$;
(b) *$B/\mathbf{J}(B)$ is projective as an R-module;*
(c) $\mathbf{J}(B)^k = 0$ *for some* $k \geq 1$.
There is a subalgebra A of B such that $B = A \oplus \mathbf{J}(B)$ as R-modules, and $A \cong B/\mathbf{J}(B)$ as algebras. If B satisfies
(a′) $\operatorname{Dim} B/\mathbf{J}(B) = 0$,
then for any two subalgebras A and A' of B that satisfy $B = A \oplus \mathbf{J}(B) = A' \oplus \mathbf{J}(B)$, there exists $w \in \mathbf{J}(B)$ such that

$$A' = (1 - w)^{-1}A(1 - w).$$

PROOF. The existence of A is established by induction on k. If $k = 1$, there is nothing to do but take $A = B$. Assume that $J^2 = 0$, where J abbreviates $\mathbf{J}(B)$. Let $\pi: B \to B/J$ be the natural projection. Since B/J is projective by (b), the exact sequence of R-modules $0 \to J \to B \to B/J \to 0$ splits. Thus, there is an R-module homomorphism $\kappa: B/J \to B$ such that $\pi\kappa = id_{B/J}$. For $x, y \in B/J$, define $\Phi(x,y) = \kappa(xy) - \kappa(x)\kappa(y)$. Thus, Φ is a measure of the degree to which κ fails to be an algebra homomorphism.

$$\Phi(x,y) \in J \quad \text{for all} \quad x, y \in B/J. \tag{1}$$

In fact, $\pi\Phi(x,y) = \pi\kappa(xy) - \pi(\kappa(x)\kappa(y)) = xy - xy = 0$, since π is an algebra homomorphism and $\pi\kappa = id$. Thus, $\Phi(x,y) \in \operatorname{Ker} \pi = J$. Use κ to define right and left scalar operations of B/J on J; explicitly, $ux = u\kappa(x)$, $xu = \kappa(x)u$.

$$J \text{ is a } B/J\text{-bimodule.} \tag{2}$$

Indeed, since κ is an R-module homomorphism, and $R \subseteq \mathbf{Z}(B)$, the only doubtful bimodule axioms are $(xy)u = x(yu)$ and $u(xy) = (ux)y$. By (1), $(xy)u - x(yu) = \kappa(xy)u - \kappa(x)\kappa(y)u = \Phi(x,y)u \in J^2 = 0$. Similarly, $u(xy) = (ux)y$.

$$\Phi \in Z_R^2(B/J,J). \tag{3}$$

Plainly, Φ is bilinear. Moreover,

$$\begin{aligned}(\delta^{(2)}\Phi)(x,y,z) &= x(\kappa(yz) - \kappa(y)\kappa(z)) - (\kappa(xyz) - \kappa(xy)\kappa(z)) \\ &\quad + (\kappa(xyz) - \kappa(x)\kappa(yz)) - (\kappa(xy) - \kappa(x)\kappa(y))z \\ &= \kappa(x)\kappa(yz) - \kappa(x)\kappa(y)\kappa(z) - \kappa(xyz) \\ &\quad + \kappa(xy)\kappa(z) + \kappa(xyz) - \kappa(x)\kappa(yz) \\ &\quad - \kappa(xy)\kappa(z) + \kappa(x)\kappa(y)\kappa(z) = 0.\end{aligned}$$

By (3) and the hypothesis (a), $\Phi \in B_R^2(B/J,J)$. That is, there exists $\phi \in \mathrm{Hom}_R(B/J,J)$ such that $\Phi(x,y) = x\phi(y) - \phi(xy) + \phi(x)y$ for all $x, y \in B/J$. Let $\psi = \kappa + \phi \in \mathrm{Hom}_R(B/J,B)$. Then $\pi\psi = \pi\kappa = id_{B/J}$, since $\pi(J) = 0$. Moreover, $\psi(xy) - \psi(x)\psi(y) = \kappa(xy) + \phi(xy) - (\kappa(x) + \phi(x))(\kappa(y) + \phi(y)) = \kappa(xy) - \kappa(x)\kappa(y) - (x\phi(y) - \phi(xy) + \phi(x)y) - \phi(x)\phi(y) = 0$, because $\kappa(xy) - \kappa(x)\kappa(y) = \Phi(x,y) = x\phi(y) - \phi(xy) + \phi(x)y$, and $\phi(x)\phi(y) \in J^2 = 0$. Also, $\psi(1_{B/J}) - 1_B \in J$ implies $0 = (\psi(1) - 1)^2 = \psi(1)^2 - 2\psi(1) + 1 = 1 - \psi(1)$. Thus, ψ is an algebra homomorphism, and $A = \mathrm{Im}\,\psi$ is a subalgebra of B that satisfies $B = A \oplus J$. Assume now that $k > 2$, and the existence portion of the theorem has been established for algebras whose radicals are nilpotent of order less than k. Let $B_1 = B/J^2$. Then $J/J^2 \lhd B_1$, $B_1/(J/J^2) \cong B/J$, and $(J/J^2)^2 = 0$. Thus, $\mathbf{J}(B_1) = J/J^2$, and B_1 satisfies the hypotheses (a), (b), and (c). By the case $k = 2$ that has been completed, there is a subalgebra A_1 of B_1 such that $B_1 = A_1 \oplus J/J^2$. Let $A_1 = C/J^2$, where C is a subalgebra of B such that $C \cap J = J^2$. As R-algebras, $C/J^2 = C/(C \cap J) \cong (C + J)/J = B/J$. Moreover, $(J^2)^{k-1} = J^{k+(k-2)} \subseteq J^k = 0$. Thus, $\mathbf{J}(C) = J^2$ and C satisfies (a), (b), and (c) with $\mathbf{J}(C)^{k-1} = 0$. By the induction hypothesis, there is a subalgebra A of C such that $C = A \oplus J^2$. Consequently, $A + J = C + J = B$, and $A \cap J = A \cap C \cap J = A \cap J^2 = 0$, that is, $B = A \oplus J$. This completes the "existence" portion of the theorem. For the proof of uniqueness, assume that (a′), (b), and (c) are satisfied, and that A and A' are subalgebras of B such that $B = A \oplus J = A' \oplus J$. There exist commutative diagrams

$$\begin{array}{ccc} B & \xrightarrow{\rho} & A \\ {\scriptstyle\pi}\downarrow & \nearrow_{\psi} & \\ B/J & & \end{array} \qquad \begin{array}{ccc} B & \xrightarrow{\rho'} & A' \\ {\scriptstyle\pi}\downarrow & \nearrow_{\psi'} & \\ B/J & & \end{array}$$

in which ρ and ρ' are the canonical projections associated with the decompositions $B = A \oplus J$ and $B = A' \oplus J$. Note that ρ and ρ' are algebra homomorphisms; therefore, so are ψ and ψ'. In fact, if $x, y \in A$ and $z, w \in J$, then $\rho((x+z)(y+w)) = \rho(xy + (xw + zy + zw)) = xy = \rho(x+z)\rho(y+w)$. Define a B/J-bimodule structure on J by $xu = \psi(x)u$ and $ux = u\psi'(x)$. The bimodule axioms are satisfied since ψ and ψ' are algebra homomorphisms. Define $\chi: B/J \to B$ by $\chi(x) = \psi(x) - \psi'(x)$. Note that $\pi\chi\pi = \pi\psi\pi - \pi\psi'\pi = \pi\rho - \pi\rho' = \pi(id_B - \rho') - \pi(id_B - \rho) = 0$, because $\operatorname{Im}(id_B - \rho) = \operatorname{Im}(id_B - \rho') = J = \operatorname{Ker}\pi$. Hence, $\operatorname{Im}\chi = \operatorname{Im}\chi\pi \subseteq J$, that is, $\chi \in \operatorname{Hom}_R(B/J, J)$. Moreover, $\chi(xy) = \psi(xy) - \psi'(xy) = \psi(x)\psi(y) - \psi'(x)\psi'(y) = \psi(x)(\psi(y) - \psi'(y)) + (\psi(x) - \psi'(x))\psi'(y) = x\chi(y) + \chi(x)y$. Thus, χ is a derivation of B/J to J. Since $\operatorname{Dim} B/J = 0$, χ is an inner derivation: there exists $w \in J$ such that $\chi(x) = xw - wx$ for all $x \in B/J$. That is, $\psi(x) - \psi'(x) = xw - wx = \psi(x)w - w\psi'(x)$, and $\psi(x)(1-w) = (1-w)\psi'(x)$. Since $w^k = 0$, $1 - w$ has an inverse, and $A' = \psi'(B/J) = (1-w)^{-1}\psi(B/J)(1-w) = (1-w)^{-1}A(1-w)$. □

Corollary. *If F is a perfect field, and B is a finite dimensional F-algebra, then there is a subalgebra A of B such that $B = A \bigoplus \mathbf{J}(B)$. Moreover, A is unique up to conjugation by units of the form $1 - w$, where $w \in \mathbf{J}(B)$.*

PROOF. $B/\mathbf{J}(B)$ is a finite dimensional semisimple algebra over a perfect field. By Corollary 10.7b, $B/\mathbf{J}(B)$ is separable, and therefore $\operatorname{Dim} B/\mathbf{J}(B) = 0$ by Corollary 11.5a. Since F is a field, all F-modules are projective. Finally, $(\mathbf{J}(B))^k = 0$ for some $k \geq 1$ by Proposition 4.4. Since the hypotheses (a′), (b), and (c) of the theorem are satisfied, the corollary is proved. □

EXERCISES

1. (a) Let E be a field of characteristic 2, $F = E(\mathbf{x})$, and define $A = F(c)$, where $c^2 = \mathbf{x}$. Show that $Z_F^n(A, M) \cong M^{(A)}$ under the mapping $\Phi \to \Phi(c, c, \ldots, c)$. Deduce that $\operatorname{Dim} A = \infty$.

   (b) Let B be the four dimensional F-algebra (where F is the field that was defined in (a)) with the basis $1_B, d, y, z$, and the multiplication defined by $d^2 = 1_B\mathbf{x} + y + z$, $dy = yd = z$, $dz = zd = y\mathbf{x}$, and $y^2 = z^2 = yz = zy = 0$. Prove that B is an associative algebra with $\mathbf{J}(B) = yF + zF$, $B/\mathbf{J}(B) \cong A$, and no subalgebra of B is isomorphic to A.

2. Let B be the three dimensional F-algebra with basis 1_B, e, x that was defined in Exercise 1 of Section 11.5. Prove that $\mathbf{J}(B) = xF$ and $B/\mathbf{J}(B) \cong F \dotplus F$. Show that $A = eF + (1-e)F$ and $A' = (e+x)F + (1-e-x)F$ are subalgebras of B such that $B = A \bigoplus \mathbf{J}(B) = A' \bigoplus \mathbf{J}(B)$, and there is no unit $u \in B$ such that $A' = u^{-1}Au$.

11.7. Split Extensions of Algebras

The Wedderburn Principal Theorem can be viewed as a result concerning algebra extensions. In this section we will introduce split extensions of algebras and formulate Theorem 11.6 as a statement about such extensions.

Definition. Let A be an R-algebra. A *multiplicative A-bimodule* is an A_R-bimodule N on which an R-bilinear, associative multiplication $(u,v) \mapsto uv$ is defined, and

$$(uv)x = u(vx), \quad (ux)v = u(xv), \quad (xu)v = x(uv) \tag{1}$$

for all $u, v \in N$, and $x \in A$.

It is not assumed that a multiplicative A-bimodule N has a unity element for multiplication, so that N may not be an R-algebra in the sense of Definition 1.1. Nevertheless, the terminology of algebra theory can be applied to multiplicative bimodules. In particular, if $k \in \mathbb{N}$, we will call a multiplicative A-bimodule N *k-nilpotent* if $N^k = 0$; that is, $u_1 u_2 \cdots u_k = 0$ for every sequence $u_1, u_2, \ldots, u_k$ of elements in N. A homomorphism from a multiplicative A-bimodule M to a multiplicative A-bimodule N is a bimodule homomorphism $\phi\colon M \to N$ such that $\phi(uv) = \phi(u)\phi(v)$ for all $u, v \in M$. If ϕ is also bijective, then it is called an isomorphism.

The equations (1) and the bimodule identities $(ux)a = u(xa) = (ua)x$, $(xu)a = x(ua) = (xa)u$ for $u \in N$, $x \in A$, $a \in R$ imply that the multiplication mapping $(u,v) \mapsto uv$ on a multiplicative A-bimodule N induces $\mu \in \mathrm{Hom}_{A^e}(N \otimes_A N, N)$ such that the associativity condition

$$\mu(u \otimes \mu(v \otimes w)) = \mu(\mu(u \otimes v) \otimes w) \tag{2}$$

is satisfied for all u, v, and w in N. Conversely, any $\mu \in \mathrm{Hom}_{A^e}(N \otimes_A N, N)$ that satisfies (2) defines a multiplicative structure on the A-bimodule N. In particular, the zero mapping $N \otimes_A N \to 0 \in N$ defines a 2-nilpotent multiplicative structure on N.

There is another source of multiplicative A-bimodules. Let B be an R-algebra, A a subalgebra of B, and N an ideal of B. The product in B imposes an A-bimodule structure on N ($(u,x) \mapsto ux$, $(x,u) \mapsto xu$), and a multiplication ($(u,v) \mapsto uv$) that make N a multiplicative A-bimodule. The next result shows that this example is universal.

Lemma a. *Let N be a multiplicative A-bimodule. Define $N \rtimes A = N \oplus A$ as R-modules, and $(u,x)(v,y) = (uv + xv + uy, xy)$ for $u, v \in N$; $x, y \in A$.*

(i) *$N \rtimes A$ is an R-algebra with unity* $(0,1)$.
(ii) *$A' = \{(0,x)\colon x \in A\}$ is a subalgebra of $N \rtimes A$ that is isomorphic to A by the mapping* $(0,x) \mapsto x$.
(iii) *$N' = \{(u,0)\colon u \in N\}$ is an ideal of $N \rtimes A$ that is isomorphic (as a multiplicative A-bimodule) to N by the mapping* $(u,0) \mapsto u$.
(iv) $N \rtimes A = N' \oplus A'$.

The statements (i), (ii), (iii), and (iv) can be verified by routine computation. The converse result is more interesting.

Lemma b. *If B is an R-algebra, A is a subalgebra of B, and N is an ideal of B such that $B = N \oplus A$ as R-modules, then N is a multiplicative A-bimodule with the bimodule and multiplication operations inherited from B, and $B \cong N \rtimes A$ as R-algebras.*

PROOF. The fact that N is a multiplicative A-bimodule is a special case of earlier remarks. The mapping $\theta: (u,x) \mapsto u + x$ is an R-module isomorphism of $N \rtimes A$ to B because $B = N \oplus A$. Moreover, $\theta((u,x)(v,y)) = \theta(uv + xv + uy, xy) = uv + xv + uy + xy = (u + x)(v + y) = \theta(u,x)\theta(v,y)$, so that θ is an R-algebra isomorphism. □

If the hypotheses of Lemma b are satisfied, then B is called a *split extension* of N by A. This terminology and the notation $N \rtimes A$ is motivated by the analogy of split extensions of groups. However, a warning should be given. The product $N \dotplus A$ of R-algebras is generally not an instance of a split extension because A is not a subalgebra of $N \dotplus A$: $1_{N+A} \neq 1_A$ if $N \neq 0$. (See Exercises 4 and 5.)

The principal result of this section is essentially a reformulation of (a special case of) the Wedderburn Principal Theorem, using Lemmas a and b.

Proposition. *If B is a finite dimensional F-algebra such that $A = B/\mathbf{J}(B)$ is separable, then $\mathbf{J}(B)$ is a nilpotent, multiplicative A-bimodule, and $B \cong \mathbf{J}(B) \rtimes A$. Conversely, if A is a semisimple F-algebra, N is a nilpotent, multiplicative A-bimodule, and $B = N \rtimes A$, then $N \cong \mathbf{J}(B)$ and $A \cong B/\mathbf{J}(B)$.*

There is a uniqueness statement that accompanies the proposition. Roughly, it states that $N \rtimes A \cong N' \rtimes A'$ if and only if A and A' can be identified in such a way that N and N' are isomorphic as multiplicative A-bimodules.

Corollary. *Let A and A' be separable F-algebras, and suppose that N and N' are multiplicative A- and A'-bimodules respectively. If $N \rtimes A \cong N' \rtimes A'$ as F-algebras, then there is an F-algebra isomorphism $\theta: A \to A'$ and an F-space isomorphism $\psi: N \to N'$ such that*

$$\psi(uv) = \psi(u)\psi(v), \quad \psi(xu) = \theta(x)\psi(u), \quad \psi(ux) = \psi(u)\theta(x) \tag{3}$$

for all $u, v \in N$ and $x \in A$. Conversely, if $\theta: A \to A'$ is an F-algebra isomorphism and $\psi: N \to N'$ is an F-space isomorphism such that (3) is satisfied, then $N \rtimes A \cong N' \rtimes A'$.

PROOF. Let $\phi: N \rtimes A \to N' \rtimes A'$ be an F-algebra isomorphism. By the proposition, $\phi(N) = \phi(\mathbf{J}(N \rtimes A)) = \mathbf{J}(N' \rtimes A') = N'$, and $N' \rtimes A' = \mathbf{J}(N' \rtimes A') \oplus \phi(A)$. By Theorem 11.6, $A' = (1 - w)^{-1}\phi(A)(1 - w)$ for some $w \in N'$. Define $\theta: A \to A'$ by $\theta(x) = (1 - w)^{-1}\phi(x)(1 - w)$ and

$\psi\colon N \to N'$ by $\psi(u) = (1 - w)^{-1}\phi(u)(1 - w)$. Plainly, θ is an algebra isomorphism, and ψ is an F-space isomorphism. The equations (3) are obtained by a routine calculation. For the converse, define $\phi\colon N \rtimes A \to N' \rtimes A'$ by $\phi(u,x) = (\psi(u),\theta(x))$. A straightforward check shows that ϕ is an F-algebra isomorphism. □

EXERCISES

1. Prove the statements (i), (ii), and (iii) in Lemma a.
2. Complete the proof of the corollary.
3. Prove that if A is an R-algebra and N is an A-bimodule, then there is an R-algebra B that contains an ideal I such that $B/I \cong A$, $I^2 = 0$, and I is isomorphic to N as an A-bimodule (with the bimodule operations induced by the isomorphism $A \cong B/I$).
4. Let N be a multiplicative A-bimodule such that there is a unity element for the multiplication of N. Prove that $N \rtimes A \cong N \dotplus A$. Hint. Show that $(1_N,0)$ and $(-1_N,1_A)$ are central idempotent elements in $N \rtimes A$.
5. Prove that the $\mathbb{Z}$-algebra $\mathbb{Z}/2\mathbb{Z} \dotplus \mathbb{Z}/3\mathbb{Z}$ cannot be written as a split extension of a non-zero multiplicative bimodule by a non-zero algebra.

11.8. Algebras with 2-nilpotent Radicals

Classification is still the fundamental problem in the theory of algebras. For many algebras the results of Section 11.7 shift the classification problem to the study of nilpotent, multiplicative bimodules. In this section, we will use this approach to study finite dimensional algebras B over an algebraically closed field, that satisfy $\mathbf{J}(B)^2 = 0$. By combining the results of previous chapters, it is possible to give a complete classification of these algebras, and to characterize the algebras in this class that have finite representation types.

In this section, assume that F is an algebraically closed field. The algebras under consideration are finite dimensional F-algebras. Since F is perfect, such an algebra is separable if and only if it is semisimple. By Proposition 11.7, we can limit our attention to the algebras of the form $N \rtimes A$, where A is semisimple and N is a nilpotent multiplicative A-bimodule. Since $\mathbf{J}(N \rtimes A) \cong N$, the assumption that $N \rtimes A$ is 2-nilpotent is the same as the condition $N^2 = 0$, that is, N has the trivial multiplication. Thus, the isomorphism classes of finite dimensional F-algebras B such that $\mathbf{J}(B)^2 = 0$ are in one-to-one correspondence with the isomorphism classes of pairs (A,N) such that A is finite dimensional, semisimple, and N is a finite dimensional A-bimodule, or equivalently a right A^e-bimodule.

Since F is algebraically closed, it follows from the Wedderburn Structure

Theorem that every finite dimensional, semisimple F-algebra is isomorphic to a product $M_{n_1}(F) \dot{+} M_{n_2}(F) \dot{+} \cdots \dot{+} M_{n_r}(F)$ of full matrix algebras with the ordering of the factors chosen so that $1 \leq n_1 \leq n_2 \leq \cdots \leq n_r$. The non-decreasing sequence $(n_1,n_2,\ldots,n_r)$ of natural numbers constitutes a complete set of invariants for the finite dimensional, semisimple F-algebras.

To simplify notation, write $A = A_1 \dot{+} A_2 \dot{+} \cdots \dot{+} A_r$, where $A_i \cong M_{n_i}(F)$. The enveloping algebra $A^e = A^* \bigotimes A$ is the product over all pairs (i,j) with $1 \leq i,j \leq r$ of the algebras $A_{ij} = A_i^* \bigotimes A_j \cong M_{n_i}(F)^* \bigotimes M_{n_j}(F) \cong M_{n_in_j}(F)$ (since $M_{n_i}(F)^* \cong M_{n_i}(F)$ by the transpose mapping). In particular, A^e is semisimple, so that every right A^e-module is isomorphic to a direct sum of simple modules N_{ij}, where N_{ij} is a minimal right ideal of A_{ij} (considered as a right A^e-module). To within isomorphism, N_{ij} depends only on (i,j). Any finite dimensional A-bimodule is isomorphic to a unique direct sum $\bigoplus_{1 \leq i,j \leq r} \bigoplus m_{ij} N_{ij}$, in which the m_{ij} are non-negative integers.

The pairs that consist of a non-decreasing sequence $(n_1,n_2,\ldots,n_r)$ of natural numbers and an r by r matrix $[m_{ij}]$ of non-negative integers determine a set of representatives of all isomorphism classes of finite dimensional F-algebras B such that $\mathbf{J}(B)^2 = 0$. If the n's are not distinct, then different matrices may correspond to isomorphic algebras. For example, if $n_1 = n_2$, then interchanging the first and second rows and columns of $[m_{ij}]$ gives a new invariant for the same isomorphism class of algebras. A genuine invariant can be obtained by defining a suitable equivalence relation on the matrices $[m_{ij}]$. The details of this procedure are sketched in Exercise 1.

We now take a final look at the representation types of algebras.

Theorem. *Let F be an algebraically closed field, and suppose that B is a finite dimensional F-algebra such that $\mathbf{J}(B)^2 = 0$. For B to have finite representation type, it is necessary and sufficient that $\mathbf{I}(B)$ is distributive, and the separated quiver $\Gamma^s(B)$ has a diagram that is a disjoint union of Dynkin diagrams of types, A_n, D_n, E_6, E_7, or E_8.*

The proof of this result consists of assembling a few of the deep theorems that we have proved concerning algebras and their representations. We will sketch a map of the path that leads to the conclusion of the theorem.

By Corollary 9.6, B has finite representation type if and only if the basic algebra of B has finite representation type. Proposition 6.6b therefore permits us to assume that B is reduced, that is, $A = B/\mathbf{J}(B)$ is a product of copies of F (since F is algebraically closed). In other words, the sequence associated with A is $(1,1,\ldots,1)$. For B to have finite representation type, it is necessary by Theorem 6.7 that the lattice $\mathbf{I}(B)$ of ideals in B be distributive. By Proposition 4.8, $\mathbf{I}(B)$ is distributive if and only if the lattice of sub-bimodules of $\mathbf{J}(B)$ is distributive, where $\mathbf{J}(B)$ can be viewed as an A-bimodule because $\mathbf{J}(B)^2 = 0$. The distributivity of $\mathbf{I}(B)$ therefore translates via Corollary 2.4c to the condition that the entries m_{ij} in the matrix $[m_{ij}]$ associated with B are all 0 or 1. An easy check shows that $m_{ij} \neq 0$

exactly when (i,j) belongs to the edge set of the quiver $\Gamma(B)$. Consequently, if B is reduced and $\mathbf{I}(B)$ is distributive, then B is isomorphic to the algebra $B_{\Gamma(B)}$ that was defined in Section 8.1. The theorem follows from Proposition 8.3.

The hypothesis $\mathbf{J}(B)^2 = 0$ severely restricts the usefulness of the theorem. However, if B has finite representation type, then so does $B/\mathbf{J}(B)^2$. Thus, for a finite dimensional algebra over an algebraically closed field, the theorem imposes a condition on $\Gamma(B/\mathbf{J}(B)^2)$ that is necessary for B to have finite representation type.

Exercises

1. Let F be a field. For each pair (ν,μ) consisting of a sequence $\nu = (n_1, n_2, \ldots, n_r)$ of natural numbers with $n_1 \leq n_2 \leq \cdots \leq n_r$, and a matrix $\mu = [m_{ij}]$ of non-negative integers, associate the finite dimensional F-algebra $B(\nu,\mu) = N \rtimes A$, where $A = A_1 \times A_2 \times \cdots \times A_r$ with $A_i = M_{n_i}(F)$, and N is the multiplicative A-bimodule $\bigoplus_{1 \leq i,j \leq r} \bigoplus m_{ij} N_{ij}$, where the N_{ij} are simple right ideals in $A_i^* \otimes A_j$, and $N^2 = 0$. Thus, if F is algebraically closed, then the discussion of Section 11.8 shows that every finite dimensional F-algebra B such that $\mathbf{J}(B)^2 = 0$ is isomorphic to an algebra of the form $B(\nu,\mu)$. Prove the following statements:

   (a) If $\nu = (n_1, \ldots, n_r)$ and $G_\nu = \{\pi \in S_r : \pi(i) = j \text{ implies } n_i = n_j\}$, then G is a subgroup of the symmetric group S_r.

   (b) With ν and G_ν as in (a), define $[m'_{ij}] \sim_\nu [m_{ij}]$ if there exists $\pi \in G_\nu$ such that $m'_{ij} = m_{\pi(i)\pi(j)}$, for all i and j. Then $\sim_\nu$ is an equivalence relation on the set of r by r matrices over ω.

   (c) $B(\nu',\mu') \cong B(\nu,\mu)$ if and only if $\nu' = \nu$ and $\mu' \sim_\nu \mu$.

   (d) $\dim_F B(\nu,\mu) = \nu(\mu + \iota_r)\nu^t$.

   (e) For $\nu = (n_1, n_2, \ldots, n_r)$ and $\nu_1 = (1, 1, \ldots, 1)$ (a sequence of length r), $B(\nu_1,\mu)$ is a basic subalgebra of $B(\nu,\mu)$.

2. Let F be an algebraically closed field. Enumerate the isomorphism classes of four dimensional F-algebras B such that $\mathbf{J}(B)^2 = 0$.

Notes on Chapter 11

The cohomology groups of algebras were introduced by G. Hochschild. It was also Hochschild who found the connection between separable algebras and the theorems of Wedderburn and Malcev. Our treatment of these results follows Hochschild's paper [44]. There are more sophisticated ways to obtain the cohomology modules of algebras than Hochschild's original construction, but his approach efficiently produces the needed machinery, using fairly primitive tools. The proof of the Snake Lemma given in Section 11.3 is due to J. B. Leicht. The reduction of the structure problem for algebras to the case of nilpotent algebras (as in Section 11.7) is enlightening, but it does not come close to solving the problem. In his 1939 book [3],

Albert remarked on page 172 that only fragmentary results were known about the structure of nilpotent algebras. That situation has apparently not changed. Theorem 11.8 first appeared in the paper [34] by Gabriel. This work initiated a flurry of activity by numerous mathematicians. The dust that this work stirred up has not yet settled. The best current summaries of progress in the theory of algebra representations can be found in [29]. These notes also contain a very complete bibliography of papers on the subject.

CHAPTER 12
Simple Algebras

This chapter is the beginning of a systematic study of simple algebras over a field. We will concentrate our attention on finite dimensional algebras. The problems encountered in the study of infinite dimensional simple algebras are formidable; they lead to a theory that bears little resemblance to the subject of finite dimensional, simple algebras.

The highlights of the chapter are four classical theorems: the Jacobson Density Theorem, the Jacobson–Bourbaki Theorem, the Noether–Skolem Theorem, and the Double Centralizer Theorem. These results, together with Wedderburn's Structure Theorem, comprise the foundation of the theory of simple algebras.

The center plays a fundamental role in the study of simple algebras. Especially important are the central simple algebras, that is, simple F-algebras A such that $\mathbf{Z}(A) = F$. Some basic properties of these algebras are established in Section 12.4. The Brauer group of a field is introduced in Section 12.5. These goups are of fundamental importance in the theory of central simple algebras. The computation of the Brauer groups of various fields is our principal theme in the remaining chapters of this book. The last two sections of this chapter present proofs of the Noether–Skolem Theorem and the Double Centralizer Theorem.

12.1. Centers of Simple Algebras

If A is an R-algebra, the *center* of A is the set $\mathbf{Z}(A) = \{y \in A : xy = yx$ for all $x \in A\}$. A two line check shows that $\mathbf{Z}(A)$ is a subalgebra of A. In particular, $R \subseteq \mathbf{Z}(A)$. By definition, $\mathbf{Z}(A)$ is commutative.

The center plays an important part in the study of simple algebras. We will increasingly appreciate its usefulness as this chapter evolves.

Lemma. *Let A be an R-algebra.*

(i) *A is simple if and only if A is a simple right A^e-module.*
(ii) *$\mathbf{Z}(A) \cong \mathbf{E}_{A^e}(A)$ under the mapping $y \mapsto \lambda_y$, where $\lambda_y(x) = yx$.*

PROOF. The assertion (i) is a consequence of the observation that a subset I of A is an algebra ideal if and only if I is an A^e-submodule of A. The statement (ii) follows from Proposition 1.3 and the observation that $\mathbf{Z}(\mathbf{E}_A(A)) = \mathbf{E}_{A^e}(A)$. □

This lemma and Schur's Lemma give the main result of this section.

Proposition. *If A is a simple algebra, then $\mathbf{Z}(A)$ is a field.*

Because of this proposition, the study of simple algebras over fields is as general as the investigation of simple algebras over an arbitrary commutative ring.

Since $\mathbf{Z}(A \dotplus B) = \mathbf{Z}(A) \dotplus \mathbf{Z}(B)$, it follows from the proposition and Wedderburn's Structure Theorem that the center of a semisimple algebra is a product of fields. The structure theorem also reduces the calculation of the center of a simple algebra to the determination of the center of a division algebra. This assertion is based on a more general observation.

EXAMPLE. For any algebra A, $\mathbf{Z}(M_n(A)) = \iota_n \mathbf{Z}(A)$.

PROOF. Let $\alpha = \sum_{i,j=1}^n \varepsilon_{ij} x_{ij} \in \mathbf{Z}(M_n(A))$, where $x_{ij} \in A$. For $1 \leq k, l \leq n$, $\sum_{i=1}^n \varepsilon_{il} x_{ik} = \alpha \varepsilon_{kl} = \varepsilon_{kl} \alpha = \sum_{j=1}^n \varepsilon_{kj} x_{lj}$. Thus, $x_{kj} = 0$ for $k \neq j$, and $x_{kk} = x_{ll}$. That is, $\alpha = \iota_n x$ for some $x \in A$. Since $\iota_n xy = (\iota_n x)(\iota_n y) = (\iota_n y)(\iota_n x) = \iota_n yx$ for all $y \in A$, it follows that $x \in \mathbf{Z}(A)$. Hence, $\mathbf{Z}(M_n(A)) \subseteq \iota_n \mathbf{Z}(A)$. The reverse inclusion is obvious. □

This example can be reformulated as a statement about the center of certain endomorphism algebras: if P is a finitely generated, free A-module, then $\mathbf{Z}(\mathbf{E}(P)) = id_P \mathbf{Z}(A)$.

EXERCISES

1. Prove that the following F-algebras are central, that is, have F as their centers:
   (a) the algebra of Exercise 2, Section 3.1 (where $F = \mathbb{Q}$);
   (b) the algebra of Exercise 1, Section 3.3;
   (c) the algebra of Exercise 5, Section 3.3 (where $F = \mathbb{R}$);
   (d) the algebra of Exercise 1, Section 11.5.

2. Let $\theta: A \to B$ be an algebra homomorphism.

(a) Prove that if θ is surjective, then $\theta(\mathbf{Z}(A)) \subseteq \mathbf{Z}(B)$.

(b) Prove that this inclusion can be proper, even when A is a finite dimensional F-algebra. Hint. Let A be the algebra of Exercise 5, Section 11.5, and $B = A/\mathbf{J}(A)$.

(c) Prove that if θ is not assumed to be surjective, then the inclusion of (a) can fail.

3. (a) Prove that if A is an algebra such that $\mathbf{J}(A)$ and $\mathbf{J}(\mathbf{Z}(A))$ are nilpotent, then

$$\mathbf{J}(\mathbf{Z}(A)) = \mathbf{Z}(A) \cap \mathbf{J}(A).$$

(b) Let R be the localization of $\mathbb{Z}$ at the prime p, that is, $R = \{m/n: m \in \mathbb{Z}, n \in \mathbb{Z} - p\mathbb{Z}\}$. Define

$$A = \begin{bmatrix} R & \mathbb{Q} \\ 0 & \mathbb{Q} \end{bmatrix} = \left\{ \begin{bmatrix} a & x \\ 0 & y \end{bmatrix} : a \in R, x, y \in \mathbb{Q} \right\}.$$

Prove that A is a $\mathbb{Z}$-algebra for which $\mathbf{J}(\mathbf{Z}(A)) = \iota_2(pR) \neq 0 = \mathbf{Z}(A) \cap \mathbf{J}(A)$.

12.2. The Density Theorem

Jacobson's Density Theorem is often viewed as a generalization of Wedderburn's Structure Theorem to infinite dimensional algebras. However, it is also an extremely useful tool for treating questions about finite dimensional simple algebras. In this section we will prove one variant of the Density Theorem. The traditional version of the Density Theorem is outlined in Exercise 2.

Lemma. *Let M be a semisimple right A-module. Denote $D = \mathbf{E}_A(M)$. Consider M as a D-$\mathbf{E}_D(M)$-bimodule. If $\phi \in \mathbf{E}_D(M)$ and $u_1, \ldots, u_n \in M$, then $x \in A$ exists so that $u_i\phi = u_i x$ for $1 \leq i \leq n$.*

PROOF. By Corollary 2.4b, $N = \bigoplus n M$ is a semisimple A-module. Let $w = (u_1, \ldots, u_n) \in N$. By Proposition 2.4, there is a submodule P of N such that $N = wA \bigoplus P$. Let $\pi \in \mathbf{E}_A(N)$ be the corresponding projection of N to wA. By Corollary 3.4b, we can identify $\mathbf{E}_A(N)$ with $M_n(\mathbf{E}_A(M)) = M_n(D)$. Plainly, $\iota_n\phi$ is an endomorphism of N, considered as a left $M_n(D)$-module. Thus, $(u_1\phi, \ldots, u_n\phi) = w(\iota_n\phi) = (\pi w)(\iota_n\phi) = \pi(w(\iota_n\phi)) \in wA$. That is, there exists $x \in A$ such that $(u_1\phi, \ldots, u_n\phi) = wx = (u_1x, \ldots, u_nx)$. Equivalently, $u_i\phi = u_ix$ for $1 \leq i \leq n$. □

Density Theorem (Jacobson). *Assume that M is a simple right A-module. Consider M as a left D-space, where D is the division algebra $\mathbf{E}_A(M)$. If $u_1, \ldots, u_n \in M$ are linearly independent over D and $w_1, \ldots, w_n \in M$ are any elements, then $x \in A$ exists so that $u_ix = w_i$ for $1 \leq i \leq n$.*

PROOF. Since M is simple, D is a division algebra by Schur's Lemma. In particular, ${}_DM$ is a D-space, that is, semisimple and free as a D-module. By Proposition 2.4 and the assumption that $u_1, \ldots, u_n$ are linearly independent, there exists a D-subspace N of M such that $M = Du_1 \oplus \cdots \oplus Du_n \oplus N$. Define $\phi \in \mathbf{E}_D(M)$ by the conditions $u_i\phi = w_i$ for $1 \leq i \leq n$ and $N\phi = 0$. By the lemma, there is an element $x \in A$ such that $w_i = u_i\phi = u_ix$ for $1 \leq i \leq n$. □

EXERCISES

1. The various parts of this exercise outline an alternative proof of the density theorem. This argument is essentially the original proof that was given by Jacobson. Let P be a simple right A-module over an R-algebra A. Denote by D the division algebra $\mathbf{E}_A(P)$. Consider P as a left D-space. Let $u_1, \ldots, u_n \in P$ be D-linearly independent. Use induction on n to prove that $(u_1, \ldots, u_n)A = \bigoplus n P$. Hint. For $1 \leq i \leq n$, let $w_i = (u_1, \ldots, \hat{u}_i, \ldots, u_n) \in \bigoplus(n-1)P$. Define $\phi_i \in \operatorname{Hom}_A(A, \bigoplus(n-1)P)$ by $\phi_i(x) = w_ix$. By the induction hypothesis, ϕ_i is surjective. Define $\psi_i \in \operatorname{Hom}_A(A,P)$ by $\psi_i(x) = u_ix$. Use the linear independence of $u_1, \ldots, u_n$ to prove that $\operatorname{Ker}\phi_i \nsubseteq \operatorname{Ker}\psi_i$. Let $x_i \in \operatorname{Ker}\phi_i - \operatorname{Ker}\psi_i$. Use the hypothesis that P is simple to show that $(u_1, \ldots, u_n)A \supseteq (u_1x_1, 0, \ldots, 0)A + \cdots + (0, \ldots, 0, u_nx_n)A = \bigoplus n P$.

2. Let D be an R-algebra, and suppose that M is a left D-module. Consider M as a right $\mathbf{E}_D(M)$-module. The *finite topology* on $\mathbf{E}_D(M)$ is defined by taking an open basis consisting of the sets

$$N_\phi(u_1, \ldots, u_n) = \{\psi \in \mathbf{E}_D(M)\colon u_i\psi = u_i\phi \quad \text{for} \quad 1 \leq i \leq n\}.$$

A subalgebra of $\mathbf{E}_D(M)$ is called a *dense algebra* of endomorphisms of M if it is dense in the finite topology.

(a) Use the Density Theorem to prove that an R-algebra A is primitive (as in Exercise 4, Section 4.3) if and only if A is isomorphic to a dense algebra of endomorphisms of a D-space, where D is a division algebra over R. (Note that if endomorphisms operate on the right, then the right regular representation is an algebra isomorphism rather than an anti-isomorphism.)

(b) Prove that the Jacobson radical of an R-algebra A is zero if and only if A is isomorphic to a subdirect product of dense algebras of endomorphisms of vector spaces over division algebras.

(c) Use the result (a) to give an alternative proof of the Wedderburn Structure Theorem for simple Artinian algebras.

3. Let A be a primitive R-algebra that is finitely generated as an R-module. Prove that A is a simple Artinian algebra.

12.3. The Jacobson–Bourbaki Theorem

If A is a simple F-algebra, then $\mathbf{Z}(A)/F$ is a field extension. If $\dim_F A < \infty$, then this extension is, of course, finite. The Jacobson–Bourbaki Theorem establishes a Galois connection between the fields K satisfying $F \subseteq K \subseteq \mathbf{Z}(A)$ and certain subalgebras of $\mathbf{E}_F(A)$. The most important application of this result is in Galois Theory, but it is also useful in the study of central simple algebras.

Let A be an R-algebra. We will consider A as a left $\mathbf{E}_R(A)$-module and a right $\mathbf{E}_R(A)^*$-module. In general, A is not an $\mathbf{E}_R(A)$-$\mathbf{E}_R(A)^*$ bimodule. By restriction, A becomes a left module over the subalgebras of $\mathbf{E}_R(A)$ and a right module over the subalgebras of $\mathbf{E}_R(A)^*$. In particular, the left A-module structure of A coincides with the module structure defined by the subalgebra $\lambda(A)$ of $\mathbf{E}_R(A)$, since the left regular representation λ is injective. It is convenient to be somewhat careless with notation, and identify A with its image $\lambda(A)$ in $\mathbf{E}_R(A)$.

Since A is a right A^e-module, the right regular representation ρ defines a homomorphism of A^e to $\mathbf{E}_R(A)^*$. Specifically, $x\rho_w = xw$ for $x \in A$, $w \in A^e$. The image of A^e under ρ is a subalgebra of $\mathbf{E}_R(A)^*$ (hence also of $\mathbf{E}_R(A)$) that is called the *multiplication algebra* of A. We will denote this algebra by $\mathbf{M}(A)$.

The right regular representation ρ of A can be viewed as an injective homomorphism of A to $\mathbf{E}_R(A)^*$, acting on the right of A. In some cases it is also useful to view the left regular representation λ as a homomorphism from A^* to $\mathbf{E}_R(A)^*$, again operating on the right side of A. Since $A^* \cup A$ generates $A^* \otimes A$, it follows that $\mathbf{M}(A)$ is the subalgebra of $\mathbf{E}_R(A)^*$ that is generated by $\lambda(A^*) \cup \rho(A)$. This characterization of $\mathbf{M}(A)$ is often elevated to the status of a definition.

Lemma. *For a subalgebra B of $\mathbf{E}_R(A)^*$ and a subalgebra D of $\mathbf{E}_R(A)$, define $\kappa(B) = \mathbf{E}(A_B)$ and $\beta(D) = \mathbf{E}({}_DA)$. Then $\kappa(B)$ is a subalgebra of $\mathbf{E}_R(A)$, $\beta(D)$ is a subalgebra of $\mathbf{E}_R(A)^*$, and*

(i) $B_1 \subseteq B_2$ *implies* $\kappa(B_1) \supseteq \kappa(B_2)$,
(i′) $D_1 \subseteq D_2$ *implies* $\beta(D_1) \supseteq \beta(D_2)$,
(ii) $B \subseteq \beta(\kappa(B))$,
(ii′) $D \subseteq \kappa(\beta(D))$.

Moreover, $\kappa(\mathbf{M}(A)) = \mathbf{Z}(A)$, *and* $\mathbf{M}(A) \subseteq \beta(\mathbf{Z}(A))$.

PROOF. The statements (i), (i′), (ii), and (ii′) are routine consequences of the definitions of κ and β. It is clear from the definition of $\mathbf{M}(A)$ that $\kappa(\mathbf{M}(A)) = \mathbf{E}(A_{\mathbf{M}(A)}) = \mathbf{E}(A_{A^e}) = \lambda(\mathbf{Z}(A)) = \mathbf{Z}(A)$ by Lemma 12.1. Consequently, $\mathbf{M}(A) \subseteq \beta(\kappa(\mathbf{M}(A))) = \beta(\mathbf{Z}(A))$. □

Jacobson–Bourbaki Theorem. *Let A be a finite dimensional, simple F-algebra. The mappings κ and β define mutually inverse, one-to-one correspondences between the set $\mathfrak{B}$ of subalgebras of $\mathbf{E}_F(A)^*$ that contain $\mathbf{M}(A)$, and the set $\mathfrak{K}$ of subfields of $\mathbf{Z}(A)$ that include F.*

PROOF. If $B \in \mathfrak{B}$, then $\mathbf{M}(A) \subseteq B$ implies $\kappa(B) \subseteq \kappa(\mathbf{M}(A)) = \mathbf{Z}(A)$ by the lemma. Since $\kappa(B)$ is a finite dimensional subalgebra of a finite field extension of F, $\kappa(B)$ is itself a field. (This fact also follows easily from the definition of $\kappa(B)$.) Hence, $\kappa(B) \in \mathfrak{K}$. Similarly, if $K \in \mathfrak{K}$, then $\mathbf{M}(A) \subseteq \beta(\mathbf{Z}(A)) \subseteq \beta(K)$; hence, $\beta(K) \in \mathfrak{B}$. It remains to show that $\beta(\kappa(B)) = B$ for all $B \in \mathfrak{B}$ and $\kappa(\beta(K)) = K$ for all $K \in \mathfrak{K}$. Let $B \in \mathfrak{B}$. Since A is a simple algebra, $A_{\mathbf{M}(A)}$ is a simple module by the first part of Lemma 12.1. Indeed, $\mathbf{S}(A_{\mathbf{M}(A)}) = \mathbf{S}(A_{A^e})$ by the definition of $\mathbf{M}(A)$. Therefore, A_B is simple because $\mathbf{M}(A) \subseteq B$. Moreover, if $K = \kappa(B)$, then $\dim_K A \leq \dim_F A < \infty$ because $F \subseteq K$. In particular, A is a finitely generated left module over $K = \mathbf{E}(A_B)$. By the Density Theorem, every $\phi \in \mathbf{E}({}_K A)$ can be matched on a generating set by an element of B. That is, $\beta(\kappa(B)) = \mathbf{E}({}_K A) = B$. The proof that $\kappa(\beta(K)) = K$ for $K \in \mathfrak{K}$ begins with a useful observation: if $B \in \mathfrak{B}$, then $\kappa(B) = \{x \in A: \lambda_x \in \mathbf{Z}(B)\}$. Indeed, $y(\lambda_x \phi) = (xy)\phi$ and $y(\phi\lambda_x) = x(y\phi)$ for $x, y \in A$ and $\phi \in \mathbf{E}_F(A)^*$. Thus, $x \in \mathbf{E}(A_B) = \kappa(B)$ if and only if $\lambda_x \in \mathbf{Z}(B)$. Using this comment in the case $B = \beta(K) = \mathbf{E}({}_K A)$ gives $\kappa(\beta(K)) = \lambda^{-1}(\mathbf{Z}(\mathbf{E}({}_K A)) = \lambda^{-1}(id_A K) = K$ by Example 12.1. □

Corollary. *If A is a finite dimensional, simple F-algebra such that $\mathbf{Z}(A) = F$, then* $\mathbf{M}(A) = \mathbf{E}_F(A)^*$.

EXERCISE

This exercise shows how the Jacobson–Bourbaki Theorem can be used to prove the Fundamental Theorem of Galois Theory. Let E/F be a finite field extension. In particular, E is a finite dimensional, simple F-algebra.

(a) Prove that the multiplication algebra $\mathbf{M}(E)$ is $\rho(E)$.

(b) Deduce from the Jacobson–Bourbaki Theorem that the mappings $\beta: K \mapsto \mathbf{E}_K(E)$, $\kappa: B \mapsto \{x \in E: \rho_x \in \mathbf{Z}(B)\}$ are mutually inverse, inclusion reversing bijections between the set $\mathfrak{K}$ of fields K between F and E, and the set $\mathfrak{B}$ of subalgebras B of $\mathbf{E}_F(E)^*$ such that $\rho(E) \subseteq B$.

(c) For each $K \in \mathfrak{K}$, let $\mathbf{G}(E/K)$ be the Galois group of E/K, that is, the group of automorphisms σ of E such that $\sigma(x) = x$ for all $x \in K$. For each subgroup H of $\mathbf{G}(E/F)$, let E^H be the fixed field of H, that is, $E^H = \{x \in E: \sigma(x) = x \text{ for all } \sigma \in H\}$. Show that $K \mapsto \mathbf{G}(E/K)$ and $H \mapsto E^H$ form a Galois connection: $F \subseteq K_1 \subseteq K_2 \subseteq E$ implies $\mathbf{G}(E/K_1) \supseteq \mathbf{G}(E/K_2)$, $H_1 \subseteq H_2 \subseteq \mathbf{G}(E/F)$ implies $E^{H_1} \supseteq E^{H_2}$, $H \subseteq \mathbf{G}(E/E^H)$, and $K \subseteq E^{\mathbf{G}(E/F)}$.

(d) Show that if $\sigma \in \mathbf{G}(E/F)$ and $x \in E$, then $\rho_x \sigma = \sigma \rho_{\sigma(x)}$.

(e) For each subgroup H of $\mathbf{G}(E/F)$, define $B_H = \{\sum_{\sigma \in H} \rho_x \sigma: x_\sigma \in E\}$. Prove that $B_H \in \mathfrak{B}$.

(f) Prove that $\kappa(B_H) = E^H$ for every subgroup H of $\mathbf{G}(E/F)$, and $\beta(K) \cap \mathbf{G}(E/F) = \mathbf{G}(E/K)$ for every $K \in \mathfrak{K}$.

(g) Deduce from (f) that for each subgroup H of $\mathbf{G}(E/F)$, $B_H = \mathbf{E}_K(E)^*$, where $K = E^H$.

The results (a) through (g) provide a factorization of the Galois connection through the class $\mathfrak{B}$ of F-algebras. To make this point explicit, let $\mathfrak{H}$ be the class of all subgroups of $\mathbf{G}(E/F)$. We have constructed mappings

$$\mathfrak{K} \underset{\kappa}{\overset{\beta}{\rightleftarrows}} \mathfrak{B} \underset{B_H \leftarrow H}{\overset{B \mapsto B \cap \mathbf{G}(E/F)}{\rightleftarrows}} \mathfrak{H}$$

with $\kappa = \beta^{-1}$, $\beta(K) \cap \mathbf{G}(E/F) = \mathbf{G}(E/K)$, and $\kappa(B_H) = E^H$.

We need a variant of Dedekind's theorem on the independence of automorphisms.

(h) Let $B \in \mathfrak{B}$. Assume that $\sigma_1, \ldots, \sigma_k \in \mathbf{G}(E/F)$ are distinct. Prove that if $x_1, \ldots, x_k \in E^\circ$ exist so that $\rho_{x_1}\sigma_1 + \cdots + \rho_{x_k}\sigma_k \in B$, then $\sigma_1, \ldots, \sigma_k \in B$. Hint. Deny the assertion, and let k be minimal so that there is a counterexample: $\rho_{x_1}\sigma_1 + \cdots + \rho_{x_k}\sigma_k \in B$, not all $\sigma_i \in B$. Minimality implies that $\sigma_i \notin B$ for all i. Also, $k > 1$ because $\rho_x\sigma \in B$ yields $\sigma = \rho_{x^{-1}}\rho_x\sigma \in B$. Derive a contradiction to the minimality of k using (d) and the observation that for all $y \in E$,

$$\rho_y(\rho_{x_1}\sigma_1 + \cdots + \rho_{x_k}\sigma_k) - (\rho_{x_1}\sigma_1 + \cdots + \rho_{x_k}\sigma_k)\rho_{\sigma_k(y)} \in B.$$

(i) Deduce from (h) that $B_H \cap \mathbf{G}(E/F) = H$ for all $H \in \mathfrak{H}$, and if $B \in \mathfrak{B}$ satisfies $B \subseteq B_{\mathbf{G}(E/F)}$, then $B = B_H$, where $H = B \cap \mathbf{G}(E/F)$.

(j) Prove the *Fundamental Theorem of Galois Theory*: (1) If H is a subgroup of $\mathbf{G}(E/F)$, then $\mathbf{G}(E/E^H) = H$. (2) If E/F is Galois, that is, $E^{\mathbf{G}(E/F)} = F$, then $E^{\mathbf{G}(E/K)} = K$ for all fields K between F and E. Hint: For the proof of (2), use (g) and the hypothesis that E/F is Galois to show that $B_{\mathbf{G}(E/K)} = \mathbf{E}_F(E)$, so that (i) applies to all $B \in \mathfrak{B}$.

12.4. Central Simple Algebras

An F-algebra A is *central simple* if A is simple and $\mathbf{Z}(A) = F$. Every simple algebra is central simple over its center, so that the study of simple algebras can be factored into two parts: scalar extensions, central simple algebras. In this section we will concentrate on central simple algebras with emphasis on the tensor products of these algebras. Exercises 3 and 4 at the end of the section give some results on the tensor products of simple algebras that are not central.

We begin by looking at tensor products in which one factor is central simple. In this case, the characterization of the tensor product can be simplified.

Lemma a. *Let B and C be subalgebras of the F-algebra A such that $C \subseteq \mathbf{C}_A(B)$. Assume that B is central simple. If $x_1, \ldots, x_n$ is a linearly independent sequence of elements in B and $y_1, \ldots, y_n \in C$, then $y_1x_1 + \cdots + y_nx_n = 0$ implies $y_1 = \cdots = y_n = 0$.*

Proof. The assumption $C \subseteq \mathbf{C}_A(B)$ implies that A is a C-B^e bimodule: $(yz)(x \otimes x') = x(yz)x' = y(xzx') = y(z(x \otimes x'))$ for all $y \in C$, $z \in A$, and $x, x' \in B$. By Lemma 12.1, B is a simple B^e-module, and $\mathbf{E}(B_{B^e}) \cong \mathbf{Z}(B) = F$. Since $x_1, \ldots, x_n$ is an independent sequence, it follows from the Density Theorem that B^e contains elements w_j with the property that $x_iw_j = 0$ for $i \neq j$ and $x_jw_j = 1_B = 1_A$. Thus, $y_1x_1 + \cdots + y_nx_n = 0$ implies $0 = (y_1x_1 + \cdots + y_nx_n)w_j = y_1(x_1w_j) + \cdots + y_n(x_nw_j) = y_j$. □

Proposition a. *Let B and C be subalgebras of the finite dimensional F-algebra A such that $C \subseteq \mathbf{C}_A(B)$ and B is central simple. The following conditions are equivalent.*

(i) $A = BC$.
(ii) $\dim_F A = (\dim_F B)(\dim_F C)$.
(iii) *The inclusion mappings of B and C into A induce an isomorphism*
$$B \otimes C \cong A.$$

PROOF. Let $x_1, \ldots, x_n$ be an F-space basis of B, and suppose that $y_1, \ldots, y_m$ is a basis of C. If $a_{ij} \in F$ satisfy $\sum_{i=1}^n \sum_{j=1}^m x_i y_j a_{ij} = 0$, then $\sum_{j=1}^n y_j a_{ij} = 0$ for $1 \leq i \leq n$ by Lemma a. Hence, $a_{ij} = 0$ for all i and j. This argument proves that $\{x_i y_j: 1 \leq i \leq n, 1 \leq j \leq m\}$ is linearly independent. Either of the hypotheses (i) or (ii) implies that this set is a basis of A, so that $A \cong B \otimes C$ by Proposition 9.2c. The same proposition shows that (iii) implies both (i) and (ii). □

When the subalgebras B and C of an algebra A are such that the inclusion mappings induce an isomorphism of $B \otimes C$ to A, we will write $A = B \otimes C$ and call A the inner tensor product of B and C. If B and C are F-algebras, then the homomorphisms $x \mapsto x \otimes 1_C$, $y \mapsto 1_B \otimes y$ map B and C isomorphically to the respective subalgebras $B \otimes F$ and $F \otimes C$ of $B \otimes C$. It is usually permissible to identify B with $B \otimes F$ and C with $F \otimes C$.

Lemma b. *Let B and C be F-algebras.*

(i) *If $B \otimes C$ is simple, then B and C are simple.*
(ii) *If B is central simple and C is simple, then $B \otimes C$ is simple.*

PROOF. (i) If B is not simple, then either $B = 0$ and $B \otimes C = 0$, or there is a non-zero homomorphism $\phi: B \to B'$ such that $\operatorname{Ker} \phi \neq 0$. In the second case, $\phi \otimes id_C: B \otimes C \to B' \otimes C$ is a non-zero homomorphism with non-zero kernel, so that $B \otimes C$ is not simple. The same conclusion is obtained if C is not simple.

(ii) Since B and C are simple, they are not 0, and therefore $B \otimes C \neq 0$. Let $\phi: B \otimes C \to A$ be a non-zero homomorphism. Since $B \otimes F \cong B$ and $F \otimes C \cong C$ are simple, the restrictions $\phi | B \otimes F$ and $\phi | F \otimes C$ are injective. Thus, $\phi(B \otimes F)$ is central simple and $\phi(F \otimes C) \subseteq \mathbf{C}_A(\phi(B \otimes F))$. Let $z \in \operatorname{Ker} \phi$. It is possible to write $z = x_1 \otimes y_1 + \cdots + x_n \otimes y_n$ with $x_1, \ldots, x_n$ linearly independent in B. Since $\phi(1 \otimes y_1)\phi(x_1 \otimes 1) + \cdots + \phi(1 \otimes y_n)\phi(x_n \otimes 1) = \phi(z) = 0$, it follows from Lemma a that $\phi(1 \otimes y_1) = \cdots = \phi(1 \otimes y_n) = 0$. Therefore, $1 \otimes y_1 = \cdots = 1 \otimes y_n = 0$ because $\phi | F \otimes C$ is injective, and $z = 0$. The argument proves that every non-zero homomorphism of $B \otimes C$ is injective. Hence $B \otimes C$ is simple. □

Lemma c. *Let B and C be F-algebras. Denote $B \otimes C$ by A.*

(i) $\mathbf{C}_A(B \otimes F) = \mathbf{Z}(B) \otimes C$.
(ii) $\mathbf{Z}(A) = \mathbf{Z}(B) \otimes \mathbf{Z}(C)$.

PROOF. Let $\{y_j : j \in J\}$ be an F-space basis of C. By Proposition 9.1c, every element of A has the form $w = \sum_j x_j \otimes y_j$ with the $x_j \in B$ uniquely determined by w. If $w \in \mathbf{C}_A(B \otimes F)$, then $0 = (x \otimes 1)w - w(x \otimes 1) = \sum_{j \in J}(xx_j - x_j x) \otimes y_j$ for all $x \in B$. Thus, $xx_j = x_j x$ for all $x \in B$ and $j \in J$. That is, every x_j is in the center of B, and $w \in \mathbf{Z}(B) \otimes C$. This discussion shows that $\mathbf{C}_A(B \otimes F) \subseteq \mathbf{Z}(B) \otimes C$. The reverse inclusion is obvious. By symmetry, $\mathbf{C}_A(F \otimes C) = B \otimes \mathbf{Z}(C)$. Therefore, $\mathbf{Z}(A) = \mathbf{C}_A(B \otimes F) \cap \mathbf{C}_A(F \otimes C) = (\mathbf{Z}(B) \otimes C) \cap (B \otimes \mathbf{Z}(C)) = \mathbf{Z}(B) \otimes \mathbf{Z}(C)$. □

Proposition b. *Let B and C be central simple F-algebras, and suppose that E is a field and an F-algebra.*

(i) *$B \otimes C$ is central simple.*
(ii) *$B \otimes E$ is a central simple E-algebra.*
(iii) *B^* is central simple.*
(iv) *If $\dim_F B = n < \infty$, then $B^* \otimes B \cong M_n(F)$.*

PROOF. The properties (i) and (ii) are special cases of Lemmas b and c. The assertion (iii) is an obvious consequence of the observations that $\mathbf{I}(B^*) = \mathbf{I}(B)$ and $\mathbf{Z}(B^*) = \mathbf{Z}(B)$. It follows from (i) and (iii) that $B^e = B^* \otimes B$ is simple. Therefore, $B^e \cong \mathbf{M}(B)$; and if $\dim_F B = n < \infty$, then $B^* \otimes B \cong \mathbf{E}_F(B)^* \cong M_n(F)$ by Corollary 12.3. □

The corollary of Proposition b pays a debt that was incurred in the proof of Proposition 10.7.

Corollary. *Every finite dimensional central simple F-algebra is separable.*

This result follows from Corollary 10.6, using part (ii) of the proposition.

EXERCISES

1. Let F be a field of characteristic p. Suppose that $a \in F - F^p$. Consider the field $A = F(a^{1/p})$ as an F-algebra. Prove that $A \otimes A$ is not simple.

2. Let A be the $\mathbb{Z}$-algebra that is a free $\mathbb{Z}$-module on the basis $1_A, e, x$ with multiplication defined by $e^2 = e$, $ex = x$, $xe = 0$, $x^2 = 0$. Define $B = A^{\mathrm{o}}$. still considered as a $\mathbb{Z}$-algebra. Let $C = A/I$, where I is the ideal $x(2\mathbb{Z})$. Prove that $\mathbf{Z}(B) \otimes \mathbf{Z}(C)$ is properly contained in $\mathbf{Z}(B \otimes C)$.

3. Let A be a finite dimensional F-algebra. Assume that $E = \mathbf{Z}(A)$ is a field. Suppose that B and C are simple subalgebras of A such that $\mathbf{Z}(B) \subseteq E$, $\mathbf{Z}(C) \subseteq E$, $C \subseteq \mathbf{C}_A(B)$, and $A = BC$. Prove that $A \cong (B \otimes_{\mathbf{Z}(B)} E) \otimes_E (C \otimes_{\mathbf{Z}(C)} E)$. Hint. Let $B' = BE$ and $C' = CE$. Use Proposition a to show that $B' \cong B \otimes_{\mathbf{Z}(B)} E$, $C' \cong C \otimes_{\mathbf{Z}(C)} E$, B' and C' are central simple E-algebras, and $A = B' \otimes_E C'$.

4. Let B and C be simple, separable F-algebras. Note that B and C are finite dimensional, and $\mathbf{Z}(B)/F$, $\mathbf{Z}(C)/F$ are separable field extensions by the results of Chapter 10. Hence, $\mathbf{Z}(B) \otimes \mathbf{Z}(C) = E_1 \dotplus \cdots \dotplus E_r$, where each E_i is a field that contains both $\mathbf{Z}(B)$ and $\mathbf{Z}(C)$. Use the result of Exercise 3 to prove that $B \otimes C = (B^{E_1} \otimes_{E_1} C^{E_1}) \dotplus \cdots \dotplus (B^{E_r} \otimes_{E_r} C^{E_r})$.

12.5. The Brauer Group

The results of Proposition 12.4b can be put in a very interesting form by considering the isomorphism classes of central simple algebras modulo a suitable equivalence relation. The relation is the Morita equivalence that was discussed in Section 9.6. Up to equivalence, the central simple F-algebras form an abelian group that is called the *Brauer Group* of F. The term "Brauer group" honors Richard Brauer who made the first systematic study of this fundamental invariant. The importance of the Brauer Groups in the theory of rings and fields is now firmly established. Much of the rest of this book will be concerned with various properties of Brauer groups.

It will be economical to introduce notation for classes of central simple algebras. Let F be a field. We will denote by $\mathfrak{S}(F)$ the class of all finite dimensional, simple F-algebras A such that $\mathbf{Z}(A) = F$, that is, central simple F-algebras. A simple F-algebra can fail to be in $\mathfrak{S}(F)$ by being infinite dimensional over F, or by having a center that is properly larger than F.

Morita equivalence for central simple algebras takes a simple form.

Lemma. *Let A and B be members of $\mathfrak{S}(F)$. The following conditions are equivalent.*

(i) *The basic algebras of A and B are isomorphic.*
(ii) *There is a division algebra $D \in \mathfrak{S}(F)$ and positive integers m and n such that $A \cong M_n(D)$ and $B \cong M_m(D)$.*
(iii) *There exist positive integers r and s such that $A \otimes M_r(F) \cong B \otimes M_s(F)$.*

PROOF. By the Wedderburn Structure Theorem $A \cong M_n(D_1)$ and $B \cong M_m(D_2)$, where D_1 and D_2 are finite dimensional division algebras over F. Moreover. D_1 and D_2 are central by Example 12.1. Therefore, the fact that (i) implies (ii) is a consequence of the observation in Example 6.6 that D_1 is the basic algebra of A and D_2 is the basic algebra of B. If (ii) is satisfied, then $A \otimes M_m(F) \cong M_{mn}(D) \cong B \otimes M_n(F)$, which is (iii) with $r = m$ and $s = n$. On the other hand, if $A \otimes M_r(F) \cong B \otimes M_s(F)$, then $M_{rn}(D_1) \cong M_{sm}(D_2)$. By the uniqueness statement in Wedderburn's theorem, this isomorphism implies that $D_1 \cong D_2$. Hence, A and B have isomorphic basic algebras. □

We will say that algebras A and B in $\mathfrak{S}(F)$ are *equivalent* if they satisfy the conditions of the lemma. The notation $A \sim B$ will abbreviate the statement that A and B are equivalent. It is clear from property (i) of the lemma that $\sim$ is an equivalence relation on $\mathfrak{S}(F)$. The equivalence class of A in $\mathfrak{S}(F)$ will be denoted by $[A]$.

Proposition a. *For a field F, the set $\mathbf{B}(F) = \{[A]: A \in \mathfrak{S}(F)\}$ is an abelian group with the product $[A][B] = [A \otimes B]$, the unity element $[F]$, and the inverse operation $[A]^{-1} = [A^*]$.*

PROOF. If $A \cong B$, then $A \sim B$, Thus, $\mathbf{B}(F)$ is a set by Proposition 1.5. In fact, $|\mathbf{B}(F)| \leq \aleph_0 |F|$. If $A, B \in \mathfrak{S}(F)$, then $A \otimes B \in \mathfrak{S}(F)$ by Proposition 12.4b. Moreover, $A \sim A'$ and $B \sim B'$ implies $A \otimes B \sim A' \otimes B'$. Indeed, if $A \otimes M_r(F) \cong A' \otimes M_s(F)$ and $B \otimes M_k(F) \cong B' \otimes M_l(F)$, then

$$\begin{aligned} A \otimes B \otimes M_{rk}(F) &\cong (A \otimes M_r(F)) \otimes (B \otimes M_k(F)) \\ &\cong (A' \otimes M_s(F)) \otimes (B' \otimes M_l(F)) \\ &= A' \otimes B' \otimes M_{sl}(F). \end{aligned}$$

Thus, the tensor product on $\mathfrak{S}(F)$ induces a product on $\mathbf{B}(F)$ by the rule $[A][B] = [A \otimes B]$. The commutativity and associativity of $\otimes$ translate to corresponding identities in $\mathbf{B}(F)$; and $A \otimes F \cong A$ implies $[F] = 1$. Finally, if $A \in \mathfrak{S}(F)$, then $A^* \in \mathfrak{S}(F)$ and $A^* \otimes A \cong M_n(F) \sim F$ by Proposition 12.4b. Therefore $\mathbf{B}(F)$ is a group in which $[A]^{-1} = [A^*]$. □

The group $\mathbf{B}(F)$ is called the Brauer group of the field F.

Our next result is an easy corollary of Wedderburn's Structure Theorem, but it is very useful.

Proposition b. *Let F be a field.*

(i) *If $A, B \in \mathfrak{S}(F)$, then $A \cong B$ if and only if $[A] = [B]$ in $\mathbf{B}(F)$ and $\dim_F A = \dim_F B$.*

(ii) *Every class in $\mathbf{B}(F)$ is represented by a division algebra that is unique to within isomorphism.*

PROOF. If $[A] = [B]$, that is, $A \sim B$, then $A \cong M_n(D)$ and $B \cong M_m(D)$ by the lemma; and $\dim_F A = \dim_F B$ implies $n = m$, so that $A \cong B$. Conversely, $A \cong B$ plainly implies $[A] = [B]$ and $\dim_F A = \dim_F B$. By the Wedderburn Structure Theorem, every $A \in \mathfrak{S}(F)$ satisfies $A \cong M_n(D)$, where $D \in \mathfrak{S}(F)$ is a division algebra. Thus, $[A] = [D]$. If D_1 and D_2 are equivalent division algebras, say $M_n(D_1) \cong M_m(D_2)$, then the uniqueness statement in Wedderburn's Structure Theorem yields $D_1 \cong D_2$. □

Corollary. *If F is algebraically closed, then $\mathbf{B}(F) = \{1\}$.*

Indeed, the only finite dimensional division algebra over F is F itself by Lemma 3.5.

The content of Proposition b is that $\mathbf{B}(F)$ classifies the finite dimensional division algebras over F. However, the group structure of $\mathbf{B}(F)$ cannot be defined within the class of division algebras. In general, the tensor product of two division algebras is not a division algebra.

We conclude this section by showing that the Brauer group is the object map of a functor.

Proposition c. *If $\phi: F \to E$ is a homomorphism of fields, then ϕ induces a group homomorphism $\phi_*: \mathbf{B}(F) \to \mathbf{B}(E)$ by $\phi_*([A]) = [A \otimes {}_\phi E]$. The correspondences $F \mapsto \mathbf{B}(F)$ and $\phi \mapsto \phi_*$ define a functor from the category of fields to the category of abelian groups.*

The notation ${}_\phi E$ in this statement has the meaning that was described in Section 2.1: ${}_\phi E$ is an F-algebra with the scalar operation defined by $ab = \phi(a)b$ for $a \in F$ and $b \in E$.

By Proposition 12.4b, $A \otimes {}_\phi E \in \mathfrak{S}(E)$. Moreover, $A \otimes B \otimes {}_\phi E \cong A \otimes {}_\phi E \otimes B \cong A \otimes ({}_\phi E \otimes_E {}_\phi E) \otimes B \cong (A \otimes {}_\phi E) \otimes_E (B \otimes {}_\phi E)$. In particular, if $A \sim A'$, say $A \otimes M_r(F) \cong A' \otimes M_s(F)$, then

$$\begin{aligned}(A \otimes {}_\phi E) \otimes_E M_r(E) &\cong (A \otimes {}_\phi E) \otimes_E (M_r(F) \otimes {}_\phi E)\\ &\cong (A \otimes M_r(F)) \otimes {}_\phi E\\ &\cong (A' \otimes M_s(F)) \otimes {}_\phi E\\ &\cong (A' \otimes {}_\phi E) \otimes_E M_s(E).\end{aligned}$$

Thus, $\phi_*([A]) = [A \otimes {}_\phi E]$ is a well defined group homomorphism of $\mathbf{B}(F)$ to $\mathbf{B}(E)$. If $\psi: E \to K$ is another homomorphism of fields, then it is easy to see that $(A \otimes {}_\phi E) \otimes {}_\psi K \cong A \otimes {}_{\psi\phi} K$ as K-algebras. Hence, $\psi_* \phi_* = (\psi\phi)_*$. This completes the proof of Proposition c.

In general, distinct embeddings of F into a field E give rise to different homomorphisms of $\mathbf{B}(F)$ to $\mathbf{B}(E)$. Exercise 1 provides an example of this phenomenon.

Exercises

1. Let $F = \mathbb{Q}(2^{1/2})$, $E = \mathbb{Q}(2^{1/4})$. Define homomorphisms $\phi: F \to E$ and $\psi: F \to E$ by $\phi(2^{1/2}) = 2^{1/2}$, $\psi(2^{1/2}) = -2^{1/2}$. Let D be the quaternion algebra $\left(\frac{-1, -2^{1/2}}{F}\right)$. Prove that $[D] \in \operatorname{Ker} \psi_*$ and $[D] \notin \operatorname{Ker} \phi_*$ in $\mathbf{B}(F)$. Hint. Let $K = \mathbb{Q}(i2^{1/4})$. Show that $K \cong {}_\psi E$ as F-algebras. Use Proposition 1.6 to show that $A \otimes K \cong \left(\frac{-1, -2^{1/2}}{K}\right) \cong M_2(K)$ and $A \otimes E \cong \left(\frac{-1, -2^{1/2}}{E}\right)$ is a division algebra.

2. Let $\phi: F \to E$ and $\psi: F \to E$ be homomorphisms of fields. Assume that $E/\phi(L)$ is a (finite) Galois extension, where $L = \{x \in F: \phi(x) = \psi(x)\}$. Prove that $\phi_* = \psi_*$. Hint. Use the assumption that $E/\phi(L)$ is Galois to show that there is an automorphism χ of E such that $\chi\phi = \psi$. Deduce that $\chi: {}_\phi E \to {}_\psi E$ is an F-algebra isomorphism, and therefore, $A \otimes {}_\phi E \cong A \otimes {}_\psi E$ as E-algebras.

3. Involutions of F-algebras were defined in Exercise 4 of Section 10.1. Prove that if there is an involution τ of $A \in \mathfrak{S}(F)$, then $[A]^2 = 1$ in $\mathbf{B}(F)$. Deduce that $[A]^2 = 1$ for every quaternion algebra.

4. Use the result of Exercise 4, Section 1.7 to prove that $\mathbf{B}(\mathbb{Q})$ is an infinite group.

12.6. The Noether–Skolem Theorem

The purpose of this section is to prove the Noether–Skolem Theorem for algebras in the class $\mathfrak{S}(F)$. A more general result is outlined in Exercise 1.

We begin with a special case of the Noether–Skolem Theorem from which the full result will then be deduced.

Lemma. *Let B be a finite dimensional, simple F-algebra, and suppose that M is an F-space. If ϕ and ψ are F-algebra homomorphisms of B to $\mathbf{E}_F(M)$, then there exists $\theta \in \mathbf{E}_F(M)^\circ$ such that $\phi(x) = \theta^{-1}\psi(x)\theta$ for all $x \in B$.*

The idea of the proof is that ϕ and ψ impose B^*-module structures on M. The resulting modules must be isomorphic because B is simple and the modules over finite dimensional simple algebras are classified by their dimension according to Corollary 3.3b. The required linear transformation θ is just the isomorphism between these modules. In detail, define M_ϕ to be the right B^*-module on M with the scalar operation $ux = \phi(x)(u)$, and let M_ψ have scalar operation defined by $u \circ x = \psi(x)(u)$. Routine calculations show that M_ϕ and M_ψ satisfy the module axioms. Let $\theta: M_\phi \to M_\psi$ be the B^*-module isomorphism whose existence is guaranteed by Corollary 3.3b. Then $\theta \in \mathbf{E}_F(M)^\circ$, and $\theta(\phi(x)(u)) = \theta(ux) = \theta(u) \circ x = \psi(x)(\theta(u))$. That is, $\phi(x) = \theta^{-1}\psi(x)\theta$ for all $x \in B$.

Noether–Skolem Theorem. *Let $A \in \mathfrak{S}(F)$, and suppose that B is a simple subalgebra of A. If χ is an algebra homomorphism of B to A, then there exists $u \in A^\circ$ such that $\chi(y) = u^{-1}yu$ for all $y \in B$.*

PROOF. By Proposition 12.4b, there is an algebra isomorphism $\rho: A^e = A^* \otimes A \to \mathbf{E}_F(A)$. Define $\phi = \rho(id \otimes \chi): A^* \otimes B \to \mathbf{E}_F(A)$ and $\psi = \rho(id \otimes \kappa): A^* \otimes B \to \mathbf{E}_F(A)$, where $\kappa: B \to A$ is the inclusion homomorphism. Since $A^* \otimes B$ is simple by Lemma 12.4b, it follows from the lemma that there exists $\theta \in \mathbf{E}_F(A)^\circ$ such that $\phi(x \otimes y) = \theta^{-1}\psi(x \otimes y)\theta$ for all $x \in A^*, y \in B$. Let $z = \rho^{-1}(\theta) \in A^e$. Since θ is a unit, so is z, and $\theta^{-1} = \rho(z^{-1})$.

Moreover,

$$\rho(z(x \otimes \chi(y))) = \rho(z)\rho(x \otimes \chi(y)) = \theta\phi(x \otimes y)$$
$$= \psi(x \otimes y)\theta = \rho(x \otimes y)\rho(z) = \rho((x \otimes y)z),$$

because ρ is an algebra homomorphism. Since ρ is injective,

$$x \otimes \chi(y) = z^{-1}(x \otimes y)z \quad \text{for all} \quad x \in A^*, y \in B. \tag{1}$$

By letting $y = 1$ in (1), we obtain $z(x \otimes 1) = (x \otimes 1)z$; that is, $z \in \mathbf{C}_{A^e}(A^* \otimes F) = F \otimes A$ by Lemma 12.4c. Similarly, $z^{-1} \in F \otimes A$. Therefore, $z = 1 \otimes u$ and $z^{-1} = 1 \otimes v$ with $u, v \in A$. Hence, $uv = 1$, $u \in A^\circ$, and $v = u^{-1}$. Finally, if $x = 1$ in (1), then $1 \otimes \chi(y) = 1 \otimes u^{-1}yu$ for all $y \in B$; therefore, $\chi(y) = u^{-1}yu$. □

Exercises

1. Show that the hypothesis $\dim_F A < \infty$ in the Noether–Skolem Theorem can be replaced by the condition that B is finite dimensional. Hint. Note that the image of $\chi: A^e \to \mathbf{E}_F(A)$ is a dense subalgebra, and $\operatorname{Ker} \chi = 0$ because A^e is simple. (It is still assumed that A is central simple.) This remark makes it possible to extend the proof of the Noether–Skolem Theorem that is given in this section.

2. (a) Let $\alpha \in M_n(F)$ be such that the minimum polynomial $\Phi(\mathbf{x})$ of α (over F) is irreducible. Prove that $\beta \in M_n(F)$ is similar to α (that is, $\beta = \gamma^{-1}\alpha\gamma$ for some $\gamma \in M_n(F)^\circ$) if and only if $\Phi(\mathbf{x})$ is also the minimum polynomial of β over F.
   (b) Prove that

$$\alpha = \begin{bmatrix} 1 & 0 & 0 \\ 0 & 1 & 0 \\ 0 & 0 & 2 \end{bmatrix}$$

and

$$\beta = \begin{bmatrix} 1 & 0 & 0 \\ 0 & 2 & 0 \\ 0 & 0 & 2 \end{bmatrix}$$

in $M_3(\mathbb{Q})$ have the same minimum polynomial $(\mathbf{x} - 1)(\mathbf{x} - 2)$ over $\mathbb{Q}$, but α is not similar to β.

3. Prove that every automorphism of a finite dimensional, central simple algebra A is an inner automorphism, that is, a mapping $x \mapsto u^{-1}xu$ for a fixed $u \in A^\circ$.

12.7. The Double Centralizer Theorem

The term "Double Centralizer Theorem" (abbreviated D.C.T.) is the generic name for a class of theorems that relate subalgebras B of an algebra A to their second centralizers $\mathbf{C}_A(\mathbf{C}_A(B))$. It is evident from the definition of the centralizer that $B \subseteq \mathbf{C}_A(\mathbf{C}_A(B))$ in all cases. Generally this inclusion is proper. Double Centralizer Theorems deal with conditions in which $B = \mathbf{C}_A(\mathbf{C}_A(B))$.

In this section we will prove a classical D.C.T. A more general result (with a different proof) is sketched in Exercise 5. Our discussion begins with a modest generalization of the second part of Lemma 12.1.

Lemma. *Let A be an F-algebra, and B a subalgebra of A. Consider A as a right $B^* \otimes A$-module, by way of the homomorphism $B^* \otimes A \to A^* \otimes A = A^e$. The left regular representation of A maps $\mathbf{C}_A(B)$ isomorphically to $\mathbf{E}_{B^* \otimes A}(A)$.*

Proof. If $x \in B$, $y \in \mathbf{C}_A(B)$, and z, $w \in A$, then $\lambda_y(w(x \otimes z)) = yxwz = xywz = \lambda_y(w)(x \otimes z)$. Thus, $\lambda(\mathbf{C}_A(B)) \subseteq \mathbf{E}_{B^* \otimes A}(A)$. If $\lambda_y \in \mathbf{E}_{B^* \otimes A}(A)$ and $x \in B$, then $yx = \lambda_y(1(x \otimes 1)) = \lambda_y(1)(x \otimes 1) = xy$. Therefore, $y \in \mathbf{C}_A(B)$. It follows from Proposition 1.3 that λ maps $\mathbf{C}_A(B)$ isomorphically to $\mathbf{E}_{B^* \otimes A}(A)$. □

Theorem. *Let $A \in \mathfrak{S}(F)$, and suppose that B is a simple subalgebra of A.*

(i) *$\mathbf{C}_A(B)$ is simple.*
(ii) *$(\dim_F B)(\dim_F \mathbf{C}_A(B)) = \dim_F A$.*
(iii) *$\mathbf{C}_A(\mathbf{C}_A(B)) = B$.*
(iv) *If B is central simple, then $\mathbf{C}_A(B)$ is central simple, and $A = B \otimes \mathbf{C}_A(B)$.*

Proof. By Lemma 12.4b, $B^* \otimes A$ is simple. This algebra is also Artinian because it is finite dimensional over the field F. Let P be a minimal right ideal of $B^* \otimes A$. By the Wedderburn Structure Theorem in the form of Corollary 3.5a, $B^* \otimes A \cong M_n(D)$, where D is the division algebra $\mathbf{E}_{B^* \otimes A}(P)$. Hence, $B^* \otimes A \cong \bigoplus n\, P$ and $P \cong \bigoplus n\, D$, as in Example 3.3. In particular,

$$(\dim A)(\dim B) = n^2(\dim D). \tag{1}$$

Since A is a finite dimensional $B^* \otimes A$-module, it follows from Proposition 3.3b that $A = \bigoplus k\; P$ for a suitable $k \in \mathbb{N}$. Thus by the lemma, $\mathbf{C}_A(B) \cong \mathbf{E}_{B^* \otimes A}(\bigoplus k\, P) \cong M_k(D)$. Consequently, $\mathbf{C}_A(B)$ is simple, and

$$\dim A = k(\dim P) = kn(\dim D) \tag{2}$$

$$\dim \mathbf{C}_A(B) = k^2(\dim D). \tag{3}$$

By eliminating k, n, and $\dim D$ from (1), (2), and (3), we obtain (ii): $(\dim B)(\dim \mathbf{C}_A(B)) = \dim A$. Since $\mathbf{C}_A(B)$ is simple, it is permissible to replace B by $\mathbf{C}_A(B)$ in (ii) to obtain $(\dim \mathbf{C}_A(B))(\dim \mathbf{C}_A(\mathbf{C}_A(B))) = \dim A = (\dim B)(\dim \mathbf{C}_A(B))$. Therefore, $\dim B = \dim \mathbf{C}_A(\mathbf{C}_A(B))$; and $B = \mathbf{C}_A(\mathbf{C}_A(B))$ because $B \subseteq \mathbf{C}_A(\mathbf{C}_A(B))$. If B is central simple, then $A = B \otimes \mathbf{C}_A(B)$ by virtue of Proposition 12.4a and (ii). Moreover, $F = \mathbf{Z}(B \otimes \mathbf{C}_A(B)) = F \otimes \mathbf{Z}(\mathbf{C}_A(B))$, so that $\mathbf{C}_A(B)$ is central simple. □

Exercises

1. Let $A = M_n(F)$, and take B to be the set of all lower triangular matrices $[a_{ij}] \in A$, that is, $a_{ij} = 0$ for $i > j$. Prove that B is a subalgebra of A such that $\mathbf{C}_A(B) = \iota_n F$; hence $\mathbf{C}_A(\mathbf{C}_A(B)) = A \supset B$.

2. Let B be an F-algebra. Denote $A = \mathbf{E}_F(B)$. Prove that $\mathbf{C}_A(\lambda(B)) = \rho(B)$ and $\mathbf{C}_A(\rho(B)) = \lambda(B)$.

3. Generalize Lemma 12.4c: if A and B are F-algebras, and if C and D are respectively subalgebras of A and B, then $\mathbf{C}_{A \otimes B}(C \otimes D) = \mathbf{C}_A(C) \otimes \mathbf{C}_B(D)$.

4. Let B be a subalgebra of the F-algebra A. Prove that if $z \in A^\circ$, then $\mathbf{C}_A(z^{-1}Bz) = z^{-1}\mathbf{C}_A(B)z$.

5. Generalize the D.C.T.: if A is a central simple F-algebra (not necessarily finite dimensional) and B is a finite dimensional simple subalgebra of A, then $\mathbf{C}_A(B)$ is simple, and $\mathbf{C}_A(\mathbf{C}_A(B)) = B$. Hint. Denote $C = \mathbf{E}_F(B)$ and $D = A \otimes C$. Note that $\lambda(B)$ and $\rho(B)$ are subalgebras of C with $\mathbf{C}_C(\lambda(B)) = \rho(B)$, $\mathbf{C}_C(\rho(B)) = \lambda(B)$ (Exercise 2), and D is simple by Lemma 12.4b. Use the version of the Noether–Skolem Theorem in Exercise 1, Section 12.6 to obtain $z \in D^\circ$ such that $z^{-1}(F \otimes \lambda(B))z = B \otimes F$. Use the results of Exercises 3 and 4 to obtain $\mathbf{C}_A(B) \otimes C = z^{-1}(A \otimes \rho(B))z$, so that $\mathbf{C}_A(B)$ is simple. Repeat this strategy to get $\mathbf{C}_A\mathbf{C}_A(B) \otimes F = B \otimes F$.

Notes on Chapter 12

The material in this chapter is classical; so is our exposition of it. In the interest of simplicity, we have added finite dimensionality to the hypotheses of the Jacobson–Bourbaki Theorem, the Noether–Skolem Theorem, and the Double Centralizer Theorem. More general results can be found in most expositions of non-commutative ring theory (for example, in [41], [46], and [55]), and some generalizations are outlined in the exercises of Sections 12.6 and 12.7.

CHAPTER 13

Subfields of Simple Algebras

In this chapter we set the stage for a systematic study of the Brauer group. The details of the program will be worked out in the next chapter, using two main tools: the cohomology theory that was introduced in Chapter 11, and the properties of subfields of central simple algebras which will be established in the first five sections of this chapter. The last section of the chapter gives applications of the theory, including Wedderburn's Theorem on Finite Division Algebras, and the Cartan–Brauer–Hua Theorem.

13.1. Maximal Subfields

A *subfield* of an F-algebra A is a subalgebra E of A such that E is a field. In particular, E contains $1_A F$, so that E can be viewed as an extension of F. As usual, $[E:F]$ is the dimension of E as an F-space. Plainly, $[E:F] \leq \dim_F A$. If there is no subfield K of A such that $E \subset K$, then E is called a *maximal subfield* of A.

Lemma a. *If B is an F-algebra with* $\dim_F B = k < \infty$, *and if $n \in \mathbb{N}$ is divisible by k, then B is isomorphic to a subalgebra of $M_n(F)$.*

Proof. If $k = n$, then the lemma restates Corollary 5.5b. The general case follows from this special situation because the diagonal map $x \mapsto (x, \ldots, x)$ is an injective algebra homomorphism of B to a product A of n/k copies of B, and $\dim_F A = n$. □

We will often use the following simple property of division algebras.

Lemma b. *Let D be a division algebra over F. If $x \in D$, then there is a subfield E of D such that $x \in E$. If $\dim_F D < \infty$, then the subalgebra $F[x] = \{\Phi(x)\colon \Phi \in F[\mathbf{x}]\}$ is a subfield of D.*

PROOF. Since $F \subseteq \mathbf{Z}(D)$, the set $F[x]$ is a commutative subalgebra of D and $\theta\colon \Phi \mapsto \Phi(x)$ is an algebra homomorphism of $F[\mathbf{x}]$ to $F[x]$. Since D has no proper zero divisors, $F[x]$ is an integral domain. Thus, $\operatorname{Ker}\theta$ is a prime ideal of $F[\mathbf{x}]$. If $\operatorname{Ker}\theta \neq 0$ (which must be the case if $\dim_F D < \infty$), then $\operatorname{Ker}\theta$ is maximal and $F[x]$ is a field. If $\operatorname{Ker}\theta = 0$, then $E = \{\Phi(x)\Psi(x)^{-1}\colon \Phi, \Psi \in F[\mathbf{x}], \Psi \neq 0\}$ is a subfield of D that includes x. □

It follows from Lemma b that if $D \in \mathfrak{S}(F)$ and D is a division algebra, then every subalgebra B of D is also a division algebra. Indeed, if $0 \neq x \in B$, then $x^{-1} \in F[x] \subseteq B$.

For a natural number n, we will say that the field F is *n-closed* if there is no proper extension E of F such that $[E:F]$ divides n. Every field is 1 closed since no proper extension has degree 1. At the opposite extreme, F is n-closed for all $n \in \mathbb{N}$ if and only if F is algebraically closed. The field $\mathbb{R}$ is n-closed for all odd n, but $\mathbb{R}$ is not 2-closed. It is obvious from the definition that if F is n-closed, then F is k-closed for every divisor k of n.

Lemma c. *If A is a simple, finite dimensional F-algebra such that F is a maximal subfield of A, then $A \cong M_n(F)$ and F is n-closed, where $n \in \mathbb{N}$ is $(\dim_F A)^{1/2}$.*

PROOF. Since A is simple and finite dimensional, the Wedderburn Structure Theorem yields $A \cong M_n(D)$, where D is a division algebra over F. In fact, $D = F$. Otherwise, by Lemma b there is a subfield E of D that properly contains F. The assumption that F is a maximal subfield of A excludes this possibility. If F is not n-closed, then there is a proper extension E/F such that $[E:F]$ divides n. In this case, $M_n(F) \cong A$ contains a subfield that is isomorphic to E by Lemma a, which again contradicts the maximality of F. □

The converse of this lemma follows from the next result by taking $A = M_n(F)$ and $E = F$.

Proposition. *Let $A \in \mathfrak{S}(F)$, and suppose that E is a subfield of A with $[E:F] = k$. The following conditions are equivalent.*

(i) *E is a maximal subfield of A.*
(ii) *$\mathbf{C}_A(E) \cong M_n(E)$ and E is n-closed.*

If (i) *and* (ii) *are satisfied, then $\dim_F A = (kn)^2$.*

PROOF. Assume that E is a maximal subfield of A. Since E is simple, so is $\mathbf{C}_A(E)$ by the Double Centralizer Theorem. Moreover, $E \subseteq \mathbf{Z}(\mathbf{C}_A(E))$ because E is commutative. Thus, $\mathbf{C}_A(E)$ is a simple E-algebra, and since E is maximal in A, it is also maximal in $\mathbf{C}_A(E)$. By Lemma c, there exists $n \in \mathbb{N}$ such that $\mathbf{C}_A(E) \cong M_n(E)$ and E is n-closed. The D.C.T. also gives $\dim_F A = (\dim_F E)(\dim_F \mathbf{C}_A(E)) = [E:F](\dim_F M_n(E)) = n^2k^2$. Conversely, suppose that (ii) is satisfied. Let $E \subseteq K$, where K is a maximal subfield of A. Then $K \subseteq \mathbf{C}_A(E) \cong M_n(E)$. Hence, K is a maximal subfield of $B = \mathbf{C}_A(E) \in \mathfrak{S}(E)$. The first part of the proof gives $\mathbf{C}_B(K) \cong M_m(K)$ and $n^2 = \dim_E B = (m[K:E])^2$. In particular $[K:E]$ divides n. However, E is n-closed by assumption. Thus, $E = K$ is a maximal subfield of A. □

Corollary a. *If $A \in \mathfrak{S}(F)$, then $\dim_F A = m^2$ for some $m \in \mathbb{N}$. For a subfield E of A, $[E:F]$ divides m.*

These statements reformulate the last part of the proposition, since every subfield of A can be enlarged to a maximal subfield. The natural number m is called the *degree* of A. It will be denoted by $\operatorname{Deg} A$. Explicitly $\operatorname{Deg} A = (\dim_F A)^{1/2}$ for $A \in \mathfrak{S}(F)$.

If E is a subfield of $A \in \mathfrak{S}(F)$, then $[E:F] \leq \operatorname{Deg} A$ by Corollary a. Hence, when $[E:F] = \operatorname{Deg} A$, the field E is necessarily a maximal subfield of A. The converse is not true in general. For example, if F is n-closed, then F is maximal in $M_n(F)$, and $\operatorname{Deg} M_n(F) = n$. We will say that a subfield E of $A \in \mathfrak{S}(F)$ is *strictly maximal* if $[E:F] = \operatorname{Deg} A$. Exercise 2 provides an example of an algebra that contains a strictly maximal subfield and also a maximal subfield that is not strictly maximal.

Corollary b. *A subfield E of $A \in \mathfrak{S}(F)$ is strictly maximal if and only if $\mathbf{C}_A(E) = E$. If A is a division algebra, then every maximal subfield of A is strictly maximal.*

Proof. The first assertion is a consequence of the D.C.T. because $E \subseteq \mathbf{C}_A(E)$, and $(\operatorname{Deg} A)^2 = \dim_F A = [E:F](\dim_F \mathbf{C}_A(E))$. If E is a maximal subfield of the division algebra A, so that $M_n(E) \cong \mathbf{C}_A(E) \subseteq A$ by the proposition, then $n = 1$ since A has no non-zero nilpotent elements. Thus, $\mathbf{C}_A(E) = E$, and E is strictly maximal. □

We conclude this section with an application of maximal subfields that proves an assertion that was made in Section 1.6.

Theorem. *Let F be a field whose characteristic is not 2. If $A \in \mathfrak{S}(F)$ has degree 2, then A is isomorphic to a quaternion algebra.*

Proof. Let E be a maximal subfield of A. If $E = F$, then $A \cong M_2(F) \cong \left(\frac{1,1}{F}\right)$ by Lemma c. If $E \neq F$, then E/F is a quadratic extension; and since $\operatorname{char} F \neq 2$, it is possible to write $E = F(x)$, where $x^2 = a \in F^\circ$, $x \notin F$. The mapping $x \mapsto -x$ defines an automorphism of E, so that by the Noether–

Skolem Theorem, there exists $y \in A^\circ$ such that $y^{-1}xy = -x$. Plainly, $y \in A - E$. Therefore, $\dim_F(F + xF + yF + xyF) = 4 = \dim_F A$. Note that $xy = -yx$ implies $xy^2 = -yxy = y^2x$. Hence, $y^2 \in \mathbf{Z}(A) = F$, say $y^2 = b \in F^\circ$. Our discussion shows that the correspondences $1 \mapsto 1$, $x \mapsto \mathbf{i}$, $y \mapsto \mathbf{j}$, $xy \mapsto \mathbf{k}$ extend to an F-algebra isomorphism of A to $\left(\frac{a,b}{F}\right)$. □

Corollary c (Frobenius). *Up to isomorphism, the only finite dimensional, non-commutative division algebra over* $\mathbb{R}$ *is* $\mathbb{H} = \left(\frac{-1,-1}{\mathbb{R}}\right)$. *Hence,* $\mathbf{B}(\mathbb{R}) = \mathbb{Z}/2\mathbb{Z}$.

Proof. Let D be a finite dimensional, non-commutative division algebra over $\mathbb{R}$. Since $\mathbb{C}$ is the only non-trivial algebraic extension of $\mathbb{R}$, either $\mathbf{Z}(D) = \mathbb{R}$ or $\mathbf{Z}(D) = \mathbb{C}$. The second possibility is excluded because $\mathbf{B}(\mathbb{C}) = \{1\}$. Thus, $D \in \mathfrak{S}(\mathbb{R})$. Let E be a maximal subfield of D. By Corollary b, $\operatorname{Deg} D = [E:\mathbb{R}] = [\mathbb{C}:\mathbb{R}] = 2$. Thus, D is a quaternion algebra, so that $D \cong \mathbb{H}$ according to the remark after Corollary 1.7. □

Exercises

1. Assume that F is a field of characteristic zero. Prove the following statements.

   (a) F is n-closed if and only if there is no irreducible $\Phi \in F[\mathbf{x}]$ such that $\deg \Phi > 1$ and $\deg \Phi$ divides n.

   (b) Let E/F be a finite extension of degree m. If E is n-closed and m is relatively prime to n, then F is n-closed.

   (c) Let E and K be field extensions of F with K/F finite. If $n \in \mathbb{N}$ is such that E is m-closed for all $m \le n$ and $[K:F] \le n$, then $K \subseteq E$. Hint. Consider KE/E.

   (d) For each $n \in \mathbb{N}$, there is a smallest field E between F and its algebraic closure with the property that E is m-closed for all $m \le n$.

   (e) If F is n-closed, then $(F^\circ)^p = F^\circ$ for all primes p that divide n.

   (f) F is 2-closed if and only if $(F^\circ)^2 = F^\circ$.

   (g) F is both 2-closed and 3-closed if and only if $(F^\circ)^3 = (F^\circ)^2 = F^\circ$. Hint. Use Cardano's formula for the solution of a cubic equation.

   (h) If F is an algebraic number field, that is, a finite extension of $\mathbb{Q}$, then F is not n-closed for any $n > 1$.

2. Let K be the maximal solvable extension of $\mathbb{Q}$, that is, the compositum of all finite Galois extensions $E/\mathbb{Q}$ such that $\mathbf{G}(E/\mathbb{Q})$ is a solvable group. Let $F = K \cap \mathbb{R}$. Prove the following statements.

   (a) $[K:F] = 2$.

   (b) K is 2-closed and 3-closed.

   (c) K is not 6-closed. Hint. Let $\Phi \in \mathbb{Q}[\mathbf{x}]$ be a polynomial of degree 6 such that the Galois group of the splitting field of Φ is the symmetric group on 6-letters. For a proof that such polynomials exists, see [73], p. 201. An explicit example is given on p. 109 of [47]. Show that Φ is irreducible over K.

   (d) $M_6(F)$ contains a maximal subfield that is isomorphic to K, and also a strictly maximal subfield that is isomorphic to $F(y)$, where y is a root of Φ.

3. Let F be a field of characteristic 2 such that F is not 2-closed. Assume that $A \in \mathfrak{S}(F)$ has degree 2. Prove that there exist elements x and $y \in A$ such that $A = F + xF + yF + xyF$, where $yx = x(y + 1)$, $x^2 = a$, $y^2 + y + b = 0$, and the polynomials $\mathbf{x}^2 + a$, $\mathbf{x}^2 + \mathbf{x} + b$ are irreducible over F. Conversely, show that an F-algebra A that is defined by this recipe is central simple, and A is a division algebra if and only if there are no elements c and d in F such that $c^2 + cd + d^2b = a$.

4. Let $D = \left(\frac{a,b}{F}\right)$ be a quaternion division algebra. Prove that if $0 \neq x \in D$ is a pure quaternion (that is, $x = \mathbf{i}c_1 + \mathbf{j}c_2 + \mathbf{k}c_3$ for some c_1, c_2, c_3 in F), then $F((-\nu(x))^{1/2})$ is isomorphic to a maximal subfield of D, where $\nu\colon D \to F$ is the quaternion norm. Conversely, show that every maximal subfield of D has the form $F(c^{1/2})$, where $c = -\nu(x)$ and x is a pure quaternion in D.

13.2. Splitting Fields

The division algebra $\mathbb{H} = \left(\frac{-1,-1}{\mathbb{R}}\right)$ of real quaternions loses its glamour when the coefficient domain is extended to $\mathbb{C}$: $H \otimes \mathbb{C} \cong \left(\frac{-1,-1}{\mathbb{C}}\right) \cong M_2(\mathbb{C})$ by Proposition 1.6. This phenomenon is of fundamental importance in the theory of central simple algebras. It is the key to the construction of all such algebras.

Definition. Let $A \in \mathfrak{S}(F)$. An extension field E of F is a *splitting field* for A if $A^E \cong M_n(E)$ as E-algebras, where $n = \operatorname{Deg} A^E = \operatorname{Deg} A$.

Recall that A^E is our notation for $A \otimes E$, that is, A^E is the E-algebra obtained from A by extending the coefficient domain from F to E. We will often say that *E splits A* if E is a splitting field for A.

It will be useful to have alternative characterizations of splitting fields.

Proposition a. *Let $A \in \mathfrak{S}(F)$ have degree n. The following conditions are equivalent for a field extension E of F.*

(i) *E is a splitting field for A.*
(ii) *There is an F-algebra homomorphism $\phi\colon A \to M_n(E)$.*
(iii) *There is an F-algebra homomorphism $\phi\colon A \to M_n(E)$ such that $\phi(A)E = M_n(E)$ (that is, $\phi(A)$ spans the E-space $M_n(E)$).*
(iv) *For some $m \in \mathbb{N}$, there is an F-algebra homomorphism $\phi\colon A \to M_m(E)$ such that $\phi(A)E = M_m(E)$.*

Proof. If E is a splitting field for A, then the composite mapping $A \to A \otimes E \to M_n(E)$ is an F-algebra homomorphism. Assume that $\phi\colon A \to M_n(E)$ is an F-algebra homomorphism. Write B for $M_n(E)$, viewed as an F-algebra.

Then $E = \mathbf{Z}(B) \subseteq \mathbf{C}_B(\phi(A))$, $\phi(A)$ is central simple, and $\dim_F B = n^2[E:F] = (\dim_F \phi(A))(\dim_F E)$. By Proposition 12.4a, $B = \phi(A) \otimes E$. In particular, $B = \phi(A)E$. Obviously, (iii) implies (iv). Finally, if (iv) is satisfied, then there is an F-algebra isomorphism $M_m(E) \cong \phi(A) \otimes E \cong A^E$ by Proposition 12.4a; and $m^2 = \dim_E M_m(E) = \dim_E A^E = n^2$. Hence, E is a splitting field for A. □

The splitting fields of a central simple F-algebra have an important relation with the Brauer group of F. It is convenient to introduce some notation that will be used extensively in the next chapter.

Let E be a field extension of F. Denote the inclusion homomorphism of F to E by κ. Then κ induces a group homomorphism $\kappa_*: \mathbf{B}(F) \to \mathbf{B}(E)$. The kernel of κ_* is called the relative Brauer group of E/F; it will be denoted by $\mathbf{B}(E/F)$.

The following lemma reformulates the definition of a splitting field in the terminology of Brauer groups.

Lemma. *Let $A \in \mathfrak{S}(F)$. If E/F is a field extension, then E is a splitting field for A if and only if $[A] \in \mathbf{B}(E/F)$.*

If E/F and K/E are field extensions with corresponding inclusions κ_1 and κ_2, then $\kappa_2\kappa_1$ is the inclusion of F in K. Since $\operatorname{Ker}\kappa_{1*} \subseteq \operatorname{Ker}\kappa_{2*}\kappa_{1*} = \operatorname{Ker}(\kappa_2\kappa_1)_*$, it follows that $\mathbf{B}(E/F) \subseteq \mathbf{B}(K/F)$. Thus, the lemma has the following useful consequence.

Corollary. *If E is a splitting field for $A \in \mathfrak{S}(F)$, then every field extension of E splits A.*

Results on the subfields of splitting fields are rare. The last proposition of this section is one of the few specimens of such theorems.

Proposition b. *Let L be a splitting field for $A \in \mathfrak{S}(F)$. There is a subfield E of L that is finitely generated over F and splits A.*

PROOF. Proposition a allows us to assume that A is a subalgebra of $M_n(L)$, where $n = \operatorname{Deg} A$, and $M_n(L) = AL$. Let $\{\xi_{ij}: 1 \le i, j \le n\}$ be an F-basis of A. Write $\xi_{ij} = \sum_{1 \le k,l \le n} \varepsilon_{kl} c_{klij}$, where $c_{klij} \in L$. On the other hand, since $AL = M_n(L)$, there exist $d_{ijkl} \in L$ $(1 \le i, j, k, l \le n)$ such that $\varepsilon_{kl} = \sum_{1 \le i,j \le n} \xi_{ij} d_{ijkl}$. It follows that the $n^2 \times n^2$ matrices $[c_{klij}]$ and $[d_{ijkl}]$ are inverses of each other. If E is the field generated over F by $\{c_{klij}: 1 \le i, j, k, l \le n\}$, then $d_{ijkl} \in E$ for all i, j, k, l. Therefore, $A \subseteq M_n(E) = \sum_{1 \le k,l \le n} \varepsilon_{ij} E$, and $AE = M_n(E)$. By Proposition a, E splits A. □

One might suspect that the field E in Proposition b could be chosen to be algebraic over F. Exercise 2 shows that this is not the case.

EXERCISES

1. Let $D = \left(\frac{a,b}{F}\right)$ be a quaternion division algebra, where char $F \neq 2$ and $a, b \in F^\circ$. Prove that $E = F(\sqrt{a})$ is a splitting field for D by showing explicitly that the mapping $\phi: D \to M_2(E)$ defined by

$$\phi(c_0 + \mathbf{i}c_1 + \mathbf{j}c_2 + \mathbf{k}c_3) = \begin{bmatrix} e & f \\ b\bar{f} & \bar{e} \end{bmatrix},$$

$e = c_0 + \sqrt{a}c_1, \bar{e} = c_0 - \sqrt{a}c_1, f = c_2 + \sqrt{a}c_3, \bar{f} = c_2 - \sqrt{a}c_3$ is an F-algebra homomorphism.

2. Let $D = \left(\frac{a,b}{F}\right)$ be a quaternion division algebra, where char $F \neq 2$, and $a, b \in F^\circ$.
   (a) Show that if t is transcendental over F, then $\mathbf{x}^2 - b(t^2 - a)$ is irreducible in $F(t)[\mathbf{x}]$.
   Define $E = F(t,y)$, where y is a root of $\mathbf{x}^2 - b(t^2 - a)$.
   (b) Use Proposition 1.6 to prove that E splits D. Hint. Note that $(bt)^2 = ab^2 + by^2$.
   (c) Show that F is algebraically closed in E. Hint. Prove that if $c \in E - F$ is algebraic over F, then $F(t,c) = E$. Show that $y \in F(c)(t)$ is impossible.

13.3. Algebraic Splitting Fields

In this section our attention is on the finite algebraic extensions of a field F that split a given central simple F-algebra A. We will see that these extensions are closely related to maximal subfields of A.

Lemma. *Let $A \in \mathfrak{S}(F)$. If E is a subfield of A, then $\mathbf{C}_A(E) \in \mathfrak{S}(E)$, and $\mathbf{C}_A(E) \sim A^E$ as E-algebras.*

The proof of this result is the same as the first part of the proof of the D.C.T. Since E is a subfield of A, we can view A as a right A^E-module, and from this standpoint $\mathbf{C}_A(E) \cong \mathbf{E}_{A^E}(A)$ by Lemma 12.7. Let P be a representative of the unique (because A^E is simple) isomorphism class of right A^E-modules. If D is the division algebra $\mathbf{E}_{A^E}(P)$, then for suitable natural numbers k and n, $\mathbf{C}_A(E) \cong \mathbf{E}_{A^E}(A) \cong \mathbf{E}_{A^E}(\bigoplus k P) \cong M_k(D) \sim M_n(D) \cong \mathbf{E}_{A^E}(\bigoplus n P) \cong \mathbf{E}_{A^E}(A^E) \cong A^E$ as E-algebras. In particular, $\mathbf{C}_A(E) \in \mathfrak{S}(E)$.

Proposition. *Let $A \in \mathfrak{S}(F)$. For a subfield E of A, the following conditions are equivalent.*

(i) *E is a splitting field for A.*
(ii) *$\mathbf{C}_A(E) \cong M_k(E)$, where $k[E:F] =$ Deg A.*
(iii) *$A = B \otimes C$, where $B \in \mathfrak{S}(F)$, $C \cong M_k(F)$, and E is a strictly maximal subfield of B.*

PROOF. If E is a splitting field for A, then $\mathbf{C}_A(E) \cong M_k(E)$ by the lemma, and the D.C.T. yields $\operatorname{Deg} A = k[E: F]$. It follows from (ii) that $\mathbf{C}_A(E) = E \otimes C$, where $C \cong M_k(F)$. Let $B = \mathbf{C}_A(C)$. Clearly, E is a subfield of B. Since $C \in \mathfrak{S}(F)$, it follows from the D.C.T. that $B \in \mathfrak{S}(F)$, $A = B \otimes C$, and $\operatorname{Deg} B = (\operatorname{Deg} A)/k = [E:F]$, according to (ii). Thus, E is a strictly maximal subfield of B. Assume that (iii) is satisfied. By Corollary 13.1b, $\mathbf{C}_B(E) = E$. Thus, E splits B by the lemma. Hence, $[A] = [B] \in \mathbf{B}(E/F)$, and E splits A by Lemma 13.2. □

It follows from the proposition and Proposition 13.1 that every maximal subfield of $A \in \mathfrak{S}(F)$ splits A. Up to equivalence, the converse is true.

Theorem. *Let $A \in \mathfrak{S}(F)$. For a finite field extension E/F, the following conditions are equivalent.*

(i) *E is a splitting field for A.*
(ii) *There exists $B \in \mathfrak{S}(F)$ such that $B \sim A$ and E is a strictly maximal subfield of B.*
(iii) *There exists $B \in \mathfrak{S}(F)$ such that $B \sim A$ and E is a maximal subfield of B.*

PROOF. Plainly, (ii) implies (iii); and (i) follows from (iii) by the proposition, Proposition 13.1, and Lemma 13.2. Assume that E splits A. By Lemma 13.1a it can be assumed that E is a subfield of $M_n(F)$, where $n = [E:F]$. In this case, E is a subfield of $A \otimes M_n(F)$ and E splits $A \otimes M_n(F)$. By the proposition, $A \sim A \otimes M_n(F) = B \otimes C \cong B \otimes M_k(F) \sim B$, where $B \in \mathfrak{S}(F)$ and E is a strictly maximal subfield of B. □

Corollary. *If $A \in \mathfrak{S}(F)$ and E/F is a finite field extension such that $[E:F] = \operatorname{Deg} A$, then E splits A if and only if E is isomorphic as an F-algebra to a strictly maximal subfield of A.*

The corollary follows from the theorem and Proposition 12.5b.

EXERCISE

In this problem, D is the division algebra of rational (Hamiltonian) quaternions $\left(\frac{-1,-1}{\mathbb{Q}}\right)$. Prove the following assertions for subfields F and E of $\mathbb{C}$.

(a) E is a splitting field for D if and only if there are elements a and b in E such that $a^2 + b^2 = -1$.

(b) If p is a prime divisior of $2^r + 1$, where $r \geq 1$, z is a primitive p'th root of unity, and $E = \mathbb{Q}(z)$, then E is a splitting field for D.

Hint. Let $X = \{c^2 + d^2 : c, d \in E\}$. Show that X is closed under multiplication and includes $1 + z^m$ for all $m \in \mathbb{N}$ (in fact, $z^m \in (E^\circ)^2$). Let $2^r + 1 = sp$, and deduce that

$$0 = \left(\sum_{i=0}^{p-1} z^i\right)\left(\sum_{j=0}^{s-1} z^{jp}\right) = z^{-1} + \prod_{k=0}^{r-1}(1 + z^{2^k}),$$

so that

$$-1 = z\prod_{k=0}^{r-1}(1 + z^{2^k}) \in X.$$

Apply (a).

(c) Suppose that $[E:\mathbb{Q}] = 2^n m$ with m odd, where E is one of the fields that was defined in (b). Then there is a unique subfield F of E such that $[F:\mathbb{Q}] = 2^n$, and F is a splitting field for D.

Hint. $E/\mathbb{Q}$ is Galois and $\mathbf{G}(E/\mathbb{Q})$ is cyclic. Use this observation together with Proposition 13.4 below.

(d) If $F/\mathbb{Q}$ is Galois with $\mathbf{G}(F/\mathbb{Q})$ cyclic of order 2^n, then no proper subfield of F splits D.

Hint. Show that every proper subfield of F is contained in $\mathbb{R}$ by proving that $[F: F \cap \mathbb{R}] \leq 2$.

(e) If p is a prime divisor of the Fermat number $2^{2^k} + 1$, then 2^{k+1} divides $p - 1$.

Hint. Show that the multiplicative order of 2 in $\mathbb{F}_p$ is 2^{k+1}.

(f) For infinitely many natural numbers n, there exist splitting fields F for D such that $[F:\mathbb{Q}] = 2^n$, and no proper subfield of F splits D.

13.4. The Schur Index

The degree mapping is plainly not invariant under the Morita equivalence. Because of this fact, and for other reasons, it is useful to define a different numerical function on central simple algebras.

If $A \in \mathfrak{S}(F)$, then by Proposition 12.5b there is a division algebra D such that $A \sim D$, and D is unique to within isomorphism. Explicitly, D is determined by the conditions: $A \cong M_n(D)$ for a suitable $n \in \mathbb{N}$; and D is a division algebra. The *Schur index* of A is

$$\operatorname{Ind} A = \operatorname{Deg} D.$$

For simplicity, we will usually refer to Ind A as the index of A, since Schur's fame is sufficiently honored by the expression "Schur's Lemma."

In this section, we will use the results concerning splitting fields to prove some basic facts about the index.

Lemma. *Let $A \in \mathfrak{S}(F)$. If E is a finite field extension of F that splits A, then* Ind A *divides $[E:F]$. Conversely, A contains a subfield E that splits A, and $[E:F] =$* Ind A.

PROOF. If E splits A, then by Theorem 13.3 there is an algebra $B \in \mathfrak{S}(F)$ that contains E as a strictly maximal subfield, and $B \sim A$. If $B \cong M_n(D)$, where D is a division algebra, then $A \sim D$ and $[E:F] = \operatorname{Deg} B =$

$n(\text{Deg}\, D) = n(\text{Ind}\, A)$. To prove the converse, write $A = D \otimes B$, where D is a division algebra and $B \cong M_k(F)$ for some $k \in \mathbb{N}$. By Corollary 13.1b, a maximal subfield E of D is strictly maximal. That is, $[E:F] = \text{Deg}\, D = \text{Ind}\, A$. By Theorem 13.3, any maximal subfield of D is a splitting field for A. □

Proposition. *Let $A, B \in \mathfrak{S}(F)$, and suppose that E/F is a finite field extension.*

(i) *If* $[A] = [B]$, *then* $\text{Ind}\, A = \text{Ind}\, B$.
(ii) $\text{Ind}\, A$ *divides* $\text{Deg}\, A$, *and* $\text{Ind}\, A = \text{Deg}\, A$ *if and only if A is a division algebra.*
(iii) $\text{Ind}\, A = \min\{[K:F]: K \text{ splits } A\}$.
(iv) $\text{Ind}\, A^E$ *divides* $\text{Ind}\, A$.
(v) $\text{Ind}\, A$ *divides* $[E:F](\text{Ind}\, A^E)$.
(vi) *If* $\text{Ind}\, A$ *is relatively prime to* $[E:F]$, *then* $\text{Ind}\, A^E = \text{Ind}\, A$; *in this case, if A is a division algebra, then so is A^E.*
(vii) $\text{Ind}(A \otimes B)$ *divides* $(\text{Ind}\, A)(\text{Ind}\, B)$.
(viii) *For* $m \geq 1$, $\text{Ind}\, A^{\otimes m}$ *divides* $\text{Ind}\, A$, *where $A^{\otimes m}$ is the tensor product of m copies of A.*

PROOF. The statements (i) and (ii) are easy consequences of the definitions of the index and the degree. The formula (iii) is an obvious consequence of the lemma. For the proofs of the remaining statements, it can be assumed that A and B are division algebras. Properties (iv) and (vii) then follow from (ii). To prove (v), let K/E be a field extension such that K splits A^E and $[K:E] = \text{Ind}\, A^E$. Clearly, K is a splitting field for A, so that $\text{Ind}\, A$ divides $[K:F] = [K:E][E:F] = [E:F](\text{Ind}\, A^E)$. The combination of (iv) and (v) plainly implies (vi). To prove (viii), let K/F be a field extension such that K splits A and $[K:F] = \text{Ind}\, A$. In the Brauer group of K, $[(A^{\otimes m})^K] = [A^K]^m = 1$. By the lemma, $\text{Ind}\, A^{\otimes m}$ divides $[K:F] = \text{Ind}\, A$. □

As an application of the index, we will prove a useful variant of Corollary 13.3.

Corollary. *Assume that $D \in \mathfrak{S}(F)$ is a division algebra, and K/F is a field extension such that $[K:F]$ is a prime divisor of* $\text{Deg}\, D$. *The following properties are equivalent.*

(i) *K is isomorphic to a subfield of D.*
(ii) *D^K is not a division algebra.*
(iii) $\text{Deg}\, D = [K:F](\text{Ind}\, D^K)$.

PROOF. For the proof that (i) implies (ii) it can be assumed that K is a subfield of D. Extend K to a maximal subfield E of D. By Theorem 13.3, E is a splitting field for D; therefore E also splits D^K. The lemma and Proposition 13.1 yield $\text{Ind}\, D^K \leq [E:K] < [E:F] \leq \text{Deg}\, D = \text{Deg}\, D^K$, so that D^K is not a division algebra by part (ii) of the proposition. Conversely, if D^K is not a division algebra, then $\text{Ind } D^K < \text{Deg } D$. In this case, it follows from parts (iv) and (v) of the proposition and the hypothesis "$[K: F]$ is prime" that $\text{Deg } D = [K : F]$ $(\text{Ind } D^K)$. Finally, by the lemma there is an extension E/K such that E splits D^K and $[E : K] = \text{Ind } D^E$. Thus, E splits D, and (iii) implies that $[E : F] = \text{Deg } D$. By Corollary 13.3, E is isomorphic to a subfield of D. Therefore, so is K. □

Exercises

1. Let $A \in \mathfrak{S}(F)$ have index k. Assume that E is a splitting field for A. Prove that if P is a minimal right ideal of A and Q is a minimal right ideal of A^E, then $P^E = P \otimes E$ (considered as an A^E-module) is isomorphic to a direct sum of k copies of Q. Hint. Compute the dimensions of P^E and Q.

2. The Schur index can be defined for arbitrary separable F-algebras in the following way. Let A be a separable F-algebra. In particular, A is semisimple and finite dimensional. Let $A = A_1 \dotplus \cdots \dotplus A_r$ with each A_i simple. Hence $A_i \in \mathfrak{S}(\mathbf{Z}(A_i))$. The index of A_i is defined to be its index as a central simple algebra over its center. The index of A is the sequence $(\text{Ind } A_1, \ldots, \text{Ind } A_r)$. It is this notion of the Schur index that occurs in the theory of group algebras.

   Let A be a finite dimensional F-algebra. An extension field E of F is a splitting field for A if A^E is a product of matrix rings over E, that is, $A^E \cong M_{n_1}(E) \dotplus \cdots \dotplus M_{n_r}(E)$ for suitable $n_i \geq 1$. Prove the following results.

   (a) A has a splitting field if and only if A is separable.

   (b) If $A = F(x)$, where x is separable and algebraic over F with minimum polynomial $\Phi \in F[\mathbf{x}]$, then E is a splitting field for A if and only if Φ decomposes into a product of linear factors in E; that is, E is a splitting field for Φ in the usual field theoretic sense. In particular, F is isomorphic to a subfield of E.

   (c) If A is a separable F-algebra, and $A = A_1 \dotplus \cdots \dotplus A_r$ with each A_i simple, then E is a splitting field for A if and only if E is a splitting field for each $\mathbf{Z}(A_i)$ and also a splitting field for each A_i, considered as a $\mathbf{Z}(A_i)$-algebra. Hint. Use the result of Exercise 4, Section 12.4.

   (d) With the hypotheses of (c) and the notation $m_i = [\mathbf{Z}(A_i) : F]$, $n_i = \text{Ind } A_i$, prove that there is a splitting field E for A such that $[E : F] \leq (m_1!) \cdots (m_r!)$ $n_1 \cdots n_r$.

13.5. Separable Splitting Fields

The connection between the Brauer group of a field F and cohomology groups is based on the existence for each $A \in \mathfrak{S}(F)$ of a finite Galois extension E/F such that E splits A. Since every separable extension can be enlarged to a Galois extension, it is sufficient to prove the existence of a splitting field that is separable over F.

Lemma. *Let $D \in \mathfrak{S}(F)$ be a division algebra. If every subfield of D is purely inseparable over F, then $D = F$.*

PROOF. By Corollary 13.1b, D contains a strictly maximal subfield K; say $\operatorname{Deg} D = [K:F] = n$. If $\operatorname{char} F = 0$, then K/F is separable and purely inseparable, so that $F = K = D$. Assume that $\operatorname{char} F = p > 0$. In this case, $n = [K:F] = p^m$ for some $m \geq 0$, since K/F is purely inseparable. The algebraic closure E of F splits D, so that there is an F-algebra homomorphism $\phi: D \to M_n(E)$, with $M_n(E) = \phi(D)E$. If $x \in D$, then $F(x)/F$ is purely inseparable. Hence, $x^{p^k} = a$ for some $k \leq m$ and $a \in F$. Thus,

$$(\phi(x) - \iota_n a^{1/p^k})^{p^k} = 0.$$

Consequently,

$$\operatorname{tr}(\phi(x)) - na^{1/p^k} = \operatorname{tr}(\phi(x) - \iota_n a^{1/p^k}) = 0,$$

because the trace of a nilpotent matrix is 0. If $n > 1$, then p divides n, $\operatorname{tr}(\phi(x)) = 0$ for all $x \in D$, and $\operatorname{tr} \alpha = 0$ for all $\alpha \in \phi(D)E = M_n(E)$. This contradiction proves that $n = 1$ and $F = D$ in the prime characteristic case. □

Proposition. *If $D \in \mathfrak{S}(F)$ is a division algebra, and K is a subfield of D that is maximal with the property that K/F is separable, then K is a strictly maximal subfield of D.*

PROOF. By Lemma 13.3 and the remark that was made after Lemma 13.1b, $\mathbf{C}_D(K) \in \mathfrak{S}(K)$ is a division algebra. Since K is maximal with the property that K/F is separable, and a separable extension of a separable extension is separable, it follows that every subfield of $\mathbf{C}_D(K)$ is purely inseparable over K. By the lemma, $\mathbf{C}_D(K) = K$. Therefore, K is strictly maximal in D by Corollary 13.1b. □

In this book, the term "Galois extension" is used to describe a finite, separable, normal, field extension. Infinite Galois extensions will appear only in a few exercises of Chapter 14. By a standard result of Galois theory, E/F is Galois if and only if E is the splitting field of a separable polynomial in $F[\mathbf{x}]$. Therefore, any finite separable extension K/F can be enlarged to a Galois extension E/F: let E be the splitting field of a separable polynomial $\Phi \in F[\mathbf{x}]$ such that $K \cong F[\mathbf{x}]/\Phi F[\mathbf{x}]$.

Theorem. *If $A \in \mathfrak{S}(F)$, then there exists $B \in \mathfrak{S}(F)$ and a strictly maximal subfield E of B such that $B \sim A$ and E/F is a Galois extension.*

PROOF. Let $A \sim D \in \mathfrak{S}(F)$, where D is a division algebra. By the proposition, D has a strictly maximal subfield K such that K/F is separable. Let E/F be a Galois extension with $K \subseteq E$. Since K splits D by Proposition 13.3, so does E. The theorem therefore follows from Theorem 13.3. □

Corollary. *For a field F, the Brauer group $\mathbf{B}(F)$ is the union of the subgroups $\mathbf{B}(E/F)$, where E/F ranges over all Galois extensions. Every element of $\mathbf{B}(E/F)$*

has the form $[A]$, *where* $A \in \mathfrak{S}(F)$ *contains* E *as a strictly maximal subfield; the algebra* A *with this property is unique up to isomorphism.*

The uniqueness statement in the corollary is a consequence of Proposition 12.5b because the strict maximality of E in A implies that $\dim_F A = [E:F]^2$.

EXERCISES

1. The existence of a separable maximal subfield of certain infinite dimensional division algebras can be proved using the result of this exercise. Let D be a non-commutative division algebra whose center is a field F of prime characteristic p. Assume that every $x \in D$ is algebraic over F. Prove that there exists $x \in D - F$ such that $F(x)/F$ is separable. Hint. Deny the assertion. Show that some $a \in D - F$ satisfies $a^p \in F$. Fix such an a, and define $\phi: D \to D$ by $\phi(x) = xa - ax$. Prove that $\phi \neq 0 = \phi^p$. Let $y \in D$ and $k > 1$ be such that $\phi^k(y) = 0 \neq \phi^{k-1}(y)$. Define $w = \phi^{k-2}(y)$, $x = \phi^{k-1}(y) = \phi(w) \neq 0$. Write $x = au$, $v = wu^{-1}$. Show that $v^{p^m} = 1 + av^{p^m}a^{-1}$ for all $m \in \mathbb{N}$, and obtain a contradiction when $v^{p^m} \in F$.

2. Let D be a division algebra over a field of prime characteristic. Prove that every finite subgroup of D° is abelian. Hint. Show that a finite subgroup of D generates a finite subalgebra, and use the fact (to be proved) that every finite division algebra is a field.

3. An algebra A is *strongly regular* if for every $x \in A$, there exists $y \in A$ such that $x^2y = x$. For example, division algebras are strongly regular. Prove the following facts concerning a strongly regular algebra A.
   (a) A has no non-zero nilpotent elements.
   (b) If $x^2y = x$, then $xyx = yx^2 = x$, so that xy is idempotent.
   (c) If $e \in A$ is idempotent, then $e \in \mathbf{Z}(A)$. Hint. Compute $(exe - xe)^2$ and $(exe - ex)^2$.
   (d) $\mathbf{J}(A) = 0$.
   (e) Every (left) ideal of A is a two sided ideal in A.
   (f) Every homomorphic image of A is strongly regular.
   (g) A is a subdirect product of division algebras.

4. Prove *Jacobson's Commutativity Theorem*: If A is an algebra such that for each $x \in A$ there exists $n \geq 2$ (depending on x) such that $x^n = x$, then A is commutative. Hint. Use Exercise 3 to reduce the proof to the case in which A is a division algebra. Let $F = \mathbf{Z}(A)$, and show that the characteristic of F is prime. Assume the existence of $x \in A - F$, and show that $\sigma \in \mathbf{G}(F(x)/F)$ exists such that $\sigma(x) = x^i \neq x$. Deduce from the generalized Noether–Skolem Theorem (Exercise 1, Section 12.6) that $u^{-1}xu = x^i \neq x$ for some $u \in A^\circ$. Obtain a contradiction by applying the result of Exercise 2 to the group generated by u and x.

13.6. The Cartan–Brauer–Hua Theorem

There are more questions about the subfields of central simple algebras than there are theorems on this subject. Some of the most important problems concern the relation between the group of units of $A \in \mathfrak{S}(F)$ and the multiplicative subgroups of the subfields of A. The Cartan–Brauer–Hua Theorem deals with one aspect of this relationship.

Lemma. *Assume that F is an infinite field and A is a finite dimensional F-algebra. If $x \in A$, then there exists $a \in F^\circ$ such that $x - a$ is a unit in A. In particular, every element of A is a sum of two units in A.*

The proof of this result is outlined in Exercise 1.

For a subset X of an algebra A, define the *normalizer* of X in A by

$$\mathbf{N}_A(X) = \{u \in A^\circ : u^{-1}Xu = X\}.$$

Obviously, $\mathbf{N}_A(X)$ is a subgroup of A°.

Cartan–Brauer–Hua Theorem. *Let $A \in \mathfrak{S}(F)$, where F is an infinite field. If D is a subalgebra of A such that D is a division algebra and D° is a normal subgroup of A°, then either $D = F$ or $D = A$.*

Proof. The hypothesis that D° is normal in A° implies

$$\mathbf{N}_A(D) = A^\circ. \tag{1}$$

With the aim of getting a contradiction, suppose that $F \subset D \subset A$. By the last statement of the lemma, $D^\circ \subset A^\circ$. Moreover, $\mathbf{C}_A(D)^\circ \subset A^\circ$, since $F \subset D$ implies $\mathbf{C}_A(D) \subset A$ by the D.C.T. Thus, $D^\circ \cup \mathbf{C}_A(D)^\circ \subset A^\circ$. (See Exercise 2.) Let $w \in A^\circ - (D^\circ \cup \mathbf{C}_A(D)^\circ)$. Since $w \notin \mathbf{C}_A(D)$, it follows from (1) that $x, y \in D$ exist such that

$$x \neq y \quad \text{and} \quad wx = yw. \tag{2}$$

By the lemma, $w - a \in A^\circ$ for some $a \in F^\circ$. Consequently, $(w - a)x = z(w - a)$ for some $z \in D$ by (1). This equation and (2) give

$$(z - x)a = (z - y)w. \tag{3}$$

Since $x \neq y$ in D and $a \in F^\circ$, it follows that $z - y \in D^\circ$. Hence, $w = (z - y)^{-1}(z - x)a \in D^\circ$, which contradicts the choice of w. □

Corollary a. *Let $D \in \mathfrak{S}(F)$ be a division algebra. If E is a subfield of D that properly contains F, then $\bigcup_{x \in D^\circ} x^{-1}Ex$ generates D as an F-algebra.*

Proof. The existence of a division algebra $D \in \mathfrak{S}(F)$ with $\operatorname{Deg} D > 1$ implies that F is infinite, as we will soon prove. Let A be the subalgebra of D that is generated by $\bigcup_{x \in D^\circ} x^{-1}Ex$. By Lemma 13.1b, A is a division algebra, and $A \supseteq E \supset F$. The definition of A implies that A° is a normal subgroup of D°. Hence, $A = D$. □

Rougly speaking, this corollary tells us that if a proper extension E of the field F can be embedded in the division algebra D, then the conjugates of E are "dense" in D. A related property characterizes the division algebras with a unique maximal subfield.

Proposition. *Let $D \in \mathfrak{S}(F)$ be a division algebra, and suppose that E is a maximal subfield of D. All of the maximal subfields of D are isomorphic to E if and only if E/F is separable and $D = \bigcup_{x \in D^\circ} x^{-1}Ex$.*

PROOF. If all maximal subfields of D are isomorphic, then E/F is separable by Proposition 13.5. Moreover, every $y \in D$ is in some maximal subfield K of D, and $K \cong E$ implies that $K = x^{-1}Ex$ for some $x \in D^\circ$ by the Noether–Skolem Theorem. Thus, $D = \bigcup_{x \in D^\circ} x^{-1}Ex$. Conversely, assume that E/F is separable and $D = \bigcup_{x \in D^\circ} x^{-1}Ex$. If $y \in D$, then $y \in x^{-1}Ex$ for a suitable $x \in D^\circ$. Consequently, y is separable over F. It follows that every maximal subfield of D has the form $F(y)$ for some $y \in D$; and $y \in x^{-1}Ex$ implies that $F(y) \cong xF(y)x^{-1} = E$ by maximality. □

The condition that $D = \bigcup_{x \in D^\circ} x^{-1}Ex$ for a maximal subfield E of D implies that the multiplicative group D° contains an abelian subgroup E° such that D° is the union of the conjugates of E°. An easy counting argument that is sketched in Exercise 3 shows that no finite, non-abelian group can satisfy this condition. Thus, the proposition implies another one of Wedderburn's celebrated results.

Wedderburn's Finite Division Algebra Theorem. *Every finite division algebra is a field.*

PROOF. If D is a finite division algebra with center F, then $D \in \mathfrak{S}(F)$. All maximal subfields of D are finite fields with $|F|^n$ elements, where $n = \operatorname{Deg} D$. Since finite fields with the same number of elements are isomorphic, the proposition implies that $D^\circ = \bigcup_{x \in D^\circ} x^{-1}E^\circ x$, where E is a maximal subfield of D. Because D° is finite, this can only happen if D° is abelian, that is, $D = F$. □

EXERCISES

1. Prove the lemma. Hint. Since $\dim_F A < \infty$, every $x \in A$ satisfies $\Phi(x) = 0$ for some $\Phi \in F[\mathbf{x}] - \{0\}$. Show that since F is infinite, there is an $a \in F^\circ$ such that $\Phi(\mathbf{x} + a)$ has a non-zero constant term.

2. Let H_1 and H_2 be proper subgroups of the group G. Prove that $H_1 \cup H_2 \subset G$. Hint. Reduce the proof to the case in which $H_1 \nsubseteq H_2$ and $H_2 \nsubseteq H_1$. Show that if $x \in H_1 - H_2$ and $y \in H_2 - H_1$, then $xy \in G - (H_1 \cup H_2)$.

3. Prove that if G is a finite group, and H is a proper subgroup of G, then $G \neq \bigcup_{x \in G} x^{-1}Hx$. Hint. Note that $|\{x^{-1}Hx : x \in G\}|$ is at most equal to the index $[G:H]$ of H in G, and $1 \in x^{-1}Hx$ for all $x \in G$. Use these observations and the assumption that $[G:H] \geq 2$ to show that $|\bigcup_{x \in G} x^{-1}Hx| < |G|$.

4. A field F is *Pythagorean* if any sum of squares of elements in F is in F^2. Thus, F is Pythagorean if and only if $a^2 + b^2 \in F^2$ for all a, $b \in F$. A field F (necessarily of characteristic zero) is *formally real* if -1 cannot be written as a sum of squares in F. In particular, a Pythagorean field is formally real if and only if $-1 \notin F^2$. Prove that a quaternion division algebra D over F has the property that all maximal subfields of D are isomorphic if and only if F is formally real, Pythagorean, and $D \cong \left(\frac{-1,-1}{F}\right)$. Hint. Show that for a, $b \in F - F^2$, the fields $F(\sqrt{a})$ and $F(\sqrt{b})$ are isomorphic if and only if $a/b \in F^2$. Use the result of Exercise 4, Section 13.1 to show that if all maximal subfields of D are isomorphic, then $D \cong \left(\frac{-1,-1}{F}\right)$. Moreover, since $c^2 + d^2 = \nu(\mathbf{i}c + \mathbf{j}d)$ in $\left(\frac{-1,-1}{F}\right)$, the hypothesis that maximal subfields are isomorphic yields $F(\sqrt{-(c^2 + d^2)}) \cong F(\sqrt{-1})$. Deduce that F is formally real and Pythagorean.

5. The purpose of this exercise is to show that there are formally real, Pythagorean fields E such that $E^\circ/(E^\circ)^2$ has arbitrarily large (finite) order. Of course, $\mathbb{R}$ is formally real and Pythagorean, but $|\mathbb{R}^\circ/(\mathbb{R}^\circ)^2| = 2$. Let F be a formally real, Pythagorean field such that $|F^\circ/(F^\circ)^2| = 2^m$. Let E be the field of formal Laurant series $\sum_{k \geq n} \mathbf{x}^k a_k$, $n \in \mathbb{Z}$, $a_k \in F$ with componentwise addition and the convolution product, that is, $\mathbf{x}^i\mathbf{x}^j = \mathbf{x}^{i+j}$. Prove the following statements.

   (a) $1 + \mathbf{x}a_1 + \mathbf{x}^2 a_2 + \cdots \in (E^\circ)^2$ for all $a_1, a_2, \ldots$ in F.
   (b) E is Pythagorean.
   (c) E is formally real.
   (d) $|E^\circ/(E^\circ)^2| = 2^{m+1}$. Hint. $\mathbf{x}$ and $-\mathbf{x}$ are in different cosets of $(E^\circ)^2$.

Notes on Chapter 13

Our discussion of subfields and splitting fields for simple algebras follows the classical line in two respects: only finite dimensional algebras are considered; all subfields are required to contain the center. In discussing maximal subfields of arbitrary central simple algebras rather than just division algebras, we have been slightly unorthodox. This extra generality makes it necessary to fuss over the pathology of n-closed fields. However, such fields do exist, and the property of n-closure shows up in the structure of the Brauer group. The Cartan–Brauer–Hua Theorem is usually stated for subalgebras of division algebras. The extra generality in our statement of this result falls short of what is known about the normal subgroup structure of $M_n(D)^\circ = GL_n(D)$ when $n \geq 2$. An elegant treatment of this topic is given in Chapter IV of Artin's book [6]. The example that is outlined in the exercise of Section 13.3 is due to Brauer and Noether. (See [19].)

CHAPTER 14
Galois Cohomology

The explicit calculation of the Brauer group of a field is usually a formidable task. In this chapter, we develop the machinery that in principle will compute $\mathbf{B}(F)$ for any field F. The key results in this program are: (1) $\mathbf{B}(F) = \bigcup \mathbf{B}(E/F)$, where the union is taken over all Galois extensions E/F (Corollary 13.5); (2) $\mathbf{B}(E/F) \cong H^2(\mathbf{G}(E/F), E^\circ)$, the second cohomology group of E°, considered as a $\mathbb{Z}\mathbf{G}(E/F)$-bimodule in a suitable way (Theorem 14.2); (3) the isomorphism of (2) lifts to an isomorphism of $\mathbf{B}(F) = \bigcup \mathbf{B}(E/F)$ with the direct limit $H^2(\mathbf{G}(F_s/F), F_s^\circ) = \varinjlim H^2(\mathbf{G}(E/F), E^\circ)$, where F_s is the maximal separable extension of F (Theorem 14.6). The group $H^2(\mathbf{G}(F_s/F), F_s)$ is one of the *Galois cohomology groups* of F. To provide some relief from this morass of formalism, we have inserted an application of Theorem 14.2: a proof that $\mathbf{B}(F)$ is a torsion group.

Like Chapters 9 and 11, this chapter is heavily oriented toward technical constructions: the cohomology of groups, direct and inverse limits, and Galois cohomology. It would be an unusual reader indeed who developed great enthusiasm for the results that are given here. The theorems in this chapter are not that exciting. However, they are absolutely fundamental tools for modern research in the theory of central simple algebras. The only available way to construct the Brauer groups of arbitrary fields is by using these techniques. Moreover, Galois cohomology provides the bridge between central simple algebras and class field theory that leads to the fundamental theorems on the Brauer groups of local fields and algebraic number fields. These are among the most profound results in modern algebra.

14.1. Crossed Products

If E/F is a finite Galois extension, then there is a construction that produces the central simple F-algebras that contain E as a strictly maximal subfield. By Corollary 13.5, every element of the relative Brauer group $\mathbf{B}(E/F)$ is represented uniquely by such an algebra. In this section we describe the construction.

Throughout this section, E/F denotes a Galois extension with Galois group $\mathbf{G}(E/F) = G$. It is convenient to represent the action of G on E by exponential notation $\sigma: c \mapsto c^\sigma$.

Lemma. *Assume that $A \in \mathfrak{S}(F)$ contains E as a strictly maximal subfield.*

(i) *There is a set $\{u_\sigma : \sigma \in G\} \subseteq A^\circ$ such that for all $c \in E$ and $\sigma \in G$,*

$$c^\sigma = u_\sigma^{-1} c u_\sigma. \tag{1}$$

(ii) *If $\{u_\sigma : \sigma \in G\}$ satisfies* (i), *then this set is an E-space basis of A_E; moreover, if $\sigma, \tau \in G$, then $\Phi(\sigma,\tau) = (u_{\sigma\tau})^{-1} u_\sigma u_\tau \in E^\circ$, and*

$$\Phi(\sigma,\tau)\Phi(\rho\sigma,\tau)^{-1}\Phi(\rho,\sigma\tau)(\Phi(\rho,\sigma)^\tau)^{-1} = 1 \quad \text{for all } \rho, \sigma, \tau \in G. \tag{2}$$

(iii) *If $u_1 = 1$ and Φ is defined as in* (ii), *then $\Phi(\sigma,1) = \Phi(1,\sigma) = 1$ for all $\sigma \in G$.*

Proof. The statement (i) is a special case of the Noether–Skolem Theorem. To prove the first part of (ii), it is sufficient to show that $\{u_\sigma : \sigma \in G\}$ is linearly independent, since the strict maximality of E in A implies that $\dim_E A_E = [E:F] = |G|$. If the u_σ are linearly dependent over E, then there is a relation $\sum_{\sigma \in X} u_\sigma c_\sigma = 0$ in which $c_\sigma \in E^\circ$ for $\sigma \in X$ and $\varnothing \neq X \subseteq G$ is minimal. Since each $u_\sigma \neq 0$, the set X includes at least two elements of G. For each $d \in E^\circ$, the equation (1) yields $\sum_{\sigma \in X} u_\sigma d^\sigma c_\sigma = d(\sum_{\sigma \in X} u_\sigma c_\sigma) = 0$. By the minimality of X, the sequences $(\ldots d^\sigma c_\sigma \ldots)_{\sigma \in X}$ and $(\ldots c_\sigma \ldots)_{\sigma \in X}$ must be proportional; that is, $d^\sigma = d^\tau$ for all $d \in E^\circ$ and $\sigma, \tau \in X$. This conclusion contradicts $|X| > 1$, and proves that $\{u_\sigma : \sigma \in G\}$ is a basis of A_E. We remark for future reference that the proof of the linear independence of $\{u_\sigma : \sigma \in G\}$ has used only (1). An easy computation based on the property (1) shows that $(u_{\sigma\tau})^{-1} u_\sigma u_\tau c = c(u_{\sigma\tau})^{-1} u_\sigma u_\tau$ for all $c \in E$. Hence, $(u_{\sigma\tau})^{-1} u_\sigma u_\tau \in \mathbf{C}_A(E) = E$ by Corollary 13.1b. Clearly, $\Phi(\sigma,\tau) = (u_{\sigma\tau})^{-1} u_\sigma u_\tau \neq 0$. Finally,

$$\begin{aligned}\Phi(\rho\sigma,\tau)^{-1}\Phi(\rho,\sigma\tau)\Phi(\sigma,\tau) &= u_\tau^{-1} u_{\rho\sigma}^{-1} u_{\rho\sigma\tau} u_{\rho\sigma\tau}^{-1} u_\rho u_{\sigma\tau} u_{\sigma\tau}^{-1} u_\sigma u_\tau \\ &= u_\tau^{-1} u_{\rho\sigma}^{-1} u_\rho u_\sigma u_\tau = \Phi(\rho,\sigma)^\tau.\end{aligned}$$

The statement (iii) is obvious. □

The mapping $\Phi: G^2 \to E^\circ$ defined in (ii) determines a cochain, and the equation (2) is the *cocycle condition* for suitable cohomology groups. The details of this identification will be given in the next section.

Proposition. *Let E/F be a Galois extension with $\mathbf{G}(E/F) = G$, and suppose that $\Phi: G^2 \to E^\circ$ satisfies the cocycle condition. Let $\{u_\sigma : \sigma \in G\}$ be a basis for the E-space $A = \bigoplus_{\sigma \in G} u_\sigma E$. Define $\mu: A \times A \to A$ by*

$$\mu\left(\sum_{\sigma \in G} u_\sigma c_\sigma, \sum_{\tau \in G} u_\tau d_\tau\right) = \sum_{\sigma,\tau} u_{\sigma\tau}\Phi(\sigma,\tau)c_\sigma^\tau d_\tau. \tag{3}$$

The mapping μ is F-bilinear, and it defines a product on A such that $A \in \mathfrak{S}(F)$, $1_A E$ is a strictly maximal subfield of A, the basis $\{u_\sigma : \sigma \in G\}$ satisfies (i) *of the lemma, and $(u_{\sigma\tau})^{-1}u_\sigma u_\tau = 1_A\Phi(\sigma,\tau)$.*

Proof. The F-bilinearity of μ is a consequence of the fact that G fixes the elements of F. By a routine calculation, the form $\Phi(\rho,\sigma\tau)\Phi(\sigma,\tau) = \Phi(\rho\sigma,\tau)\Phi(\rho,\sigma)^\tau$ of the cocycle condition implies that μ defines an associative multiplication. It is also easy to check that $1_A = u_1\Phi(1,1)^{-1}$; and $(1_A c)(1_A d) = 1_A cd$, so that $1_A E$ is a subfield of A that is isomorphic to E. Moreover,

$$u_\sigma u_{\sigma^{-1}} = u_1\Phi(\sigma,\sigma^{-1}) = 1_A\Phi(1,1)\Phi(\sigma,\sigma^{-1}) \in A^\circ,$$

$u_\sigma^{-1}1_A c u_\sigma = u_\sigma^{-1}u_\sigma c^\sigma = 1_A c^\sigma$ for $\sigma \in G$, $c \in E$, and $(u_{\sigma\tau})^{-1}u_\sigma u_\tau = u_{\sigma\tau}^{-1}u_{\sigma\tau}\Phi(\sigma,\tau) = 1_A\Phi(\sigma,\tau)$. If $\theta: A \to B$ is a surjective F-algebra homomorphism, then $\theta|(1_A E)$ is injective, and $\{\theta u_\sigma : \sigma \in G\}$ satisfies (1). As we noted in the proof of the lemma, this fact implies that $\{\theta u_\sigma : \sigma \in G\}$ is linearly independent over E. Thus, $\dim_F A \geq \dim_F B \geq |G|[E:F] = \dim_F A$, and θ is an isomorphism. Hence, A is simple. If $x = \sum_{\sigma \in G} u_\sigma c_\sigma \in \mathbf{Z}(A)$, then for all $d \in E$, $0 = (1_A d)x - x(1_A d) = \sum_{\sigma \in G} u_\sigma(d^\sigma - d)c_\sigma$. Thus, $(d^\sigma - d)c_\sigma = 0$ for all $\sigma \in G$ and $d \in E$, which implies that $c_\sigma = 0$ for all $\sigma \neq 1$. That is, $x = 1_A c$ for some $c \in E$. In fact, since $1_A c^\sigma = u_\sigma^{-1}(1_A c)u_\sigma = 1_A c$ for all $\sigma \in G$, the assumption that E/F is Galois puts c in F. □

The algebra A that is obtained from the construction of the proposition is called the *crossed product* of E and G relative to Φ. We will denote A by (E,G,Φ). As usual, E will be identified with the subfield $1_A E$ of A, and for $c \in E$, the element $1_A c$ will be designated by c.

Corollary. *The mapping $\Phi \to [(E,G,\Phi)]$ is surjective from the mappings $\Phi: G^2 \to E^\circ$ that satisfy the cocycle condition to $\mathbf{B}(E/F)$.*

The construction of (E,G,Φ) is somewhat neater if Φ is *normalized*, that is, $\Phi(1,1) = 1$. In this case $1_A = u_1$, and the cocycle condition yields $\Phi(\sigma,1) = \Phi(1,\sigma) = 1$ for all $\sigma \in G$. The lemma implies that for any Ψ there exists a normalized Φ such that $(E,G,\Psi) = (E,G,\Phi)$.

Example. Let E/F be Galois, with $[E:F] = n$ and $\mathbf{G}(E/F) = G$. Denote $A = \mathbf{E}_F(E) \cong M_n(F)$. The left regular representation maps E to a strictly maximal subfield of A. For $\sigma \in G$ and $c \in E$, define $\phi_\sigma \in A$ by $\phi_\sigma(c) = c^{\sigma^{-1}}$. If $b \in E$, then $(\phi_\sigma^{-1}\lambda_b\phi_\sigma)(c) = b^\sigma c = \lambda_{b^\sigma}(c)$. That is, $\phi_\sigma^{-1}\lambda_b\phi_\sigma = \lambda_{b^\sigma}$. Moreover, $\phi_\sigma\phi_\tau = \phi_{\sigma\tau}$ for $\sigma, \tau \in G$. Thus, $A \cong (E,G,I)$, where $I(\sigma,\tau) = 1$ for all $\sigma, \tau \in G$.

EXERCISES

1. Prove that if $\Phi: G^2 \to E^\circ$ is a mapping that satisfies the cocycle condition (2) and $\Phi(1,1) = 1$, then $\Phi(\sigma,1) = \Phi(1,\sigma)$ for all $\sigma \in G$.

2. Fill in the details that were omitted from the following parts of the proof of the proposition.
   (a) The multiplication defined by (3) is associative.
   (b) $u_1\Phi(1,1)^{-1}$ is the unity element of A.
   (c) If $c, d \in E$, then $(1_A c)(1_A d) = 1_A cd$.

3. Let F be a field whose characteristic is not 2, and suppose that $A = \left(\frac{a,b}{F}\right)$ is a quaternion algebra with $a \in F - F^2, b \in F^\circ$. Let $E = F(\sqrt{a})$, $G = \mathbf{G}(E/F) = \{1,\sigma\}$. Prove that $A \cong (E,G,\Phi)$, where $\Phi(1,1) = \Phi(1,\sigma) = \Phi(\sigma,1) = 1$ and $\Phi(\sigma,\sigma) = b$.

4. Let $A \in \mathfrak{S}(F)$, and suppose that B and C are simple subalgebras of A such that $C = \mathbf{C}_A(B)$. Denote $G = \mathrm{Aut}_F B$ and $H = \mathrm{Aut}_F C$. Prove the following statements.
   (a) There exist elements $u_\sigma \in \mathbf{N}_A(B)$ for all $\sigma \in G$ and $v_\tau \in \mathbf{N}_A(C)$ for all $\tau \in H$ such that $\sigma(x) = u_\sigma^{-1}xu_\sigma$, $x \in B$, and $\tau(y) = v_\tau^{-1}yv_\tau$, $y \in C$.
   (b) With the notation of (a), $u_{\sigma\tau}^{-1}u_\sigma u_\tau \in C^\circ$ for $\sigma, \tau \in G$, and $v_{\sigma\tau}^{-1}v_\sigma v_\tau \in B^\circ$ for $\sigma, \tau \in H$.
   (c) $\mathbf{N}_A(B)$ is the disjoint union of the cosets $C^\circ u_\sigma$, $\sigma \in G$, and $\mathbf{N}_A(C)$ is the disjoint union of the cosets $B^\circ v_\tau$, $\tau \in H$.
   (d) $\sigma \in G$ is an inner automorphism if and only if $u_\sigma \in B^\circ C^\circ$, and $\tau \in H$ is an inner automorphism if and only if $v_\tau \in B^\circ C^\circ$. Let X be a set of representatives of the right cosets in $\mathrm{Aut}_F B$ of the group $\mathrm{Inn}_F B$ of inner automorphisms of B, and let Y be a set of representatives of the cosets of $\mathrm{Inn}_F C$ in $\mathrm{Aut}_F C$.
   (e) $\mathbf{N}_A(B) = \bigcup_{\sigma\in X} B^\circ C^\circ u_\sigma$ and $\mathbf{N}_A(C) = \bigcup_{\tau \in Y} B^\circ C^\circ v_\tau$.
   (f) If B is a division algebra, then $\{u_\sigma : \sigma \in X\}$ is linearly independent over B, and $\sum_{\sigma\in X} BCu_\sigma$ is a subalgebra of A.

14.2. Cohomology and Brauer Groups

The results of the last section lead to a cohomological interpretation of relative Brauer groups. The correspondence will be described in this section.

Lemma. *Let E/F be a Galois extension with Galois group G. If Φ and Ψ are mappings from G^2 to E° that satisfy the cocycle condition, then $(E,G,\Phi) \cong (E,G,\Psi)$ if and only if there is a mapping $\Theta: G \to E^\circ$ such that*

$$\Phi(\sigma,\tau)\Psi(\sigma,\tau)^{-1} = \Theta(\tau)\Theta(\sigma\tau)^{-1}\Theta(\sigma)^\tau \quad \textit{for all} \quad \sigma, \tau \in G. \tag{1}$$

PROOF. Let $A = (E,G,\Phi) = \bigoplus_{\sigma\in G} u_\sigma E$, and $B = (E,G,\Psi) = \bigoplus_{\sigma\in G} v_\sigma E$, as in Proposition 14.1. Suppose that there is an isomorphism $\phi: A \to B$. There is no harm in assuming that $\phi(1_A c) = 1_B c$ for all $c \in E$. In fact, by the Noether–Skolem Theorem, the homomorphism $1_B c \mapsto \phi(1_A c)$ extends to an automorphism α of B, and $\alpha^{-1}\phi$ is an F-algebra and E-space isomorphism of A to B. By applying ϕ to the equation $u_\sigma^{-1}(1_A c)u_\sigma = 1_A c^\sigma$, we obtain $\phi(u_\sigma)^{-1}(1_B c)\phi(u_\sigma) = 1_B c^\sigma = v_\sigma^{-1}(1_B c)v_\sigma$ for all $c \in E$. Thus, $\phi(u_\sigma)v_\sigma^{-1} \in \mathbf{C}_B(1_B E) = 1_B E$, so that there exists $\Theta(\sigma) \in E^\circ$ satisfying

$$\phi(u_\sigma) = v_\sigma \Theta(\sigma). \tag{2}$$

As σ ranges over G, the equation (2) defines the required mapping from G to E°. In fact,

$$\begin{aligned} 1_B\Phi(\sigma,\tau) &= \phi(1_A\Phi(\sigma,\tau)) \\ &= \phi(u_{\sigma\tau}^{-1}u_\sigma u_\tau) \\ &= 1_B\Theta(\sigma\tau)^{-1}v_{\sigma\tau}^{-1}v_\sigma\Theta(\sigma)v_\tau\Theta(\tau) \\ &= 1_B\Theta(\sigma\tau)^{-1}v_{\sigma\tau}^{-1}v_\sigma v_\tau\Theta(\sigma)^\tau\Theta(\tau) \\ &= 1_B\Theta(\sigma\tau)^{-1}\Psi(\sigma,\tau)\Theta(\sigma)^\tau\Theta(\tau), \end{aligned}$$

so that (1) is satisfied. Conversely, if there is a mapping $\Theta\colon G \to E^\circ$ such that (1) holds, then the equation (2) defines an E-space isomorphism from A to B. A routine calculation using (1) and the definition of multiplication in crossed products shows that $\phi((u_\sigma c)(u_\tau d)) = \phi(u_\sigma c)\phi(u_\tau d)$ for all σ, $\tau \in G$ and c, $d \in E$. Thus, ϕ is an F-algebra isomorphism. □

The cocycle condition and condition (1) in the lemma can be interpreted as cohomology relations for a suitable bimodule. Let E/F be a Galois extension with the corresponding Galois group G. The multiplicative group E° becomes a $\mathbb{Z}G$-bimodule in which the elements of G act as the identity mapping on the left, and on the right they operate as the automorphisms that they are. Explicitly, if $z = \sum_{\sigma\in G}\sigma n_\sigma \in \mathbb{Z}G$, and $c \in E^\circ$, then

$$^z c = \prod_{\sigma\in G} c^{n_\sigma},\ c^z = \prod_{\sigma\in G}(c^\sigma)^{n_\sigma}. \tag{3}$$

A routine check shows that E° is a $\mathbb{Z}G$-bimodule with the scalar operations defined by (3).

The cohomology modules corresponding to the bimodule E° are defined as in Section 11.1, except that the bimodule addition is written multiplicatively. The role of R is played by $\mathbb{Z}$, so that the cohomology modules are just abelian groups. It is convenient to simplify the notation of Section 11.1 by writing $C^n(G,E^\circ)$, $Z^n(G,E^\circ)$, $B^n(G,E^\circ)$, and $H^n(G,E^\circ)$ for the groups of n-cochains, n-cocycles, n-coboundaries, and n-cohomology classes respectively. For each $\Phi \in Z^n(G,E^\circ)$, the cohomology class $\Phi B^n(G,E^\circ)$ will be denoted by $[\Phi]$.

Since $\mathbb{Z}G$ is a group algebra, the elements of $C^n(G,E^\circ)$ can be identified with the mappings of G^n to E°, as we noted in Section 11.1. Taking this viewpoint, the first three coboundary homomorphisms take the following forms: $(\delta^{(0)}c)(\sigma) = c(c^\sigma)^{-1} = c^{1-\sigma}$ for $c \in E^\circ = C^0(G,E^\circ)$ and $\sigma \in G$; $(\delta^{(1)}\Theta)(\sigma,\tau) = \Theta(\tau)\Theta(\sigma\tau)^{-1}\Theta(\sigma)^\tau$ for a mapping $\Theta\colon G \to E^\circ$ and $\sigma, \tau \in G$; $(\delta^{(2)}\Phi)(\rho,\sigma,\tau) = \Phi(\sigma,\tau)\Phi(\rho\sigma,\tau)^{-1}\Phi(\rho,\sigma\tau)(\Phi(\rho,\sigma)^\tau)^{-1}$ for a mapping Φ: $G^2 \to E^\circ$. Therefore, the cocycle condition on a mapping $\Phi\colon G^2 \to E^\circ$ is identical with the assumption that $\Phi \in Z^2(G,E^\circ)$. Moreover, the condition (1) of the lemma is equivalent to $\Phi\Psi^{-1} = \delta^{(1)}\Theta$, that is, $[\Phi] = [\Psi]$.

Theorem. *If E/F is a Galois extension with $G = \mathbf{G}(E/F)$, then the mapping $\theta_{E/F}\colon [\Phi] \mapsto [(E,G,\Phi)]$ is an isomorphism of $H^2(G,E^\circ)$ to $\mathbf{B}(E/F)$.*

By Corollary 14.1 and the lemma, the mapping $\theta_{E/F}$ is a well defined bijection from $H^2(G,E^\circ)$ to $\mathbf{B}(E/F)$. The proof that $\theta_{E/F}$ is a group homomorphism occupies the next section.

A special case of the fact that $\theta_{E/F}$ is a group homomorphism was outlined in Exercise 5 of Section 9.2: if char $F \neq 2$ and a, b, $c \in F^\circ$, then $\left(\frac{a,b}{F}\right) \otimes \left(\frac{a,c}{F}\right) \cong \left(\frac{a,bc}{F}\right) \otimes M_2(F)$. If $a \notin F^2$, then by Exercise 1 of Section 14.1, the algebras $\left(\frac{a,b}{F}\right)$, $\left(\frac{a,c}{F}\right)$, and $\left(\frac{a,bc}{F}\right)$ can be identified with the respective crossed products (E,G,Φ), (E,G,Ψ), and $(E,G,\Phi\Psi)$ in which $E = F(\sqrt{a})$, $G = \mathbf{G}(E/F) = \{1,\sigma\}$, and $\Phi(1,1) = \Phi(1,\sigma) = \Phi(\sigma,1) = \Psi(1,1) = \Psi(1,\sigma) = \Psi(\sigma,1) = 1$, $\Phi(\sigma,\sigma) = b$, $\Psi(\sigma,\sigma) = c$. The assertion $(E,G,\Phi) \otimes (E,G,\Psi) \sim (E,G,\Phi\Psi)$ is confirmed by the Exercise 5 in Section 9.2.

Exercises

1. Prove that the mapping ϕ that is defined in the lemma by (2) satisfies $\phi((u_\sigma c)(u_\tau d)) = \phi(u_\sigma c)\phi(u_\tau d)$ for all σ, $\tau \in G$ and c, $d \in E$.
2. Prove that E° is a $\mathbb{Z}G$-bimodule with the scalar operations defined by (3).
3. Prove *Hilbert's Theorem* 90: if E/F is a Galois extension with $\mathbf{G}(E/F) = G$, then $H^0(G,E^\circ) \cong F^\circ$ and $H^1(G,E^\circ) = \{1\}$. Hint. Let $\Theta\colon G \to E^\circ$ satisfy $(\delta^{(1)}\Theta)(\sigma,\tau) = 1$ for all σ, $\tau \in G$. Using an argument that is similar to the proof of Lemma 14.1, show that there exists $c \in E^\circ$ such that $d = \sum_{\sigma \in G} \Theta(\sigma)c^\sigma \neq 0$. Deduce that $d^\tau = \Theta(\tau)^{-1}d$ for all $\tau \in G$, so that $\Theta \in B^1(G,E^\circ)$.
4. Let E/F be a Galois extension with the Galois group G. Consider the additive group of E as a $\mathbb{Z}G$-bimodule in which the elements of G act as the identity on the left and by their standard action on the right. Consider F as a submodule of ${}_{\mathbb{Z}G}E$. Prove that $E \cong \mathrm{Hom}_{\mathbb{Z}}(\mathbb{Z}G,F)$ as $\mathbb{Z}G$-bimodules, where $\mathrm{Hom}_{\mathbb{Z}}(\mathbb{Z}G,F)$ has the $\mathbb{Z}G$-bimodule structure that was defined in the Coinduced Bimodules statement of Theorem 11.2. Conclude that $H^n(\mathbb{Z}G,E) = 0$ for all $n \geq 1$. Hint. By the normal basis theorem of Galois theory, there exists $d \in E$ such that $\{d^\sigma : \sigma \in G\}$ is a basis of E_F. Prove that the mapping $\phi\colon \mathrm{Hom}_{\mathbb{Z}}(\mathbb{Z}G,F) \to E$ defined by $\phi(\theta) = \sum_{\sigma \in G} d^\sigma\theta(\sigma^{-1})$ is a $\mathbb{Z}G$-bimodule isomorphism.

14.3. The Product Theorem

The purpose of this section is to complete the proof of Theorem 14.2. It only remains to show that the correspondence between $H^2(G,E^\circ)$ and $\mathbf{B}(E/F)$ that was given in Section 14.2 is a group homomorphism.

Proposition. *Let E/F be a Galois extension with $G = \mathbf{G}(E/F)$. If $\Phi, \Psi \in Z^2(G,E^\circ)$, then*

$$(E,G,\Phi) \otimes (E,G,\Psi) \sim (E,G,\Phi\Psi).$$

The proof of this proposition is based on two lemmas, the first of which is elementary and widely useful.

Lemma a. *Let $A \in \mathfrak{S}(F)$. If e is a non-zero idempotent element of A, then $eAe \in \mathfrak{S}(F)$ and $eAe \sim A$.*

PROOF. Let P be a minimal right ideal of A. Since A is finite dimensional and simple, it follows from Proposition 3.3b that $eA \cong \bigoplus k\,P$ as right A-modules for a suitable $k \in \mathbb{N}$. By Corollaries 6.4b and 3.4a, $eAe \cong \mathbf{E}_A(\bigoplus k\,P) \cong M_k(D)$, where $D = \mathbf{E}_A(P)$ is a division algebra. Thus, eAe is simple, and $eAe \sim D$. In particular, if $e = 1$, then $A \sim D$. Hence $eAe \sim A$, and $\mathbf{Z}(eAe) = \mathbf{Z}(A) = F$. □

We will prove the proposition by applying the lemma to $A = (E,G,\Phi) \otimes (E,G,\Psi)$, and an idempotent e that satisfies $eAe = (E,G,\Phi\Psi)$. This idempotent is found in the subalgebra $E \otimes E$ of A. The structure of $E \otimes E$ is known from Example 9.4: if $E = F(d)$, then $E \otimes E \cong E[\mathbf{x}]/\Phi E[\mathbf{x}]$, where Φ is the minimum polynomial of d over F. Since E/F is Galois, Φ is a product of distinct linear factors in $E[\mathbf{x}]$. By the Chinese Remainder Theorem, $E \otimes E$ is isomorphic to a product of $[E:F]$ copies of E. Consequently, $E \otimes E$ contains a set of $[E:F]$ pairwise orthogonal, primitive idempotents e such that $e(E \otimes E) \cong E$. We will show that one of these idempotents satisfies $eAe \cong (E,G,\Phi\Psi)$. Some information about the primitive idempotents in $E \otimes E$ is needed to obtain this isomorphism. A careful look at the decomposition of $E \otimes E$ as a product of copies of E leads to the second lemma of this section.

Lemma b. *Let E/F be a Galois extension with $G = \mathbf{G}(E/F)$. Denote $R = E \otimes E$. There exists $\{e_\sigma \in R: \sigma \in G\}$ such that:*

(i) $e_\sigma^2 = e_\sigma$, $e_\sigma e_\tau = 0$ *for* $\sigma \neq \tau$, *and* $\sum_{\sigma \in G} e_\sigma = 1_R$*;*
(ii) $e_\sigma(c \otimes 1) = e_\sigma(1 \otimes c^\sigma)$ *for all* $c \in E$*;*
(iii) $c \mapsto e_\sigma(c \otimes 1)$ *maps E isomorphically onto* $e_\sigma R$*;*

(iv) *if* $f \in R$ *is a non-zero idempotent element such that* $f(c \otimes 1) = f(1 \otimes c^\sigma)$ *for all* $c \in E$, *then* $f = e_\sigma$.

PROOF. Since every Galois extension is simple, there exists $d \in E$ such that $E = F(d)$. Let Θ be the minimum polynomial of d over F; thus, $\Theta(\mathbf{x}) = \prod_{\sigma \in G}(\mathbf{x} - d^\sigma) = \mathbf{x}^n + a_1\mathbf{x}^{n-1} + \cdots + a_n$, where $n = [E:F]$ and $a_i \in F$ for $1 \le i \le n$. For $\sigma \in G$, define

$$\Theta_\sigma(\mathbf{x}) = \prod_{\tau \neq \sigma} (d^\sigma - d^\tau)^{-1}(\mathbf{x} - d^\tau) = b_{\sigma 1}\mathbf{x}^{n-1} + b_{\sigma 2}\mathbf{x}^{n-2} + \cdots + b_{\sigma n},$$

with $b_{\sigma i} \in E$. By construction, $\Theta_\sigma(d^\tau) = 0$ if $\tau \neq \sigma$ and $\Theta_\sigma(d^\sigma) = 1$. Thus, $1 - \sum_{\sigma \in G}\Theta_\sigma$ is a polynomial of degree less than n that has n distinct roots. Therefore,

$$\sum_{\sigma \in G} \Theta_\sigma = 1. \tag{1}$$

For each $\sigma \in G$, define $1 \otimes \Theta_\sigma \in R[\mathbf{x}]$ by $(1 \otimes \Theta_\sigma)(\mathbf{x}) = (1 \otimes b_{\sigma 1})\mathbf{x}^{n-1} + (1 \otimes b_{\sigma 2})\mathbf{x}^{n-2} + \cdots + (1 \otimes b_{\sigma n})$. It follows from (1) that

$$\sum_{\sigma \in G} 1 \otimes \Theta_\sigma = 1_{R[\mathbf{x}]}. \tag{2}$$

Moreover, if $1 \otimes \Theta = (1 \otimes 1)\mathbf{x}^n + (1 \otimes a_1)\mathbf{x}^{n-1} + \cdots + (1 \otimes a_n) \in R[\mathbf{x}]$, then

$$(\mathbf{x} - 1 \otimes d^\sigma)(1 \otimes \Theta_\sigma) = (1 \otimes b_{\sigma 1})(1 \otimes \Theta), \tag{3}$$

and

$$\text{if } \sigma \neq \tau, \quad \text{then} \quad (1 \otimes \Theta_\sigma)(1 \otimes \Theta_\tau) = (1 \otimes \Theta)\Upsilon_{\sigma\tau} \tag{4}$$

for some $\Upsilon_{\sigma\tau} \in R[\mathbf{x}]$. Define $e_\sigma = (1 \otimes \Theta_\sigma)(d \otimes 1) = (d^{n-1} \otimes b_{\sigma 1}) + (d^{n-2} \otimes b_{\sigma 2}) + \cdots + (1 \otimes b_{\sigma n})$. Since $d^{n-1}, d^{n-2}, \ldots, 1$, are linearly independent, it follows from Proposition 9.1c that $e_\sigma \neq 0$. However,

$$\begin{aligned}(1 \otimes \Theta)(d \otimes 1) &= (d^n \otimes 1) + (d^{n-1} \otimes a_1) + \cdots + (1 \otimes a_n)\\ &= (d^n + a_1 d^{n-1} + \cdots + a_n) \otimes 1 = 0,\end{aligned}$$

so that $\sum_{\sigma \in G} e_\sigma = 1$, $(d \otimes 1 - 1 \otimes d^\sigma)e_\sigma = 0$, and $e_\sigma e_\tau = 0$ for $\sigma \neq \tau$, by (2), (3), and (4). Consequently, $e_\sigma = \sum_{\tau \in G} e_\sigma e_\tau = e_\sigma^2$, which proves (i). Moreover, for all $k \in \mathbb{N}$, $e_\sigma(d^k \otimes 1) = (e_\sigma(d \otimes 1))^k = (e_\sigma(1 \otimes d^\sigma))^k = e_\sigma(1 \otimes (d^k)^\sigma)$. The statement (ii) follows from this observation because the powers of d span E_F, and σ fixes the elements of F. Since E is a field, $c \mapsto e(c \otimes 1)$ is an isomorphism from E to $e(E \otimes F)$ for every non-zero idempotent $e \in R$. This observation proves (iii) because $e_\sigma(E \otimes F) = e_\sigma R$ by (ii). Suppose that f is a non-zero idempotent in R such that $f(c \otimes 1) = f(1 \otimes c^\sigma)$ for all $c \in E$. If $\tau \neq \sigma$, then $c^{\sigma\tau^{-1}} \neq c$ for some $c \in E$, and $e_\tau f(c \otimes 1) = fe_\tau(1 \otimes c^\sigma) = e_\tau f(c^{\sigma\tau^{-1}} \otimes 1)$. Thus, $e_\tau f = 0$, since $c - c^{\sigma\tau^{-1}}$ is in the kernel of $b \mapsto e_\tau f(b \otimes 1)$. By (i), $f = \sum_{\tau \in G} fe_\tau = fe_\sigma$. On the other hand, $f = e_\sigma f$ is a non-zero idempotent element in the field $e_\sigma R$, so that $f = 1_{e_\sigma R} = e_\sigma$. $\square$

We can now prove the proposition. Let $A \in \mathfrak{S}(F)$ be such that $A = B \otimes C$, where $B \cong (E,G,\Phi)$ and $C \cong (E,G,\Psi)$. By Lemma a, it suffices to find a non-zero idempotent element $e \in A$ such that $eAe \cong (E,G,\Phi\Psi)$. The assumptions $B \cong (E,G,\Phi)$ and $C \cong (E,G,\Psi)$ mean that there are F-algebra homomorphisms $\phi: E \to B$, $\psi: E \to C$, and sets $\{u_\sigma : \sigma \in G\} \subseteq B^\circ$, $\{v_\tau : \tau \in G\} \subseteq C^\circ$ such that $B = \bigoplus_\sigma u_\sigma \phi(E)$, $C = \bigoplus_\tau v_\tau \psi(E)$,

$$u_\sigma^{-1}\phi(c)u_\sigma = \phi(c^\sigma),\ v_\sigma^{-1}\psi(c)v_\sigma = \psi(c^\sigma) \quad \text{for } \sigma \in G,\ c \in E, \tag{5}$$

and

$$u_{\sigma\tau}^{-1}u_\sigma u_\tau = \phi(\Phi(\sigma,\tau)),\ v_{\sigma\tau}^{-1}v_\sigma v_\tau = \psi(\Psi(\sigma,\tau)) \quad \text{for } \sigma, \tau \in G. \tag{6}$$

Our aim is to find a non-zero idempotent element $e \in R = \phi(E) \otimes \psi(E)$, a non-zero F-algebra homomorphism $\chi: E \to eAe$, and a set $\{w_\sigma : \sigma \in G\} \subseteq (eAe)^\circ$ such that: $eAe = \bigoplus_\sigma w_\sigma \chi(E)$; $\chi(E)$ is a maximal subfield of eAe; $w_\sigma^{-1}\chi(c)w_\sigma = \chi(c^\sigma)$ for $\sigma \in G$, $c \in E$; and $w_{\sigma\tau}^{-1}w_\sigma w_\tau = \chi(\Phi(\sigma,\tau)\Psi(\sigma,\tau))$ for $\sigma, \tau \in G$. It will then follow that $(E,G,\Phi\Psi) \cong eAe \sim A$ by Lemma a. The facts that B centralizes C in A and R is a commutative subalgebra of A will be used often in the rest of the proof. Apply Lemma b to $R = \phi(E) \otimes \psi(E)$ to get non-zero idempotents $e_\sigma \in R$ such that

$$e_\sigma e_\tau = 0 \quad \text{if } \sigma \neq \tau, \tag{7}$$

$$c \mapsto e_\sigma \phi(c) \tag{8}$$

is an F-algebra isomorphism of E onto $e_\sigma R$,

$$e_\sigma \phi(c) = e_\sigma \psi(c^\sigma) \quad \text{for } \sigma \in G,\ c \in E, \tag{9}$$

and e_σ is the unique non-zero idempotent in R that satisfies this condition. Define $e = e_1$ and $\chi: E \to eR \subseteq eAe$ by $\chi(c) = e\phi(c)$. Note that $e\phi(c) = e\psi(c)$ by (9). The following equations follow from (5) and (9):

$$u_\sigma e = e_\sigma u_\sigma \quad \text{and} \quad ev_\tau = v_\tau e_\tau \quad \text{for } \sigma, \tau \in G. \tag{10}$$

Indeed,

$$\begin{aligned} u_\sigma R u_\sigma^{-1} &= u_\sigma \phi(E)\psi(E)u_\sigma^{-1} \\ &= u_\sigma \phi(E) u_\sigma^{-1}\psi(E) \\ &= \phi(E^{\sigma^{-1}})\psi(E) \\ &= \phi(E)\psi(E) = R, \end{aligned}$$

so that $u_\sigma e u_\sigma^{-1}$ is a non-zero idempotent in R; and $u_\sigma e u_\sigma^{-1}\phi(c) = u_\sigma e\phi(c^\sigma)u_\sigma^{-1} = u_\sigma e\psi(c^\sigma)u_\sigma^{-1} = u_\sigma e u_\sigma^{-1}\psi(c^\sigma)$ because B centralizes $\psi(E)$. Thus $u_\sigma e u_\sigma^{-1} = e_\sigma$. A similar calculation gives the second part of (10). It follows from (7) and (10) that if $\sigma \neq \tau$, then $ev_\tau u_\sigma e = v_\tau e_\tau e_\sigma u_\sigma = 0$. Thus, $eAe = e(\sum_{\sigma,\tau}$

$v_\tau u_\sigma R)e = \sum_\sigma w_\sigma \chi(E)$, where $w_\sigma = ev_\sigma u_\sigma e = v_\sigma e_\sigma e_\sigma u_\sigma = v_\sigma e_\sigma u_\sigma = ev_\sigma u_\sigma = v_\sigma u_\sigma e$. By Lemma a, $eAe \in \mathfrak{S}(F)$ and $\dim_F eAe \le |G|[\chi(E):F] = [\chi(E):F]^2 \le \dim_F eAe$. Thus, $eAe = \bigoplus_\sigma w_\sigma \chi(E)$ and $\chi(E)$ is a maximal subfield of eAe. Since $w_\sigma(eu_\sigma^{-1}v_\sigma^{-1}e) = e$, the elements w_σ are units in eAe. If $c \in E$, then $w_\sigma^{-1}\chi(c)w_\sigma = ev_\sigma^{-1}u_\sigma^{-1}\phi(c)u_\sigma v_\sigma e = ev_\sigma^{-1}\phi(c^\sigma)v_\sigma e = e\phi(c^\sigma) = \chi(c^\sigma)$. Finally,

$$\begin{aligned} w_\sigma w_\tau &= ev_\sigma u_\sigma ev_\tau u_\tau \\ &= ev_{\sigma\tau}\psi(\Psi(\sigma,\tau))u_{\sigma\tau}\phi(\Phi(\sigma,\tau)) \\ &= ev_{\sigma\tau}u_{\sigma\tau}\psi(\Psi(\sigma,\tau))\phi(\Phi(\sigma,\tau)) \\ &= w_{\sigma\tau}e\psi(\Psi(\sigma,\tau))\phi(\Phi(\sigma,\tau)) \\ &= w_{\sigma\tau}e\phi(\Psi(\sigma,\tau))\phi(\Phi(\sigma,\tau)) \\ &= w_{\sigma\tau}\chi(\Phi(\sigma,\tau)\Psi(\sigma,\tau)). \end{aligned}$$

Exercise

Specialize the construction that was used in Lemma b and the proposition to the quaternion algebras $B = \left(\frac{a,b}{F}\right)$, $C = \left(\frac{a,c}{F}\right)$, where $\operatorname{char} F \neq 2$, a, b, $c \in F^\circ$, $a \notin F^2$, and $E = F(\sqrt{a})$. Specifically, show that $e = (1/2)(a^{-1}(\mathbf{i} \otimes \mathbf{i}') + (1 \otimes 1))$ and $e_\sigma = (1/2)((-a^{-1})(\mathbf{i} \otimes \mathbf{i}') + (1 \otimes 1))$, where $\sigma(\sqrt{a}) = -\sqrt{a}$. Verify that eR has the F-basis e, $e(\mathbf{i} \otimes 1) = (1/2)((1 \otimes \mathbf{i}') + (\mathbf{i} \otimes 1))$, and that $w_1 = e$ and $w_\sigma = e(\mathbf{j} \otimes \mathbf{j}') = (1/2)((\mathbf{j} \otimes \mathbf{j}') + (\mathbf{k} \otimes \mathbf{k}'))$. Prove that eAe has the F-basis e, $e(\mathbf{j} \otimes \mathbf{j}')$, $e(\mathbf{i} \otimes 1)$, and $-e(\mathbf{k} \otimes \mathbf{j}')$. Compare these results with the hints for Exercise 5, Section 9.2.

14.4. Exponents

This section is a digression from the formalism of the rest of the chapter. We will use crossed products to prove that the Brauer groups have no elements of infinite order.

Lemma a. *If $A \in \mathfrak{S}(F)$ has index n, then $[A]^n = 1$ in $\mathbf{B}(F)$.*

Proof. By Proposition 13.4 and Theorem 13.5, it can be assumed that A contains a strictly maximal subfield E such that E/F is Galois. By Proposition 13.4, $n = \operatorname{Ind} A$ divides $\operatorname{Deg} A = [E:F]$, say $[E:F] = mn$. Thus, $A \cong M_m(D)$, where D is a division algebra of degree n. It follows that there is a D-A bimodule M with $\dim_D M = m$. Since E is a subfield of A, M is also a right E-space. The bimodule condition $au = ua$ for $a \in F$, $u \in M$ implies that $\dim M_F = \dim_F M$. Therefore, $(\dim M_E)mn = (\dim M_E)[E:F] = \dim M_F = \dim_F M = (\dim_D M)(\dim_F D) = mn^2$; that is, $\dim M_E = n$. Let $w_1, \ldots, w_n$ be a basis of M_E. By Lemma 14.1, A_E has a basis $\{u_\sigma : \sigma \in G\}$

(where $G = \mathbf{G}(E/F)$) such that $u_\sigma^{-1} c u_\sigma = c^\sigma$ for all $\sigma \in G$, $c \in E$, and $u_{\sigma\tau}^{-1} u_\sigma u_\tau = \Phi(\sigma,\tau)$ for all σ, $\tau \in G$ with $\Phi \in Z^2(G,E^\circ)$. That is, $A \cong (E,G,\Phi)$. Relative to the basis $\{w_1, \ldots, w_n\}$ of M, the right regular representation of A on M is described by matrices. In particular, if $\sigma \in G$ and $1 \leq j \leq n$, then

$$w_j u_\sigma = \sum_{i=1}^{n} w_i \mu_{ij}(\sigma). \tag{1}$$

Let $\mu(\sigma) = [\mu_{ij}(\sigma)] \in M_n(E)$. Since $u_\sigma \in A^\circ$, the matrix $\mu(\sigma)$ has an inverse. Moreover, $\sum_{i=1}^n w_i \mu_{ij}(\sigma\tau)\Phi(\sigma,\tau) = w_j u_{\sigma\tau}\Phi(\sigma,\tau) = w_j u_\sigma u_\tau = (\sum_{k=1}^n w_k \mu_{kj}(\sigma))u_\tau = \sum_{k=1}^n w_k u_\tau \mu_{kj}(\sigma)^\tau = \sum_{i=1}^n w_i(\sum_{k=1}^n \mu_{ik}(\tau)\mu_{kj}(\sigma)^\tau)$, so that

$$\mu(\sigma\tau)\Phi(\sigma,\tau) = \mu(\tau)\mu(\sigma)^\tau, \tag{2}$$

where $\mu(\sigma)^\tau = [\mu_{ij}(\sigma)^\tau]$. Define $\Theta: G \to E^\circ$ by $\Theta(\sigma) = \det \mu(\sigma)$. By taking the determinant each side of (2) we get $\Theta(\sigma\tau)\Phi(\sigma,\tau)^n = \Theta(\tau)\Theta(\sigma)^\tau$. Hence, $\Phi^n = \delta^{(1)}\Theta$ and $[\Phi]^n = [\Phi^n] = 1$ in $H^2(G,E)$. By Theorem 14.2, $[A]^n = \theta_{E/F}([\Phi]^n) = 1$. □

Proposition a. *If E/F is a finite field extension with $[E:F] = n$, then $[A]^n = 1$ for all $[A] \in \mathbf{B}(E/F)$. In particular $\mathbf{B}(F)$ is a torsion group.*

Proof. By Lemma 13.2, E splits any algebra $A \in \mathfrak{S}(F)$ such that $[A] \in \mathbf{B}(E/F)$. In this case, $\operatorname{Ind} A$ divides $[E:F] = n$ by Lemma 13.4. Thus, $[A]^n = 1$ by Lemma a. □

The fact that $H^2(G,E^\circ)$ is a torsion group can be proved directly, and in a more general form. An outline of this development is given in Exercise 3.

Definition. Let $A \in \mathfrak{S}(F)$. The exponent $\operatorname{Exp} A$ of A is the order of $[A]$ in $\mathbf{B}(F)$.

In other words, $\operatorname{Exp} A$ is the least $m \in \mathbb{N}$ such that the tensor product of m copies of A is a matrix algebra over F. The exponent is similar to the index in many ways, and for important classes of algebras these invariants are equal. The rest of this section is occupied with the proofs of fundamental properties of the exponent.

Lemma b. *Let $A \in \mathfrak{S}(F)$ have index $p^e m$, where p is a prime, $e \geq 1$, and p does not divide m. There is a finite, separable field extension K/F such that p does not divide $[K:F]$ and $\operatorname{Ind} A^K = p^e$.*

Proof. By Theorem 13.5 there is a Galois extension E/F such that E splits A. Let H be a Sylow p-subgroup of $\mathbf{G}(E/F)$, and define K to be the fixed field of H. This construction gives $[E:K] = |H| = p^r$ for some $r < \omega$ and $[K:F]$ is not divisible by p. Proposition 13.4 shows that $\operatorname{Ind} A^K = p^e$. In fact, $\operatorname{Ind} A^K$ divides both $\operatorname{Ind} A = p^e m$ and $[E:K] = p^r$ because E splits

A^K. Thus, $\operatorname{Ind} A^K = p^s$, where $s \leq e$. On the other hand, $p^e m = \operatorname{Ind} A$ divides $[K:F] \operatorname{Ind} A^K = [K:F]p^s$; hence, $e \leq s$. □

Proposition b. *Let $A, B \in \mathfrak{S}(F)$, and suppose that K/F is a finite field extension.*

(i) *If* $[A] = [B]$, *then* $\operatorname{Exp} A = \operatorname{Exp} B$.
(ii) $\operatorname{Exp} A$ *divides* $\operatorname{Ind} A$; *every prime divisor of* $\operatorname{Ind} A$ *divides* $\operatorname{Exp} A$.
(iii) $\operatorname{Exp} A^K$ *divides* $\operatorname{Exp} A$.
(iv) $\operatorname{Exp} A$ *divides* $[K:F] \operatorname{Exp} A^K$.
(v) *If* $\operatorname{Ind} A$ *is relatively prime to* $[K:F]$, *then* $\operatorname{Exp} A^K = \operatorname{Exp} A$.
(vi) $\operatorname{Exp}(A \otimes B)$ *divides the least common multiple of* $\operatorname{Exp} A$ *and* $\operatorname{Exp} B$.
(vii) $\operatorname{Exp} A^{\otimes m} = (\operatorname{Exp} A)/k$ *where k is the greatest common divisor of m and* $\operatorname{Exp} A$.
(viii) *If* $\operatorname{Ind} A$ *and* $\operatorname{Ind} B$ *are relatively prime, then* $\operatorname{Ind}(A \otimes B) = (\operatorname{Ind} A)(\operatorname{Ind} B)$ *and* $\operatorname{Exp}(A \otimes B) = (\operatorname{Exp} A)(\operatorname{Exp} B)$; *in this case, if A and B are division algebras, then so is $A \otimes B$.*

PROOF. The statement (i) is obvious from the definition of the exponent, and the statements (vi) and (vii) are translations of elementary group theoretic facts. The property (iii) is a consequence of the observation that $[A] \mapsto [A^K]$ is a homomorphism from $\mathbf{B}(F)$ to $\mathbf{B}(K)$. The first part of (ii) follows from Lemma a; the second half of (ii) is a consequence of the first part of (ii), Lemma b, and (iii). Indeed, if p divides $\operatorname{Ind} A$, then there is an extension K/F such that $\operatorname{Ind} A^K = p^e$ with $e \geq 1$. Thus, $\operatorname{Exp} A^K = p^f$ with $f \geq 1$ since $[A^K] \neq 1$. Therefore, p divides $\operatorname{Exp} A$. To prove (iv), let $m = \operatorname{Exp} A^K$. Then K splits $A^{\otimes m}$, so that $\operatorname{Ind} A^{\otimes m}$ divides $[K:F]$ by Proposition 13.4. By (vii) and (ii), $\operatorname{Exp} A/\operatorname{Exp} A^K = \operatorname{Exp} A^{\otimes m}$ divides $\operatorname{Ind} A^{\otimes m}$; hence, $\operatorname{Exp} A$ divides $[K:F] \operatorname{Exp} A^K$. Clearly, (v) is a consequence of (ii), (iii), and (iv). It remains to prove (viii). By the hypothesis and (ii), $\operatorname{Exp} A$ and $\operatorname{Exp} B$ are relatively prime. Thus, $m \in \mathbb{N}$ exists so that $m \equiv 1 \pmod{\operatorname{Exp} A}$ and $m \equiv 0 \pmod{\operatorname{Exp} B}$. Thus, $(A \otimes B)^{\otimes m} \cong A^{\otimes m} \otimes B^{\otimes m} \sim A$. By Proposition 13.4 and (vii), $\operatorname{Ind} A$ divides $\operatorname{Ind}(A \otimes B)$ and $\operatorname{Exp} A$ divides $\operatorname{Exp}(A \otimes B)$. Similarly, $\operatorname{Ind} B$ divides $\operatorname{Ind}(A \otimes B)$ and $\operatorname{Exp} B$ divides $\operatorname{Exp}(A \otimes B)$. Thus, $(\operatorname{Ind} A)(\operatorname{Ind} B) = \operatorname{Ind}(A \otimes B)$ and $(\operatorname{Exp} A)(\operatorname{Exp} B) = \operatorname{Exp}(A \otimes B)$ by Proposition 13.4 and (vi). If A and B are division algebras, then

$$\operatorname{Ind}(A \otimes B) = (\operatorname{Ind} A)(\operatorname{Ind} B) = (\operatorname{Deg} A)(\operatorname{Deg} B) = \operatorname{Deg}(A \otimes B),$$

so that $A \otimes B$ is a division algebra. □

It is a consequence of this result that every central simple division algebra admits a primary decomposition.

Primary Decomposition Theorem. *Let $D \in \mathfrak{S}(F)$ be a division algebra with* $\operatorname{Deg} D = p_1^{e_1} \cdots p_r^{e_r}$, *where $p_1, \ldots, p_r$ are distinct primes and $e_i \in \mathbb{N}$ for $1 \leq i \leq r$. There is a unique (to isomorphism) decomposition $D = D_1 \otimes \cdots \otimes D_r$ with $D_i \in \mathfrak{S}(F)$ a division algebra such that* $\operatorname{Deg} D_i = p_i^{e_i}$ *for $1 \leq i \leq r$.*

PROOF. By Proposition b, $\operatorname{Exp} D = p_1^{f_1} \cdots p_r^{f_r} = n$, with $1 \leq f_i \leq e_i$. If $n_i = n/p_i^{f_i}$, then the greatest common divisor of $n_1, \ldots, n_r$ is 1, that is, $1 = \sum_{i=1}^r m_i n_i$ for suitable $m_i \in \mathbb{Z}$. For $1 \leq i \leq r$, let $D_i \in \mathfrak{S}(F)$ be the unique division algebra such that $[D_i] = [D]^{m_i n_i}$. Then $[D_1 \otimes \cdots \otimes D_r] = \prod_{i=1}^r [D]^{m_i n_i} = [D]$, and $\operatorname{Exp} D_i = p_i^{f_i}$ by Proposition b. Hence, $D_1 \otimes \cdots \otimes D_r$ is a division algebra that is necessarily isomorphic to D. Moreover, $\operatorname{Deg} D_i = \operatorname{Ind} D_i = p_i^{l_i}$, where $1 \leq f_i \leq l_i$. In fact, $p_1^{e_1} \cdots p_r^{e_r} = \operatorname{Deg} D = \operatorname{Deg}(D_1 \otimes \cdots \otimes D_r) = p_1^{l_1} \cdots p_r^{l_r}$ implies $e_i = l_i$ for $1 \leq i \leq r$. The uniqueness of the D_i is clear because our construction is reversible. □

The algebras D_i in the theorem are called the primary components of D. Many questions about division algebras can be reduced to the case of algebras whose degree is a prime power, that is, primary division algebras.

EXERCISES

1. Fill in the details of the proof that the primary components of a division algebra $D \in \mathfrak{S}(F)$ are unique.

2. Let F be an n-closed field. Prove that the order of every element of $\mathbf{B}(F)$ is relatively prime to n.

3. Let E/F be a Galois extension of degree n. Prove that $[\Phi]^n = 1$ for all $\Phi \in Z^k(G, E^\circ)$ if $k \geq 1$. Hint. Write the condition $\delta^{(k)}\Phi = 1$ in the form

$$\Phi(\sigma_1, \ldots, \sigma_k) = \Phi(\sigma_0\sigma_1, \sigma_2, \ldots, \sigma_k)\Phi(\sigma_0, \sigma_1\sigma_2, \ldots, \sigma_k)^{-1} \cdots \Phi(\sigma_0, \sigma_1, \ldots, \sigma_{k-1})^{\pm\sigma_k},$$

and take the product over all $\sigma_0 \in \mathbf{G}(E/F)$.

14.5. Inflation

By Corollary 13.5, the Brauer group $\mathbf{B}(F)$ is the union over all Galois extensions E/F of the relative Brauer groups $\mathbf{B}(E/F)$. In turn, the relative Brauer groups can be identified with cohomology groups. In order to relate the full Brauer group to cohomological data, an interpretation is needed for the inclusion mappings $\mathbf{B}(K/F) \to \mathbf{B}(E/F)$ that arise when $F \subseteq K \subseteq E$ with K/F and E/F Galois extensions. Those inclusions correspond to the inflation homomorphisms. We discuss inflation in this section.

It will be useful to standardize the notation for this section. Let E/F and K/F be Galois extensions with $K \subseteq E$ and $[E:K] = r$. Denote $G = \mathbf{G}(E/F)$ and $H = \mathbf{G}(K/F)$.

The restriction mapping $\sigma \mapsto \sigma|K$ is a surjective homomorphism of G to H that induces an adjoint homomorphism $C^n(H, K^\circ) \to C^n(G, E^\circ)$ by $\Phi \mapsto \Phi^*$, where $\Phi^*(\sigma_1, \ldots, \sigma_n) = \Phi(\sigma_1|K, \ldots, \sigma_n|K)$. A simple calculation shows that this map commutes with the coboundary: $(\delta^{(n)}\Phi)^* = \delta^{(n)}(\Phi^*)$.

Thus, the adjoint map carries $Z^n(H,K^\circ)$ to $Z^n(G,E^\circ)$ and $B^n(H,K^\circ)$ to $B^n(G,E^\circ)$. Consequently, it induces a group homomorphism of $H^n(H,K^\circ)$ to $H^n(G,E^\circ)$ that is called the *inflation mapping*, and is denoted by inf (or if necessary, $\inf^{(n)}_{K/F\to E/F}$). Explicitly, $\inf[\Phi] = [\Phi^*]$ for $\Phi \in Z^n(H,K^\circ)$.

Proposition. *If* $\kappa\colon \mathbf{B}(K/F) \to \mathbf{B}(E/F)$ *is the inclusion homomorphism, then* $\kappa \circ \theta_{K/F} = \theta_{E/F} \circ \inf$. *That is,*

$$\begin{array}{ccc} H^2(\mathbf{G}(K/F),K^\circ) & \overset{\inf}{\to} & H^2(\mathbf{G}(E/F),E^\circ) \\ \theta_{K/F}\downarrow & & \downarrow\theta_{E/F} \\ \mathbf{B}(K/F) & \underset{\kappa}{\to} & \mathbf{B}(E/F) \end{array}$$

commutes.

If $\Phi \in Z^2(H,K^\circ)$, then $\theta_{E/F} \circ \inf[\Phi] = [(E,G,\Phi^*)]$ and $\kappa\theta_{K/F}[\Phi] = [(K,H,\Phi)]$. The proposition is equivalent to $(E,G,\Phi^*) \sim (K,H,\Phi)$. Denote $B = (K,H,\Phi) = \bigoplus_{\tau\in H} v_\tau K$, where $\{v_\tau : \tau \in H\}$ satisfies the conditions of Lemma 14.1 and K is a strictly maximal subfield of B. Let $A = M_r(B) \cong M_r(F) \otimes B$. Then $A \in \mathfrak{S}(F)$, $A \sim B$, and $\operatorname{Deg} A = \operatorname{Deg}(E,G,\Phi^*)$ since $r = [E:K]$. The proposition will be proved by embedding E in A and constructing a set $\{u_\sigma : \sigma \in G\} \subseteq A^\circ$ satisfying 14.1(1) and $u_{\sigma\tau}^{-1}u_\sigma u_\tau = \Phi^*(\sigma,\tau)$ for all σ, $\tau \in G$.

Lemma. *Let G act on* $M_r(K)$ *by* $[c_{ij}]^\sigma = [c_{ij}^\sigma]$. *There is an injective K-algebra homomorphism* $\lambda\colon E \to M_r(K)$ *and a mapping* $\mu\colon G \to M_r(K)^\circ$ *such that:*

(i) $\mu(\sigma\tau) = \mu(\tau)\mu(\sigma)^\tau$ *for all* σ, $\tau \in G$;
(ii) $\mu(\sigma)\lambda(d)^\sigma = \lambda(d^\sigma)\mu(\sigma)$ *for all* $\sigma \in G$, $d \in E$.

Proof. Fix a basis $\{b_1, \ldots, b_r\}$ of E_K, and define $\lambda(d) = [l_{ij}(d)]$, $\mu(\sigma) = [m_{ij}(\sigma)]$ by $db_j = \sum_{i=1}^r b_i l_{ij}(d)$, $b_j^\sigma = \sum_{i=1}^r b_i m_{ij}(\sigma)$. Routine calculations show that λ is a K-algebra homomorphism and conditions (i) and (ii) are satisfied. The computations can be simplified by a couple of general observations. First, note that $N = \bigoplus r\, E$ is an E-FG bimodule and a faithful, cyclic right $M_r(K)$-module (under matrix multiplication) with $\beta = (b_1, \ldots, b_r)$ as a generator. The mappings λ and μ are characterized by $d\beta = \beta\lambda(d)$ and $\beta^\sigma = \beta\mu(\sigma)$. Hence, $\beta\mu(\sigma\tau) = \beta^{\sigma\tau} = (\beta\mu(\sigma))^\tau = \beta^\tau\mu(\sigma)^\tau = \beta\mu(\tau)\mu(\sigma)^\tau$ gives (i); and $\beta\mu(\sigma)\lambda(d)^\sigma = \beta^\sigma\lambda(d)^\sigma = (\beta\lambda(d))^\sigma = (d\beta)^\sigma = d^\sigma\beta^\sigma = d^\sigma\beta\mu(\sigma) = \beta\lambda(d^\sigma)\mu(\phi)$ implies (ii). Since $\mu(1) = \iota_r$, it is a consequence of (i) that $\mu(G) \subseteq M_r(K)^\circ$. □

Proof of the Proposition. Let λ and μ be the mappings that were defined in the lemma. Since $K \subseteq B$, we can view λ as an embedding of E in $M_r(B) = A$. Since λ is an F-algebra homomorphism, the Galois group $\mathbf{G}(E/F)$ acts on $\lambda(E)$ by $\sigma\colon \lambda(d) \mapsto \lambda(d^\sigma)$. To avoid confusion with the notation that was introduced in the lemma, we will not write $\lambda(d)^\sigma$ for $\lambda(d^\sigma)$; these expressions have different meanings in the proof. For $\sigma \in G$, define $u_\sigma \in M_r(B)^\circ = A^\circ$ to be the scalar product of $v_{\sigma|K}$ with the matrix $\mu(\sigma)^{-1}$. That is, $u_\sigma = v_{\sigma|K}\mu(\sigma)^{-1}$. It follows from the lemma that $u_\sigma^{-1}\lambda(d)u_\sigma = \mu(\sigma)v_{\sigma|K}^{-1}\lambda(d)v_{\sigma|K}\mu(\sigma)^{-1} = \mu(\sigma)\lambda(d)^\sigma\mu(\sigma)^{-1} = \lambda(d^\sigma)\mu(\sigma)\mu(\sigma)^{-1} = \lambda(d^\sigma)$, and

$$\begin{aligned} u_\sigma u_\tau &= v_{\sigma|K}\mu(\sigma)^{-1}v_{\tau|K}\mu(\tau)^{-1} \\ &= v_{\sigma|K}v_{\tau|K}(\mu(\sigma)^\tau)^{-1}\mu(\tau)^{-1} \\ &= v_{(\sigma|K)(\tau|K)}\Phi(\sigma|K,\tau|K)\mu(\sigma\tau)^{-1} \\ &= v_{\sigma\tau|K}\mu(\sigma\tau)^{-1}\Phi^*(\sigma,\tau) \\ &= u_{\sigma\tau}\Phi^*(\sigma,\tau). \end{aligned}$$

Thus, $A \cong (E,G,\Phi^*)$. □

EXERCISES

1. Prove that the inflation mapping is functorial: if $F \subseteq K \subseteq E \subseteq L$ are fields such that K/F, E/F, and L/F are Galois, then $\inf^{(n)}_{K/F\to L/F} = \inf^{(n)}_{E/F\to L/F} \circ \inf^{(n)}_{K/F\to E/F}$.
2. With the notation that was used in the proof of the proposition, prove that $\mathbf{C}_A(K) = K \otimes C$, where $C \cong M_r(F)$, and $\mathbf{C}_A(C) \cong B$.

14.6. Direct Limits

This section gives the definition of direct limits and the basic existence and uniqueness theorems that are associated with this concept. Our main result establishes an isomorphism between the Brauer group of a field F and the second Galois cohomology group of F that is obtained as a direct limit of the groups $H^2(\mathbf{G}(E/F),E^\circ)$.

Definition a. Let I be a set that is partially ordered by $\leq$. A *direct* (*inverse*) *system* of groups over I is a pair (G,ϕ) of functions such that for each $i \in I$, G_i is a group, and for each pair (i,j) with $i \leq j$, ϕ_{ij} is a homomorphism from G_i to G_j (respectively, G_j to G_i); it is assumed that $\phi_{ii} = id_{G_i}$ for all $i \in I$, and $i \leq j \leq k$ implies $\phi_{ik} = \phi_{jk}\phi_{ij}$ (respectively, $\phi_{ik} = \phi_{ij}\phi_{jk}$). If (G,ϕ) and (G',ϕ') are direct (inverse) systems over I, then a *morphism* from (G,ϕ) to (G',ϕ') is a function θ on I such that each θ_i is a homomorphim from G_i to G_i', and if $i \leq j$, then $\theta_j\phi_{ij} = \phi_{ij}'\theta_i$ (respectively, $\phi_{ij}'\theta_j = \theta_i\phi_{ij}$). That is, the diagram

$$\begin{array}{ccc} G_i & \xrightarrow{\phi_{ij}} & G_j \\ \theta_i \downarrow & & \downarrow \theta_j \\ G_i' & \xrightarrow[\phi_{ij}']{} & G_j' \end{array}$$

commutes for all $i \leq j$.

The morphisms of direct and inverse systems of groups compose componentwise: if $\theta: (G,\phi) \to (G',\phi')$ and $\theta': (G',\phi') \to (G'',\phi'')$ are morphisms, then $\theta'\theta: (G,\phi) \to (G'',\phi'')$ is a morphism, where $(\theta'\theta)_i = \theta'_i\theta_i$ for all $i \in I$. The identity morphism of (G,ϕ) is $id_{(G,\phi)}$ whose value at i is id_{G_i}. Thus, the class of all inverse (direct) systems over I is a category. For future reference, note that a morphism θ is an isomorphism, that is, θ has an inverse in the category, if and only if all θ_i are isomorphisms.

The definitions of direct and inverse systems are categorically dual. As a result, the basic concepts in the theory of inverse systems can be obtained from their counterparts in the theory of direct systems by "reversing the arrows." This observation justifies our concentration on direct systems. Some facts about inverse systems involve new methods; these are described in Exercise 3.

In most of the direct systems that occur in applications the index set I is directed, that is, for each $i, j \in I$, there exists $k \in I$ such that $i \leq k$ and $j \leq k$. All of the examples that we will consider have this property.

Example a. For a given field F, let I be the set of all Galois extensions E/F such that E is a subfield of the algebraic closure of F. Partially order I by defining $E/F \leq K/F$ if $E \subseteq K$. The set I is directed because any two Galois extensions E/F and L/F have a compositum EL/F that is also Galois. There are two direct systems and one inverse system of groups over I that interest us.

(i) The system of relative Brauer groups $(\mathbf{B},\kappa)$ is the direct system of abelian groups $\mathbf{B}(E/F)$ and the inclusion mappings $\kappa: \mathbf{B}(E/F) \to \mathbf{B}(K/F)$ defined when $E \subseteq K$.

(ii) The system of n'th cohomology groups (H^n,inf) is the direct system of abelian groups $H^n(\mathbf{G}(E/F),E^\circ)$ and the inflation mappings

$$\text{inf}^{(n)}_{E/F\to K/F}: H^n(\mathbf{G}(E/F),E^\circ) \to H^n(\mathbf{G}(K/F),K^\circ),$$

defined when $E \subseteq K$.

(iii) The system of Galois groups $(\mathbf{G},\rho)$ is the inverse system of groups $\mathbf{G}(E/F)$ and the restriction mappings $\rho: \sigma \to \sigma|E$ from $\mathbf{G}(K/F)$ to $\mathbf{G}(E/F)$ defined when $E \subseteq K$.

Proposition 14.5 can be stated succinctly in the terminology of Definition a: θ is an isomorphism from (H^2,inf) to $(\mathbf{B},\kappa)$.

Definition b. Let (G,ϕ) be a direct system of groups over I.

(i) A *prelimit* of (G,ϕ) is a pair (H,χ) in which H is a group, χ is a function on I such that $\chi_i: G_i \to H$ is a homomorphism for all $i \in I$, and

$$i \leq j \quad \text{implies} \quad \chi_i = \chi_j\phi_{ij}; \tag{1}$$

H is generated by

$$\bigcup_{i \in I} \operatorname{Im} \chi_i, \tag{2}$$

that is, no proper subgroup of H contains this union.

(ii) If (H,χ) and (H',χ') are prelimits of (G,ϕ), then a morphism from (H,χ) to (H',χ') is a homomorphism ψ of H to H' such that $\chi_i' = \psi\chi_i$ for all $i \in I$. (It is clear from (2) that ψ is necessarily surjective.)

It is easily checked that the composition of morphisms of prelimits is a morphism, and that id_H is a morphism of (H,χ). Thus, the class of prelimits of (G,ϕ) is another category.

If the indexing set I is directed, then the condition (2) in the definition of a prelimit is equivalent to a simpler property:

$$H = \bigcup_{i \in I} \operatorname{Im} \chi_i. \tag{3}$$

In fact, if I is directed, then $\bigcup_{i \in I} \operatorname{Im} \chi_i$ is a subgroup of H.

Lemma a. *If (H,χ) and (H',χ') are prelimits of (G,ϕ), then there is at most one morphism $\psi: (H,\chi) \to (H',\chi')$.*

PROOF. If ψ_1 and ψ_2 are morphisms from (H,χ) to (H',χ'), then $\psi_1\chi_i = \chi_i' = \psi_2\chi_i$ for all $i \in I$. Thus, $\psi_1|(\bigcup_{i \in I} \operatorname{Im} \chi_i) = \psi_2|(\bigcup_{i \in I} \operatorname{Im} \chi_i)$. Since $\{x \in H: \psi_1(x) = \psi_2(x)\}$ is a subgroup of H, it follows from (2) that $\psi_1 = \psi_2$. □

Definition c. Let (G,ϕ) be a direct system over I. A *limit* of (G,ϕ) is a prelimit (H,χ) of (G,ϕ) such that for all prelimits (H',χ') of (G,ϕ), there is a morphism of (H,χ) to (H',χ').

Proposition a. *Any two limits (H,χ) and (H',χ') of (G,ϕ) are isomorphic.*

PROOF. By the definition of a limit, there exist homomorphisms $\psi: (H,\chi) \to (H',\chi')$ and $\psi': (H',\chi') \to (H,\chi)$. Since $\psi'\psi$ and $\psi\psi'$ are homomorphisms, it follows from Lemma a that $\psi'\psi = id_H$ and $\psi\psi' = id_{H'}$. □

The uniqueness of limits allows us to speak of *the* limit of (G,ϕ). The notation $(H,\chi) = \varinjlim(G,\phi)$ (or just $H = \varinjlim G$) abbreviates the assertion that (H,χ) is the limit of the direct system (G,ϕ).

Free products can be used to show that every direct system of groups has a limit. For direct systems of abelian groups, free products can be replaced by direct sums.

Proposition b. *If (G,ϕ) is a direct system of abelian groups over a directed set, then the limit of (G,ϕ) exists.*

PROOF. Let I be the directed set that indexes (G,ϕ). For each $i \in I$, denote the natural embedding of G_i in $\bigoplus_{j \in I} G_j$ by κ_i. Define N to be the subgroup of $\bigoplus_{j\in I} G_j$ that is generated by $\bigcup_{i\le j} \operatorname{Im}(\kappa_j\phi_{ij} - \kappa_i)$. Let $H = (\bigoplus_{j\in I} G_j)/N$, and $\chi_i = \pi\kappa_i\colon G_i \to H$, where $\pi\colon \bigoplus_{j\in I} G_j \to H$ is the natural projection. We will show that $(H,\chi) = \varinjlim(G,\phi)$. If $i \le j$, then $\chi_j\phi_{ij} - \chi_i = \pi(\kappa_j\phi_{ij} - \kappa_i) = 0$ by the definition of N. Since $\bigoplus_{j\in I} G_j$ is generated by $\bigcup_{j\in I} \kappa_j G_j$, it follows that $H = \pi(\bigoplus_{j\in I} G_j)$ is generated by $\pi(\bigcup_{j\in I} \kappa_j G_j) = \bigcup_{j\in I} \operatorname{Im}\chi_j$. Thus, (H,χ) is a prelimit of (G,ϕ). Let (H',χ') be another prelimit of (G,ϕ). Since I is directed and all G_j are abelian, $H' = \bigcup_{j\in I} \chi_j'(G_j)$ is abelian. Thus, there is a homomorphism $\sigma\colon \bigoplus_{j\in I} G_j \to H'$ given by $\sigma(\sum_{j\in I} x_j) = \sum_{j\in I} \chi_j' x_j$ that satisfies $\sigma\kappa_i = \chi_i'$ for all $i \in I$. Moreover, $\sigma(\kappa_j\phi_{ij} - \kappa_i) = \chi_j'\phi_{ij} - \chi_i'$, so that $N \subseteq \operatorname{Ker}\sigma$. Thus, σ factors through π, say $\sigma = \psi\pi$, where $\psi\colon H \to H'$ is a homomorphism. Because $\chi_i' = \sigma\kappa_i = \psi\pi\kappa_i = \psi\chi_i$, the mapping ψ is a morphism from (H,χ) to (H',χ'). Since (H',χ') is an arbitrary prelimit of (G,ϕ), it follows that $(H,\chi) = \varinjlim(G,\phi)$. □

The next lemma characterizes limits in a special case.

Lemma b. *Let (G,ϕ) be a direct system of groups over a direct set I. If (H,χ) is a prelimit of (G,ϕ) such that all χ_i are injective, then $(H,\chi) = \varinjlim(G,\phi)$.*

PROOF. Let (H',χ') be a prelimit of (G,ϕ). Define $\psi = \bigcup_{i\in I} \chi_i'\chi_i^{-1}$. Since each χ_i is injective and I is directed, the set ψ is a homomorphism to H' with domain $\bigcup_{i\in I} \operatorname{Im}\chi_i = H$ (by (3)). By definition, $\psi\chi_i = \chi_i'$, so that ψ is a morphism of (H,χ) to (H',χ'). Thus, $(H,\chi) = \varinjlim(G,\phi)$. □

EXAMPLE B. For any field F, $(\mathbf{B}(F),\chi) = \varinjlim(\mathbf{B},\kappa)$, where $\chi_{E/F}\colon \mathbf{B}(E/F) \to \mathbf{B}(F)$ is the inclusion homomorphism.

Lemma c. *Let $\theta\colon (G,\phi) \to (G',\phi')$ be a morphism of direct systems of groups over I. If $(H,\chi) = \varinjlim(G,\phi)$ and $(H',\chi') = \varinjlim(G',\phi')$, then there is a unique morphism $\psi\colon (H,\chi) \to (H',\chi')$ such that $\psi\chi_i = \chi_i'\theta_i$ for all $i \in I$.*

PROOF. The composition $\chi_i'' = \chi_i'\theta_i$ defines the prelimit (H',χ'') of (G,ϕ). Since (H,χ) is the limit of (G,ϕ), there is a unique morphism $\psi\colon (H,\chi) \to (H',\chi'')$; that is, $\psi\chi_i = \chi_i'' = \chi_i'\theta_i$. □

Write $\psi = \varinjlim \theta$, where θ and ψ have the same meaning as in Lemma c. If $(G,\phi) \xrightarrow{\theta} (G',\phi') \xrightarrow{\theta'} (G'',\phi'')$, then $\varinjlim \theta'\theta = (\varinjlim \theta')(\varinjlim \theta)$ by the uniqueness property in Lemma c. Also, $\varinjlim id_{(G,\phi)} = id_{\varinjlim G}$. Thus, the limit is a functor on the category of direct systems.

The principal result of this section follows directly from Proposition b, Lemma c, and Examples a and b.

Theorem. *For any field F,*

$$\varinjlim \theta\colon \varinjlim H^2(\mathbf{G}(E/F),E^\circ) \to \mathbf{B}(F)$$

is an isomorphism.

The group $\varinjlim H^2(\mathbf{G}(E/F),E^\circ)$ is the second Galois cohomology group of the field F. If F_s is the separable algebraic closure of F, then $\varinjlim H^2(\mathbf{G}(E/F), E^\circ)$ can be viewed as the second cohomology group of the $\mathbb{Z}\mathbf{G}(F_s/F)$-bimodule F_s°. However, this definition does not agree with the definition of the cohomology groups that was given in Section 11.1. The cochain groups $C^n(\mathbf{G}(F_s/F),F_s^\circ)$ must be modified to take account of the topology of the infinite Galois group $\mathbf{G}(F_s/F)$.

Exercises

1. Let I be the unordered set $\{i,j\}$. Define G_i and G_j to be the subgroups of the group H of all permutations of $\{1,2,3\}$ that are respectively generated by the cycles (1,2) and (1,2,3). If $\phi_{ii} = id_{G_i}$, $\phi_{jj} = id_{G_j}$, then (G,ϕ) is a direct system. Show that (H,χ) is a prelimit of (G,ϕ), where $\chi_i\colon G_i \to H$ and $\chi_j\colon G_j \to H$ are inclusion mappings, but (H,χ) is not a limit of (G,ϕ). It can be shown that $\varinjlim G$ is the free product of G_i and G_j, and this group is isomorphic to $PSL_2(\mathbb{Z})$.

2. (a) Let (G,ϕ) be a direct system of abelian groups over a directed partially ordered set I, and suppose that (H,χ) is a prelimit of (G,ϕ). Prove that $(H,\chi) = \varinjlim(G,\phi)$ if and only if $\operatorname{Ker}\chi_i = \bigcup_{j\geq i}\operatorname{Ker}\phi_{ij}$ for all $i \in I$. Hint. For the "only if" part of the proof, it can be assumed that (H,χ) is defined by the construction in the proof of Proposition b. To obtain the converse, let (H',χ') be a limit of (G,ϕ), so that there is a morphism $\psi\colon (H',\chi') \to (H,\chi)$. Use the hypothesis to prove that ψ is an isomorphism.

   (b) Let I be a directed partially ordered set. Suppose that $(G,\phi) \xrightarrow{\theta} (G',\phi') \xrightarrow{\theta'} (G'',\phi'')$ is an exact sequence of direct systems of abelian groups over I, that is, for all $i \in I$, $\operatorname{Ker}\theta_i' = \operatorname{Im}\theta_i$. Use the result of (a) to prove that $\operatorname{Ker}(\varinjlim\theta') = \operatorname{Im}(\varinjlim\theta)$: the sequence $\varinjlim(G,\phi) \to \varinjlim(G',\phi') \to \varinjlim(G'',\phi'')$ is exact.

3. In this exercise, (G,ϕ) is an inverse system of groups over the partially ordered set I. A *prelimit* of (G,ϕ) is a pair (H,χ) in which H is a group and χ is a function on I with each χ_i a homomorphism from H to G_i such that (1′) $\chi_i = \phi_{ij}\chi_j$ if $i \leq j$, and (2′) $\bigcap_{j\in I}\operatorname{Ker}\chi_j = \{1\}$. If (H,χ) and (H',χ') are prelimits of (G,ϕ), then a morphism from (H',χ') to (H,χ) is a homomorphism $\psi\colon H' \to H$ such that $\chi_i' = \chi_i\psi$ for all $i \in I$. Prove the following statements.

   (a) The class of prelimits of (G,ϕ) is a category.

   (b) If (H',χ') and (H,χ) are prelimits of (G,ϕ), then there is at most one morphism from (H',χ') to (H,χ).

   A prelimit (H,χ) of (G,ϕ) is a *limit* of this inverse system if, for every prelimit (H',χ') of (G,ϕ), there is a morphism of (H',χ') to (H,χ). Denote (H,χ) by $\varprojlim(G,\phi)$.

   (c) Any two limits of (G,ϕ) are isomorphic.

   (d) For $i \leq j$, define $H_{ij} = \{x \in \prod_{k\in I}G_k\colon \phi_{ij}\pi_j x = \pi_i x\}$, where $\pi_j\colon \prod_{k\in I}G_k \to G_j$ is the component projection. Let $H = \bigcap_{i\leq j}H_{ij}$. Define $\chi_i = \pi_i|H\colon H \to G_i$. Then (H,χ) is the limit of (G,ϕ).

   (e) If $\theta\colon (G',\phi') \to (G,\phi)$ is a morphism of inverse systems of groups over I, and (H,χ) and (H',χ') are the limits of (G,ϕ) and (G',ϕ') respectively, then there is a unique morphism $\psi\colon (H',\chi') \to (H,\chi)$ such that $\chi_i\psi = \theta_i\chi_i'$ for all $i \in J$.

   (f) Let $\{N_i\colon i \in I\}$ be a set of normal subgroups of the group G, where I is a directed partially ordered set. Assume that: (i) $i \leq j$ implies $N_j \subseteq N_i$; (ii) $\bigcap_{j\in I}N_j =$

$\{1\}$; and (iii) $G_i = G/N_i$ is finite for all $i \in I$. For $i \leq j$, let $\phi_{ij}: G_j \to G_i$ be the projection $\phi_{ij}(xN_j) = xN_i$. Prove that $(\{G_i\},\{\phi_{ij}\})$ is an inverse system of groups, and $(G,\{\chi_i\})$ is a prelimit of this system, where $\chi_i(x) = xN_i$. Prove that $(G,\{\chi_i\})$ is a limit of $(\{G_i\},\{\phi_{ij}\})$ if and only if G is compact in the topology that is defined by letting the set of cosets $\{xN_i: x \in G, i \in I\}$ be a neighborhood basis of G. (In this case, G is called a *profinite* group.) Hint. Apply Tychonov's theorem to the construction (d).

(g) Let F_s be the separable algebraic closure of F. Define $G = \mathbf{G}(F_s/F)$ to be the group of F-algebra automorphisms of F_s. For each Galois extension E/F, let $\chi_{E/F}$ be the restriction homomorphism $\sigma \mapsto \sigma|E$ from G to $\mathbf{G}(E/F)$. Use (f) to prove that (G,χ) is the limit of the inverse system $(\mathbf{G},\rho)$ that was defined in Example a.

4. In this exercise the notation and hypotheses are the same as they were in part (f) of Exercise 3, including the assumption that G is profinite. Let M be a right $\mathbb{Z}G$-module. Consider M as a $\mathbb{Z}G$-bimodule with the elements of G acting trivially (that is, as the identity) on the left side of M. Denote $M_j = \{u \in M: ux = u \text{ for all } x \in N_j\}$. Call M *discrete* if $\bigcup_{j \in I} M_j = M$. It is assumed that the $\mathbb{Z}G$-modules in this exercise are discrete.

(a) Prove that M_j is a $\mathbb{Z}G_j$-bimodule with the right scalar operation $u(xN_j) = ux$.

(b) For $i \leq j$, define a mapping of $C^n(\mathbb{Z}G_i,M_i) \to C^n(\mathbb{Z}G_j,M_j)$ by $\Phi \mapsto \Phi^*$, where $\Phi^*(x_1N_j, \ldots, x_nN_j) = \Phi(x_1N_i, \ldots, x_nN_i)$, that is, $\Phi \mapsto \Phi^*$ is the adjoint of ϕ_{ij}. Show that $\Phi \mapsto \Phi^*$ induces a generalized inflation homomorphism $\inf_{i,j}^{(n)}: H^n(G_i, M_i) \to H^n(G_j,M_j)$, and $(\{H^n(G_i,M_i)\},\inf^{(n)})$ is a direct system to abelian groups over I. The direct limit $H^n(G,M) = \varinjlim(\{H^n(G_i,M_i)\},\inf)$ is the n'th Galois cohomology group of G with coefficients in M.

(c) Prove that a homomorphism $\psi: M \to M'$ of discrete $\mathbb{Z}G$-modules induces a group homomorphism $\psi_*: H^n(G,M) \to H^n(G,M')$, namely, $\psi_* = \varinjlim\{(\psi|M_j)_*: j \in I\}$.

(d) Let G have the topology that is defined by the neighborhood basis $\{xN_j: x \in G, j \in I\}$. Thus G is compact. For $n \in \mathbb{N}$, endow G^n with the product topology, and give M the discrete topology. Let $C_c^n(G,M)$ be the abelian group of all continuous mappings from G^n to M, with pointwise addition. Show that the usual coboundary homomorphism maps $C_c^n(G,M)$ to $C_c^{n+1}(G,M)$. (Hint. If $\Phi \in C_c^n(G,M)$, then $\Phi(G^n)$ is discrete and compact, hence finite. Use this observation and the hypothesis $M = \bigcup_{j \in I} M_j$ to show that $\delta^{(n)}\Phi$ is continuous.) Define $Z_c^n(G,M) = \operatorname{Ker} \delta^{(n)}$, $B_c^n(G,M) = \operatorname{Im} \delta^{(n-1)}$, and $H_c^n(G,M) = Z_c^n(G,M)/B_c^n(G,M)$. Prove that the adjoint of the projection mapping $\chi_j: G \to G_j$ induces a homomorphism $\psi_j: H^n(G_j, M_j) \to H_c^n(G,M)$, by the same construction that defines the inflation mapping. Show that $(H_c^n(G,M),\psi)$ is the limit of $(\{H^n(G_j,M_j)\},\inf)$. Thus, $H^n(G,M) \cong H_c^n(G,M)$.

(e) Use the result (d) to generalize the cohomology properties described in Section 11.2 to the Galois cohomology groups of discrete $\mathbb{Z}G$-modules.

(f) Let M be a discrete $\mathbb{Z}G$-module. For a submodule N of M, denote by λ_{NM} the homomorphism of $H^n(G,N)$ to $H^n(G,M)$ that is induced by the inclusion map of N to M. Prove that $(\{H^n(G,N)\},\{\lambda_{NN'}\})$ is a direct system over the set of finitely generated submodules of M (ordered by inclusion), and $(H^n(G,M),\{\lambda_{NM}\})$ is a limit of this system. That is, $H^n(G,M) = \varinjlim H^n(G,N)$. Hint. Use Exercise 2 and the previously observed fact that continuous cochains have finite images.

5. Let F have characteristic $p > 0$. Denote the separable algebraic closure of F by F_s, and write $G = \mathbf{G}(F_s/F)$.

(a) Use the result of Exercise 4 in Section 14.2 to prove that F_s is a discrete $\mathbb{Z}G$-module (under addition and the usual action of G) such that the Galois cohomology groups $H^n(G,F_s) = 0$ for all $n \geq 1$.

(b) Define $\psi: F_s \to F_s$ by $\psi(x) = x^p - x$. Show that ψ is a surjective $\mathbb{Z}G$-module homomorphism with $\operatorname{Ker}\psi \cong \mathbb{Z}/p\mathbb{Z}$, where $\mathbb{Z}G$ acts trivially on $\mathbb{Z}/p\mathbb{Z}$. (Hint. Note that if $a \in F_s$, then $\mathbf{x}^p - \mathbf{x} - a$ is separable.) Use the Long Exact Sequence Theorem for Galois cohomology to show that $H^n(G,\mathbb{Z}/p\mathbb{Z}) = 0$ for all $n \geq 2$.

14.7. Restriction

If F is a subfield of K, then the inclusion mapping $\kappa: F \to K$ induces a homomorphism $\kappa_*: \mathbf{B}(F) \to \mathbf{B}(K)$. When these Brauer groups are represented as unions of relative Brauer groups corresponding to cohomology groups, the description of κ_* can be given in terms of certain homomorphisms that are standard tools of cohomology theory. The purpose of this section is to define these homomorphisms, and relate them to the mappings of the Brauer groups.

Let H be a subgroup of the finite group G. If M is a right $\mathbb{Z}G$-module, then M can also be viewed as a $\mathbb{Z}H$-module because $\mathbb{Z}H$ is a subalgebra of $\mathbb{Z}G$. Moreover, the trivial action of G on the left of M yields a $\mathbb{Z}G$-bimodule or a $\mathbb{Z}H$-bimodule. Let $\Phi \in C^n(G,M)$ be an n-cochain, considered as a mapping from G^n to M. The restriction $\Phi|H^n$ is then an element of $C^n(H,M)$. The coboundary homomorphism plainly satisfies $\delta^{(n)}(\Phi|H^n) = (\delta^{(n)}\Phi)|H^n$, so that $\Phi \mapsto \Phi|H^n$ maps $Z^n(G,M)$ to $Z^n(H,M)$ and $B^n(G,M)$ to $B^n(H,M)$. Therefore, $\Phi \mapsto \Phi|H^n$ induces a group homomorphism.

$$\operatorname{res}: H^n(G,M) \to H^n(H,M)$$

that is called the *restriction mapping*. Explicitly, $\operatorname{res}[\Phi] = [\Phi|H^n]$ for all $\Phi \in Z^n(G,M)$. For clarity we will sometimes denote this restriction mapping by $\operatorname{res}_{G\to H}$.

The applications of the restriction mapping that interest us occur when $n = 2$, $G = \mathbf{G}(E/F)$, and $H = \mathbf{G}(E/K)$, where E/F is Galois and K is a field between F and E. Moreover, M will generally be E° with the usual $\mathbb{Z}G$-bimodule structure. Note that $H^0(G,E^\circ) = (E^\circ)^G = F^\circ$ and $H^0(H,E^\circ) = (E^\circ)^H = K^\circ$. It is easy to check that $\operatorname{res}: H^0(G,E^\circ) \to H^0(H,E^\circ)$ is the inclusion map of F° to K°.

Lemma a. *Let E/F be a Galois extension, $G = \mathbf{G}(E/F)$, H a subgroup of G, and $K = E^H$, the fixed field of H. If $\Phi \in Z^2(G,E^\circ)$, then $\Psi = \Phi|H^2 \in Z^2(H,E^\circ)$ and $(E,H,\Psi) \cong \mathbf{C}_A(K) \sim A^K$, where $A = (E,G,\Phi)$.*

Proof. In the notation of Section 14.1, $A = \bigoplus_{\sigma\in G} u_\sigma E$, where the $u_\sigma \in A^\circ$ satisfy $u_\sigma^{-1}cu_\sigma = c^\sigma$ for all $c \in E$, and $u_{\sigma\tau}^{-1}u_\sigma u_\tau = \Phi(\sigma,\tau)$. Since H is a subgroup of G and $\Psi = \Phi|H^2$, it follows that $B = \bigoplus_{\tau\in H} u_\tau E$ is a subalgebra of A that is isomorphic to (E,H,Ψ). If $\tau \in H$ and $c \in K$, then $u_\tau^{-1} c u_\tau = c^\tau = c$. Therefore, $B \subseteq \mathbf{C}_A(K)$. The Double Centralizer Theorem implies that $\dim_F \mathbf{C}_A(K) = (\dim_F A)/[K:F] = |G|[E:F]/[K:F] = |H|[E:F] = \dim_F B$. Thus, $B = \mathbf{C}_A(K)$. By Lemma 13.3, $\mathbf{C}_A(K) \sim A^K$. □

Proposition a. *Let $F \subseteq K \subseteq E$ be fields such that E/F is a Galois extension. If $\kappa: F \to K$ is the inclusion mapping, then $\kappa_*(\mathbf{B}(E/F)) \subseteq \mathbf{B}(E/K)$ and the diagram*

$$\begin{array}{ccc} H^2(\mathbf{G}(E/F),E^\circ) & \xrightarrow{\text{res}} & H^2(\mathbf{G}(E/K),E^\circ) \\ \theta_{E/F}\downarrow & & \downarrow\theta_{E/K} \\ \mathbf{B}(E/F) & \xrightarrow[\kappa_*|B(E/F)]{} & \mathbf{B}(E/K) \end{array} \tag{1}$$

is commutative.

Proof. If $A \in \mathfrak{S}(F)$ is split by E, then E also splits A^K since $(A^K)^E \cong A^{KE} = A^E$. Thus, $\kappa_*(\mathbf{B}(E/F)) \subseteq \mathbf{B}(E/K)$. Moreover, if $\Phi \in Z^2(\mathbf{G}(E/F),E^\circ)$ and $A = (E,\mathbf{G}(E/F),\Phi)$, then $\kappa_*\theta_{E/F}[\Phi] = \kappa_*[A] = [A^K] = [(E,\mathbf{G}(E/K),\Phi|\mathbf{G}(E/K)^2)] = \theta_{E/K}\,\text{res}\,[\Phi]$ by Lemma a. □

If $F \subseteq K \subseteq E \subseteq L$ is a chain of fields such that E/F and L/F are Galois extensions, then an easy calculation shows that the following diagram commutes.

$$\begin{array}{ccc} H^n(\mathbf{G}(E/F),E^\circ) & \xrightarrow{\text{res}} & H^n(\mathbf{G}(E/K),E^\circ) \\ \text{inf}\downarrow & & \downarrow\text{inf} \\ H^n(\mathbf{G}(L/F),L^\circ) & \xrightarrow[\text{res}]{} & H^n(\mathbf{G}(L/K),L^\circ) \end{array} \tag{2}$$

In the ordering of extensions, the Galois extensions of F constitute a cofinal subset of the Galois extensions of K. This remark and the commutativity of the diagram above implies that the family of restriction homomorphism induces a restriction homomorphism of the Galois cohomology groups: res: $H^n(\mathbf{G}(F_s/F),F_s^\circ) \to H^n(\mathbf{G}(F_s/K),F_s^\circ)$. Of course, $F_s = K_s$. It is easy to check that if $\kappa: F \to K$ is the inclusion mapping, then $\kappa_*: \mathbf{B}(F) \to \mathbf{B}(K)$ is the limit of $\{\kappa_*|B(E/F): E/F \text{ Galois and } K \subseteq E\}$. The result of Proposition a therefore implies that the diagram

$$\begin{array}{ccc} H^2(\mathbf{G}(F_s/F),F_s^\circ) & \xrightarrow{\text{res}} & H^2(\mathbf{G}(F_s/K),F_s^\circ) \\ \varinjlim\theta_{E/F}\downarrow & & \downarrow\varinjlim\theta_{E/K} \\ \mathbf{B}(F) & \xrightarrow[\kappa_*]{} & \mathbf{B}(K) \end{array}$$

is commutative when K/F is a finite separable extension.

We will need a stronger version of Proposition a, based on some elementary field theory. If E and K are fields that contain a common subfield F, and at least one of $[E:F]$, $[K:F]$ is finite, then E and K are said to be *linearly disjoint* over F if $E \otimes_F K$ is a field. When E and K are subfields of a common field L, and $[E:F]$ is finite, the condition that E and K are linearly disjoint over F is equivalent to the (algebra) compositum EK being a tensor product over F of E and K. Indeed, EK is a field and a homomorphic image of $E \otimes K$ by Proposition 9.2b, so that $E \otimes K$ is a field if and only if this mapping is an isomorphism. In this case, it follows from Proposition 9.2c that E and K are linearly disjoint over F if and only if every basis of E_F is also a basis of $(EK)_K$, that is $[E:F] = [EK:K]$. If E/F is also Galois, then there is a sharper result.

Lemma b. *Let F, K, and E be subfields of the field L with $F \subseteq E \cap K$. Assume that E/F is Galois.*

(i) *E and K are linearly disjoint over $E \cap K$.*
(ii) *EK/K is Galois, and $\sigma \mapsto \sigma|E$ is an isomorphism of $\mathbf{G}(EK/K)$ to $\mathbf{G}(E/E \cap K)$.*

PROOF. For notational convenience and without loss of generality, assume that $E \cap K = F$. Since E/F is Galois, this extension is simple, say $E = F(d)$. Let Θ be the minimum polynomial of d over F. Then $EK = K(d)$ and Θ splits completely in $EK[\mathbf{x}]$. Thus, EK is the splitting field of Θ over K. In particular, EK/K is a Galois extension. If $\Phi \in K[\mathbf{x}]$ is a monic divisor of Θ, then the roots of Φ are also roots of Θ. Since Θ splits in $E[\mathbf{x}]$, it follows that $\Phi = (\mathbf{x} - c_1) \cdots (\mathbf{x} - c_r)$ with $c_i \in E$. Therefore, $\Phi \in E[\mathbf{x}] \cap K[\mathbf{x}] = (E \cap K)[\mathbf{x}] = F[\mathbf{x}]$. Consequently, $\Phi = \Theta$ because Θ is irreducible in $F[\mathbf{x}]$. This argument shows that Θ is also irreducible in $K[\mathbf{x}]$. Thus, $[EK:K] = \deg \Theta = [E:F]$, E and K are linearly disjoint over F, and $|\mathbf{G}(EK/K)| = |\mathbf{G}(E/F)|$. Since E/F is Galois, every $\sigma \in \mathbf{G}(EK/K)$ maps E to itself, that is, $\sigma|E \in \mathbf{G}(E/F)$. If $\sigma \in \mathbf{G}(EK/K)$ satisfies $\sigma|E = id_E$, then $\sigma = id_{EK}$: $\sigma \mapsto \sigma|E$ is an injective group homomorphism of $\mathbf{G}(EK/K)$ to $\mathbf{G}(E/F)$. This homomorphism is an isomorphism because the finite groups $\mathbf{G}(EK/K)$ and $\mathbf{G}(E/F)$ have the same cardinality. □

Lemma c. *Let F, K, and E be subfields of the field L such that $F = E \cap K$, $L = EK$, and E/F is Galois. Identify $\mathbf{G}(L/K)$ with $G = \mathbf{G}(E/F)$ by the restriction isomorphism $\sigma \mapsto \sigma|E$. If $\Phi \in Z^2(G,E^\circ)$, then $\Phi \in Z^2(G,L^\circ)$ and $(L,G,\Phi) \cong (E,G,\Phi)^K$.*

PROOF. The identification of $\mathbf{G}(L/K)$ with G gives the inclusion $Z^2(G,E^\circ) \subseteq Z^2(G,L^\circ)$. Let $(E,G,\Phi) = \bigoplus_{\sigma \in G} u_\sigma E$, where $u_\sigma^{-1} c u_\sigma = c^\sigma$ for $\sigma \in G$, $c \in E$, and $u_{\sigma\tau}^{-1} u_\sigma u_\tau = \Phi(\sigma,\tau)$. As K-spaces, $(E,G,\Phi)^K \cong \bigoplus_{\sigma \in G} u_\sigma EK = \bigoplus_{\sigma \in G} u_\sigma L$. If $d = \sum_{i=1}^r c_i b_i$ with $c_i \in E$, $b_i \in K$, then $u_\sigma^{-1} d u_\sigma = \sum_{i=1}^r u_\sigma^{-1} c_i u_\sigma b_i = \sum_{i=1}^r c_i^\sigma b_i = d^\sigma$. Hence, $(E,G,\Phi)^K \cong (L,G,\Phi)$. □

This lemma has a cohomological interpretation. Let $\lambda\colon E \to L$ be the inclusion mapping, and $\kappa = \lambda|F\colon F \to K$. If $A \in \mathfrak{S}(F)$ has degree n, and if E splits A, then $(A^K)^L \cong (A^E)^L \cong M_n(L)$. Hence, $\kappa_*(\mathbf{B}(E/F)) \subseteq \mathbf{B}(L/K)$. Moreover, the following diagram commutes.

$$\begin{array}{ccc} H^2(G,E^\circ) & \xrightarrow{\lambda_*} & H^2(G,L^\circ) \\ \theta_{E/F}\downarrow & & \downarrow\theta_{L/K} \\ \mathbf{B}(E/F) & \xrightarrow[\kappa_*]{} & \mathbf{B}(L/K) \end{array} \qquad (3)$$

In fact, $\kappa_*\theta_{E/F}[\Phi] = [(E,G,\Phi)^K] = [(L,G,\Phi)] = \theta_{L/K}\lambda_*[\Phi]$.

Proposition b. *Let F, K, and E be subfields of a field L such that $F \subseteq E \cap K$ and E/F is Galois. The extension EK/K is Galois with*

$$\mathbf{G}(EK/K) \cong \mathbf{G}(E/E \cap K),$$

and the diagram of group homomorphisms

$$\begin{array}{ccc} H^2(\mathbf{G}(E/F),E^\circ) & \xrightarrow{\chi} & H^2(\mathbf{G}(EK/K),(EK)^\circ) \\ \theta_{E/F}\downarrow & & \downarrow\theta_{EK/K} \\ \mathbf{B}(E/F) & \xrightarrow[\kappa_*|\mathbf{B}(E/F)]{} & \mathbf{B}(EK/K) \end{array}$$

commutes, where $\kappa\colon F \to K$ is inclusion and χ is the composition of the restriction homomorphism $\mathrm{res}_{\mathbf{G}(E/F)\to\mathbf{G}(E/E\cap K)}$ with the cohomology map that corresponds to the inclusion of E° in $(EK)^\circ$.

This result combines Proposition a with Lemmas b and c. Indeed, we have the commutative diagram

$$\begin{array}{ccccc} H^2(\mathbf{G}(E/F),E^\circ) & \xrightarrow{\mathrm{res}} & H^2(\mathbf{G}(E/E\cap K),E^\circ) & \to & H^2(\mathbf{G}(EK/K),(EK)^\circ) \\ \downarrow & & \downarrow & & \downarrow \\ \mathbf{B}(E/F) & \longrightarrow & \mathbf{B}(E/E\cap K) & \to & \mathbf{B}(EK/K) \end{array}$$

in which the left square is (1) and the right square is (3).

Exercises

1. Prove that diagram (2) is commutative.

2. Prove that if K/F and E/F are Galois extensions such that $K \subseteq E$, then the sequence $0 \to H^2(\mathbf{G}(K/F),K^\circ) \xrightarrow{\mathrm{inf}} H^2(\mathbf{G}(E/F),E^\circ) \xrightarrow{\mathrm{res}} H^2(\mathbf{G}(E/K),E^\circ)$ is exact. Hint. Pass to relative Brauer groups.

3. Let H be a subgroup of the finite group G. Let $O \to M \to N \to P \to O$ be an exact sequence of right $\mathbb{Z}G$-modules. Prove that the following diagram commutes.

$$\begin{array}{ccccccccccccc} 0 \to & H^0(G,M) & \to & H^0(G,N) & \to \cdots \to & H^n(G,M) & \to & H^n(G,N) & \to & H^n(G,P) & \to & H^{n+1}(G,M) & \to \cdots \\ & \mathrm{res}\downarrow & & \mathrm{res}\downarrow & & \mathrm{res}\downarrow & & \mathrm{res}\downarrow & & \mathrm{res}\downarrow & & \mathrm{res}\downarrow & \\ 0 \to & H^0(H,M) & \to & H^0(H,N) & \to \cdots \to & H^n(H,M) & \to & H^n(H,N) & \to & H^n(H,P) & \to & H^{n+1}(H,M) & \to \cdots \end{array}$$

4. Let H be a subgroup of index m in the finite group G. Fix a coset decomposition $G = Hx_1 \cup \cdots \cup Hx_m$ of G. Suppose that M is a right $\mathbb{Z}G$-module. For $u \in M^H$, define $\mathrm{cor}\, u = u(\sum_{k=1}^m x_k)$.

   (a) Prove that $\mathrm{cor}\, u$ does not depend on the choice $x_1, \ldots, x_m$ of coset representatives, and cor is a group homomorphism from $M^H = H^0(H,M)$ to $M^G = H^0(G,M)$.

   (b) Fix an exact sequence $0 \to M \to N \to P \to 0$ of right $\mathbb{Z}G$-modules such that $H^n(G,N) = 0$ for all $n \geq 1$. Use induction on $n \geq 1$ to get a sequence of homomorphisms cor: $H^n(H,M) \to H^n(G,M)$ such that

$$\begin{array}{ccccccc} H^{n-1}(H,N) & \to & H^{n-1}(H,P) & \to & H^n(H,M) & \to & 0 \\ \text{cor}\downarrow & & \text{cor}\downarrow & & \text{cor}\downarrow & & \\ H^{n-1}(G,N) & \to & H^{n-1}(G,P) & \to & H^n(G,M) & \to & 0 \end{array}$$

commutes. The homomorphisms that are defined in this way are called the *corestriction mappings*.

(c) Prove by induction on n that $\text{cor}(\text{res}\, u) = mu$ for all $u \in H^n(G,M)$.

(d) Specialize H to be a Sylow p-subgroup of G. Use the result of (c) to show that the restriction mapping res: $H^n(G,M) \to H^n(H,M)$ is injective on the p-primary component of $H^n(G,M)$.

5. Let K be the separable closure of the field F. For a prime p, let L be a subfield of K that is maximal with the property that $p \nmid [F(d) : F]$ for all $d \in L$. Zorn's Lemma shows that L exists.

(a) Show that L is a subfield of K with $F \subseteq L$, K is the separable closure of L, and if $L \subseteq E \subseteq K$ with E/L a Galois extension, then $\mathbf{G}(E/L)$ is a p-group.

(b) Denote $G = \mathbf{G}(K/F)$, $H = \mathbf{G}(K/L)$. Prove that H is a closed subgroup of G and if N is an open normal subgroup of G, then $H/H \cap N$ is a Sylow p-subgroup of the finite group G/N. Deduce that $H = \varprojlim H/H \cap N$, where N ranges through the open normal subgroups of G.

(c) Let M be a discrete $\mathbb{Z}G$-module. Show that the restriction mapping res: $H^n(G,M) \to H^n(H,M)$ (defined as the limit of the family $\{\text{res}_{G/N \to H/H \cap N}\}$) is injective on the p-primary component of $H^n(G,M)$.

6. The purpose of this exercise is to outline the proof of a theorem that is due to E. Witt: if F is a field of prime characteristic p, then $\mathbf{B}(F)$ is p-divisible, that is, $p\mathbf{B}(F) = \mathbf{B}(F)$. To this end, we adopt the notation and hypotheses of Exercise 5 plus the assumption that char $F = p$. Prove the following facts.

(a) If N is a finitely generated, discrete $\mathbb{Z}H$-module such that $pN = 0$, then N is finite.

(b) If P is a simple, discrete $\mathbb{Z}H$-module such that $pP = 0$, then $P \cong \mathbb{Z}/p\mathbb{Z}$ with the trivial action of H. Hint. By (a), P is a finite $\mathbb{F}_p$-space, so that $|P| = p^m$ for some $m \geq 1$. Use discreteness to show that if $v \in P$, then $|vH| = p^k$ with $0 \leq k < \omega$. By a counting argument, deduce that there exists $0 \neq v \in P$ such that $vH = \{v\}$. Hence, $P = v\mathbb{Z}H = v\mathbb{Z} \cong \mathbb{Z}/p\mathbb{Z}$.

(c) If N is a discrete, finitely generated $\mathbb{Z}H$-module such that $pN = 0$, then $H^2(H,N) = 0$. Hint. If N is simple, use (b) and Exercise 5 of Section 14.6 (with F replaced by L). The general case can be obtained by induction on $|N|$ (because of (a)), using the long exact sequence for Galois cohomology.

(d) If M is a discrete $\mathbb{Z}H$-module such that $pM = 0$, then $H^2(H,M) = 0$. Hint. Use Exercise 4(f) of Section 14.6.

(e) If M is a discrete $\mathbb{Z}G$-module such that $pM = 0$, then the p-primary component of $H^2(G,M)$ is 0. Hint. Use Exercise 5c.

(f) $\mathbf{B}(F)$ is p-divisible. Hint. Char $F = p$ implies that $1 \to K^\circ \xrightarrow{d \mapsto d^p} K^\circ \to M \to 1$ is exact with $M = K^\circ/(K^\circ)^p$. Use the long exact sequence $\cdots \to H^2(G,K^\circ) \xrightarrow{u \mapsto u^p} H^2(G,K^\circ) \to H^2(G,M) \to \cdots$ and (d) to conclude that the p-primary components of $H^2(G,K^\circ)$ and $\mathbf{B}(F)$ are p-divisible. Note that if $q \neq p$ is prime, then the q-primary component of a torsion abelian group is p-divisible.

Notes on Chapter 14

The title of this chapter is probably inappropriate. We have given only an introductory look at Galois Cohomology. The reader who wants to explore this subject more fully is advised to read Serre's monograph [71]. The exercises in Sections 14.6 and 14.7 go a bit beyond the text in these sections.

The first five sections of the chapter cover the basic connection between Brauer groups and cohomology in about the same way that this topic is handled in the books of Artin, Nesbitt and Thrall [9], Herstein [41], Jacobson [48], and Reiner [66].

CHAPTER 15

Cyclic Division Algebras

The first examples of division algebras that were found after the quaternions belong to the class of cyclic division algebras. This class still plays a major role in the theory of central simple algebras. If F is a local field, an algebraic number field, or more generally a global field, then every central division algebra over F is cyclic. This fact will be proved later; it is one of the most profound results in this book.

This chapter has two purposes. The first two sections collect basic facts about cyclic algebras that will be used later. The rest of the chapter elaborates the theory of cyclic division algebras. In particular, we prove Wedderburn's theorem that all division algebras of degree 3 are cyclic. The final section presents an example that is due to Albert of a non-cyclic division algebra of degree 4.

15.1. Cyclic Algebras

A field extension E/F is called cyclic if E/F is Galois and $\mathbf{G}(E/F)$ is a cyclic group.

Definition. An algebra $A \in \mathfrak{S}(F)$ is *cyclic* if there is a strictly maximal subfield E of A such that E/F is a cyclic extension.

In particular, a cyclic algebra is a crossed product. However, cyclic algebras are very special crossed products. The purpose of this section is to specialize the results of Chapter 14 to cyclic algebras.

Proposition a. *Let E/F be a cyclic extension such that $G = \mathbf{G}(E/F)$ is cyclic of order n with the generator σ. If $A \in \mathfrak{S}(F)$ contains E as a strictly maximal subfield, then there is an element $u \in A^\circ$ that satisfies*

(i) $A = \bigoplus_{0 \le j < n} u^j E$,
(ii) $u^{-1} d u = d^\sigma$ *for all* $d \in E$, *and*
(iii) $u^n = a \in F^\circ$.

Conversely, if A is the F-algebra that is defined by the conditions (i), (ii), *and* (iii), *then* $A \cong (E, G, \Phi_a)$, *where*

$$\Phi_a(\sigma^i, \sigma^j) = \begin{cases} 1 & \text{if } 0 \le i, j, i + j < n, \\ a & \text{if } 0 \le i, j < n \le i + j. \end{cases}$$

Proof. Assume that E is a strictly maximal subfield of $A \in \mathfrak{S}(F)$. By the Noether–Skolem Theorem, there is an element $u \in A^\circ$ such that $u^{-1} d u = d^\sigma$ for all $d \in E$. By induction, $(u^j)^{-1} d u^j = d^{\sigma^j}$ for $0 \le j \le n$. In particular, $u^n \in \mathbf{C}_A(E) = E$. Since $G = \{1, \sigma, \sigma^2, \ldots, \sigma^{n-1}\}$, Lemma 14.1 implies that $A = \bigoplus_{0 \le j < n} u^j E$. Therefore $u^n \in \mathbf{Z}(A) = F$. To obtain the converse, check by computation (Exercise 1) that $\Phi_a \in Z^2(G, E^\circ)$. Define $A = (G, E, \Phi_a) = \bigoplus_{0 \le j < n} u_\sigma E$ with $u_1 = 1_A$ (because Φ_a is normalized) and $u_\sigma = u$ (as a notational convenience). If $1 < j < n$, then $u_\sigma u_{\sigma^{j-1}} = \Phi(\sigma, \sigma^{j-1}) u_{\sigma^j} = u_{\sigma^j}$. Therefore, $u_{\sigma^j} = u^j$ for all $1 \le j < n$ by induction. Also, $u^n = u_\sigma u_{\sigma^{n-1}} = \Phi(\sigma, \sigma^{n-1}) u_1 = a$. Thus, A is the algebra that is defined by (i), (ii), and (iii). □

It is convenient to simplify the crossed product notation. In the case of cyclic algebras, we will write (E, σ, a) instead of $(E, \mathbf{G}(E/F), \Phi_a)$ when $\mathbf{G}(E/F) = \langle \sigma \rangle$. The symbols u and u_σ will denote an element of $(E, \sigma, a)^\circ$ that satisfies $u^{-1} d u = d^\sigma$ for all $d \in E$, and $u^n = a$, where n is the order of σ.

The results of Chapter 14 can be translated as three corollaries of Proposition a. In these statements, E/F is a cyclic extension of degree n, $G = \mathbf{G}(E/F) = \langle \sigma \rangle$, and a and b are elements of F°.

Corollary a. (i) $(E, \sigma, a) \otimes (E, \sigma, b) \sim (E, \sigma, ab)$. *In particular,* $(E, \sigma, 1) \sim F$.
(ii) *If $k \in \mathbb{Z}$ is relatively prime to n, then* $(E, \sigma^k, a^k) \cong (E, \sigma, a)$.

Since $\Phi_a \Phi_b = \Phi_{ab}$, (i) follows from Theorem 14.2; (ii) is clear because $G = \langle \sigma^k \rangle$, $u^{-k} d u^k = d^{\sigma^k}$, and $(u^k)^n = a^k$.

Corollary b. *If K is a subfield of E that contains F and $[K : F] = m$, then K/F is cyclic with $\mathbf{G}(K/F) = \langle \sigma | K \rangle$ and $(K, \sigma | K, a) \sim (E, \sigma, a^{n/m})$.*

Proof. This is essentially a reformulation of Proposition 14.5, but it takes some work to connect these results. Since G is cyclic, all subgroups of G are normal and the factor groups of G are cyclic. If $\tau = \sigma | K$, then $H = \mathbf{G}(K/F) = \langle \tau \rangle$, where τ has order m. Define $\Psi_a : H^2 \to F^\circ$ as in Proposition a: $\Psi_a(\tau^k,\tau^l) = 1$ for $0 \le k, l, k + l < m$, $\Psi_a(\tau^k,\tau^l) = a$ for $0 \le k, l < m \le k + l$. Thus, $(K,\tau,a) = (K,H,\Psi_a)$. By Proposition 14.5, $(K,\tau,a) \sim (E,G,\Psi_a^*)$, where $\Psi_a^*(\sigma^i,\sigma^j) = \Psi_a(\tau^i,\tau^j)$ for $0 \le i, j < n$. In order to obtain an explicit formula for Ψ_a^* we represent i and j by the division algorithm. That is, $i = rm + k$, $j = sm + l$ with $0 \le k, l < m$. Plainly,

$$\Psi_a^*(\sigma^i,\sigma^j) = \begin{cases} 1 & \text{if } k + l < m \\ a & \text{if } k + l \ge m. \end{cases} \tag{1}$$

The proof of the corollary will be completed by showing that Ψ_a^* and $\Phi_{a^{n/m}}$ belong to the same cohomology class. Define $\Theta: G \to F^\circ$ by $\Theta(\sigma^i) = a^r$ if $0 \le i = rm + k < n$, $0 \le k < m$. A computation gives

$$(\delta^{(1)}\Theta)(\sigma^i,\sigma^j) = \begin{cases} 1 & \text{if } i + j < n, k + l < m \\ a^{-1} & \text{if } i + j < n, k + l \ge m \\ a^{n/m} & \text{if } i + j \ge n, k + l < m \\ a^{(n/m)-1} & \text{if } i + j \ge n, k + l \ge m. \end{cases} \tag{2}$$

Then, (1) and (2) imply

$$((\delta^{(1)}\Theta)\Psi_a^*)(\sigma^i,\sigma^j) = \begin{cases} 1 & \text{if } i + j < n \\ a^{n/m} & \text{if } i + j \ge n, \end{cases}$$

so that $\Phi_{a^{m/n}} = (\delta^{(1)}\Theta)\Psi_a^*$ and $[\Phi_{a^{n/m}}] = [\Psi_a^*]$. □

For cyclic algebras, Proposition 14.7b takes the following form.

Corollary c. *If F, K, and E are subfields of a field L, $F \subseteq E \cap K$, and E/F is cyclic, then EK/K is cyclic and* $\mathbf{G}(EK/K) \cong \langle\sigma^r\rangle$, *where $r = [E \cap K : F]$. Moreover, $(E,\sigma,a)^K$ is cyclic and $(E,\sigma,a)^K \sim (EK,\sigma^r,a)$.*

This corollary comes directly from Proposition 14.7b when we use the restriction of automorphisms to identify $\mathbf{G}(EK/K)$ with $\mathbf{G}(E/E \cap K) = \langle\sigma^r\rangle$.

For a finite extension E/F, denote the field norm from E to F by $N_{E/F}$. The norm can be viewed as a group homomorphism from E° to F°. If E/F is Galois with $G = \mathbf{G}(E/F)$, then $N_{E/F}(d) = \prod_{\sigma \in G} d^\sigma$. In the cyclic case $G = \langle\sigma\rangle$, $N_{E/F}(d) = d^{1+\sigma+\cdots+\sigma^{n-1}}$, where $n = |G|$.

Lemma. *Let E/F be a cyclic extension of degree n with* $\mathbf{G}(E/F) = \langle\sigma\rangle$. *If $a, b \in F^\circ$, then $(E,\sigma,a) \cong (E,\sigma,b)$ if and only if $b/a \in N_{E/F}(E^\circ)$. In particular, $(E,\sigma,a) \cong M_n(F)$ if and only if $a \in N_{E/F}(E^\circ)$.*

Proof. If $u \in (E,\sigma,a)^\circ$ satisfies $u^{-1}du = d^\sigma$ for all $d \in E$ and $u^n = a$; and if $v = uc$ for some $c \in E^\circ$, then $v^2 = ucuc = u^2c^{1+\sigma}$, $v^3 = ucu^2c^{1+\sigma} = u^3c^{1+\sigma+\sigma^2}, \ldots, v^n = u^nc^{1+\sigma+\cdots+\sigma^{n-1}} = aN_{E/F}(c)$. This calculation shows that if $b/a \in N_{E/F}(E^\circ)$, then $(E,\sigma,a) \cong (E,\sigma,b)$. Conversely, suppose that ϕ:

$(E,\sigma,b) \to (E,\sigma,a)$ is an isomorphism, where $(E,\sigma,a) = \bigoplus_{j<n} u^j E$, $(E,\sigma,b) = \bigoplus_{j<n} v^j E$, $u^n = a$, and $v^n = b$. As in the proof of Lemma 14.2, it can be assumed that $\phi|E = id_E$ and $\phi(v) = uc \in uE^\circ$. In this case, $b = \phi(b) = \phi(v^n) = (uc)^n = aN_{E/F}(c)$, that is $b/a \in N_{E/F}(E^\circ)$. The last statement of the lemma is a consequence of Corollary a. □

Proposition b. *If E/F is a cyclic extension with* $\mathbf{G}(E/F) = \langle\sigma\rangle$, *then* $\mathbf{B}(E/F) \cong F^\circ/N_{E/F}(E^\circ)$ *by the mapping* $aN_{E/F}(E^\circ) \mapsto [(E,\sigma,a)]$.

PROOF. By Proposition a and Corollary a, the mapping $a \mapsto [(E,\sigma,a)]$ is a surjective homomorphism from F° to $\mathbf{B}(E/F)$. The kernel of this homomorphism is $N_{E/F}(E^\circ)$ by the lemma. □

This proposition has many applications. Here is one of them; two other consequences of the proposition are outlined in Exercises 4 and 5.

Corollary d. *If E/F is a cyclic extension of degree n, and if the order of $a \in F^\circ$ modulo $N_{E/F}(E^\circ)$ is n, then (E,σ,a) is a division algebra.*

PROOF. By Propositions b, 13.4, and 14.4b, $n = \text{Exp}(E,\sigma,a) \le \text{Ind}(E,\sigma,a) \le \text{Deg}(E,\sigma,a) = n$. Hence, $\text{Ind}(E,\sigma,a) = \text{Deg}(E,\sigma,a)$ and (E,σ,a) is a division algebra. □

EXERCISES

1. Prove that the mapping Φ_a that was defined in Proposition a satisfies the cocycle condition.

2. Prove the formula (2).

3. (a) Show that if char $F \neq 2$, then every $A \in \mathfrak{S}(F)$ of degree 2 is cyclic. Hint. Use Theorem 13.1.

   (b) Use the result of Exercise 3 in Section 13.1 to obtain the same result for fields of characteristic 2. Hint. If char $F = 2$, then the splitting field over F of $\mathbf{x}^2 + \mathbf{x} + a$, $a \in F$, is cyclic.

4. Use Proposition b or the lemma to give a new proof of Proposition 1.6.

5. Use Proposition b to give a new proof that $\mathbf{B}(F) = \{1\}$ for a finite field F. Hint. Let $|F| = q$ and suppose that $[E:F] = n$. Thus, $|E| = q^n$ and E° is cyclic of order $q^n - 1$. Let $E^\circ = \langle c\rangle$. Use the fact that $\mathbf{G}(E/F)$ is generated by the Frobenius automorphism $d \mapsto d^q$ to show that $N_{E/F}(c) = c^{1+q+\cdots+q^{n-1}}$ has order $q - 1$ in F°.

6. Give an example of a cyclic division algebra D such that $M_n(D)$ is not cyclic for all $n > 1$. Hint. Try $D = \mathbb{H}$.

15.2. Constructing Cyclic Algebras by Inflation

The main result of this section shows that a central simple algebra of index greater than one can be inflated to an algebra that is equivalent to a cyclic division algebra of prime index. This proposition is a useful tool for inductive proofs. Its most important application is in Chapter 18. In this section we will use the result of characterize fields whose finite separable extensions have trivial Brauer groups. Another application is an improvement of Proposition 14.4b(ii).

We begin with a generalization of Lemma 14.4b.

Lemma. *Let $A \in \mathfrak{S}(F)$ have index $p^e m$, where p is prime, $e \geq 1$, and p does not divide m. There is a natural number n that is not divisible by p, and a chain of fields $F \subseteq K_e \subset K_{e-1} \subset \cdots \subset K_0$ such that:*

(i) *K_i/F is a separable extension of degree $p^{e-i}n$;*
(ii) *K_i/K_{i+1} is cyclic for $0 \leq i < e$;*
(iii) *$\operatorname{Ind} A^{K_i} = p^i$ for $0 \leq i \leq e$; in particular, K_0 splits A.*

Proof. If A is replaced by A^L, where L is the field that was constructed in Lemma 14.4b, then the proof is reduced to the case in which $m = 1$. Hence, assume that $\operatorname{Ind} A = p^e$. By Proposition 13.5, there is a separable extension K/F with $[K:F] = p^e$ such that K splits A. Extend K to a field E that is a Galois extension of F. Let $|\mathbf{G}(E/F)| = p^s n$, where p does not divide n. Denote a Sylow p-subgroup of $\mathbf{G}(E/F)$ by H, and define L to be the fixed field of H. Since $[L:F] = n$ and $[KL:L][L:F] = [KL:K][K:F]$, it follows that $[KL:L] = [K:F] = p^e$. If $H_0 = \mathbf{G}(E/KL)$, then $H_0 \subseteq H$ and $[H:H_0] = [KL:L] = p^e$. Since H is a p-group, there is a chain $H_0 \subset H_1 \subset \cdots \subset H_e = H$ with H_i normal in H_{i+1} and H_{i+1}/H_i cyclic of order p for $0 \leq i < e$. Let K_i be the fixed field of H_i. Then $KL = K_0 \supset K_1 \supset \cdots \supset K_e = L$ and K_i/K_{i+1} is Galois with $\mathbf{G}(K_i/K_{i+1}) \cong H_{i+1}/H_i$ a cyclic group of order p. Thus, $[K_i:F] = p^{e-i}[L:F] = p^{e-i}n$. Since $K_0 = KL$ splits A^{K_i}, it follows from Proposition 13.4 that $\operatorname{Ind} A^{K_{i+1}}$ divides $p \operatorname{Ind} A^{K_i}$ and $\operatorname{Ind} A^{K_e} = \operatorname{Ind} A^L = \operatorname{Ind} A = p^e$. By induction, $\operatorname{Ind} A^{K_i} = p^i$ for $0 \leq i \leq e$. □

Proposition. *Let $A \in \mathfrak{S}(F)$ have index $p^e m$, where p is prime, $e \geq 1$, and p does not divide m. There is a separable extension K/F such that $[K:F] = p^{e-1}n$, p does not divide n, and $A^K \sim D$ where $D \in \mathfrak{S}(K)$ is a cyclic division algebra of degree p.*

Proof. Define $K = K_1$ where K_1 has the properties that were listed in the lemma. Let $D \in \mathfrak{S}(K)$ be a division algebra that satisfies $A^K \sim D$. Then K_0/K is cyclic with $[K_0:K] = p = \operatorname{Deg} D$ and K_0 splits D. By Corollary 13.3, D contains a strictly maximal subfield that is isomorphic to K_0 as a K-algebra. Thus, D is cyclic. □

An important application of the proposition concerns the index of tensor products.

Corollary a. *If* $A \in \mathfrak{S}(F)$ *has index* rs*, then* $\operatorname{Ind} A^{\otimes r}$ *divides* s.

PROOF. It can be assumed that $\operatorname{Ind} A = p^e m$ and $r = p$. If K is defined as in the proposition, then $\operatorname{Exp} A^K = \operatorname{Ind} A^K = p$. Thus, $(A^{\otimes p})^K \cong (A^K)^{\otimes p} \sim K$, so that $\operatorname{Ind} A^{\otimes p}$ divides $[K:F] = p^{e-1}n$. Since $\operatorname{Ind} A^{\otimes p}$ also divides $\operatorname{Ind} A$ by Proposition 13.4, the corollary is established. □

The second corollary of the proposition characterizes cohomologically trivial fields. It is an easy consequence of the proposition and Proposition 15.1b.

Corollary b. *For a field* F*, the following conditions are equivalent.*

(i) $\mathbf{B}(E) = \{1\}$ *for all finite separable extensions* E/F.
(ii) *If* E/F *is a finite separable extension, and* K/E *is a cyclic extension, then* $N_{K/E}(K^\circ) = E^\circ$.

EXERCISE

In this exercise, assume that $\operatorname{char} F = p$, $A \in \mathfrak{S}(F)$, and $\operatorname{Deg} A = p^e$, where $e \geq 1$. Our aim is to prove that there is a finite, purely inseparable extension K/F such that K splits A. Prove the following statements.

(a) If L and K are subfields of the algebraic closure of F, L/F is finite separable, and K/F is finite purely inseparable, then $[KL:K] = [L:F]$. Hint. Use Exercise 2, Section 10.7.

(b) There is a finite separable extension L/F of degree $p^{e-1}n$ (where $p \nmid n$) and a cyclic extension E/L of degree p such that $A^L \sim (E,\sigma,c)$, where $\mathbf{G}(E/L) = \langle\sigma\rangle$ and $c \in L^\circ$ has the form $c = a_1 x_1^p + \cdots + a_m x_m^p$ with $x_i \in L$, $a_i \in F$.

(c) There is a purely inseparable extension K/F of finite degree (with K contained in the algebraic closure of F) and an element $d \in KL$ such that $d^p = c$.

(d) KL splits A. Hint. Show that $c \in N_{KE/KL}(KE)$.

(e) $\operatorname{Ind} A^K = p^f$, where $f \leq e - 1$. Hint. By (d), $\operatorname{Ind} A^K$ divides $[KL:K]$. Use (a).

(f) There is a finite, purely inseparable extension of F that splits A. Hint. Use induction based on the result of (e).

(g) If F is a perfect field of characteristic $p > 0$ and $A \in \mathfrak{S}(F)$, then p does not divide $\operatorname{Ind} A$. Hence, the p-primary component of $\mathbf{B}(F)$ is zero.

15.3. The Primary Decomposition of Cyclic Algebras

In Section 14.4 it was proved that every division algebra D is uniquely a tensor product of division algebras that have prime power degrees. To use that result effectively it is necessary to relate the properties of D with corresponding properties of its factors. In this section we prove that D is cyclic if and only if its primary components are cyclic.

We need to supplement the results on linearly disjoint field extensions that were given in Section 14.7. The next lemma is more general than is necessary for this section. The extra generality will be used in Section 15.5.

Lemma a. *Let K/F and L/F be finite field extensions where K and L are subfields of the algebraic closure of F.*

(i) *If $[K:F]$ and $[L:F]$ are relatively prime, then K and L are linearly disjoint over F.*
(ii) *If K/F and L/F are Galois, and K and L are linearly disjoint over F, then $(K \otimes L)/F$ is Galois and $\mathbf{G}((K \otimes L)/F) \cong \mathbf{G}(L/F) \times \mathbf{G}(K/F)$.*
(iii) *If E/F is Galois, $\mathbf{G}(E/F) = H_1 \times H_2$, K is the fixed field of H_2 and L is the fixed field of H_1, then K and L are linearly disjoint over F, $E = K \otimes L$, K/F and L/F are Galois, and $\mathbf{G}(K/F) \cong H_1$, $\mathbf{G}(L/F) \cong H_2$.*

Proof. The statement (i) is clear from the observations $[KL:L] \le [K:F]$, $[KL:K] \le [L:F]$, and $[KL:K][K:F] = [KL:L][L:F]$. It follows from Example 9.2 and Corollary 9.3a that the mapping θ: $\mathbf{G}(K/F) \times \mathbf{G}(L/F) \to \mathbf{E}_F(K \otimes L)$ defined by $\theta(\sigma,\tau) = \sigma \otimes \tau$ is injective. An easy calculation with rank one tensors shows that $\operatorname{Im}\theta \subseteq \mathbf{G}((K \otimes L)/F)$, and θ is a group homomorphism. Thus, $|\mathbf{G}((K \otimes L)/F)| \ge |\operatorname{Im}\theta| = |\mathbf{G}(K)| \cdot |\mathbf{G}(L)| = [(K \otimes L):F] \ge |\mathbf{G}((K \otimes L)/F)|$. Hence, $(K \otimes L)/F$ is Galois and θ is an isomorphism. The statement (iii) is an easy consequence of Galois theory: $\mathbf{G}(E/L) = H_1 \lhd \mathbf{G}(E/F)$, $\mathbf{G}(E/K) = H_2 \lhd \mathbf{G}(E/F)$, so that K/F and L/F are Galois with $\mathbf{G}(K/F) \cong H_1$ and $\mathbf{G}(L/F) \cong H_2$; the F-automorphisms of E that fix all elements of KL are in $H_1 \cap H_2 = \{1\}$, so that $KL = E$; and $[E:F] = |H_1| \cdot |H_2| = [K:F][L:F]$ implies $E = K \otimes L$. □

Lemma b. *Let K/F and L/F be cyclic extensions with $\mathbf{G}(K/F) = \langle\sigma\rangle$, $\mathbf{G}(L/F) = \langle\tau\rangle$. Assume that $m = [K:F]$ is relatively prime to $n = [L:F]$, say $rm + sn = 1$, where $r, s \in \mathbb{Z}$. If $a \in F^\circ$, then $(K,\sigma,a^s) \otimes (L,\tau,a^r) \cong ((K \otimes L), (\sigma,\tau), a)$.*

Proof. By Lemma a, $(K \otimes L)/F$ is a Galois extension with Galois group $\langle\sigma\rangle \times \langle\tau\rangle$. Since the orders m and n of σ and τ are relatively prime, $\langle\sigma\rangle \times \langle\tau\rangle$ is cyclic with the generator $\langle(\sigma,\tau)\rangle$. Let $A = (K,\sigma,a^s) = \bigoplus_{i<m} u^i K$ and $B = (L,\tau,a^r) = \bigoplus_{j<n} v^j L$, as in Proposition 15.1a. Then $A \otimes B = \bigoplus_{i<m,j<n} (u^i \otimes v^j)(K \otimes L) = \bigoplus_{k<mn} (u \otimes v)^k (K \otimes L)$, $(u \otimes v)^{-1}(c \otimes d)(u \otimes v) = (u^{-1}cu) \otimes (v^{-1}dv) = c^\sigma \otimes d^\tau = (c \otimes d)^{(\sigma,\tau)}$ for $c \in K$, $d \in L$, and $(u \otimes v)^{mn} = a^{sn} \otimes a^{rm} = (1 \otimes 1)a^{sn+rm} = (1 \otimes 1)a$. Thus, $A \otimes B \cong (K \otimes L, (\sigma,\tau), a)$. □

Proposition. *If A and B in $\mathfrak{S}(F)$ have relatively prime degrees, then $A \otimes B$ is cyclic if and only if A is cyclic and B is cyclic.*

PROOF. Let $\operatorname{Deg} A = m$, $\operatorname{Deg} B = n$, where $rm + sn = 1$ for suitable integers r and s. Assume that $A \otimes B = (E,\sigma,a)$ with $a \in F^\circ$ and $\mathbf{G}(E/F) = \langle\sigma\rangle$. Note that $\langle\sigma\rangle \cong \langle\sigma^{sn}\rangle \times \langle\sigma^{rm}\rangle$ under the correspondence $\sigma \mapsto (\sigma^{sn},\sigma^{rm})$. Let K be the fixed field of σ^{rm} and let L be the fixed field of σ^{sn}. By Lemma a, $E = K \otimes L$, and $\mathbf{G}(K/F) = \langle\sigma^{sn}\rangle$, $\mathbf{G}(L/F) = \langle\sigma^{rm}\rangle$. Also, $A_1 = (K,\sigma^n,a) \cong (K,\sigma^{sn},a^s)$ and $B_1 = (L,\sigma^m,a) \cong (L,\sigma^{rm},a^r)$, using Corollary 15.1a. By Lemma b, $A_1 \otimes B_1 \cong (K \otimes L, (\sigma^{sn},\sigma^{rm}), a) = (E,\sigma,a) = A \otimes B$. Since $\operatorname{Deg} A_1 = \operatorname{Deg} A = m$ and $\operatorname{Deg} B_1 = \operatorname{Deg} B = n$, it follows from Proposition 14.4b that in $\mathbf{B}(F)$, $[A] = [A][A]^{-rm}[B]^{sn} = ([A][B])^{sn} = [A \otimes B]^{sn} = [A_1 \otimes B_1]^{sn} = [A_1]$. Similarly, $[B] = [B_1]$. By Proposition 12.5b, $A \cong A_1$ and $B \cong B_1$, so that A and B are cyclic. Conversely, if A and B are cyclic, then by Lemmas a and b so is $A \otimes B$. □

Corollary. *A central division algebra is cyclic if and only if its primary components are cyclic.*

EXERCISES

1. Verify the statement in the proof of Lemma a that θ is a homomorphism of $\mathbf{G}(K/F) \times \mathbf{G}(L/F)$ to $\mathbf{G}((K \otimes L)/F)$.
2. Let E/F be a cyclic extension, and suppose that K is a subfield of E that contains F, and $[E:K] = n$. Assume that $A \in \mathfrak{S}(F)$ contains a strictly maximal subfield that is isomorphic as an F-algebra to K. Prove that $A \otimes M_n(F)$ contains a strictly maximal subfield that is isomorphic to E.
3. Use Exercise 2 to show that if $-1 \in F^2$ and $a \in F - F^2$, then $\left(\frac{a,b}{F}\right) \otimes \left(\frac{a,c}{F}\right)$ is a cyclic algebra for all b, c in F°. In Section 15.7, an example will be given of a noncyclic division algebra that is a tensor product of two quaternion algebras.

15.4. Characterizing Cyclic Division Algebras

The problem of characterizing cyclic division algebras is interesting and important. It does not have a fully satisfactory solution, but in this section we will obtain a partial characterization of cyclic algebras in terms of the maximal subfields of the algebras.

Lemma. *If $D = (E,\sigma,a)$ is a cyclic division algebra of degree n, then $\mathbf{x}^n - a$ is irreducible in $F[\mathbf{x}]$ and D contains a strictly maximal subfield K that is isomorphic to $F(a^{1/n})$ as an F-algebra.*

PROOF. Using the notation of Section 15.1, $D = E \oplus uE \oplus \cdots \oplus u^{n-1}E$, where $u \in D^\circ$ satisfies $u^n = a$. Thus, u is a root of $\mathbf{x}^n - a$ and $[F(u):F] = n$. Hence, $\mathbf{x}^n - a$ is irreducible. □

It is worth remarking that even though the sequence $1, u, \ldots, u^{n-1}$ is linearly independent over E, the polynomial $\mathbf{x}^n - a$ may not be irreducible over E. The arithmetic of polynomials over division algebras is not a straightforward generalization of the arithmetic of polynomials over a field.

In order to prove a converse of the lemma, some additional hypothesis is needed. If enough roots of unity are in F, then the converse is easily obtained. In the next section we will show that this assumption can be dropped for division algebras of prime degree.

Proposition. *Let F be a field whose characteristic does not divide the natural number n, and suppose that there is a primitive n'th root of unity in F. Assume that $A \in \mathfrak{S}(F)$ has degree n. If $a \in F$ is such that $\mathbf{x}^n - a$ is irreducible and A contains a subfield that is isomorphic to $F(a^{1/n})$, then A is a cyclic algebra.*

PROOF. If E is a subfield of A that is isomorphic to $F(a^{1/n})$, then $[E:F] = n = \operatorname{Deg} A$ because $\mathbf{x}^n - a$ is irreducible. Thus, E is strictly maximal in A. By the theory of Kummer extensions, E/F is cyclic. Thus, A is cyclic. □

It is easy to describe the algebras that satisfy the hypotheses of the proposition. The subfield E of A has the form $F(v)$ where $v^n = a$. The Galois group of $F(v)/F$ is generated by an automorphism σ that is defined by the condition $v^\sigma = v\zeta$, where $\zeta \in F$ is a primitive n'th root of unity. If $u \in A$ is such that $u^{-1}vu = v^\sigma$, then $vu = uv\zeta$ and $u^n = b \in F^\circ$. These remarks are summarized by the equations

$$A = \bigoplus\nolimits_{0 \le i,j < n} u^i v^j F, \tag{1}$$

$$vu = uv\zeta, \tag{2}$$

$$u^n = b, \quad v^n = a. \tag{3}$$

Conversely, if $\mathbf{x}^n - a$ is irreducible, then the F-algebra that is defined by (1), (2), and (3) is central simple and cyclic.

The algebras that are defined by (1), (2), and (3) generalize the quaternion algebras. Indeed, if $n = 2$ and $\zeta = -1$, then $A = \left(\frac{a,b}{F}\right)$. This observation motivates the notation $\left(\frac{a,b}{F,\zeta}\right)$ that is used to denote the algebra that is defined by the conditions (1), (2), and (3).

If $\mathbf{x}^n - a$ and $\mathbf{x}^n - b$ are both irreducible, then $\left(\frac{a,b}{F,\zeta}\right) = \left(\frac{b,a}{F,\zeta^{-1}}\right)$. In the notation of cyclic algebras, $\left(\frac{a,b}{F,\zeta}\right) = (F(a^{1/n}),\sigma,b) = (F(b^{1/n}),\sigma^{-1},a) = (F(b^{1/n}),\sigma,a^{-1}) = \left(\frac{b,a^{-1}}{F,\zeta}\right) = \left(\frac{b,a}{F,\zeta}\right)^*$. If $\left(\frac{a,b}{F,\zeta}\right)$ is a division algebra, then $\mathbf{x}^n - b$ is necessarily irreducible by the Lemma.

EXERCISES

1. Give an example of a cyclic division algebra (E,σ,a) such that $\mathbf{x}^n - a$ is reducible over E. Hint. Try $\mathbb{H}$.

2. Let F be a field whose characteristic does not divide $n \in \mathbb{N}$. Assume that $\zeta \in F$ is a primitive n'th root of unity. Prove the following equivalences for a, b, $c \in F^\circ$.
   (a) $\left(\dfrac{a,bc}{F,\zeta}\right) \sim \left(\dfrac{a,b}{F,\zeta}\right) \otimes \left(\dfrac{a,c}{F,\zeta}\right)$.
   (b) $\left(\dfrac{ab,c}{F,\zeta}\right) \sim \left(\dfrac{a,c}{F,\zeta}\right) \otimes \left(\dfrac{b,c}{F,\zeta}\right)$.
   (c) $\left(\dfrac{a, 1-a}{F,\zeta}\right) \sim F$. Hint. Compute $N_{F(v)/F}(1 - v)$, where v is a root of $\mathbf{x}^n - a$.

   It can be assumed that $\mathbf{x}^n - a$ is irreducible.

   Remark. The mapping $F^\circ \times F^\circ \to \mathbf{B}(F)$, that is defined by $(a,b) \mapsto \left[\left(\dfrac{a,b}{F,\zeta}\right)\right] = \{a,b\}$, satisfies $\{a,bc\} = \{a,b\}\{a,c\}$, $\{ab,c\} = \{a,c\}\{b,c\}$, and $\{a, 1-a\} = 1$. Any mapping from $F^\circ \times F^\circ$ to an abelian group with these properties is called a *Steinberg symbol* on F. By a theorem of Matsumoto, every Steinberg symbol on a field F induces a homomorphism from K_2F, the image of F under the second algebraic K-theory functor. A more complete discussion of this subject can be found in Milnor's book [57].

15.5. Division Algebras of Prime Degree

The assumption that the field F contains a primitive n'th root of unity limits the usefulness of Proposition 15.4. However, if this hypothesis is omitted, the proposition isn't true. Albert has given an example of a non-cyclic division algebra of degree four that contains a subfield of the form $F(a^{1/4})$.

In this section we will prove that the converse of Lemma 15.4 is true for division algebras of prime degree p without any extra hypotheses. The strategy of the proof is to extend the field F by a p'th root of unity ζ, apply Proposition 15.4, and then use Corollary 15.1c to cancel $F(\zeta)$. The fact that $[F(\zeta):F]$ is prime to p is used several times in the proof; the failure of this property when ζ is a p^e'th root of unity is the main reason that the characterization of cyclic division algebras does not extend to algebras of prime power degree.

Throughout this section, p is a prime that is different from the characteristic of the field F. (Fields of characteristic p are treated in Exercise 3.) Let ζ be a primitive p'th root of unity in the algebraic closure of F. Denote $L = F(\zeta)$, and $m = [L:F]$. Since p is prime and $m < p$, the greatest common divisor of m and p is 1. The extension L/F is cyclic, say $\mathbf{G}(L/F) = \langle\tau\rangle$. The conjugates of ζ are powers of ζ, so that $\zeta^\tau = \zeta^k$ for a $k \in \mathbb{N}$ such that $1 \le k < p$. Since the order of τ is m, it follows that $k^m \equiv 1 \pmod{p}$ and $k^j \not\equiv 1 \pmod{p}$ for $1 \le j < m$.

Lemma. *If $b \in L - L^p$ satisfies $b^\tau/b^k \in L^p$ and $E = L(b^{1/p})$, then E/F is cyclic and $E = K \otimes L$, where K/F is cyclic of degree p.*

PROOF. By assumption, $b^\tau = b^k c^p$ for a suitable $c \in L$. Let $E = L(u)$, where $u^p = b$. Extend τ to E by the condition $u^\tau = u^k c$. This prescription defines an F-algebra automorphism of E because $(u^k c)^p = b^k c^p = b^\tau$. The assumption $b \notin L^p$ implies that $u \notin L$, so that there is an L-algebra automorphism σ of E that satisfies $u^\sigma = \zeta u$. Thus, $[E:L] = p$ and $\mathbf{G}(E/L) = \langle\sigma\rangle$. Since $u^{\tau\sigma} = (u^k c)^\sigma = \zeta^k u^k c = (\zeta u)^\tau = u^{\sigma\tau}$ and $\zeta^{\tau\sigma} = \zeta^k = \zeta^{\sigma\tau}$, the mappings σ and τ generate a commutative group G of F-algebra automorphisms of E whose fixed field is F. Thus, E/F is Galois with $\mathbf{G}(E/F) = G$. The order m of $\tau|L$ divides the order of τ; hence, τ^l has order m for some $l \in \mathbb{N}$. Therefore, $\langle\sigma\rangle \cap \langle\tau^l\rangle = \{1\}$ and $\langle\sigma,\tau^l\rangle = \langle\sigma\rangle \times \langle\tau^l\rangle$ has order mp. On the other hand, $|\mathbf{G}(E/F)| = [E:L][L:F] \leq pm$. Consequently, $\mathbf{G}(E/F) = \langle\sigma\rangle \times \langle\tau^l\rangle$ is cyclic, and by Lemma 15.3a, $E = K \otimes L$, where K is the fixed field of τ^l. Thus, K/F is cyclic of degree p. □

The converse of this lemma is true: if $b \in L - L^p$ is such that $L(b^{1/p})/F$ is a cyclic extension, then $b^\tau/b^k \in L^p$. (See Exercise 1.)

Proposition. *Let $D \in \mathfrak{S}(F)$ be a division algebra of prime degree p, where* char $F \neq p$. *The algebra D is cyclic if and only if D contains a subfield that is isomorphic to $F(a^{1/p})$ for some $a \in F - F^p$.*

PROOF. If D is cyclic, then D contains a subfield of the form $F(a^{1/p})$ by Lemma 15.4. In the proof of the converse result, we retain the notational conventions that preceded the lemma. Since $[L:F]$ is relatively prime to p, it follows from Proposition 13.4 that D^L is a division algebra of degree p over L. Moreover, $a \in F - F^p$ implies that $\mathbf{x}^p - a$ is irreducible over L. (See Exercise 2.) By assumption, $v^p = a$ for some $v \in D$. Thus, by Proposition 15.4, $D^L = \bigoplus_{j<p} u^j L(v)$, where $vu = uv\zeta$ and $u^p = b \in L^\circ$. It suffices to prove: b can be chosen so that

$$b^\tau/b^k \in L^p. \tag{1}$$

Indeed, by the lemma and Corollary 15.1c it will follow that $D^L = \left(\frac{a,b}{L,\zeta}\right) = \left(\frac{b,a^{-1}}{L,\zeta}\right) = (L(u),\sigma,a^{-1}) = (K \otimes L, \sigma, a^{-1}) \cong (K,\sigma,a^{-1})^L$ with K/L cyclic. If A is the cyclic algebra (K,σ,a^{-1}), then $[D][A]^{-1} \in \mathbf{B}(L/F)$ implies $[D]^m = [A]^m$ by Proposition 14.4a. Also, $[D]^p = 1 = [A]^p$ because $\operatorname{Deg} D = \operatorname{Deg} A = p$. Thus, $[D] = [A]$, and $D \cong A$ is cyclic by Proposition 12.5b. To simplify the notation, write $E = L(v)$. By Lemma 15.1, $(E,\sigma,b) \cong (E,\sigma,c)$ if $c/b \in N_{E/L}(E^\circ)$. Thus, the proof can be completed by producing $x \in E^\circ$ such that $c = bN_{E/L}(x)$ satisfies (1), that is, $c^\tau/c^k \in L^p$. Define $\rho = id_D \otimes \tau \in \operatorname{Aut}_F D^L$. Note that $\rho(v) = v$ because $v \in D$, and $\rho(d) = d^\tau$ for all

$d \in L$. In particular, $\rho(\zeta) = \zeta^\tau = \zeta^k$. Thus, $(\rho u)^{-1}v(\rho u) = \rho(u^{-1}vu) = \rho(v^\sigma) = \rho(\zeta v) = \zeta^k v = u^{-k}vu^k$, so that $(\rho u)u^{-k} = y \in \mathbf{C}_{D^L}(E) = E$. Consequently, $b^\tau = (\rho u)^p = (yu^k)^p = u^{kp}y^{\sigma^k + \sigma^{2k} + \cdots + \sigma^{pk}} = b^k N_{E/L}(y)$, since $\{k, 2k, \ldots, pk\}$ is a complete system of residues modulo p. A similar calculation gives the more general result

$$b^{\tau^i} = b^{k^i} N_{E/L}(y_i) \quad \text{for } 0 \le i < m, \tag{2}$$

where $y_i \in E^\circ$. Choose $l, n \in \mathbb{N}$ so that $kl \equiv 1 \equiv mn \pmod p$, and define $c = (\prod_{0 \le i < m} (b^{\tau^i})^{l^i})^n$. By (2), $c = (\prod_i b^{k^i l^i})^n N_{E/L}(z) = be^p N_{E/L}(z) = bN_{E/L}(ez)$, where $e \in L^\circ$, $z \in E^\circ$. Moreover, since $l^m \equiv l^m k^m \equiv 1^m = 1 \pmod p$, there exist $r, s \in \mathbb{Z}$ such that $l^m = 1 + rp$ and $lk = 1 + sp$. Consequently,

$$b^{\tau^m l^m} = b^{l^m} = bb^{rp},$$

$$(c^\tau)^l = (\textstyle\prod_{0 \le i < m} (b^{\tau^{i+1}})^{l^{i+1}})^n = cb^{rnp},$$

and

$$c^\tau / c^k = (b^{rnk}/c^{\tau s})^p \in L^p. \qquad \square$$

The proposition can be reformulated and generalized somewhat.

Corollary. *If n is a square-free natural number that is not divisible by* char F, *and if $D \in \mathfrak{S}(F)$ is a division algebra of degree n, then D is cyclic if and only if there exists $a \in F$ such that $F(a^{1/n})$ splits D.*

Proof. If D is cyclic, then such a splitting field exists by Lemma 15.4 and Corollary 13.3. For the proof of the converse, let $n = p_1 \cdots p_r$, where the p_i are distinct primes. The Primary Decomposition Theorem yields $D = D_1 \otimes \cdots \otimes D_r$, where each D_i is a division algebra of degree p_i. The field $E_i = F(a^{1/p_i})$ must split D_i; otherwise $\operatorname{Ind} D_i^{E_i} = p_i$, in which case $F(a^{1/n})$ would not split D_i since p_i doesn't divide $[F(a^{1/n}) : E_i]$. Thus, $[E_i : F] = p_i$. By Corollary 13.3, D_i contains a subfield that is isomorphic to E_i. Therefore, D_i is cyclic according to the proposition. Corollary 15.3 shows that D is also cyclic. $\square$

Exercises

1. Prove the converse of the lemma: if $L = F(\zeta)$ and $E = L(b^{1/p})$ is such that E/F is cyclic, then $b^\tau/b^k \in L^p$. Hint. Let $v \in E$ satisfy $v^p = b$. Use $\tau\sigma = \sigma\tau$ to show that $v^\tau/v^k \in L$.

2. Prove that if $a \in F - F^p$ and L/F is an extension such that $p \nmid [L : F]$, then $\mathbf{x}^p - a$ is irreducible over L. Hint. Otherwise, there is an extension K/L of degree $m < p$ such that $x^p = a$ for some $x \in K$. Obtain a contradiction from $N_{K/F}(a) = N_{K/F}(x)^p$.

3. The purpose of this problem is to show that the proposition is also true when char $F = p$. Throughout the problem assume that F is a field of prime characteristic p.
   (a) Suppose that $\alpha \in M_p(F)$ satisfies $\alpha^p = \iota_p a$, where $a \in F - F^p$. Prove that

$\beta \in M_p(F)$ exists such that $\beta\alpha = \alpha(\beta + 1)$. Hint. Use Exercise 2 of Section 12.6 to reduce the problem to the case

$$\alpha = \begin{bmatrix} 0 & 0 & \cdots & 0 & a \\ 1 & 0 & \cdots & 0 & 0 \\ 0 & 1 & \cdots & 0 & 0 \\ & & \cdots & & \\ 0 & 0 & \cdots & 1 & 0 \end{bmatrix}. \quad \text{Try } \beta = \begin{bmatrix} 1 & 0 & \cdots & 0 & 0 \\ 0 & 2 & \cdots & 0 & 0 \\ & & \cdots & & \\ 0 & 0 & \cdots & p-1 & 0 \\ 0 & 0 & \cdots & 0 & 0 \end{bmatrix}.$$

(b) Suppose that $D \in \mathfrak{S}(F)$ is a division algebra of degree p that contains an element x such that $x^p = a \in F$, where $\mathbf{x}^p - a$ is irreducible in $F[\mathbf{x}]$. Prove that there exists $y \in D$ such that $yx = x(y + 1)$.

Hint. Let K be a separable extension of F that splits D. Show that $\mathbf{x}^p - a$ is irreducible in $K[\mathbf{x}]$. Let $\phi: D \to M_p(K)$ be an injective F-algebra homomorphism such that $M_p(K) = \phi(D)K$. Use the result of (b) to find $\beta \in M_p(K)$ such that $\beta\phi(x) = \phi(x)(\beta + 1)$. Write $\beta = \phi(y) + \phi(y_2)c_2 + \cdots + \phi(y_p)c_p$, where $1, c_2, \ldots, c_p$ is an F-basis of K. Show that y does the job.

(c) Suppose that $D \in \mathfrak{S}(F)$ is a division algebra of degree p. Prove that D is cyclic if and only if D contains a maximal subfield of the form $F(a^{1/p})$, $a \in F$.

Hint. Let $x \in D$ satisfy $x^p = a$. Let $y \in D$ satisfy $yx = x(y + 1)$ as in (b). Show that $F(y)/F$ is cyclic of degree p.

15.6. Division Algebras of Degree Three

A division algebra of degree two is necessarily cyclic because separable quadratic extensions are cyclic. In this section we will prove that every division algebra of degree three is cyclic. This result is another one of Wedderburn's fundamental contributions to the theory of associative algebras.

Theorem. *If $D \in \mathfrak{S}(F)$ is a division algebra of degree three, then D is cyclic.*

We will prove this result for the fields F with char $F \neq 3$. The proof for fields of characteristic 3 is outlined in the Exercise 2.

By Corollary 15.5 it is sufficient to show that D has a splitting field of the form $F(a^{1/3})$, where $a \in F - F^3$. This splitting field will be found among the subfields of D^L, where L is a quadratic extension of F such that D^L is cyclic. Here are the details of the proof.

Let K be a maximal separable subfield of D. By Proposition 13.5, K is strictly maximal in D, that is, $[K:F] = 3$. Since K/F is separable, there is an element $v \in D$ such that $K = F(v)$. Moreover, v can be chosen so that its minimum polynomial over F has the form $\Phi(\mathbf{x}) = \mathbf{x}^3 + b_1\mathbf{x} + b$ because char $F \neq 3$. Let E be a splitting field of Φ over F with $K \subseteq E$; define $G = \mathbf{G}(E/F)$. We can assume that G is not abelian; otherwise K/F is cyclic and the proof is finished. Thus, $[E:F] = 6$ and G is the group of all permutations of the three roots of Φ. If we let $\mathbf{G}(E/K) = \langle\tau\rangle$ and choose σ to be a generator of the (normal) Sylow 3-subgroup of G, then $G = \langle\sigma,\tau\rangle$, τ has order 2, σ has order 3, and $\tau\sigma\tau = \sigma^2$. Let L be the fixed field of σ; L/F is Galois of degree 2, and $\mathbf{G}(L/F) = \langle\tau|L\rangle$, $\mathbf{G}(E/L) = \langle\sigma\rangle$. Plainly, $v \in E - L$, so that $E = L(v)$ and the three roots of Φ are v, v^σ, and v^{σ^2}. In particular,

$$v + v^\sigma + v^{\sigma^2} = 0, \quad \text{and} \tag{1}$$

$$vv^\sigma v^{\sigma^2} = b \in F. \tag{2}$$

The algebra D^L is cyclic because it contains the strictly maximal subfield $E = L(v)$ and E/L is cyclic. Thus, $D^L = E \oplus uE \oplus u^2E$, where $u^{-1}yu = y^\sigma$ for all $y \in E$. Define $\rho = id_D \otimes (\tau|L) \in \mathrm{Aut}_F D^L$. Since $v^\tau = v$, it is clear that $\rho(y) = y^\tau$ for all $y \in E$. Hence, $v\rho(u) = \rho(v)\rho(u) = \rho(vu) = \rho(uv^\sigma) = \rho(u)\rho(v^\sigma) = \rho(u)v^{\sigma\tau} = \rho(u)v^{\tau\sigma\tau} = \rho(u)v^{\sigma^2}$, and $c\rho(u) = \rho(u)c$ for all $c \in L = \mathbf{Z}(D^L)$. Therefore, $w = \rho(u)u^{-1}$ satisfies $w^{-1}yw = y^\sigma$ for all $y \in E$; and $\rho(w) = \rho^2(u)\rho(u)^{-1} = u\rho(u)^{-1} = (\rho(u)u^{-1})^{-1} = w^{-1}$, since $\tau^2 = id_E$ implies $\rho^2 = id$. These observations and Proposition 15.1a give

$$vw = wv^\sigma, \quad v^\sigma w = wv^{\sigma^2}, \quad v^{\sigma^2}w = wv, \quad \text{and} \tag{3}$$

$$w^3 = d \in L, \quad \text{where } d^\tau = d^{-1}. \tag{4}$$

Define $z = (1 + w + w^{-1})v$. A straightforward computation using (1), (2), (3), and (4) leads to the result $z^3 = a$, where $a = b(d + d^\tau - 2) \in F$. The cubic polynomial $\mathbf{x}^3 - a$ is irreducible over L. Otherwise, $z \in E$ because E/L is Galois. Then (3) and (4) yield $w = d(v - v^\sigma)(v^{\sigma^2} - v)^{-1} \in E$. However, this conclusion contradicts (3) because the roots of Φ are distinct. Thus, $L(z)$ is strictly maximal in D^L, so that $L(z)$ splits D^L. Since $[L:F] = 2$, it is clear from Proposition 13.4 that $F(z)$ splits D. Thus, D is cyclic by Corollary 15.5.

Corollary. *If $D \in \mathfrak{S}(F)$ is a division algebra of degree six, then D is cyclic.*

The corollary follows from the proposition and Corollary 15.3 when $\operatorname{char} F \neq 3$. The case in which $\operatorname{char} F = 3$ is handled by Exercise 2.

Exercises

1. If $z = (1 + w + w^{-1})v$ is defined as it was in the proof of the proposition, show that $z^2 = v^\sigma v^{\sigma^3}(w(d^{-1} - 1) + w^{-1}(d - 1))$, and deduce that $z^3 = b(d^{-1} + d - 2) = b(d + d^\tau - 2) \in F$.

2. Prove that the proposition is also true when $\operatorname{char} F = 3$. Hint. Follow the general pattern of the proof for the case $\operatorname{char} F \neq 3$ with one change: define $z = w + w^{-1}$. Then $z^3 = d + d^\tau \in F$, so that the result of Exercise 3, Section 15.5 is applicable.

15.7. A Non-cyclic Division Algebra

The first division algebras of degree greater than 2 were found by L. E. Dickson in 1914. They were cyclic algebras of the form $\left(\frac{a,b}{F,\zeta}\right)$ that were defined in Section 15.4. For eighteen years after Dickson's paper, the existence of division algebras that are not cyclic was an open problem. In a 1932 paper, A. A. Albert gave an example of such an algebra. He followed this work with constructions of non-cyclic algebras that have various other properties. In this section, we will describe Albert's first example.

Let $F = K(u,v)$, where u and v are algebraically independent over a totally ordered field K. For instance, K can be any subfield of $\mathbb{R}$. Define $D_1 = \left(\frac{u,-1}{F}\right)$, $D_2 = \left(\frac{-u,v}{F}\right)$, and $A = D_1 \otimes D_2$.

Theorem. *A is a division algebra that is not cyclic.*

There are four ingredients in the proof of this statement. The first two of these technical preliminaries are general facts about cyclic extensions of degree four. They don't depend on the special form of F.

If -1 is not a sum of squares in F, and E/F is a cyclic extension of degree four, then $-1 \notin E^2$. (1)

Let E/F be a cyclic extension with $[E:F] = 4$. Suppose that $-1 \notin E^2$. If L is the unique subfield of E such that $[L:F] = 2$, then $L \cong F((r^2 + s^2)^{1/2})$ for some $r, s \in F$. (2)

The other two facts that we need are special properties of $F = K(u,v)$. In particular, they depend on the assumption that K is totally ordered. If $p \in K[u,v]$, we will denote by $\deg_u p$ and $\deg_v p$ the degrees of p considered as a polynomial in u and v respectively.

If $p = r_1^2 + \cdots + r_k^2$ and $q = s_1^2 + \cdots + s_l^2$, where the r_i and s_j are non-zero members of $K[u,v]$, then $p \neq 0$, $p \pm uq \neq 0$, and the degrees $\deg_u p$, $\deg_v p$, $\deg_v p + uq$, and $\deg_v p - uq$ are even. (3)

If $L = F((r^2 + s^2)^{1/2})$ with $r, s \in K[u,v]$ and $r^2 + s^2 \notin F^2$, then A^L is a division algebra. (4)

In particular, (4) implies that A is a division algebra. If A is cyclic, then by (1), (2), and (3), there is a subfield L of A with $[L:F] = 2$, such that $L = F((r^2 + s^2)^{1/2})$ for suitable $r, s \in K[u,v]$. However, for such an extension L/F, (4) implies that A^L is a division algebra. By Corollary 13.4, A^L

cannot be a division algebra if L is a subfield of A that properly contains F. Hence, A is not cyclic.

The proofs of (1), (2), and (3) are fairly easy. They are left as exercises. The rest of this section is devoted to the proof of (4).

Lemma. *Let F be a field with char $F \neq 2$, and suppose that $a_1, b_1, a_2, b_2 \in F^\circ$. Define $D_1 = \left(\frac{a_1, b_1}{F}\right)$, $D_2 = \left(\frac{a_2, b_2}{F}\right)$, and $A = D_1 \otimes D_2$. The following properties are equivalent.*

(i) *A is a division algebra.*
(ii) *If E_i is a maximal subfield of D_i for $i = 1, 2$, then $E_1 \not\cong E_2$.*
(iii) *If $x_i, y_i, z_i \in F$ satisfy $a_1x_1^2 + b_1y_1^2 - a_1b_1z_1^2 = a_2x_2^2 + b_2y_2^2 - a_2b_2z_2^2$, then $x_1 = y_1 = z_1 = x_2 = y_2 = z_2 = 0$.*

Proof. The equivalence of (ii) and (iii) is a consequence of two observations: $F(\sqrt{c}) \cong F(\sqrt{d})$ if and only if $c/d \in F^2$; the maximal subfields of a quaternion algebra D have the form $F((-v(z))^{1/2})$ where $z \in D$ is a pure quaternion and v is the norm of D. (See Exercise 4, Section 13.1.) If (ii) fails, then c, d, and e can be found in F so that $D_1 = \left(\frac{c, d}{F}\right)$, $D_2 = \left(\frac{c, e}{F}\right)$, and $A \sim \left(\frac{c, de}{F}\right)$ has index 2 by Corollary 15.1b. Thus, (i) implies (ii). (A more general form of this implication is given in Exercise 1.) Assume that (ii) is satisfied. By Corollary 13.4, this hypothesis implies that if E_i is a subfield of D_i for $i = 1, 2$, then $E_1 \otimes D_2$ and $D_1 \otimes E_2$ are division algebras. To prove that A is a division algebra, it will therefore suffice to show that if $z \neq 0$ in A, then there exists $u \in A$ such that zu is a non-zero element of $E_1 \otimes D_2$ or $D_1 \otimes E_2$ for suitable subfields E_i of D_i. Let 1, **i**, **j**, and **ij** be the quaternion units of D_1. Thus, $A = D_2 \oplus \mathbf{i}D_2 \oplus \mathbf{j}D_2 \oplus \mathbf{ij}D_2 = F(\mathbf{i})D_2 \oplus \mathbf{j}(F(\mathbf{i})D_2)$. Write $z = w_0 + \mathbf{j}w_1$ with $w_0, w_1 \in F(\mathbf{i})D_2$. If $w_1 = 0$, then $z \in F(\mathbf{i})D_2$ is a unit. Otherwise, $zw_1^{-1} = w_0w_1^{-1} + \mathbf{j}$. Thus, we can assume that $z = w + \mathbf{j}$ with $w = x + \mathbf{i}y \in F(\mathbf{i})D_2, x, y \in D_2$. Since $(w + \mathbf{j})(\mathbf{j}^{-1}w\mathbf{j} - \mathbf{j}) = w\mathbf{j}^{-1}w\mathbf{j} - \mathbf{j}^2 = x^2 - ay^2 - b + \mathbf{i}(yx - xy) \in F(\mathbf{i})D_2$, the proof is finished unless $x^2 - ay^2 - b = 0$ and $yx = xy$. If $y \in F$, then $z = x + \mathbf{i}y + \mathbf{j} \in F(\mathbf{i}y + \mathbf{j}) \otimes D_2 \subseteq A^\circ$. If $y \notin F$, then $yx = xy$ implies $x \in \mathbf{C}_{D_2}(F(y)) = F(y)$. In this case, $z \in D_1 \otimes F(y) \subseteq A^\circ$. □

We now prove (4). Let $\Phi = \Phi_1 - \Phi_2$, where Φ_1 and Φ_2 are the quadratic forms associated with D_1 and D_2 in Section 1.7. Explicitly, $\Phi(\mathbf{x}_1, \ldots, \mathbf{x}_6) = u\mathbf{x}_1^2 - \mathbf{x}_2^2 + u\mathbf{x}_3^2 + u\mathbf{x}_4^2 - v\mathbf{x}_5^2 - uv\mathbf{x}_6^2$. By the lemma, it is sufficient to prove that if $\vec{x} = (x_1, \ldots, x_6) \in F^6$ and $\vec{y} = (y_1, \ldots, y_6) \in F^6$ satisfy $\Phi(\vec{x} + (r^2 + s^2)^{1/2}\vec{y}) = 0$, then $\vec{x} = \vec{y} = (0, \ldots, 0)$. By homogeneity, it can be assumed that the x_i and y_j are in $K[u,v]$. If Θ is the bilinear form obtained by polarizing Φ, then $\Phi(\vec{x}) + (r^2 + s^2)\Phi(\vec{y}) + 2(r^2 + s^2)^{1/2}\Theta(\vec{x},\vec{y}) = \Phi(\vec{x} + (r^2 + s^2)^{1/2}\vec{y}) = 0$. The hypothesis $r^2 + s^2 \notin F^2$ yields $\Phi(\vec{x}) +$

$(r^2 + s^2)\Phi(\vec{y}) = 0$; that is, $up_1 - p_2 + up_3 = -up_4 + vp_5 + uvp_6$, where $p_i = x_i^2 + (ry_i)^2 + (sy_i)^2$. If both sides of the equation are 0, that is, $p_2 = u(p_1 + p_3)$ and $up_4 = v(p_5 + up_6)$, then $p_1 = p_2 = p_3 = p_4 = p_5 = p_6 = 0$ by (3). Otherwise, $v(p_5 + up_6) = u(p_1 + p_3 + p_4) - p_2$; and (3) implies $p_5 = p_6 = 0$ and $p_2 = u(p_1 + p_3 + p_4)$, since $\deg_v v(p_5 + up_6)$ is odd and $\deg_v u(p_1 + p_3 + p_4) - p_2$ is even unless $p_5 = p_6 = 0$. Similarly, $p_2 = u(p_1 + p_3 + p_4)$ yields $p_2 = p_1 = p_3 = p_4 = 0$. Since r and s are not zero, one more application of (3) gives the required conclusion that $\vec{x} = \vec{y} = (0, \ldots, 0)$.

Exercises

1. Let $D_1, D_2 \in \mathfrak{S}(F)$ be division algebras that respectively contain maximal subfields E_1 and E_2 such that E_1 and E_2 are not linearly disjoint. Prove that $D_1 \otimes D_2$ is not a division algebra. Hint. The assumption that E_1 and E_2 are not linearly disjoint implies the existence of a compositum K of E_1 and E_2 such that $[K:F] < [E_1:F][E_2:F] = \operatorname{Deg} D_1 \otimes D_2$. Show that K splits $D_1 \otimes D_2$.

2. Prove the statement (3).

3. Prove the statement (1). Hint. Otherwise, $E = F(i,u)$, where $u^2 = x \in F(i)$, $i^2 = -1$, $\mathbf{G}(E/F(i)) = \langle\sigma\rangle$ with $u^\sigma = -u$, $\mathbf{G}(E/F) = \langle\tau\rangle$, and $\sigma = \tau^2$. Show that $u^\tau u \in F(i)$ and $(u^\tau u)^\tau = -u^\tau u$, hence $u^\tau u = ci$ for some $c \in F$. If $x = a + ib$, derive the contradiction $-1 = (a/c)^2 + (b/c)^2$.

4. Prove the statement (2). Hint. Let $K = E \otimes F(i) = E(i)$, where $i^2 = -1$. Show that K is a field, K/F is Galois, $\mathbf{G}(K/F) = \mathbf{G}(K/E) \times \mathbf{G}(K/F(i)) = \langle\tau\rangle \times \langle\sigma\rangle$, $K = F(i,w)$ with $w^4 = x \in F(i)$, $w^\sigma = iw$, $i^\tau = -i$. Deduce from $\sigma\tau = \tau\sigma$ that $w^\tau w \in F$, say $w^\tau = aw^{-1}$. Show that L is the fixed field of $\langle\sigma^2,\tau\rangle$, and deduce that the elements of L have the form $y_0 + y_2 w^2$, where $y_0^\tau = y_0 \in F$ and $y_2^\tau = y_2 x a^{-2}$. Prove that $(y_2 w^2)^2 = r^2 + s^2$ for suitable $r, s \in F$. Finally, note that if $y_2 w^2, z_2 w^2 \in L$, then $y_2/z_2 \in F$.

5. Let D_1 and D_2 be quaternion algebras that satisfy the hypotheses of the lemma. Prove that if K/F is a quadratic extension, then $D_1^K \cong D_2^K$. This fact can also be proved using some fairly deep properties of quadratic forms. It is then possible to give an easy proof that (ii) implies (i) in the lemma.

Notes on Chapter 15

The material in this chapter is about 50 or 60 years old. It seems surprisingly youthful. Most of the results that are presented here are due to Albert, Dickson, and Wedderburn. Except for the example in Section 15.7, the exposition is based on Albert's book [3]. Section 15.7 is a rewrite of Albert's paper [2].

We conclude these notes with a survey of what is known about cyclic division algebras. In this chapter it has been shown that all division algebras of degrees two, three, and six are cyclic, but there are non-cyclic algebras of degree four. Albert has proved (in [1], for instance) that every division algebra of degree four over F is a crossed product. (See Exercise 7, Section 20.8.) Little is known about division algebras of degree p for primes p greater than three. Brauer proved that if $D \in \mathfrak{S}(F)$ is a division algebra with $\operatorname{Deg} D = 5$, then there is a solvable extension E/F of degree 12 such that D^E is cyclic; it is still not known if such a D is cyclic, or even a crossed product. Amitsur has proved that if n is divisible by 8 or the square of an odd prime, then there are division algebras of degree n that are not crossed products. We will prove Amitsur's Theorem in Chapter 20. Much more is known about the division algebras over special kinds of fields. Trivially, if $\mathbf{B}(F) = \{1\}$, then all questions about the division algebras in $\mathfrak{S}(F)$ evaporate. However, there are important classes of fields such that $\mathbf{B}(F)$ is not trivial and all of the division algebras in $\mathfrak{S}(F)$ are cyclic. We will prove (in Chapters 17 and 18) that this is the case when F is a local field or a number field. It would be interesting to have a description of the fields F such that all the division algebras in $\mathfrak{S}(F)$ are cyclic, but such an objective is now out of sight. There is a related problem that seems more tractable: for which fields F is $\mathbf{B}(F)$ generated by the equivalence classes of cyclic algebras? It has been conjectured that all fields have this property.

CHAPTER 16
Norms

A finite dimensional central simple F-algebra can be thought of as a non-commutative analogue of a finite field extension. If we adopt that viewpoint, it is natural to look for analogues of the useful ideas in field theory. This chapter is concerned with the counterpart for central simple algebras of the norm of a field extension. These mappings are called reduced norms.

The definition and the basic properties of the reduced norm are presented in the first half of the chapter. The last three sections use the norm to get some information about the multiplicative structure of central simple algebras. This development leads us to one of the active frontiers of research on central simple algebras: the investigation of the Reduced Whitehead Groups of algebras.

The reduced norm will reappear in later chapters. The most impressive application of the norm will be its use in Chapter 19 to prove Tsen's Theorem.

16.1. The Characteristic Polynomial

This section presents an array of candidates for the role of the norm of an associative algebra. It is shown that for the class of central simple algebras, all of the norms can be obtained as powers of a single one, the reduced norm.

It is useful to generalize the definition in Section 5.5 of a representation of an F-algebra A. If K is an extension of the field F, n is a natural number, and ϕ is an F-algebra homomorphism of A to $M_n(K)$, then ϕ will be called a *representation* of A. For $K = F$ this concept agrees with Definition 5.5. The advantage of adopting a more liberal definition stems from our results on splitting fields in Section 13.2. On the basis of Proposition 13.2a, we will call an F-algebra homomorphism $\phi: A \to M_n(K)$ a *splitting representation* of the finite dimensional central simple F-algebra A if $n = \operatorname{Deg} A$.

Definition. Let $\phi: A \to M_n(K)$ be a representation of the F-algebra A. The *ϕ-characteristic polynomial* of an element $y \in A$ is

$$X_\phi(y,\mathbf{x}) = \det(\iota_n\mathbf{x} - \phi(y)).$$

The *ϕ-trace* and *ϕ-norm* of y are $\tau_\phi(y) = \operatorname{tr}\phi(y)$ and $\nu_\phi(y) = \det\phi(y)$.

Two special cases of this definition are familiar. If $A = M_n(F)$ and ϕ is the identity homomorphism, then $X_\phi(\alpha,\mathbf{x})$ is the characteristic polynomial of the matrix α, $\nu_\phi(\alpha) = \det\alpha$, and $\tau_\phi(\alpha) = \operatorname{tr}\alpha$. In this case, ϕ is a splitting representation because $\operatorname{Deg} M_n(F) = n$. The other well known case of the definition occurs when A is a finite field extension of F and ϕ is a matrix representation that corresponds to the left regular representation of A. In this situation ν_ϕ is the field norm $N_{A/F}$ and τ_ϕ is the trace mapping $T_{A/F}$.

It is clear from the definitions that

$$X_\phi(y,\mathbf{x}) = \mathbf{x}^n - \tau_\phi(y)\mathbf{x}^{n-1} + \cdots + (-1)^n\nu_\phi(y). \tag{1}$$

This identity enables us to deduce many facts about the trace and the norm from the properties of characteristic polynomials.

Another useful observation is a familiar property of determinants.

$$\text{If } \alpha \in M_n(K) \text{ and } \gamma \in M_n(K)^\circ, \text{ then } \det(\iota_n\mathbf{x} - \gamma^{-1}\alpha\gamma) = \det(\iota_n\mathbf{x} - \alpha). \tag{2}$$

In fact, $\det(\iota_n\mathbf{x} - \gamma^{-1}\alpha\gamma) = \det(\gamma^{-1}(\iota_n\mathbf{x} - \alpha)\gamma) = (\det\gamma)^{-1}(\det(\iota_n\mathbf{x} - \alpha))(\det\gamma) = \det(\iota_n\mathbf{x} - \alpha)$.

Lemma. *Let $A \in \mathfrak{S}(F)$. If $\phi: A \to M_n(K)$ is a splitting representation of A and $\psi: A \to M_m(K)$ is an arbitrary representation of A, then $m = nk$ for some $k \in \mathbb{N}$ and $X_\psi = X_\phi^k$.*

Proof. By Proposition 9.2b, ϕ and ψ extend to K-algebra homomorphisms $\phi: A^K \to M_n(K)$, $\psi: A^K \to M_m(K)$ such that $\phi(y \otimes c) = \phi(y)c$, $\psi(y \otimes c) = \psi(y)c$; and the extension of ϕ is an isomorphism by Proposition 13.2a. Since $A^K \in \mathfrak{S}(K)$, ψ is injective, and the Double Centralizer Theorem implies that $M_m(K) = \psi(A^K) \otimes B$ for a suitable $B \in \mathfrak{S}(K)$. Therefore, $m = (\operatorname{Deg}\psi(A^K))(\operatorname{Deg} B) = nk$, where $k = \operatorname{Deg} B \in \mathbb{N}$. Define $\theta: A^K \to M_m(K)$ by $\theta(z) = \phi(z) \otimes \iota_k$. By the Noether–Skolem Theorem, there is an element $\gamma \in M_m(K)^\circ$ such that $\psi(z) = \gamma^{-1}\theta(z)\gamma$ for all $z \in A^K$. In particular, if $y \in A$, then $X_\psi(y,\mathbf{x}) = \det(\iota_m\mathbf{x} - \psi(y)) = \det(\iota_m\mathbf{x} - \theta(y)) = \det((\iota_n\mathbf{x} - \phi(y)) \otimes \iota_k) = X_\phi(y,\mathbf{x})^k$ by (2). □

Proposition. *Let $A \in \mathfrak{S}(F)$. If $\phi: A \to M_n(E)$ is a splitting representation of A, then $X_\phi \in F[\mathbf{x}]$. Moreover, if $\psi: A \to M_n(L)$ is another splitting representation of A, then $X_\psi = X_\phi$.*

Proof. Assume that E/F is Galois. If $\sigma \in \mathbf{G}(E/F)$, then σ determines an automorphism of $M_n(E)$ by $\sigma([c_{ij}]) = [c_{ij}^\sigma]$, and $\sigma\phi$ is a splitting representation of A. The lemma yields $X_\phi = X_{\sigma\phi} = \sigma(X_\phi)$, where σ operates on the coefficients of polynomials in $E[\mathbf{x}]$. Since E/F is Galois, it follows that $X_\phi \in F[\mathbf{x}]$. Let $\psi: A \to M_n(L)$ be another splitting representation. There is a field extension K/F and F-algebra homomorphisms $\chi: E \to K$ and $\theta: L \to K$. (See Exercise 1.) These mappings extend to F-algebra homomorphisms $\chi: M_n(E) \to M_n(K)$ and $\theta: M_n(L) \to M_n(K)$ by $\chi([c_{ij}]) = [\chi(c_{ij})]$ and $\theta([d_{ij}]) = [\theta(d_{ij})]$. The lemma implies $\theta(X_\psi) = X_{\theta\psi} = X_{\chi\phi} = \chi(X_\phi) = X_\phi$ because $X_\phi \in F[\mathbf{x}]$ and χ is an F-algebra homomorphism. Thus, $X_\psi = X_\phi \in F[\mathbf{x}]$. By Theorem 13.5, splitting representations $\phi: A \to M_n(E)$ such that E/F is Galois exist. This observation finishes the proof. □

The proposition enables us to simplify our notation. We will write $X_{A/F}$, $\tau_{A/F}$, and $\nu_{A/F}$ (or just X, τ, and ν when there is no danger of confusion) for X_ϕ, τ_ϕ, and ν_ϕ respectively, where ϕ is any splitting representation of A. If $y \in A$, then $X(y,\mathbf{x})$, $\tau(y)$, and $\nu(y)$ are called the *characteristic polynomial*, *trace*, and *reduced norm* of y.

Corollary a. *Let $A \in \mathfrak{S}(F)$ have degree n. If $\phi: A \to M_m(K)$ is a representation of A, then n divides m and $X_\phi = X^{m/n}$, $\tau_\phi = (m/n)\tau$, and $\nu_\phi = \nu^{m/n}$. In particular, $\tau_\phi(y)$, $\nu_\phi(y) \in F$ for all $y \in A$.*

Corollary b. *Let E/F be a field extension, and suppose that $A \in \mathfrak{S}(F)$, $B \in \mathfrak{S}(E)$. If $\theta: A \to B$ is an F-algebra homomorphism, then* $\operatorname{Deg} B = k \operatorname{Deg} A$ *for some $k \in \mathbb{N}$ and $X_{B/E}(\theta(y),\mathbf{x}) = X_{A/F}(y,\mathbf{x})^k$ for all $y \in A$. In particular, $X_{A^E/E}(y,\mathbf{x}) = X_{A/F}(y,\mathbf{x})$, $\tau_{A^E/E}(y) = \tau_{A/F}(y)$, and $\nu_{A^E/E}(y) = \nu_{A/F}(y)$ for all $y \in A$.*

Proof. Let $\operatorname{Deg} B = m$, $\operatorname{Deg} A = n$. Suppose that $\phi: B \to M_m(K)$ is a splitting representation of B. Since $\phi\theta: A \to M_m(K)$ is a representation of A, it follows from the lemma that $m = kn$ and $X_{B/E}(\theta(y),\mathbf{x}) = X_{\phi\theta}(y,\mathbf{x}) = X_{A/F}(y,\mathbf{x})^k$ for some $k \in \mathbb{N}$. □

Exercises

1. Let E/F and L/F be field extensions. Prove that there is an extension K/F such that both E and L are isomorphic as F-algebras to subfields of K that contain F. Hint. Let M be a maximal ideal of $E \otimes L$. Consider $(E \otimes L)/M$.

2. Let A be a finite dimensional F-algebra. For a fixed basis $\{w_1, w_2, \ldots, w_n\}$ of A and $y \in A$, define $\phi(y) \in M_n(F)$ by $[yw_1, yw_2, \ldots, yw_n] = [w_1, w_2, \ldots, w_n]\phi(y)$. That is, $\phi(y)$ is the matrix of λ_y relative to the basis $\{w_1, w_2, \ldots, w_n\}$. Prove the following statements.

   (a) $\phi: A \to M_n(F)$ is a representation of A, and X_ϕ does not depend on the choice of the basis $\{w_1, w_2, \ldots, w_n\}$. Thus, there is no ambiguity when we write X_λ for the ϕ-characteristic polynomial that is defined in this way.

(b) If E/F is a field extension and $y \in A$, then $X_\lambda(y,\mathbf{x})$ is the same polymomial when y is viewed as an element of A^E as it is for $y \in A$.

(c) If $A \in \mathfrak{S}(F)$ has degree n, then $X_\lambda = X^n$. As a challenge, prove this statement without using Corollary a, by appling the result (b) in the case that E splits A. In this situation, A^E can be identified with $M_n(E)$.

16.2. Computations

Explicit calculations of the characteristic polynomials, norms, and traces in central simple algebras are usually difficult. In this section we exhibit some cases in which formulas can be given for these quantities.

By 16.1(1), the norm can be recovered from the characteristic polynomial. However, it is sometimes easier to work directly with the norm. We begin this section by showing that the characteristic polynomial can be viewed as a norm.

Lemma a. *Let $\phi: A \to M_n(K)$ be a representation of the F-algebra A. If $E = F(\mathbf{x})$ and $\psi = \phi \otimes id_E: A^E \to M_n(K(\mathbf{x}))$, where $M_n(K) \otimes F(\mathbf{x})$ is identified with a subalgebra of $M_n(K(\mathbf{x}))$, then $X_\phi(y,\mathbf{x}) = \nu_\psi(\mathbf{x} - y)$ for all $y \in A$. In particular, if $A \in \mathfrak{S}(F)$, then $X_{A/F}(y,\mathbf{x}) = \nu_{A^E/E}(\mathbf{x} - y)$.*

Proof. By definition, $\psi(y) = \phi(y)$ for all $y \in A$, and $\psi(\mathbf{x}) = \phi(1) \otimes \mathbf{x} = \iota_n\mathbf{x}$. Thus, $\nu_\psi(\mathbf{x} - y) = \det(\psi(\mathbf{x} - y)) = \det(\iota_n\mathbf{x} - \phi(y)) = X_\phi(y,\mathbf{x})$. □

If K/F is a finite field extension, then it follows from the lemma that the characteristic polynomial of an element $d \in K$ over F is $\mathrm{N}_{K(\mathbf{x})/F(\mathbf{x})}(\mathbf{x} - d)$. We will need a more general version of this observation.

Lemma b. *Assume that K/F is a field extension of degree r. If $\Phi(\mathbf{x}) = \mathbf{x}^m + d_1\mathbf{x}^{m-1} + \cdots + d_m \in K[\mathbf{x}]$, then $\Psi(\mathbf{x}) = N_{K(\mathbf{x})/F(\mathbf{x})}(\Phi(\mathbf{x}))$ is a monic polynomial of degree mr in $F[\mathbf{x}]$, and $\Psi(0) = N_{K/F}(d_m)$.*

Proof. Let $w_1, \ldots, w_r$ be a basis of K_F. Define $\phi: K \to M_r(F)$ by $d[w_1, \ldots, w_r] = [w_1, \ldots, w_r]\phi(d)$, that is, ϕ is the left regular matrix representation of K relative to $w_1, \ldots, w_r$. Since $w_1, \ldots, w_r$ is also a basis of $K(\mathbf{x})_{F(\mathbf{x})}$, it follows that $\Phi(\mathbf{x})[w_1, \ldots, w_r] = [w_1, \ldots, w_r](\iota_r\mathbf{x}^m + \phi(d_1)\mathbf{x}^{m-1} + \cdots + \phi(d_m))$ and $\Psi(\mathbf{x}) = \det(\iota_r\mathbf{x}^m + \phi(d_1)\mathbf{x}^{m-1} + \cdots + \phi(d_m)) = \mathbf{x}^{mr} + \cdots + N_{K/F}(d_m)$. □

Proposition a. *Let $A \in \mathfrak{S}(F)$. If K is a subfield of A, then $k \in \mathbb{N}$ exists such that* $\operatorname{Deg} A = k[K:F]$, *$\tau(y) = kT_{K/F}(y)$, and $\nu(y) = N_{K/F}(y)^k$ for all $y \in K$.*

Proof. It was pointed out in Corollary 13.1a that $[K:F]$ divides $\operatorname{Deg} A$. Let $v_1, \ldots, v_m$ be a basis of ${}_KA$. If $u_1, \ldots, u_r$ is a basis of ${}_FK$, then $\{u_iv_j: 1 \le i \le r, 1 \le j \le m\}$ is a basis of ${}_FA$. Let ψ be the left regular matrix representation of K relative to $u_1, \ldots, u_r$; that is, $y[u_1, \ldots, u_r] = [u_1, \ldots,$

$u_r]\psi(y)$ for $y \in K$. Since $\psi(y) \in M_r(F)$ and $F = \mathbf{Z}(A)$, it follows that $y[u_1v_j, \ldots, u_rv_j] = [u_1v_j, \ldots, u_rv_j]\psi(y)$ for $1 \le j \le m$. Thus, the left regular matrix representation of A relative to $\{u_iv_j\}$ has the form of a matrix tensor product $\phi(y) = \psi(y) \otimes \iota_m$ if $y \in K$. By Corollary 16.1a, $kr = \operatorname{Deg} A$ divides mr, and $\nu(y)^{m/k} = \nu_\phi(y) = (\det \psi(y))^m = N_{K/F}(y)^m$. Up to a root of unity, this is the last statement in the proposition. When A, K, and F are replaced by $B = A^{F(\mathbf{x})}$, $K(\mathbf{x})$, and $F(\mathbf{x})$ respectively, Lemma a gives $X_{A/F}(y,\mathbf{x})^{m/k} = \nu_{B/F(\mathbf{x})}(\mathbf{x} - y)^{m/k} = N_{K(\mathbf{x})/F(\mathbf{x})}(\mathbf{x} - y)^m = X_\psi(y,\mathbf{x})^m$. Since $X_{A/F}(y,\mathbf{x})$ and $X_\psi(y,\mathbf{x})$ are monic polynomials in $\mathbf{x}$, it follows that $X_{A/F}(y,\mathbf{x}) = X_\psi(y,\mathbf{x})^k$. In particular, $\tau_{A/F}(y) = kT_{K/F}(y)$ and $\nu_{A/F}(y) = N_{K/F}(y)^k$ by 16.1(1). □

Corollary. *If $A \in \mathfrak{S}(F)$ is a division algebra of degree n, and if $y \in A$ has the minimum polynomial $\Phi(\mathbf{x})$ over F, then $\nu(y) = (-1)^n\Phi(0)^{n/r}$, where $r = \deg \Phi$.*

In fact, $K = F(y)$ is a subfield of the division algebra A, and $[K:F] = r$. The same proof works when the hypothesis that A is a division algebra is replaced by the assumption that the minimum polynomial of y is irreducible. Indeed, this is exactly the case in which y belongs to a subfield of A.

When A is a crossed product, then it is possible to give an explicit splitting representation of A and hence a formula for the norm.

Proposition b. *Let $A = (E,G,\Phi)$ be a crossed product of degree n, where $G = \mathbf{G}(E/F)$, $A = \bigoplus_{\sigma \in G} u_\sigma E$, $u_\sigma^{-1} d u_\sigma = d^\sigma$ for all $\sigma \in G$, $d \in E$, and $u_{\sigma\tau}^{-1}u_\sigma u_\tau = \Phi(\sigma,\tau)$. The formula*

$$\phi\left(\sum_{\rho \in G} u_\rho c_\rho\right) = [d_{\sigma\tau}] \quad \text{with} \quad d_{\sigma\tau} = \Phi(\sigma\tau^{-1},\tau)c^\tau_{\sigma\tau^{-1}} \tag{1}$$

defines a splitting representation of A. In particular, if $A = (E,\sigma,a)$ is cyclic and $A = \bigoplus_{i<n} u^i E$ with $u^{-1}du = d^\sigma$ for all $d \in E$ and $u^n = a$, then

$$\phi\left(\sum_{i<n} u^i c_i\right) = \begin{bmatrix} c_0 & ac^\sigma_{n-1} & ac^{\sigma^2}_{n-2} & \cdots & ac^{\sigma^{n-1}}_1 \\ c_1 & c^\sigma_0 & ac^{\sigma^2}_{n-1} & \cdots & ac^{\sigma^{n-1}}_2 \\ \cdot & \cdot & \cdot & \cdots & \cdot \\ c_{n-1} & c^\sigma_{n-2} & c^{\sigma^2}_{n-3} & \cdots & c^{\sigma^{n-1}}_0 \end{bmatrix}, \tag{2}$$

and

$$\nu\left(\sum_{i<n} u^i c_i\right) = (-1)^{n-1}N_{E/F}(c_{n-1})a^{n-1} + \cdots + N_{E/F}(c_0). \tag{3}$$

Proof. If $y = \sum_{\rho \in G} u_\rho c_\rho$, then $\phi(y)$ is the matrix of λ_y relative to the basis $\{u_\sigma : \sigma \in G\}$ of A_E. The calculation that proves this assertion is left as Exercise 1. □

If $A = \left(\frac{a,b}{F}\right) = (F(\sqrt{a}),\sigma,b)$ is a quaternion algebra, then (3) takes the form $\nu(c_0 + \mathbf{i}c_1 + \mathbf{j}c_2 + \mathbf{ij}c_3) = N_{(F\sqrt{a})/F}(c_0 + c_1\sqrt{a}) - N_{F(\sqrt{a})/F}(c_2 -$

$c_3\sqrt{a})b = c_0^2 - ac_1^2 - bc_2^2 + abc_3^2$. That is, the reduced norm for a quaternion algebra is the norm that was introduced in Section 1.6.

A word of caution is needed: if $n > 2$, then the coefficients of a^i in (3) for $1 \leq i < n - 1$ are fairly complicated homogeneous polynomials in the various c_i and their conjugates. They are not simple expressions like $(-1)^i N_{E/F}(c_i)$. Exercise 2 illustrates this fact.

EXERCISES

1. Complete the proof of Proposition b.
2. Let F be a field with char $F \neq 3$ such that there is a primitive third root of unity ζ in F. Let $A = \left(\frac{a,b}{F,\zeta}\right)$ be the cyclic algebra of degree 3 that was defined in Section 15.4. Thus, A has an F-basis $\{u^i v^j : 0 \leq i,j < 3\}$ such that $vu = uv\zeta$, $u^3 = a \in F^\circ$, $v^3 = b \in F^\circ$. Prove that if $z = \sum_{0 \leq i,j < 3} u^i v^j c_{ij}$, then $\nu(z) = a^2(c_{20}^3 + bc_{21}^3 + b^2 c_{22}^3 - 3bc_{20}c_{21}c_{22}) + a[(c_{10}^3 + bc_{11}^3 + b^2 c_{12}^3 - 3bc_{10}c_{11}c_{12}) - 3(c_{00}c_{10}c_{20} + bc_{01}c_{11}c_{21} + b^2 c_{02}c_{12}c_{22}) - 3b\zeta(c_{00}c_{12}c_{21} + c_{01}c_{10}c_{22} + c_{02}c_{11}c_{20}) - 3b\zeta^2(c_{00}c_{11}c_{22} + c_{02}c_{10}c_{21} + c_{01}c_{12}c_{20})] + (c_{00}^3 + bc_{01}^3 + b^2 c_{02}^3 - 3bc_{00}c_{01}c_{02})$.

16.3. The Reduced Norm

In this section we will translate standard facts about matrices and determinants into statements about the reduced norm. The main result of this program is a norm criterion for an element of a central simple algebra to be a unit.

Lemma a. *Let $\phi\colon A \to M_n(K)$ be a representation of the F-algebra A. If $x, y \in A$ and $a, b \in F$, then*

(i) $\tau_\phi(xa + yb) = \tau_\phi(x)a + \tau_\phi(y)b$,
(ii) $\nu_\phi(xy) = \nu_\phi(x)\nu_\phi(y)$,
(iii) $\nu_\phi(a) = a^n$.

If $y_1, \ldots, y_m$ is a basis of A_F, then there is a homogeneous polynomial Φ of degree n in $\mathbf{x}_1, \ldots, \mathbf{x}_m$ with coefficients in K such that $\nu_\phi(\sum_{i=1}^m y_i a_i) = \Phi(a_1, \ldots, a_m)$.

PROOF. Equations (i), (ii), and (iii) reflect corresponding properties of the trace and determinant mappings of matrices. For example, $\nu_\phi(xy) = \det(\phi(xy)) = \det(\phi(x)\phi(y)) = \det\phi(x)\det\phi(y) = \nu_\phi(x)\nu_\phi(y)$. Let $\phi(y_i) = [b_{jk}^i] \in M_n(K)$. Then

$$\nu_\phi\left(\sum_{i=1}^m y_i a_i\right) = \det\left[\sum_{i=1}^m b_{jk}^i a_i\right]$$

$$= \sum_{\pi} \operatorname{sgn} \pi \left(\prod_{j=1}^{n} \left(\sum_{i=1}^{m} b^i_{j\pi(j)} a_i \right) \right)$$

$$= \Phi(a_1, \ldots, a_m),$$

where $\Phi \in K[\mathbf{x}_1, \ldots, \mathbf{x}_m]$ is homogeneous of degree n. □

Lemma b. *If $A \in \mathfrak{S}(F)$, and if $\phi: A \to M_n(K)$ is a representation of A, then every element of A is a root of its ϕ-characteristic polynomial.*

PROOF. By Corollary 16.1a, $X_\phi(y,\mathbf{x})$ has coefficients in F, so that

$$\phi(X_\phi(y,y)) = X_\phi(y,\phi(y))$$

because ϕ is an F-algebra homomorphism. By definition, $X_\phi(y,\mathbf{x})$ is the characteristic polynomial of the matrix $\phi(y)$. Thus, $X_\phi(y,\phi(y)) = 0$ by the Cayley–Hamilton Theorem for matrices. Finally, since A is simple, ϕ is injective. Hence, $X_\phi(y,y) = 0$. □

Proposition a. *Let $A \in \mathfrak{S}(F)$. An element $y \in A$ is a unit if and only if $\nu(y) \neq 0$.*

PROOF. If $y \in A^\circ$, then $\nu(y^{-1})\nu(y) = \nu(1) = 1$ by Lemma a. Thus, $\nu(y) \neq 0$. Assume that $\nu(y) \neq 0$. Then $X(y,\mathbf{x}) = \mathbf{x}^n + a_1\mathbf{x}^{n-1} + \cdots + a_n$ with $a_n = (-1)^n\nu(y) \neq 0$ by 16.1(1). Since $y^n + a_1 y^{n-1} + \cdots + a_n = 0$ by Lemma b, it follows that $-(y^{n-1} + a_1 y^{n-2} + \cdots + a_{n-1})a_n^{-1}$ is the inverse of y. □

Corollary a. *If ϕ is a representation of $A \in \mathfrak{S}(F)$, then A is a division algebra if and only if $\nu_\phi(y) \neq 0$ for all $y \in A - \{0\}$.*

This corollary follows from the proposition and Corollary 16.1a.

It follows from Lemma a and the proposition that the reduced norm is a group homomorphism from A° to F°. Since F° is commutative, the kernel of ν contains the commutator subgroup $A' = [A^\circ, A^\circ]$ of A°. Therefore, ν induces a homomorphism ν^{ab} from $A^{ab} = A^\circ/A'$ to F°: $\nu^{ab}(xA') = \nu(x)$.

Corollary b. *If $A \in \mathfrak{S}(F)$, then ν is a group homomorphism of A° to F° that induces a homomorphism $\nu^{ab}: A^{ab} \to F^\circ$.*

The groups $\operatorname{Ker} \nu^{ab} = \operatorname{Ker} \nu/A'$ and $\operatorname{Coker} \nu^{ab} = \operatorname{Coker} \nu$ are important invariants of central simple algebras. The kernel of ν^{ab} is encountered in algebraic K-theory; it is called the *Reduced Whitehead Group* of A, and it is denoted by $SK_1(A)$. Corollary b yields an exact sequence

$$1 \to SK_1(A) \to A^{ab} \to F^\circ \to \operatorname{Coker} \nu \to 1. \tag{1}$$

It is clear from Lemma a that if $\operatorname{Deg} A = n$, then $(F^\circ)^n \subseteq \operatorname{Im} \nu$. Therefore, $\operatorname{Coker} \nu$ is a homomorphic image of $F^\circ/(F^\circ)^n$. In particular, the exponent of

Coker ν divides n. The same statement will be proved for $SK_1(A)$ in Section 16.6.

We conclude this section with a proof that SK_1 and Coker ν are the object maps of functors.

Proposition b. *Let K/F be a field extension. Suppose that $A \in \mathfrak{S}(F)$, $B \in \mathfrak{S}(K)$, $\operatorname{Deg} A = m$, and $\operatorname{Deg} B = n$. An F-algebra homomorphism $\theta: A \to B$ induces group homomorphisms $\theta^{ab}: A^{ab} \to B^{ab}$, $\theta_1: SK_1(A) \to SK_1(B)$, and θ_2: Coker $\nu_{A/F} \to$ Coker $\nu_{B/K}$ such that the diagram*

$$\begin{array}{ccccccccc} 1 \to & SK_1(A) & \to & A^{ab} & \to & F^\circ & \longrightarrow & \operatorname{Coker}\nu_{A/F} & \to 1 \\ & \downarrow{\scriptstyle\theta_1} & & \downarrow{\scriptstyle\theta^{ab}} & & \downarrow{\scriptstyle\eta_{n/m}} & & \downarrow{\scriptstyle\theta_2} & \\ 1 \to & SK_1(B) & \to & B^{ab} & \to & K^\circ & \longrightarrow & \operatorname{Coker}\nu_{B/K} & \to 1 \end{array} \tag{2}$$

commutes, where η_k is the exponential mapping $a \mapsto a^k$.

PROOF. The composite homomorphism $A^\circ \xrightarrow{\theta} B^\circ \to B^{ab}$ has the kernel $\theta^{-1}(B')$. Thus, $A^\circ/\theta^{-1}(B')$ is isomorphic to a subgroup of the commutative group B^{ab}. Therefore, $A' \subseteq \theta^{-1}(B')$, and θ induces $\theta^{ab}: A^{ab} \to B^{ab}$ such that

$$\theta^{ab}(xA') = \theta(x)B'.$$

By Corollary 16.1b, $\nu_{B/K}^{ab}\theta^{ab}(xA') = \nu_{B/K}\theta(x) = \nu_{A/F}(x)^{n/m} = \eta_{n/m}\nu_{A/F}^{ab}(xA')$ for all $x \in A^\circ$. Thus, the middle square of (2) commutes. From this fact and the exactness of the rows in (2) it follows easily that there are unique homomorphisms θ_1 and θ_2 such that the whole diagram commutes. □

If θ: $A \to B$ and ψ: $B \to C$ are respectively F-algebra and K-algebra homomorphisms, then the construction in the proof of Proposition b shows that

$$(\psi\theta)^{ab} = \psi^{ab}\theta^{ab}, \quad (\psi\theta)_1 = \psi_1\theta_1, \quad \text{and} \quad (\psi\theta)_2 = \psi_2\theta_2. \tag{3}$$

In particular, SK_1 is a functor from the category $\mathfrak{S}(F)$ to the category of abelian groups.

EXERCISES

1. Complete the proof of Proposition b. That is, show that the exactness of rows and the commutativity of the center square in (2) implies the existence and the uniqueness of homomorphisms θ_1 and θ_2 so that the diagram commutes.

2. Compute Coker $\nu_{A/F}$ in the following cases.
   (a) F arbitrary, $A = M_n(F)$.
   (b) $F = \mathbb{R}$, $A = \mathbb{H}$.
   (c) $F = \mathbb{Q}$, $A = \left(\frac{-1,-1}{\mathbb{Q}}\right)$.
   (d) $F = \mathbb{Q}(\mathbf{x})$, $A = \left(\frac{-1,-1}{F}\right)$.

3. Let $A \in \mathfrak{S}(F)$. If $y \in A$, then the minimum polynomial of y over F is the monic polynomial Φ of least degree in $F[\mathbf{x}]$ such that $\Phi(y) = 0$. Prove that every $y \in A$ has a unique minimum polynomial Φ, and Φ divides $X(y,\mathbf{x})$ in $F[\mathbf{x}]$. Show that y is contained in a subfield K of A if and only if Φ is irreducible in $F[\mathbf{x}]$.

4. Let $D \in \mathfrak{S}(F)$ be a division algebra. Denote the F-algebra of polynomials with coefficients in D by $D[\mathbf{x}]$. (By definition, $\mathbf{x}y = y\mathbf{x}$ for all $y \in D$.) Define the degree $\deg \Phi$ of $\Phi \in D[\mathbf{x}] - \{0\}$ in the usual way. For $y \in D$ and $\Phi = x_0 + \mathbf{x}x_1 + \cdots + \mathbf{x}^n x_n \in D[\mathbf{x}]$, define $\Phi_l(y) = x_0 + yx_1 + \cdots + y^n x_n$. Prove the following statements.

(a) If $\Phi, \Psi \in D[\mathbf{x}] - \{0\}$, then $\deg \Phi\Psi = \deg \Phi + \deg \Psi$.

(b) If $\Phi \in D[\mathbf{x}]$ has degree $n \geq 1$ and $y \in D$, then there exists $\Psi \in D[\mathbf{x}]$ such that $\deg \Psi = n - 1$ and $\Phi = (\mathbf{x} - y)\Psi + \Phi_l(y)$. Deduce that $\Phi_l(y) = 0$ if and only if $\Phi = (\mathbf{x} - y)\Psi$ for some $\Psi \in D[\mathbf{x}]$.

(c) If $\Phi = \Psi X$ and $y \in D$ satisfies $\Phi_l(y) = 0 \neq \Psi_l(y)$, then $X_l(z^{-1}yz) = 0$ for some $z \in D^\circ$. Hint. Apply (b) to Φ and Ψ. Take $z = \Psi_l(y)$.

(d) If $\Psi \in D[\mathbf{x}] - \{0\}$ is monic of minimal degree with the property $\Psi_l(z^{-1}yz) = 0$ for all $z \in D$, then $\Psi \in F[\mathbf{x}]$; hence $\deg \Psi \geq [F(y) : F]$.

(e) If $\Phi \in F[\mathbf{x}] \subseteq D[\mathbf{x}]$ is the minimum polynomial over F of $y \in D$, then $\Phi = (\mathbf{x} - y)(\mathbf{x} - y_2) \cdots (\mathbf{x} - y_n)$, where the y_i are conjugates of y in D. Hint. Use (b), (c), and (d).

(f) If $\operatorname{Deg} D \geq 2$, then there is no total ordering of D such that sums and products of positive elements are positive. Hint. Find $y \in D$ such that the minimum polynomial of y has the form $\mathbf{x}^n + a_2\mathbf{x}^{n-2} + \cdots + a_n$. Show that if $y > 0$, then the conjugates of y are positive, and if $y < 0$, then the conjugates of y are negative. Obtain a contradiction.

16.4. Transvections and Dilatations

If D is a division algebra in $\mathfrak{S}(F)$ and $A = M_n(D)$, then the reduced norm $\nu_{A/F}$ is analogous to a determinant mapping. In this section and the next one we will construct a determinant for matrices over a division algebra that is closer in spirit to the usual determinant than the norm. The difference between the reduced norm and the determinant is measured to a large extent by the reduced Whitehead group and the cokernel of the norm.

The letter D denotes a division algebra that is not necessarily finite dimensional over the field F. Denote $M_n(D)^\circ$ by $GL_n(D)$. Thus, $GL_n(D)$ is the group of invertible n by n matrices with entries in D. In this section we tacitly assume that $n \geq 2$; most of the statements that make sense for $GL_1(D) = D^\circ$ are trivially true.

It is useful to introduce notation for two classes of n by n matrices. Let $1 \leq i \neq j \leq n$; define $\tau_{ij}(x) = \iota + \varepsilon_{ij}x$ for $x \in D$, and $\delta_i(x) = \sum_{k \neq i} \varepsilon_{kk} + \varepsilon_{ii}x = \iota + \varepsilon_{ii}(x - 1)$ if $x \in D^\circ$. If it is necessary to incorporate n into this notation we will write $\tau_{ij}^n(x)$ and $\delta_i^n(x)$ instead of $\tau_{ij}(x)$ and $\delta_i(x)$. The matrices $\delta_n(x)$ play a special part in the theory, and it is convenient to abbreviate $\delta_n(x)$ by $\delta(x)$.

Every $\tau_{ij}(x)$ and $\delta_i(x)$ belongs to $GL_n(D)$. In fact $\tau_{ij}(x)^{-1} = \tau_{ij}(-x)$ and $\delta_i(x)^{-1} = \delta_i(x^{-1})$, since $\tau_{ij}(x + y) = \tau_{ij}(x)\tau_{ij}(y)$ and $\delta_i(xy) = \delta_i(x)\delta_i(y)$. In particular,

$$\delta\colon D^\circ \to GL_n(D) \text{ is an injective group homomorphism.} \tag{1}$$

The matrices $\tau_{ij}(x)$ are called *transvections*; the $\delta_i(x)$ are *dilatations*. There are geometrical definitions of transvections and dilatations which assign these titles to matrices that don't have the forms $\tau_{ij}(x)$ and $\delta_i(x)$, but this matter won't concern us.

Lemma a. *If $D \neq \mathbb{F}_2$, then the transvections $\tau_{ij}(x)$ are elements of the commutator subgroup $GL_n(D)'$ of $GL_n(D)$.*

PROOF. Since $\tau_{ij}(x)$ has only one non-zero entry off the diagonal, it suffices to prove the lemma when $n = 2$. Choose $y \in D - \{0, -x\}$. This choice is possible because $D \neq \mathbb{F}_2$. Let $z = y(x + y)^{-1}$. A routine calculation shows that $\begin{bmatrix} 1 & 0 \\ x & 1 \end{bmatrix} = \alpha^{-1}\beta^{-1}\alpha\beta \in GL_n(D)'$, where $\alpha = \begin{bmatrix} 1 & 0 \\ y & 1 \end{bmatrix}$ and $\beta = \begin{bmatrix} 1 & 0 \\ 0 & z \end{bmatrix}$. Also, $\begin{bmatrix} 1 & x \\ 0 & 1 \end{bmatrix} \in GL_n(D)'$, since $GL_n(D)'$ is closed under transposition. □

Denote the subgroup of $GL_n(D)'$ that is generated by $\{\tau_{ij}(x)\colon i \neq j, x \in D^\circ\}$ by H or H_n. In the next section we will see that H is the commutator subgroup of $GL_n(D)$. Since $\tau_{ij}(x)^{-1} = \tau_{ij}(-x)$, every element of H is a product of transvections.

If $\alpha, \beta \in GL_n(D)$ with $n \geq 2$, write $\alpha \simeq \beta$ when α and β are in the same right coset of H; that is $\beta = \gamma\alpha$ for some $\gamma \in H$. Plainly, $\simeq$ is an equivalence relation on $GL_n(D)$.

The relation $\simeq$ has a familiar characterization:

$\alpha \simeq \beta$ if and only if β can be obtained from α by a sequence of elementary row transformations of the first kind, that is, transformations that add a left multiple of one row to another row. (2)

Indeed, if $\alpha = [y_{kl}] = \sum_{k,l}\varepsilon_{kl}y_{kl}$, then $\tau_{ij}(x)\alpha = (\iota + \varepsilon_{ij}x)(\sum_{k,l}\varepsilon_{kl}y_{kl}) = \sum_{k,l}\varepsilon_{kl}y_{kl} + \sum_l \varepsilon_{il}xy_{jl} = [z_{kl}]$, where $z_{kl} = y_{kl}$ if $k \neq i$ and $z_{il} = y_{il} + xy_{jl}$. That is, $\tau_{ij}(x)\alpha$ is obtained from α by adding the left multiple by x of row j to row i.

Lemma b. *If $\alpha \in GL_n(D)$, then $\alpha \simeq \delta(x)$ for some $x \in D^\circ$.*

Using only elementary row transformations of the first kind, the Gauss elimination process carries an invertible matrix α to a matrix of the form $\delta(x)$. Thus, $\alpha \simeq \delta(x)$ by (2). The details of this construction are outlined in Exercise 1.

Corollary a. *H is a normal subgroup of $GL_n(D)$. In particular, $\alpha_1 \simeq \beta_1$ and $\alpha_2 \simeq \beta_2$ imply $\alpha_1\alpha_2 \simeq \beta_1\beta_2$.*

PROOF. If $i, j \neq n$, then $\delta(x)^{-1}\tau_{ij}(y)\delta(x) = \tau_{ij}(y)$; and $\delta(x)^{-1}\tau_{in}(y)\delta(x) = \tau_{in}(yx)$, $\delta^{-1}(x)\tau_{nj}(y)\delta(x) = \tau_{nj}(x^{-1}y)$. Thus, $H \lhd GL_n(D)$ by Lemma b. Moreover, $\alpha_1 = \gamma_1\beta_1$ and $\alpha_2 = \gamma_2\beta_2$ with $\gamma_1, \gamma_2 \in H$ implies $\alpha_1\alpha_2 = \gamma_1\beta_1\gamma_2\beta_1^{-1}\beta_1\beta_2 \simeq \beta_1\beta_2$. □

Lemma c. *If $x_1, \ldots, x_r \in D^\circ$ and $1 \leq i_1, \ldots, i_r \leq n$, then $\delta_{i_1}(x_1)\cdots\delta_{i_r}(x_r) \simeq \delta(x_1\cdots x_r)$. In particular, $\delta_k(x) \simeq \delta_l(x)$ for all k and l.*

PROOF. By induction on r, it is sufficient to prove that $\delta_i(x)\delta(y) \simeq \delta(xy)$. Moreover, it can be assumed that $n = 2$ and (by (1)) $i = 1$. By (2), $\delta_1(x)\delta(y)$

$$= \begin{bmatrix} x & 0 \\ 0 & y \end{bmatrix} \simeq \begin{bmatrix} x & 0 \\ x & y \end{bmatrix} \simeq \begin{bmatrix} 1 & (x^{-1} - 1)y \\ x & y \end{bmatrix} \simeq \begin{bmatrix} 1 & (x^{-1} - 1)y \\ 0 & xy \end{bmatrix} \simeq \begin{bmatrix} 1 & 0 \\ 0 & xy \end{bmatrix}$$

$= \delta(xy)$. □

Corollary b. *If $\alpha \in GL_n(D)$, then $\alpha^n = \gamma(\iota x)$ for some $\gamma \in H$ and $x \in D^\circ$.*

By Lemma b there is an $x \in D^\circ$ such that $\alpha \simeq \delta(x)$. Therefore, $\alpha^n \simeq \delta(x)^n \simeq \delta_1(x)\cdots\delta_n(x) = \iota x$ by Corollary a and Lemma c.

This corollary enables us to prove an important analogue of Corollary 16.1b in which the roles of E and F are reversed.

Proposition. *Let E/F be a finite field extension, $A \in \mathfrak{S}(E)$, and $B \in \mathfrak{S}(F)$. Assume that $[E:F] = l$, $\operatorname{Deg} A = m$, and $\operatorname{Deg} B = n$. If $\theta: A \to B$ is an F-algebra homomorphism, then $n = mlk$ for some $k \in \mathbb{N}$ and $\nu_{B/F}(\theta(y)) = N_{E/F}(\nu_{A/E}(y))^k$ for all $y \in A$.*

PROOF. Since A is simple, we can assume that $A \subseteq B$ and θ is the inclusion homomorphism. In this case, E is a subfield of B and $F \subseteq E \subseteq A$. By the D.C.T. and Corollary 16.1b, $n^2 = l^2(\operatorname{Deg} \mathbf{C}_B(E))^2 = (lmk)^2$. Fix an element $y \in A$. If $y \in B^\circ$, then $y^{-1} \in F[y] \subseteq A$ by Lemma 16.3b. Hence, $\nu_{A/E}(y) = 0$ implies $\nu_{B/F}(y) = 0$. Assume that $\nu_{A/E}(y) \neq 0$, that is, $y \in A^\circ$. By the Wedderburn Structure Theorem, $A \cong M_s(D)$, where D is a division algebra in $\mathfrak{S}(E)$. It can be assumed for convenience of notation that $D \subseteq A$. The assumption that $y \in A^\circ$ implies by Corollary b that $y^s = zw$ for some $z \in A' \subseteq B'$ and $w \in D^\circ$. It follows from Corollary 16.3b that $\nu_{B/F}(y)^s = \nu_{B/F}(z)\nu_{B/F}(w) = \nu_{B/F}(w)$ and $\nu_{A/E}(y)^s = \nu_{A/E}(w)$. Let $K = E[w]$. Since $w \in D$ and D is a finite dimensional division algebra over E, K/E is a finite field extension, say $[K:E] = t$. Then $[K:F] = lt$ and $\nu_{B/F}(w) = N_{K/F}(w)^{n/lt} = N_{K/F}(w)^{km/t} = N_{E/F}(\nu_{A/E}(w))^k$ by Proposition 16.2a. Hence, $\nu_{B/F}(y)^s = (N_{E/F}(\nu_{A/E}(y))^k)^s$. The device that was introduced at the end of the proof of Proposition 16.2a can again be used to conclude that $\nu_{B/F}(y) = N_{E/F}(\nu_{A/E}(y))^k$. The details of the argument constitute Exercise 2. □

Corollary c. *If the hypotheses of the proposition are satisfied, then* θ *induces group homomorphisms* θ^{ab}, θ_1, *and* θ_2 *such that the diagram*

$$\begin{array}{ccccccccc} 1 \to & SK_1(A) & \to & A^{ab} & \to & E^\circ & \to & \mathrm{Coker}\, \nu_{A/E} & \to 1 \\ & \downarrow \theta_1 & & \downarrow \theta^{ab} & & \downarrow \psi & & \downarrow \theta_2 & \\ 1 \to & SK_1(B) & \to & B^{ab} & \to & F^\circ & \to & \mathrm{Coker}\, \nu_{B/F} & \to 1 \end{array}$$

commutes, where $\psi = \eta_k N_{E/F}$.

This corollary follows from the proposition by the same reasoning that was used to deduce Proposition 16.3b from Corollary 16.1b.

EXERCISES

1. Prove Lemma b. Hint. For $0 \le k < n$, denote by G_k the set of all $\alpha \in GL_n(D)$ that have the form $\begin{bmatrix} \iota_k & * \\ 0 & \beta \end{bmatrix}$ with $\beta \in M_{n-k}(D)$. Show that $\beta \in GL_{n-k}(D)$. Prove that if $\alpha \in G_k$ and $k+1 < n$, then $\alpha \simeq \begin{bmatrix} \iota_k & * \\ 0 & \gamma \end{bmatrix}$, where $\gamma = [y_{ij}] \in M_{n-k}(D)$ satisfies $y_{11} = 1$. Deduce that $\alpha \simeq \alpha' \in G_{k+1}$. Finally, show that if $\alpha \in G_{n-1}$, then $\alpha \simeq \delta(x)$ for some $x \in D^\circ$.
2. Complete the proof of the proposition by applying the result obtained in the first part of the argument to $F(\mathbf{x})$, $E(\mathbf{x})$, $A^{E(\mathbf{x})}$, $B^{F(\mathbf{x})}$, and $\mathbf{x} - y$. It is necessary to use both Lemma 16.2a and Lemma 16.2b.
3. Prove that Lemma a is true for $D = \mathbb{F}_2$ provided $n \ge 3$, but it is false for $GL_2(\mathbb{F}_2)$.

16.5. Non-commutative Determinants

The construction of the determinant mapping for $GL_n(D)$ is completed in this section. We also complete the preparations for the study of SK_1 in Section 6.

The notation that was introduced in Section 4 has the same meaning in this section. In particular, H is the subgroup of $GL_n(D)$ that is generated by the transvections $\tau_{ij}(x)$. We have shown that $H \lhd GL_n(D)$, $H \subseteq GL_n(D)'$, and $\delta(D^\circ)$ is a subgroup of $GL_n(D)$ such that $H\delta(D^\circ) = GL_n(D)$. To complete this picture, it is necessary to determine $H \cap \delta(D^\circ)$.

Lemma a. *If* $n \ge 2$ *and* $x \in D^\circ$, *then* $\delta(x) \in H$ *if and only if* $x \in D'$, *the commutator subgroup of* D°.

PROOF. Assume that $x = y^{-1}z^{-1}yz$. We will prove that $\delta(x) \in H$. It is sufficient to consider the case $n = 2$:

$$\begin{bmatrix} 1 & 0 \\ 0 & x \end{bmatrix} \simeq \begin{bmatrix} 1 & 0 \\ 1 & x \end{bmatrix} \simeq \begin{bmatrix} 0 & -x \\ 1 & x \end{bmatrix} \simeq \begin{bmatrix} 0 & -x \\ 1 & z \end{bmatrix} \simeq \begin{bmatrix} y^{-1}z^{-1}y & 0 \\ 1 & z \end{bmatrix}$$

$$\simeq \begin{bmatrix} y^{-1}z^{-1}y & 0 \\ y & z \end{bmatrix} \simeq \begin{bmatrix} 0 & -y^{-1} \\ y & z \end{bmatrix} \simeq \begin{bmatrix} 0 & -y^{-1} \\ y & 1 \end{bmatrix} \simeq \begin{bmatrix} 1 & 0 \\ y & 1 \end{bmatrix}$$
$$\simeq \begin{bmatrix} 1 & 0 \\ 0 & 1 \end{bmatrix}.$$

Since every element of D' is a product of commutators, it follows from 16.4(1) that $\delta(D') \subseteq H$. The proof that $\delta(x) \in H$ implies $x \in D'$ is more difficult. This implication is equivalent to the existence of a determinant homomorphism from $GL_n(D)$ to D^{ab}. It is convenient to extend our notation to cover the case $n = 1$. Write H_1 for D', $x \simeq y$ if $y^{-1}x \in H_1$, and $\delta_1^1(x) = x$. For $n \geq 1$, denote the natural projection homomorphism from $GL_n(D)$ to $GL_n(D)/H_n$ by π. The proof of the lemma will be completed by constructing a mapping $\theta\colon GL_n(D) \to GL_{n-1}(D)/H_{n-1}$ for $n \geq 2$ such that

$$\alpha \simeq \beta \quad \text{implies} \quad \theta(\alpha) = \theta(\beta), \tag{1}$$

and

$$\theta(\delta_n^n(x)) = \pi(\delta_{n-1}^{n-1}(x)). \tag{2}$$

This will do the job because $\delta_n^n(x) = \delta(x) \in H_n$ implies $\delta_{n-1}^{n-1}(x) \in H_{n-1}$, and by induction $x = \delta_1^1(x) \in H_1 = D'$. If $\alpha = [x_{ij}] \in GL_n(D)$, $n \geq 2$, and $1 \leq i \leq n$, define the row vectors $\xi_i = [x_{i1}, x_{i2}, \ldots, x_{in}]$ and $\eta_i = [x_{i2}, \ldots, x_{in}]$. Let $[y_1, y_2, \ldots, y_n]$ be the first row of α^{-1}. This vector is characterized by

$$y_1\xi_1 + y_2\xi_2 + \cdots + y_n\xi_n = [1,0,\ldots,0]. \tag{3}$$

In particular,

$$y_1\eta_1 + y_2\eta_2 + \cdots + y_n\eta_n = 0. \tag{4}$$

If $y_i \neq 0$, then $[y_i^{-1},0,\ldots,0] = y_i^{-1}y_1\xi_1 + \cdots + \xi_i + \cdots + y_i^{-1}y_n\xi_n$ by (3). It follows from Lemma 16.4c that

$$\alpha \simeq \begin{bmatrix} \xi_1 \\ \cdot \\ \cdot \\ \cdot \\ 1,0,\ldots,0 \\ \cdot \\ \cdot \\ \cdot \\ y_i^{-1} \quad \xi_k \\ \cdot \\ \cdot \\ \cdot \\ \xi_n \end{bmatrix}$$

for every $k \neq i$. Define

$$\alpha_{ik} = \begin{bmatrix} \eta_1 \\ \cdot \\ \cdot \\ \cdot \\ \hat{\eta}_i \\ \cdot \\ \cdot \\ \cdot \\ (-1)^{i+1} y_i^{-1} \eta_k \\ \cdot \\ \cdot \\ \cdot \\ \eta_n \end{bmatrix}$$

where the symbol $\hat{\eta}_i$ indicates that the row i is omitted. If $D = F$, so that the determinant mapping is defined, then $\det \alpha = \det \alpha_{ik}$. Our definition of θ is motivated by this observation: $\theta(\alpha) = \pi(\alpha_{ik})$, where i and k satisfy $y_i \neq 0$ and $k \neq i$.

$$\text{If } y_i \neq 0 \text{ and } y_j \neq 0, \quad \text{then } \alpha_{ik} \simeq \alpha_{jl} \text{ for all } k \neq i \text{ and } l \neq j. \tag{5}$$

By Lemma 16.4c the choices of k and l are irrelevant; it is sufficient to prove that $\alpha_{ij} \simeq \alpha_{ji}$ under the assumptions $y_i \neq 0$, $y_j \neq 0$, and $i < j$. By (4), $-y_i^{-1}\eta_j = \sum_{k \neq j} y_i^{-1} y_j^{-1} y_k \eta_k$. Thus, elementary row transformations of the first kind yield

$$\alpha_{ij} = \begin{bmatrix} \cdot \\ \cdot \\ \cdot \\ \hat{\eta}_i \\ \cdot \\ \cdot \\ \cdot \\ \sum_{k \neq j} (-1)^i y_i^{-1} y_j^{-1} y_k \eta_k \\ \cdot \\ \cdot \\ \cdot \end{bmatrix} \simeq \begin{bmatrix} \cdot \\ \cdot \\ \cdot \\ \hat{\eta}_i \\ \cdot \\ \cdot \\ \cdot \\ (-1)^i y_i^{-1} y_j^{-1} y_i \eta_i \\ \cdot \\ \cdot \\ \cdot \end{bmatrix}.$$

If the matrix γ is obtained from β by interchanging rows k and $k + 1$ and multiplying the new row k by -1, then $\gamma = \tau_{k\,k+1}(1)\tau_{k+1\,k}(-1)\tau_{k\,k+1}(1)\beta \simeq \beta$. Consequently,

$$\alpha_{ij} \simeq \begin{bmatrix} \cdot \\ \cdot \\ \cdot \\ (-1)^{j+1} y_i^{-1} y_j^{-1} y_i \eta_i \\ \cdot \\ \cdot \\ \cdot \\ \hat{\eta}_j \\ \cdot \\ \cdot \\ \cdot \end{bmatrix} = \delta_i^{n-1}(y_i^{-1} y_j^{-1} y_i y_j)\alpha_{ji} \simeq \alpha_{ji}$$

by the first part of the proof. Thus, (5) is proved. It is clear from the definition of θ that (2) is satisfied. For the proof of (1), it can be assumed that $\beta = \tau_{ij}(z)\alpha$. Thus, row i of β is $\xi_i + z\xi_j$, and the other rows of β are the same as the corresponding rows of α. By (3), $y_1\xi_1 + \cdots + y_i(\xi_i + z\xi_j) + \cdots + (y_j - y_i z)\xi_j + \cdots + y_n\xi_n = [1,0,\ldots,0]$, so that the first row of β^{-1} is $[y_1,\ldots,y_i,\ldots,y_j - y_i z,\ldots,y_n]$. If $y_k \neq 0$ for some $k \neq j$, then $\beta_{kj} = \tau_{ij}((-1)^{k+1} z y_k)\alpha_{kj} \simeq \alpha_{kj}$. In this case, $\theta(\beta) = \pi(\beta_{kj}) = \pi(\alpha_{kj}) = \theta(\alpha)$ by (5). (If $k = i$, then in fact $\beta_{ij} = \alpha_{ij}$.) If $y_k = 0$ for all $k \neq j$, then by (4), $\eta_j = 0$ and $y_j - y_i z = y_j$. In this case, $\beta_{ji} = \alpha_{ji}$ and $\theta(\beta) = \theta(\alpha)$ by (5). □

Proposition a. *If D is a division algebra, $D \ncong \mathbb{F}_2$, and $n \geq 2$, then the commutator subgroup of $GL_n(D)$ is generated by the transvections $\tau_{ij}(x)$. Moreover, $GL_n(D)^{ab} \cong D^{ab}$.*

Proof. By 16.4(1) and Lemma 16.4b, $\delta(D^\circ)$ is a subgroup of $GL_n(D)$ such that $H\delta(D^\circ) = GL_n(D)$. Lemma a is equivalent to the equation $H \cap \delta(D^\circ) = \delta(D')$. Therefore, since $H \lhd GL_n(D)$ by Corollary 16.4a, the Noether isomorphism yields $GL_n(D)/H = H\delta(D^\circ)/H \cong \delta(D^\circ)/(H \cap \delta(D^\circ)) = \delta(D^\circ)/\delta(D') \cong D^{ab}$. Thus, $H \supseteq GL_n(D)'$ because D^{ab} is commutative. By Lemma 16.4a, $H = GL_n(D)'$. □

It is a consequence of Exercise 3 in Section 16.4 that the proposition is true for $D = \mathbb{F}_2$, provided $n \geq 3$. The second statement of the proposition is valid with no restrictions on D or n.

Corollary a. *If D is a division algebra and $n \in \mathbb{N}$, then there is a unique group homomorphism* Det: $GL_n(D) \to D^{ab}$ *such that* $\operatorname{Det}\delta(x) = \pi(x)$ *for $x \in D^\circ$, where $\pi: D^\circ \to D^{ab}$ is the projection homomorphism.*

The explicit definition of $\operatorname{Det}\alpha$ is this: write $\alpha = \gamma\delta(x)$ where $\gamma \in H$ and $x \in D^\circ$; $\operatorname{Det}\alpha = \pi(x)$. By Lemma 14.4b and Lemma a, this recipe furnishes the unique homomorphism that satisfies the condition $\operatorname{Det}\delta(x) = \pi(x)$.

The mapping Det was introduced by J. Dieudonné in his study of the classical groups. If D is a field, then Det is the ordinary determinant. The kernel of Det is called the *Special Linear Group*, and it is denoted by $SL_n(D)$. If $D \neq \mathbb{F}_2$, then $SL_n(D) = GL_n(D)'$ by the proposition. In particular, $SL_1(D) = D'$.

Lemma b. *If $A = M_n(D)$, where $D \in \mathfrak{S}(F)$ is a division algebra, then $\nu_{A/F} \circ \delta = \nu_{D/F}|D^\circ$.*

PROOF. Let $\phi: D \to M_m(K)$ be a splitting representation of D. Define ψ: $A \to M_{nm}(K)$ by $\psi([x_{ij}]) = [\phi(x_{ij})]$, considered as a matrix of blocks. Plainly, ψ is a splitting representation of A. Thus, if $x \in D^\circ$, then $\nu_{A/F}(\delta(x)) = \det(\psi(\delta(x))) = \det\begin{bmatrix} \iota_{(n-1)m} & 0 \\ 0 & \phi(x) \end{bmatrix} = \det \phi(x) = \nu_{D/F}(x)$. □

Proposition b. *Let $A = M_n(D)$, where $D \in \mathfrak{S}(F)$ is a division algebra such that $|D| > 2$. There are isomorphisms δ^{ab}: $D^{ab} \to A^{ab}$ and $\delta_1 : SK_1(D) \to SK_1(A)$ such that the following diagram commutes.*

$$\begin{array}{ccccccccc} 1 \to & SK_1(D) & \to & D^{ab} & \to & F^\circ & \to & \operatorname{Coker} \nu_{D/F} & \to 1 \\ & \downarrow \delta_1 & & \downarrow \delta^{ab} & & \downarrow id & & \downarrow id & \\ 1 \to & SK_1(A) & \to & A^{ab} & \to & F^\circ & \to & \operatorname{Coker} \nu_{A/F} & \to 1 \end{array}$$

PROOF. By Lemma a, $\delta^{ab}(xD') = \delta(x)A'$ is a well defined, injective group homomorphism from D^{ab} to A^{ab}. By Lemma 16.4b, δ^{ab} is surjective. It is a consequence of Lemma b that the middle square of the diagram commutes. Therefore, the other squares commute, and $\delta_1 = \delta^{ab}|SK_1(D)$ is an isomorphism. □

Corollary b. *If $A = M_n(D)$, where $D \in \mathfrak{S}(F)$ is a division algebra, then* $\nu_{A/F}|A^\circ = \nu^{ab} \circ \delta^{ab} \circ \operatorname{Det}$.

PROOF. Assume that $D \neq \mathbb{F}_2$. Let $\alpha \in A$. Write $\alpha = \gamma\delta(x)$ with $\gamma \in H = A'$ and $x \in D^\circ$. Then $\nu^{ab}\delta^{ab}\operatorname{Det}\alpha = \nu^{ab}\delta^{ab}\pi(x) = \nu^{ab}\delta(x)A' = \nu^{ab}(\alpha A') = \nu(\alpha)$. The case $D = \mathbb{F}_2$ is trivial. □

Corollary c. *Assume that $|F| > 2$. If A, $B \in \mathfrak{S}(F)$ satisfy $A \sim B$, then* $\operatorname{Coker} \nu_{A/F} = \operatorname{Coker} \nu_{B/F}$ *and $SK_1(A) \cong SK_1(B)$. In particular,* $|SK_1(M_n(F))| = |\operatorname{Coker} \nu_{M_n(F)/F}| = 1$.

EXERCISES

1. Prove the assertion that the second statement of Proposition a is true for all D and all $n \in \mathbb{N}$. Show that $SK_1(M_2(\mathbb{F}_2)) \simeq \mathbb{Z}/2\mathbb{Z}$. Hint. Show that $GL_2(\mathbb{F}_2)$ is isomorphic to the symmetric group of order 6.
2. Let $\Delta: GL_n(D) \to D^\circ/D'$ be a mapping that satisfies:

(i) if β is obtained from $\alpha \in GL_n(D)$ by adding a left multiple of one row to another row, then $\Delta(\beta) = \Delta(\alpha)$;
(ii) if β is obtained from $\alpha \in GL_n(D)$ by multiplying one row of α on the left by $x \in D^\circ$, then $\Delta(\beta) = (xD')\Delta(\alpha)$;
(iii) $\Delta(\iota_n) = D'$.

Prove that $\Delta = \mathrm{Det}$.

3. Prove by induction on n that there is a mapping Δ that satisfies conditions (i), (ii) and (iii) of Exercise 2. Hint. Use the technique that was developed in the proof of Lemma a.

4. Let D be a division algebra. Let $\pi\colon D^\circ \to D^{ab}$ be the projection homomorphism. Derive the following formulas for determinants of the elements of $M_2(D)$.

(a) If $y, z \in D^\circ$, then $\mathrm{Det}\begin{bmatrix} 0 & y \\ z & w \end{bmatrix} = \pi(-zy)$.

(b) If $x \in D^\circ$ and $w \neq zx^{-1}y$ then $\mathrm{Det}\begin{bmatrix} x & y \\ z & w \end{bmatrix} = \pi(xw - xzx^{-1}y)$.

(c) If $yz \neq zy$, then $\mathrm{Det}\begin{bmatrix} 1 & y \\ z & yz \end{bmatrix} = \pi(yz - zy)$.

16.6. The Reduced Whitehead Group

This section presents some of the elementary results concerning the structure of $SK_1(A)$, where $A \in \mathfrak{S}(F)$. It is shown that the exponent of $SK_1(A)$ divides the index of A, the primary decomposition of a division algebra A induces a corresponding decomposition of $SK_1(A)$, and $SK_1(A)$ is trivial if $\mathrm{Ind}\, A$ is square-free. Deeper properties of the Reduced Whitehead Groups are described in the Notes. To avoid the anomaly $SK_1(M_2(\mathbb{F}_2)) \neq \{1\}$, we assume in this section that $|F| > 2$.

Lemma. *Let K/F be a finite field extension. If $D \in \mathfrak{S}(F)$ is a division algebra, $A \in \mathfrak{S}(K)$, $B \cong M_k(F)$, and $\chi\colon A \to B$ is an F-algebra homomorphism, then there is a sequence $SK_1(D) \xrightarrow{\phi} SK_1(D \otimes A) \xrightarrow{\psi} SK_1(D \otimes B) \xrightarrow{\theta} SK_1(D)$ such that θ is an isomorphism and $\theta\psi\phi = \eta_k$, where $\eta_k(x) = x^k$.*

Proof. Proposition 16.3b, Corollary 16.4c, and Proposition 16.5b give the commutative diagram

$$\begin{array}{ccccccccc} SK_1(D) & \xrightarrow{\phi} & SK_1(D \otimes A) & \xrightarrow{\psi} & SK_1(D \otimes B) & \xrightarrow{\tau} & SK_1(M_k(D)) & \xleftarrow{\delta_1} & SK_1(D) \\ \downarrow & & \downarrow & & \downarrow & & \downarrow & & \downarrow \\ D^{ab} & \to & (D \otimes A)^{ab} & \to & (D \otimes B)^{ab} & \to & M_k(D)^{ab} & \leftarrow & D^{ab} \end{array}$$

in which the vertical mappings are inclusions, ϕ is induced by $x \mapsto x \otimes 1$, ψ is induced by $x \otimes y \mapsto x \otimes \chi(y)$, and τ is induced by $x \otimes z \mapsto \rho(z)x$ (where $\rho\colon B \to M_k(F)$ is the isomorphism that is assumed to exist). Since

ρ is an isomorphism, so is τ; and δ_1 is an isomorphism by Proposition 16.5b. Thus, $\theta = \delta_1^{-1}\tau$: $SK_1(D \otimes B) \to SK_1(D)$ is an isomorphism. If $xD' \in SK_1(D)$, then $\theta\psi\phi(xD')$ is the coset (mod D') of the image of $x \mapsto x \otimes 1 \mapsto x \otimes \chi(1) \mapsto \iota_k x \mapsto \delta^{-1}(\iota_k x)$. By Corollary 16.4b, $\delta^{-1}(\iota_k x)D' = x^k D'$. Hence, $\theta\psi\phi = \eta_k$. □

Proposition a. *If $A \in \mathfrak{S}(F)$, then the exponent of $SK_1(A)$ divides* Ind A.

Proof. By Corollary 16.5c, we can assume that $A = D$ is a division algebra. Let K be a maximal subfield of D, so that $[K:F] = \operatorname{Deg} D = \operatorname{Ind} D = k$ and $D \otimes K \cong M_k(K)$. Thus, $SK_1(D \otimes K) = \{1\}$ by Corollary 16.5c. Apply the lemma with $A = K$ and χ: $K \to M_k(F) = B$ is a left regular representation. If $xD' \in SK_1(D)$, then $(xD')^k = \eta_k(xD') = \theta\psi\phi(xD') = \theta\psi(1) = 1$. □

Proposition b. *If the division algebra $D \in \mathfrak{S}(F)$ has the primary decomposition $D = D_1 \otimes \cdots \otimes D_r$, then $SK_1(D)$ has the primary decomposition $SK_1(D) \cong SK_1(D_1) \times \cdots \times SK_1(D_r)$.*

Proof. By Proposition 14.4b, it will be sufficient to prove that if A and B are division algebras of relatively prime degrees m and n, then $SK_1(A \otimes B) \cong SK_1(A) \times SK_1(B)$. By Proposition a, the exponent of $SK_1(A \otimes B)$ divides mn, so that $SK_1(A \otimes B) = G \times H$, where the exponent of G divides m and the exponent of H divides n. In particular, $\eta_n(G) = G$ and $\eta_n(H) = \{1\}$. By symmetry, it suffices to show that $G \cong SK_1(A)$. Define χ: $B \to B \otimes B^* \cong M_{n^2}(F)$ by $\chi(x) = x \otimes 1$. The lemma provides a sequence of homomorphisms $SK_1(A) \xrightarrow{\phi} SK_1(A \otimes B) \xrightarrow{\psi} SK_1(A \otimes B \otimes B^*) \xrightarrow{\theta} SK_1(A)$ with θ an isomorphism, and $\theta\psi\phi = \eta_{n^2}$. Thus, $SK_1(A) = \eta_n\eta_{n^2}(SK_1(A)) = \eta_n\theta\psi\phi(SK_1(A)) \subseteq \theta\psi\eta_n(G \times H) = \theta\psi(G) \subseteq SK_1(A)$. Hence, $\theta\psi$: $G \to SK_1(A)$ is a surjective homomorphism. The proof is completed by showing that $\psi|G$ is injective. To do so, we use the lemma with the left regular representation $B^* \to M_{n^2}(F)$ to obtain $SK_1(A \otimes B) \xrightarrow{\phi'} SK_1(A \otimes B \otimes B^*) \xrightarrow{\psi'} SK_1(A \otimes B \otimes M_{n^2}(F)) \xrightarrow{\theta'} SK_1(A \otimes B)$ with $\theta'\psi'\phi' = \eta_{n^2}$. A check of the definitions shows that $\phi' = \psi$. If $w = x(A \otimes B)' \in G - \{1\}$, then $\theta'\psi'\psi(w) = \theta'\psi'\phi'(w) = \eta_{n^2}(w) = w^{n^2} \neq 1$. Thus, $\psi|G$ is injective. □

Proposition c. *If $D \in \mathfrak{S}(F)$ is a division algebra and L/F is a finite field extension such that $[L:F]$ is relatively prime to* Deg D, *then the inclusion homomorphism $D \to D^L$ induces a split injection $SK_1(D) \to SK_1(D^L)$.*

Proof. Let $n = [L:F]$. By the lemma, the left regular representation $L \to M_n(F)$ leads to the sequence $SK_1(D) \xrightarrow{\phi} SK_1(D^L) \xrightarrow{\psi} SK_1(D \otimes M_n(F)) \xrightarrow{\theta} SK_1(D)$ with $\theta\psi\phi = \eta_n$. Since $(n, \operatorname{Deg} D) = 1$, it follows from Proposition a that η_n is an automorphism of $SK_1(D)$. Therefore, ϕ is a split injection. □

Theorem. *If the index of $A \in \mathfrak{S}(F)$ is square-free, then $SK_1(A) = \{1\}$.*

PROOF. By Corollary 16.5c and Proposition b, it can be assumed that $A = D$ is a division algebra of prime degree p. Let $x \in D$ satisfy $\nu(x) = 1$. We must prove that $x \in D'$. This conclusion follows from

there is a finite extension L/F such that p does not divide $[L:F]$, and D^L contains a maximal subfield E that is a cyclic extension of L with $x \in E$. (1)

In fact, (1) implies that $N_{E/L}(x) = \nu_{D^L/L}(x) = \nu_{D/F}(x) = 1$ by Proposition 16.2a and Corollary 16.1b. Since E/L is cyclic, it follows from Hilbert's Theorem 90 and the Noether–Skolem Theorem that $x \in (D^L)'$. (See Exercises 1 and 2.) Thus, $x \in D'$ by Proposition c. The proof of (1) is a simple version of the proof of Lemma 15.2 with a minor twist: $F(x)/F$ is separable, so that there is a maximal subfield K of D such that K/F is separable and $x \in K$. (If $x \in F$, then $F(x)/F$ is obviously separable. Otherwise, $N_{F(x)/F}(x) = \nu_{D/F}(x) = 1$ by Proposition 16.2a and the hypothesis; since $[F(x):F] = p$, it follows that $F(x)/F$ must be separable.) The proof of Lemma 15.2 yields a finite extension L/F such that p does not divide $[L:F]$, $[KL:L] = p$, and KL/L is Galois. Thus, $E = KL$ is a maximal subfield of D^L, E/L is cyclic, and $x \in E$. □

EXERCISES

1. Use the results of Exercises 3 and 4 of Section 14.2 to prove that if K/F is a cyclic extension with $\mathbf{G}(K/F) = \langle\sigma\rangle$, then:
   (a) if $x \in K$ satisfies $T_{K/F}(x) = 0$, then $x = y - y^\sigma$ for some $y \in K$;
   (b) if $x \in K$ satisfies $N_{K/F}(x) = 1$, then $x = (y^\sigma)^{-1}y$ for some $y \in K^\circ$.
   In fact, (a) and (b) are the original statements of Hilbert's Theorem 90.
2. Let $A \in \mathfrak{S}(F)$. Prove the following statements.
   (a) If $x \in A$ satisfies $\nu(x) = 1$, and if there is a strictly maximal subfield K of A such that $x \in K$ and K/F is cyclic, then $x \in A'$. Hint. Use the result of Exercise 1 and the Noether–Skolem Theorem.
   (b) Prove that if A is a quaternion algebra, and $x \in A$, then $\nu(x) = 1$ if and only if x is a commutator in A°. This result strengthens the theorem for the case of algebras of degree two.
3. Complete the proof of the theorem. In particular, show that $F(x)/F$ is separable, and give a detailed proof of (1).

Notes on Chapter 16

The characteristic polynomials, traces, and reduced norms of central simple algebras are treated in most expositions of associative algebras. The discussion of these topics in the first three sections of this chapter follows the traditional line. The connection between Dieudonné's non-commutative determinant and the reduced norm was first exploited by S. Wang in [75]. Most of the results on $SK_1(A)$ in Section 16.6 were first proved in [75]. One of the deepest results on the Reduced Whitehead Groups is Wang's theorem that if F is an algebraic number field and $A \in \mathfrak{S}(F)$, then $SK_1(A) = \{1\}$. Early results suggested that $SK_1(A)$ might be trivial for all central simple algebras A. This question was called the "Tannaka–Artin Problem." In 1975, V. P. Platonov showed that there are algebras A such that $SK_1(A) \neq \{1\}$. It is now known that virtually all bounded torsion abelian groups occur as Reduced Whitehead Groups. An extensive discussion of the Tannaka–Artin Problem is given in the monograph [32]. There is also an interesting and concise exposition of this topic in Platonov's survey paper [63].

CHAPTER 17
Division Algebras over Local Fields

This chapter gives a fairly complete description of the finite dimensional division algebras over fields that are locally compact in the topology of a discrete valuation, that is, local fields. The most important property of these algebras is that they contain maximal subfields that are unramified extensions of their centers. It follows that all such algebras are cyclic. Moreover, the classification of the unramified extensions of local fields gives a characterization of the Brauer groups of such fields; they are all isomorphic to $\mathbb{Q}/\mathbb{Z}$.

The theory of field valuations can be extended in a straightforward way to division algebras. The first half of this chapter gives a self-contained development of this subject. No prior knowledge of valuation theory is assumed.

17.1. Valuations of Division Algebras

The basic definitions in the theory of valuations are given in this section. Our main result relates the valuations of a division algebra $D \in \mathfrak{S}(F)$ to the reduced norm $\nu_{D/F}$.

Definition. A *valuation* of a division algebra D is a mapping $v\colon D \to \mathbb{R}$ such that

$$v(x) \geq 0 \text{ for all } x \in D \text{ and } v(x) = 0 \text{ if and only if } x = 0, \tag{1}$$

$$v(xy) = v(x)v(y) \text{ for all } x, y \in D, \tag{2}$$

$$\text{there is a positive real number } a \text{ such that } v(x + y) \leq a \max\{v(x), v(y)\} \text{ for all } x, y \in D. \tag{3}$$

If v is a valuation of D. then $v|D^\circ$ is a homomorphism of D° to the multiplicative group $\mathbb{R}^+$ of positive real numbers. Conversely, a homomorphism $v\colon D^\circ \to \mathbb{R}^+$ can be extended to a valuation by $v(0) = 0$ if (3) is satisfied. For example, the homomorphism of D° to $\{1\}$ gives a valuation v such that $v(x) = 1$ for all $x \neq 0$ and $v(0) = 0$. This v is called the *trivial valuation* of D. Other valuations of D are called non-trivial.

The uniqueness of roots in $\mathbb{R}^+$ implies that if $v(x^n) = v(y^n)$, then $v(x) = v(y)$. In particular, if x is a root of unity in D, then $v(x) = v(1_D) = 1$. Two special cases of this observation are used frequently: $v(-x) = v(x)$; $v(-1_D) = 1$.

By (3), the set $\{v(1 + x)\colon x \in D, v(x) \leq 1\}$ is bounded. Define

$$a(v) = \sup\{v(1 + x)\colon x \in D, v(x) \leq 1\}. \tag{4}$$

Lemma a. *If v is a valuation of the division algebra D and $x_1, x_2, \ldots, x_n \in D$, then*

$$v(x_1 + x_2 + \cdots + x_n) \leq a(v)^m \max\{v(x_i)\colon 1 \leq i \leq n\},$$

where m is the least integer greater than $\log_2 n$.

An easy argument gives this inequality for $n = 2$. Induction extends it to powers of 2. The final form of the lemma is obtained by adjoining $2^m - n$ zeros to the sum $x_1 + \cdots + x_n$.

If v is a valuation of the division algebra D and $e \in \mathbb{R}^+$, then the mapping $v^e\colon D \to \mathbb{R}$ defined by $v^e(x) = v(x)^e$ is also a valuation of D. Moreover, since $v^e(x) \leq 1$ if and only if $v(x) \leq 1$, it follows from (4) that

$$a(v^e) = a(v)^e. \tag{5}$$

Two valuations v and w of the division algebra D are *equivalent* if $w = v^e$ for some $e \in \mathbb{R}^+$. Since $(v^e)^f = v^{ef}$, the concept of equivalence for valuations is an equivalence relation on the set of valuations of D.

Lemma b. *Let v be a valuation of the division algebra D.*

(i) $a(v) \geq 1$.
(ii) *If $a(v) = 1$, then $a(w) = 1$ for all valuations w that are equivalent to v.*
(iii) *If $a(v) > 1$ and $1 < a \in \mathbb{R}$, then there is a unique valuation w such that w is equivalent to v and $a(w) = a$.*

The property (i) is clear from (4) because $v(0) = 0$ and $v(1) = 1$; the statements (ii) and (iii) follow easily from (5).

Proposition. *If v is a valuation of the division algebra D, then $a(v) \leq 2$ if and only if v satisfies the* triangle inequality*: $v(x + y) \leq v(x) + v(y)$ for all $x, y \in D$. Every valuation of D is equivalent to a valuation that satisfies the triangle inequality.*

Proof. If v satisfies the triangle inequality, then $v(1 + x) \le v(1) + v(x) \le 2$ for all x such that $v(x) \le 1$. Hence, $a(v) \le 2$. Conversely, if $a(v) \le 2$, then Lemma a implies that $v(x_1 + \cdots + x_n) \le (2n) \max\{v(x_i): 1 \le i \le n\}$ since $a(v)^m \le 2^m < 2n$. Thus, $v(x + y)^n = v((x + y)^n) = v(z_0 + \cdots + z_n) \le 2(n + 1) \max\{v(z_k): 0 \le k \le n\}$, where z_k is the sum of the $\binom{n}{k}$ monomials $x^{i_1}y^{j_1} \cdots x^{i_r}y^{j_r}$ with $i_1 + \cdots + i_r = k$, $j_1 + \cdots + j_r = n - k$. The summands of z_k satisfy $v(x^{i_1}y^{j_1} \cdots x^{i_r}y^{j_r}) = v(x)^k v(y)^{n-k}$, so that $v(z_k) \le 2\binom{n}{k}v(x)^k v(y)^{n-k}$. Therefore, $v(x + y)^n \le 4(n + 1)\sum_{k=0}^{n}\binom{n}{k}v(x)^k v(y)^{n-k} = 4(n + 1)(v(x) + v(y))^n$. Taking n'th roots and letting $n \to \infty$ gives the triangle inequality because $\lim_{n\to\infty}(4(n + 1))^{1/n} = 1$. The last statement of the proposition follows from Lemma b. □

There is an obvious property of valuations that is worth mentioning. If v is a valuation of the division algebra D and A is a subalgebra of D that is also a division algebra, then $v|A$ is a valuation of A. If D is finite dimensional, then every subalgebra of D is a division algebra.

Theorem. *Let $D \in \mathfrak{S}(F)$ be a division algebra. If v is a valuation of D, then $v = w \circ \nu_{D/F}$, where w is the valuation of F that is defined by $w = (v|F)^{1/n}$, with $n = \operatorname{Deg} D$.*

Proof. If $x \in D^\circ$, then $y = x^n \nu(x)^{-1}$ satisfies $v(y) = v(x^n)v(\nu(x)^{-1}) = v(x)^n v(x)^{-n} = 1$. By Proposition 16.6a, $y^n \in D' \subseteq \operatorname{Ker} \nu$. Thus, $v(y)^n = 1$. That is, $v(x^n \nu(x)^{-1}) = v(y) = 1$, and $v(x) = v(\nu(x))^{1/n} = w(\nu(x))$. □

The converse of this theorem is false: if w is a valuation of F, then the mapping $w \circ \nu_{D/F}$ may not satisfy (3). (See Exericse 4 in Section 17.2.)

Exercises

1. Give the details of the proof of Lemma a.
2. Show that if D is a division algebra and $v: D^\circ \to \mathbb{R}^+$ is a group homomorphism such that $\{v(1 + x): x \in D - \{0, -1\}, v(x) \le 1\}$ is bounded, then v can be extended to a valuation of D.
3. (a) Prove that the mapping $v: \mathbb{R} \to \mathbb{R}$ that is defined by the absolute value $v(b) = |b|$ is a valuation of $\mathbb{R}$ such that $a(v) = 2$.
   (b) Prove that the mapping $v: \mathbb{C} \to \mathbb{R}$ that is defined by $v(c + id) = c^2 + d^2$ is a valuation of $\mathbb{C}$ such that $a(v) = 4$.
   (c) Prove that the reduced norm $\nu_{\mathbb{H}/\mathbb{R}}$ is a valuation v of $\mathbb{H}$ such that $a(v) = 4$.
4. Prove that the only valuation of a finite field is the trivial valuation.

17.2. Non-archimedean Valuations

A valuation v of the division algebra D is called *non-archimedean* if the constant $a(v)$ defined by 17.1(4) is equal to 1. In other words, v satisfies

$$v(x + y) \leq \max\{v(x), v(y)\} \tag{1}$$

for all $x, y \in D$. If $a(v) > 1$, then v is an *archimedean valuation.* It is clear from Lemma 17.1b that the dichotomy between archimedean and non-archimedean valuations is respected by the equivalence relation. The main purpose of this section is to obtain an effective characterization of non-archimedean valuations. We begin by recording an important consequence of the inequality (1).

Domination principle. *If v is a non-archimedean valuation of the division algebra D, and $x_1, x_2, \ldots, x_n$ are elements of D such that $v(x_1) > v(x_i)$ for $2 \leq i \leq n$, then $v(x_1 + x_2 + \cdots + x_n) = v(x_1)$.*

PROOF. Denote $y = x_2 + \cdots + x_n$. By (1),

$$v(y) \leq \max\{v(x_2), \ldots, v(x_n)\} < v(x_1).$$

Hence,

$$\begin{aligned} v(x_1 + y) \leq \max\{v(x_1), v(y)\} = v(x_1) &\leq \max\{v(x_1 + y), v(-y)\} \\ &= \max\{v(x_1 + y), v(y)\}. \end{aligned}$$

Thus,

$$v(x_1 + y) = v(x_1). \qquad \square$$

Proposition. *For a valuation v of the division algebra D, the following conditions are equivalent.*

(i) *v is non-archimedean.*
(ii) *$v(m1_D) \leq 1$ for all $m \in \mathbb{Z}$.*
(iii) *$\{v(m1_D): m \in \mathbb{N}\}$ is bounded.*

PROOF. If v is non-archimedean, then $v(-m1_D) = v(m1_D) \leq 1$ for all $m \in \mathbb{N}$, and $v(0) = 0$. Plainly, (ii) implies (iii). Assume that $v(m1_D) \leq b$ for all $m \in \mathbb{N}$. If $x \in D$ satisfies $v(x) \leq 1$, then for all $k \in \mathbb{N}$, $v(1 + x)^k = v(\sum_{j=0}^{k} \binom{k}{j} x^j) \leq a(v)^l \max\{v(\binom{k}{j})v(x)^j: 0 \leq j \leq k\}$, where $\log_2 k \leq l < \log_2 k + 1$. Consequently, $v(1 + x)^k \leq a(v)^l b$; and $v(1 + x) \leq \lim_{k\to\infty} a(v)^{l/k} b^{1/k} = 1$, since $(\log_2 k)/k \to 0$ as $k \to \infty$. Hence, $a(v) = 1$ and v is non-archimedean. $\square$

Corollary. *Let D be a division algebra over the field F. Denote the prime field of F by L. A valuation v of D is non-archimedean if and only if $v|L$ is non-archimedean. In particular, if the characteristic of F is a prime, then every valuation of D is non-archimedean.*

The last statement of the corollary is a consequence of the observation that if L is finite then $\{v(m1_D): m \in \mathbb{N}\}$ is finite, hence bounded.

If F is a subfield of $\mathbb{C}$, then the ordinary absolute value $v(z) = |z| = (z\bar{z})^{1/2}$ is an archimedean valuation of F. (See Exercise 3, Section 17.1.) One example of a non-archimedean valuation is always on hand: the trivial valuation is plainly non-archimedean. We concluded this section with a fairly general construction of non-archimedean valuations.

Example. Let R be a principal ideal domain with the fraction field F. Fix a real number d that satisfies $0 < d < 1$. Corresponding to each irreducible $p \in R$, define $v_p\colon F^\circ \to \mathbb{R}^+$ by $v_p(x) = d^k$, where k is the unique integer such that $x = p^k(a/b)$ with $a, b \in R - pR$. Easy calculations show that v_p extends to a non-archimedean valuation of F.

Two cases of this example are especially interesting: $R = \mathbb{Z}$ and $F = \mathbb{Q}$; $R = K[\mathbf{x}]$ and $F = K(\mathbf{x})$, where K is any field. In the next section we will show that all non-archimedean valuations of $\mathbb{Q}$ have the form v_p.

Exercises

1. Prove that the mapping v_p of the example is a non-archimedean valuation.
2. For each positive rational prime p, define $v_p\colon \mathbb{Q}^\circ \to \mathbb{R}^+$ as in the example with $d = 1/p$. Define $v_\infty(x) = |x|$ for $x \in \mathbb{Q}^\circ$. Prove that for each $x \in \mathbb{Q}^\circ$, $v_p(x) = 1$ for almost all p, and $(\prod_p v_p(x))v_\infty(x) = 1$.
3. Let $F = K(\mathbf{x})$ be the fraction field of the polynomial algebra $K[\mathbf{x}]$, where K is any field. Fix $d \in \mathbb{R}$ with $0 < d < 1$. Define $v_\infty : K(\mathbf{x})^\circ \to \mathbb{R}^+$ by $v_\infty(\Phi/\Psi) = d^k$, where $\Phi, \Psi \in K[\mathbf{x}]$ and $k = \deg \Psi - \deg \Phi$. Prove that v_∞ extends to a non-trivial, non-archimedean valuation of F such that $v_\infty | K$ is trivial. Show that v_∞ is not equivalent to a valuation of F that is obtained by the construction in the example.
4. Let p be a rational prime with $p \equiv 3 \pmod 4$. Thus, $D = \left(\frac{-1, p}{\mathbb{Q}}\right)$ is a division algebra by Exercise 4, Section 1.7. Let q be an odd prime for which p is a quadratic residue. Prove that there is no valuation v of D such that $v|\mathbb{Q}$ is equivalent to v_q. Hint. For each $n \in \mathbb{N}$, let $c_n \in \mathbb{N}$ be such that $(c_n)^2 \equiv p \pmod{q^n}$. Prove that if $v|\mathbb{Q}$ is equivalent to v_q, then $v(c_n \pm \mathbf{j}) \to 0$ as $n \to \infty$. Derive the contradiction $v(2\mathbf{j}) = 0$.

17.3. Valuation Rings

As Example 17.2 correctly suggests, there is a close relation between non-archimedean valuations of a division algebra D and certain subrings of D. These subrings are called valuation rings. They play a major role in the theory of valuations.

Lemma. *Let v be a non-archimedan valuation of the division algebra D. If $O(v) = \{x \in D\colon v(x) \le 1\}$ and $P(v) = \{x \in D\colon v(x) < 1\}$, then $O(v)$ is a* *subring of D that is a local $\mathbb{Z}$-algebra with $\mathbf{J}(O(v)) = P(v)$, and $E(v) = O(v)/P(v)$ is a division algebra.*

PROOF. If $x, y \in O(v)$, then $v(x - y) \leq \max\{v(x),v(y)\} \leq 1$; thus, $x - y \in O(v)$. Similarly, $x, y \in P(v)$ implies $x - y \in P(v)$. The equation $v(xy) = v(x)v(y)$ implies that $O(v)$ is a subring of D and $P(v)$ is an ideal of $O(v)$. If $x \in O(v) - P(v)$, then $v(x) = 1$. Thus, $v(x^{-1}) = 1$, $x^{-1} \in O(v)$, and $x \in O(v)^\circ$. Conversely, if $x \in O(v)^\circ$, then $v(x) \leq 1$ and $v(x^{-1}) \leq 1$. Hence, $v(x) = 1$, that is, $x \in O(v) - P(v)$. In particular, $O(v) - O(v)^\circ = P(v)$ is closed under addition, $O(v)$ is a local ring, and $E(v) = O(v)/P(v)$ is a division algebra. □

If v is a valuation of the division algebra D, denote

$$O(D,v) = \{x \in D\colon v(x) \leq 1\}$$

and

$$P(D,v) = \{x \in D\colon v(x) < 1\}.$$

When v is non-archimedean we will call these sets the *valuation ring* of v and the *valuation ideal* of v. By the lemma this terminology is accurate. If we are considering only one division algebra or valuation then the notation $O(D,v)$ and $P(D,v)$ will be shortened to $O(D)$ or $O(v)$ and $P(D)$ or $P(v)$.

If v is a non-archimedean valuation of the division algebra D, then the algebra $E(D,v) = O(D,v)/P(D,v)$ is called the *residue class field* of D. (In most cases that interest us, $E(D,v)$ turns out to be commutative.) Unless it causes confusion the notation $E(D,v)$ will be shortened to $E(D)$ or $E(v)$. It is convenient and custormary to write $\bar{x}$ for the image in $E(v)$ of an element $x \in O(v)$, that is, $\bar{x} = x + P(v)$. Moreover, if $\mathbf{\Phi}(\mathbf{x}) = a_0\mathbf{x}^n + a_1\mathbf{x}^{n-1} + \cdots + a_n \in O(v)[\mathbf{x}]$, then we will write $\overline{\mathbf{\Phi}}(\mathbf{x})$ for $\bar{a}_0\mathbf{x}^n + \bar{a}_1\mathbf{x}^{n-1} + \cdots + \bar{a}_n$.

Our next result shows that $O(v)$, $P(v)$, and $E(v)$ are unchanged when v is replaced by an equivalent valuation.

Proposition. *Let v and w be non-trivial valuations of the division algebra D. The following conditions are equivalent.*

(i) *v and w are equivalent valuations.*
(ii) *$O(v) = O(w)$ and $P(v) = P(w)$.*
(iii) *$O(v) \subseteq O(w)$.*
(iv) *$P(v) \subseteq P(w)$.*

PROOF. If $w = v^e$ for some $e \in R^+$, then it is evident that $v(x) \leq 1$ if and only if $w(x) \leq 1$ and $v(x) < 1$ if and only if $w(x) < 1$. It is therefore sufficient to show that both (iii) and (iv) imply that v is equivalent to w. Assume that $O(v) \subseteq O(w)$. We will first prove that $O(v) = O(w)$. If $x \in O(w)$, then $w(x^{-n}) \geq 1$ for all $n \in \mathbb{N}$. Since w is non-trivial, there exists $y \in D^\circ$ such that $w(y) > 1$. Consequently, $w(x^{-n}y) > 1$ for all $n \in \mathbb{N}$. Since $O(v) \subseteq O(w)$, it follows that $v(x^{-n}y) > 1$ for all $n \in \mathbb{N}$. Hence $v(x) \leq 1$, that is, $x \in O(v)$. Let $x, y \in D^\circ$ satisfy $w(x) > 1$ and $w(y) > 1$. Then $v(x) > 1$ and $v(y) > 1$. For $m, n \in \mathbb{N}$, the inequality $\log w(x)/\log w(y) \leq m/n$ is equivalent to $\log w(x^n) \leq \log w(y^m)$ or $w(x^n y^{-m}) \leq 1$. Thus, $\log w(x)/\log w(y) \leq m/n$ if and only if $x^n y^{-m} \in O(w) = O(v)$. Hence, $\{m/n: m, n \in \mathbb{N}, \log w(x)/\log w(y) \leq m/n\} = \{m/n: m, n \in \mathbb{N}, \log v(x)/\log v(y) \leq m/n\}$, and $\log w(x)/\log w(y) = \log v(x)/\log v(y)$. Therefore, $w(x) = v(x)^e$ for all $x \in D$ such that $w(x) > 1$, where $e = (\log w(y))/(\log v(y)) \in \mathbb{R}^+$. It follows easily that $w = v^e$. Finally, if $P(v) \subseteq P(w)$, then $O(w) - \{0\} = \{x \in D^\circ: x^{-1} \notin P(w)\} \subseteq \{x \in D^\circ: x^{-1} \notin P(v)\} = O(v) - \{0\}$. Therefore, v and w are equivalent. □

Corollary. *Let R be a principal ideal domain whose fraction field is F. If v is a non-trivial, non-archimedean valuation of F such that $v(x) \leq 1$ for all $x \in R$, then there is an irreducible element $p \in R$ such that v is equivalent to v_p, and $E(v) \cong R/pR$. Moreover, v_p is equivalent to v_q if and only if the irreducible elements p and q are associates in R.*

Proof. By assumption, $O(v) \supseteq R$. The lemma implies that $R \cap P(v)$ is a non-zero prime ideal of R. Since R is a principal ideal domain, there is an irreducible $p \in R$ such that $R \cap P(v) = pR$. If $x \in F$ satisfies $v_p(x) \leq 1$, then $x = a/b$ with $a \in R$, $b \in R - pR = R - P(v)$. Thus, $v(x) = v(a) \leq 1$. This remark shows that $O(v_p) \subseteq O(v)$. Hence, v is equivalent to v_p by the proposition. Clearly, $R \cap P(v_p) = pR$ and $R + P(v_p) \subseteq O(v_p)$. If $x \in O(v_p)$, say $x = a/b$ with $a \in R$ and $b \in R - pR$, then $1 = cb + dp\,(c, d \in R)$ implies that $x = ac + p(ad/b) \in R + P(v_p)$. Thus, $E(v) = O(v)/P(v) = (R + P(v))/P(v) \cong R/R \cap P(v) = R/pR$. If v_p is equivalent to v_q, then $pR = R \cap O(v_p) = R \cap O(v_q) = qR$; that is, p and q are associates in R. Conversely, if p and q are associates in R, then v_p and v_q are equivalent by definition. □

Theorem. (i) *If v is a non-archimedean valuation of $\mathbb{Q}$, then v is equivalent to v_p for some rational prime p. In this case, $E(v) \cong \mathbb{Z}/p\mathbb{Z}$.*

(ii) *If v is an archimedean valuation of $\mathbb{Q}$, then v is equivalent to v_∞, the absolute value on $\mathbb{Q}$.*

Proof. If v is non-archimedean, then $v(n) \leq 1$ for all $n \in \mathbb{Z}$ by Proposition 17.2. Therefore (i) is a special case of the corollary. Let v be archimedean. By Proposition 17.1, it can be assumed that v satisfies the triangle inequality. By the proposition, it is sufficient to show that $v(x) < 1$ implies $|x| < 1$. We can assume that $x > 0$, say $x = m/n$, where m and n are relatively prime natural numbers. Suppose that $v(x) < 1$ and $x > 1$, that is, $m > n$. Every $k \in \mathbb{N}$ has an m/n-ary representation: $k = \sum_{i=0}^{r} a_i(m/n)^i$, $a_i \in \{0, 1, \ldots, m - 1\}$. (See Exercise 1.) By the triangle inequality, $v(k) \leq \sum_{i=0}^{r} a_i v(m/n)^i \leq (m - 1)\sum_{i=0}^{\infty} v(x)^i = (m - 1)/(1 - v(x))$. It follows from Proposition 17.2 that v is non-archimedean, which is contrary to hypothesis. □

EXERCISES

1. Let $m, n \in \mathbb{N}$ with $m > n$. Prove that every $k \in \mathbb{N}$ can be written in the form $k = \sum_{i=0}^{r} a_i(m/n)^i$ with $a_i \in \{0, 1, \ldots, m-1\}$. Hint. Choose $a_0 \in \{0, 1, \ldots, m-1\}$ so that $k - a_0 = ml$ with $l \geq 0$ in $\mathbb{Z}$, and apply induction to $k' = ln$.

2. Let $F = K(\mathbf{x})$ be the fraction field of $K[\mathbf{x}]$. Let v be a valuation of F such that $v|K$ is trivial and $v(\mathbf{x}) > 1$. Prove that v is equivalent to the valuation v_∞ that was defined in Exercise 3 of Section 17.2. Deduce that every valuation of F is equivalent to exactly one of the valuations v_∞ or v_Φ, where Φ is a monic irreducible polynomial in $K[\mathbf{x}]$.

3. Let v be a non-archimedean valuation of the field F. Use the Domination Principle to prove that $O(v)$ is integrally closed in F, that is, if $x \in F$ satisfies $x_n + a_1 x^{n-1} + \cdots + a_n = 0$, where all a_i are elements of $O(v)$, then $x \in O(v)$.

17.4. The Topology of a Valuation

If v is a valuation of the division algebra D and v satisfies the triangle inequality, then v determines a distance function $\delta: D \times D \to \mathbb{R}$ by

$$\delta(x, y) = v(x - y). \tag{1}$$

Thus, v defines a metric topology on D in which the neighborhoods of $x \in D$ are defined for $e \in \mathbb{R}^+$ by

$$N(x,e,v) = \{y \in D: v(x - y) < e\}. \tag{2}$$

It is a standard consequence of the triangle inequality that the sets defined by (2) have the properties of a neighborhood basis for a Hausdorff topology on D. (In this chapter, a few standard results of set topology are assumed to be known.) Even if v does not satisfy the triangle inequality, the sets described by (2) define a topology because $N(x,e,v) = N(x,e^f,v^f)$ for all $f \in \mathbb{R}^+$; and if f is sufficiently small, then v^f does satisfy the triangle inequality by 17.1(5) and Proposition 17.1.

For a valuation v of the division algebra D, the *v-topology* of D is the topology that is defined by taking the family of sets $\{N(x,e,v): x \in D, e \in \mathbb{R}^+\}$ as a neighborhood basis for the open subsets of D. Thus, a subset of D is open in the v-topology if and only if it is a union of sets of the form $N(x,e,v)$.

Lemma a. *If v and w are valuations of D, then the v-topology has the same open sets as the w-topology if and only if v and w are equivalent.*

PROOF. We have already observed that equivalent valuations generate the same neighborhood bases, hence the same topologies. Conversely, if the v and w topologies coincide, then there exists $e \in \mathbb{R}^+$ such that $N(0,e,w) \subseteq N(0,1,v)$. If $x \in D$ satisfies $w(x) < 1$, then $w(x^n) = w(x)^n < e$ for a sufficiently large n. Thus, $x^n \in N(0,e,w) \subseteq N(0,1,v)$, that is, $v(x)^n = v(x^n) < 1$. Consequently, $v(x) < 1$. By Proposition 17.3, v is equivalent to w. □

In the rest of this section, assume that v is a valuation of the division algebra D, and v satisfies the triangle inequality. Let δ be the distance function on D that is defined by (1).

Lemma b. *Addition and subtraction in D are uniformly continuous; multiplication is uniformly continuous on bounded subsets of D; the inverse operation is uniformly continuous on sets that are bounded away from* 0.

These facts follow from the estimates $\delta((x+z)+(y+w), x+y) \le v(z)+v(w)$, $\delta((x+z)(y+w), xy) \le v(z)v(y)+v(w)v(x)+v(z)v(w)$, and $\delta((x+z)^{-1}, x^{-1}) = v((x+z)((x+z)^{-1}-x^{-1})x)/v(x+z)v(x) = v(z)/v(x+z)v(x)$.

A metric space is *complete* if every Cauchy sequence has a limit in the space. It is a basic fact of topology that every metric space X can be embedded in a complete metric space $\hat{X}$ such that X is dense in $\hat{X}$, and the restriction to X of the distance function on $\hat{X}$ coincides with the original distance function on X. The completion $\hat{X}$ of X is unique: if Y is a complete metric space that contains X as a dense subspace, then there is a unique distance preserving homeomorphism ϕ of $\hat{X}$ to Y such that $\phi|X = id_X$.

Proposition. *Let v be a valuation of the division algebra D. There is a division algebra $\hat{D}$ that contains D as a subalgebra and a valuation $\hat{v}$ of $\hat{D}$ such that:*

(i) *$\hat{D}$ is complete in the $\hat{v}$-topology;*
(ii) *D is dense in $\hat{D}$;*
(iii) $\hat{v}|D = v$;
(iv) $a(\hat{v}) = a(v)$;
(v) *if v is non-archimedean, then* $\hat{v}(\hat{D}^\circ) = v(D^\circ)$;
(vi) *if D is a field, then so is $\hat{D}$.*

The division algebra $\hat{D}$ with properties (i), (ii), *and* (iii) *is unique to within an isomorphism that is the identity on D.*

Proof. Let $\hat{D}$ be the metric space completion of D. The uniform continuity of the ring operations of D implies that these operations extend uniquely to $\hat{D}$. For example, if $\hat{x}, \hat{y} \in \hat{D}$, then there exist sequences $\{x_n : n < \omega\} \subseteq D$ and $\{y_n : n < \omega\} \subseteq D$ such that $\lim_{n\to\infty} x_n = \hat{x}$ and $\lim_{n\to\infty} y_n = \hat{y}$ because D is dense in $\hat{D}$. These Cauchy sequences are necessarily bounded, so that by Lemma b, $\{x_n y_n : n < \omega\}$ is a Cauchy sequence in D and $\hat{D}$. Since $\hat{D}$ is complete, $\lim_{n\to\infty} x_n y_n$ exists. It is easy to check that this limit does not depend on the choice of the sequences converging to $\hat{x}$ and $\hat{y}$. Therefore, $\hat{x}\hat{y} = \lim_{n\to\infty} x_n y_n$ provides a definition of the product in $\hat{D}$. Addition, subtraction, and the inverse operation are similarly derived from the corresponding operations of D by a limit process. The identities that define a division algebra or field follow by continuity from the corresponding laws in D. For example, if $\hat{x} \neq 0$, then $\hat{x} = \lim_{n\to\infty} x_n$, where $\{x_n : n < \omega\} \subseteq D^\circ$,

$\hat{x}^{-1} = \lim_{n\to\infty} x_n^{-1}$, and $\hat{x}\hat{x}^{-1} = \lim_{n\to\infty} x_n x_n^{-1} = 1_D = 1_{\hat{D}}$. If $x \in D$, then $x = \lim_{n\to\infty} x_n$, where $x_n = x$ for all n. It follows that D is a subalgebra of $\hat{D}$. For $\hat{x} \in \hat{D}$, define $\hat{v}(\hat{x}) = \hat{\delta}(\hat{x},0)$. By the definition of the metric topology, $\hat{v}$ is a continuous mapping of $\hat{D}^\circ$ to $\mathbb{R}$; and $\hat{v}|D = v$ since $\hat{\delta}|D^2 = \delta$. It follows easily that $\hat{v}$ is a valuation of $\hat{D}$ with $a(\hat{v}) = a(v)$, and $\hat{\delta}$ is the distance function induced by $\hat{v}$. Thus, $\hat{D}$ is complete in the $\hat{v}$-topology. If v is non-archimedean, then so is $\hat{v}$ by (iv). In this case, if $\hat{x} \neq 0$ in $\hat{D}$, then $x \in D$ exists so that $\hat{v}(\hat{x} - x) < \hat{v}(\hat{x})$. The Domination Principle then implies that $\hat{v}(\hat{x}) = \hat{v}(\hat{x} - (\hat{x} - x)) = \hat{v}(x) = v(x)$. Therefore, $\hat{v}(\hat{D}^\circ) = v(D^\circ)$. The uniqueness of $\hat{D}$ is clear from Lemma b and the assumption that D is dense in $\hat{D}$. □

The division algebra $\hat{D}$ is called the *completion of* D in the v-topology. We will generally use the symbol v instead of $\hat{v}$ to represent the valuation on $\hat{D}$ that is the extension of the valuation on D.

Corollary a. *Let v be a valuation of the division algebra D such that D is complete in the v-topology. If A is a sub-division algebra of D, then the closure of A in D is isomorphic to $\hat{A}$.*

PROOF. By the continuity of addition, subtraction, multiplication and division, the closure B of A is a division algebra that contains A as a dense subalgebra. Since B is complete in the $v|B$-topology, it follows from the uniqueness statement of the proposition that $B \cong \hat{A}$. □

Corollary b. *If v is a non-archimedean valuation of the division algebra D, then $O(\hat{D},v) = P(\hat{D},v) + O(D,v)$ and $P(D,v) = P(\hat{D},v) \cap O(D,v)$. Thus, the inclusion mapping of $O(D,v)$ to $O(\hat{D},v)$ induces an isomorphism of $E(D,v)$ to $E(\hat{D},v)$.*

PROOF. If $0 \neq \hat{x} \in O(\hat{D},v)$, then (since D is dense in $\hat{D}$) there is an element $x \in D$ such that $v(\hat{x} - x) < v(\hat{x}) \le 1$. By the Domination Principle, $v(x) = v(\hat{x}) \le 1$. Hence, $\hat{x} = x + (\hat{x} - x) \in O(D,v) + P(\hat{D},v)$. Obviously, $P(D,v) = P(\hat{D},v) \cap O(D,v)$. Thus, $E(\hat{D},v) = O(\hat{D},v)/P(\hat{D},v) = (O(D,v) + P(\hat{D},v))/P(\hat{D},v) \cong O(D,v)/(P(\hat{D},v) \cap O(D,v)) = O(D,v)/P(D,v) = E(D,v)$. □

If $v = v_\infty$ is the absolute value of the field of rational numbers, then the completion $\hat{\mathbb{Q}}$ is the real field $\mathbb{R}$. Indeed, $\mathbb{R}$ is complete with respect to v_∞ and $\mathbb{Q}$ is dense in $\mathbb{R}$. The completion of $\mathbb{Q}$ in the v_p-topology defined by the prime p is called the field of *p-adic numbers*. We will denote this field by $\hat{\mathbb{Q}}_p$.

The final result of this section demonstrates the usefulness of the completeness property.

Hensel's Lemma. *Let D be a division algebra over the field F, and suppose that D is complete in the topology of a non-archimedean valuation v. Assume that $\Phi \in O(F,v)[\mathbf{x}]$ is such that $\overline{\Phi}$ and $\overline{\Phi}'$ (the derivative of $\overline{\Phi}$) are relatively prime in $E(F,v)[\mathbf{x}]$. If $\bar{x} \in E(D,v)$ is a root of $\overline{\Phi}$, then $y \in O(D,v)$ exists so that $\Phi(y) = 0$ and $\bar{y} = \bar{x}$.*

PROOF. Since the coefficients of Φ are in $\mathbf{Z}(D)$, the expressions $\Phi(y)$ and $\overline{\Phi}(\bar{x})$ are not ambiguous. The same remark applies to similar notation in the proof. The hypotheses that $\overline{\Phi}$ and $\overline{\Phi}'$ are relatively prime and $\overline{\Phi}(\bar{x}) = 0$ imply that $\overline{\Phi}'(\bar{x}) \neq 0$. Thus, if $\bar{x}$ is the coset of $x \in O(D,v)$, then $v(\Phi(x)) < 1 = v(\Phi'(x))$. This inequality enables us to use x as a first approximation of the required y. The construction is a non-archimedean analogue of Newton's method for obtaining the real roots of a polynomial. We use the Taylor expansions

$$\Phi(\mathbf{x} + \mathbf{z}) = \Phi(\mathbf{x}) + \Phi'(\mathbf{x})\mathbf{z} + \Phi_0(\mathbf{x},\mathbf{z})\mathbf{z}^2 \tag{3}$$

and

$$\Phi'(\mathbf{x} + \mathbf{z}) = \Phi'(\mathbf{x}) + \Phi_1(\mathbf{x},\mathbf{z})\mathbf{z} \tag{4}$$

with $\Phi_0, \Phi_1 \in O(F,v)[\mathbf{x},\mathbf{z}]$. Define recursively $x_0 = x$, $x_{n+1} = x_n + z_n$, where $z_n = -\Phi(x_n)/\Phi'(x_n)$. The choice of z_n is such that when $\mathbf{x}$ is replaced by x_n and z_n is substituted for $\mathbf{z}$ in (3), we get $\Phi(x_{n+1}) = \Phi_0(x_n,z_n)(z_n)^2$. From this observation and (4), it follows by induction that $v(\Phi'(x_n)) = 1$ and $v(\Phi(x_n)) \leq v(\Phi(x))^{2^n}$. Consequently,

$$v(x_{n+1} - x_n) = v(z_n) = v(\Phi(x_n)) \leq v(\Phi(x))^{2^n}.$$

Therefore, $y = \lim_{n\to\infty} x_n$ exists by the completeness of D. Moreover, $\Phi(y) = \lim_{n\to\infty} \Phi(x_n) = 0$, and $v(x - y) \leq \max\{v(x_{n+1} - x_n) : n < \infty\} \leq v(\Phi(x)) < 1$ implies $\bar{y} = \bar{x}$. □

There are generalized versions of Hensel's Lemma, but we won't need them in this chapter.

EXERCISES

1. Fill in the details of the proof of the proposition. In particular, show that the definitions of addition, subtraction, multiplication, and division do not depend on the choices of sequences converging to $\hat{x}$ and $\hat{y}$. Verify the identities for division algebras (and the commutative law when D is a field). Show that $\hat{v}$ is a valuation and $a(\hat{v}) = a(v)$.

2. Let K be a field. Define $F = K((\mathbf{x}))$ to be the set of all formal Laurent series $\sum_{n\geq k} \mathbf{x}^n a_n$, with $a_n \in K$, $k \in \mathbb{Z}$. Define the K-space operations componentwise on F, and let multiplication be given by convolution: $(\sum_{n\geq k} \mathbf{x}^n a_n)(\sum_{n\geq l} \mathbf{x}^n b_n) = \sum_{n\geq k+l} \mathbf{x}^n c_n$, where $c_n = \sum_{r+s=n} a_r b_s$. Show that F is a field. Fix $d \in \mathbb{R}^+$ with $d < 1$, and define $v: F^\circ \to \mathbb{R}^+$ by $v(\sum_{n\geq k} \mathbf{x}^n a_n) = d^k$ when $a_k \neq 0$. Prove that v extends to a non-archimedean

valuation of F and F is complete in the v-topology. Show that F is isomorphic to the completion in the $v_{\mathbf{x}}$-topology of the fraction field $K(\mathbf{x})$ of the principal ideal domain $K[\mathbf{x}]$.

3. Show that the polynomial $\Phi(\mathbf{x}) = \mathbf{x}^2 - 2 \in O(\hat{\mathbb{Q}}_2)[\mathbf{x}]$ has no root in $\hat{\mathbb{Q}}_2$, even though $\overline{\Phi}(\mathbf{x}) \in E(\hat{\mathbb{Q}}_2)[\mathbf{x}]$ has a root in $E(\hat{\mathbb{Q}}_2)$. Reconcile this observation with Hensel's Lemma.

17.5. Local Fields

This section introduces the class of fields that occupies our attention in the rest of the chapter. A *local field* is a field F with a non-archimedean valuation v such that $O(F,v)$ is compact in the v-topology. The main result of this section is a practical characterization of local fields.

A valuation v of the division algebra D is *discrete* if $v(D^\circ)$ is a cyclic subgroup of $\mathbb{R}^+$. Thus, v is discrete if and only if there is an element $z \in D^\circ$ such that $v(D^\circ) = \{v(z)^n : n \in \mathbb{Z}\}$. If also $v(z) < 1$, then z is called a *uniformizer* at v. The historical roots of the terminology are in the theory of Riemann surfaces. It is obvious that if v is equivalent to a discrete valuation, then v is discrete.

Proposition a. *Let v be a valuation of the division algebra D.*

(i) *If v is discrete, then v is non-archimedean, and every uniformizer at v generates $P(v)$, both as a right ideal and as a left ideal.*
(ii) *If v is non-archimedean and $P(v)$ is principal as either a left or right ideal of $O(v)$, then v is discrete.*

PROOF. An archimedean valuation v cannot be discrete because $v(D^\circ) \supseteq v(\mathbb{Q}^\circ)$, and $v(\mathbb{Q}^\circ)$ is dense in $\mathbb{R}^+$ by Corollary 17.2 and Theorem 17.3. If z is a uniformizer at the discrete valuation v and $x \in P(v)$, then $v(x) = v(z)^n$ for some $n \in \mathbb{N}$. Hence, $v(z^{-n}x) = 1$, $z^{-n}x \in O(v)$, and $x = z^n(z^{-n}x) \in zO(v)$. Similarly, $x \in O(v)z$. Therefore, z generates $P(v)$ as a left or right ideal. Conversely, assume that v is non-archimedean and $P(v) = zO(v)$. It is then easy to see that $v(z) = \max\{v(x): x \in D, v(x) < 1\}$ and $v(z) < 1$. In particular, if $x \in P(v) - \{0\}$ then $v(z)^n \geq v(x) > v(z)^{n+1}$ for some $n \in \mathbb{N}$. Consequently, $1 \geq v(z^{-n}x) > v(z)$, so that $v(z^{-n}x) = 1$. That is $v(x) = v(z)^n$. It follows easily that $v(D^\circ) = \{v(z)^n : n \in \mathbb{Z}\}$. A similar proof shows that v is discrete if $P(v)$ is a principal left ideal of $O(v)$. □

Lemma. *Let v be a non-trivial discrete valuation of the division algebra D. If z is a uniformizer at v, then $z^nO(v)/z^{n+1}O(v) \cong E(v)$ as abelian groups for all $n \in \mathbb{N}$. In particular, if $E(v)$ is finite, then $|O(v)/z^nO(v)| = |E(v)|^n$ for all $n \in \mathbb{N}$.*

PROOF. The mapping $x \mapsto zx$ is a surjective group homomorphism ϕ of $z^kO(v)$ to $z^{k+1}O(v)$, and $\phi^{-1}(z^{k+2}O(v)) = z^{k+1}O(v)$, since $z \neq 0$ (because v is non-trivial). Hence,

$$z^{k+1}O(v)/z^{k+2}O(v) \cong z^kO(v)/z^{k+1}O(v) \cong \cdots \cong O(v)/zO(v) = E(v). \quad \square$$

Proposition b. *Let v be a non-archimedean valuation of the division algebra D. The ring $O(v)$ is compact in the v-topology if and only if v is discrete, D is complete, and $E(v)$ is finite.*

PROOF. Assume that $O(v)$ is compact. If $\{x_n : n < \omega\}$ is a Cauchy sequence in D, then there exists $m < \omega$ such that $v(x_m - x_{m+k}) \leq 1$ for all $k < \omega$. Therefore, $\{x_m - x_{m+k} : k < \omega\}$ is a Cauchy sequence in the compact metric space $O(v)$. Since compact spaces are complete, it follows that $y = \lim_{k\to\infty} (x_m - x_{m+k})$ exists. Therefore, $\lim_{n\to\infty} x_n = x_m - y$ exists. Hence D is complete. Let $X \subseteq O(v)$ be a set of representatives of the cosets of $P(v)$, that is, $E(v) = \{x + P(v)\colon\ x \in X\}$. Plainly, $O(v) = \bigcup_{x\in X}(x + P(v))$, each set $x + P(v)$ is open in $O(v)$, and $(x + P(v)) \cap (y + P(v)) = \varnothing$ if $x \neq y$ in X. By compactness, X is finite. Therefore, $E(v)$ is finite. Moreover, $P(v) = O(v) - \bigcup\{x + P(v) \colon x \in X - P(v)\}$ is closed in $O(v)$. In particular, $P(v)$ is compact. Thus, $\max\{v(x)\colon x \in P(v)\}$ is attained: there exists $z \in P(v)$ such that $v(x) \leq v(z)$ for all $x \in P(v)$. As in the proof of Proposition a, it follows that v is discrete. Conversely, assume that v is discrete, D is complete, and $E(v)$ is finite. Since $O(v)$ is closed in D, it is also complete. To prove that this set is compact, it suffices (by an elementary theorem of metric topology) to prove that $O(v)$ is totally bounded, that is, for each $e > 0$, $O(v)$ is a finite union of sets with diameter less than e. Let z be a uniformizer at the discrete valuation v. Choose n so that $v(z^n) < e$. It follows easily from the inequality 17.2 (1) that the cosets $x + z^nO(v)$, $x \in O(v)$, have diameter less than e. By the lemma, $O(v)/z^nO(v)$ is finite. Hence, $O(v)$ is a finite union of sets of diameter less than e, that is, $O(v)$ is totally bounded and therefore compact. $\square$

If $O(v)$ is compact in the v-topology of the division algebra D, then D is clearly a locally compact space in the v-topology. The converse statement is false because D is locally compact in the discrete topology, which is the topology of the trivial valuation. However, this is the only exception to the converse. (See Exercise 1.) It is therefore safe to call a division algebra locally compact in the v-topology if $O(v)$ is compact.

It can be proved that the only fields that are locally compact in the topology of an archimedean valuation are $\mathbb{R}$ and $\mathbb{C}$ (Ostrowski's Theorem). By definition, the local fields are the fields that are locally compact in the topology of a non-archimedean valuation.

If v is a non-trivial valuation of the division algebra D such that $O(v)$ is compact in the v-topology, then every closed, bounded subset X of D is compact. In fact, if $0 < v(z) < 1$, then $z^nX \subseteq O(v)$ for a sufficiently large

$n \in \mathbb{N}$, and $x \mapsto z^n x$ is a homeomorphism from X to $z^n X$. In particular, every closed, bounded subset of a local field is compact, since a field with the discrete topology is a local field only if it is finite.

Corollary a. *If v is a discrete valuation of the field F such that $E(F,v)$ is a finite field, then the completion $\hat{F}$ of F in the v-topology is a local field.*

The corollary follows directly from Proposition a, Proposition b, Proposition 17.4, and Corollary 17.4b.

Corollary b. *If p is a rational prime, then $\hat{\mathbb{Q}}_p$ is a local field.*

If F is the fraction field of a principal ideal domain R and $p \in R$ is irreducible, then v_p is a discrete valuation by definition. In the case that $F = \mathbb{Q}$, $R = \mathbb{Z}$, and p is a prime, it follows that $E(v_p) = \mathbb{F}_p$ is finite by Theorem 17.3. Thus, Corollary b follows from Corollary a.

It is convenient to refer to *the* valuation of the local field F, meaning the valuation v such that $O(v)$ is compact. In fact, it can be shown that the valuation with this property is unique up to equivalence.

Exercises

1. Let v be a non-trivial valuation of the division algebra D such that D is a locally compact space in the v-topology. Prove that $O(v)$ is compact. Hint. Local compactness implies that $\{x \in E: v(x) < e\}$ has compact closure for some $e > 0$. Since v is non-trivial, there exists $y \in D^\circ$ that satisfies $v(y) < 1$. Conclude that $y^n O(v)$ is compact for a suitable n.

2. Prove that if v is an archimedean valuation of the division algebra D, and D is locally compact in the v-topology, then D is complete and $v(D^\circ) = \mathbb{R}^+$.

3. Let v be a non-trivial discrete valuation of the field F such that F is complete in the v-topology. Let z be a uniformizer at v. Suppose that X is a set of representatives in $O(v)$ of the cosets of $P(v)$. Prove that every element of F has a unique "Laurant series" representation
$$\sum_{n \geq k} z^n a_n = \lim_{m \to \infty} \sum_{k \leq n \leq m} z^n a_n,$$
where $a_n \in X$ for all n and $k \in \mathbb{Z}$. Deduce that if X can be chosen to be a subfield K of $O(v)$, then F is isomorphic to $K((\mathbf{x}))$.

4. Prove the following statements.
   (a) If K is a finite field, then $K((\mathbf{x}))$ is a local field.
   (b) If K and L are finite fields such that $K((\mathbf{x})) \cong L((\mathbf{x}))$ as fields, then $K \cong L$.
   Hint. Prove that K is the algebraic closure in $K((\mathbf{x}))$ of the prime field of $K((\mathbf{x}))$.
5. Let F be a local field of prime characteristic p. Prove that $F \cong K((\mathbf{x}))$, where $K = E(F)$. Hint. Let $|K| = q$, where q is a power of p. Use Hensel's Lemma to find

$x \in O(F)$ such that $x^{q-1} = 1$. Show that $\mathbb{F}_p(x) \subseteq O(F)$ is a set of representatives of the cosets of $P(F)$. Use the result of Exercise 3.

6. Prove that if F is a field of prime characteristic such that F is locally compact in the topologies of the discrete valuations v and w, then v is equivalent to w.

7. Prove that if F is a local field in the topology of the discrete valuation v, and $z \in P(v)$ is a uniformizer, then every proper non-zero ideal of $O(v)$ has the form $z^n O(v)$ for some $n \in \mathbb{N}$. In particular, $O(v)$ is a principal ideal domain.

17.6. Extension of Valuations

In this section we will prove that if D is a finite dimensional division algebra over a local field F, then the valuation of F extends uniquely to a valuation of D. The following hypotheses and notation are in effect throughout the section: F is a field, v is a valuation of F that satisfies the triangle inequality, and D is a finite dimensional division algebra over F.

Lemma a. *Let M be a finite dimensional F-space with the basis $x_1, \ldots, x_n$. Define the mapping $x \mapsto \|x\|$ from M to $\mathbb{R}$ by $\|x_1 a_1 + \cdots + x_n a_n\| = \max\{v(a_j): 1 \leq j \leq n\}$.*

(i) *$\|x\| \geq 0$ for all $x \in M$, and $\|x\| = 0$ only if $x = 0$.*
(ii) *$\|x + y\| \leq \|x\| + \|y\|$ for all $x, y \in M$.*
(iii) *If $x \in M$ and $a \in F$, then $\|xa\| = \|x\| v(a)$.*

These statements follow routinely from the triangle inequality.

The mapping $x \mapsto \|x\|$ is called the *uniform norm* on M relative to the basis $x_1, \ldots, x_n$. The uniform norm determines a distance function $(x,y) \mapsto \|x - y\|$ which gives a metric topology on M that is called the *uniform topology*.

Lemma b. *Assume that $O(F,v)$ is compact. Let M be a finite dimensional F-space. Define the uniform norm $\|\cdot\|$ relative to a basis $x_1, \ldots, x_n$. If $\theta: M \to \mathbb{R}$ is a mapping that satisfies*

(i) *$\theta(x) \geq 0$ for all $x \in M$, and $\theta(x) = 0$ only if $x = 0$,*
(ii) *$\theta(xa) = \theta(x)v(a)$ for $x \in M$, $a \in F$, and*
(iii) *θ is continuous in the uniform topology of M,*
then c and d exist in $\mathbb{R}^+$ such that $c\|x\| \leq \theta(x) \leq d\|x\|$ for all $x \in M$.

Proof. The set $U = \{y \in M: \|y\| = 1\}$ is compact in the uniform topology. Indeed, U is a closed subset of $V = \{y \in M: \|y\| \leq 1\}$, and $(a_1, \ldots, a_n) \mapsto \sum_{i=1}^n x_i a_i$ is a homeomorphism of the compact space $O(v) \times \cdots \times O(v)$ to V. By (i), $\theta(y) > 0$ for all $y \in U$. Since θ is continuous and U is compact, there exist $c, d \in \mathbb{R}^+$ such that $c \leq \theta(y) \leq d$ for all $y \in U$. Let $0 \neq x =$

$x_1a_1 + \cdots + x_na_n \in M$. Choose i so that $v(a_i) = \max\{v(a_j): 1 \le j \le n\} = \|x\|$. Then $y = xa_i^{-1}$ satisfies $\|y\| = 1$, that is, $y \in U$. Hence, $c \le \theta(y) \le d$ and $c\|x\| \le \theta(y)v(a_i) = \theta(x) \le d\|x\|$. □

This lemma shows in particular that the uniform topology on a finite dimensional space over a local field does not depend on the basis that is used to define it.

Proposition. *Let D be a finite dimensional division algebra over the locally compact field F with the valuation v.*

(i) *v extends uniquely to a valuation w of D.*
(ii) *D is locally compact in the w-topology.*
(iii) *If v is discrete, then so is w and $[w(D^\circ) : v(F^\circ)]$ divides $[\mathbf{Z}(D) : F]$ Deg D.*

PROOF. If v is trivial, then F and D are finite fields. In this case there is nothing to prove. Assume that v is non-trivial. Denote $K = \mathbf{Z}(D)$, $r = [K:F]$, $s = \operatorname{Deg} D$, and $m = rs$. Define $w: D \to \mathbb{R}$ by

$$w(x) = v(N_{K/F}(\nu_{D/K}(x)))^{1/m}.$$

Clearly, w is a group homomorphism of D° to $\mathbb{R}^+$. If $a \in F$, then $w(a) = v(N_{K/F}(\nu_{D/K}(a)))^{1/m} = v(N_{K/F}(a^s))^{1/m} = v(a^{sr})^{1/m} = v(a)$. Thus, w extends v. In particular, $w(xa) = w(x)v(a)$ if $x \in D$ and $a \in F$. We will use Lemma b to prove that w is a valuation. It is necessary to know that w is continuous in the uniform topology of D. Let $x_1, \ldots, x_n$ be a basis of D_F. By Lemma 16.3a, there is a polynomial $\Phi \in F[\mathbf{x}_1, \ldots, \mathbf{x}_n]$ such that

$$N_{K/F}\nu_{D/K}(x_1a_1 + \cdots + x_na_n) = \Phi(a_1, \ldots, a_n).$$

It follows from Lemma 17.4b that the mapping $(a_1, \ldots, a_n) \mapsto \Phi(a_1, \ldots, a_n)$ from F^n to F is continuous; and $v: F \mapsto \mathbb{R}$ is continuous by the definition of the topology of F. Therefore, w is continuous. By Lemma b there are numbers $c, d \in \mathbb{R}^+$ such that $c\|x\| \le w(x) \le d\|x\|$ for all $x \in D$. Thus, if $x, y \in D$, then $w(x + y) \le d\|x + y\| \le d(\|x\| + \|y\|) \le dc^{-1}(w(x) + w(y)) \le 2dc^{-1}\max\{w(x), w(y)\}$. Hence, w is a valuation of D. By Lemma b, any two extensions of v to D determine the same topology and are therefore equivalent by Lemma 17.4a. However, two equivalent extensions of a non-trivial valuation are obviously identical. Thus, w is unique. By Lemma b, the w-topology of D coincides with the uniform topology. Hence, $O(D,w)$ is closed and bounded in the uniform topology. Since F is locally compact, the closed, bounded subsets of F are compact. This property extends to the uniform topology on D because D is homeomorphic to a finite product of copies of F. Thus, $O(D,w)$ is compact; that is, D is locally compact. The definition of w implies that $w(x) \mapsto w(x)^m$ is an injective group homomorphism from $w(D^\circ)$ to $v(F^\circ)$. If v is discrete, then $v(F^\circ)$ is cyclic, and it follows that the index of $v(F^\circ)$ in $w(D^\circ)$ divides $m = [K:F]\operatorname{Deg} D$. □

Henceforth we will use v to denote both the valuation of F and the extension of this valuation to a finite dimensional division algebra over F.

Corollary. *If F is a local field and K/F is a finite field extension, then K is a local field.*

Exercises

1. Prove Lemma a.
2. By embedding $\mathbb{Q}(i)$ in $\hat{\mathbb{Q}}_p(i)$, prove:
   (a) if $\mathbf{x}^2 + 1$ is irreducible in $\hat{\mathbb{Q}}_p[\mathbf{x}]$, then v_p has a unique extension to $\mathbb{Q}(i)$;
   (b) if $\mathbf{x}^2 + 1$ factors in $\hat{\mathbb{Q}}_p[\mathbf{x}]$ and $p \neq 2$, then v_p has two extensions to $\mathbb{Q}(i)$.
3. Let $E/\mathbb{Q}$ be an algebraic extension. Suppose that $\phi: E \to \mathbb{C}$ is a $\mathbb{Q}$-algebra homomorphism. Define $v_\phi(x) = |\phi(x)|$ for $x \in E$. Prove that v_ϕ is a valuation of E that extends the absolute value on $\mathbb{Q}$. Apply this construction with $E = \mathbb{Q}(\sqrt{2})$ to obtain two inequivalent extensions of the absolute value on $\mathbb{Q}$.

17.7. Ramification

Let v be a discrete valuation of the division algebra D. If K is a subfield of D such that $v|K$ is non-trivial, then $v(K^\circ)$ is a subgroup of finite index in $v(D^\circ)$. The order of the finite cyclic group $v(D^\circ)/v(K^\circ)$ is called the *ramification index* of K in D (at v), and this number is denoted by $e(D/K)$ (or $e_v(D/K)$). When $e(D/K) = 1$, D is said to be an *unramified extension* of K. If $\pi: O(D,v) \to E(D,v)$ is the natural projection, then $\operatorname{Ker}(\pi|O(K,v)) = K \cap P(D,v) = P(K,v)$. Thus, the inclusion mapping $O(K,v) \to O(D,v)$ induces an injective ring homomorphism $E(K,v) \to E(D,v)$. In this situation it will be convenient to identify $E(K,v)$ with a subfield of $E(D,v)$. In particular, $E(D,v)$ can be viewed as a right $E(K,v)$-space whose dimension is called the *relative degree* of D/K (at v). The relative degree of D/K is denoted by $f(D/K)$ (or $f_v(D/K)$). By definition, $f(D/K)$ is a non-negative integer or ∞. The purpose of this section is to establish some fundamental properties of $e(D/F)$ and $f(D/F)$ in the case that F is a local field and D is a finite dimensional division algebra over F. In the lemmas, local compactness is not needed.

Lemma a. *Let D be a finite dimensional division algebra over the field F. If v is a non-trivial discrete valuation of D, then $v|F$ is non-trivial, $E(D,v)$ is a finite dimensional $E(F,v)$-algebra, and $e(D/F)f(D/F) \leq \dim_F D$.*

Proof. If $v|F$ is trivial, then $\sup v(D) = \sup v(X)$, where X is a basis of D_F. The assumption that $\dim_F D < \infty$ implies that v is bounded. However, non-trivial valuations are obviously unbounded. Therefore, $v|F$ is non-trivial. We have noted that $E(F,v)$ is a subfield of the division algebra $E(D,v)$

under the identification of $E(F,v)$ with the subring $(O(F,v) + P(D,v))/P(D,v)$ of $O(D,v)/P(D,v) = E(D,v)$. Since $F \subseteq \mathbf{Z}(D)$, it follows that $E(F,v) \subseteq \mathbf{Z}(E(D,v))$, that is, $E(D,v)$ is an $E(F,v)$-algebra. It remains to show that $e(D/F)f(D/F) \le \dim_F D$. To simplify our notation, write $e = e(D/F)$, $B = O(D,v)$, $A = O(F,v) = B \cap F$, $Q = P(D,v)$, and $P = P(F,v) = Q \cap F$. Let $z \in Q$ be a uniformizer at v, and suppose that $c \in P$ is a uniformizer at $v|F$. It follows that $v(z)^e = v(c)$, $z^e B = cB$, and $P = cA = z^e B \cap A$. In particular, if $\pi\colon B \to B/z^e B$ is the projection mapping, then $\pi(A) \cong A/(z^e B \cap A) = A/P = E(F,v)$, and $B/z^e B$ is a $\pi(A)$-space. By Lemma 17.5, $E(D,v) = B/zB \cong zB/z^2B \cong \cdots \cong z^{e-1}B/z^e B$ as $\pi(A)$-spaces. Thus, $\dim_{\pi(A)} B/z^e B = e \dim_{\pi(A)} E(D,v)$. If $x_1, \ldots, x_m \in B$ are such that $\pi(x_1), \ldots, \pi(x_m)$ are linearly independent over $\pi(A)$, then $x_1, \ldots, x_m$ are linearly independent over F. Otherwise, there is an equation of the form $x_i = \sum_{j \ne i} x_j a_j$ with $a_j \in A$; and this expression yields the contradiction $\pi(x_i) = \sum_{j \ne i} \pi(x_j)\pi(a_j)$ to the independence of $\pi(x_1), \ldots, \pi(x_m)$. Thus, $ef(D/F) = e \dim_{E(F,v)} E(D,v) = \dim_{\pi(A)} B/z^e B \le \dim_F D$. □

Lemma b. *Let v be a discrete valuation of the division algebra D. Assume that K is a subfield of D such that $v|K$ is non-trivial and $f(D/K) = 1$. If $z \in P(D,v)$ is a uniformizer and $e = e(D/K)$, then the K-space $N = K + zK + \cdots + z^{e-1}K$ is dense in D.*

PROOF. Let $c \in K \cap P(D,v) = P(K,v)$ be a uniformizer at $v|K$; thus, $v(c) = v(z)^e$. Since v is discrete (hence 0 is the only limit point of $v(D^\circ)$), it is sufficient to show that for each $x \in D^\circ$, there exists $y \in N$ such that $v(x - y) < v(x)$. Let $v(x) = v(z)^n$, where $n \in \mathbb{Z}$. If $n = ke + r$ with $k, r \in \mathbb{Z}$, $0 \le r < e$, then $v(z^{-r}xc^{-k}) = 1$. Thus, $z^{-r}xc^{-k} \in O(v)$. The hypothesis $f(D/K) = 1$ implies $E(D,v) = E(K,v)$, that is, $O(v) = (K \cap O(D,v)) + P(D,v)$. Consequently, $z^{-r}xc^{-k} = d + w$ for a suitable $d \in K$ and $w \in P(v)$. Let $y = z^r dc^k$. Then $y \in z^r K \subseteq N$ and $v(x - y) = v(z^r wc^k) = v(z)^{ke+r}v(w) < v(z)^n = v(x)$. □

We can now prove the main result of this section.

Proposition. *Let D be a division algebra over the infinite field F. Assume that v is a discrete valuation of D such that F is closed in D and D is locally compact in the v-topology.*

(i) *F is a local field and* $\dim_F D = e(D/F)f(D/F) < \infty$.
(ii) *There is a subfield K of D such that K/F is unramified, $f(D/K) = 1$, and $f(D/F) = [K:F]$.*
(iii) *If $F = \mathbf{Z}(D)$, then $e(D/F) = f(D/F)$ and K is a maximal subfield of D.*

PROOF. The hypothesis that F is closed in D implies that $O(F,v) = F \cap O(D,v)$ is closed in the compact space $O(D,v)$. Therefore, F is a local field, and $v|F$ is nontrivial because F is infinite. By Proposition 17.5b, $E(D,v)$ and $E(F,v)$ are finite fields. In particular, $E(D,v) = E(F,v)(\bar{x})$ for some $\bar{x}$. Let $\Phi \in O(F,v)[\mathbf{x}]$ be a monic polynomial of degree $f(D/F)$ such that $\bar{\Phi}$ is the minimum polymial of $\bar{x}$ over $E(F,v)$. Any such Φ is irreducible over F; otherwise, Gauss's Lemma applied to the principal ideal domain $O(F,v)$ (Exercise 7, Section 17.5) gives $\Phi = \Psi\Theta$ in $O(F,v)[\mathbf{x}]$ and $\bar{\Phi} = \bar{\Psi}\bar{\Theta}$ in $E(F,v)[\mathbf{x}]$, which contradicts the irreducibility of $\bar{\Phi}$. Since $E(F,v)$ is perfect and $\bar{\Phi}$ is irreducible, the polynomials $\bar{\Phi}$ and $\bar{\Phi}'$ are relatively prime. Therefore, Hensel's Lemma provides an element $y \in D$ such that $\Phi(y) = 0$ and $\bar{y} = \bar{x}$. If $K = F(y)$, then K is a subfield of D that satisfies $[K:F] = \deg \Phi = f(D/F)$ and $E(K,v) = E(D,v)$, that is $f(D/K) = 1$. By Lemma b, there is a subspace N of D_K that is dense in D, $\dim_K N \le e(D/K) \le e(D/F)$, and $\dim_F N \le e(D/F)f(D/F)$. Since the v-topology on N coincides with the uniform topology by Lemma 17.6b, N is complete. Consequently, $N = D$, and $\dim_F D \le e(D/F)f(D/F) < \infty$. By Lemma a, $\dim_F D = e(D/F)f(D/F)$. Moreover, $e(D/K) = e(D/F)$, $v(K^\circ) = v(F^\circ)$, and K/F is unramified. Assume that $\mathbf{Z}(D) = F$. By Proposition 17.6, $e(D/F) \le \operatorname{Deg} D$; and $f(D/F) = [K:F] \le \operatorname{Deg} D$ because K is a subfield of D. Therefore, $[K:F] = f(D/F) = e(D/F) = \operatorname{Deg} D$, since $e(D/F)f(D/F) = \dim_F D = (\operatorname{Deg} D)^2$. □

The first corollary of the proposition plays a fundamental role in the theory of division algebras over local fields.

Corollary a. *Let F be a local field. If $D \in \mathfrak{S}(F)$ is a division algebra, then there is a maximal subfield K of D such that K/F is unramified.*

Corollary b. *Let $F \subseteq K \subseteq E$ be a chain of local fields such that E/F is a finite extension.*

(i) *$e(E/F) = e(E/K)e(K/F)$ and $f(E/F) = f(E/K)f(K/F)$.*
(ii) *It is possible to choose K so that K/F is unramified and $f(E/F) = f(K/F) = [K:F]$.*

The index of $v(F^\circ)$ in $v(E^\circ)$ is the product of the index of $v(F^\circ)$ in $v(K^\circ)$ and the index of $v(K^\circ)$ in $v(E^\circ)$, so that $e(E/F) = e(E/K)e(K/F)$ whether or not the fields F, K, and E are local. The other statements of the corollary follow from the proposition.

The last corollary provides a description of the local fields that have characteristic zero.

Corollary c. *If F is a field of characteristic zero, then F is local if and only if F is a finite extension of $\hat{\mathbb{Q}}_p$ for some prime p.*

Proof. If $F/\hat{\mathbb{Q}}_p$ is finite, then F is local by Corollaries 17.5b and 17.6. Let F be a local field with $\mathbb{Q} \subseteq F$. The restriction $v|\mathbb{Q}$ is non-trivial; otherwise, the projection $O(F,v) \to E(F,v)$ would embed $\mathbb{Q}$ in the finite field $E(F,v)$. By

Theorem 17.3, $v|\mathbb{Q}$ is equivalent to v_p for some prime p, and the closure of $\mathbb{Q}$ in F is isomorphic to $\hat{\mathbb{Q}}_p$ by Corollary 17.4a. The proposition implies that F is a finite extension of $\hat{\mathbb{Q}}_p$. □

Another characterization of the local fields of characteristic zero will be given in Exercise 3 of Section 17.8.

Exercises

1. Let v be a discrete valuation of the division algebra D such that D is locally compact in the v-topology. Prove that $\mathbf{Z}(D)$ is a local field and D is finite dimensional over $\mathbf{Z}(D)$. Hint. Prove that $\mathbf{Z}(D)$ is closed in D. Apply the proposition.

2. Let K be a finite field, $E = K((\mathbf{x}))$, and $F = K((\mathbf{x}^n))$, that is, F is the field of formal Laurent series in $\mathbf{x}^n$. Note that $E \cong F$ as K-algebras by $\mathbf{x} \mapsto \mathbf{x}^n$. Prove that if L/F is any field extension of degree n such that $e(L/F) = n$, then $L \cong E$ as F-algebras.

3. Let p be an odd prime. Prove the following statements.

   (a) If $a \in \mathbb{Z}$ is not divisible by p, then $a \in \hat{\mathbb{Q}}_p^2$ if and only if a is a quadratic residue modulo p.

   (b) If $a \in \mathbb{Z}$ is not divisible by p and a is a non-residue modulo p, then $\hat{\mathbb{Q}}_p(\sqrt{a})$ is an unramified quadratic extension of $\mathbb{Q}_p$. Moreover, if $b \in \mathbb{Z}$ is also not divisible by p and a non-residue modulo p, then $\hat{\mathbb{Q}}_p(\sqrt{b}) \cong \hat{\mathbb{Q}}_p(\sqrt{a})$ as $\hat{\mathbb{Q}}_p$ algebras.

   (c) $p \notin \hat{\mathbb{Q}}_p^2$ and $e(\hat{\mathbb{Q}}_p(\sqrt{p})/\hat{\mathbb{Q}}_p) = 2$.

4. (a) Use Hensel's Lemma to prove that if p is a prime, $n \in \mathbb{N}$ is relatively prime to $p - 1$, and $x \in \hat{\mathbb{Q}}_p$ satisfies $v_p(x) = 1$, then $x \in (\hat{\mathbb{Q}}_p)^n$.

   (b) Deduce from (a) that if p and q are distinct primes, then v_q cannot be extended to a discrete valuation of $\hat{\mathbb{Q}}_p$.

   (c) Use (b) to show that if v and w are discrete valuations of the field F, char $F = 0$, and F is locally compact in the v-topology and the w-topology, then v is equivalent to w.

5. Let v be a discrete valuation of the finite dimensional division algebra D over the field F. The extension D/F is *totally ramified* (at v) if $e(D/F) = \dim_F D$. In the following statements it is assumed that F is a local field. Prove these assertions.

   (a) If E/F is a finite field extension, then there is a field K between F and E such that K/F is unramified and E/K is totally ramified.

   (b) If x is a root of a polynomial $\mathbf{x}^n + a_1\mathbf{x}^{n-1} + \cdots + a_n$, where $a_i \in P(F,v)$ for all i and a_n is a uniformizer at v, then $F(x)/F$ is a totally ramified extension of degree n. Any polynomial of this form is called an *Eisenstein polynomial.*

   (c) If K/F is a totally ramified extension of degree n, then $K = F(x)$, where x is a uniformizer in $P(K,v)$ at v. Moreover, the minimum polynomial of x over F is an Eisenstein polynomial.

17.8. Unramified Extensions

Corollary 17.7a is the key to the structure of Brauer groups of local fields. The application of this result is based on the properties of unramified extensions of local fields. The purpose of this section is to prove the needed facts about such unramified extensions.

It is convenient to fix our hypotheses and introduce simplified notation to be used throughout this section. Let K/F be a finite field extension of degree n. Assume that K and F are local fields relative to the discrete valuation v. For our purposes, the case in which K and F are finite is uninteresting. Therefore, it can be assumed that v is non-trivial. Denote $A = O(F,v)$, $P = P(F,v)$, $B = O(K,v)$, and $Q = P(K,v)$. We will also write $\bar{F}$ for $E(F,v)$ and $\bar{K}$ for $E(K,v)$. As usual, $\bar{F}$ is identified with the image of A under the residue class mapping $y \mapsto \bar{y}$ of B to $\bar{K}$. Since K and F are local fields, $\bar{K}$ and $\bar{F}$ are finite. Denote the order of $\bar{F}$ by q, where q is a power of the characteristic of $\bar{F}$.

Lemma a. *If K/F is a Galois extension and $\sigma \in \mathbf{G}(K/F)$, then $v(y^\sigma) = v(y)$ for all $y \in K$.*

Proof. The mapping $v: K \to \mathbb{R}$ defined by $v^\sigma(y) = v(y^\sigma)$ is clearly a valuation of K such that $v^\sigma | F = v | F$. By the uniqueness part of Proposition 17.6, $v^\sigma = v$. That is, $v(y^\sigma) = v(y)$ for all $y \in K$. □

Lemma b. *If K/F is a Galois extension, then there is a homomorphism $\phi: \mathbf{G}(K/F) \to \mathbf{G}(\bar{K}/\bar{F})$ such that $\overline{y^\sigma} = \bar{y}^{\phi(\sigma)}$ for all $y \in B$ and $\sigma \in \mathbf{G}(K/F)$.*

Proof By Lemma a, $\sigma(B) = B$, $\sigma(Q) = Q$, and $\sigma|A = id_A$. Thus, σ induces an automorphism $\phi(\sigma)$ of $B/Q = \bar{K}$ such that $\bar{y}^{\phi(\sigma)} = \overline{y^\sigma}$. If $b \in A$, then $\bar{b}^{\phi(\sigma)} = \overline{b^\sigma} = \bar{b}$. Thus, $\phi(\sigma) \in \mathbf{G}(\bar{K}/\bar{F})$. Clearly, $\phi(\sigma\tau) = \phi(\sigma)\phi(\tau)$; that is, ϕ is a group homomorphism. □

Proposition. *The following properties of K/F are equivalent.*

(i) *K is the splitting field over F of $\mathbf{x}^{l-1} - 1$, where $l = q^n$.*
(ii) *K/F is Galois, and $K = F(y)$, where y is a root of a monic polynomial $\Phi \in A[\mathbf{x}]$ such that $\bar{\Phi}$ has distinct roots in an extension of $\bar{F}$.*
(iii) *K/F is Galois, and $K = F(y)$, where y is a root of a monic, irreducible polynomial $\Psi \in A[\mathbf{x}]$ such that $\bar{\Psi}$ has distinct roots in $\bar{K}$.*
(iv) *K/F is Galois, and the homomorphism $\phi: \mathbf{G}(K/F) \to \mathbf{G}(\bar{K}/\bar{F})$ of Lemma b is an isomorphism.*
(v) *K/F is unramified.*

Proof. If (i) is true, then K/F is Galois, and there is a primitive l'th root of unity $y \in K$. Hence, $K = F(y)$, so that (ii) is satisfied. Assume that (ii) holds. Write $\Phi = \Psi_1 \cdots \Psi_r$ with $\Psi_1, \ldots, \Psi_r$ monic and irreducible in $F[\mathbf{x}]$. Since

A is a principal ideal domain, Gauss's Lemma implies that $\Psi_i \in A[\mathbf{x}]$ for all i. For some i, $\overline{\Psi}_i(y) = 0$, since $\Phi(y) = 0$. The roots of $\overline{\Psi}_i$ are distinct because the roots of $\overline{\Phi}$ are distinct. By Lemma a, Ψ_i splits completely in $B[\mathbf{x}]$ because K/F is Galois; hence, all the roots of $\overline{\Psi}_i$ are in $\bar{K}$. If (iii) is satisfied, then $\Psi = \prod_{\sigma \in G}(\mathbf{x} - y^\sigma)$ and $\overline{\Psi} = \prod_{\sigma \in G}(\mathbf{x} - \bar{y}^{\phi(\sigma)})$, where G abbreviates $\mathbf{G}(K/F)$. The assumption that $\overline{\Psi}$ has distinct roots therefore implies that ϕ is injective. Since $\bar{K}$ is finite, $\bar{K}/\bar{F}$ is Galois, and therefore $|\mathbf{G}(\bar{K}/\bar{F})| = [\bar{K}:\bar{F}] = [\bar{F}(\bar{y}):\bar{F}] \leq \deg \overline{\Psi} = \deg \Psi = [K:F] = |\mathbf{G}(K/F)|$. It follows that ϕ is an isomorphism. If (iv) holds, then so does (v) because $e(K/F)f(K/F) = [K:F] = |\mathbf{G}(K/F)| = |\mathbf{G}(\bar{K}/\bar{F})| = [\bar{K}:\bar{F}] = f(K/F)$ by Proposition 17.7. Assume that K/F is unramified. By Proposition 17.7, $[\bar{K}:\bar{F}] = [K:F] = n$. Therefore, $|\bar{K}| = |\bar{F}|^n = q^n = l$ and $\bar{K}$ is the splitting field over $\bar{F}$ of $\mathbf{x}^{l-1} - 1$. By Hensel's Lemma, there is an element $y \in B$ such that $y^{l-1} = 1$ and $\bar{y}$ is a primitive l'th root of unity in $\bar{K}$. Thus, y is a primitive l'th root of unity in K, $\bar{K} = \bar{F}(\bar{y})$, and $n = [\bar{K}:\bar{F}] = [\bar{F}(\bar{y}):\bar{F}] \leq [F(y):F] \leq [K:F] = n$ by Lemma 17.7a. Consequently, $K = F(y)$ is the splitting field over F of $\mathbf{x}^{l-1} - 1$. □

Let F_s denote the separable algebraic closure of the field F.

Corolloary a. *If F is a local field, then for each $n \in \mathbb{N}$, there is a unique field K between F and F_s such that $[K:F] = n$ and K/F is unramified.*

This corollary is a direct consequence of the equivalence of (i), (iv), and (v) in the proposition.

For a local field F, let $K_n(F)$ denote the subfield of F_s such that $K_n(F)/F$ is an unramified extension of degree n. By Corollary 17.6, $K_n(F)$ is a local field and $v(K_n(F)) = v(F)$. The extension $K_n(F)/F$ is Galois with $\mathbf{G}(K_n(F)/F) \cong \mathbf{G}(\overline{K_n(F)}/\bar{F})$. Since $\bar{F}$ is finite of order q, the group $\mathbf{G}(\overline{K_n(F)}/\bar{F})$ is cyclic of order n. It has a canonical generator: the mapping $\bar{x} \mapsto \bar{x}^q$. The corresponding element of $\mathbf{G}(K_n(F)/F)$ is called the *Frobenius automorphism* (or Frobenius substitution) of $K_n(F)/F$. We will denote this generator of $\mathbf{G}(K_n(F)/F)$ by σ_F (or just σ when this abbreviation is permissible). Explicitly, σ is defined to be the F-algebra automorphism of $K_n(F)$ that satisfies $v(x^\sigma - x^q) < 1$ for all $x \in O(K_n(F))$.

From our viewpoint, the next corollary is the most important consequence of the proposition.

Corollary b. *If F is a local field and $D \in \mathfrak{S}(F)$ is a division algebra, then D is cyclic.*

This result follows from Corollary 17.7a and the proposition.

Exercises

1. Let F_s be the separable algebraic closure of the local field F. Define $F_{nr} = \bigcup_{n \in \mathbb{N}} K_n(F)$. Prove the following statements.

   (a) F_{nr} is a subfield of F_s such that F_{nr}/F is an infinite Galois extension with $\mathbf{G}(F_{nr}/F) \cong \varprojlim \mathbb{Z}/n\mathbb{Z} = \hat{\mathbb{Z}}$.

   (b) There exists $\sigma \in \mathbf{G}(F_{nr}/F)$ such that $\sigma|K_n(F)$ is the Frobenius automorphism of $K_n(F)/F$.

   (c) If E is a finite extension of F with $E \subseteq F_s$, then $(E \cap F_{nr})/F$ is an unramified extension and $E/(E \cap F_{nr})$ is a totally ramified extension.

2. (a) Prove *Krasner's Lemma*: let F, K, and E be subfields of the local field L such that $F \subseteq E \cap K$, E/F is Galois, and K is an infinite, closed subfield of L; if $x \in E$ and $y \in K$ satisfy $v(x - y) < v(x^\sigma - x)$ for all $\sigma \in \mathbf{G}(E/F)$ such that $x^\sigma \neq x$, then $x \in K$. Hint. Use Lemma a and the Domination Principle to prove that $x^\sigma = x$ for all $\sigma \in \mathbf{G}(EK/K)$.

   (b) Deduce from (a) that if K is a local field, E/K is a Galois extension, and $x, y \in E$ satisfy $v(x - y) < v(x^\sigma - x)$ for all $\sigma \in \mathbf{G}(E/K)$ such that $x^\sigma \neq x$ and $v(x - y) < v(y^\tau - y)$ for all $\tau \in \mathbf{G}(E/K)$ such that $y^\tau \neq y$, then $F(y) = F(x)$.

3. (a) Let K be a local field of characteristic zero, and suppose that L is the splitting field over K of a monic polynomial $\Phi \in K[\mathbf{x}]$. Write $\Phi(\mathbf{x}) = (\mathbf{x} - x_1) \cdots (\mathbf{x} - x_n) = \mathbf{x}^n + a_1\mathbf{x}^{n-1} + \cdots + a_n$ with $x_i \in L$ and $a_i \in K$. Denote $r = \min\{v(x_i - x_j): i \neq j\}$ and $s = \max\{v(x_i)^k: 0 \leq k < n, 1 \leq i \leq n\}$. Prove that $r > 0, s \geq 1$, and if $b_1, \ldots, b_n \in K$ satisfy $v(b_k - a_k) < r^n/s$ for $1 \leq k \leq n$, then $\Psi(\mathbf{x}) = \mathbf{x}^n + b_1\mathbf{x}^{n-1} + \cdots + b_n$ is irreducible in $K[\mathbf{x}]$, and there is a root y of Ψ such that $K(y) = K(x_1)$. Hint. Write $\Psi(\mathbf{x}) = (\mathbf{x} - y_1) \cdots (\mathbf{x} - y_n)$, with the y_j in a splitting field of Ψ over L. Note that for $1 \leq i \leq n, \prod_{j=1}^n v(x_i - y_j) = v(\Psi(x_i) - \Phi(x_i)) \leq \max\{v(b_k - a_k)v(x_i)^{n-k}: 1 \leq k \leq n\} < r^n$. Deduce that there exists $j(i)$ such that $v(x_i - y_{j(i)}) < r$. Use the Domination Principle to show that $v(y_{j(i)} - y_{j(k)}) \geq r$ if $i \neq k$. Apply Exercise 2(b)

   (b) Prove that if E is a local field of characteristic zero, then there is a finite extension F of $\mathbb{Q}$ and a discrete valuation v of F such that E is isomorphic to the completion of F in the v-topology. Hint. Use Corollary 17.7c and the result in (a).

17.9. Norm Factor Groups

We are close to the description of $\mathbf{B}(F)$ when F is a local field. By Corollary 17.7a, $\mathbf{B}(F) = \bigcup \mathbf{B}(K/F)$, where the union is over the finite, unramified extensions of F. Every unramified extension K/F is cyclic by Proposition 17.8, so that $\mathbf{B}(K/F) \cong F^\circ/N_{K/F}(K^\circ)$. It remains to describe the norm factor groups $F^\circ/N_{K/F}(K^\circ)$ of unramified extensions of the local field F. It will be shown in this section that if $[K:F] = n$, then $F^\circ/N_{K/F}(K^\circ)$ is cyclic of order n.

The hypotheses and notation that were introduced in Section 17.8 will be used in this section. We also assume that $K = K_n(F)$, that is, the unramified extension K/F (of local fields) has degree n. Therefore, $[\bar{K}:\bar{F}] = n$ and the Galois groups $G = \mathbf{G}(K/F)$ and $\bar{G} = \mathbf{G}(\bar{K}/\bar{F})$ are cyclic of order n. The Frobenius automorphism σ generates G, and the image $\bar{\sigma}$ of σ under the isomorphism $\phi: G \to \bar{G}$ satisfies $\bar{x}^{\bar{\sigma}} = \bar{x}^q$, where $q = |\bar{F}|$. It is convenient to abbreviate the norm and trace mappings from K to F by N and T. Thus,

$N(y) = \prod_{j<n} y^{\sigma^j}$ and $T(y) = \sum_{j<n} y^{\sigma^j}$ for all $y \in K$. Similarly, $\bar{N}: \bar{K} \to \bar{F}$ and $\bar{T}: \bar{K} \to \bar{F}$ will denote the norm and trace mappings for $\bar{K}/\bar{F}$.

We record a standard fact, leaving its proof as Exercise 1.

Lemma a. $\bar{N}(\bar{K}) = \bar{F}$; $\bar{T}(\bar{K}) = \bar{F}$.

The heart of the proof of our main result is in the next lemma.

Lemma b. *If* $V = B^\circ = \{y \in K: v(y) = 1\}$ *and* $U = A^\circ = \{a \in F: v(a) = 1\}$, *then* $N(V) = U$.

Proof. If $y \in K$, then $v(N(y)) = \prod_{j<n} v(y^{\sigma^j}) = v(y)^n$, and $v(T(y)) = v(\sum_{j<n} d^{\sigma^j}) \le v(d)$ by Lemma 17.8a and the fact that v is non-archimedean. In particular, N maps the compact set V continuously to U. Hence, $N(V)$ is a closed subset of U, and the lemma can be proved by showing that $N(V)$ is dense in U. If $y \in B$, then

$$\overline{N(y)} = \bar{N}(\bar{y}) \quad \text{and} \quad \overline{T(y)} = \bar{T}(\bar{y}). \tag{1}$$

Indeed, $\overline{N(y)} = \overline{\prod_{j<n} y^{\sigma^j}} = \prod_{j<n} \bar{y}^{\bar{\sigma}^j} = \bar{N}(\bar{y})$. A similar calculation gives $\overline{T(y)} = \bar{T}(\bar{y})$. It follows from Lemma a and (1) that

$$A = T(B) + P, \tag{2}$$

and

$$U = N(V)(1 + P), \tag{3}$$

that is, every element of U can be written in the form $N(y)(1 + b)$, where $y \in V$ and $b \in P$. Let $a \in P$ be a uniformizer: $P = aA$. Since K/F is unramified, we also have $Q = aB$. By (2), $P^k = a^k A = a^k T(B) + a^k P = T(a^k B) + P^{k+1} = T(Q^k) + P^{k+1}$ for $k \ge 1$. Thus, $1 + P^k = 1 + T(Q^k) + P^{k+1}$. Moreover, if $y \in Q^k$, then $N(1 + y) = \prod_{j<n}(1 + y^{\sigma^j}) = 1 + \sum_{j<n} y^{\sigma^j} + c = 1 + T(y) + c$, where $c \in Q^{2k} \cap A = P^{2k}$. That is, $1 + T(Q^k) \subseteq N(1 + Q^k) + P^{2k}$. Therefore, $1 + P^k \subseteq N(1 + Q^k) + P^{k+1} \subseteq N(1 + Q^k)(1 + P^{k+1})$ for $k \ge 1$. It follows by induction from (3) that $U \subseteq N(V)N(1 + Q^k)(1 + P^{k+1}) = N(V)(1 + P^{k+1}) \subseteq U$ for all k. Hence, $N(V)$ is dense in U. □

Proposition. *If K is an unramified extension of the local field F and* $[K:F] = n$, *then* $F^\circ/N_{K/F}(K^\circ)$ *is a cyclic group of order n. Moreover, if* $a \in F^\circ$ *is a uniformizer, then* $aN_{K/F}(K^\circ)$ *generates* $F^\circ/N_{N/F}(K^\circ)$.

Proof. The Snake Lemma gives the following commutative diagram with exact rows and columns.

$$\begin{array}{ccccccc}
1 \to & V & \to & K^\circ & \overset{v}{\to} \langle v(a)\rangle \to 1 \\
& \downarrow N & & \downarrow N & \downarrow \eta_n \\
1 \to & U & \to & F^\circ & \overset{v}{\to} \langle v(a)\rangle \to 1 \\
& \downarrow & & \downarrow & \downarrow \\
& U/N(V) & \to & F^\circ/N(K^\circ) & \to \mathbb{Z}/n\mathbb{Z} \to 1
\end{array}$$

As in Chapter 16, η_n is the exponential mapping $t \mapsto t^n$. By Lemma b, $U/N(V) = 1$. Thus, v induces an isomorphism of $F^\circ/N(K^\circ)$ to $\mathbb{Z}/n\mathbb{Z}$ with $aN(K^\circ)$ mapped to a generator of $\mathbb{Z}/n\mathbb{Z}$. □

EXERCISES

1. Prove Lemma a. Hint. Use Proposition 15.1b to show that $\bar{N}(\bar{K}) = \bar{F}$. By linearity, the trace map is surjective if it is not zero.

2. Let v be a non-trivial, discrete valuation of the field F. Denote $A = O(v)$, $P = P(v)$, $\bar{F} = E(v) = A/P$, $U_0 = U = A^\circ = O(v) - P(v)$, and for $n \in \mathbb{N}$, $U_n = 1 + P^n = \{1 + b : b \in P^n\}$. Show that U_n is a subgroup of U, and prove the following statements.
   (a) $F^\circ \cong U \times \mathbb{Z}$. Hint. Show that the exact sequence $1 \to U \to F^\circ \to v(F^\circ) \to 1$ splits.
   (b) $U_0/U_1 \cong \bar{F}^\circ$. Hint. Map $U_0 \to \bar{F}^\circ$ by $c \mapsto \bar{c}$.
   (c) For $n \in \mathbb{N}$, U_n/U_{n+1} is isomorphic to the additive group of $\bar{F}$. Hint. Let a be a uniformizer at v. Map $1 + a^n c \in U_n$ to $\bar{c} \in \bar{F}$.

17.10. Brauer Groups of Local Fields

We are ready to assemble parts from the previous sections to obtain a complete description of the Brauer groups of local fields.

Theorem. *If F is a local field with the uniformizer a, then $\mathbf{B}(F) \cong \mathbb{Q}/\mathbb{Z}$ via the mapping*

$$\theta_F : k/n + \mathbb{Z} \mapsto [(K_n(F), \sigma, a^k)],$$

where $n \in \mathbb{N}$ and $0 \le k < n$.

Recall that $K_n(F)$ denotes the unique subfield K of the separable closure of F such that K/F is unramified of degree n; and σ is the Frobenius automorphism of K/F. The theorem is a combination of three statements.

(i) For a fixed $n \in \mathbb{N}$, the mapping $\theta_n(k/n + \mathbb{Z}) = [(K_n(F), \sigma, a^k)]$, $0 \le k < n$, is a well defined isomorphism of the groups $n^{-1}\mathbb{Z}/\mathbb{Z}$ and $\mathbf{B}(K_n(F)/F)$.

(ii) If $m, n \in \mathbb{N}$, then the diagram

$$\begin{array}{ccc} n^{-1}\mathbb{Z}/\mathbb{Z} & \to & (nm)^{-1}\mathbb{Z}/\mathbb{Z} \\ {\scriptstyle \theta_n}\downarrow & & \downarrow{\scriptstyle \theta_{nm}} \\ \mathbf{B}(K_n(F)/F) & \to & \mathbf{B}(K_{nm}(F)/F) \end{array}$$

is commutative (where the horizontal mappings are inclusions).

(iii) $\mathbb{Q}/\mathbb{Z} = \bigcup_{n \in \mathbb{N}} n^{-1}\mathbb{Z}/\mathbb{Z}$ and $\mathbf{B}(F) = \bigcup_{n \in \mathbb{N}} \mathbf{B}(K_n(F)/F)$.

PROOFS OF (i), (ii), AND (iii). The mapping $k \mapsto [(K_n(F), \sigma, a^k)]$ is a group homomorphism from $\mathbb{Z}$ to $\mathbf{B}(K_n(F)/F)$ by Corollary 15.1a. By Propositions 17.9 and 15.1b, this homomorphism is surjective and has the kernel $n\mathbb{Z}$. The statement (i) follows, because multiplication by n induces an isomorphism of $n^{-1}\mathbb{Z}/\mathbb{Z}$ to $\mathbb{Z}/n\mathbb{Z}$. The commutativity property in (ii) follows from Corollary 15.1b: $(K_{nm}(F), \sigma, a^{km}) \sim (K_n(F), \sigma, a^k)$. The first statement in (iii) is obvious. If $D \in \mathfrak{S}(F)$ is a division algebra of degree n, then $[D] \in \mathbf{B}(K_n(F)/F)$ by Corollary 17.7a and Proposition 17.8. Thus, $\mathbf{B}(F) = \bigcup_{n \in \mathbb{N}} \mathbf{B}(K_n(F)/F)$. □

It is useful to define a new invariant for the algebras in $\mathfrak{S}(F)$, where F is a local field. For $A \in \mathfrak{S}(F)$, denote

$$\mathrm{INV}_F A = \theta^{-1}([A]),$$

where θ is the isomorphism of $\mathbb{Q}/\mathbb{Z}$ to $\mathbf{B}(F)$ that was defined in the theorem. When only one field F is under consideration, we will write INV instead of INV_F. Plainly, INV can be viewed as an invariant of the elements of the Brauer group. It is used this way in local class field theory. However, for our purposes, it is more convenient to consider INV as an invariant of central simple algebras.

Corollary a. *Let F be a local field; suppose that $A, B \in \mathfrak{S}(F)$ and $m \in \mathbb{N}$.*

(i) $A \sim B$ *if and only if* $\mathrm{INV}\, A = \mathrm{INV}\, B$.
(ii) $A \sim F$ *if and only if* $\mathrm{INV}\, A = 0$.
(iii) $\mathrm{Ind}\, A$ *is the order of* $\mathrm{INV}\, A$ *in* $\mathbb{Q}/\mathbb{Z}$.
(iv) $\mathrm{INV}(A \otimes B) = \mathrm{INV}\, A + \mathrm{INV}\, B$.
(v) $\mathrm{INV}\, A^{\otimes m} = m\, \mathrm{INV}\, A$.

PROOF. The assertions (i), (ii), (iv), and (v) follow directly from the theorem and Corollary 15.1a. For the proof of (iii), let $\mathrm{INV}\, A = k/n + \mathbb{Z}$, where $n \in \mathbb{N}$, $0 \le k < n$, and $(n,k) = 1$. Thus, n is the order of $\mathrm{INV}\, A$. The order of a^k modulo $N_{K_n(F)/F}(K_n(F)^\circ)$ is n by Proposition 17.9. Thus, $\mathrm{Ind}\, A = \mathrm{Ind}(K_n(F), \sigma, a^k) = n$, according to Corollary 15.1d. □

The properties (ii), (iii), and (v) of the corollary imply one of the fundamental facts about the central simple algebras over a local field.

Corollary b. *If F is a local field and $A \in \mathfrak{S}(F)$, then* $\mathrm{Exp}\, A = \mathrm{Ind}\, A$.

The last result of this chapter relates INV_E to INV_F, when E/F is a finite extension of local fields. It is based on the work that was done in Section 14.7.

Proposition. *If E/F is a finite extension of degree m, where F is a local field, then* $\mathrm{INV}_E A^E = m(\mathrm{INV}_F A)$.

The proof is based on a diagram of field extensions.

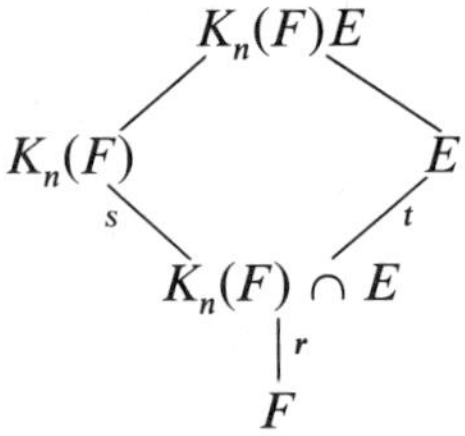

Let $[K_n(F) \cap E : F] = r$, $[K_n(F) : K_n(F) \cap E] = s$, and $[E : K_n(F) \cap E] = t$. By Lemma 14.7b, $[K_n(F)E : E] = s$ and $[K_n(F)E : K_n(F)] = t$. It follows from Corollary 17.7b that the fields between F and $K_n(F)$ are unramified. Therefore, $r = (n, f(E/F))$ and $f(E/K_n(F) \cap E)$ is relatively prime to s. Since $f(K_n(F)E/E)f(E/K_n(F) \cap E) = f(K_n(F)E/K_n(F))s$, it follows that s divides $f(K_n(F)E/E)$. Thus, $f(K_n(F)E/E) = s = [K_n(F)E/E]$ and $K_n(F)E/E$ is unramified. The uniqueness of unramified extensions implies that $K_n(F) \cap E = K_r(F)$ and $K_n(F)E = K_s(E)$. This discussion is summarized by:

$$m = rt, \quad n = rs; \tag{1}$$

$$\text{if } l = f(E/K_n(F) \cap E), \quad \text{then } f(E/F) = lr,\ t = le(E/F), \quad \text{and } (l,s) = 1. \tag{2}$$

The Frobenius automorphisms of $K_s(E)$ and $K_n(F)$ are related by

$$\sigma_E | K_n(F) = \sigma_F^{f(E/F)}. \tag{3}$$

Moreover, if a is a uniformizer for E, then $b = a^{e(E/F)}$ is a uniformizer for F. We can use these data to prove that the diagram

$$\begin{array}{ccc} \mathbb{Q}/\mathbb{Z} & \xrightarrow{m} & \mathbb{Q}/\mathbb{Z} \\ {\scriptstyle \theta_F}\downarrow & & \downarrow{\scriptstyle \theta_E} \\ \mathbf{B}(F) & \xrightarrow[\kappa_*]{} & \mathbf{B}(E) \end{array}$$

commutes, where κ_* is induced by the inclusion of F in E. This commutativity is obviously equivalent to the statement of the proposition. If $n \in \mathbb{N}$ and $1 \leq k < n$, then $\kappa_*\theta_F(k/n + \mathbb{Z}) = \kappa_*[(K_n(F), \sigma_F, b^k)] = [(K_n(F)E, \sigma_F^r, b^k)] = [(K_s(E), \sigma_F^{lr}, a^{lke(E/F)})] = [(K_s(E), \sigma_E, a^{kt})] = [(K_n(E), \sigma_E, a^{km})] = \theta_E(m(k/n + \mathbb{Z}))$. In addition to (1), (2), and (3) we have used Corollary 15.1c in this calculation.

Exercises

1. Let F be a local field. Prove that for each $n \in \mathbb{N}$, the number of isomorphism classes of division algebras $D \in \mathfrak{S}(F)$ of degree n is $\phi(n)$ (where ϕ is the Euler Totient).
2. Let A and B be division algebras in $\mathfrak{S}(F)$, where F is a local field. Prove that $A \otimes B$ is a division algebra if and only if the degrees of A and B are relatively prime.
3. Let p be an odd prime, and suppose that $a \in \mathbb{N}$ is not a quadratic residue modulo p.

Prove that the quaternion algebra $D = \left(\frac{a,p}{\hat{\mathbb{Q}}_p}\right)$ is a division algebra, and show that every division algebra of degree 2 in $\mathfrak{S}(\hat{\mathbb{Q}}_p)$ is isomorphic to D.

4. Let F be a local field, and suppose that $A \in \mathfrak{S}(F)$ has degree $n > 1$. Prove the following statements.

(a) If K/F is a field extension of degree n, then A contains a maximal subfield that is isomorphic to K as an F-algebra. Hint. Use the proposition and Corollary 13.3.

(b) If $x \in F^\circ$, then there is an extension K/F of degree n and $y \in K$ such that $N_{K/F}(y) = x$. Hint. Let $x = z^r u$, where z is a uniformizer in F and $v(u) = 1$. If n divides r, let $K = K_n(F)$ and $y = z^{r/n} w$ with $w \in K$ chosen according to Lemma 17.9b so that $N_{K/F}(w) = u$. If n is relatively prime to r, let $K = F(y)$, where y is a root of $\Phi(\mathbf{x}) = \mathbf{x}^n + z^{|r|+1}\mathbf{x} + (-1)^n x$. Thus, $v(z)^r = v(x) \leq \max\{v(y)^n, v(y)v(z)^{|r|+1}\}$. Use the Domination Principle to show that in fact $v(x) = v(y)^n$, and deduce from the hypothesis $(n,r) = 1$ that n divides $e(K/F)$. Thus, Φ is irreducible over F and $x = N_{K/F}(y)$. Combine the special cases to obtain the general result.

(c) $\nu_{A/F}(A) = F$. Hint. Use (a), (b), and Proposition 16.2a.

Notes on Chapter 17

This chapter gives a brief introduction to valuation theory, following the traditional development of this subject. Its coverage is limited to topics that are needed for the study of division algebras. The results in this chapter provide the foundation for the study of division algebras over number fields, our subject in the next chapter.

In the interest of keeping the exposition finite, it has been necessary to trim off some of the most interesting topics in the theory of valuations. The reader who wants to probe this subject more deeply can find many references that are less goal oriented. The most complete discussion of local fields is given in Serre's book [71]. Also recommended are Artin's books [7] and [8], and Chapters 1, 2, and 6 of [22]. In Chapter 19 we will take a brief look at fields that are complete under a discrete valuation, but are not local.

CHAPTER 18
Division Algebras over Number Fields

In this chapter we come to some of the deepest and most beautiful results in modern algebra. These are the theorems that classify and describe the central simple algebras over algebraic number fields. This work is associated with the names of several of the greatest heroes of mathematics: Hasse, Brauer, Noether, and Albert. It is based on developments in number theory that are due to Kronecker, Weber, Hilbert, Minkowski, Furtwangler, Artin, Takagi, Hasse, Witt, and many others.

It will no longer be possible for us to give self-contained proofs of the basic theorems. Instead, we will quote some results from class field theory and derive the classification and structure theory of algebras from these deep number theoretic facts. Some theorems on rational division algebras that can be derived in an elementary way are outlined in the exercises. However, even in this simple case, the best results require the use of number theoretical tools that cannot be called elementary.

18.1. Field Composita

In the last chapter we studied two cases of the question "when can a valuation v of a field F be extended to a larger field K?" It was shown that such extensions exist and are unique when K is the completion of F in the v-topology and also when F is a local field and K/F is finite. Our work in this chapter requires an answer to the extension question in the case that K/F is a finite extension and no hypothesis is made on F. By using field composites we will reduce this problem to the cases that were treated in the last chapter. This section is concerned with those topics in the theory of field composita that are needed to solve the valuation extension problem.

Let K and L be fields that contain F as a subfield. *A compositum* of K and L over F is a triple (E,ϕ,ψ) in which E is a field that contains F, and $\phi: K \to E$, $\psi: L \to E$ are F-algebra homomorphisms such that $E = \phi(K)\psi(L)$. Two such composita (E,ϕ,ψ) and (E',ϕ',ψ') are equivalent if there is an F-algebra homomorphism $\theta: E \to E'$ such that $\phi' = \theta\phi$ and $\psi' = \theta\psi$. In this case, $\theta(E) = \theta(\phi(K)\psi(L)) = \phi'(K)\psi'(L) = E'$, so that θ is a field isomorphism. It follows that equivalence of composita is a symmetric relation; plainly, it is also reflexive and transitive.

Lemma a. *If K/F is a finite, separable field extension and L/F is a field extension, then $K \otimes L = K^L = E_1 \dotplus \cdots \dotplus E_r$, where E_i/L is a field extension such that $[K:F] = \sum_{i=1}^r [E_i : L]$. Write $1_{K\otimes L} = e_1 + \cdots + e_r$, where $e_i = 1_{E_i}$, and define $\phi_i: K \to E_i$, $\psi_i: L \to E_i$ by $\phi_i(x) = e_i(x \otimes 1)$, $\psi_i(y) = e_i(1 \otimes y)$. The triples (E_i,ϕ_i,ψ_i) are pairwise inequivalent composita of K and L over F, and every compositum of K and L over F is equivalent to one of the (E_i,ϕ_i,ψ_i).*

PROOF. By Lemma 10.7b and Proposition 10.6a, $K \otimes L = K^L$ is a separable L-algebra. In particular, $K \otimes L$ is semisimple. Since $K \otimes L$ is commutative, the Wedderburn Structure Theorem takes the form $K \otimes L = E_1 \dotplus \cdots \dotplus E_r$ in which each E_i is a field that contains F. The mappings ϕ_i and ψ_i are clearly F-algebra homomorphisms of K and L to E_i, and $\phi_i(K)\psi_i(L) = e_i(K \otimes L) = E_i$. Thus, (E_i,ϕ_i,ψ_i) is a compositum of K and L over F. If E_i is given the L-space structure that is induced by ψ_i, then $\sum_{i=1}^r [E_i : L] = \dim_L K^L = [K:F]$. Suppose that there is an isomorphism $\theta: E_i \to E_j$ such that $\theta\phi_i = \phi_j$, $\theta\psi_i = \psi_j$ with $i \neq j$. If $e_i = \sum_k (x_k \otimes y_k)$, then $e_j = \theta(e_i) = \theta(\sum_k e_i(x_k \otimes y_k)) = \theta(\sum_k \phi_i(x_k)\psi_i(y_k)) = \sum_k \phi_j(x_k)\psi_j(y_k) = \sum_k e_j(x_k \otimes y_k) = e_j e_i = 0$, which is a contradiction. Thus, (E_i,ϕ_i,ψ_i) is not equivalent to (E_j,ϕ_j,ψ_j) if $i \neq j$. Suppose that (E,ϕ,ψ) is a compositum of K and L. The mapping $\phi \otimes \psi: K \otimes L \to E$ is an F-algebra homomorphism such that $(\phi \otimes \psi)(K \otimes L) = \phi(K)\psi(L) = E$. In particular, there is an index i such that $\theta = (\phi \otimes \psi)|E_i$ is a non-zero F-algebra homomorphism of E_i to E. Thus, $\theta(e_i) = 1_E$, and if $x \in K$, then $\theta\phi_i(x) = \theta(e_i(x \otimes 1)) = \theta(e_i)\phi(x)\psi(1) = \phi(x)$. Similarly, $\theta\psi_i = \psi$. Consequently, (E,ϕ,ψ) is equivalent to (E_i,ϕ_i,ψ_i). □

Lemma b. *Let K/F be a Galois extension with the Galois group G, and suppose that L/F is an arbitrary field extension. Assume that (E,ϕ,ψ) is a compositum of K and L over F. If $\sigma \in G$ then $(E,\phi\sigma^{-1},\psi)$ is a compositum of K and L over F. Every compositum of K and L over F is equivalent to $(E,\phi\sigma^{-1},\psi)$ for some $\sigma \in G$.*

PROOF. It is obvious that if $\sigma \in G$ then $(E,\phi\sigma^{-1},\psi)$ is a compositum. For the proof of the last statement, we adopt the notation that was introduced in Lemma a. Assume that $[E_1 : L] \leq \cdots \leq [E_r : L]$. The mapping $\sigma \mapsto \theta_\sigma =$

$\sigma \otimes id_L$ is an injective group homomorphism of G to $\mathrm{Aut}_F K \otimes L$. If $\chi = \theta_\sigma$, then χ permutes the minimal ideals of $K \otimes L$. Thus, there is a unique j such that $\chi(E_1) = E_j$. In particular, $\chi(e_1) = e_j$, so that $\chi\phi_1(x) = \chi(e_1(x \otimes 1)) = \chi(e_1)((\sigma \otimes id_L)(x \otimes 1)) = e_j(x^\sigma \otimes 1) = (\phi_j\sigma)(x)$ for all $x \in K$. Similarly, $\chi\psi_1(y) = e_j(1^\sigma \otimes y) = \psi_j(y)$ for all $y \in L : \chi$ is an L-space homomorphism. Let $H = \{\sigma \in G\colon \theta_\sigma E_1 = E_1\}$. Since $\sigma \mapsto \theta_\sigma$ is a homomorphism, H is a subgroup of G and $\theta_\sigma E_1 = \theta_\tau E_1$ if and only if $\tau^{-1}\sigma \in H$. Therefore, the left cosets of H are in one-to-one correspondence with the fields $\{\theta_\sigma E_1 : \sigma \in G\}$. If $\sigma \in H$, then $\theta_\sigma | E_1 \in \mathbf{G}(E_1/L)$. When $\theta_\sigma | E_1$ is the identity automorphism, $\phi_1(x) = \theta_\sigma\phi_1(x) = \phi_1(x^\sigma)$ for all $x \in K$, so that $\sigma = id_K$. Consequently, $|H| \leq |\mathbf{G}(E_1/L)| \leq [E_1 : L]$, and $r|H| \leq r[E_1 : L] \leq \sum_{i=1}^r [E_i : L] = [K:F] = |G| = [G:H]|H|$. Thus, $r \geq |\{\theta_\sigma E_1 : \sigma \in G\}| = [G:H] \geq r$. Therefore, every E_j has the form $\theta_\sigma E_1$ for some $\sigma \in G$. This conclusion, together with Lemma a proves the last statement of the lemma. Indeed, we can assume that $(E,\phi,\psi) = (E_1,\phi_1,\psi_1)$. If $\theta_\sigma E_1 = E_j$, then $\phi\sigma^{-1} = \theta_\sigma^{-1}\phi_j$ and $\psi = \theta_\sigma^{-1}\psi_j$; hence (E_j,ϕ_j,ψ_j) is equivalent to $(E,\phi\sigma^{-1},\psi)$. □

The composita $(E,\phi\sigma^{-1},\psi)$ are not in general inequivalent. In fact, $(E,\phi\sigma^{-1},\psi)$ is equivalent to (E,ϕ,ψ) if and only if $\phi\sigma\phi^{-1} \in \mathbf{G}(\phi K/\phi K \cap \psi L)$.

Exercises

1. Prove the last statement of the section: if K/F is Galois, (E,ϕ,ψ) is a compositum of K and L over F, and $\sigma \in \mathbf{G}(K/F)$, then $(E,\phi\sigma^{-1},\psi)$ is equivalent to (E,ϕ,ψ) if and only if $(\phi\sigma\phi^{-1})(z) = z$ for all $z \in \phi(K) \cap \psi(L)$.
2. Prove that if K/F and L/F are Galois extensions with the compositum (E,ϕ,ψ), then E/F is Galois.
3. Let F, K, and L be fields with $F \subseteq K \subseteq L$. Show that if $\phi\colon K \to L$ is an F-algebra homomorphism, then (L,ϕ,id_L) is a compositum of K and L over F. Prove that if $\phi'\colon K \to L$ is another F-algebra homomorphism, then (L,ϕ',id_L) is equivalent to (L,ϕ,id_L) if and only if $\phi' = \phi$. Deduce that the number of equivalence classes of composita of K and L over F may be infinite unless K/F is a finite extension.

18.2. More Extensions of Valuations

Our aim in this section is to survey the extensions to K of a valuation v of the field F when K/F is a finite separable field extension. The approach to this problem is: extend v to the valuation $\hat{v}$ of $\hat{F}$; extend $\hat{v}$ to E, where (E,ϕ,ψ) is a compositum of K and $\hat{F}$ over F; restrict the valuation on E back to K.

We will write $w \supseteq v$ or $v \subseteq w$ if w is an extension of v to a field K that contains F as a subfield. If $w|F$ is equivalent to v, though not necessarily equal to v, then w *divides* v. This property is indicated by writing $w|v$. If v

and w are non-archimedean valuations, then it follows from Proposition 17.3 that $w|v$ if and only if $P(K,w) \supseteq P(F,v)$. In other words, the ideal $P(K,w)$ divides the ideal $P(F,v)O(K,w)$.

We will need a generalization of Corollary 17.8b.

Lemma a. *Let $F \subseteq K \subseteq E$ be a chain of fields. If w is a discrete valuation of E, and $v = w|K$, then*

(i) $e_w(E/F) = e_w(E/K)e_v(K/F)$,
(ii) $f_w(E/F) = f_w(E/K)f_v(K/F)$.

The first part of the proof of Corollary 17.8b gives (i). In the argument, the assumption that F is a local field was not used. We leave the proof of (ii) as an easy exercise.

A technical lemma leads to the main result of this section. In the statement of this lemma the topological terms refer to the relevant valuation metrics and topologies.

Lemma b. *Let F, K, L, and E be fields with $F \subseteq L \subseteq E$, $F \subseteq K$, and $[K:F] < \infty$. Suppose that v is a valuation of F, and u and w are respectively extensions of v to E and K. Assume that L is locally compact (that is, $O(L,u)$ is compact), F is dense in L, and $\phi: K \to E$ is an isometric F-algebra homomorphism.*

(i) *There is an isometric F-algebra isomorphism $\psi: \hat{F} \to L$.*
(ii) *$\phi(K)L$ is closed in E and $\phi(K)$ is dense in $\phi(K)L$.*
(iii) *There is an extension of ϕ to an isometric F-algebra isomorphism $\theta: \hat{K} \to \phi(K)L$.*

PROOF. The assumptions that L is locally compact (hence complete) and F is dense in L imply $L \cong \hat{F}$ algebraically and topologically, by Corollary 17.4a. The isometric isomorphism ψ is defined by letting $\psi(x)$ be the limit in L of a Cauchy sequence $\{a_n\} \subseteq F$ whose limit in $\hat{F}$ is x. Since $\dim_L \phi(K)L \leq [K:F]$, it follows from Lemma 17.6b that the uniform topology of the L-space $\phi(K)L$ coincides with the u-topology. Moreover, $\phi(K)L$ is locally compact in the uniform topology, hence also in the u-topology. In particular, $\phi(K)L$ is complete and therefore closed in E. Since F is dense in L, it is clear that $\phi(K) = \phi(K)F$ is dense in $\phi(K)L$. If $\{y_n\} \subseteq K$ is a Cauchy sequence that converges to $y \in \hat{K}$, then $\{\phi(y_n)\}$ is a Cauchy sequence in $\phi(K)L$ that converges to an element $\theta(y)$. A routine check shows that θ is a well defined, isometric F-algebra isomorphism of $\hat{K}$ to $\phi(K)L$. □

Proposition. *Let v be a non-trivial valuation of the field F such that $\hat{F}$ is locally compact. Assume that K/F is a finite, separable field extension. If w is a valuation of K that extends v, let $\hat{K}_w$ denote the completion of K in the w-topology, ϕ_w the standard embedding of K in $\hat{K}_w$, and ψ_w the isometric F-algebra homomorphism of $\hat{F}$ to the closure of F in $\hat{K}_w$ (defined in Lemma b). The* *mapping* $w \mapsto (\hat{K}_w, \phi_w, \psi_w)$ *induces a bijection between the extensions* w *of* v *to* K *and the equivalence classes of composita of* K *and* $\hat{F}$ *over* F. *If* v *is discrete, then all extensions* w *of* v *are discrete,* $e_w(K/F) = e_w(\hat{K}_w/\hat{F})$, $f_w(K/F) = f_w(\hat{K}_w/\hat{F})$, *and* $e_w(K/F)f_w(K/F) = [\hat{K}_w : \hat{F}]$.

PROOF. If w is an extension of v to K, then $(\hat{K}_w, \phi_w, \psi_w)$ is a compositum of K and $\hat{F}$ over F. In fact, $(\phi_w K)(\psi_w \hat{F}) = K(\psi_w \hat{F})$ is closed in $\hat{K}_w$ by Lemma b; hence $(\phi_w K)(\psi_w \hat{F}) = \hat{K}_w$. Let (E,ϕ,ψ) be a compositum of K and $\hat{F}$ over F. We wish to associate an extension of v with this compositum. Since $\hat{F}$ is locally compact by assumption, and $\dim_{\psi(\hat{F})} E \leq \dim_F K < \infty$, it follows from Proposition 17.6 that there is a unique valuation $\hat{w}$ of E such that $\hat{w}(\psi(y)) = v(y)$ for all $y \in \hat{F}$. Define the valuation w of K by $w(x) = \hat{w}(\phi(x))$. If $a \in F$, then $w(a) = \hat{w}(a) = v(a)$, so that $w \supseteq v$. We must prove that these constructions are mutually inverse to within equivalence of composita. One way is clear: w is the extension of v that corresponds to $(\hat{K}_w, \phi_w, \psi_w)$. Indeed, the extension $\hat{w}$ of w to $\hat{K}_w$ does satisfy $\hat{w}(\psi_w(y)) = v(y)$ for all $y \in \hat{F}$ because ψ_w is an isometry; and obviously $\hat{w}(\phi_w(x)) = w(x)$ for all $x \in K$. Suppose that w is obtained from the compositum (E,ϕ,ψ) by our construction. We will show that $(\hat{K}_w, \phi_w, \psi_w)$ is equivalent to (E,ϕ,ψ). By the definition of w and $\hat{w}$, the mappings ϕ and ψ are isometries. Proposition 17.6 implies that E is locally compact, hence complete in the $\hat{w}$-topology. By lemma b, $\phi(K)$ is dense in E. Therefore, ϕ extends to an isometric isomorphism $\theta: \hat{K}_w \to E$. By the definition of ϕ_w, $\theta\phi_w = \theta | K = \phi$. To prove that $\theta\psi_w = \psi$, let $y = \lim x_n \in \hat{F}$ with $\{x_n\} \subseteq F$. Since θ and ψ are isometric mappings, we have $\theta\psi_w(y) = \theta\psi_w(\lim^{(\hat{F})} x_n) = \theta(\lim^{(\hat{K})} x_n) = \lim^{(E)} x_n = \psi(\lim^{(\hat{F})} x_n) = \psi(y)$. (The superscripts on lim keep track of the space in which the limits are taken.) We next prove that equivalent composita give rise to the same extension of v. Let (E',ϕ',ψ') and (E,ϕ,ψ) be equivalent composita, say $\theta: E' \to E$ is an isomorphism that satisfies $\phi = \theta\phi'$ and $\psi = \theta\psi'$. Denote the valuations associated with (E,ϕ,ψ) and (E',ϕ',ψ') by w and w' respectively. The mapping $z \mapsto \hat{w}(\theta(z))$ is a valuation of E' that satisfies $\hat{w}(\theta\psi'(y)) = \hat{w}(\psi(y)) = v(y)$ for all $y \in \hat{F}$. The uniqueness statement in Proposition 17.6 implies that $\hat{w}(\theta(z)) = \hat{w}'(z)$ when $z \in E'$. In particular, if $x \in K$, then $w'(x) = \hat{w}'(\phi'(x)) = \hat{w}(\theta\phi'(x)) = \hat{w}(\phi(x)) = w(x)$. It remains to consider the consequences of discreteness. Let v be a discrete valuation of F. If (E,ϕ,ψ) is a compositum of K and $\hat{F}$, then the extension of v to $\hat{F}$ is discrete by Proposition 17.4, and the further extension to the valuation $\hat{w}$ of E is also discrete according to Proposition 17.6. Hence, the valuation w of K corresponding to (E,ϕ,ψ) is discrete. Proposition 17.4 and its Corollary 17.4b yield $e_w(\hat{K}_w/K) = e_v(\hat{F}/F) = 1 = f_w(\hat{K}_w/K) = f_v(\hat{F}/F)$. Thus, $e_w(K/F) = e_w(\hat{K}_w/\hat{F})$ and $f_w(K/F) = f_w(\hat{K}_w/\hat{F})$ by Lemma a. It follows from Proposition 17.7 that $e_w(K/F)f_w(K/F) = [\hat{K}_w : \hat{F}]$. □

Corollary a. *Let* v *be a non-trivial valuation of the field* F *such that* $\hat{F}$ *is locally compact. If* K/F *is a finite, separable field extension, then there are finitely many* *distinct extensions $w_1, \ldots, w_r$ of v to K, and $\sum_{i=1}^r [\hat{K}_{w_i} : \hat{F}] = [K:F]$. If v is discrete, then all of the w_i are discrete and $\sum_{i=1}^r e_{w_i}(K/F)f_{w_i}(K/F) = [K:F]$.*

This corollary is a consequence of the proposition and Lemma 18.1a.

Corollary b. *Let v be a non-trivial valuation of the field F such that $\hat{F}$ is locally compact. Assume that K/F is a Galois extension. If w is a valuation of K that extends v, then for all $\sigma \in \mathbf{G}(K/F)$, w^σ is an extension of v to K, where $w^\sigma(x) = w(x^{\sigma^{-1}})$ for all $x \in K$. Every extension of v to K has the form w^σ for some $\sigma \in \mathbf{G}(K/F)$.*

PROOF. If $\sigma \in \mathbf{G}(K/F)$, then $(\hat{K}_w, \phi_w\sigma^{-1}, \psi_w)$ is a compositum of K and $\hat{F}$ over F, and the corresponding valuation of K is defined by $\hat{w}(\phi_w\sigma^{-1}(x)) = \hat{w}(\phi_w(x^{\sigma^{-1}})) = w^\sigma(x)$. Thus, w^σ is an extension of v to K. By Lemma 18.1b and the proposition, every extension of v to K has the form w^σ for some $\sigma \in \mathbf{G}(K/F)$. □

When the extension K/F is Galois, then Corollary b gives a nice classification of the valuations of K that extend the valuation v of F. Our last corollary lists some consequences of this classification.

Corollary c. *Let the hypotheses and notation be as they were in Corollary b. In particular, K/F is Galois.*

(i) *As an $\hat{F}$-algebra, $\hat{K}_w$ is independent of the choice of w.*
(ii) *$\hat{K}_w/\hat{F}$ is Galois.*
(iii) *$\sigma \mapsto \phi_w^{-1}\sigma\phi_w$ is an injective homomorphism from $\mathbf{G}(\hat{K}_w/\hat{F})$ to $\mathbf{G}(K/F)$; the image of this mapping is $G_w = \mathbf{G}(K/\phi_w^{-1}(\phi_w(K) \cap \psi_w(\hat{F})))$.*
(iv) *If $\sigma, \tau \in \mathbf{G}(K/F)$, then $w^\sigma = w^\tau$ if and only if $\sigma\tau^{-1} \in G_w$; in particular, $G_w = \{\sigma \in \mathbf{G}(K/F) : w^\sigma = w\}$, and $G_{w^\tau} = \tau^{-1}G_w\tau$.*
(v) *If v is discrete, then $e_{w^\sigma}(K/F) = e_w(K/F)$, $f_{w^\sigma}(K/F) = f_w(K/F)$, and $e_w(K/F)f_w(K/F)g_w(K/F) = [K:F]$, where $g_w(K/F)$ is the index of G_w in $\mathbf{G}(K/F)$.*

PROOF. The statement (i) is clear from the proposition and Corollary b. The properties (ii) and (iii) restate Lemma 14.7b. To prove (iv), note that $w^\sigma = w^\tau$ if and only if $(\hat{K}_w, \phi_w\sigma^{-1}, \psi_w)$ is equivalent to $(\hat{K}_w, \phi_w\tau^{-1}, \psi_w)$, that is, there is an automorphism ρ of $\hat{K}_w$ satisfying $\rho\psi_w = \psi_w$ and $\rho\phi_w\sigma^{-1} = \phi_w\tau^{-1}$. The equation $\rho\psi_w = \psi_w$ is equivalent to $\rho \in \mathbf{G}(\hat{K}_w/\hat{F})$; the second equation translates to $\sigma\tau^{-1} = \phi_w^{-1}\rho\phi_w \in G_w$ (noting that the switch from composition of maps to exponentiation reverses the order of the product of τ^{-1} and σ). The last assertion of (iv) comes from the easily checked observation that $w^{(\sigma\tau)} = (w^\sigma)^\tau$. Finally, if v is discrete, then $e_{w^\sigma}(K/F) = e_w(\hat{K}_w/\hat{F}) = e_w(K/F)$ and $f_{w^\sigma}(K/F) = f_w(\hat{K}_w/\hat{F}) = f_w(K/F)$ by (i). □

The group G_w is called the *decomposition group* of w. When $\mathbf{G}(K/F)$ is abelian, then G_w is the same subgroup of $\mathbf{G}(K/F)$ for all extensions w of v. In this case, it is customary to write G_v instead of G_w, and identify this group with $\mathbf{G}(\hat{K}_w/\hat{F})$.

Exercises

1. Prove the statement (ii) of Lemma a.
2. Prove the assertion $w^{(\sigma\tau)} = (w^\sigma)^\tau$ that was made in the proof of Corollary c.
3. Let a be a square free integer, and denote $K = \mathbb{Q}(\sqrt{a})$. Assume that p is an odd prime that does not divide a. Prove that the p-adic valuation v_p of $\mathbb{Q}$ has two extensions to valuations of K if a is a quadratic residue modulo p, and v_p has one extension to K if a is a non-residue modulo p. Hint. Note that $\mathbf{x}^2 - a$ is irreducible over $\hat{\mathbb{Q}}_p$ if and only if a is a non-residue modulo p. Prove that $e_{v_p}(K/\mathbb{Q}) = 1$ in all of these cases.

18.3. Valuations of Algebraic Number Fields

An *algebraic number field* is a subfield F of $\mathbb{C}$ such that $[F:\mathbb{Q}] < \infty$. We will use the results of the previous section to survey the valuations of algebraic number fields.

Proposition. *Let F be an algebraic number field.*

(i) *Every non-trivial valuation of F divides a non-trivial valuation of $\mathbb{Q}$.*
(ii) *If v is a non-trivial, non-archimedean valuation of F, then v divides a valuation v_p of $\mathbb{Q}$ for a unique prime p, v is discrete, $E(F,v)$ is a finite field, and $\hat{F}_v$ is a local field. For each prime p, there are at most $[F:\mathbb{Q}]$ extensions of v_p to F; in fact, $\sum_{v \supseteq v_p} e_v(F/\mathbb{Q}) f_v(F/\mathbb{Q}) = [F:\mathbb{Q}]$.*
(iii) *Every archimedean valuation v of F is equivalent to a valuation $w(x) = |\phi(x)|$, where $\phi\colon F \to \mathbb{C}$ is a non-zero field homomorphism. There are at most $[F:\mathbb{Q}]$ equivalence classes of archimedean valuations of F.*

Proof. If v is non-trivial, then so is $v|\mathbb{Q}$ by (the proof of) Lemma 17.7a. If v is non-trivial and non-archimedean, then $v|\mathbb{Q}$ is also non-archimedean. Hence, $v|\mathbb{Q}$ is equivalent to v_p for a unique prime p. The remaining assertions in (ii) follow from Proposition 18.2: the separability hypothesis is automatically satisfied because char $F = 0$. If v is archimedean, then $v|\mathbb{Q}$ is equivalent to the absolute value v_∞. Since $\hat{\mathbb{Q}}_{v_\infty} = \mathbb{R}$ and $\hat{F}_v/\hat{\mathbb{Q}}_{v_\infty}$ is finite, either $\hat{F}_v \cong \mathbb{R}$ or $\hat{F}_v \cong \mathbb{C}$. Thus, (iii) also follows from Proposition 18.2. □

If K/F is an extension of algebraic number fields, v is a valuation of F, and w is a valuation of K that divides v, then $\hat{K}_w/\hat{F}_v$ is a finite field extension whose degree is called the *local degree* of K/F at w. When K/F is Galois, $\hat{K}_w$ is the same $\hat{F}_v$-algebra for all extensions w of v by Corollary 18.2c. In particular, the local degrees at the extensions of v are the same. In this case, we will write $\hat{K}_v$ instead of $\hat{K}_w$, and denote the local degree "at v" by $[\hat{K}_v : \hat{F}_v]$.

Corollary. *Let K/F be a Galois extension of algebraic number fields. Assume that v is a non-trivial valuation of F and w is a valuation of K that extends v.*

(i) $[\hat{K}_v : \hat{F}_v]$ *divides* $[K : F]$.

(ii) *If v is discrete and $e_w(K/F) = 1$, then the decomposition group G_w of w is cyclic. There is a unique generator σ_w of G_w such that $w(x^{\sigma_w} - x^q) < 1$ for all $x \in O(K,w)$, where $q = |E(F,v)|$. If $\tau \in \mathbf{G}(K/F)$, then $\sigma_{w^\tau} = \tau^{-1}\sigma_w\tau$.*

Proof. By Corollary 18.2c, $[\hat{K}_w : \hat{F}_v] = |\mathbf{G}(\hat{K}_w : \hat{F}_v)| = |G_w|$ divides $|\mathbf{G}(K/F)| = [K : F]$ for all extensions w of v. Since $[\hat{K}_v : \hat{F}_v] = [\hat{K}_w : \hat{F}_v]$, this observation proves (i). Assume that v is discrete and $e_w(K/F) = 1$ for one (hence every) extension w of v. By Proposition 18.2, $e_w(\hat{K}_w/\hat{F}_v) = 1$; that is, $\hat{K}_w/\hat{F}_v$ is an unramified extension of local fields. Consequently, $\mathbf{G}(\hat{K}_w/\hat{F}_v)$ is cyclic with the Frobenius automorphism as a generator, according to the results of Section 17.8. The generator σ_w of G_w is the image of the Frobenius automorphism under the isomorphism of $\mathbf{G}(\hat{K}_w/\hat{F}_v)$ to G_w. The characterization of σ_w by the condition $w(x^{\sigma_w} - x^q) < 1$ for all $x \in O(K,w)$ is a consequence of the analogous characterization of the Frobenius automorphism. If $\tau \in \mathbf{G}(K/F)$ and $x \in O(K,w^\tau)$, then $x^{\tau^{-1}} \in O(K,w)$ and $w^\tau(x^{\tau^{-1}\sigma_w\tau} - x^q) = w(x^{\tau^{-1}\sigma_w} - (x^{\tau^{-1}})^q) < 1$. Therefore, $\sigma_{w^\tau} = \tau^{-1}\sigma_w\tau$. □

The element σ_w of $\mathbf{G}(K/F)$ is called the *Frobenius automorphism* of K/F at w. When K/F is Galois, these automorphisms are defined for all discrete valuations w of K such that $e_w(K/F) = 1$. As we will see, almost all valuations of K have these properties. It is clear that $\sigma_w = \sigma_{w'}$ when w and w' are equivalent.

If K/F is an abelian extension, that is, K/F is Galois and $\mathbf{G}(K/F)$ is an abelian group, then by part (ii) of the corollary, $\sigma_w = \sigma_{w'}$ whenever w and w' divide the same valuation v. In this case, $w \mapsto \sigma_w$ can be viewed as a mapping from a certain set of discrete valuations of F to $\mathbf{G}(K/F)$, and it is natural to write σ_v instead of σ_w if $w \mid v$. These remarks apply in particular when K/F is a cyclic extension.

The inconvenience of dealing with equivalence classes of valuations can be avoided in the study of algebraic number fields. There are several ways to select a canonical representative from each equivalence class of valuations. The standardization that we will adopt leads to an elegant product formula, due to Artin and Nesbitt.

Let v be a non-trivial valuation of the algebraic number field F. By the proposition, v divides a non-trivial valuation v_p of $\mathbb{Q}$, where p is a prime or $p = \infty$; that is, v_p is either a p-adic valuation of $\mathbb{Q}$ or the absolute value. If $v \mid \mathbb{Q} = v_p^{n(v)}$, where $n(v) = [\hat{F}_v : \hat{\mathbb{Q}}_p]$ ($[\hat{F}_v : \mathbb{R}]$ if $p = \infty$), then v is called a *normalized valuation* of F. Every non-trivial valuation of F is equivalent to a unique normalized valuation. Moreover, if p is a prime or ∞, then the number of normalized valuations that divide v_p is finite. Denote by $\mathbf{S}(F)$ the set of all (non-trivial) normalized valuations of F; if p is a prime or infinity, let $\mathbf{S}_p(F) = \{v \in \mathbf{S}(F) : v \text{ divides } v_p\}$.

Lemma. *Let F and K be algebraic number fields with $F \subseteq K$.*

(i) *If $v \in \mathbf{S}(F)$, $w \in \mathbf{S}(K)$ and $w \mid v$, then $w \mid F = v^k$, where $k = [\hat{K}_w : \hat{F}_v]$.*
(ii) *If K/F is Galois, $\sigma \in \mathbf{G}(K/F)$, $v \in \mathbf{S}(F)$, $w \in \mathbf{S}(K)$, and $w \mid v$, then $w^\sigma \in \mathbf{S}(K)$ and $w^\sigma \mid v$.*
(iii) *If K/F is Galois, $v \in \mathbf{S}(F)$, and $y \in K$, then $\prod_{w \in \mathbf{S}(K), w \mid v} w(y) = v(N_{K/F}(y))$.*
(iv) *If p is a prime or ∞ and $x \in F$, then $\prod_{v \in \mathbf{S}_p(F)} v(x) = v_p(N_{F/\mathbb{Q}}(x))$.*

PROOF. Let $v \in \mathbf{S}_p(F)$ with p a prime or ∞. Note that if $w \in \mathbf{S}(K)$ and $w \mid v$, then $w \in \mathbf{S}_p(K)$. Moreover, $w \mid F = v^k$ for some k. If $n(w) = [\hat{K}_w : \hat{\mathbb{Q}}_p]$ and $n(v) = [\hat{F}_v : \hat{\mathbb{Q}}_p]$, then $v_p^{n(w)} = w \mid \mathbb{Q} = v^k \mid \mathbb{Q} = (v \mid \mathbb{Q})^k = v_p^{n(v)k}$. Hence, $k = n(w)/n(v) = [\hat{K}_w : \hat{\mathbb{Q}}_p]/[\hat{F}_v : \hat{\mathbb{Q}}_p] = [\hat{K}_w : \hat{F}_v]$. Assume that K/F is Galois. Denote $G = \mathbf{G}(K/F)$. Plainly, if $\sigma \in G$, then $w^\sigma \mid F = w \mid F$. Therefore, $w^\sigma \mid v$. Moreover, $w^\sigma \mid \mathbb{Q} = w \mid \mathbb{Q} = v_p^{n(w)} = v_p^{n(w^\sigma)}$ by Corollary 18.2c. Thus, $w^\sigma \in \mathbf{S}_p(K)$. Let $G_w = \{\rho \in G : w^\rho = w\}$ be the decomposition group of w. By Corollary 18.2c, the order k of G_w is $[\hat{K}_w : \hat{F}_v]$; and $\sigma, \tau \in G$ satisfy $w^\sigma = w^\tau$ if and only if $\sigma\tau^{-1} \in G_w$. If $G = G_w\sigma_1 \cup \cdots \cup G_w\sigma_g$ is a coset decomposition of G, then

$$\left(\prod_{w \in \mathbf{S}(K), w \mid v} w(y)\right)^k = \prod_{i=1}^{g} \prod_{\rho \in G_w} w^{\rho\sigma_i}(y) = \prod_{\sigma \in G} w^\sigma(y) = w\left(\prod_{\sigma \in G} y^{\sigma^{-1}}\right)$$
$$= w(N_{K/F}(y)) = v(N_{K/F}(y))^k \quad \text{for all } y \in K.$$

Since k'th roots are unique in $\mathbb{R}^+$, this calculation proves (iii). For the proof of (iv), choose the algebraic number field K so that $K/\mathbb{Q}$ is Galois and $F \subseteq K$. Thus, K/F is Galois. If $s = [K : F]$, $v \in \mathbf{S}_p(F)$, and $x \in F$, then $v(x)^s = v(N_{K/F}(x)) = \prod_{w \in \mathbf{S}(K), w \mid v} w(x)$. Note that if $w \in \mathbf{S}_p(K)$, then there is a unique $v \in \mathbf{S}_p(F)$ such that $w \mid v$. Conversely, $w \in \mathbf{S}(K)$, $v \in \mathbf{S}_p(F)$ and $w \mid v$ implies $w \in \mathbf{S}_p(K)$. Hence, if $x \in F$, then $(\prod_{v \in \mathbf{S}_p(F)} v(x))^s = \prod_{w \in \mathbf{S}_p(K)} w(x) = v_p(N_{K/\mathbb{Q}}(x)) = v_p(N_{F/\mathbb{Q}}(x))^s$, which yields (iv). □

The lemma leads to an important property of normalized valuations.

The Product Formula. *If F is an algebraic number field and $x \in F^\circ$, then $v(x) = 1$ for almost all $v \in \mathbf{S}(F)$ and $\prod_{v \in \mathbf{S}(F)} v(x) = 1$.*

PROOF. If $x \in F$, then x is algebraic over $\mathbb{Q}$, say $x^n + a_1x^{n-1} + \cdots + a_n = 0$, where $a_i \in \mathbb{Q}$, $a_n \neq 0$, and $n \geq 1$. The set X of all prime factors of the numerators and denominators of the a_i is finite; and if p is a prime that is not in X, then $v_p(a_i) = 1$ whenever $a_i \neq 0$. Let p be a prime, $p \notin X$, and $v \in \mathbf{S}_p(F)$. Then v is non-archimedean and $v(a_ix^{n-i}) = v(x)^{n-i}$ if $a_i \neq 0$. The Domination Principle implies that $v(x) = 1$. Indeed, if $v(x) > 1$, then $v(x)^n > v(a_i x^{n-i})$ for $1 \leq i \leq n$ yields $0 = v(0) = v(x)^n > 0$. If $v(x) < 1$, then $1 = v(a_n) > v(a_i x^{n-i})$ for $0 \leq i < n$; hence, $0 = v(0) = 1$. Therefore, $v(x) = 1$ unless v is in the finite set $\mathbf{S}_\infty(F) \cup \bigcup_{p \in X} \mathbf{S}_p(F)$. In particular, the product $\prod_{v \in \mathbf{S}(F)} v(x)$ is finite. By the lemma, $\prod_{v \in \mathbf{S}(F)} v(x) = v_\infty(N_{F/\mathbb{Q}}(x))(\prod_{p\ prime} v_p(N_{F/\mathbb{Q}}(x)))$. It is easy to see from the definition of the p-adic valuations that for any rational number c, the product $v_\infty(c)(\prod_p v_p(c))$ is equal to 1. (See Exercise 2, Section 17.2.) □

We will frequently use the part of this result that states $v(x) = 1$ for almost all v. The actual product formula is less important for us. However, this equation has an important role in class field theory, and it appears implicitly in many results that will be stated in the next few sections.

Exercises

1. An *algebraic function field* is a finite separable extension of a field of the form $K(\mathbf{x})$, where K is a finite field.

   (a) Prove that all valuations of an algebraic function field are discrete.

   (b) Prove the analogue of Proposition a for algebraic function fields.

   (c) Using the same definition of normalized valuation that was given for algebraic number fields, prove that the product formula is valid for algebraic function fields. Hint. See Exercise 3, Section 17.2 and Exercise 2, Section 17.3.

   Artin and Nesbitt proved that if the product formula holds for a field F, then F is either isomorphic to an algebraic number field or an algebraic function field. The members of the union of these classes of fields are called *global fields*. The basic results of class field theory are valid for all global fields. All of the results that we will describe in the next few sections generalize to global fields. This assertion will be completely obvious if the claim that class field theory applies to global fields is accepted.

2. The purpose of this problem is to outline a proof of the *Weak Approximation Theorem*: if F is an algebraic number field, $v_1, \ldots, v_m \in \mathbf{S}(F)$ are distinct, $x_1, \ldots, x_m \in F^\circ$, and $e \in \mathbb{R}^+$, then there exists $y \in F^\circ$ such that $v_i(y - x_i) < e$ for $1 \leq i \leq m$.

   (a) Prove that if $v \in \mathbf{S}(F)$ and $x, y \in F^\circ$, then in the v-topology $\lim_{k\to\infty} x^k y/(1 + x^k) = y$ if $v(x) > 1$, and $\lim_{k\to\infty} x^k y/(1 + x^k) = 0$ if $v(x) < 1$.

   (b) Use (a) and induction on m to show that there exists $z \in K^\circ$ satisfying $v_1(z) > 1$ and $v_i(z) < 1$ for $2 \leq i \leq m$.

   (c) Choose $z_1, \ldots, z_m \in K^\circ$ so that $v_i(z_j) < 1$ if $i \neq j$ and $v_i(z_i) > 1$. Prove that if $k \in \mathbb{N}$ is sufficiently large, then $y = \sum_{i=1}^m (z_i^k x_i/(1 + z_i^k))$ satisfies $v_i(y - x_i) < e$ for $1 \leq i \leq m$.

3. Let F be an algebraic number field. For each homomorphism $\phi: F \to \mathbb{C}$, denote by v the archimedean valuation $v_\phi(x) = |\phi(x)|$. If $\operatorname{Im} \phi \subseteq \mathbb{R}$, then ϕ is a *real embedding*, otherwise a *complex embedding*. Prove the following statements.

   (a) v_ϕ is equivalent to v_ψ if and only if $\phi = \psi$ or $\phi = \bar{\psi}$, where $\bar{\psi}(x)$ is the complex conjugate of $\psi(x)$. Hint. Use Proposition 18.2, noting that the only isometric homomorphisms of $\mathbb{R}$ to $\mathbb{R}$ or $\mathbb{R}$ to $\mathbb{C}$ are the embedding maps.

(b) If ϕ is a real embedding, then v_ϕ is normalized. If ϕ is a complex embedding, then v_ϕ^2 is normalized.

(c) The number of complex embeddings of F in $\mathbb{C}$ is even, say $2r_2$, and $r_1 + 2r_2 = [F:\mathbb{Q}]$, where r_1 is the number of real embeddings of F in $\mathbb{C}$. In this case, the number of normalized archimedean valuations of F is $r_1 + r_2$.

(d) With the notation of (c), denote the real embeddings of F by $\phi_1, \ldots, \phi_{r_1}$ and the complex embeddings of F by $\psi_1, \ldots, \psi_{r_2}, \bar{\psi}_1, \ldots, \bar{\psi}_{r_2}$. Let $x_1, \ldots, x_{r_1}$ be arbitrary real numbers, $z_1, \ldots, z_{r_2}$ arbitrary complex numbers, and $e \in \mathbb{R}^+$. There exists $y \in F$ such that $|\phi_i(y) - x_i| < e$ for $1 \le i \le r_1$ and $|\psi_j(y) - z_j| < e$ for $1 \le j \le r_2$. Hint. Use the weak approximation theorem.

18.4. The Albert–Hasse–Brauer–Noether Theorem

The most profound result in the theory of central simple algebras is the Albert–Hasse–Brauer–Noether Theorem. It was proved independently by Hasse, Brauer, and Noether in [40] and by Albert and Hasse in [4].

Albert–Hasse–Brauer–Noether Theorem. *Let F be an algebraic number field. If $A \in \mathfrak{S}(F)$ satisfies $A \otimes \hat{F}_v \sim \hat{F}_v$ for all $v \in \mathbf{S}(F)$, then $A \sim F$.*

For convenience, we will refer to this result as the "Basic Theorem" throughout the rest of this section.

The Basic Theorem is closely related to one of the deep results of class field theory, the Hasse Norm Theorem. As we will show, the Basic Theorem can be deduced fairly easily from the norm theorem. On the other hand, the norm theorem is essentially the statement that a certain cohomology group in class field theory vanishes, and the triviality of this cohomology group is an easy consequence of the Basic Theorem. A direct proof of the Basic Theorem can be obtained from the analysis of generalized zeta functions. Expositions of this proof are given in the books by Deuring [26] and Weil [78].

The Hasse Norm Theorem. *Let K/F be a cyclic extension of algebraic number fields. An element a in F is the norm of an element of K if and only if $a \in N_{\hat{K}_v/\hat{F}_v}(\hat{K}_v)$ for all $v \in \mathbf{S}(F)$.*

All proofs of the norm theroem are long. We will use the result without proving it. An algebraic treatment of class field theory, including the norm theorem, is given in the article by Tate in [22] and in Artin–Tate [10]. Proofs of the norm theorem that use some analytic machinery can be found in the books by Janusz [52] and Lang [56].

Our statement of The Hasse Norm Theorem merits some explanation.

Since K/F is cyclic (hence Galois), all extensions w of v to K produce the same completion $\hat{K}_w$. In particular, $N_{\hat{K}_w/\hat{F}_v}(\hat{K}_w)$ does not depend on w. Therefore, the notation $\hat{K}_v$ instead of $\hat{K}_w$ is justified.

If K/F is Galois and $v \in \mathbf{S}(F)$, then it can (and will) be assumed that K and $\hat{F}_v$ are subfields of $\hat{K}_v$ such that $\hat{K}_v = K\hat{F}_v$. There may be different embeddings of K in $\hat{K}_v$, but they all have the same image. Let G_v be one of the various conjugate decomposition groups associated with an extension of v to K. If $\mathbf{G}(K/F) = \sigma_1 G_v \cup \cdots \cup \sigma_g G_v$ is a coset decomposition and $y \in K$, then

$$N_{K/F}(y) = \prod_{i=1}^{g} \prod_{\tau \in G_v} y^{\sigma_i \tau} = \prod_{i=1}^{g} N_{\hat{K}_v/\hat{F}_v}(y^{\sigma_i}) = N_{\hat{K}_v/\hat{F}_v}\left(\prod_{i=1}^{g} y^{\sigma_i}\right).$$

This observation shows that if $x \in F$ is the norm of an element in K, then $x \in N_{\hat{K}_v/\hat{F}_v}(\hat{K}_v)$ for all $v \in \mathbf{S}(F)$, that is, one implication of the norm theorem is true and easy to prove for Galois extensions. The difficult reverse implication is a special property of cyclic extensions. (See [22], p. 360.)

Proof of the Basic Theorem. We begin the proof by treating the case in which A is a cyclic algebra: $A = (K,\sigma,a)$, where K/F is a cyclic extension of algebraic number fields, $\mathbf{G}(K/F) = \langle\sigma\rangle$, and $a \in F^\circ$. By Corollary 15.1c and our hypothesis, if $v \in \mathbf{S}(F)$, then $\hat{F}_v \sim A \otimes \hat{F}_v \sim (K\hat{F}_v,\sigma^g,a) = (\hat{K}_v,\sigma^g,a)$, where $g[\hat{K}_v : \hat{F}_v] = [K : F]$. It follows from Lemma 15.1 that $a \in N_{\hat{K}_v/\hat{F}_v}(\hat{K}_v)$ for all $v \in \mathbf{S}(F)$. The Hasse Norm Theorem implies that $a \in N_{K/F}(K)$, so that $A \sim F$ by Lemma 15.1 again. Consider the general case. With the aim of getting a contradiction, assume that $\operatorname{Ind} A > 1$. If p is a prime divisor of $\operatorname{Ind} A$, then by Proposition 15.2 there is an algebraic number field E containing F such that $A^E \sim D$, where $D \in \mathfrak{S}(E)$ is a cyclic division algebra of degree p. In particular, $D \nsim E$. Let $w \in \mathbf{S}(E)$ divide $v \in \mathbf{S}(F)$. Identify $\hat{F}_v$ with a subfield of $\hat{E}_w$. Our hypothesis gives $D \otimes_E \hat{E}_w \sim A^E \otimes_E \hat{E}_w \cong (A \otimes \hat{F}_v) \otimes_{\hat{F}_v} \hat{E}_w \sim \hat{F}_v \otimes_{\hat{F}_v} \hat{E}_w \cong \hat{E}_w$. Since D is cyclic, the first part of the proof leads to the required contradiction $D \sim E$. □

The Basic Theorem can be put in a convenient form by using the mappings INV_F for central simple algebras over local fields that were introduced in Section 17.10. First, we must extend the definition of INV_F to the cases $F = \mathbb{R}$ and $F = \mathbb{C}$. If $A \in \mathfrak{S}(\mathbb{R})$, then either $A \sim \mathbb{R}$ or $A \sim \mathbb{H}$. Define $\mathrm{INV}_{\mathbb{R}}: \mathfrak{S}(\mathbb{R}) \to \mathbb{Q}/\mathbb{Z}$ by $\mathrm{INV}_{\mathbb{R}} A = 0$ if $A \sim \mathbb{R}$ and $\mathrm{INV}_{\mathbb{R}} A = 1/2 + \mathbb{Z}$ if $A \sim \mathbb{H}$. Plainly, $\mathrm{INV}_{\mathbb{R}} A = \mathrm{INV}_{\mathbb{R}} B$ if and only if $A \sim B$. Thus, $\mathrm{INV}_{\mathbb{R}}$ can be viewed as a bijective mapping from $\mathbf{B}(\mathbb{R})$ to $(1/2)\mathbb{Z}/\mathbb{Z}$. Since $\mathbb{H} \otimes \mathbb{H} = \left(\frac{-1,-1}{\mathbb{R}}\right) \otimes \left(\frac{-1,-1}{\mathbb{R}}\right) \sim \left(\frac{-1,1}{\mathbb{R}}\right) \sim \mathbb{R}$ by the results of Section 15.4, it is clear that $\mathrm{INV}_{\mathbb{R}}$ is a group homomorphism. Define $\mathrm{INV}_{\mathbb{C}}$ to be the zero homomorphism from $\mathfrak{S}(\mathbb{C})$ to $\mathbb{Q}/\mathbb{Z}$. In this case it is a trivial observation that $\mathrm{INV}_{\mathbb{C}}$ is an injective group homomorphism of $\mathbf{B}(\mathbb{C})$ to $\mathbb{Q}/\mathbb{Z}$ because $\mathbf{B}(\mathbb{C}) = \{1\}$.

If v is an archimedean valuation of the algebraic number field F, then $\hat{F}_v$ is either $\mathbb{R}$ or $\mathbb{C}$ by Proposition 18.3. In these respective cases v is called a *real* or *complex valuation* of F. The invariants $\mathrm{INV}_{\hat{F}_v}$ are therefore defined for all $v \in \mathbf{S}(F)$; they induce injective group homomorphisms from the multiplicative groups $\mathbf{B}(\hat{F}_v)$ to the additive group $\mathbb{Q}/\mathbb{Z}$. To simplify notation we will write INV_v instead of $\mathrm{INV}_{\hat{F}_v}$. The same expression will be used to designate the corresponding homomorphism $\mathbf{B}(\hat{F}_v) \to \mathbb{Q}/\mathbb{Z}$.

If $A \in \mathfrak{S}(F)$ and $v \in \mathbf{S}(F)$, then $\mathrm{INV}_v(A \otimes \hat{F}_v)$ is called the *local invariant* of A at v. By collecting the local invariants we get the *global invariant* of A. It is the mapping $\mathrm{INV} = \mathrm{INV}^{(F)}: \mathfrak{S}(F) \to \prod_{v \in \mathbf{S}(F)} \mathbb{Q}/\mathbb{Z}$ that is defined by $\mathrm{INV}\, A = (\ldots \mathrm{INV}_v(A \otimes \hat{F}_v) \ldots)$. Clearly, $\mathrm{INV}\, A = \mathrm{INV}\, B$ if $A \sim B$. Thus, INV can be viewed as a mapping from $\mathbf{B}(F)$ to a product of copies of $\mathbb{Q}/\mathbb{Z}$. The same notation INV or $\mathrm{INV}^{(F)}$ will be used in both contexts.

Proposition. *If F is an algebraic number field, then $\mathrm{INV}^{(F)}$ is an injective group homomorphism from $\mathbf{B}(F)$ to $\prod_{v \in \mathbf{S}(F)} \mathbb{Q}/\mathbb{Z}$.*

Since $(A \otimes B) \otimes \hat{F}_v \cong (A \otimes \hat{F}_v) \otimes_{\hat{F}_v} (B \otimes \hat{F}_v)$, it follows from Corollary 17.10a that $\mathrm{INV}^{(F)}$ is a group homomorphism; the injectivity of this homomorphism is a restatement of the Basic Theorem.

Corollary a. *Let $A, B \in \mathfrak{S}(F)$.*
(i) *$A \sim B$ if and only if $\mathrm{INV}^{(F)} A = \mathrm{INV}^{(F)} B$.*
(ii) *$A \cong B$ if and only if $\mathrm{INV}^{(F)} A = \mathrm{INV}^{(F)} B$ and $\mathrm{Deg}\, A = \mathrm{Deg}\, B$.*

Lemma. *If K/F is an extension of algebraic number fields, $w \in \mathbf{S}(K)$, $v \in \mathbf{S}(F)$, and $w|v$, then $\mathrm{INV}_w(A^K \otimes_K \hat{K}_w) = [\hat{K}_w : \hat{F}_v]\,\mathrm{INV}_v(A \otimes \hat{F}_v)$.*

PROOF. If v is discrete, then the lemma restates Proposition 17.10. Assume that v is archimedean. If $\mathrm{INV}_v(A \otimes \hat{F}_v) = 0$, then $A \otimes \hat{F}_v \sim \hat{F}_v$. Consequently, $A^K \otimes_K \hat{K}_w \sim \hat{K}_w$, as in the last part of the proof of the Basic Theorem. If $\mathrm{INV}_v(A \otimes \hat{F}_v) \neq 0$, then $\hat{F}_v = \mathbb{R}$ and $A \otimes \hat{F}_v \sim \mathbb{H}$. Either $\hat{K}_w = \mathbb{C}$ or $\hat{K}_w = \mathbb{R} = \hat{F}_v$. In the former case, $A^K \otimes_K \hat{K}_w \sim \mathbb{H} \otimes_{\mathbb{R}} \mathbb{C} \sim \mathbb{C}$ and $0 = \mathrm{INV}_w(A^K \otimes_K \hat{K}_w) = 2(1/2 + \mathbb{Z}) = [\hat{K}_w : \hat{F}_v]\,\mathrm{INV}_v(A \otimes \hat{F}_v)$. If $\hat{K}_w = \mathbb{R}$, then $A^K \otimes_K \hat{K}_w \cong A \otimes \mathbb{R} \sim \mathbb{H}$, and $\mathrm{INV}_w(A^K \otimes_K \hat{K}_w) = 1/2 + \mathbb{Z} = \mathrm{INV}_v(A \otimes \hat{F}_v) = [\hat{K}_w : \hat{F}_v]\mathrm{INV}_v(A \otimes \hat{F}_v)$. □

Corollary b. *Let K/F be an extension of algebraic number fields, and suppose that $A \in \mathfrak{S}(F)$.*
(i) *K splits A if and only if $[\hat{K}_w : \hat{F}_v]\,\mathrm{INV}_v(A \otimes \hat{F}_v) = 0$ for all $v \in \mathbf{S}(F)$ and $w \in \mathbf{S}(K)$ such that $w|v$.*
(ii) *K is isomorphic to a strictly maximal subfield of A if and only if $\mathrm{Deg}\, A = [K:F]$ and $[\hat{K}_w : \hat{F}_v]\,\mathrm{INV}_v(A \otimes \hat{F}_v) = 0$ for all $v \in \mathbf{S}(F)$ and $w \in \mathbf{S}(K)$ such that $w|v$.*

This corollary is a consequence of the lemma, the Basic Theorem, and Corollary 13.3.

Exercises

1. Prove that if $A \in \mathfrak{S}(F)$ has odd index, then $\text{INV}_v(A \otimes \hat{F}_v) = 0$ for all $v \in \mathbf{S}_\infty(F)$.

2. The purpose of this exercise is to outline a proof of a special case of Hasse's Norm Theorem.

   Theorem. If $a, b \in \mathbb{Q}$ are such that $b \in N_{\hat{\mathbb{Q}}_p(\sqrt{a})/\hat{\mathbb{Q}}_p}(\hat{\mathbb{Q}}_p(\sqrt{a}))$ for all primes p and $p = \infty$, then $b \in N_{\mathbb{Q}(\sqrt{a})/\mathbb{Q}}(\mathbb{Q}(\sqrt{a}))$.

   (a) Prove that it suffices to establish the theorem with the added hypotheses: $a, b \in \mathbb{Z}$ and a, b are square free.

   Henceforth, assume that $a, b \in \mathbb{Z}$ are square free. The proof is by induction on $|a| + |b|$. Denote $N_a = N_{\mathbb{Q}(\sqrt{a})/\mathbb{Q}}(\mathbb{Q}(\sqrt{a}))$ and for p a prime or ∞, $N_{p,a} = N_{\hat{\mathbb{Q}}_p(\sqrt{a})/\hat{\mathbb{Q}}_p}(\hat{\mathbb{Q}}_p(\sqrt{a}))$, with similar meanings assigned to N_b and $N_{p,b}$.

   (b) Prove that $b \in N_a$ if and only if $a \in N_b$ and for all p, $b \in N_{p,a}$ if and only if $a \in N_{p,b}$. Our hypothesis is that $b \in N_{p,a}$ for all p. Hence, $a \in N_{p,b}$ for all p.

   (c) Use the hypothesis with $p = \infty$ to prove that if $|a| = |b| = 1$, then $a = 1$ or $b = 1$, hence $a \in N_b$ or $b \in N_a$ and the theorem is true.

   (d) Show that if $a = x^2 - by^2 \in N_{p,b}$ with $x, y \in \hat{\mathbb{Q}}_p$, and if p divides b but p does not divide a, then $v_p(x) = 1$.

   (e) Deduce from (d) that if p is a prime divisor of b, then a is a quadratic residue mod p. Use the Chinese Remainder Theorem and the assumption that b is square free to conclude that a is a quadratic residue mod b.

   (f) Assume (as we may by (b) and (c)) that $|a| \leq |b|$ and $|b| \geq 2$. Conclude from (e) that there are integers c, d, e with c square free such that $bcd^2 = e^2 - a = \mathbb{N}_{\mathbb{Q}(\sqrt{a})/\mathbb{Q}}(e + \sqrt{a})$, where $|c| \leq |cd^2| < |b|$. Prove that $b \in N_a$ if and only if $c \in N_a$, and $c \in N_{p,a}$ for all p.

   (g) Use induction to complete the proof of the theorem.

3. Let $A \in \mathfrak{S}(\mathbb{Q})$ be a quaternion algebra. Use the result of Exercise 2 to prove that $A \sim \mathbb{Q}$ if and only if $A \otimes \hat{\mathbb{Q}}_p \sim \hat{\mathbb{Q}}_p$ for all primes p and for $p = \infty$.

4. This exercise develops machinery that can be used to compute the local invariants of quaternion algebras over $\mathbb{Q}$. It also introduces a concept of classical number theory that is an ancestor of the Artin Reciprocity Law. Assume throughout the exercise that F is a field with char $F \neq 2$. For $a, b \in F^\circ$, the *norm residue symbol* of a and b relative to F is defined by $(a,b)_F = -1$ if $a\mathbf{x}^2 + b\mathbf{y}^2 - \mathbf{z}^2 = 0$ has no solution except the trivial one $\mathbf{x} = \mathbf{y} = \mathbf{z} = 0$, and $(a,b)_F = 1$ if there is a non-trivial solution of this equation in F. The norm residue symbol was introduced by Hilbert. Prove the following results for elements $a, b, c \in F^\circ$.

   (a) The following three conditions are equivalent: $(a,b)_F = 1$; $\left(\dfrac{a,b}{F}\right) \sim F$; $b \in N_{F(\sqrt{a})/F}(F(\sqrt{a}))$. Hint. Use Proposition 1.6.

   (b) $(a,b)_F = (b,a)_F$; $(a,1)_F = (a, -a)_F = (a, 1 - a)_F = 1$; $(a,bc^2)_F = (a,b)_F$; $(a,b)_F = (a, -ab)_F = (a, (1 - a)b)_F$.

   In the remaining parts of this exercise, it is assumed that $F = \hat{\mathbb{Q}}_p$ with p a prime, or

$F = \hat{\mathbb{Q}}_\infty = \mathbb{R}$, and $a, b \in \mathbb{Z} - \{0\}$ are square free. The notation $(a,b)\hat{\mathbb{Q}}_p$ is shortened to $(a,b)_p$ where p is a prime or ∞.

(c) If $A = \left(\frac{a,b}{\mathbb{Q}}\right) \otimes \hat{\mathbb{Q}}_p \cong \left(\frac{a,b}{\hat{\mathbb{Q}}_p}\right)$, then $(a,b)_p = e^{2\pi i(\mathrm{INV}_{v_p} A)}$.

(d) $(a,bc)_p = (a,b)_p(a,c)_p$. Hint. Use (c) and the fact that $\left(\frac{a,b}{\mathbb{Q}}\right) \otimes \left(\frac{a,c}{\mathbb{Q}}\right) \sim \left(\frac{a,bc}{\mathbb{Q}}\right)$.

(e) $(a,b)_\infty = -1$ if and only if $a < 0$ and $b < 0$.

(f) If p is a prime that does not divide a, and $a\mathbf{x}^2 + p\mathbf{y}^2 = \mathbf{z}^2$ has a non-trivial solution in $\hat{\mathbb{Q}}_p$, then this equation has a solution (x,y,z) such that $v_p(x) = v_p(z) = 1$ and $v_p(y) \leq 1$.

(g) If p is an odd prime that does not divide a, then $(a,p)_p = \left(\frac{a}{p}\right)$, the Legendre Symbol. Hint. By Hensel's Lemma, $\left(\frac{a}{p}\right) = 1$ implies that $a \in \mathbb{Q}_p^2$. For the converse, use (f).

(h) If p is an odd prime, then $(p,p)_p = \left(\frac{-1}{p}\right) = (-1)^{(p-1)/2}$. Hint. $(p,p)_p = (-1,p)_p(-p,p)_p$.

(i) If p is an odd prime that does not divide either a or b, then $(a,b)_p = 1$. Hint. Use Lemma 17.9b.

(j) If a is an odd integer, then $a \in (\hat{\mathbb{Q}}_2)^2$ if and only if $a \equiv 1 \pmod 8$. Hint. For the implication that $a \equiv 1 \pmod 8$ implies $a \in (\hat{\mathbb{Q}}_2)^2$, show that if $z_n^2 \equiv a \pmod{2^{n+2}}$ for $n \geq 1$, then a suitable $y_n \in \mathbb{Z}$ satisfies $(z_n + 2^{n+1}y_n)^2 \equiv a \pmod{2^{n+3}}$.

(k) $\hat{\mathbb{Q}}_2^\circ/(\hat{\mathbb{Q}}_2^\circ)^2$ is a group of order 8 whose elements are the (multiplicative) cosets of ± 1, ± 3, ± 2, and ± 6.

(l) If a is odd, then $a\mathbf{x}^2 + b\mathbf{y}^2 = \mathbf{z}^2$ has a non-trivial solution in $\hat{\mathbb{Q}}_2$ if and only if it has a non-nilpotent solution in $\mathbb{Z}/8\mathbb{Z}$. Hint. Assume that (x,y,z) is a solution in $\mathbb{Z}^3$ modulo 8 with not all of x, y, z even. If one of a or b is even, then $ax^2 + by^2$ is odd, and by (j) there exists $w \in \hat{\mathbb{Q}}_2$ such that $w^2 = ax^2 + by^2$. If a and b are both odd, then one of ax^2, by^2 is odd, and a similar argument applies, using $a(z^2 - by^2)$ or $b(z^2 - ax^2)$ as the constant. Note that $a^2 \equiv b^2 \equiv 1 \pmod 8$, and a^{-1}, b^{-1} exist in $\hat{\mathbb{Q}}_2$.

(m) If p and q are odd primes, then $(-1,-1)_2 = -1$, $(2,2)_2 = 1$, $(-1,p)_2 = (-1)^{(p-1)/2}$, $(2,p)_2 = \left(\frac{p}{2}\right) = (-1)^{(p^2-1)/8}$, $(p,q)_2 = (-1)^{(p-1)(q-1)/4}$. Hint. Use (k), (l), and a lot of paper.

(n) If p is an odd prime, $a = p^e a'$, and $b = p^f b'$, where p does not divide a' or b', then $(a,b)_p = (-1)^{ef(p-1)/2}\left(\frac{a'}{p}\right)^f\left(\frac{b'}{p}\right)^e$. If $a = 2^e a'$ and $b = 2^f b'$, where a' and b' are odd, then $(a,b)_2 = (-1)^{(a'-1)(b'-1)/4}\left(\frac{a'}{2}\right)^f\left(\frac{b'}{2}\right)^e$.

5. Use the Basic Theorem and the results of Exercise 4 to determine which pairs of the following quaternion algebras are isomorphic, that is, classify the algebras by isomorphism types: $M_2(\mathbb{Q}) = \left(\frac{1,1}{\mathbb{Q}}\right)$, $\left(\frac{2,3}{\mathbb{Q}}\right)$, $\left(\frac{-1,2}{\mathbb{Q}}\right)$, $\left(\frac{-1,3}{\mathbb{Q}}\right)$, $\left(\frac{-1,6}{\mathbb{Q}}\right)$, $\left(\frac{-2,3}{\mathbb{Q}}\right)$, $\left(\frac{2,-3}{\mathbb{Q}}\right)$, $\left(\frac{-1,-2}{\mathbb{Q}}\right)$, $\left(\frac{-1,-3}{\mathbb{Q}}\right)$, $\left(\frac{-1,-6}{\mathbb{Q}}\right)$, $\left(\frac{-2,-3}{\mathbb{Q}}\right)$.

18.5. The Brauer Groups of Algebraic Number Fields

The Albert–Hasse–Brauer–Noether Theorem shows that the Brauer group of an algebraic number field can be embedded in a product of copies of $\mathbb{Q}/\mathbb{Z}$. The main theorem of this section gives a precise description of the embedding.

Theorem. *If F is an algebraic number field, then there is an exact sequence*

$$1 \to \mathbf{B}(F) \to \mathbf{I}(F) \to \mathbb{Q}/\mathbb{Z} \to 1,$$

where $\mathbf{I}(F) = \bigoplus_{v \in \mathbf{S}(F)} I_v(F)$, $I_v(F) = \mathbb{Q}/\mathbb{Z}$ *if v is discrete,* $I_v(F) = (1/2)\mathbb{Z}/\mathbb{Z}$ *if v is real, and* $I_v(F) = 0$ *if v is complex.*

The homomorphism $\mathbf{B}(F) \to \mathbf{I}(F)$ is the invariant mapping $\mathrm{INV}^{(F)}$. The homomorphism $\mathbf{I}(F) \to \mathbb{Q}/\mathbb{Z}$ is the coproduct γ of the inclusion mappings $I_v(F) \to \mathbb{Q}/\mathbb{Z}$, that is, $\gamma: (\ldots t_v \ldots) \mapsto \sum_v t_v$.

The proof of the theorem occupies the rest of this chapter. We have noted that the injectivity of $\mathrm{INV}^{(F)}$ is equivalent to the Albert–Hasse–Brauer–Noether Theorem. Most of this section is devoted to the proof that the image of $\mathrm{INV}^{(F)}$ is a subgroup of $\mathbf{I}(F)$. This fact is obtained by an elementary argument. The deepest part of the proof is the exactness of the sequence at $\mathbf{I}(F)$. We will prove this fact in Section 18.7, using preliminary results from Section 18.6 and two basic theorems of class field theory. It is obvious that γ is surjective: if v is discrete, so that $I_v(F) = \mathbb{Q}/\mathbb{Z}$, then $\gamma(I_v(F)) = \mathbb{Q}/\mathbb{Z}$.

Lemma. *If F and K are algebraic number fields and $F \subseteq K$, then $e_w(K/F) = 1$ for almost all $w \in \mathbf{S}(K)$.*

PROOF. We can assume that K/F is Galois. Otherwise, enlarge K to an algebraic number field L such that L/F is Galois, and note that if u is an extension to L of $w \in \mathbf{S}(K)$, then $e_w(K/F)$ divides $e_u(L/F)$. In particular, if $e_u(L/F) = 1$ for almost all $u \in \mathbf{S}(L)$, then $e_w(K/F) = 1$ for almost all $w \in \mathbf{S}(K)$. The assumption that K/F is Galois implies that $\hat{K}_w/\hat{F}_v$ is Galois and $\hat{K}_w$ doesn't depend on the choice of the extension w of v. Moreover, $e_w(K/F) = e_w(\hat{K}_w/\hat{F}_v)$, so that it suffices to show that the field extension $K\hat{F}_v/\hat{F}_v$ is unramified for almost all $v \in \mathbf{S}(F)$. Let $K = F(y)$. If $\Phi \in F[\mathbf{x}]$ is the minimum polynomial of y over F, then Φ is monic, irreducible, and the roots of Φ are distinct. Thus, Φ is relatively prime to its derivative Φ'. Since $F[\mathbf{x}]$ is a principal ideal domain, $\Theta\Phi + \Psi\Phi' = 1$ for suitable Θ and Ψ in $F[\mathbf{x}]$. The Product Formula guarantees the existence of a finite set X of valuations of F with $\mathbf{S}_\infty(F) \subseteq X$ such that $\Phi, \Phi', \Theta, \Psi \in O(F,v)[\mathbf{x}] \subseteq O(\hat{F}_v)[\mathbf{x}]$ for all $v \in \mathbf{S}(F) - X$. Thus, if $v \notin X$, then the residue field mapping of $O(\hat{F}_v)[\mathbf{x}]$ to $E(\hat{F}_v)[\mathbf{x}]$ is defined on Φ, Φ', Θ, and Ψ. It gives $\overline{\Theta}\overline{\Phi} + \overline{\Psi}\overline{\Phi}' =$

1. Consequently, $\bar{\Phi}$ has distinct roots in an extension of $E(\hat{F}_v)$. By Proposition 17.8, $\hat{F}_v(y)/\hat{F}_v = K\hat{F}_v/\hat{F}_v$ is unramified for all $v \in \mathbf{S}(F) - X$, that is, for almost all $v \in \mathbf{S}(F)$. □

Proposition. *Let F be an algebraic number field. If $A \in \mathfrak{S}(F)$, then* $\mathrm{INV}_v(A \otimes \hat{F}_v) = 0$ *for almost all* $v \in \mathbf{S}(F)$.

PROOF. $\mathrm{INV}^{(F)}$ is constant over equivalence classes of algebras. Thus, we can assume that A is a crossed product, say $A = (K,G,\Phi)$, where K/F is a Galois extension, $G = \mathbf{G}(K/F)$, and $\Phi \in Z^2(G,K^\circ)$. By the Product Formula and the lemma, there is a finite set $X \subseteq \mathbf{S}(F)$ such that if $v \in \mathbf{S}(F) - X$ and $w \in \mathbf{S}(K)$ divides v, then v is discrete, $\hat{K}_w/\hat{F}_v$ is unramified, and $w(\Phi(\rho,\tau)) = 1$ for all ρ, $\tau \in G$. We will show that $\mathrm{INV}_v(A \otimes \hat{F}_v) = 0$ in this case. The choice of X guarantees that $\hat{F}_v$ is a local field and $\hat{K}_w/\hat{F}_v = K\hat{F}_v/\hat{F}_v$ is unramified for all $v \in \mathbf{S}(F) - X$. It follows from Proposition 17.8 that $\hat{K}_w/\hat{F}_v$ is cyclic, say $H = \mathbf{G}(\hat{K}_w/\hat{F}_v) = \langle\sigma\rangle$. Denote $k = |H| = [\hat{K}_w : \hat{F}_v]$. The definition of crossed products implies that $A = \bigoplus_{\tau \in G} u_\tau K$, $u_\tau^{-1} d u_\tau = d^\tau$ for all $d \in K$, $\tau \in G$, and $u_\rho u_\tau = u_{\rho\tau}\Phi(\rho,\tau)$. By Proposition 14.7b, $A \otimes \hat{F}_v \sim (\hat{K}_w, H, \Phi|H^2) = \bigoplus_{i \in k} u_{\sigma^i}\hat{K}_w = \bigoplus_{i<k} (u_\sigma)^i \hat{K}_w$. In fact, $u_\sigma^2 = u_{\sigma^2}\Phi(\sigma,\sigma)$, $u_\sigma^3 = u_{\sigma^3}\Phi(\sigma,\sigma^2)\Phi(\sigma,\sigma), \ldots, u_\sigma^{k-1} = u_{\sigma^{k-1}}\,\Phi(\sigma,\sigma^{k-2}) \cdots \Phi(\sigma,\sigma)$ and finally, $u_\sigma^k = u_1\Phi(\sigma,\sigma^{k-1}) \cdots \Phi(\sigma,\sigma) = a$, where $a = \Phi(1,1)\Phi(\sigma,\sigma^{k-1}) \cdots \Phi(\sigma,\sigma) \in \hat{F}_v \cap K$. In the notation of Section 15.1, $A \otimes \hat{F}_v \sim (\hat{K}_w,\sigma,a)$. Since $v \notin X$, it follows that $w(\Phi(\rho,\tau)) = 1$ for all ρ, $\tau \in G$. Therefore, $v(a) = w(a) = 1$. By Lemma 17.9b, $a \in N_{\hat{K}_w/\hat{F}_v}(\hat{K}_w)$. Consequently, $(\hat{K}_w,\sigma,a) \sim \hat{F}_v$ by Lemma 15.1b. That is, $\mathrm{INV}_v(A \otimes \hat{F}_v) = 0$. □

Corollary. *If F is an algebraic number field, then the image of* $\mathrm{INV}^{(F)}$ *is a subgroup of* $\mathbf{I}(F)$.

EXERCISES

1. Prove that if the algebraic number field F contains a primitive n'th root of unity, $n > 2$, then all non-archimedean valuations of F are complex, and $\mathbf{I}(F)$ is isomorphic to a direct sum of copies of $\mathbb{Q}/\mathbb{Z}$.

2. Give a simplified proof of the proposition in the case that A is the cyclic algebra (K,σ,a) with K/F cyclic, $\mathbf{G}(K/F) = \langle\sigma\rangle$, $a \in F^\circ$. Show that $\mathrm{INV}_v(A \otimes \hat{F}_v) = 0$ if v is discrete, $v(a) = 1$, and $e_w(K/F) = 1$ for all extensions w of v to K. Hint. Use Corollary 15.1c.

3. Use the theorem to show that if $A \in \mathfrak{S}(F)$, where F is an algebraic number field, and if $[A] \neq 1$, then $\mathrm{INV}_v(A \otimes \hat{F}_v) \neq 0$ for at least two $v \in \mathbf{S}(F)$.

4. Let $a \in \mathbb{Z}$ be square free. Prove that $e_w(\mathbb{Q}(\sqrt{a})/\mathbb{Q}) = 2$ in the following cases.
   (a) $w \supseteq v_p$, where p is a prime divisor of a.
   (b) $w \supseteq v_2$ and $a \equiv 3 \pmod 4$. Hint. Show that $w(1 + \sqrt{a}) = v_2(2)^{1/2}$.
   Prove that $e_w(\mathbb{Q}(\sqrt{a})/\mathbb{Q}) = 1$ for all other discrete valuations w of $\mathbb{Q}(\sqrt{a})$. Hint.

Show that if $w \supseteq v_2$ and $a \equiv 1 \pmod 4$, then the assumption that $1 > w(c + d\sqrt{a}) > v_2(2)$, $c, d \in \mathbb{Q}$, leads to a contradiction.

5. Prove that the lemma and the proposition are also true when F is an algebraic function field.

18.6. Cyclic Algebras over Number Fields

One remarkable consequence of the Albert–Hasse–Brauer–Noether Theorem is that every central simple algebra over an algebraic number field is cyclic. In this section we will prove this fact, using a fundamental existence theorem for cyclic extensions of algebraic number fields.

The Grunwald–Wang Theorem. *Let F be an algebraic number field. Assume that $\{(v_1,n_1), \ldots, (v_r,n_r)\}$ is a finite set of pairs such that $v_i \in \mathbf{S}(F)$, $n_i \in \mathbb{N}$, $n_i = 1$ if v_i is complex, and $n_i \leq 2$ if v_i is real. Let m be the least common multiple of $\{n_1, \ldots, n_r\}$. If $n \in \mathbb{N}$ is divisible by m, then there is a cyclic extension K/F of degree n such that n_i divides $[\hat{K}_{v_i} : \hat{F}_{v_i}]$ for $1 \leq i \leq r$. Moreover, if $2m$ divides n, then K can be chosen so that $\hat{K}_{v_i}/\hat{F}_{v_i}$ is unramified for all of the v_i that are discrete.*

We won't prove the Grunwald–Wang Theorem. It is treated fully in Chapter 10 of the Artin–Tate notes on class field theory [10]. A related weaker theorem is outlined in Exercise 3.

Theorem. *Let F be an algebraic number field. If $A \in \mathfrak{S}(F)$, then A is cyclic and* $\operatorname{Ind} A = \operatorname{Exp} A$.

PROOF. By Proposition 18.5, the set of valuations v of F such that $\mathrm{INV}_v(A \otimes \hat{F}_v) \neq 0$ is finite, and no such v is complex. Let $v_1, \ldots, v_r$ be a listing of these valuations. For $1 \leq i \leq r$, define $n_i = \operatorname{Ind}(A \otimes \hat{F}_{v_i})$. If v_i is real, then $A \otimes \hat{F}_{v_i} \sim \mathbb{H}$ and $n_i = 2$. Denote the least common multiple of $\{n_1, \ldots, n_r\}$ by m. According to Proposition 13.4, the degree n of A is divisible by each n_i; thus, m divides n. By the Grunwald–Wang Theorem, there are cyclic extensions K/F and L/F of degrees n and m respectively such that n_i divides $[\hat{K}_{v_i} : \hat{F}_{v_i}]$ and $[\hat{L}_{v_i} : \hat{F}_{v_i}]$ for $1 \leq i \leq r$. Since n_i is the order of $\mathrm{INV}_{v_i}(A \otimes \hat{F}_{v_i})$ by Corollary 17.10a, it is a consequence of Corollary 18.4b that K and L split A. By construction $\operatorname{Deg} A = [K : F]$, so that K is isomorphic to a strictly maximal subfield of A according to Corollary 13.3. Hence, A is cyclic. The fact that L splits A implies that $\operatorname{Ind} A \leq m$ by Proposition 13.4. If k is the exponent of A, then $k\,\mathrm{INV}[A] = \mathrm{INV}[A]^k = 0$; that is, $k\,\mathrm{INV}_v(A \otimes \hat{F}_v) = 0$ for all $v \in \mathbf{S}(F)$. Thus, $n_i | k$ for $1 \leq i \leq r$. Equivalently, $m|k$. In particular, $\operatorname{Ind} A \leq \operatorname{Exp} A$. The reverse inequality is valid for central simple algebras over any field by Proposition 14.4b. $\square$

The proof of the theorem gives an explicit computation of the index.

Corollary. *If F is an algebraic number field and $A \in \mathfrak{S}(F)$, then the index of A is the least common multiple of the local indices* $\mathrm{Ind}(A \otimes \hat{F}_v)$ *of* A.

EXERCISES

1. Prove the Grunwald–Wang Theorem in the case that $F = \mathbb{Q}$ and $n = 2$. Explicitly, show that if $n_\infty, n_2, n_3, n_5, \ldots, n_p$ are natural numbers, and each n_i is either 1 or 2, then there is a number $a \in \mathbb{Z}$ such that $[\mathbb{R}(\sqrt{a}) : \mathbb{R}] = n_\infty$, $[\hat{\mathbb{Q}}_2(\sqrt{a}) : \hat{\mathbb{Q}}_2] = n_2$, $[\hat{\mathbb{Q}}_3(\sqrt{a}) : \hat{\mathbb{Q}}_3] = n_3$, $[\hat{\mathbb{Q}}_5(\sqrt{a}) : \hat{\mathbb{Q}}_5] = n_5, \ldots, [\hat{\mathbb{Q}}_p(\sqrt{a}) : \hat{\mathbb{Q}}_p] = n_p$, and $\hat{\mathbb{Q}}_q(\sqrt{a})/\hat{\mathbb{Q}}_q$ is unramified for all primes $q \leq p$. Hint. By part (j) of Exercise 4 in Section 18.4, if $a \in \mathbb{Z}$ satisfies $a \equiv 1 \pmod 8$, then $a \in (\hat{\mathbb{Q}}_2)^2$ and if $a \equiv 5 \pmod 8$, then $a \notin (\hat{\mathbb{Q}}_2)^2$. Prove that if q is an odd prime, then $a \in (\hat{\mathbb{Q}}_q)^2$ if $a \equiv 1 \pmod q$; and there exists $c_q \in \mathbb{N}$ with $1 < c_q < q$ such that $a \equiv c_q \pmod q$ implies that $a \notin (\hat{\mathbb{Q}}_q)^2$. Use these observations together with the Chinese Remainder Theorem and the result of Exercise 4, Section 18.5 to prove the desired result.

2. The result of this exercise is needed to prove the weak version of the Grunwald–Wang Theorem that will be given in the next exercise. Prove that if K is a local field, then $W = \bigcup_n \{x \in K : x^n = 1\}$ is finite. Hint. Let v be the valuation of K. Denote $W_1 = \{x \in W : v(x - 1) < 1\}$. Note that W_1 is the kernel of the restriction to W of the residue class mapping, so that W/W_1 is finite. Use the result of Exercise 2, Section 17.9 to show that the multiplicative order of every element in W_1 is a power of $p = \operatorname{char} E(K,v)$. Prove that if $y \in W_1$, then $v(y^p - 1) < v(y - 1)$. Use this result to show that if $x \in W_1$ is such that $v(x - 1) \geq v(y - 1)$ for all $y \in W_1$, then $W_1 = \langle x \rangle$. (Note that the finite subgroups of the multiplicative group of a field are cyclic.)

3. This exercise outlines the proof of a weak form of the Grunwald–Wang Theorem.

   Proposition. If F is an algebraic number field, $m \in \mathbb{N}$, and X is a finite set of discrete valuations of F, then there is a cyclic extension K/F such that all archimedean valuations of K are complex and m divides the local degree $[\hat{K}_v : \hat{F}_v]$ for all $v \in X$.

   (a) Prove that if the proposition is true in the case that $F = \mathbb{Q}$, then it is true for all algebraic number fields F.

   (b) Let p be an odd prime, $k \in \mathbb{N}$, $\zeta = e^{2\pi i/p^{k+1}} \in \mathbb{C}$, and $L_p = \mathbb{Q}(\zeta)$. Prove that $L_p/\mathbb{Q}$ is cyclic of degree $p^k(p-1)$. Deduce that L_p contains a subfield K_p such that $K_p/\mathbb{Q}$ is cyclic of degree p^k.

   (c) Let $\zeta = e^{2\pi i/2^{k+2}} \in \mathbb{C}$ and $L_2 = \mathbb{Q}(\zeta)$. Prove that $L_2/\mathbb{Q}$ is Galois with $\mathbf{G}(L_2/\mathbb{Q}) = \{\tau_n : n \text{ odd}\} \cong \mathbb{Z}/2\mathbb{Z} \oplus \mathbb{Z}/2^k\mathbb{Z}$, where $\zeta^{\tau} = \zeta^n$. Denote $x = \zeta - \zeta^{-1}$, $K_2 = \mathbb{Q}(x)$. Show that $x^{\tau_r} = x^{\tau_s}$ if and only if $r \equiv s \pmod{2^{k+2}}$ or $r \equiv 2^{k+1} - s \pmod{2^{k+2}}$. Deduce that $\mathbf{G}(K_2/\mathbb{Q}) = \{\tau_n|K_2 : n \equiv 1 \pmod 4\} \cong \mathbb{Z}/2^k\mathbb{Z}$. Prove that if $y = x^{\tau_t}$ with $t = 2^k + 1$, then $y \in K_2$ and $(x/2)^2 + (y/2)^2 = -1$.

   (d) Let K and L be the respective composita in $\mathbb{C}$ of the fields K_p and L_p, where p ranges over the prime divisors of $2m$. Note that K and L depend on k, which has yet to be specified. Prove that $K/\mathbb{Q}$ is cyclic and $L/\mathbb{Q}$ is Galois. Deduce from the last statement of (c) that the archimedean valuations of K are complex.

   (e) Let p be a prime divisor of m, and suppose that $v_q \in X$. Denote by $p^{\mu(k)}$ the largest power of p that divides $[L_p\hat{\mathbb{Q}}_q : \hat{\mathbb{Q}}_q]$. Prove that $p^{\mu(k)-1}$ divides the local degree of $K/\mathbb{Q}$ at v_q. (In fact, $p^{\mu(k)}$ divides this local degree if p is odd.) Deduce that the proposition will follow from the conclusion that $\mu(k) \to \infty$ as $k \to \infty$.

(f) Assume that p is an odd prime divisor of m. Thus, $L_p/\mathbb{Q}$ is cyclic, say $\mathbf{G}(L_p/\mathbb{Q}) = \langle \rho \rangle$. Let F be the fixed field of ρ^{p-1}. Prove that $F = \mathbb{Q}(w)$, where $w = e^{2\pi i/p}$. Establish the equality $[(L_p \cap \hat{\mathbb{Q}}_q)F : \mathbb{Q}] = p^{k-\mu(k)}(p-1)$. Deduce from the fact that the subfields of L_p are totally ordered by inclusion that $\zeta^{p^{\mu(k)}} \in (L_p \cap \hat{\mathbb{Q}}_q)F \subseteq \hat{\mathbb{Q}}_q(w)$. Use the result of Exercise 2 to show that $\mu(k) \to \infty$ as $k \to \infty$. Hint. $[L_p\hat{\mathbb{Q}}_q : \hat{\mathbb{Q}}_q] = [L_p : L_p \cap \hat{\mathbb{Q}}_q]$.

(g) Complete the proof of the proposition with a similar argument in the case $p = 2$. Hint. $L_2/\mathbb{Q}(i)$ is cyclic of degree 2^k.

4. Deduce from the proposition of Exercise 3 that if F is an algebraic number field and $A \in \mathfrak{S}(F)$, then there is a cyclic extension of F that splits A. This result gives a weak version of the theorem: every $A \in \mathfrak{S}(F)$ is equivalent to a cyclic algebra.

5. Let F be an algebraic number field, $D_1 = \left(\dfrac{a,b}{F}\right)$, $D_2 = \left(\dfrac{c,d}{F}\right)$, where $a, b, c, d \in F^\circ$.

(a) Prove that $\operatorname{Ind}(D_1 \otimes D_2) \leq 2$.

(b) Deduce from (a) that D_1 and D_2 contain maximal subfields E_1 and E_2 such that $E_1 \cong E_2$ as F-algebras. Hint. See Lemma 15.7.

18.7. The Image of INV

In this section we will complete the proof of Theorem 18.5. It remains to show that the image of INV is equal to the kernel of γ. This will be done in two steps: if $A \in \mathfrak{S}(F)$, then $\sum_{v \in \mathbf{S}(F)} \mathrm{INV}_v(A \otimes \hat{F}_v) = 0$; if $(\ldots t_v \ldots) \in \mathbf{I}(F)$ satisfies $\sum_{v \in \mathbf{S}(F)} t_v = 0$, then $A \in \mathfrak{S}(F)$ exists such that $\mathrm{INV}_v(A \otimes \hat{F}_v) = t_v$ for all $v \in \mathbf{S}(F)$. The proofs of these results use the Artin Reciprocity Law and the Tchebotarev Density Theorem.

We begin this section with a description of Artin's Reciprocity Law. If F is an algebraic number field, then the *idele group* of F is the subgroup J_F of $\prod_{v \in \mathbf{S}(F)} \hat{F}_v^\circ$ that consists of all sequences $(\ldots y_v \ldots)$ such that $v(y_v) = 1$ for almost all $v \in \mathbf{S}(F)$. There is a topology on J_F that is defined in terms of the v-topology of the factors $\hat{F}_v^\circ$. We won't give this definition because the topology of J_F does not occur in the statement of the reciprocity law. If $x \in F^\circ$, then $v(x) = 1$ for almost all $v \in \mathbf{S}(F)$ by the Product Formula. Therefore, the diagonal mapping $x \mapsto (\ldots x \ldots)$ embeds F° as a subgroup of J_F. Our notation for this subgroup will also be F°.

Let K be an algebraic number field that contains F. If $v \in \mathbf{S}(F)$ and $w \in \mathbf{S}(K)$ with $w|v$, then $\hat{F}_v$ can be viewed as a subfield of $\hat{K}_w$. The norm mapping from $\hat{K}_w^\circ$ to $\hat{F}_v^\circ$ will be abbreviated by N_w. These mappings induce a homomorphism $N: J_K \to J_F$ by the rule

$$N: (\ldots z_w \ldots) \to (\ldots \textstyle\prod_{w|v} N_w(z_w) \ldots).$$

Indeed, if $w(z_w) = 1$ for all $w \in \mathbf{S}(K)$ that divide v, then $v(\prod_{w|v} N_w(z_w)) = 1$ by 17.9(1). We have seen that if K/F is Galois, then the completions $\hat{K}_w$

are the same for all divisors w of v. Therefore, $N_w(\hat{K}_w^\circ)$ also depends only on v. We will usually write $\hat{K}_v$ instead of $\hat{K}_w$ and $N_v(K_v^\circ)$ for $N_w(\hat{K}_w^\circ)$ when K/F is Galois.

Artin Reciprocity Law. *Let K/F be an abelian extension of algebraic number fields. There is a homomorphism $\alpha\colon J_F \to \mathbf{G}(K/F)$ such that the sequence*

$$1 \to N(J_K)F^\circ \to J_F \xrightarrow{\alpha} \mathbf{G}(K/F) \to 1$$

is exact.

There are several reasons why this theorem is called a "reciprocity law." Exercise 1 shows that the classical quadratic reciprocity theorem is a special case of Artin's Reciprocity Law.

The homomorphism $\alpha\colon J_F \to \mathbf{G}(K/F)$ is called the *Artin mapping*. We will need information about the Artin mapping in order to compute it in special cases. The results that we need come directly from the definition of α.

The restrictions of α to the components $\hat{F}_v^\circ$ of J_F give local Artin mappings $\alpha_v\colon \hat{F}_v^\circ \to \mathbf{G}(K/F)$ such that $\alpha_v(x_v) = 1$ if $x_v \in N_v(\hat{K}_v^\circ)$. If $\hat{K}_v/\hat{F}_v$ is unramified and $v(x_v) = 1$, then $x_v \in N_v(\hat{K}_v^\circ)$ by Lemma 17.9b. This observation and the assumption that $\mathbf{G}(K/F)$ is abelian imply that $(\ldots x_v \ldots) \mapsto \prod_{v \in \mathbf{S}(F)} \alpha_v(x_v)$ is a well defined homomorphism of J_F to $\mathbf{G}(K/F)$. The fact is that this definition reproduces α: the Artin mapping is the coproduct of the local Artin mappings. Thus, to determine α, it is sufficient to describe all α_v. We will do so in two cases: v is discrete and K/F is unramified at v; v is archimedean, K/F is cyclic, and $[\hat{K}_v : \hat{F}_v] = 2$.

Assume that v is discrete and $e_v(K/F) = 1$. Let $x_v \in \hat{F}_v^\circ$ have the exponential value l, that is, $v(x_v) = v(a_v)^l$, where $a_v \in \hat{F}_v$ is a uniformizer. The local Artin mapping at v is defined by $\alpha_v(x_v) = \sigma_v^l$. As in Section 18.3, σ_v denotes the Frobenius automorphism of K/F corresponding to v.

If K/F is cyclic of degree n, v is archimedean, and $[\hat{K}_v : \hat{F}_v] = 2$, then n is even because $[\hat{K}_v : \hat{F}_v]$ divides $[K : F]$. In this case, α_v is the unique, nontrivial homomorphism from $\hat{F}_v^\circ = \mathbb{R}$ to $\mathbf{G}(K/F)$ that has the kernel $N_v(\hat{K}_v^\circ) = N_v(\mathbb{C}) = \mathbb{R}^2$. If $\mathbf{G}(K/F) = \langle\tau\rangle$, then $\alpha_v(x_v) = \tau^{n/2}$ for $x_v < 0$ and $\alpha_v(x_v) = 1$ for $x_v > 0$.

Lemma. *Let K/F be a cyclic extension of algebraic number fields with $\mathbf{G}(K/F) = \langle\tau\rangle$ and $[K : F] = n$. The elements of $(1/n)\mathbb{Z}/\mathbb{Z}$ act as endomorphisms of $\mathbf{G}(K/F)$ by the rule $\tau^{(k/n+\mathbb{Z})} = \tau^k$. If $A = (K,\tau,a)$ and $v \in \mathbf{S}(F)$, then $\alpha_v(a) = \tau^{\mathrm{INV}_v(A \otimes \hat{F}_v)}$ in the following cases: v is discrete and $e_v(K/F) = 1$; v is archimedean.*

Proof. Assume that v is discrete and $e_v(K/F) = 1$. By Corollary 15.1c, $A \otimes \hat{F}_v \sim (\hat{K}_v, \tau^{k(v)}, a)$, where $k(v) = n/n(v)$ with $n(v) = [\hat{K}_v : \hat{F}_v]$. Both $\tau^{k(v)}$ and σ_v are generators of the decomposition group $G_v \cong \mathbf{G}(\hat{K}_v/\hat{F}_v)$.

Thus, $\sigma_v = \tau^{k(v)m(v)}$ for a suitable integer $m(v)$ that is relatively prime to $n(v)$. Corollary 15.1a yields $A \otimes \hat{F}_v \sim (\hat{K}_v,\tau^{k(v)m(v)},a^{m(v)}) = (\hat{K}_v,\sigma_v,a^{m(v)})$. Let l be the exponential value of a in $\hat{F}_v$. If $a_v \in \hat{F}_v$ is a uniformizer, then by Lemma 17.9b, $a \in a_v^l N_v(\hat{K}_v^\circ)$; and $A \otimes \hat{F}_v \sim (\hat{K}_v,\sigma_v,a_v^{lm(v)})$ according to Lemma 15.1. By comparing this expression with the definition of INV_v in Section 17.10, we conclude that $\mathrm{INV}_v(A \otimes \hat{F}_v) = lm(v)/n(v) + \mathbb{Z} = lk(v)m(v)/n + \mathbb{Z}$. The description of the local Artin mapping at a discrete unramified valuation gives $\alpha_v(a) = \sigma_v^l = \tau^{lk(v)m(v)} = \tau^{\mathrm{INV}_v(A \otimes \hat{F}_v)}$. Suppose that $v \in \mathbf{S}_\infty(F)$. If $\hat{K}_v = \hat{F}_v$, then $\mathrm{INV}_v(A \otimes \hat{F}_v) = 0$ and $\alpha_v(a) = 1 = \tau^{\mathrm{INV}_v(A \otimes \hat{F}_v)}$. If $\hat{K}_v = \mathbb{C} \supset \mathbb{R} = \hat{F}_v$, then $\mathrm{INV}_v(A \otimes \hat{F}_v) = 0$ if and only if $a \in N_{\mathbb{C}/\mathbb{R}}(\mathbb{C}^\circ) = \mathbb{R}^+$. If $a < 0$, then $\mathrm{INV}_v(A \otimes \hat{F}_v) = 1/2 + \mathbb{Z}$. The description of α_v for non-archimedean v implies that $\alpha_v(a) = 1 = \tau^{\mathrm{INV}_v(A \otimes \hat{F}_v)}$ if $a > 0$, and $\alpha_v(a) = \tau^{n/2} = \tau^{\mathrm{INV}_v(A \otimes \hat{F}_v)}$ if $a < 0$. □

Proposition a. *If $A \in \mathfrak{S}(F)$, where F is an algebraic number field, then* $\sum_{v \in \mathbf{S}(F)} \mathrm{INV}_v(A \otimes \hat{F}_v) = 0$.

Proof. By the Grunwald–Wang Theorem and Corollary 18.4b, there is a cyclic extension K/F such that K splits A and $e_v(K/F) = 1$ for all discrete $v \in \mathbf{S}(F)$ such that $\mathrm{INV}_v(A \otimes \hat{F}_v) \neq 0$. We can assume on the basis of Theorem 13.3 that $A = (K,\tau,a)$, where $\mathbf{G}(K/F) = \langle\tau\rangle$ and $a \in F^\circ$. If $\mathrm{INV}_v A \otimes \hat{F}_v = 0$, then $a \in N_v(\hat{K}_v^\circ)$ by Lemma 15.1. In this case, $\alpha_v(a) = 1$. By the lemma and the Artin Reciprocity Law, $\tau^{\sum_v \mathrm{INV}(A \otimes \hat{F}_v)} = \prod_v \alpha_v(a) = \alpha(a) = 1$. Thus, $\sum_v \mathrm{INV}_v(A \otimes \hat{F}_v) = 0$. □

To complete the proof of Theorem 18.5, we need a weak version of the Tchebotarev Density Theorem. The result generalizes Dirichlet's theorem on the existence of primes in arithmetic progressions. (See Exercise 2.) It also sharpens the statement in the Reciprocity Law that α is surjective.

Tchebotarev Density Theorem. *Let K/F be an abelian extension of algebraic number fields. If $\rho \in \mathbf{G}(K/F)$, then there are infinitely many discrete valuations v of F with $e_v(K/F) = 1$ such that $\sigma_v = \rho$.*

Proposition b. *Let F be an algebraic number field. If $\xi = (\ldots t_v \ldots) \in \mathbf{I}(F)$ satisfies $\sum_v t_v = 0$, then $\xi = \mathrm{INV}^{(F)} A$ for some $A \in \mathfrak{S}(F)$.*

Proof. Let $X = \{v \in \mathbf{S}(F)\colon t_v \neq 0\}$. For $v \in X$, write $t_v = k_v/n_v + \mathbb{Z}$, where $k_v, n_v \in \mathbb{N}$, $n_v > 1$, and $(k_v,n_v) = 1$. If $v \in X \cap \mathbf{S}_\infty(F)$, then v is real and $t_v = 1/2 + \mathbb{Z}$. By the Grunwald–Wang Theorem, there is a cyclic extension K/F of algebraic number fields such that n_v divides $n(v) = [\hat{K}_v : \hat{F}_v]$ for all $v \in X$ and $e_v(\hat{K}_v/\hat{F}_v) = 1$ for all discrete $v \in X$. Let $\mathbf{G}(K/F) = \langle\tau\rangle$ have order n. If v is discrete, then Theorem 17.10 produces cyclic $\hat{F}_v$-algebras with the required invariant t_v. That is, there exists $z_v \in \hat{F}_v^\circ$ such that $\mathrm{INV}_v(\hat{K}_v,\sigma_v,z_v) = t_v$. As in the proof of Proposition a, $\sigma_v = \tau^{k(v)m(v)}$ where

$k(v) = n/n(v)$ and $(m(v),n(v)) = 1$. The fact that $m(v)$ is relatively prime to $n(v)$ implies the existence of $y_v \in \hat{F}_v^\circ$ such that $z_v/y_v^{m(v)} \in \hat{F}_v^{n(v)} \subseteq N_v(\hat{K}_v^\circ)$. Therefore, $(\hat{K}_v,\sigma_v,z_v) \cong (\hat{K}_v,\tau^{k(v)m(v)},y_v^{m(v)}) \cong (\hat{K}_v,\tau^{k(v)},y_v)$, and

$$\mathrm{INV}_v(\hat{K}_v,\tau^{k(v)},y_v) = t_v. \tag{1}$$

When $v \in X$ is archimedean, we obtain (1) directly by letting $y_v = -1$. In this case, $\hat{K}_v = \mathbb{C}$, $\hat{F}_v = \mathbb{R}$, $t_v = 1/2 + \mathbb{Z}$, and $k(v) = n/2$. By the Tchebotarev Density Theorem, there is a discrete valuation $v_0 \in \mathbf{S}(F) - X$ with $e_{v_0}(K/F) = 1$ and a uniformizer $y_{v_0} \in \hat{F}_{v_0}$ such that $\alpha_{v_0}(y_{v_0}) = (\prod_{v \in X} \alpha_v(y_v))^{-1}$. Define $(\ldots x_v \ldots) \in J_F$ by the conditions $x_v = y_v$ for $v \in X \cup \{v_0\}$ and $x_v = 1$ otherwise. By construction, $(\ldots x_v \ldots) \in \operatorname{Ker} \alpha = N(J_K)F^\circ$. Thus, $a \in F^\circ$ exists with the property that $x_v a^{-1} \in N_v(\hat{K}_v)$ for all $v \in \mathbf{S}(F)$. Let $A = (K,\tau,a) \in \mathfrak{S}(F)$. If $v \in \mathbf{S}(F)$, then

$$A \otimes \hat{F}_v \sim (\hat{K}_v,\tau^{k(v)},a) \cong (\hat{K}_v,\tau^{k(v)},x_v) \tag{2}$$

by Lemma 15.1. For $v \in X$, (1) and (2) yield $\mathrm{INV}_v(A \otimes \hat{F}_v) = t_v$. When $v \in \mathbf{S}(F) - (X \cup \{v_0\})$ we have $A \otimes \hat{F}_v \sim (\hat{K}_v,\tau^{k(v)},1) \sim \hat{F}_v$, and $\mathrm{INV}_v(A \otimes \hat{F}_v) = 0 = t_v$. Finally, from Proposition a and the hypothesis $\sum_v t_v = 0$, it follows that $\mathrm{INV}_{v_0}(A \otimes \hat{F}_{v_0}) = -\sum_{v \neq v_0} \mathrm{INV}_v(A \otimes \hat{F}_v) = -\sum_{v \neq v_0} t_v = t_{v_0}$. Hence, $\mathrm{INV}^{(F)} A = (\ldots t_v \ldots) = \xi$ as required. □

Exercises

1. Let a be a square free integer, $b \in \mathbb{Z}$. Let the norm residue symbol $(a,b)_p$ be defined as in Exercise 4, Section 18.4.

   (a) Use the results of Exercise 4, Section 18.4 and the law of quadratic reciprocity to prove $\prod_p (a,b)_p = 1$, where the product is over all primes and ∞. Hint. By part (d) of Exercise 4, Section 18.4, it is sufficient to prove this formula in the cases: $a = b = -1$; $a = -1$, $b = 2$; $a = -1$, $b =$ odd prime; $a = b = 2$; $a = 2$, $b =$ odd prime; $a = b =$ odd prime; a and b are distinct odd primes.

   (b) Show that the formula $\prod_p (a,b)_p = 1$ yields the quadratic reciprocity law in the case that a and b are distinct odd primes.

   (c) Prove that $(a,b)_p = \alpha_{v_p}(b)$, where α_{v_p} is the local Artin mapping that corresponds to the extension $\mathbb{Q}(\sqrt{a})/\mathbb{Q}$, provided p is an odd prime that does not divide a or $p = \infty$. The result is true for all p, but this fact cannot be proved on the basis of our incomplete discussion of the local Artin mappings.

2. Let $m > 1$ be a natural number, $\zeta = e^{2\pi i/m}$ a primitive m'th root of unity, $K = \mathbb{Q}(\zeta)$.

   (a) Prove that $\mathbf{G}(K/\mathbb{Q}) = \{\tau_k\colon (m,k) = 1\}$, where $\zeta^{\tau_k} = \zeta^k$. Hence, $K/\mathbb{Q}$ is abelian.

   (b) Let p be a prime such that p does not divide m, and $e_{v_p}(K/\mathbb{Q}) = 1$. (These conditions are actually equivalent.) Prove that the Frobenius automorphism σ_{v_p} is τ^p.

   (c) Deduce from the Tchebotarev Density Theorem that if $(m,k) = 1$, then there are infinitely many primes p of the form $rm + k$, $r \in \mathbb{N}$. This result is a major corollary of Dirichlet's Density Theorem.

3. Let X be a finite subset of $\{\infty, 2, 3, \ldots, p, \ldots\}$ (the rational primes and ∞) such that $|X|$ is even. Prove that there are square free integers a and b such that $(a,b)_p = -1$ for all $p \in X$ and $(a,b)_p = 1$ for all $p \notin X$. Hint. Assume that $\infty \notin X$. Let $a = \prod\{p : p \in X \cup \{2\}\}$. For each odd prime p in X, let $c_p \in \mathbb{Z}$ be a non-residue modulo p. In particular, p does not divide c_p. Define $c_2 = 1$ if $2 \notin X$, $c_2 = 5$ if $2 \in X$. Use the Chinese Remainder Theorem to find $m \in \mathbb{N}$ such that $m \equiv c_p \pmod{p}$ for the odd primes $p \in X$ and $m \equiv c_2 \pmod{8}$. Note that $4a$ is relatively prime to m. Use the Dirichlet Density Theorem to find an odd prime b of the form $4an + m$. Use the result of Exercise 4, Section 18.4 to show that $(a,b)_p = -1$ if p is an odd prime in X, $(a,b)_2 = -1$ if $2 \in X$, $(a,b)_2 = 1$ if $2 \notin X$, and $(a,b)_p = 1$ if p is a prime that is not in $X \cup \{b\}$. Deduce that $(a,b)_b = 1$ from Exercise 1 and the assumption that $|X|$ is even. If $\infty \in X$, then a and b must be negative. The construction is similar.

4. Use the results of Exercises 3 and 4 of Section 18.4, and Exercises 1 and 3 to show that there is an exact sequence

$$1 \to \mathbf{B}(\mathbb{Q})_2 \to \mathbf{I}(\mathbb{Q})_2 \to (1/2)\mathbb{Z}/\mathbb{Z} \to 1,$$

where $\mathbf{B}(\mathbb{Q})_2$ is the subgroup of $\mathbf{B}(\mathbb{Q})$ consisting of elements whose order is 2 (that is, the group of classes of rational quaternion algebras), and $\mathbf{I}(\mathbb{Q})_2 = \bigoplus_p I_p(\mathbb{Q})_2$ with $I_p(\mathbb{Q})_2 = (1/2)\mathbb{Z}/\mathbb{Z}$ and the sum is over all primes and ∞.

Notes on Chapter 18

The primary source of the results in this chapter is the classical paper [40] of Hasse, Brauer, and Noether. An amplified version of this paper is given in Deuring's book [26]. The work [3] by Albert presents the material of Sections 18.4 and 18.6 in about the same way that we have. Our discussion of the Brauer groups of algebraic number fields is similar to Reiner's in [66]. We have tried to give a fuller description than Reiner or Deuring of the role of class field theory in the proofs of the fundamental theorems on division algebras over fields.

A very accessible exposition of class field theory for algebraic number fields is available in the book [56] by Lang. The cohomological approach to class field theory is stressed in the book [22] and in the Artin–Tate notes [10]. Our Theorem 18.5 can be obtained easily from the structure of certain cohomology groups. Anyone who is comfortable with class field theory will find Artin and Tate's discussion in [10] of the Grunwald–Wang Theorem easy to follow; most other proofs of this result are less lucid.

CHAPTER 19

Division Algebras over Transcendental Fields

In this chapter we consider the central simple algebras over fields that are transcendental extensions of their prime fields. In contrast with the abundant information about algebras over local and global fields that was presented in the last two chapters, our knowledge about division algebras over transcendental fields is sparse. The most important result on this subject is Tsen's Theorem. It will be proved in Section 4. Tsen's Theorem is a generalization of Wedderburn's Theorem on finite division algebras. Using properties of the reduced norm we will prove a result that includes Tsen's Theorem and Wedderburn's Theorem as special cases. Tsen's Theorem is the basis of most work on the Brauer groups of transcendental extensions. In Section 5 we will use it to prove a relativized version of the Auslander–Brumer–Faddeev Theorem that describes $\mathbf{B}(F(\mathbf{x}))$. The study of $\mathbf{B}(F(\mathbf{x}))$ leads to a construction of division algebras in Section 6 that clarifies the relation between the Schur index and the exponent of a central simple algebra. The last three sections of the chapter examine algebras over Laurent series fields.

19.1. The Norm Form

It was noted in Lemma 16.3a that the reduced norm of a central simple F-algebra can be computed by evaluating a homogeneous form. We will now examine that observation with more care. In particular, it will be shown that the coefficients of the form are in F. For convenience, we review some simple properties of homogeneous forms.

Let $\mathbf{x}_1, \ldots, \mathbf{x}_m$ be independent commuting variables. A non-zero polynomial $\Phi \in F[\mathbf{x}_1, \ldots, \mathbf{x}_m]$ is a *homogeneous form* of degree k in m variables if $\Phi = \sum \mathbf{x}_1^{r_1} \cdots \mathbf{x}_m^{r_m} a_{r_1 \cdots r_m}$, where the sum is over all sequences $(r_1, \ldots, r_m)$ of non-negative integers such that $\sum_{i=1}^m r_i = k$. The symbols $\mathbf{x}_1, \ldots, \mathbf{x}_m$ that denote the variables in a form will often be replaced by other bold faced letters. When it isn't necessary to specify the variables in a form, we will write Φ instead of $\Phi(\mathbf{x}_1, \ldots, \mathbf{x}_m)$. Throughout the section, F is an arbitrary field.

Lemma. (i) *If $\Delta \in F[\mathbf{x}_{11}, \mathbf{x}_{12}, \ldots, \mathbf{x}_{nn}]$ is defined by $\Delta = \det[\mathbf{x}_{ij}]$, then Δ is homogeneous of degree n.*

(ii) *If $\Phi, \Psi \in F[\mathbf{x}_1, \ldots, \mathbf{x}_m] - \{0\}$, then $\Phi\Psi$ is homogeneous of degree k if and only if Φ is homogeneous of degree r, Ψ is homogeneous of degree s, and $r + s = k$.*

(iii) *If $\Phi(\mathbf{x}_1, \ldots, \mathbf{x}_m)$ is homogeneous of degree k, and $\Psi_1, \ldots, \Psi_m \in F[\mathbf{y}_1, \ldots, \mathbf{y}_n]$ are homogeneous of degree l, then $\Phi(\Psi_1, \ldots, \Psi_m)$ is either zero or homogeneous of degree kl.*

Proof. By definition, $\det[\mathbf{x}_{ij}] = \sum_\pi (\operatorname{sgn} \pi)\mathbf{x}_{1\pi(1)} \cdots \mathbf{x}_{n\pi(n)}$, where the sum is over all permutations π of $\{1, \ldots, n\}$. Hence, Δ is homogeneous of degree n. An easy calculation shows that if Φ is homogeneous of degree r and Ψ is homogeneous of degree s, then $\Phi\Psi$ is homogeneous of degree $r + s$. To prove the converse, write $\Phi = \Phi_1 + \cdots + \Phi_k$, $\Psi = \Psi_1 + \cdots + \Psi_l$ with $\Phi_1, \ldots, \Phi_k$, $\Psi_1, \ldots, \Psi_l$ non-zero, homogeneous polynomials such that $\deg \Phi_1 < \cdots < \deg \Phi_k$, $\deg \Psi_1 < \cdots < \deg \Psi_l$. The product $\Phi\Psi$ is equal to $\Phi_1\Psi_1 + \cdots + \Phi_k\Psi_l$, where $\deg \Phi_1\Psi_1 < \deg \Phi_i\Psi_j$ if $1 < i$ or $1 < j$, and $\deg \Phi_i\Psi_j < \deg \Phi_k\Psi_l$ if $i < k$ or $j < l$. If $\Phi\Psi$ is homogeneous, then necessarily $k = 1$ and $l = 1$. That is, Φ and Ψ are homogeneous. The last assertion of the lemma follows by induction from (ii) and the observation that a non-zero sum of polynomials that are homogeneous of degree kl is itself homogeneous of degree kl. □

Two homogeneous forms $\Phi, \Psi \in F[\mathbf{x}_1, \ldots, \mathbf{x}_m]$ are *equivalent* (over $F[\mathbf{x}_1, \ldots, \mathbf{x}_m]$) if there is an invertible matrix $\alpha = [a_{ij}] \in M_m(F)$ such that

$$\Psi(\mathbf{x}_1, \ldots, \mathbf{x}_m) = \Phi([\mathbf{x}_1, \ldots, \mathbf{x}_m]\alpha) = \Phi\left(\sum_{i=1}^m \mathbf{x}_i a_{i1}, \ldots, \sum_{i=1}^m \mathbf{x}_i a_{im}\right).$$

In this case we will write $\Phi \simeq \Psi$. If $\Psi(\mathbf{x}_1, \ldots, \mathbf{x}_m) = \Phi([\mathbf{x}_1, \ldots, \mathbf{x}_m]\alpha)$, then $\Phi(\mathbf{x}_1, \ldots, \mathbf{x}_m) = \Psi([\mathbf{x}_1, \ldots, \mathbf{x}_m]\alpha^{-1})$. Thus, the relation $\simeq$ is symmetric; it is also transitive and reflexive, that is, an equivalence relation.

Let A be a finite dimensional F-algebra with $\dim A_F = m$. Denote $R = F[\mathbf{x}_1, \ldots, \mathbf{x}_m]$ and $K = F(\mathbf{x}_1, \ldots, \mathbf{x}_m)$, where $\mathbf{x}_1, \ldots, \mathbf{x}_m$ are independent commuting variables. The inclusion mapping of R to K induces an injective F-algebra homomorphism of A^R to A^K. It is convenient to view A^R as an F-subalgebra of A^K. An element $z \in A^R$ is *generic for A* if $z = \sum_{i=1}^m w_i\mathbf{x}_i$, where $\{w_1, \ldots, w_m\}$ is a basis of A_F. The justification for this terminology is the observation that an F-algebra homomorphism of R to F defined by

$\mathbf{x}_i \mapsto a_i$ induces an F-algebra homomorphism $A^R \to A$ which sends z to $\sum_{i=1}^m w_i a_i$. Therefore, every element of A is the image of z under a suitable homomorphism of A^R to A.

Proposition. *Let $A \in \mathfrak{S}(F)$ have degree n. Denote $R = F[\mathbf{x}_{11}, \mathbf{x}_{12}, \ldots, \mathbf{x}_{nn}]$ and $K = F(\mathbf{x}_{11}, \mathbf{x}_{12}, \ldots, \mathbf{x}_{nn})$, where the $\mathbf{x}_{ij}$ are commuting variables. Corresponding to a generic $z \in A^R$, define Δ^z to be $v_{A^K/K}(z)$.*

(i) *$\Delta^z \in F[\mathbf{x}_{11}, \mathbf{x}_{12}, \ldots, \mathbf{x}_{nn}]$ is homogeneous of degree n and irreducible.*
(ii) *If $x \in A$, then there exist $a_{11}, a_{12}, \ldots, a_{nn}$ in F such that $v_{A/F}(x) = \Delta^z(a_{11}, a_{12}, \ldots, a_{nn})$.*
(iii) *If $y \in A^R$ is generic, then Δ^y is equivalent to Δ^x. Conversely, if Φ is a form such that $\Phi \simeq \Delta^x$, then $\Phi = \Delta^y$ for a generic $y \in A^R$.*

To simplify the proof of this proposition and various statements in later parts of the chapter, we will use the vector notation $\vec{a}$ for a row vector $[a_1, \ldots, a_m] \in F^m$ and $\vec{\mathbf{x}}$ for a row $[\mathbf{x}_1, \ldots, \mathbf{x}_m]$ of variables. The zero vector $[0, \ldots, 0] \in F^m$ is denoted by $\vec{0}$. In the proposition $m = n^2$. As usual, $\vec{a}^t$ and $\vec{\mathbf{x}}^t$ denote the column vectors obtained from $\vec{a}$ and $\vec{\mathbf{x}}$ by transposition.

Proof of the Proposition. Let E/F be a field extension such that E splits A, and suppose that $\phi: A \to M_n(E)$ is a splitting representation of A. The tensor product $\psi = \phi \otimes id_K: A^K \to M_n(E) \otimes K \subseteq M_n(E(\vec{\mathbf{x}}))$ is a splitting representation of A^K that can be used to compute Δ^z. If $z = \sum_{i,j} w_{ij}\mathbf{x}_{ij} = \vec{w}\vec{\mathbf{x}}^t$ with $\{w_{ij}: 1 \leq i, j \leq n\}$ a basis of A_F, then $\Delta^z = v_{A^K/K}(z) = \det \psi(z) = \det(\sum_{i,j} \phi(w_{ij})\mathbf{x}_{ij})$ is either homogeneous of degree n in $E[\vec{\mathbf{x}}]$ or identically zero. By Corollary 16.1a, $v_{A^K/K}(z) \in K = F(\vec{\mathbf{x}})$. Thus, $\Delta^z \in F[\vec{\mathbf{x}}]$. If $x \in A$, then $x = \vec{w}\vec{a}^t$ for a unique $\vec{a} \in F^{n^2}$. Hence, $v(x) = \det \phi(x) = \det(\sum_{i,j} \phi(w_{ij})a_{ij}) = \Delta^z(\vec{a})$. This calculation proves (ii); it also shows that $\Delta^z \neq 0$ because $v(1_A) = 1$. Hence, Δ^z is homogeneous of degree n. Let $y \in A^R$ be generic, say $y = \sum_{i,j} w'_{ij}\mathbf{x}_{ij}$, where $\{w'_{ij}: 1 \leq i, j \leq n\}$ is a basis of A_F. There exists $\alpha \in GL_{n^2}(F)$ such that $\vec{w}' = \vec{w}\alpha$. Thus, $\Delta^y(\vec{\mathbf{x}}) = v(y) = \det(\sum_{i,j} \phi(w'_{ij})\mathbf{x}_{ij}) = \det(\phi(\vec{w})\alpha\vec{\mathbf{x}}^t) = \Delta^x(\vec{\mathbf{x}}\alpha^t)$, so that $\Delta^y \simeq \Delta^x$. Conversely, if $\Phi(\vec{\mathbf{x}}) = \Delta^x(\vec{\mathbf{x}}\alpha^t)$, then $\Phi = \Delta^y$, where $y = \vec{w}\alpha\vec{\mathbf{x}}^t$ is generic. It remains to show that Δ^z is irreducible. By Proposition 13.2a, $\psi(z) = \sum_{i,j} \phi(w_{ij})\mathbf{x}_{ij}$ is generic for $M_n(E)$, and $\Delta^{\psi(z)} = \Delta^z$ according to the definition of Δ^z. The matrix $y = \sum_{ij} \varepsilon_{ij}\mathbf{x}_{ij}$ is also generic for $M_n(E)$. Consequently, $\Delta^z \simeq \Delta^y$ over $E[\vec{\mathbf{x}}]$. Clearly, $\Delta^y(\vec{\mathbf{x}}) = \det[\mathbf{x}_{ij}]$. An elementary degree argument that is outlined in Exercise 2 shows that Δ^y is irreducible in $E[\vec{\mathbf{x}}]$. It follows easily that Δ^z is also irreducible. □

The homogeneous polynomials Δ^z are called the *norm forms* of A. They constitute one class of equivalent forms in $F[\vec{\mathbf{x}}]$. When we use a norm form, it usually doesn't matter which representative of the class is employed.

Thus, it is customary to refer to *the* norm form of A, and to denote any representative of this class of forms by $\Delta_{A/F}$.

Corollary. *If $A \in \mathfrak{S}(F)$ has degree n, then A is a division algebra if and only if $\Delta_{A/F}$ has only the trivial zero. That is, $\Delta_{A/F}(\vec{a}) = 0$ for $\vec{a} \in F^{n^2}$ implies $\vec{a} = \vec{0}$.*

This corollary restates Corollary 16.3a in terms of the norm form of A.

Exercises

1. Give a detailed proof of part (iii) of the Lemma.

2. (a) Prove that if $\{\mathbf{x}_{ij}: 1 \le i, j \le n\}$ are independent commuting variables over a field E, then $\det[\mathbf{x}_{ij}]$ is irreducible in $E[\vec{\mathbf{x}}]$. Hint. As a polynomial in $\mathbf{x}_{ij}$, $\det[\mathbf{x}_{ij}]$ has degree one. Thus, if $\det[\mathbf{x}_{ij}] = \Phi\Psi$, then either $\deg_{\mathbf{x}_{ij}}\Phi = 1$, $\deg_{\mathbf{x}_{ij}}\Psi = 0$ or vice versa. Show that if $\deg_{\mathbf{x}_{ij}}\Phi = 1$, then $\deg_{\mathbf{x}_{kj}}\Phi = 1$ for all k, and $\deg_{\mathbf{x}_{ik}}\Phi = 1$ for all k. Deduce that Ψ is a constant in this case.

   (b) Prove that if y and z are generic elements for $A \in \mathfrak{S}(F)$, and if Δ^y is irreducible, then Δ^z is irreducible. Hint. Use part (iii) of the proposition and part (iii) of the lemma.

3. Let $A \in \mathfrak{S}(F)$ have degree n. If $z = \sum_{i,j=1}^{n} w_{ij}\mathbf{x}_{ij} \in A^R$ is generic for A, define the generic polynomial of A corresponding to z to be the characteristic polynomial $X(z,\mathbf{x})$ of z. Prove the following statements.

   (a) $X(z,\mathbf{x}) \in F[\mathbf{x}, \mathbf{x}_{11}, \mathbf{x}_{12}, \ldots, \mathbf{x}_{nn}]$ is homogeneous of degree n.

   (b) If $y \in A$, then there is a homomorphism ϕ of $F[\mathbf{x}, \mathbf{x}_1, \ldots, \mathbf{x}_{nn}]$ to $F[\mathbf{x}]$ such that $\phi(X(z,\mathbf{x})) = X(y,\mathbf{x})$.

   (c) Considered as an element of $F(\mathbf{x}_{11}, \ldots, \mathbf{x}_{nn})[\mathbf{x}]$, $X(z,\mathbf{x})$ is irreducible, and the Galois group of this polynomial is isomorphic to the symmetric group on n symbols.

4. Assume that char $F \ne 2$ and $A = \left(\frac{a,b}{F}\right)$, where $a, b \in F^\circ$.

   (a) Prove that $a\mathbf{x}^2 + b\mathbf{y}^2 - 1$ is irreducible in $F[\mathbf{x},\mathbf{y}]$, so that $R = F[\mathbf{x},\mathbf{y}]/(a\mathbf{x}^2 + b\mathbf{y}^2 - 1)$ is an integral domain. Denote the fraction field of R by E.

   (b) Show that the following conditions are equivalent for an extension K of F.

   (i) K splits A.
   (ii) $ac^2 + bd^2 = 1$ for some $c, d \in K$.
   (iii) There is an F-algebra homomorphism $\phi: R \to K$.

   Hint. For the equivalence of (i) and (ii), use Proposition 1.6 and the observation (that has to be proved) that if $a\mathbf{x}^2 + b\mathbf{y}^2 = 0$ has a non-trivial solution in K, then $a\mathbf{x}^2 + b\mathbf{y}^2 = 1$ has a solution in K.

   (c) Prove that $A \sim F$ if and only if $E \cong F(\mathbf{t})$ for some variable $\mathbf{t}$. Hint. Assume that $c, d \in F$ satisfy $ac^2 + bd^2 = 1$. Let x and y be the images in R of $\mathbf{x}$ and $\mathbf{y}$. Define $\mathbf{t} = (y - d)/(x - c)$. Show that $\mathbf{t}$ is transcendental over F and $x, y \in F(\mathbf{t})$. For the converse, note that if $x, y \in F(\mathbf{t})$, then a solution of $a\mathbf{x}^2 + b\mathbf{y}^2 = 1$ can be found in F by assigning suitable values to $\mathbf{t}$.

It is clear from (b) that the field E splits A. This field is called a *generic splitting field for* A. It can be shown that two quaternion algebras over F are isomorphic if and only if their associated generic splitting fields are isomorphic as F-algebras. The construction given in this exercise is due to E. Witt. (See [79].) It has been generalized to arbitrary central simple algebras by Amitsur.

19.2. Quasi-algebraically Closed Fields

A field F is *quasi-algebraically closed* (abbreviated QAC) if every homogeneous form $\Phi \in F[\mathbf{x}_1, \ldots, \mathbf{x}_m]$ in m variables whose degree n satisfies $1 \leq n < m$ has a non-trivial zero in F^m. The alternative term "C_1-field" is often used in the literature of field theory. Our interest in QAC fields is based on the following consequence of Corollary 19.1.

Theorem. *If F is a quasi-algebraically closed field, then* $\mathbf{B}(F) = \{1\}$.

Proof. If $A \in \mathfrak{S}(F)$ has degree $n > 1$, then $\Delta_{A/F}$ is homogeneous of degree n in $n^2 > n$ variables. Since F is QAC, $\Delta_{A/F}$ has a non-trivial zero, and A is not a division algebra by Corollary 19.1. It follows that $\mathbf{B}(F) = \{1\}$. □

It is clear that an algebraically closed field is QAC. All finite fields are QAC; the proof of this fact is outlined in Exercise 2. Tsen's Theorem is the statement that if F is algebraically closed, then every algebraic extension of $\mathbf{F}(x)$ is QAC. Using this result and some algebraic geometry it can be proved that if F is algebraically closed, then the field $F((\mathbf{x}))$ of formal Laurant series is QAC. In short, QAC fields are fairly ubiquitous.

The rest of this section is devoted to proving that the class of QAC fields is closed under algebraic extensions. The proof uses a new concept. A *normic form* of degree k over a field F is a homogeneous polynomial $\Lambda \in F[\mathbf{x}_1, \ldots, \mathbf{x}_k]$ with $\deg \Lambda = k$ such that Λ has only the trivial zero in F^k: $\Lambda(\vec{a}) = 0$ implies $\vec{a} = \vec{0}$. The following example shows that if F is not k-closed, then normic forms of degree k over F exist.

Example. Let K/F be a field extension of degree k. By Lemma 16.3a, the left regular matrix representation ϕ of K relative to a basis $y_1, \ldots, y_k$ of K_F yields a homogeneous form $\Lambda \in F[\mathbf{x}_1, \ldots, \mathbf{x}_k]$ of degree k such that

$$\Lambda(a_1, \ldots, a_k) = \det \phi(y_1 a_1 + \cdots + y_k a_k) = N_{K/F}(y_1 a_1 + \cdots + y_k a_k).$$

Since $N_{K/F}$ maps K° to F°, Λ is normic.

The example motivates the terminology "normic form;" it also shows that the condition $n < m$ in the definition of QAC fields is natural.

Lemma. *Let F be a QAC field such that there is a normic form of degree k over F. Let $\Omega_1, \ldots, \Omega_k \in F[\mathbf{y}_1, \ldots, \mathbf{y}_m]$ be homogeneous of degree $n \geq 1$. If $m > nk$, then $\vec{a} \in F^m - \{\vec{0}\}$ exists such that $\Omega_1(\vec{a}) = \cdots = \Omega_k(\vec{a}) = 0$.*

Proof. Assume that $\Lambda \in F[\mathbf{x}_1, \ldots, \mathbf{x}_k]$ is a normic form. By Lemma 19.1, $\Psi = \Lambda(\Omega_1, \ldots, \Omega_k)$ is either identically 0 or homogeneous of degree $nk < m$. Since F is QAC, there exists $\vec{a} \in F^m - \{\vec{0}\}$ such that $\Psi(\vec{a}) = 0$. Therefore, $\Omega_1(\vec{a}) = \cdots = \Omega_k(\vec{a}) = 0$ because Λ is normic. □

Proposition. *If F is quasi-algebraically closed, then every algebraic extension of F is quasi-algebraically closed.*

Proof. Let K/F be an algebraic extension. We must show that if $\Phi \in K[\mathbf{x}_1, \ldots, \mathbf{x}_m]$ is homogeneous of degree n, where $1 \leq n < m$, then $\vec{b} \in K^m - \{\vec{0}\}$ exists satisfying $\Phi(\vec{b}) = 0$. Since the coefficients of Φ generate a finite extension of F, it can be assumed that $[K:F] = k < \infty$. By the example, there is a normic form of degree k over F, so that the hypotheses of the lemma are satisfied. Let $y_1, \ldots, y_k$ be a basis of K_F. Define linear forms

$$\Psi_j = y_1\mathbf{x}_{1j} + \cdots + y_k\mathbf{x}_{kj}, \quad 1 \leq j \leq m \tag{1}$$

in the new variables $\mathbf{x}_{ij}$. Since the elements of K are linear combinations of $y_1, \ldots, y_k$ with coefficients in F, it follows from Lemma 19.1 that there are homogeneous forms $\Omega_l \in F[\ldots\mathbf{x}_{ij}\ldots]$ of degree n such that

$$\Phi(\Psi_1, \ldots, \Psi_m) = y_1\Omega_1 + \cdots + y_k\Omega_k. \tag{2}$$

The number mk of variables $\mathbf{x}_{ij}$ exceeds nk, so that $\Omega_1(\vec{a}) = \cdots = \Omega_k(\vec{a}) = 0$ for some $\vec{a} \in F^{mk} - \{\vec{0}\}$ by the lemma. If $b_j = \Psi_j(\vec{a}) = y_1a_{1j} + \cdots + y_ka_{kj}$ and $\vec{b} = (b_1, \ldots, b_m) \in K^m$, then by (1) and (2), $\Phi(\vec{b}) = 0$. Moreover, $\vec{b} \neq \vec{0}$ because $\vec{a} \neq \vec{0}$ and $y_1, \ldots, y_k$ is a basis of K_F. □

Exercises

1. Prove that if the field F is not algebraically closed, then there are normic forms over F that have arbitrarily large degrees. Hint. Show that if $\Lambda(\mathbf{x}_1, \ldots, \mathbf{x}_k)$ is normic of degree $k > 1$, then $\Lambda(\Lambda(\mathbf{x}_{11}, \ldots, \mathbf{x}_{1k}), \ldots, \Lambda(\mathbf{x}_{k1}, \ldots, \mathbf{x}_{kk}))$ is normic of degree $k^2 > k$.

2. This exercise outlines a proof of the *Chevalley–Warning Theorem*: every finite field is *QAC*. This result was conjectured by Artin and proved independently by Chevalley and Warning in 1936. By the proposition, it suffices to prove the result in the case that $F = \mathbb{F}_p$, where p is prime. Let Φ be a form of degree n in m variables, where $1 \leq n < m$. Prove the following statements.
   (a) If $0 \leq k < p - 1$, then $\sum_{a \in F} a^k = 0$ (with the convention that $0^0 = 1$). Hint. Choose $b \in F$ so that $F^\circ = \langle b \rangle$. Observe that $\sum_{a \in F} a^k = \sum_{a \in F} (ab)^k = b^k \sum_{a \in F} a^k$ for all $k \in \mathbb{N}$.

(b) If $0 \leq k_j < p - 1$ for some j, then $\sum_{(a_1,\ldots,a_m)} a_1^{k_1} \ldots a_m^{k_m} = 0$.

(c) $$\sum_{(a_1,\ldots,a_m)\in F^m - \{0\}} \Phi(a_1, \ldots, a_m)^{p-1} = 0.$$

Hint. Write $\Phi(a_1, \ldots, a_m)^{p-1}$ as a sum $\sum c_k a_1^{k_1} \cdots a_m^{k_m}$, reverse the order of summation, and apply (b). Note that $\Phi(0, \ldots, 0) = 0$ because Φ is a form of degree $n \geq 1$.

(d) Deduce from (c) that $\Phi(a_1, \ldots, a_m) = 0$ for some $(a_1, \ldots, a_m) \in F^m - \{\vec{0}\}$. Hint. Otherwise, the sum in (c) would be $p^m - 1 = -1$ by Fermat's Theorem.

19.3. Krull's Theorem

The next section presents the proof of Tsen's Theorem, the main result of this chapter. The key step of the proof uses a standard theorem of algebraic geometry.

Proposition. *Assume that F is an algebraically closed field. Let $\Phi_1, \ldots, \Phi_s \in F[\mathbf{x}_1, \ldots, \mathbf{x}_k]$ be homogeneous forms of positive degree. If $s < k$, then there exists $\vec{a} \in F^k - \{\vec{0}\}$ such that $\Phi_1(\vec{a}) = \cdots = \Phi_s(\vec{a}) = 0$.*

The geometrical proof of this proposition uses the dimension of algebraic sets: the set of common zeros of $\Phi_1, \ldots, \Phi_s$ is either empty or it has a component of dimension $\geq k - s$; and the zero set contains $\vec{0}$ when the $\Phi_1, \ldots, \Phi_s$ are forms of positive degree. To make this argument rigorous requires the introduction of a substantial number of geometrical definitions and theorems. An alternative path to the proposition goes by way of the Krull Height Theorem. This is the approach that we will take.

Principal Ideal Theorem (*Krull*). *Let d be a non-zero element of the commutative, Noetherian, integral domain R. If P is minimal in the set of prime ideals of R that contain d, then there is no non-zero prime ideal Q of R such that $\{0\} \subset Q \subset P$.*

Proof. Let K be the fraction field of R. Define $S = \{a^{-1}b \in K : b \in R, a \in R - P\}$. A standard calculation shows that S is a Noetherian local subring of K with the unique maximal ideal $SP = \{a^{-1}b : b \in P, a \in R - P\}$. In the terminology of commutative ring theory, $S = R_P$ is the localization of R at the prime ideal P. The minimality of P over d implies that SP is the only prime ideal of S that includes d. Assume that Q is a prime ideal of R such that $\{0\} \subset Q \subset P$. It is easy to check that $\{0\} \subset SQ \subset SP$. This construction puts us in the context of a local ring. It is convenient to revert to our original notation, adding the hypothesis that P is the unique maximal ideal of R. Thus, $\mathbf{J}(R) = P$. Since P is minimal over d, the factor ring R/Rd has exactly one prime ideal P/Rd which is maximal. It follows that

$$\bar{R} = R/Rd \text{ is Artinian.} \tag{1}$$

(See Part (j) of the exercise in Section 4.5.) The proof will be completed by obtaining a contradiction to the assumption that there is a prime ideal Q such that $\{0\} \subset Q \subset P$. Localize R at Q: denote $T = R_Q = \{b^{-1}c \colon c \in R, b \in R - Q\}$. Thus, T is local with $\mathbf{J}(T) = TQ$. Denote $I_n = (TQ)^n \cap R$. Plainly, $I_1 \supseteq I_2 \supseteq I_3 \supseteq \cdots$ is a chain of ideals in R. Moreover,

$$TI_n = (TQ)^n. \tag{2}$$

In fact, $TI_n \subseteq (TQ)^n$ because $(TQ)^n$ is an ideal of T. On the other hand, if $c \in (TQ)^n$, then $bc \in R$ for some $b \in R - Q$; hence $bc \in I_n$ and $c = b^{-1}(bc) \in TI_n$. We also have

$$Rd \cap I_n = dI_n. \tag{3}$$

Indeed,

$$\begin{aligned} Rd \cap I_n &= Rd \cap R \cap (TQ)^n = Rd \cap (TQ)^n \\ &= Rd \cap d(TQ)^n = d(R \cap (TQ)^n) = dI_n, \end{aligned}$$

because $d \notin Q$ implies $d^{-1} \in T$. Apply the Artinian property (1) to the chain $Rd + I_1 \supseteq Rd + I_2 \supseteq Rd + I_3 \supseteq \cdots$: there exists $n \geq 1$ such that $Rd + I_n = Rd + I_{n+1}$. The modular law yields $I_n = I_n \cap (Rd + I_{n+1}) = (I_n \cap Rd) + I_{n+1} = dI_n + I_{n+1}$ by (3). Since $d \in P = \mathbf{J}(R)$ and R is Noetherian, it follows from Nakayama's Lemma that $I_n = I_{n+1}$. By (2), $(TQ)^n = TI_n = TI_{n+1} = (TQ)^{n+1} = \mathbf{J}(T)(TQ)^n$. Another application of Nakayama's Lemma gives $(TQ)^n = 0$. This equality leads to the contradication $Q = 0$, because T is an integral domain. □

The result that we need is an inductive generalization of the Principal Ideal Theorem.

Krull Height Theorem. *Let $d_1, \ldots, d_n$ be elements of the commutative, Noetherian ring R. Assume that P is minimal among the prime ideals of R that contain $\{d_1, \ldots, d_n\}$. If $P_0 \subset P_1 \subset \cdots \subset P_k = P$ is a strictly ascending chain of prime ideals in R, then $k \leq n$.*

The maximum length of such a chain of prime ideals is called the height of P.

Proof. It can be assumed that R is an integral domain and $P_0 = 0$; otherwise, replace R by R/P_0 to reach this situation. Also, by passing to the localization R_P it can be assumed that R is local with $\mathbf{J}(R) = P$. If $n = 1$, then the principal ideal theorem implies that $k = 1$. Proceed inductively, assuming that $1 < n \leq k$. Since R is Noetherian, it can be assumed that there is no prime ideal of R between P_{k-1} and P. By the minimality of P over $\{d_1, \ldots, d_n\}$, some d_i is not a member of P_{k-1}, say $d_n \notin P_{k-1}$. Thus, $P_{k-1} \subset d_nR + P_{k-1} \subseteq P$, and $P/(d_nR + P_{k-1})$ is the unique prime ideal in $R/(d_nR + P_{k-1})$. By the exercise of Section 4.5, $R/(d_nR + P_{k-1})$ is Artinian.

Since $\mathbf{J}(R/(d_nR + P_{k-1})) = P/(d_nR + P_{k-1})$, it follows from Proposition 4.4 that $P^m \subseteq d_nR + P_{k-1}$ for some $m \geq 1$. In particular, if $1 \leq i < n$, then

$$d_i^m = d_n a_i + c_i, \quad a_i \in R, \quad c_i \in P_{k-1}. \tag{4}$$

Since R is Noetherian, there is a prime ideal $Q \subseteq P_{k-1}$ that is minimal with the property $\{c_1, \ldots, c_{n-1}\} \subseteq Q$. (See Part (g) of the exercise in Section 4.5.) We will show that $Q = P_{k-1}$, that is, P_{k-1} is minimal over $\{c_1, \ldots, c_{n-1}\}$. The induction hypothesis will then yield $k - 1 \leq n - 1$ and $k \leq n$. Since P_{k-1}/Q is a prime ideal of R/Q and $P_{k-1}/Q \subset P/Q$, the equality $P_{k-1} = Q$ will follow from the Principal Ideal Theorem by showing that P/Q is minimal over $d_n + Q$. If Q_1 is a prime ideal of R such that $d_n \in Q_1$ and $\{c_1, \ldots, c_{n-1}\} \subseteq Q \subseteq Q_1 \subseteq P$, then $d_i^m \in Q_1$ for $1 \leq i < n$ by (4). Hence, $d_i \in Q_1$ for $1 \leq i \leq n$ since Q_1 is prime. The minimality of P over $\{d_1, \ldots, d_n\}$ implies that $Q_1 = P$. Therefore, P/Q is minimal among the prime ideals of R/Q that include $d_n + Q$. □

The proof of the proposition also uses Hilbert's Nullstellensatz. Since this theorem is standard fare in most graduate algebra texts, we won't prove it. (However, see Exercise 3 for the proof of a special case.)

Nullstellensatz. *Let $R = F[\mathbf{x}_1, \ldots, \mathbf{x}_k]$, where the field F is algebraically closed.*

(i) *Every maximal ideal of R has the form $(\mathbf{x}_1 - a_1)R + \cdots + (\mathbf{x}_k - a_k)R$ for suitable $a_1, \ldots, a_k \in F$,*
(ii) *Every prime ideal of R is an intersection of maximal ideals.*

Proof of the Proposition. Denote $R = F[\mathbf{x}_1, \ldots, \mathbf{x}_k]$. The Hilbert Basis Theorem guarantees that R is Noetherian. Since each Φ_i is homogeneous, $\{\Phi_1, \ldots, \Phi_s\} \subseteq P = \mathbf{x}_1R + \cdots + \mathbf{x}_kR$. Note that $\{0\} \subset \mathbf{x}_1R \subset \mathbf{x}_1R + \mathbf{x}_2R \subset \cdots \subset \mathbf{x}_1R + \cdots + \mathbf{x}_kR = P$ is a strictly ascending chain of prime ideals. Since $s < k$ by assumption, it follows from Krull's Height Theorem that P is not minimal among the prime ideals that contain $\{\Phi_1, \ldots, \Phi_s\}$. Therefore, the Nullstellensatz yields $\vec{a} = (a_1, \ldots, a_k) \neq (0, \ldots, 0)$ in F^k such that $\{\Phi_1, \ldots, \Phi_s\} \subseteq (\mathbf{x}_1 - a_1)R + \cdots + (\mathbf{x}_k - a_k)R$. Consequently, $\Phi_1(\vec{a}) = \cdots = \Phi_s(\vec{a}) = 0$. □

Exercises

1. Let $R = F[\mathbf{x}_1, \ldots, \mathbf{x}_k]$, where F is any field. Show that $\{0\} \subset \mathbf{x}_1R \subset \mathbf{x}_1R + \mathbf{x}_2R \subset \cdots \subset \mathbf{x}_1R + \cdots + \mathbf{x}_kR$ is a strictly ascending chain of prime ideals in R.

2. Let $\Phi_1 = \mathbf{x}^2 + \mathbf{y}^2 - \mathbf{z}^2$ and $\Phi_2 = \mathbf{x}^2 + \mathbf{y}^2 - 2\mathbf{z}^2$ in $\mathbb{R}[\mathbf{x},\mathbf{y},\mathbf{z}]$. Show that there is no non-trivial common solution $\vec{a} = (a,b,c) \in \mathbb{R}^3$ of $\Phi_1 = 0$ and $\Phi_2 = 0$. Find all $a \in \mathbb{C}^3$ such that $\Phi_1(\vec{a}) = \Phi_2(\vec{a}) = 0$.

3. In this exercise, F denotes an algebraically closed field, $R = F[\mathbf{x}_1, \ldots, \mathbf{x}_k]$, and if

$\vec{a} = (a_1, \ldots, a_k) \in F^k$, write $I(\vec{a})$ for the ideal $(\mathbf{x}_1 - a_1)R + \cdots + (\mathbf{x}_k - a_k)R$. Prove the following statements.

(a) $I(\vec{a})$ is a maximal ideal of R.

(b) The following statements are equivalent.

(i) If I is a maximal ideal of R, then R/I is an algebraic extension of F.

(ii) Every maximal ideal of R has the form $I(\vec{a})$ for some $\vec{a} \in F^k$.

(iii) If J is a proper ideal of R, then there exists $\vec{a} \in F^k$ such that $\Phi(\vec{a}) = 0$ for all $\Phi \in J$.

(c) If the conditions of (b) are satisfied for all k, P is a prime ideal of R, and $\Phi \in R - P$, then there exists $\vec{a} \in F^k$ such that $P \subseteq I(\vec{a})$ and $\Phi(\vec{a}) \neq 0$. Hint. Let $S = F[\mathbf{x}_0, \mathbf{x}_1, \ldots, \mathbf{x}_k] \supset R$. Assuming that the statement of (c) fails, deduce from (b) (iii) that $PS + (1 - \mathbf{x}_0\Phi)S = S$, say $1 = \sum_i \Psi_i\Theta_i + (1 - \mathbf{x}_0\Phi)X$ with Θ_i, $X \in S$, $\Psi_i \in P$. Substitute Φ^{-1} for $\mathbf{x}_0$ in this identity and clear fractions to get $\Phi^n \in P$ for some $n \geq 1$. Since P is prime, this contradicts $\Phi \notin P$.

(d) Let I be a maximal ideal of R, and $K = R/I$, viewed as a field extension of F. If F is uncountable, then for each $x \in K - F$, the set $\{(x - a)^{-1}: a \in F\} \subseteq F(x)$ is linearly dependent over F. Thus x is algebraic over F. Combining (b), (c), and (d) gives a proof of the Nullstellensatz in the case that F is uncountable.

19.4. Tsen's Theorem

We are now ready to prove the main result of this chapter.

Theorem. *If F is an algebraically closed field and K is an algebraic extension of $F(\mathbf{x})$, then K is quasi-algebraically closed.*

PROOF. By Proposition 19.2, it is sufficient to prove that $K = F(\mathbf{x})$ is QAC. Let $\Phi \in K[\mathbf{x}_1, \ldots, \mathbf{x}_m]$ be homogeneous of positive degree n with $n < m$. It can be assumed that the coefficients of Φ are in $F[\mathbf{x}]$; otherwise replace Φ by $c\Phi$ where c is the product of the denominators of the coefficients of Φ. Let $t \geq 0$ be the maximum degree of the coefficients of Φ. Introduce new variables $\mathbf{y}_{ij}$, $1 \leq i \leq m$, $0 \leq j \leq r$ with r to be chosen presently. Denote

$$\Psi_i = \mathbf{y}_{i0} + \mathbf{y}_{i1}\mathbf{x} + \cdots + \mathbf{y}_{ir}\mathbf{x}^r. \tag{1}$$

Substitute Ψ_i for $\mathbf{x}_i$ in Φ and expand this form as a polynomial in $\mathbf{x}$:

$$\Phi(\Psi_1, \ldots, \Psi_m) = \Phi_0 + \Phi_1\mathbf{x} + \cdots + \Phi_{s-1}\mathbf{x}^{s-1}, \tag{2}$$

where each Φ_l is homogeneous of degree n in $F[\ldots \mathbf{y}_{ij} \ldots]$ by Lemma 19.1, and $s - 1 \leq nr + t$. The number of variables $\mathbf{y}_{ij}$ is $m(r + 1)$. If $m(r + 1) > nr + t + 1 \geq s$, then by Proposition 19.3 there exists $\vec{a} = (a_{10}, a_{11}, \ldots, a_{mr}) \in F^{m(r+1)} - \{\vec{0}\}$ such that

$$\Phi_0(\vec{a}) = \cdots = \Phi_{s-1}(\vec{a}) = 0. \tag{3}$$

Since $n < m$, the condition $m(r + 1) > s$ can be met by taking

$$r > (t + 1 - m)/(m - n).$$

Let $b_i = a_{i0} + a_{i1}\mathbf{x} + \cdots + a_{ir}\mathbf{x}^r \in F[\mathbf{x}] \subseteq K$. By (1), (2), and (3), $\Phi(\vec{b}) = 0$, and $\vec{b} \neq \vec{0}$ because $\vec{a} \neq \vec{0}$. □

Corollary a. *If K is an algebraic extension of $F(\mathbf{x})$, where the field F is algebraically closed, then $\mathbf{B}(K) = \{1\}$.*

The corollary combines the results of the theorem and Theorem 19.2.

Corollary b. *For any field F, $\mathbf{B}(F(\mathbf{x})) = \bigcup_{K/F \text{ finite}} \mathbf{B}(K(\mathbf{x})/F(\mathbf{x}))$.*

PROOF. Let E be the algebraic closure of F. If $A \in \mathfrak{S}(F(\mathbf{x}))$, then $A^{E(\mathbf{x})} \sim E(\mathbf{x})$ by Corollary a. By Proposition 13.2b, it is possible to find $\Phi_1, \ldots, \Phi_k$ in $E[\mathbf{x}]$ such that $F(\mathbf{x}, \Phi_1, \ldots, \Phi_k)$ splits A. Let K be the field that is generated over F by the coefficients of $\Phi_1, \Phi_2, \ldots$, and Φ_k. The extension K/F is finite because E is algebraic over F, and $F(\mathbf{x}, \Phi_1, \ldots, \Phi_k) \subseteq K(\mathbf{x})$ implies that $K(\mathbf{x})$ splits A. □

EXERCISES

1. Assume that F is a QAC field. Let $\Phi_1, \ldots, \Phi_r \in F[\mathbf{x}_1, \ldots, \mathbf{x}_m]$ be forms of degree n. Prove that if $1 \leq nr < m$, then there exists $\vec{a} \in F^m - \{\vec{0}\}$ such that $\Phi_1(\vec{a}) = \cdots = \Phi_r(\vec{a}) = 0$. Hint. Let Λ be a normic form of degree k, where $0 < rs \leq k < r(s + 1)$. Define $\Psi(\ldots\mathbf{x}_{ij}\ldots) = \Lambda(\Phi_1(\mathbf{x}_{11}, \ldots, \mathbf{x}_{1m}), \ldots, \Phi_r(\mathbf{x}_{11}, \ldots, \mathbf{x}_{1m}), \Phi_1(\mathbf{x}_{21}, \ldots, \mathbf{x}_{2m}), \ldots, \Phi_r(\mathbf{x}_{21}, \ldots, \mathbf{x}_{2m}), \ldots, \Phi_r(\mathbf{x}_{s1}, \ldots, \mathbf{x}_{sm}), 0, \ldots, 0)$. Prove that if $s > nr/(m - nr)$, then Ψ has a non-trivial zero, and conclude that $\Phi_1, \ldots, \Phi_r$ have a common (non-trivial) zero.

2. Let F be a QAC field. Denote $K = F(\mathbf{x})$. Use the result of Exercise 1 to show that if $\Phi \in K[\mathbf{x}_1, \ldots, \mathbf{x}_m]$ is a form of degree n such that $1 \leq n^2 < m$, then Φ has a non-trivial zero in K^m. A field K with this property is called a C_2-field.

19.5. The Structure of $\mathbf{B}(K(\mathbf{x})/F(\mathbf{x}))$

If F is a perfect field, then every finite extension K/F is separable, so that K can be embedded in a field E such that E/F is Galois. In this case, Corollary 19.4b can be sharpened: $\mathbf{B}(F(\mathbf{x})) = \bigcup_{E/F \text{ Galois}} \mathbf{B}(E(\mathbf{x})/F(\mathbf{x}))$. In this section we will determine the structure of the relative Brauer groups $\mathbf{B}(K(\mathbf{x})/F(\mathbf{x}))$ when K/F is a Galois extension.

Lemma a. *If K/F is a Galois extension, then $K(\mathbf{x})/F(\mathbf{x})$ is Galois and the mapping $\sigma \mapsto \sigma|K$ is an isomorphism of $\mathbf{G}(K(\mathbf{x})/F(\mathbf{x}))$ to $\mathbf{G}(K/F)$.*

Proof. The fields K and $F(\mathbf{x})$ can be identified with subfields of $K(\mathbf{x})$. Since the elements of $F(\mathbf{x}) - F$ are transcendental over F while every member of K is algebraic over F, it follows that $F(\mathbf{x}) \cap K = F$. Moreover, $F(\mathbf{x})K$ is a finite dimensional $F(\mathbf{x})$-algebra contained in $K(\mathbf{x})$. Thus, $F(\mathbf{x})K$ is a field that contains K and includes the element $\mathbf{x}$. This implies that $F(\mathbf{x})K = K(\mathbf{x})$. The lemma is therefore a consequence of Lemma 14.7b. □

It is convenient to identify the Galois groups $\mathbf{G}(K/F)$ and $\mathbf{G}(K(\mathbf{x})/F(\mathbf{x}))$. If $\Phi(\mathbf{x}) = a_0\mathbf{x}^n + a_1\mathbf{x}^{n-1} + \cdots + a_n \in K[\mathbf{x}]$ and $\sigma \in \mathbf{G}(K/F)$, then the action of σ on $\Phi(\mathbf{x})$ is described by the formula $\Phi^\sigma(\mathbf{x}) = a_0^\sigma\mathbf{x}^n + a_1^\sigma\mathbf{x}^{n-1} + \cdots + a_n^\sigma$.

The proof of the proposition in this section uses three standard theorems of group cohomology. We will state these results; their proofs are outlined in the exercises.

Shapiro's Lemma. *Let H be a subgroup of the finite group G. For each right $\mathbb{Z}H$-module P, define $T(P) = \mathrm{Hom}_{\mathbb{Z}H}(\mathbb{Z}G,P)$, considered as a right $\mathbb{Z}G$-module with the scalar operation $(\phi\sigma)(\sigma') = \phi(\sigma\sigma')$ for all σ, $\sigma' \in G$. If λ: $T(P) \to P$ is defined by $\lambda\phi = \phi(1)$, then λ is a homomorphism of right $\mathbb{Z}H$-modules, and the composition*

$$H^n(G,T(P)) \overset{\text{res}}{\to} H^n(H,T(P)) \overset{\lambda_*}{\to} H^n(H,P)$$

of λ_* with the restriction mapping is an isomorphism for all $n \geq 0$. ($T(P)$ and P are bimodules with the trivial actions of G and H on the left.)

Lemma b. *Let X be a finite set on which the finite group G acts transitively by $x \mapsto x^\sigma$. Define $M = \bigoplus_{x\in X} x\mathbb{Z}$ with the right $\mathbb{Z}G$-module structure induced by the action of G on X. Fix $x_0 \in X$, and let $H = \{\tau \in G: x_0^\tau = x_0\}$ be the subgroup of G that stabilizes x_0. If $\mathbb{Z}$ is considered as a right $\mathbb{Z}H$-module under the trivial action of H, then $M \cong T(\mathbb{Z}) = \mathrm{Hom}_{\mathbb{Z}H}(\mathbb{Z}G,\mathbb{Z})$ as right $\mathbb{Z}G$-modules.*

Lemma c. *If H is a finite group and $\mathbb{Q}$ is given the $\mathbb{Z}H$-bimodule structure that is induced by the trivial action of H on the left and right, then $H^n(H,\mathbb{Q}) = 0$ for all $n \geq 1$.*

The statement and proof of our main result can be simplified by the use of special notation. For a field F, denote the set of all monic irreducible polynomials in $F[\mathbf{x}]$ by $P(F)$.

Proposition (Auslander–Brumer, Faddeev). *Let K/F be a Galois extension with $\mathbf{G}(K/F) = G$. For each $\Phi \in P(F)$, choose $\Psi \in P(K)$ such that Ψ divides Φ in $K[\mathbf{x}]$, and denote $H(\Phi) = \{\tau \in G: \Psi^\tau = \Psi\}$. Then*

$$\mathbf{B}(K(\mathbf{x})/F(\mathbf{x})) \cong \mathbf{B}(K/F) \oplus \bigoplus_{\Phi\in P(F)} \mathrm{Hom}_{\mathbb{Z}}(H(\Phi),\mathbb{Q}/\mathbb{Z}).$$

Proof. By Lemma a and Theorem 14.2, $\mathbf{B}(K(\mathbf{x})/F(\mathbf{x})) \cong H^2(G,K(\mathbf{x})^\circ)$. The fundamental theorem of arithmetic for polynomial algebras takes the form $K(\mathbf{x})^\circ \cong K^\circ \oplus \bigoplus_{\Psi\in P(K)} \Psi\mathbb{Z}$, where we have shifted from multiplicative to additive notation. The direct sum $\bigoplus_{\Psi\in P(K)} \Psi\mathbb{Z}$ breaks into a sum $\bigoplus_{\Phi\in P(F)} M(\Phi)$, where $M(\Phi) = \bigoplus_{\Psi\in P(K),\, \Psi|\Phi} \Psi\mathbb{Z}$ is a $\mathbb{Z}G$-submodule. Thus, $H^2(G,K(\mathbf{x})^\circ) \cong H^2(G,K^\circ) \oplus \bigoplus_{\Phi\in P(F)} H^2(G,M(\Phi))$. (See Exercise 2, Section 11.2.) According to Lemma b, $M(\Phi) \cong \mathrm{Hom}_{\mathbb{Z}H(\Phi)}(\mathbb{Z}G,\mathbb{Z})$. Thus, Shapiro's Lemma yields $H^2(G,M(\Phi)) \cong H^2(H(\Phi),\mathbb{Z})$. The exact sequence $0 \to \mathbb{Z} \to \mathbb{Q} \to \mathbb{Q}/\mathbb{Z} \to 0$ of trivial $\mathbb{Z}H(\Phi)$-bimodules gives rise to the following segment of the Long Exact Sequence of Cohomology: $H^1(H(\Phi),\mathbb{Q}) \to H^1(H(\Phi),\mathbb{Q}/\mathbb{Z}) \to H^2(H(\Phi),\mathbb{Z}) \to H^2(H(\Phi),\mathbb{Q})$. Since $H^1(H(\Phi),\mathbb{Q}) = H^2(H(\Phi),\mathbb{Q}) = 0$ by Lemma c, we conclude that $H^2(H(\Phi),\mathbb{Z}) \cong H^1(H(\Phi),\mathbb{Q}/\mathbb{Z})$. Finally, $H^1(H(\Phi),\mathbb{Q}/\mathbb{Z}) \cong \mathrm{Hom}_{\mathbb{Z}}(H(\Phi),\mathbb{Q}/\mathbb{Z})$ because the action of $H(\Phi)$ on $\mathbb{Q}/\mathbb{Z}$ is trivial. Indeed, the derivations of $\mathbb{Z}H(\Phi)$ to $\mathbb{Q}/\mathbb{Z}$ are just group homomorphisms and 0 is the only inner derivation. By combining these isomorphisms, we get $\mathbf{B}(K(\mathbf{x})/F(\mathbf{x})) \cong H^2(G,K(\mathbf{x})^\circ) \cong \mathbf{B}(K/F) \oplus \bigoplus_{\Phi\in P(F)} \mathrm{Hom}_{\mathbb{Z}}(H(\Phi),\mathbb{Q}/\mathbb{Z})$. □

The groups $H(\Phi)$ in the proposition depend on the choice of an irreducible factor of Φ in $K[\mathbf{x}]$. However, since K/F is Galois, different choices lead to conjugate subgroups of G. Thus, the isomorphism type of $\mathrm{Hom}_{\mathbb{Z}}(H(\Phi),\mathbb{Q}/\mathbb{Z})$ depends only on Φ. Using the duality theorem for finite abelian groups, it is easy to show that in fact $\mathrm{Hom}_{\mathbb{Z}}(H(\Phi),\mathbb{Q}/\mathbb{Z}) \cong H(\Phi)/H(\Phi)'$, where $H(\Phi)'$ is the commutator of $H(\Phi)$.

Exercises

1. Let F be an infinite field with char $F \neq 2$. Prove that if K is a quadratic extension of F, then $\mathbf{B}(K(\mathbf{x})/F(\mathbf{x})) \cong \mathbf{B}(K/F) \oplus \bigoplus |F|\ \mathbb{Z}/2\mathbb{Z}$.

2. Prove Shapiro's Lemma. Hint. Denote $P^* = \mathrm{Hom}_{\mathbb{Z}}(\mathbb{Z}H,P)$. Map $\theta\colon T(P^*) \to \mathrm{Hom}_{\mathbb{Z}}(\mathbb{Z}G,P)$ by $(\theta\phi)(\sigma) = \phi(\sigma)(1)$ and $\chi\colon \mathrm{Hom}_{\mathbb{Z}}(\mathbb{Z}G,P) \to T(P^*)$ by $(\chi\psi)(\sigma)(\tau) = \psi(\sigma\tau)$. Verify that θ and χ are inverse $\mathbb{Z}G$-module homomorphisms. Embed P in P^* by $u \mapsto \lambda_u$, where $\lambda_u(\tau) = u\tau$ for $\tau \in H$. Show that the exact sequence $0 \to P \to P^* \to P^*/P \to 0$ induces an exact sequence $0 \to T(P) \to T(P^*) \to T(P^*/P) \to 0$. Since T is the Hom functor, it suffices to show that $T(P^*) \to T(P^*/P)$ is surjective, that is, for all $\phi \in T(P^*/P) = \mathrm{Hom}_{\mathbb{Z}H}(\mathbb{Z}G,P^*/P)$, there exists $\psi \in T(P^*) = \mathrm{Hom}_{\mathbb{Z}H}(\mathbb{Z}G,P^*)$ such that $\phi = \pi\psi$, where $\pi\colon P^* \to P^*/P$ is the natural projection. Write $G = \sigma_1 H \cup \cdots \cup \sigma_r H$; choose $\chi_i \in P^*$ such that $\phi(\sigma_i) = \pi\chi_i$; verify that $\psi(\sigma_i\tau) = \chi_i\tau$ for $\tau \in H$ does the job. Next, use the interpretation of zero dimensional cohomology in Section 11.2 to show that the composition $H^0(G,T(P)) \overset{\text{res}}{\to} H^0(H,T(P)) \overset{\lambda_*}{\to} H^0(H,P)$ is (essentially) the identity mapping of $P^{(H)} = \{u \in P\colon u\tau = u \text{ for all } \tau \in H\}$. Finally, use dimension shifting to prove the lemma for all $n \geq 1$. Note that this involves the properties of coinduced bimodules together with the results of Exercise 3 in Section 14.7 and Exercise 1 in Section 11.3. Moreover, the passage from $n = 0$ to $n = 1$ requires a special "diagram chasing" argument.

3. Prove Lemma b. Hint. Write $G = H\sigma_0 \cup H\sigma_1 \cup \cdots \cup H\sigma_r$, where $\sigma_0 = 1$. Let $x_i = (x_0)^{\sigma_i}$. Use the transitivity of G to show that $X = \{x_0, x_1, \ldots, x_r\}$. Prove that $\theta\colon T(\mathbb{Z}) = \operatorname{Hom}_{\mathbb{Z}H}(\mathbb{Z}G,\mathbb{Z}) \to M$, defined by $\theta(\phi) = \sum_{i=0}^{r} x_i\phi(\sigma_i^{-1})$, is a $\mathbb{Z}G$-module isomorphism.

4. Prove Lemma c. Hint. Let $\mu\colon \mathbb{Q}H \to \mathbb{Q}$ be the augmentation homomorphism $\sum \tau a_\tau \mapsto \sum a_\tau$. Deduce from the semisimplicity of $\mathbb{Q}H$ that μ is split surjective, that is, $\mathbb{Q}$ is isomorphic to a direct summand of $\mathbb{Q}H$ as a $\mathbb{Z}H$-module. Use the result of Exercise 3 to show that $\mathbb{Q}H \cong \mathbb{Q}^* = \operatorname{Hom}_{\mathbb{Z}}(\mathbb{Z}H,\mathbb{Q})$ as $\mathbb{Z}H$-modules. Thus, $H^n(H,\mathbb{Q}H) = 0$ for $n \geq 1$, and the result follows by the additivity of H^n (Exercise 2, Section 11.2).

5. The purpose of this problem is to outline a global version of the proposition. This result is known as the *Brumer–Auslander–Faddeev Theorem*. Assume that F is a field of characteristic 0, and let K be the algebraic closure of F. Let G be the (profinite) Galois group of K/F. For each $\Phi \in P(F)$, choose a root $a \in K$ of Φ, and denote by $H(\Phi)$ the subgroup of $\sigma \in G$ that fixes a. Then

$$\mathbf{B}(F(\mathbf{x})) \cong \mathbf{B}(F) \oplus \bigoplus\nolimits_{\Phi \in P(F)} \operatorname{Hom}_c(H(\Phi),\mathbb{Q}/\mathbb{Z}).$$

Hint. By Exercise 4, Section 14.6, $\mathbf{B}(F(\mathbf{x}))$ can be identified with $H^2(G,K(\mathbf{x})^\circ)$, the Galois cohomology group that is defined by continuous cocyles. Generalize Shapiro's Lemma and Lemmas b and c to the case in which G is profinite and H is a subgroup of finite index in G. The proof of the Brumer–Auslander–Faddeev Theorem is then practically identical with the proof of the proposition (using suitable versions of the cohomology properties for Galois cohomology).

19.6. Exponents of Division Algebras

The results on the structure of $\mathbf{B}(F(\mathbf{x}))$ that were obtained in Section 19.5 can be related to the construction of division algebras over $F(\mathbf{x})$. In this section we will prove a simple proposition of that kind which was obtained by Nakayama in [58]. Nakayama's theorem will then be used to establish a result of Brauer concerning the relation between the indices and exponents of central simple algebras.

Proposition. *Let $D \in \mathfrak{S}(F)$ be a division algebra with* $\operatorname{Exp} D = l$. *Assume that K/F is a cyclic extension of degree n with* $\mathbf{G}(K/F) = \langle\sigma\rangle$. *If D^K is a division algebra, then $A = D^{F(\mathbf{x})} \bigotimes_{F(\mathbf{x})}(K(\mathbf{x}),\sigma,\mathbf{x})$ is a division algebra of degree ln in* $\mathfrak{S}(F(\mathbf{x}))$, *and* $\operatorname{Exp} A$ *is the least common multiple of l and n.*

Proof. Note that by Lemma 19.5a, $K(\mathbf{x})/F(\mathbf{x})$ is cyclic with a Galois group that can be identified with $\langle\sigma\rangle$. The proof that A is a division algebra is straightforward. We will construct a convenient representation ϕ of A and show that $v_\phi(z) \neq 0$ for all $z \in A - \{0\}$. It will then follow from Corollary 16.3a that A is a division algebra. For notational purposes, let $\operatorname{Deg} D = k$, $m = k^2$, and fix an F-space basis $w_1, w_2, \ldots, w_m$ of D. Of course, w_1,

$w_2, \ldots, w_m$ is also an $F(\mathbf{x})$-space basis of $D^{F(\mathbf{x})}$. If $\psi\colon D \to M_m(F)$ is a left regular representation of D relative to any basis, then $\psi \otimes id_{F(\mathbf{x})}\colon D^{F(\mathbf{x})} \to M_m(F(\mathbf{x}))$ is a representation of $D^{F(\mathbf{x})}$. Since $[K:F] = n$, there is a unit $u \in (K(\mathbf{x}),\sigma,\mathbf{x})$ such that $(K(\mathbf{x}),\sigma,\mathbf{x}) = \bigoplus_{j<n} u^j K(\mathbf{x})$, where $u^{-1}\Phi u = \Phi^\sigma$ for $\Phi \in K(\mathbf{x})$ and $u^n = \mathbf{x}$. When $D^{F(\mathbf{x})}$ and $(K(\mathbf{x}),\sigma,\mathbf{x})$ are identified with subalgebras of A, every $z \in A$ has a unique representation.

$$z = \sum_{j<n} u^j \left(\sum_{i=1}^{m} w_i \Phi_{ij} \right), \quad \Phi_{ij} \in K(\mathbf{x}). \tag{1}$$

Let $\phi\colon A \to M_{mn}(K(\mathbf{x}))$ be the representation $(\psi \otimes id_{F(\mathbf{x})}) \otimes_{F(\mathbf{x})} \chi$, where $\chi\colon (K(\mathbf{x}),\sigma,\mathbf{x}) \to M_n(K(\mathbf{x}))$ is defined by 16.2(2). Thus, if z is given by (1), then in block form

$$\phi(z) = \begin{bmatrix} \gamma_0 & \mathbf{x}\gamma_{n-1}^{\sigma} & \mathbf{x}\gamma_{n-2}^{\sigma^2} & \cdots & \mathbf{x}\gamma_1^{\sigma^{n-1}} \\ \gamma_1 & \gamma_0^{\sigma} & \mathbf{x}\gamma_{n-1}^{\sigma^2} & \cdots & \mathbf{x}\gamma_2^{\sigma^{n-1}} \\ & & \cdot \quad \cdot \quad \cdot & & \\ \gamma_{n-1} & \gamma_{n-2}^{\sigma} & \gamma_{n-3}^{\sigma^2} & \cdots & \gamma_0^{\sigma^{n-1}} \end{bmatrix} \tag{2}$$

where $\gamma_j^{\sigma^k} = \sum_{i=1}^m \psi(w_i)\Phi_{ij}^{\sigma^k}$. Our aim is to prove that $\det \phi(z) = v_\phi(z) \neq 0$ if $z \in A - \{0\}$. Since v_ϕ is multiplicative, z can be replaced by vz or zv provided v is known to be a unit of A. Thus, we can assume that the Φ_{ij} in (1) are all polynomials in $\mathbf{x}$; otherwise multiply by the least common multiple of the denominators of these fractions. It can also be supposed that

$$\Phi_{i0}(0) \neq 0 \quad \text{for some } i \text{ between 1 and } m. \tag{3}$$

Indeed, if $\mathbf{x}^r$ is the highest power of $\mathbf{x}$ that divides all Φ_{ij} in $K[\mathbf{x}]$ and l is the smallest subscript such that $\mathbf{x}^{r+1}$ does not divide Φ_{il} for all i, then an easy calculation shows that $u^{n-l}z\mathbf{x}^{-(r+1)}$ has coefficients in $K[\mathbf{x}]$ and satisfies (3). Let $\Psi = \det \phi(z) \in K[\mathbf{x}]$. By (2)

$$\begin{aligned} \Psi(0) &= \det \begin{bmatrix} \gamma_0(0) & 0 & \cdots & 0 \\ \gamma_1(0) & \gamma_0(0)^{\sigma} & \cdots & 0 \\ & \cdot \quad \cdot \quad \cdot & & \\ \gamma_{n-1}(0) & \gamma_{n-2}(0)^{\sigma} & \cdots & \gamma_0(0)^{\sigma^{n-1}} \end{bmatrix} \\ &= (\det \gamma_0(0))(\det \gamma_0(0)^{\sigma}) \cdots (\det \gamma_0(0)^{\sigma^{n-1}}). \end{aligned}$$

Since $\psi(w_i) \in M_m(F)$, we have $\det(\gamma_0(0)^{\sigma^k}) = \det(\sum_{i=1}^m \psi(w_i)\Phi_{i0}(0)^{\sigma^k}) = (\det \sum_{i=1}^m \psi(w_i)\Phi_{i0}(0))^{\sigma^k} = (\det \gamma_0(0))^{\sigma^k}$. Thus, $\Psi(0) = N_{K/F}(\det \gamma_0(0))$. Moreover, $\det \gamma_0(0) = \det(\sum_{i=1}^m \psi(w_i)\Phi_{i0}(0)) = v_\psi(y)$, where $y = \sum_{i=1}^m w_i\Phi_{i0}(0) \in D^K$ is not zero by (3). The hypothesis that D^K is a division algebra implies that $\det \gamma_0(0) = v_\psi(y) \neq 0$; therefore $\Psi(0) \neq 0$ because $N_{K/F}$ maps K° to F°. In particular, $v_\phi(z) = \Psi \neq 0$. Hence, A is a division algebra. To determine the exponent of A, we first prove that $B = (K(\mathbf{x}),\sigma,\mathbf{x})$ has exponent n. By Corollary 15.1a and Lemma 15.1, it suffices to note that $\mathbf{x}^k \in N_{K(\mathbf{x})/F(\mathbf{x})}(K(\mathbf{x}))$ if and only if n divides k. If $\mathbf{x}^k$ is the norm of Φ/Ψ where $\Phi, \Psi \in K[\mathbf{x}]$, then $\mathbf{x}^k N_{K(\mathbf{x})/F(\mathbf{x})}(\Psi) = N_{K(\mathbf{x})/F(\mathbf{x})}(\Phi)$. Clearly, n divides the degrees of $N_{K(\mathbf{x})/F(\mathbf{x})}(\Psi)$ and

$N_{K(\mathbf{x})/F(\mathbf{x})}(\Phi)$; hence n also divides k. The converse is obvious since the norm of $\mathbf{x}$ is $\mathbf{x}^n$. Next observe that $\operatorname{Exp} D^{F(\mathbf{x})} = \operatorname{Exp} D = l$. Indeed, the first part of the proof (with $K = F$) shows that the inclusion $F \to F(\mathbf{x})$ induces an injective homomorphism $\mathbf{B}(F) \to \mathbf{B}(F(\mathbf{x}))$, so that the orders of $[D]$ and $[D^{F(\mathbf{x})}]$ are equal. To prove that $\operatorname{Exp} A = \text{l.c.m.}\{l,n\} = \text{l.c.m.}\{\operatorname{Exp}[D^{F(\mathbf{x})}], \operatorname{Exp}[B]\}$ it is sufficient to prove that $\langle[D^{F(\mathbf{x})}]\rangle \cap \langle[B]\rangle = \{1\}$ in $\mathbf{B}(F(\mathbf{x}))$. Let E/F be a Galois extension such that E splits D and $K \subseteq E$. Thus, $[A] \in \mathbf{B}(E(\mathbf{x})/F(\mathbf{x}))$. As we showed in the proof of Proposition 19.5, $\mathbf{B}(E(\mathbf{x})/F(\mathbf{x})) \cong H^2(\mathbf{G}(E/F), E(\mathbf{x})^\circ) \cong H^2(\mathbf{G}(E/F), E^\circ) \oplus \bigoplus_{\Phi \in P(F)} H^2(\mathbf{G}(E/F), M(\Phi))$. An examination of the definitions shows that under this isomorphism $[D^{F(\mathbf{x})}]$ is mapped into $H^2(\mathbf{G}(E/F), E^\circ)$ and $[B]$ goes into $H^2(\mathbf{G}(E/F), M(\mathbf{x}))$. This proves that $\langle[D^{F(\mathbf{x})}]\rangle$ and $\langle[B]\rangle$ are disjoint in $\mathbf{B}(F(\mathbf{x}))$, as we wished to show. □

Corollary a. *The inclusion mapping $F \to F(\mathbf{x})$ induces an injective homomorphism $\mathbf{B}(F) \to \mathbf{B}(F(\mathbf{x}))$ such that* $\operatorname{Ind}[A] = \operatorname{Ind}[A^{F(\mathbf{x})}]$.

Corollary b. *If K/F is a cyclic field extension of degree n with $\mathbf{G}(K/F) = \langle\sigma\rangle$, then $(K(\mathbf{x}),\sigma,\mathbf{x})$ is a cyclic division algebra of degree n and exponent n.*

Corollary c. *Let $E_1/F, \ldots, E_r/F$ be cyclic extensions with $\mathbf{G}(E_i/F) = \langle\sigma_i\rangle$ of order n_i. Assume that the E_i are linearly disjoint, that is $E_1 \otimes \cdots \otimes E_r$ is a field. If $\mathbf{x}_1, \ldots, \mathbf{x}_r$ are independent variables, then $E_i(\mathbf{x}_1, \ldots, \mathbf{x}_r)/F(\mathbf{x}_1, \ldots, \mathbf{x}_r)$ is cyclic with Galois group $\langle\sigma_i\rangle$. Denote*

$$B_i = (E_i(\mathbf{x}_1, \ldots, \mathbf{x}_r), \sigma_i, \mathbf{x}_i) \in \mathfrak{S}(F(\mathbf{x}_1, \ldots, \mathbf{x}_r)).$$

The tensor product $A = B_1 \otimes \cdots \otimes B_r$ over $F(\mathbf{x}_1, \ldots, \mathbf{x}_r)$ is a division algebra of degree $n_1 \cdots n_r$ with $\operatorname{Exp} A = \text{l.c.m.}\ \{n_1, \ldots, n_r\}$.

This corollary is obtained by induction from the proposition. The detailed proof is left as an exercise.

We will now use Corollary c to prove a theorem due to Brauer that establishes the precise relation between the exponent and Schur index. If $A \in \mathfrak{S}(F)$ has index m and exponent n, then by Proposition 14.4b n divides m and every prime divisor of m is also a factor of n. Brauer's theorem shows that in general no more can be said about the relation between $\operatorname{Ind} A$ and $\operatorname{Exp} A$.

Theorem. *If m and n are natural numbers such that n divides m and every prime divisor of m is a factor of n, then there is a field L and a division algebra $D \in \mathfrak{S}(L)$ such that* $\operatorname{Deg} D = m$ *and* $\operatorname{Exp} D = n$.

Proof. Define inductively $n_1 = n$, $n_2 = \text{g.c.d.}\ \{n,m/n_1\}, \ldots, n_{i+1} = \text{g.c.d.}\ \{n, m/(n_1 n_2 \cdots n_i)\}, \ldots$. There is a least $r \in \mathbb{N}$ such that $n_{r+1} = 1$. Since the prime divisors of m and n are the same, $n_{r+1} = 1$ implies $m =$

$n_1 n_2 \cdots n_r$. Moreover, n is clearly the least common multiple of $\{n_1, n_2, \ldots, n_r\}$. The theorem will follow from Corollary c by constructing a field F that admits cyclic extensions E_i of degree n_i for $1 \le i \le r$, such that $E_1 \otimes \cdots \otimes E_r$ is a field. Let K be a field of characteristic zero that includes a primitive n'th root of unity. Define $F = K(\mathbf{y}_1, \ldots, \mathbf{y}_r)$, where $\mathbf{y}_1, \ldots, \mathbf{y}_r$ are independent variables. If $E_i = F(\mathbf{y}_i^{1/n_i})$, then E_i/F is cyclic of degree n_i by Kummer theory, and $[F(\mathbf{y}_1^{1/n_1}, \ldots, \mathbf{y}_r^{1/n_r})\colon F] = n_1 \cdots n_r$. (See Exercise 1.) Thus, $E_1 \otimes \cdots \otimes E_r \cong F(\mathbf{y}_1^{1/n_1}, \ldots, \mathbf{y}_r^{1/n_r})$ is a field by Proposition 9.2c. According to Corollary c, the field $L = K(\mathbf{y}_1, \ldots, \mathbf{y}_r, \mathbf{x}_1, \ldots, \mathbf{x}_r)$ has the required property: there is a division algebra $D \in \mathfrak{S}(L)$ with degree m and exponent n. □

By using the results of class field theory, it can be shown that every algebraic number field F admits linearly disjoint cyclic extensions E_i of degree n_i, so that $L = F(\mathbf{x}_1, \ldots, \mathbf{x}_r)$ satisfies the requirements of the theorem.

EXERCISES

1. Using the notation that was introduced in the proof of the theorem, prove by induction on r that $[F(\mathbf{y}_1^{1/n_1}, \ldots, \mathbf{y}_r^{1/n_r})\colon F] = n_1 \cdots n_r$. Hint. Use Eisenstein's Criterion to prove that $\mathbf{x}^{n_r} - \mathbf{y}_r$ is irreducible in $F(\mathbf{y}_1^{1/n_1}, \ldots, \mathbf{y}_{r-1}^{1/n_{r-1}})[\mathbf{x}]$.

2. Prove Corollary c. Hint. Induce on r. Corollary b is the case $r = 1$. Denote $L_i = E_i(\mathbf{x}_1, \ldots, \mathbf{x}_{r-1})$, $K = F(\mathbf{x}_1, \ldots, \mathbf{x}_{r-1})$. By Lemma 19.5a, L_i/K is cyclic of degree n_i with Galois group $\langle \sigma_i \rangle$. Let $C_i = (L_i, \sigma_i, \mathbf{x}_i) \in \mathfrak{S}(K)$. Verify that $C_i^{K(\mathbf{x}_r)} \cong (L_i(\mathbf{x}_r), \sigma_i, \mathbf{x}_i) = B_i$. Show that if $1 \le i < r$, then $(E_i \otimes E_r)/E_r$ is cyclic of degree n_i with the Galois group $\langle \sigma_i \rangle$. Deduce from the induction hypothesis that $D = C_1 \otimes \cdots \otimes C_{r-1} \in \mathfrak{S}(K)$ is a division algebra of degree $n_1 \cdots n_{r-1}$ and exponent l.c.m.$\{n_1, \ldots, n_{r-1}\}$. Also use the induction hypothesis to show that $D^{L_r} \cong C_1^{L_r} \otimes \cdots \otimes C_{r-1}^{L_r}$ is a division algebra. (Note that $C_i^{L_r} \cong ((E_i \otimes E_r)(\mathbf{x}_1, \ldots, \mathbf{x}_{r-1}), \sigma_i, \mathbf{x}_i)$.) Use the proposition to complete the induction.

19.7. Twisted Laurent Series

This section introduces another construction of division algebras. The general method is due to Hilbert. It first appeared in the 1930 edition of his book on the foundations of geometry. A special case of the construction was described in Exercise 2 of Section 17.4.

Let D be a division algebra over the field F. It is not assumed that D is a finite dimensional F-algebra. Suppose that σ is an F-algebra automorphism $a \mapsto a^\sigma$ of D. As usual, we denote a variable by the symbol $\mathbf{x}$. Define $D((\mathbf{x}, \sigma))$ to be the set of all formal Laurent series $\sum_{k \ge n} \mathbf{x}^k a_k$ in which $n \in \mathbb{Z}$ and $a_k \in D$ for all $k \ge n$. In these expressions the variable $\mathbf{x}$ plays the role of a useful notational device. Equality of formal sums is defined by the rule that $\sum_{k\geq n}\mathbf{x}^k a_k = \sum_{k\geq m}\mathbf{x}^k b_k$ if $a_k = b_k$ for $k \geq \max\{m,n\}$, and $a_k = 0$ if $n \leq k < \max\{m,n\}$, whereas $b_k = 0$ if $m \leq k < \max\{m,n\}$. Thus, the notations for any finite set of elements in $D((\mathbf{x},\sigma))$ can take the form of sums that start with the same n. If almost all a_k in a formal sum $\sum_{k\geq n}\mathbf{x}^k a_k$ are zero, then we will often write the sum as a polynomial in $\mathbf{x}$ and $\mathbf{x}^{-1}$. In particular, the monomials $\mathbf{x}^k a$ are considered as elements of $D((\mathbf{x},\sigma))$, and D is identified with the constants $a_0 = \mathbf{x}^0 a_0$ in $D((\mathbf{x},\sigma))$. Give $D((\mathbf{x},\sigma))$ the structure of an F-algebra by defining addition and scalar multiplication componentwise: $(\sum_{k\geq n}\mathbf{x}^k a_k) + (\sum_{k\geq n}\mathbf{x}^k b_k) = \sum_{k\geq n}\mathbf{x}^k(a_k + b_k)$, $(\sum_{k\geq n}\mathbf{x}^k a_k)d = \sum_{k\geq n}\mathbf{x}^k(a_k d)$. Define multiplication by the convention that conjugation by $\mathbf{x}$ is the same as the action of σ on D. That is, $(\mathbf{x}^k a)(\mathbf{x}^l b) = \mathbf{x}^{k+l}a^{\sigma^l}b$. Thus,

$$\left(\sum_{k\geq m}\mathbf{x}^k a_k\right)\left(\sum_{k\geq n}\mathbf{x}^k b_k\right) = \sum_{k\geq m+n}\mathbf{x}^k c_k,$$

where $c_k = \sum_{i+j=k} a_i^{\sigma^j} b_j$. It is a routine matter to check that $D((\mathbf{x},\sigma))$ is an F-algebra with these operations, and $D = \{\mathbf{x}^0 a : a \in D\}$ is a subalgebra of $D((\mathbf{x},\sigma))$. The fact that σ is an F-algebra homomorphism is needed to prove that the associative, distributive, and scalar associativity laws are satisfied.

The algebra $D((\mathbf{x},\sigma))$ has a built in valuation v. Choose $d \in \mathbb{R}^+$ with $d < 1$. Define $v\colon D((\mathbf{x},\sigma)) - \{0\} \to \mathbb{R}^+$ by $v(\sum_{k\geq n}\mathbf{x}^k a_k) = d^m$, where $m = \min\{k : a_k \neq 0\}$. As usual, let $v(0) = 0$. It is obvious that $v(z + w) \leq \max\{v(z),v(w)\}$, and $v(zw) = v(z)v(w)$ because D has no zero divisors. Thus v is a valuation. Note that a different d would produce an equivalent valuation. Since $v(D((\mathbf{x},\sigma)) - \{0\})$ consists of the powers of d, the valuation v is discrete. Clearly, $O(v) = \{\sum_{k\geq 0}\mathbf{x}^k a_k\}$, $P(v) = \{\sum_{k>0}\mathbf{x}^k a_k\}$, and $E(v) \cong D$. In particular, $\mathbf{x}$ is a uniformizer, and $v|D$ is trivial.

Proposition a. *$D((\mathbf{x},\sigma))$ is a division algebra that is complete in the v-topology.*

PROOF. We first prove that $D((\mathbf{x},\sigma))$ is complete. Let $\{z_r : r < \omega\}$ be a Cauchy sequence in $D((\mathbf{x},\sigma))$, say $z_r = \sum_{k\geq n(r)}\mathbf{x}^k a_{kr}$. By passing to a subsequence, we can assume that $v(z_r - z_{r+1}) < d^{r+1}$ for all $r < \omega$. In this case, $a_{k\,r+1} = a_{kr}$ for all $k \leq r + 1$, and by induction $a_{k\,r+1} = a_{kk}$ for all $k \leq r + 1$. Thus, $\lim_{r\to\infty} z_r = \sum_{k\geq n(0)}\mathbf{x}^k a_{kk}$. To show that $D((\mathbf{x},\sigma))$ is a division algebra, it is sufficient to prove that every non-zero $z \in D((\mathbf{x},\sigma))$ has an inverse. Let $z = \sum_{k\geq n}\mathbf{x}^k a_k$, where $a_n \neq 0$. Write $z = \mathbf{x}^n a_n(1 - w)$ with

$$w = -\sum_{k\geq 1}\mathbf{x}^k a_n^{-\sigma^k} a_{k+n}.$$

Since $v(w) \leq d < 1$, the series $1 + w + w^2 + \cdots$ converges in $D((\mathbf{x},\sigma))$, and $z^{-1} = (1 + w + w^2 + \cdots)\mathbf{x}^{-n} a_n^{-\sigma^{-n}}$. □

The algebra $D((\mathbf{x},\sigma))$ is called a *twisted Laurent series algebra* over D. When σ is the identity automorphism, we will denote this algebra by $D((\mathbf{x}))$ and refer to it as the Laurent series algebra over D.

Our description of the center of $D((\mathbf{x},\sigma))$ uses a property of the automorphism σ. The group $\operatorname{Aut} D$ of F-algebra automorphisms of D contains the subgroup $\operatorname{Inn} D$ consisting of the inner automorphisms σ_x, where σ_x is defined for $x \in D^\circ$ by $y^{\sigma_x} = x^{-1}yx$. A simple calculation shows that $\tau^{-1}\sigma_x\tau = \sigma_{x^\tau}$. Thus, $\operatorname{Inn} D$ is a normal subgroup of $\operatorname{Aut} D$. Define the inner order of an element $\sigma \in \operatorname{Aut} D$ to be the order of the coset $\sigma \operatorname{Inn} D$ in $\operatorname{Aut} D/\operatorname{Inn} D$. If D is a field, then the inner order of σ is the same as its order in $\operatorname{Aut} D$, since $\operatorname{Inn} D = \{1\}$.

Lemma. *Let D be a division algebra over the field F such that $\dim_{\mathbf{Z}(D)} D$ is finite. Assume that σ is an F-algebra automorphism of D.*

(i) *The inner order of σ is the order of $\sigma|\mathbf{Z}(D)$.*
(ii) *If the inner order e of σ is finite, then $y \in D^\circ$ exists satisfying $y^\sigma = y$ and $\sigma^e = \sigma_y$.*

PROOF. By the definition of an inner automorphism, $\sigma_x|\mathbf{Z}(D) = id$ for each $x \in D^\circ$. Therefore, $\sigma^m \in \operatorname{Inn} D$ implies $(\sigma|\mathbf{Z}(D))^m = \sigma^m|\mathbf{Z}(D) = id$. Conversely, suppose that $\sigma^m|\mathbf{Z}(D) = id$. The Noether–Skolem Theorem applies to D considered as a $\mathbf{Z}(D)$-algebra because $\dim_{\mathbf{Z}(D)} D < \infty$; it yields $\sigma^m \in \operatorname{Inn} D$. Thus, the inner order of σ coincides with the order of $\sigma|\mathbf{Z}(D)$. Assume that this order is a finite natural number e. If $K = \{d \in \mathbf{Z}(D): d^\sigma = d\}$ is the fixed field of σ, then $\mathbf{Z}(D)/K$ is a cyclic extension of degree e with Galois group $\langle \sigma|\mathbf{Z}(D)\rangle$. On the other hand, since e is the inner order of σ, there exists $x \in D^\circ$ such that $\sigma^e = \sigma_x$. Our aim is to show that x can be chosen so that $x^\sigma = x$. Note that $\sigma_x = \sigma_y$ if and only if $\sigma_{xy^{-1}} = id$, that is, $xy^{-1} \in \mathbf{Z}(D)$. In particular, $\sigma_x = \sigma^e = \sigma^{-1}\sigma^e\sigma = \sigma^{-1}\sigma_x\sigma = \sigma_{x^\sigma}$ implies $x^\sigma = bx$ for some $b \in \mathbf{Z}(D)$. Consequently, $x^{\sigma^2} = b^\sigma x^\sigma = b^{\sigma+1}x$, $x^{\sigma^3} = b^{\sigma^2+\sigma}x^\sigma = b^{\sigma^2+\sigma+1}x, \ldots,$ and $x^{\sigma^e} = b^{\sigma^{e-1}+\cdots+\sigma+1}x = N_{\mathbf{Z}(D)/K}(b)x$. However, $x^{\sigma^e} = x^{\sigma_x} = x^{-1}xx = x$. Thus, $N_{\mathbf{Z}(D)/K}(b) = 1$. By Hilbert's Theorem 90 (Exercise 1, Section 16.6), $b = a/a^\sigma$ for some $a \in \mathbf{Z}(D)$. If $y = ax$, then $\sigma_y = \sigma_x = \sigma^e$ and $y^\sigma = a^\sigma x^\sigma = a^\sigma bx = ax = y$. □

Proposition b. *Let D be a division algebra over the field F such that $\dim_{\mathbf{Z}(D)} D$ is finite. Assume that $\sigma \in \operatorname{Aut} D$ has inner order e. Denote $K = \{a \in \mathbf{Z}(D): a^\sigma = a\}$ and $A = D((\mathbf{x},\sigma))$.*

(i) *If e is infinite, then $\mathbf{Z}(A) = K$.*
(ii) *If e is finite, then there exists $y \in D^\circ$ such that $\sigma^e = \sigma_y$, $y^\sigma = y$, and $\mathbf{Z}(A) = \{\sum_{l \geq m} (\mathbf{x}^e y^{-1})^l c_l : c_l \in K \text{ for } l \geq m\}$.*

PROOF. Let $z = \sum_{k \geq n} \mathbf{x}^k a_k \in A$. Clearly, $z \in \mathbf{Z}(A)$ if and only if $(\mathbf{x}^k a_k)(\mathbf{x}^l b) = (\mathbf{x}^l b)(\mathbf{x}^k a_k)$, that is,

$$a_k^{\sigma^l} b = b^{\sigma^k} a_k \text{ for all } k \geq n, l \in \mathbb{Z}, b \in D. \tag{1}$$

The condition (1) implies

$$a_k = a_k^\sigma \text{ for all } k \geq n, \text{ and} \tag{2}$$

$$a_k \neq 0 \text{ implies } b^{\sigma^k} = a_k b a_k^{-1} \text{ for all } b \in D. \tag{3}$$

Thus $\sigma^k \in \operatorname{Inn} D$ if $a_k \neq 0$. In particular, $z = a_0$ when e is infinite. In this case, $z \in K$ by (2). Conversely, it is obvious that $K \subseteq \mathbf{Z}(A)$. Assume that $e < \infty$. If $a_k \neq 0$, then $\sigma^k \in \operatorname{Inn} D$ and e divides k, say $k = el$. By the lemma, there exists $y \in D^\circ$ satisfying $y^\sigma = y$ and $\sigma_y = \sigma^e$. Thus,

$$\sigma_{y^l} = \sigma^{el} = \sigma^k = \sigma_{(a_k)^{-1}}$$

by (3). It follows that $a_k = y^{-l}c_k$ for some $c_k \in \mathbf{Z}(D)$. Moreover, $c_k^\sigma = c_k$ by (2) and the fact that $y^\sigma = y$. Hence, if $z \in \mathbf{Z}(A)$, then $z = \sum_{l \geq m} (\mathbf{x}^e y^{-1})^l c_l$, where $y^\sigma = y$, $\sigma^e = \sigma_y$, and $c_l \in K$ for all $l \geq m$. A routine computation shows the converse: every such z is in $\mathbf{Z}(A)$. □

Corollary. *If D is a division algebra over F such that $\dim_{\mathbf{Z}(D)} D$ is finite, and if σ is an F-algebra automorphism of D with finite inner order e, then $D((\mathbf{x},\sigma))$ is finite dimensional over its center. Moreover, as central simple algebras,* $\operatorname{Deg} D((\mathbf{x},\sigma)) = e(\operatorname{Deg} D)$.

PROOF. In the notation of Proposition b, $(\operatorname{Deg} A)^2 = \dim_{\mathbf{Z}(A)} A = e(\dim_{\mathbf{Z}(D)} D)[\mathbf{Z}(D) : K] = e^2(\operatorname{Deg} D)^2$. □

EXAMPLE. Assume that F is an algebraically closed field of characteristic zero, and let D be a division algebra over F that is finite dimensional over its center. If $D_1 = D((\mathbf{x}))$, then $\mathbf{Z}(D_1) = \mathbf{Z}(D)((\mathbf{x}))$ by Proposition b. Let $n \in \mathbb{N}$; choose a primitive n'th root of unity $\zeta \in F$. Define $\sigma \in \operatorname{Aut} D_1$ by $(\sum_{k \geq n} \mathbf{x}^k a_k)^\sigma = \sum_{k \geq n} (\mathbf{x}\zeta)^k a_k$. The inner order and the order of σ are both n. Moreover, the fixed field of $\sigma | \mathbf{Z}(D_1)$ is $\mathbf{Z}(D)((\mathbf{x}^n))$. Define $D_2 = D_1((\mathbf{y},\sigma))$. The center of D_2 is given by Proposition b using $y = 1$: $\mathbf{Z}(D_2) = \{\sum_{l \geq m} \mathbf{y}^{nl} a_l : a_l \in \mathbf{Z}(D)((\mathbf{x}^n))\} = \mathbf{Z}(D)((\mathbf{x}^n))((\mathbf{y}^n))$. By the corollary, the degree of D_2 over its center is $n(\operatorname{Deg} D)$.

Beginning with F, this construction can be iterated to obtain central simple division algebras with very interesting properties. This is the technique that we will use in Section 19.9 to prove an important theorem due to Amitsur.

EXERCISES

1. Verify that $D((\mathbf{x},\sigma))$ is an F-algebra.
2. Carry out the computation that was suggested in the last sentence of the proof of Proposition b.
3. This exercise describes Hilbert's original example of a twisted Laurent series algebra. Let F be a totally ordered field, for example $\mathbb{Q}$ or $\mathbb{R}$. Let $K = F((\mathbf{x}))$. Define $P = \{\sum_{k \geq n} \mathbf{x}^k a_k \in K^\circ : a_n > 0\}$, and for $y, z \in K$, write $z > y$ if $z - y \in P$.

(a) Prove that the binary relation $>$ is a total ordering of K such that: $z > y$ implies $z + w > y + w$ for all $y, z, w \in K$; and $y > 0$, $z > 0$, implies $yz > 0$ for all $y, z \in K$.

(b) Define $\sigma: K \to K$ by $(\sum_{k \geq m} \mathbf{x}^k a_k)^\sigma = \sum_{k\ \ m} x^k 2^k a_k$. Prove that σ is an F-algebra automorphism that preserves the ordering of K, that is, $z > w$ implies $z^\sigma > w^\sigma$. Show that σ has infinite (inner) order.

(c) Let $D = K((\mathbf{y},\sigma))$. Define $Q = \{\sum_{k \geq n} \mathbf{y}^k z_k \in D: z_n > 0\}$, $u > v$ if $u - v \in Q$. Prove that D is a totally ordered division algebra with $\mathbf{Z}(D) = F$. This result should be compared with Exercise 4 in Section 16.3, where it is proved that no division algebra that is finite dimensional over its center can be a totally ordered algebra.

4. Let $K = \mathbb{F}_2((\mathbf{x}))$. Prove that there is an $\mathbb{F}_2$-algebra automorphism σ of K such that $\mathbf{x}^\sigma = \mathbf{x} + \mathbf{x}^2$, and the fixed field of σ is $\mathbb{F}_2$. Define $D = K((\mathbf{y},\sigma))$. Show that D is a division algebra with $\mathbf{Z}(D) = \mathbb{F}_2$. Prove that $D^{ab} = D^\circ/D'$ is infinite, so that the determinant mapping of D cannot take its values in $\mathbf{Z}(D)^\circ$. Hint. D' is a subgroup of the kernel of the valuation mapping of D. (The reduced norm is not defined for D since D is not finite dimensional over its center.)

5. Let σ be an automorphism of the field K that has finite order e. Denote the fixed field of σ by F. Prove that the division algebra $D = K((\mathbf{x},\sigma))$ has center $F((\mathbf{x}^e))$ and $K((\mathbf{x}^e))$ is a maximal subfield of D with $K((\mathbf{x}^e))/F((\mathbf{x}^e))$ cyclic. Hence D is a cyclic division algebra. Show that $F((\mathbf{x}))$ is also a maximal subfield of D. Note that $K((\mathbf{x}^e))/F((\mathbf{x}^e))$ is an unramified extension, whereas $F((\mathbf{x}))/F((\mathbf{x}^e))$ is totally ramified.

19.8. Laurent Series Fields

This section lays the foundation of the proof of a major result in the next section. The material that we present here is interesting in its own right, but our treatment of it is brief. Exercise 4 gives one result on division algebras over Laurent series fields.

When F is a field, the Laurent series algebra $F((\mathbf{x}))$ constructed from the identity automorphism of F is a field. Clearly, $F((\mathbf{x}))$ has uncountable dimension over F. Hence the transcendence degree of $F((\mathbf{x}))$ over F is also uncountable. The discrete valuation of $F((\mathbf{x}))$ that was defined in Section 19.7 is a basic tool in the study of $F((\mathbf{x}))$. We will denote this valuation by v. Since v is the only valuation of $F((\mathbf{x}))$ that we will use, it can be safely called *the* valuation of $F((\mathbf{x}))$.

We will be interested in finite extensions of Laurent series fields. Some results from Chapter 17 will be used. In particular, it will be taken for granted that the valuation of $F((\mathbf{x}))$ has a unique extension to any field K such that $K/F((\mathbf{x}))$ is finite, and K is complete under the extended valuation. This fact was proved only for extensions of local fields. The proof was based on Lemma 17.6b. A different approach is needed when the compactness hypothesis is replaced by completeness. Exercise 2 outlines a proof of the generalization of Proposition 17.6 that we need.

It will be convenient to have an intrinsic characterization of Laurent series fields.

Lemma a. *Let K be a field that is complete in the topology of a discrete valuation v. If F is a subfield of K such that $v|F$ is trivial and if $t \in K^\circ$ satisfies $v(t) < 1$, then the closure of $F(t)$ in K is algebraically and topologically isomorphic to $F((\mathbf{x}))$.*

PROOF. Since v is trivial on F and $v(t) < 1$, it follows from Lemma 17.7a that t is not algebraic over F. It is therefore an easy consequence of the Domination Principle that the algebras of finite Laurent series in t and $\mathbf{x}$ are isometrically isomorphic, provided $v(\mathbf{x})$ is chosen to coincide with $v(t)$. The lemma is then a consequence of Corollary 17.4a. □

This lemma justifies the notation $F((t))$ to denote a field that is complete under a discrete valuation v such that $v|F$ is trivial and t is a uniformizer. The elements of $F((t))$ are the Laurent series $\sum_{k\geq n} t^k a_k$, all of which converge in the v-topology.

Up to now we have avoided ramified extensions of fields with a valuation. It is no longer possible to sweep ramification under the rug. However, only the tamely and totally ramified extensions require our attention.

Lemma b. *Let K be a field that is complete in the topology of a discrete valuation v. Let F be a closed subfield of K such that K/F is finite, and the characteristic of the residue class field $E(F)$ does not divide the natural number e. Denote the residue class mapping of $O(K)$ to $E(K)$ by $x \mapsto \bar{x}$.*

(i) *If $x \in O(K)$ satisfies $\bar{x} = 1$, then $x = y^e$ for some $y \in O(K)$ such that $\bar{y} = 1$.*
(ii) *Assume that K/F is a totally ramified extension, $e = e(K/F)$, t is a uniformizer for F, and $X \subseteq O(F) - P(F)$ maps onto $E(F)^\circ$ under the residue class mapping. There exists $c \in X$ and $z \in O(K)$ such that $K = F(z)$ and $z^e = ct$.*

PROOF. Let $\Phi(\mathbf{x}) = \mathbf{x}^e - x$. The characteristic of $E(F)$ does not divide e, so that the polynomial $\bar{\Phi}(\mathbf{x}) = \mathbf{x}^e - \bar{x}$ and its derivative $e\mathbf{x}^{e-1}$ are relatively prime. Since 1 is a root of Φ modulo $P(K)$, the assertion (i) follows from Hensel's Lemma. To prove (ii), let w be a uniformizer for K. If $u = w^e t^{-1}$, then $v(u) = 1$ and $\bar{u} \in E(F)^\circ$, because $e = e(K/F)$ and $f(K/F) = 1$. Therefore $c \in X$ exists satisfying $\bar{u} = \bar{c}$. Since the element $x = u^{-1}c$ satisfies $\bar{x} = \bar{1}$, it follows from (i) that $x = y^e$ for some $y \in O(K)$ such that $\bar{y} = \bar{1}$. If $z = yw$, then $z^e = ct$; and $v(z) = v(w)$ because $y \in O(K) - P(K)$. Hence $K = F(z)$ by Lemma 17.7b and the extended version of Lemma 17.6b. (See Exercise 2 (a).) □

A corollary of Lemma b will be used in Section 20.8. Although the result has nothing to do with Laurent series fields, this is an appropriate place to record it.

Corollary. *If $K/\hat{\mathbb{Q}}_p$ is a totally ramified, Galois extension of degree e and p does not divide e, then e divides $p - 1$.*

PROOF. The extension of v_p to K will be denoted by v. By Lemma b, $K = \hat{\mathbb{Q}}_p(z)$, where $z^e = cp$ with $1 \le c < p$. The polynomial $\mathbf{x}^e - cp$ is irreducible in $\hat{\mathbb{Q}}_p[\mathbf{x}]$ since $[K : \hat{\mathbb{Q}}_p] = e$. Therefore, all the roots of this polynomial are in K. It follows that there is a primitive e'th root of unity ζ in K. Note that $v(\zeta) = 1$ because $\zeta^e = 1$, so that $1 - \zeta^k \in O(K)$ and $v(1 - \zeta^k) \le 1$ for all k. The standard identity (Exercise 1) $e = \prod_{0<k<e}(1 - \zeta^k)$ yields $1 = v(e) = \prod_{0<k<e} v(1 - \zeta^k)$. Thus, $v(1 - \zeta^k) = 1$ and $\bar{\zeta}^k \ne 1$ for $0 < k < e$. That is, $\bar{\zeta}$ has order e in $E(K)^\circ = E(\hat{\mathbb{Q}}_p)^\circ = F_p^\circ$. In particular, e divides $p - 1$. □

Proposition a. *Let K be a finite field extension of $F((t))$ such that the characteristic of F does not divide $[E(K): F)]$. Denote $e = e(K/F((t)))$ and $f = f(K/F((t)))$.*

(i) *There is a finite extension E/F with $E \subseteq O(K)$ such that the residue class mapping $y \mapsto \bar{y}$ is an isomorphism of E to $E(K)$, $E((t))/F((t))$ is unramified, and $K/E((t))$ is totally ramified.*
(ii) *$K = E((z))$, where the minimum polynomial of z over $E((t))$ is $\mathbf{x}^e - yt$ with $y \in E^\circ$.*
(iii) *$[K: F((t))] = ef$.*

PROOF. Identify $E(F((t)))$ with F. The degree f of $E(K)/F$ is not divisible by char F, so that the extension is separable. Thus, $E(K) = F(\bar{x})$ for some $x \in O(K)$. The minimum polynomial Φ of $\bar{x}$ over F is irreducible of degree f and $(\Phi, \Phi') = 1$. Since it is also the case that $\Phi \in O(F((t)))[\mathbf{x}]$, Hensel's Lemma implies that $\bar{x}$ can be realized as the image of an $x \in O(K)$ that satisfies $\Phi(x) = 0$. Let $E = F(x) \subseteq O(K)$. By construction, $y \mapsto \bar{y}$ is an isomorphism of E to $E(K)$. Since E/F is finite, $v|E$ is trivial by Lemma 17.7a. By Lemma a, the closure of $E(t)$ is a Laurent series field $E((t))$ and $[E((t)) : F((t))] = [E : F] = [E(K) : F] = f$. Thus, $E((t))/F((t))$ is unramified and $K/E((t))$ is totally ramified. By Lemma b, $K = E((t))(z)$, where z is a root of $\mathbf{x}^e - yt$ with $y \in E^\circ$. This polynomial is irreducible because $[K: E((t))] \ge e$. Thus, $ef = [K: F((t))]$. Moreover, if $u = \sum_{k \ge n} t^k x_k \in E((t))$, then $u = \sum_{k \ge n} z^{ek} y^{-k} x_k \in E((z))$. Thus, $K = E((t))(z) \subseteq E((z)) \subseteq K$. □

We will extend this result to the fields that are obtained by iterating the Laurent series construction. Another lemma is needed.

Lemma c. *Assume that F is an algebraically closed field of characteristic zero. Denote $K = F((\mathbf{x}_1))((\mathbf{x}_2)) \cdots ((\mathbf{x}_s))$. If $e \in \mathbb{N}$, then every element of K° has a representation in the form $y^e \mathbf{x}_1^{k(1)} \mathbf{x}_2^{k(2)} \cdots \mathbf{x}_s^{k(s)}$, where $y \in K^\circ$ and $0 \le k(j) < e$ for $1 \le j \le s$.*

PROOF. Define $L = F((x_1)) \cdots ((x_{s-1}))$ ($L = F$ if $s = 1$). We can assume by way of induction that the lemma is true for L. If $s = 1$, then this is the case because F is algebraically closed. Since $K = L((\mathbf{x}_s))$, every element w of K° can be written in the form $z\mathbf{x}_s^{je+k}(1 + \sum_{l \ge 1} \mathbf{x}_s^l x_l)$, where $x_l \in L$, $z \in L^\circ$, and $0 \le k < e$. By induction, we have $z = y_1^e \mathbf{x}_1^{k(1)} \cdots \mathbf{x}_{s-1}^{k(s-1)}$, where $y_1 \in L^\circ$; and $1 + \sum_{l \ge 1} \mathbf{x}_s^l x_l = y_2^e$ for some $y_2 \in L((\mathbf{x}_s))^\circ$ by Lemma b. Hence, $w = y^e \mathbf{x}_1^{k(1)} \cdots \mathbf{x}_{s-1}^{k(s-1)} \mathbf{x}_s^{k(s)}$, where $k(s) = k$ and $y = y_1 y_2 \mathbf{x}_s^j$. □

Proposition b. *Let F be an algebraically closed field of characteristic zero. Define $L = F((\mathbf{x}_1)) \cdots ((\mathbf{x}_s))$. If K/L is a field extension of degree n, then $K = F((u_1)) \cdots ((u_s))$, where $u_1, \ldots, u_s \in K$ satisfy*

$$u_i^{e(i)} = u_1^{k(1,i)} \cdots u_{i-1}^{k(i-1,i)} \mathbf{x}_i, \tag{1}$$

$e(1), \ldots, e(s) \in \mathbb{N}$, $e(1) \cdots e(s) = [K : L]$, and $k(i,j) \in \mathbb{Z}$ satisfy $0 \le k(i,j) < e(j)$ for $1 \le j \le s$, $1 \le i < j$.

PROOF. If $s = 1$, then $K = E((z_1))$ with E/F finite, $z_1^e = a\mathbf{x}_1$ with $a \in E^\circ$ by Proposition a. Since F is algebraically closed, $E = F$, and $u_1 = z_1 a^{-1/e}$ satisfies (1) and $K = F((u_1))$. Assume that the result holds for $s - 1$, where $s \ge 2$. Define $L_1 = F((\mathbf{x}_1)) \cdots ((\mathbf{x}_{s-1}))$, so that $L = L_1((\mathbf{x}_s))$. Proposition a yields a finite extension K_1/L_1 of degree $f(K/L)$ and $z_s \in K$ such that $K = K_1((z_s))$, $z_s^{e(s)} = y\mathbf{x}_s$, where $e(s) = e(K/L)$ and $y \in K_1^\circ$. We can assume by way of induction that $K_1 = F((u_1)) \cdots ((u_{s-1}))$, the u_i satisfy (1) for all $i < s$, and $e(1) \cdots e(s-1) = f(K/L)$. By Lemma c, $y = b_s^{e(s)} u_1^{k(1,s)} \cdots u_{s-1}^{k(s-1,s)}$ with $b_s \in K_1^\circ$ and $0 \le k(i,s) < e(s)$ for $1 \le i < s$. We complete the induction by putting $u_s = z_s b_s^{-1}$. □

EXERCISES

1. Let K be a field of characteristic zero that includes a primitive e'th root of unity ζ. Prove that $e = \prod_{0<k<e}(1 - \zeta^k)$. Hint. Differentiate the identity $\mathbf{x}^e - 1 = \prod_{0 \le k < e} (\mathbf{x} - \zeta^k)$, and put $\mathbf{x} = 1$.

2. (a) Let F be a field that is complete in the v-topology of a valuation v that satisfies the triangle inequality. Let $u_1, \ldots, u_n$ be a basis of the n dimensional F-space M, and define the uniform norm on M by $\|\sum_{i=1}^n u_i a_i\| = \max\{v(a_i): 1 \le i \le n\}$. Let $\theta: M \to \mathbb{R}$ be a mapping that is continuous relative to the uniform topology and satisfies $\theta(x) \ge 0$ with equality only if $x = 0$, $\theta(xa) = \theta(x)v(a)$, $\theta(x + y) \le \theta(x) + \theta(y)$. Prove that there exist c, $d \in \mathbb{R}^+$ such that $c\|x\| \le \theta(x) \le d\|x\|$ for all $x \in M$. Hint. The existence of d follows from the continuity of θ. Induce on n to get

the lower bound. The case $n = 1$ follows from the completeness of F. Show that if no $c \in R^+$ exists satisfying $c\|x\| \leq \theta(x)$, then there is a sequence $\{x_k\}$ in M with the properties $\lim \theta(x_k) = 0$, $\|x_k\| = 1$ for all k, and (for some j) $x_k = \sum_{i=1}^n u_i a_{ik}$ with all $a_{jk} = 1$. Use the triangle inequality for θ and the induction hypothesis that $\theta|M_j$ satisfies (a) (where $M_j = \{\sum u_i c_i : c_j = 0\}$) to conclude that $\{x_k - x_0\}$ is a Cauchy sequence in M_j with $\lim x_k - x_0 = y \in M_j$. Hence, $\lim x_k = x_0 + y \neq 0$, so that $0 = \lim \theta(x_k) = \theta(x_0 + y) \neq 0$.

(b) Let F be a field that is complete in the v-topology of a valuation v. Let K/F be a finite field extension. Prove that if v can be extended to a valuation w of K, then w is unique and K is complete in the w-topology. Hint. Use the result of (a).

(c) Let F be a field that is complete in the topology of a discrete valuation v. Let $\Phi(\mathbf{x}) = \mathbf{x}^n + a_1\mathbf{x}^{n-1} + \cdots + a_n \in F[\mathbf{x}]$ be irreducible. Use the following generalized form of Hensel's Lemma to show that if $v(a_n) \leq 1$, then $v(a_i) \leq 1$ for all coefficients a_i of Φ. The form of Hensel's Lemma that is needed for this problem is: if $\Phi \in F[\mathbf{x}]$ is monic and $\overline{\Phi} = \Psi_1\Theta_1$ in $E(F)[\mathbf{x}]$ with $(\Psi_1, \Theta_1) = 1$, then $\Phi = \Psi\Theta$ in $O(v)[\mathbf{x}]$, where $\overline{\Psi} = \Psi_1$, $\overline{\Theta} = \Theta_1$, and $\deg \Psi = \deg \Psi_1$. For a proof of this result, see van der Waerden's book [73].

(d) Let F be a field that is complete in the topology of a discrete valuation v. Assume that K/F is a finite extension of degree n. For $x \in K$, define $w(x) = v(N_{K/F}(x))^{1/n}$. Prove that w is a discrete valuation of K that extends v. Hint. Show that $w(x) \leq 1$ implies $w(1 + x) \leq 1$ as follows. Let $\Phi(\mathbf{x}) = \mathbf{x}^m + a_1\mathbf{x}^{m-1} + \cdots + a_m$ be the minimum polynomial of x over F. Use (c) and the hypothesis $w(x) \leq 1$ to prove that $v(a_i) \leq 1$ for all i. Then note that $\Phi(\mathbf{x} - 1) \in O(F)[\mathbf{x}]$ is the minimum polynomial of $1 + x$ and $N_{K/F}(1 + x)$ is $\pm\Phi(-1)^{n/m}$.

3. Prove *Puiseaux's Theorem*: if F is algebraically closed, then for each $e \in \mathbb{N}$ there is a unique extension K_e of $F((\mathbf{x}))$ with $[K_e : F((\mathbf{x}))] = e$. Moreover, $K_e = F((z_e))$, where $(z_e)^e = \mathbf{x}$, that is $z_e = \mathbf{x}^{1/e}$.

4. Prove that if F is an algebraically closed field, then $\mathbf{B}(F((\mathbf{x}))) = \{1\}$. Hint. Use the result of Exercise 3. Note that $F((\mathbf{x}^{1/e}))/F((\mathbf{x}))$ is cyclic. Deduce from Lemma c that every element of $F((\mathbf{x}))$ is the norm of an element in $F((\mathbf{x}^{1/e}))$. In particular, the norm of $\mathbf{x}^{1/e}$ is $\mathbf{x}$. It can be shown that $F((\mathbf{x}))$ is QAC, so that the result of this exercise is a corollary of Theorem 19.2.

19.9. Amitsur's Example

The results and constructions of the last two sections will now be assembled to produce a division algebra D with a remarkable property: all maximal subfields of D are Galois extensions of $\mathbf{Z}(D)$ and their Galois groups are isomorphic. The construction is due to Amitsur. The algebra D will be used in Section 20.8 to prove the existence of division algebras that are not crossed products.

Theorem. *Assume that F is an algebraically closed field of characteristic zero. Let $(p_1, p_2, \ldots, p_r)$ be a sequence of primes that are not necessarily distinct. Define $D = F((\mathbf{x}_1))((\mathbf{y}_1, \sigma_1))((\mathbf{x}_2))((\mathbf{y}_2, \sigma_2)) \cdots ((\mathbf{x}_r))((\mathbf{y}_r, \sigma_r))$, where σ_i is* *the automorphism of* $F((\mathbf{x}_1))((\mathbf{y}_1,\sigma_1))\cdots((\mathbf{x}_i))$ *that is define by* $\mathbf{x}_i^{\sigma_i}=\zeta_i\mathbf{x}_i$ *with* ζ_i *a primitive* p_i*'th root and* $\mathbf{x}_j^{\sigma_i}=\mathbf{x}_j$, $\mathbf{y}_j^{\sigma_i}=\mathbf{y}_j$ *for* $j<i$. *The F-algebra D is a division algebra with center* $L=F((\mathbf{x}_1^{p_1}))((\mathbf{y}_1^{p_1}))\cdots((\mathbf{x}_r^{p_r}))((\mathbf{y}_r^{p_r}))$ *and degree* $p_1p_2\cdots p_r$. *If K is a maximal subfield of D, then K/L is a Galois extension and* $\mathbf{G}(K/L)\cong(\mathbb{Z}/p_1\mathbb{Z})\times\cdots\times(\mathbb{Z}/p_r\mathbb{Z})$.

The facts that $\mathbf{Z}(D)=L$ and $\operatorname{Deg} D=p_1\cdots p_r$ follow from Example 19.7 by induction. The rest of this section is occupied with the proof that K/L is Galois and $\mathbf{G}(K/L)=(\mathbb{Z}/p_1\mathbb{Z})\times\cdots\times(\mathbb{Z}/p_r\mathbb{Z})$. This conclusion is obtained by an intricate, but straightforward analysis of the extension K/L, using the result of Proposition 19.8b. We will use an analogue of the logarithm mapping to do our bookkeeping.

Lemma. *There is group homomorphism* $\lambda\colon D^\circ\to\bigoplus_{i=1}^{2r}w_i\mathbb{Z}\cong\bigoplus 2r\ \mathbb{Z}$ *such that* $\lambda(\mathbf{x}_i)=w_{2i-1}$, $\lambda(\mathbf{y}_i)=w_{2i}$, *and* $\lambda(F^\circ)=0$.

PROOF. Define λ by induction on r. If $r=0$, let $\lambda(F^\circ)=0$. If $r>0$, $D_1=F((\mathbf{x}_1))((\mathbf{y}_1,\sigma_1))\cdots((\mathbf{x}_{r-1}))((\mathbf{y}_{r-1},\sigma_{r-1}))$, and $\lambda_1\colon D_1^\circ\to\bigoplus_{i=1}^{2r-2}w_i\mathbb{Z}$ has been constructed with the required properties, define $\lambda(z)$ for

$$z=\mathbf{y}_r^n\left(\mathbf{x}_r^m z_{1m}+\sum_{l>m}\mathbf{x}_r^l z_{1l}\right)+\sum_{k>n}\mathbf{y}_r^k z_{2k}\quad(z_{1m}\in D_1^\circ,\ z_{1l}\in D_1,\ z_{2k}\in D_1((\mathbf{x}_r)))$$

by

$$\lambda(z)=\lambda_1(z_{1m})+w_{2r-1}m+w_{2r}n.$$

According to this definition, $\lambda(\mathbf{x}_i)=w_{2i-1}$, $\lambda(\mathbf{y}_i)=w_{2i}$, and $\lambda(F^\circ)=0$. If

$$z'=\mathbf{y}_r^{n'}\left(\mathbf{x}_r^{m'}z'_{1m'}+\sum_{l>m'}\cdots\right)+\sum_{k>n'}\cdots,$$

then

$$zz'=\mathbf{y}_r^{n+n'}\left(\mathbf{x}_r^{m+m'}\zeta_r^{n'm}z_{1m}z'_{1m'}+\sum_{l>m+m'}\cdots\right)+\sum_{k>n+n'}\cdots.$$

Hence,

$$\lambda(zz')=\lambda_1(\zeta_r^{n'm}z_{1m}z'_{1m'})+w_{2r-1}(m+m')+w_{2r}(n+n')=\lambda(z)+\lambda(z'),$$

since $\lambda_1(\zeta_r^{n'm})=0$. □

It will simplify formulas if we modify our notation. Denote $x_1=\mathbf{x}_1$, $x_2=\mathbf{y}_1$, $x_3=\mathbf{x}_2,\ldots,x_{2r}=\mathbf{y}_r$. For $1\le i\le 2r$, define $p(i)=p_{i/2}$ if i is even and $p(i)=p_{(i+1)/2}$ if i is odd. Let $t_i=x_i^{p(i)}$, so that

$$L=F((t_1))((t_2))\cdots((t_{2r})).$$

By Proposition 19.8c, any maximal subfield K of D° is an extension of L that has the form

$$K=F((u_1))((u_2))\cdots((u_{2r})),\tag{1}$$

where the u_j are related to the x_j by

$$t_j = x_j^{p(j)} = u_1^{-k(1,j)} \cdots u_{j-1}^{-k(j-1,j)} u_j^{e(j)} \tag{2}$$

for $1 \le j \le 2r$. The exponents in (2) satisfy $e(j) \in \mathbb{N}$ and $0 \le k(i,j) < e(j)$. Moreover,

$$e(1)e(2) \cdots e(2r) = [K:L] = p_1 \cdots p_r. \tag{3}$$

The homomorphism λ transforms (2) into a matrix equation

$$\xi\pi = \xi\alpha\beta,$$

where $\xi = [w_1, \ldots, w_{2r}]$, π is the diagonal matrix $\mathrm{diag}(p(1), \ldots, p(2r))$,

$$\beta = \begin{bmatrix} e(1) & -k(1,2) & \cdots & -k(1,2r) \\ 0 & e(2) & \cdots & -k(2,2r) \\ \cdot & \cdot & \cdots & \cdot \\ 0 & 0 & \cdots & e(2r) \end{bmatrix}$$

and $\alpha \in M_{2r}(\mathbb{Z})$ is defined by the condition $[\lambda u_1, \ldots, \lambda u_{2r}] = \xi\alpha$. Since the entries of ξ are independent, it follows that $\pi = \alpha\beta$. Therefore, α is an upper triangular matrix

$$\alpha = \begin{bmatrix} f(1) & l(1,2) & \cdots & l(1,2r) \\ 0 & f(2) & \cdots & l(2,2r) \\ \cdot & \cdot & \cdots & \cdot \\ 0 & 0 & \cdots & f(2r) \end{bmatrix}$$

and

$$e(j)f(j) = p(j) \quad \text{for} \quad 1 \le j \le 2r. \tag{4}$$

Since $e(j)$ and $f(j)$ are integers with $e(j) > 0$, and $p(j)$ is a prime, it follows from (4) that $e(j) = 1$ or $p(j)$ for $1 \le j \le 2r$.

Let m be the least common multiple of $\{p_1, \ldots, p_r\}$. Hence,

$$m = \mathrm{l.c.m.}\{p(1), \ldots, p(2r)\}$$

also. Denote $n(i) = m/p(i) \in \mathbb{N}$, and $\nu = \mathrm{diag}(n(1), \ldots, n(2r))$. According to these definitions,

$$(p(j), n(j)) = 1 \quad \text{for} \quad 1 \le j \le 2r. \tag{5}$$

Moreover, $\nu\pi = \pi\nu = \iota m$. Thus, $\nu\alpha\beta = \iota m$, and $\beta\nu\alpha = \iota m$. This last system of $2r$ equations with integral coefficients can be used to invert the product (2). The result is

$$u_j^m = t_1^{n(1)l(1,j)} \cdots t_{j-1}^{n(j-1)l(j-1,j)} t_j^{n(j)f(j)}. \tag{6}$$

(Formally, $[m\lambda u_1, \ldots, m\lambda u_{2r}] = \xi\alpha\beta\nu\alpha = \xi\pi\nu\alpha = [\lambda t_1, \ldots, \lambda t_{2r}]\nu\alpha$; the formalism can be justified by applying the identity $\beta\nu\alpha = \iota m$ directly to (2).)

Denote $v_i = u_i^{n(i)}$. Since $p(i)n(i) = m$, it follows from (6) that

$$v_i^{p(i)} \in L. \tag{7}$$

We will show by induction that

$$F((u_1)) \cdots ((u_i)) = F((t_1)) \cdots ((t_i))(v_1, \ldots, v_i). \tag{8}$$

The basis of the induction is the trivial case $F = F$. Assume that (8) is satisfied when $i = j - 1$. It is clear from (2) and the definition of v_i that $F((t_1)) \cdots ((t_j))(v_1, \ldots, v_j) \subseteq F((u_1)) \cdots ((u_j))$. To reverse the inclusion, we must prove that $u_j \in L_1((t_j))(v_j)$, where $L_1 = F((t_1)) \cdots ((t_{j-1}))(v_1, \ldots, v_{j-1}) = F((u_1)) \cdots ((u_{j-1}))$. By (2), $u_j^{e(j)} \in L_1((t_j)) \subseteq L_1((t_j))(v_j)$. If $e(j) = 1$, then we are through. Otherwise, $e(j) = p(j)$ by (4). In this case $u_j^{n(j)} = v_j \in L_1((t_j))(v_j)$. It follows from (5) that $u_j \in L_1((t_j))(v_j)$. Lemma 19.8a and (6) yield (8).

Define $J = \{i: e(i) = p(i)\}$. It is clear from (3) and (4) that $J = \{i_1, \ldots, i_r\}$ with $i_1 < \cdots < i_r$, and $e(i_1), \ldots, e(i_r)$ is a permutation of the sequence $p_1, \ldots, p_r$. By using (8),

$$K = L(v_{i_1}, \ldots, v_{i_r}). \tag{9}$$

Indeed, $K = L(v_1, \ldots, v_{2r})$ is the case $i = 2r$ of (8). Moreover, if $j \notin J$, then $e(j) = 1$ and $v_j = u_j^{n(j)} = t_j^{n(j)} \in L$ by (2) (since $0 \leq k(i,j) < e(j) = 1$).

We can now complete the proof of the theorem. By (9), $K = \prod_{i \in J} L(v_i)$. It follows from (7) that $[L(v_i) : L] = p(i)$ or 1. (See Exercise 2, Section 15.5.) Since

$$\begin{aligned} p_1 \cdots p_r &= [K : L] \\ &= [(\textstyle\prod_{i \in J} L(v_i)) : L] \leq \textstyle\prod_{i \in J} [L(v_i) : L] \leq \textstyle\prod_{i \in J} p(i) \\ &= p_1 \cdots p_r, \end{aligned}$$

we must have $[L(v_i) : L] = p(i)$ for all $i \in J$ and $[(\prod_{i \in J} L(v_i)) : L] = \prod_{i \in J} [L(v_i) : L]$. Thus, $K = \bigotimes_{i \in J} L(v_i)$ by Proposition 9.2(c). Since $v_i^{p(i)} \in L$ and there is a primitive $p(i)$'th root of unity in L, the extension $L(v_i)/L$ is cyclic of degree $p(i)$ for all $i \in J$. By Lemma 15.3a, K/L is Galois with $\mathbf{G}(K/L) \cong \prod_{i \in J} \mathbf{G}(L(v_i)/L) \cong \prod_{i=1}^r \mathbb{Z}/p_i\mathbb{Z}$.

Exercises

1. Give the details of the proof of (6).
2. Let q be a function on $\{1, 2, \ldots, 2r\}$ such that for each $i \leq r$, one of the numbers $q(2i - 1)$, $q(2i)$ is p_i and the other number is 1. Prove that $K = F((\mathbf{x}_1^{q(1)}))((\mathbf{y}_1^{q(2)}))((\mathbf{x}_2^{q(3)})) \cdots ((\mathbf{y}_r^{q(2r)}))$ is a maximal subfield of the division algebra D in the theorem. Give a direct proof that K/L is Galois and $\mathbf{G}(K/L) \cong (\mathbb{Z}/p_1\mathbb{Z}) \times \cdots \times (\mathbb{Z}/p_r\mathbb{Z})$ in all cases.

Notes on Chapter 19

The material in Section 19.1 is routine housekeeping. In Exercise 4 of this section we gave a hint of the work of Witt and Amitsur on generic splitting fields. The results in Section 19.3 are standard items of commutative ring theory. Our presentation of them was influenced by lectures of Professor Roger Wiegand. The exposition in Sections 19.2 and 19.4 follows the similar discussion in Greenberg [38]. Section 19.5 amplifies a small part of the paper [12] by Auslander and Brumer. The material in Section 19.6 is a modified version of Nakayama's article [58]. Finally, the last three sections of the chapter closely parallel Jacobson's exposition [48] of Amitsur's 1972 paper [5].

CHAPTER 20
Varieties of Algebras

This final chapter returns to the general theory of algebras over a field. It provides a brief introduction to the theory of polynomial identities for algebras. Our main goal is to prove Amitsur's Theorem which establishes the existence of finite dimensional central simple division algebras that are not crossed products. The choice of topics in the chapter is motivated by this objective. Fortunately, many interesting results on polynomial identities are encountered in the proof of Amitsur's Theorem: the Amitsur–Levitzki Theorem, the existence of central polynomials, and the Kaplansky–Amitsur Theorem on primitive *PI*-algebras.

20.1. Polynomial Identities and Varieties

An F-algebra A is commutative if $xy - yx$ is zero for all $x, y \in A$. Another way to state this definition is: the polynomial $\mathbf{y}_0\mathbf{y}_1 - \mathbf{y}_1\mathbf{y}_0$ in the free algebra $F\{\mathbf{y}_0,\mathbf{y}_1\}$ (with $\mathbf{y}_0$ and $\mathbf{y}_1$ non-commutative) belongs to the kernel of every homomorphism of $F\{\mathbf{y}_0,\mathbf{y}_1\}$ to A. The description of polynomial identities is obtained by generalizing this idea.

Let $Y = \{\mathbf{y}_i : i \in J\}$ be a set of non-commuting variables. The *word algebra* over F in the symbols of Y is the convolution algebra $F\{Y\}$ on the free groupoid generated by Y. The construction of $F\{Y\}$ is given in Section 1.2, where it is also shown that $F\{Y\}$ is the free F-algebra on the set Y: any mapping from Y to an F-algebra A extends uniquely to an F-algebra homomorphism of $F\{Y\}$ to A. The elements of $F\{Y\}$ are non-commuting polynomials in the variables of Y with coefficients in F. As usual, we will denote these polynomials by capital Greek letters Φ, Ψ, Θ, and so on. In order to distinguish commuting and non-commuting variables, we will use the following convention throughout this chapter: bold faced **x**'s (with subscripts) denote commuting variables, while bold faced **y**'s are non-commuting variables. If $\psi: F\{Y\} \to A$ is an F-algebra homomorphism and $\Phi(\mathbf{y}_{i(1)}, \ldots, \mathbf{y}_{i(m)}) \in F\{Y\}$, then the image of Φ under ψ is denoted by $\Phi(y_1, \ldots, y_m)$ if $\psi(\mathbf{y}_{i(j)}) = y_j$. Intuitively, $\psi\Phi$ is obtained by substituting y_j for $\mathbf{y}_{i(j)}$ in the expression $\Phi(\mathbf{y}_{i(1)}, \ldots, \mathbf{y}_{i(m)})$.

Definition. A polynomial $\Phi \in F\{Y\}$, or the equation $\Phi = 0$, is a *polynomial identity* of the F-algebra A if $\psi\Phi = 0$ for all F-algebra homomorphisms $\psi: F\{Y\} \to A$.

In other words, $\Phi(y_1, \ldots, y_m) = 0$ for all $y_1, \ldots, y_m \in A$. In this case, we will say that A satisfies the polynomial identity Φ or $\Phi = 0$.

Proposition a. *For $W \subseteq F\{Y\}$, define $\mathfrak{V}(W)$ to be the class of all F-algebras that satisfy $\Phi = 0$ for all $\Phi \in W$.*

(i) *If $A \in \mathfrak{V}(W)$ and $\phi: B \to A$ is an injective homomorphism of F-algebras, then $B \in \mathfrak{V}(W)$.*
(ii) *If $A \in \mathfrak{V}(W)$ and $\phi: A \to B$ is a surjective homomorphism of F-algebras, then $B \in \mathfrak{V}(W)$.*
(iii) *If $\{A_j: j \in J\} \subseteq \mathfrak{V}(W)$, then $\prod_{j \in J} A_j \in \mathfrak{V}(W)$.*

Proof. Assume that $A \in \mathfrak{V}(W)$. If $\phi: B \to A$ is injective and $\psi: F\{Y\} \to B$ is any homomorphism, then $\phi\psi\Phi = 0$ for all $\Phi \in W$ because A satisfies $\Phi = 0$. Hence, $\psi\Phi = 0$ since ϕ is injective. Thus, B also satisfies $\Phi = 0$. Therefore, $B \in \mathfrak{V}(W)$. Let $\phi: A \to B$ be surjective. Since $F\{Y\}$ is free, every homomorphism from $F\{Y\}$ to B can be written in the form $\chi = \phi\psi$, where $\psi: F\{Y\} \to A$ is a homomorphism. It follows that $\chi\Phi = \phi\psi\Phi = \phi(0) = 0$ and B satisfies $\Phi = 0$ for all $\Phi \in W$, that is, $B \in \mathfrak{V}(W)$. Finally, let $\psi: F\{Y\} \to \prod_{j \in J} A_j$, where $\{A_j: j \in J\} \subseteq \mathfrak{V}(W)$. If $\pi_i: \prod_{j \in J} A_j \to A_i$ is the projection homomorphism and $\Phi \in W$, then $\pi_i\psi\Phi = 0$. Thus,

$$\psi\Phi \in \bigcap_{i \in J} \operatorname{Ker} \pi_i = 0,$$

which proves (iii). □

A class of algebras with the properties (i), (ii), and (iii) of Proposition a is called a *variety* of algebras. The class of all F-algebras is the largest variety of algebras; the class consisting of trivial algebras is the smallest variety. A variety $\mathfrak{V}$ is called non-trivial if it includes a non-trivial algebra.

The proposition shows that every set of polynomials determines a variety. Our main objective in this section is to make the connection between varieties and polynomial identities more precise.

Proposition b. *Let Y be a set of non-commuting variables. For a class $\mathfrak{U}$ of F-algebras, define $I_Y(\mathfrak{U})$ to be the set of all $\Phi \in F\{Y\}$ that are identities of all $A \in \mathfrak{U}$.*

(i) *$I_Y(\mathfrak{U})$ is an ideal of $F\{Y\}$.*
(ii) *If ϕ is an endomorphism of $F\{Y\}$, then $\phi(I_Y(\mathfrak{U})) \subseteq I_Y(\mathfrak{U})$.*

The proof of this proposition consists of routine applications of definitions. It is left as Exercise 1.

An ideal I of an F-algebra B is called a *T-ideal* if $\phi(I) \subseteq I$ for all endomorphisms ϕ of B. The correspondence between varieties and T-ideals is bijective, as we will see. Some useful preliminary facts are needed to prove this assertion.

Proposition c. *Let $\mathfrak{B}$ be a non-trivial variety of F-algebras. For each set Y of non-commuting variables, define $C_Y(\mathfrak{B}) = F\{Y\}/I_Y(\mathfrak{B})$. Let $\pi: F\{Y\} \to C_Y(\mathfrak{B})$ be the projection homomorphism.*

(i) *$C_Y(\mathfrak{B})$ is a non-trivial F-algebra in $\mathfrak{B}$.*
(ii) *$\pi|Y$ is injective and $\pi(Y)$ generates $C_Y(\mathfrak{B})$ as an F-algebra.*
(iii) *If $A \in \mathfrak{B}$ and $\phi: \pi(Y) \to A$ is a mapping, then ϕ extends uniquely to an F-algebra homomorphism of $C_Y(\mathfrak{B})$ to A.*

Proof. If $A \in \mathfrak{B}$ is non-trivial, then $\mathbf{y} = 0$ and $\mathbf{y}_i = \mathbf{y}_j$ are not identities of A if $i \neq j$. It follows that $\pi|Y$ is injective. Moreover, $C_Y(\mathfrak{B})$ is not the zero algebra if $Y \neq \varnothing$. If $Y = \varnothing$, then $I_Y(\mathfrak{B}) = 0$ and $C_Y(\mathfrak{B}) = F\{Y\} = F$. Since Y generates $F\{Y\}$ and π is surjective, it follows that $\pi(Y)$ generates $C_Y(\mathfrak{B})$. By the definition of $I_Y(\mathfrak{B})$, there is a set $\{\phi_j: j \in J\}$ of homomorphisms of $F\{Y\}$ to various algebras $A_j \in \mathfrak{B}$ such that $I_Y(\mathfrak{B}) = \bigcap_{j \in J} \operatorname{Ker} \phi_j$. The homomorphism $\phi: F\{Y\} \to \prod_{j \in J} A_j$ defined by $\phi(x) = (\dots \phi_j(x) \dots)$ has the kernel $I_Y(\mathfrak{B})$, so that ϕ induces an injective homomorphism of $C_Y(\mathfrak{B})$ to $\prod_{j \in J} A_j$. Thus, $C_Y(\mathfrak{B}) \in \mathfrak{B}$ since $\mathfrak{B}$ is a variety. Let $A \in \mathfrak{B}$ and suppose that $\phi: \pi Y \to A$ is a mapping. The mapping $\psi = \phi\pi: Y \to A$ extends to a homomorphism of $F\{Y\}$ to A, and $I_Y(\mathfrak{B})$ is contained in the kernel of this homomorphism since $A \in \mathfrak{B}$. Thus, ψ factors through π giving a homomorphism $\phi: C_Y(\mathfrak{B}) \to A$ such that $\phi\pi = \psi$. In particular, ϕ is an extension of the given mapping of πY to A. □

The algebra $C_Y(\mathfrak{B})$ is called the *free $\mathfrak{B}$-algebra* on the generating set Y. We will often identify Y with its image $\pi(Y)$ in $C_Y(\mathfrak{B})$.

Corollary a. *Let $\mathfrak{B}$ be a variety of F-algebras. If A is an F-algebra such that every finitely generated subalgebra of A is a member of $\mathfrak{B}$, then $A \in \mathfrak{B}$.*

Proof. Let Y be a set of variables such that there is a surjective homomorphism $\psi: F\{Y\} \to A$. If $\Phi = \Phi(\mathbf{y}_{i(1)}, \dots, \mathbf{y}_{i(m)}) \in I_Y(\mathfrak{B})$ and $y_1, \dots,$

$y_m \in A$, then the subalgebra of A that is generated by $\{y_1, \ldots, y_m\}$ satisfies Φ: $\Phi(y_1, \ldots, y_m) = 0$. Hence A satisfies Φ. This argument shows that $I_Y(\mathfrak{V}) \subseteq \operatorname{Ker} \psi$. Thus, ψ factors through the projection $\pi\colon F\{Y\} \to C_Y(\mathfrak{V})$. In particular, there is a surjective homomorphism of $C_Y(\mathfrak{V})$ to A. Since $\mathfrak{V}$ is a variety and $C_Y(\mathfrak{V}) \in \mathfrak{V}$ by the proposition, it follows that $A \in \mathfrak{V}$. □

Theorem. *Let Y be a set of variables. The correspondences $W \mapsto \mathfrak{V}(W)$ and $\mathfrak{U} \mapsto I_Y(\mathfrak{U})$ are inclusion reversing mappings from subsets of $F\{Y\}$ to varieties and from classes of F-algebras to T-ideals of $F\{Y\}$.*

(i) *$I_Y(\mathfrak{V}(W)) \supseteq W$ and equality holds if and only if W is a T-ideal.*
(ii) *$\mathfrak{V}(I_Y(\mathfrak{U})) \supseteq \mathfrak{U}$ and if Y is infinite, then equality holds if and only if $\mathfrak{U}$ is a variety.*

PROOF. By virtue of Propositions a and b, the only parts of the theorem that need proofs are the "if" statements in (i) and (ii). Assume that W is a T-ideal in $F\{Y\}$ and $\Phi \in I_Y(\mathfrak{V}(W))$. We want to show that $\Phi \in W$, that is, $\pi(\Phi) = 0$, where $\pi\colon F\{Y\} \to F\{Y\}/W$ is the projection. This conclusion can be obtained by proving that $F\{Y\}/W \in \mathfrak{V}(W)$ because $\Phi \in I_Y(\mathfrak{V}(W))$. Let $\Psi(\mathbf{y}_{i(1)}, \ldots, \mathbf{y}_{i(m)}) \in W$, and suppose that $\psi\colon F\{Y\} \to F\{Y\}/W$. For each $\mathbf{y}_i \in Y$, choose $\Phi_i \in F\{Y\}$ so that $\psi(\mathbf{y}_i) = \pi(\Phi_i)$. Extend the mapping $\mathbf{y}_i \mapsto \Phi_i$ to an endomorphism ϕ of $F\{Y\}$. By construction, $\psi = \pi\phi$. Since W is a T-ideal, $\Theta = \Psi(\Phi_{i(1)}, \ldots, \Phi_{i(m)}) \in W$. Therefore,

$$\psi(\Psi(\mathbf{y}_{i(1)}, \ldots, \mathbf{y}_{i(m)})) = \pi\Psi(\phi(\mathbf{y}_{i(1)}), \ldots, \phi(\mathbf{y}_{i(m)})) = \pi\Theta = 0.$$

This shows that $I_Y(\mathfrak{V}(W)) = W$ if W is a T-ideal. Assume that $\mathfrak{U}$ is a variety of F-algebras. We will use Corollary a and Proposition c to show that $\mathfrak{V}(I_Y(\mathfrak{U})) = \mathfrak{U}$, provided Y is infinite. It can be assumed that $\mathfrak{U}$ is non-trivial; otherwise $I_Y(\mathfrak{U}) = F\{Y\}$ and $\mathfrak{V}(I_Y(\mathfrak{U}))$ is trivial. Let $A \in \mathfrak{V}(I_Y(\mathfrak{U}))$. If B is a finitely generated subalgebra of A, then there is a surjective homomorphism ψ of $F\{Y\}$ to B because Y is infinite. The hypothesis $A \in \mathfrak{V}(I_Y(\mathfrak{U}))$ guarantees that $\psi = \phi\pi$, where $\pi\colon F\{Y\} \to C_Y(\mathfrak{U})$ is the projection mapping and $\phi\colon C_Y(\mathfrak{U}) \to B$ is a surjective homomorphism. Since $\mathfrak{U}$ is a variety and $C_Y(\mathfrak{U}) \in \mathfrak{U}$ by Proposition c, it follows that $B \in \mathfrak{U}$. Therefore, $A \in \mathfrak{U}$ by Corollary a. □

Corollary b. *Let Y be an infinite set of variables, $W \subseteq F\{Y\}$, and suppose that $\mathfrak{U}$ is a class of F-algebras.*

(i) *$I_Y(\mathfrak{V}(W))$ is the smallest T-ideal in $F\{Y\}$ that contains W.*
(ii) *$\mathfrak{V}(I_Y(\mathfrak{U}))$ is the smallest variety of F-algebras that contains $\mathfrak{U}$.*

The proof is left as an exercise. We will call $\mathfrak{V}(I_Y(\mathfrak{U}))$ the variety generated by the class $\mathfrak{U}$. By Corollary b, this variety doesn't depend on Y, except that Y should be infinite. To simplify notation, write $\mathfrak{V}(\mathfrak{U})$ instead of $\mathfrak{V}(I_Y(\mathfrak{U}))$. An alternative characterization of $\mathfrak{V}(\mathfrak{U})$ is given in Exercise 3. If A is a member of the variety $\mathfrak{U}$, then A is called a *generic algebra* for $\mathfrak{U}$ if $\mathfrak{V}(A) = \mathfrak{U}$.

Corollary c. *If $\mathfrak{U}$ is a variety of F-algebras and Y is an infinite set of variables, then $C_Y(\mathfrak{U})$ is generic for $\mathfrak{U}$.*

Proof. If $A \in \mathfrak{U}$, then every finitely generated subalgebra of A is a homomorphic image of $C_Y(\mathfrak{U})$. Thus, $A \in \mathfrak{V}(C_Y(\mathfrak{U}))$ by Corollary a. The inclusion $\mathfrak{V}(C_Y(\mathfrak{U})) \subseteq \mathfrak{U}$ is clear from the theorem. □

Exercises

1. Prove Proposition b.
2. Prove Corollary b.
3. Let $\mathfrak{U}$ be a non-empty class of F-algebras. Prove that $A \in \mathfrak{V}(\mathfrak{U})$ if and only if there is a set $\{A_j: j \in J\} \subseteq \mathfrak{U}$, a subalgebra B of $\prod_{j \in J} A_j$, and a surjective homomorphism $\phi: B \to A$. Hint. Show that the class of all such A forms a variety that contains $\mathfrak{U}$.
4. Let F be an infinite field, $Y = \{\mathbf{y}_0\}$, and suppose $\mathfrak{U}$ is any non-trivial variety of F-algebras. Prove that $\mathfrak{V}(I_Y(\mathfrak{U}))$ is the variety of all F-algebras. Hint. Note that $F \in \mathfrak{U}$, and therefore $I_Y(\mathfrak{U}) = 0$.
5. Let Y be an infinite set of variables and suppose that $\mathfrak{U}$ is a non-trivial variety of F-algebras.
   (a) Let $A \in \mathfrak{U}$ contain a subset X such that $|X| = |Y|$, X generates A, and every mapping of X to $B \in \mathfrak{U}$ extends to an F-algebra homomorphism. Prove that $A \cong C_Y(\mathfrak{U})$.
   (b) Denote the natural projection of $F\{Y\}$ to $C_Y(\mathfrak{U})$ by π. Prove that $\Phi \in F\{Y\}$ is an identity for $B \in \mathfrak{U}$ if and only if $\psi\pi\Phi = 0$ for all homomorphisms $\psi: C_Y(\mathfrak{U}) \to B$.
6. Assume that Y is an infinite set of non-commuting variables, and suppose that F is an infinite field. Denote $\Gamma = \mathbf{y}_0\mathbf{y}_1 - \mathbf{y}_1\mathbf{y}_0$ and $\mathfrak{C} = \mathfrak{V}(\Gamma)$, the variety of all commutative F-algebras. Prove the following statements.
   (a) $C_Y(\mathfrak{C}) \cong F[X]$, where X is a set of commuting variables such that $|X| = |Y|$. Hint. Use Exercise 5(a).
   (b) $I_Y(F) = I_Y(\mathfrak{C})$. Hint. Use Exercise 5(b).
   (c) F is generic for $\mathfrak{C}$.
   (d) If $\mathfrak{U}$ is a non-trivial variety of F-algebras, then $\mathfrak{C} \subseteq \mathfrak{U}$.
   (e) Every commutative F-algebra is a homomorphic image of a subalgebra of a product of copies of F. Hint. Use Exercise 3.

20.2. Special Identities

The purpose of this section is to show that if an F-algebra A satisfies any non-trivial polynomial identity $\Phi = 0$ (where Φ is not the zero polynomial), then A satisfies a non-trivial identity that has a rather special form.

It is convenient to standardize our list of variables. The results of Section 20.1 show that all infinite sets of variables yield equivalent results on polynomial identities. Uncountable sets are only needed to construct large free algebras. Throughout the rest of this chapter, let $Y = \{\mathbf{y}_i: i < \omega\}$ with $\mathbf{y}_i \neq \mathbf{y}_j$ if $i \neq j$. Our notation will usually omit a reference to this set Y. In particular, we will write $I(\mathfrak{B})$ for $I_Y(\mathfrak{B})$ and $C(\mathfrak{B})$ for $C_Y(\mathfrak{B})$. However, $F\{Y\}$ will still denote the word algebra on the standardized variable set Y.

Every non-zero $\Phi \in F\{Y\}$ can be written uniquely as a sum of monomials $\Phi = M_1 + \cdots + M_r$, where $M_j = \mathbf{y}_{i(1)}, \ldots, \mathbf{y}_{i(k_j)}a_j$, $i(l) < \omega$, and $a_j \in F^\circ$. The non-negative integer k_j is the degree of M_j. The degree of Φ is $\deg \Phi = \max\{k_j: 1 \leq j \leq r\}$. For each index $i < \omega$, the degree of $\mathbf{y}_i$ in the monomial $M_j = \mathbf{y}_{i(1)} \cdots \mathbf{y}_{i(k_j)}a_j$ is the number of times that $\mathbf{y}_i$ appears in the product $\mathbf{y}_{i(1)} \cdots \mathbf{y}_{i(k_j)}$. We denote this number by $\deg_i M_j$. The degree of $\mathbf{y}_i$ in $\Phi = M_1 + \cdots + M_r$ is $\deg_i \Phi = \max\{\deg_i M_j: 1 \leq j \leq r\}$.

A non-zero polynomial Φ is *uniform* if every monomial summand of Φ is a product of the same variables. That is, if $\Phi = M_1 + \cdots + M_r$, then $\deg_i \Phi > 0$ implies $\deg_i M_j > 0$ for $1 \leq j \leq r$. If Φ is not uniform, then the monomial summands of Φ that involve the same variables can be grouped together. This process gives a representation $\Phi = \Phi_1 + \cdots + \Phi_s$ where each Φ_j is uniform and if $j \neq k$ then there is a variable $\mathbf{y}_i$ such that $\deg_i \Phi_j > 0$ and $\deg_i \Phi_k = 0$ or vice verse. This representation of Φ is plainly unique. The polynomials Φ_j are called the *uniform summands* of Φ.

Lemma. *If I is a T-ideal of $F\{Y\}$ and Φ is a non-zero polynomial in I, then all uniform summands of Φ belong to I.*

Proof. We can assume that the variables in Φ are $\mathbf{y}_0, \mathbf{y}_1, \ldots, \mathbf{y}_{m-1}$ and $\Phi = \Phi_1 + \cdots + \Phi_s$, $\Phi_1, \ldots, \Phi_s$ the uniform summands of Φ. Induce on s. If $s = 1$, then Φ is already uniform. We can assume that $s > 1$ and the Φ_i are ordered so that $\mathbf{y}_0$ occurs in $\Phi_1, \ldots, \Phi_r$ but not in $\Phi_{r+1}, \ldots, \Phi_s$ where $1 \leq r < s$. Write $\Phi = \Psi_1 + \Psi_2$, with

$$\Psi_1 = \Phi_1 + \cdots + \Phi_r, \quad \Psi_2 = \Phi_{r+1} + \cdots + \Phi_s.$$

Since I is a T-ideal,

$$\begin{aligned}\Psi_2(\mathbf{y}_0, \mathbf{y}_1, \ldots, \mathbf{y}_{m-1}) &= \Psi_1(0, \mathbf{y}_1, \ldots, \mathbf{y}_{m-1}) + \Psi_2(0, \mathbf{y}_1, \ldots, \mathbf{y}_{m-1}) \\ &= \Phi(0, \mathbf{y}_1, \ldots, \mathbf{y}_{m-1}) \in I,\end{aligned}$$

and $\Psi_1 = \Phi - \Psi_2 \in I$. The induction hypothesis applies to Ψ_1 and Ψ_2. Thus, $\Phi_1, \ldots, \Phi_s \in I$. □

A non-zero polynomial $\Phi \in F\{Y\}$ is *multilinear* if $\deg_i \Phi \leq 1$ for all $i < \omega$. If Φ is both uniform and multilinear, then Φ is a linear function of each of its variables, that is,

$$\begin{aligned}\Phi(x_1, \ldots, x_i a + y_i b, \ldots, x_m) = \Phi(x_1, \ldots, x_i, \ldots, x_m)a \\ + \Phi(x_1, \ldots, y_i, \ldots, x_m)b.\end{aligned}$$

In the next section we will derive a useful consequence of this linearity.

Proposition. *Let I be a T-ideal of $F\{Y\}$. If $\Phi \in I$ is a non-zero polynomial of degree n, then I contains a non-zero, multilinear polynomial of degree at most n.*

PROOF. It can be assumed that the variables in Φ are $\mathbf{y}_1, \ldots, \mathbf{y}_m$. Denote $k = k(\Phi) = \max\{\deg_i \Phi \colon 1 \le i \le m\}$. If $k = 0$, then Φ is a non-zero constant, $I = F\{Y\}$, and the result is obvious. If $k = 1$, then Φ is multilinear by definition. Assume that $k > 1$. Let $l = l(\Phi)$ be the number of i such that $\deg_i \Phi = k$. We proceed by double induction on (k,l), ordered lexicographically. Since I is a T-ideal, the variables in Φ can be permuted so that $\deg_1 \Phi = k$. Define

$$\Psi(\mathbf{y}_0, \mathbf{y}_1, \mathbf{y}_2, \ldots, \mathbf{y}_m) = \Phi(\mathbf{y}_0 + \mathbf{y}_1, \mathbf{y}_2, \ldots) - \Phi(\mathbf{y}_0, \mathbf{y}_2, \ldots) - \Phi(\mathbf{y}_1, \mathbf{y}_2, \ldots).$$

Then $\Psi \in I$ since I is a T-ideal and

$$\deg \Psi \le \deg \Phi = n, \quad \deg_i \Psi \le \deg_i \Phi \quad \text{for} \quad 0 \le i \le m. \tag{1}$$

Since $\Psi(0, \mathbf{y}_1, \mathbf{y}_2, \ldots) = -\Phi(0, \mathbf{y}_2, \ldots) = \Psi(\mathbf{y}_0, 0, \mathbf{y}_2, \ldots)$, every monomial summand M of Ψ satisfies $\deg_0 M = \deg_1 M = 0$ or $\deg_0 M > 0$ and $\deg_1 M > 0$. Since $\deg_0 M + \deg_1 M \le k$ by the definition of Ψ, it follows that

$$\deg_0 \Psi < k \quad \text{and} \quad \deg_1 \Psi < k. \tag{2}$$

If $\Psi \ne 0$, then the combination of (1) and (2) gives the conclusion that either $k(\Psi) < k(\Phi)$ or $k(\Psi) = k(\Phi)$ and $l(\Psi) < l(\Phi)$. In this case, the lemma follows from our induction hypothesis. It remains to prove that $\Psi \ne 0$. Let N be a monomial summand of Φ such that $\deg_1 N = k$. Write

$$N = M_0 \mathbf{y}_1 M_1 \mathbf{y}_1 \cdots M_{k-1} \mathbf{y}_1 M_k a,$$

where the M_i are (possibly empty) monomials in $\mathbf{y}_2, \ldots, \mathbf{y}_m$ and $a \in F^\circ$. It is clear that $N(\mathbf{y}_0 + \mathbf{y}_1, \mathbf{y}_2, \ldots) = \sum N_S$, where the sum ranges over all $S \subseteq \{1, \ldots, k\}$, and N_S is obtained from N by replacing $\mathbf{y}_1$ with $\mathbf{y}_0$ exactly in those positions that precede an M_i such that $i \in S$. Thus, $N_\varnothing = N(\mathbf{y}_1, \mathbf{y}_2, \ldots)$, $N_{\{1,\ldots,k\}} = N(\mathbf{y}_0, \mathbf{y}_2, \ldots)$, and

$$N(\mathbf{y}_0 + \mathbf{y}_1, \mathbf{y}_2, \ldots) - N(\mathbf{y}_0, \mathbf{y}_2, \ldots) - N(\mathbf{y}_1, \mathbf{y}_2, \ldots) = \sum_{\varnothing \ne S \ne \{1,\ldots,k\}} N_S.$$

The latter sum is not empty because $k \ge 2$. Moreover, the various summand N_S are distinct; and they are all different from the monomials that arise by applying this process to other monomial summands of Φ, because $N_S(\mathbf{y}_1, \mathbf{y}_1, \mathbf{y}_2, \ldots) = N$. Therefore, $\Psi \ne 0$. □

Corollary. *If the F-algebra A satisfies a non-trivial polynomial identity $\Phi = 0$ with $\deg \Phi = n$, then A satisfies a non-trivial identity $\Lambda = 0$ in which Λ is uniform, multilinear, and $\deg \Lambda \le n$.*

This result follows from the lemma and proposition, together with the observation that $I(A)$ is a T-ideal by Proposition 20.1b.

The polynomials that are uniform and multilinear have the following form:

$$\Lambda(\mathbf{y}_0, \mathbf{y}_1, \ldots, \mathbf{y}_{m-1}) = \sum \mathbf{y}_{\rho(0)}\mathbf{y}_{\rho(1)} \cdots \mathbf{y}_{\rho(m-1)}a_\rho, \tag{3}$$

where the sum runs over all permutations ρ of $\{0, 1, \ldots, m-1\}$ and $a_\rho \in F$. Conversely, (3) defines a uniform, multilinear polynomial.

Exercises

1. Prove that every proper T-ideal of $F\{Y\}$ is contained in the T-ideal that is generated by $\mathbf{y}_0\mathbf{y}_1 - \mathbf{y}_1\mathbf{y}_0$. Hint. Use Exercise 6, Section 20.1.

2. (a) Prove that if F does not satisfy $\Phi = 0$, then the T-ideal generated by Φ is $F\{Y\}$. In particular, this is the case if Φ has a non-zero constant term.

   (b) Prove that the uniform, multilinear polynomial Λ defined by (3) generates a proper T-ideal if and only if $\sum a_\sigma = 0$.

3. $\Phi \in F\{Y\}$ is $\mathbf{y}_i$-homogeneous if the degrees of $\mathbf{y}_i$ in the monomial summands of Φ are equal.

   (a) Prove that if $\Phi \in F\{Y\}$ is not zero then for each $i < \omega$, $\Phi = \Psi_1 + \cdots + \Psi_s$, where all Ψ_j are $\mathbf{y}_i$-homogeneous and $\deg_i \Psi_1 < \cdots < \deg_i \Psi_s$.

   (b) Prove that if F is infinite, I is a T-ideal in $F\{Y\}$, and $\Phi = \Psi_1 + \cdots + \Psi_s$ as in (a), then $\Psi_1, \ldots, \Psi_s \in I$. Hint. Choose $a_1, \ldots, a_s$ distinct in F, and invert the system of equations $\sum_{j=1}^{s} \Psi_j(\mathbf{y}_0, \ldots, \mathbf{y}_i, \ldots)a_k^{n(j)} = \Phi(\mathbf{y}_0, \ldots, a_k\mathbf{y}_i, \ldots)$, where $n(j) = \deg_i \Psi_j$.

   (c) Prove that if F is infinite, then every T-ideal is generated (as a T-ideal) by polynomials that are homogeneous in all variables.

4. Prove that if $\operatorname{char} F = 0$, then every T-ideal is generated by uniform, multilinear polynomials. Hint. Use the result of Exercise 3(c) and the process that was introduced in the proof of the proposition.

20.3. Identities for Central Simple Algebras

For $n \in \mathbb{N}$, denote by $\mathfrak{S}_n(F)$ the class of all central simple F-algebras A such that $\operatorname{Deg} A = n$. Our interest in this section focuses on the variety $\mathfrak{V}(\mathfrak{S}_n(F))$ that is generated by $\mathfrak{S}_n(F)$. It will be convenient to denote this variety by $\mathfrak{V}_n$, or $\mathfrak{V}_n(F)$ when the field F has to be identified. Our main result is that every $A \in \mathfrak{S}_n(F)$ is generic for $\mathfrak{V}_n$. In particular, $M_n(F)$ is generic. This fact provides a powerful tool for investigating the polynomial identities of $\mathfrak{V}_n$.

The proof of this result is based on the survival of polynomial identities under scalar extension. The simplest case of this phenomenon occurs for multilinear identities.

Lemma a. *Let X be a subset of the F-algebra A such that $XF = A$. If $\Lambda(\mathbf{y}_0, \mathbf{y}_1, \ldots, \mathbf{y}_{m-1})$ is a uniform, multilinear polynomial such that $\Lambda(x_0,x_1, \ldots, x_{m-1}) = 0$ for all $x_0,x_1, \ldots, x_{m-1} \in X$, then A satisfies $\Lambda = 0$.*

Proof. If $(u_0,u_1, \ldots, u_{m-1}) \in A^m$, then since $XF = A$ there exist $x_1, \ldots, x_n \in X$ and $a_{ij} \in F$ such that $u_j = \sum_{i=1}^n x_i a_{ij}$ for all $j < m$. The assumption that Λ is uniform and multilinear yields $\Lambda(u_0,u_1, \ldots, u_{m-1}) = \sum \Lambda(x_{i(0)}, \ldots, x_{i(m-1)})a_{i(0)0} \cdots a_{i(m-1)m-1} = 0$. Thus, A satisfies $\Lambda = 0$. □

There is a similar result for arbitrary polynomials, but it imposes stronger algebraic hypotheses.

Lemma b. *Let K/F be a field extension, where F is infinite. Assume that B is a K-algebra and A is an F-subalgebra of B such that $AK = B$. If $\Phi \in F\{Y\}$, then A satisfies $\Phi = 0$ if and only if B satisfies $\Phi = 0$.*

Proof. Since A is an F-subalgebra of B, the assumption that B satisfies Φ implies that A satisfies Φ by Proposition 20.1a. Assume that A satisfies $\Phi(\mathbf{y}_0,\mathbf{y}_1, \ldots, \mathbf{y}_{m-1})$. For an arbitrary natural number n, reindex a subset of Y by $\{\mathbf{y}_{ij}: i < m, 1 \le j \le n\}$, and let $\{\mathbf{x}_1, \ldots, \mathbf{x}_n\}$ be a set of variables that commute with each other and with all of the $\mathbf{y}_{ij}$. Substitute $\sum_{j=1}^n \mathbf{y}_{ij}\mathbf{x}_j$ for $\mathbf{y}_i$ in Φ, and expand the resulting polynomial as a sum of monomials in the $\mathbf{x}$'s with coefficients in $F\{Y\}$, say

$$\Phi = \sum \mathbf{x}_1^{k(1)} \cdots \mathbf{x}_n^{k(n)} \Phi_k(\ldots \mathbf{y}_{ij} \ldots),$$

summed over all sequences $k = (k(1), \ldots, k(n))$ of non-negative integers such that $\sum k(i) \le \deg \Phi$. We will show that $\Phi_k = 0$ is an identity for A. Let x_{ij} be arbitrary elements of A. For all choices $(a_1, \ldots, a_n) \in F^n$, $\sum x_{ij}a_j \in A$. Therefore,

$$0 = \Phi(\sum x_{0,j}a_j, \ldots, \sum x_{m-1,j}a_j) = \sum a_1^{k(1)} \cdots a_n^{k(n)} \Phi_k(\ldots x_{ij} \ldots).$$

Since F is infinite, it follows that $\Phi_k(\ldots x_{ij} \ldots) = 0$. (See Exercise 1.) This formal result yields the conclusion that B satisfies Φ. In fact, if $z_0, \ldots, z_{m-1} \in B$, then for a suitable $n \in \mathbb{N}$, $x_{ij} \in A$ $(i < m, 1 \le j \le n)$, and $b_1, \ldots, b_n \in K$, we have $z_i = \sum x_{ij}b_j$. Consequently, $\Phi(z_0, \ldots, z_{m-1}) = \sum b_1^{k(1)} \cdots b_n^{k(n)} \Phi_k(\ldots x_{ij} \ldots) = 0$. □

Proposition. *If $A \in \mathfrak{S}_n(F)$, then $\mathfrak{B}(A) = \mathfrak{B}_n$.*

Proof. If F is finite, then all algebras in $\mathfrak{S}_n(F)$ are isomorphic to $M_n(F)$. In this case, the proposition is clear from the definition of $\mathfrak{B}_n$. Assume that F is infinite. Let K be the algebraic closure of F. By Lemma b, $I(A) = I(A^K) \cap F\{Y\} = I(M_n(K)) \cap F\{Y\} = I(M_n(F)^K) \cap F\{Y\} = I(M_n(F))$. Thus, $I(A) = I(\mathfrak{S}_n(F))$ and $\mathfrak{B}(A) = \mathfrak{B}(I(A)) = \mathfrak{B}(I(\mathfrak{S}_n(F)) = \mathfrak{B}_n$. □

Corollary a. *If* $\Phi \in I(\mathfrak{V}_n)$, *then* $\deg \Phi \geq 2n$.

PROOF. Since $I(\mathfrak{V}_n) = I(M_n(F))$, it suffices to prove that $M_n(F)$ does not satisfy Φ if $\deg \Phi < 2n$. If this is not the case, then by Corollary 20.2, $M_n(F)$ satisfies a uniform, multilinear identity Λ of degree $m < 2n$, $\Lambda = \sum \mathbf{y}_{\rho(0)} \mathbf{y}_{\rho(1)} \cdots \mathbf{y}_{\rho(m-1)} a_\rho$, summed over the permutations ρ of $\{0, 1, \ldots, m-1\}$. We can assume that $a_1 \neq 0$. Define a homomorphism $\psi: F\{Y\} \to M_n(F)$ such that $\psi(\mathbf{y}_0) = \varepsilon_{11}$, $\psi(\mathbf{y}_1) = \varepsilon_{12}$, $\psi(\mathbf{y}_2) = \varepsilon_{22}, \ldots, \psi(\mathbf{y}_{m-1}) = \varepsilon_{rr}$ if $m = 2r - 1$, and $\psi(\mathbf{y}_{m-1}) = \varepsilon_{rr+1}$ if $m = 2r$. Since $m < 2n$, this prescription can be filled. Plainly, $\psi(\mathbf{y}_0 \mathbf{y}_1 \cdots \mathbf{y}_{m-1}) = \varepsilon_{1r}$ or ε_{1r+1}. On the other hand, if $i > j$, then $\psi(\mathbf{y}_i \mathbf{y}_j) = \varepsilon_{rs}\varepsilon_{tu}$ with $t < s$. Hence, ψ maps all monomial summands of Λ to zero, except $\mathbf{y}_0 \mathbf{y}_1 \cdots \mathbf{y}_{m-1} a_1$. This gives the contradiction $0 = \psi\Lambda = \varepsilon_{1r} a_1$ (or $\varepsilon_{1r+1} a_1) \neq 0$. □

Corollary b. $\mathfrak{V}_1 \subset \mathfrak{V}_2 \subset \mathfrak{V}_3 \subset \cdots$.

PROOF. If $m \leq n$, then $I(M_m(F)) \supseteq I(M_n(F))$ because $M_m(F)$ is a homomorphic image of a subalgebra of $M_n(F)$. (See Exercise 2.) Hence, $\mathfrak{V}_1 \subseteq \mathfrak{V}_2 \subseteq \mathfrak{V}_3 \subseteq \cdots$ by the proposition. The fact that these inclusions are strict will follow from the Amitsur-Levitzki Theorem and Corollary a. □

EXERCISES

1. Prove that if F is an infinite field, and $\Theta \in F[\mathbf{x}_1, \ldots, \mathbf{x}_n]$ is such that $\Theta(a_1, \ldots, a_n) = 0$ for all $(a_1, \ldots, a_n) \in F^n$, then Θ is the zero polynomial.
2. Let $m < n$. Define B to be the set of matrices in $M_n(F)$ that have the form $\alpha = \begin{bmatrix} \beta & 0 \\ 0 & \gamma \end{bmatrix}$ with $\beta \in M_m(F)$ and $\gamma \in M_{n-m}(F)$. Prove that B is a subalgebra of $M_n(F)$, and the mapping $\alpha \mapsto \beta$ is a surjective homomorphism of B to $M_m(F)$.
3. Show that the conclusion of Lemma b can be false if F is finite. Hint. $\mathbb{F}_p$ satisfies the identity $\mathbf{y}_0^p - \mathbf{y}_0 = 0$.

20.4. Standard Identities

The *standard polynomial* of degree n is

$$\Gamma_n(\mathbf{y}_1, \ldots, \mathbf{y}_n) = \sum (\operatorname{sgn} \rho) \mathbf{y}_{\rho(1)} \cdots \mathbf{y}_{\rho(n)},$$

where the sum is over all permutations ρ of $\{1, \ldots, n\}$. The equation $\Gamma_n = 0$ is called the *standard identity* of degree n.

The standard identities can be viewed as a sequence of progressively weaker versions of the commutative law $\mathbf{y}_1\mathbf{y}_2 - \mathbf{y}_2\mathbf{y}_1 = 0$. Indeed, $\Gamma_2(\mathbf{y}_1, \mathbf{y}_2) =$

$\mathbf{y}_1\mathbf{y}_2 - \mathbf{y}_2\mathbf{y}_1$. Our principal aim in this section is to prove that $M_n(F)$ satisfies the identity $\Gamma_{2n} = 0$. This result leads to a useful characterization of the degree of a central simple algebra. For future reference, we collect some obvious properties of the standard identities.

Lemma a. (i) *The standard polynomials are uniform and multilinear.*

(ii) *If ρ is a permutation of $\{1, \ldots, n\}$, then $\Gamma_n(\mathbf{y}_{\rho(1)}, \ldots, \mathbf{y}_{\rho(n)}) = (\operatorname{sgn} \rho)\Gamma_n(\mathbf{y}_1, \ldots, \mathbf{y}_n)$.*

(iii) *If $\phi: \{1, \ldots, n\} \to \omega$ is a mapping that is not injective, then $\Gamma_n(\mathbf{y}_{\phi(1)}, \ldots, \mathbf{y}_{\phi(n)})$ is the zero polynomial.*

(iv) *If $n \geq 2$, then $\Gamma_n(\mathbf{y}_1, \ldots, \mathbf{y}_n) = \sum_{j=1}^{n} (-1)^{j+1}\mathbf{y}_j\Gamma_{n-1}(\mathbf{y}_1, \ldots, \hat{\mathbf{y}}_j, \ldots, \mathbf{y}_n) = \sum_{j=1}^{n} (-1)^{n+j}\Gamma_{n-1}(\mathbf{y}_1, \ldots, \hat{\mathbf{y}}_j, \ldots, \mathbf{y}_n)\mathbf{y}_j$.*

It follows from Corollary 20.3a that $M_n(F)$ does not satisfy $\Gamma_m = 0$ if $m < 2n$. On the other hand, it is easy to show that any k dimensional F-algebra satisfies $\Gamma_m = 0$ for all $m > k$. (See Exercise 2.) In particular, $M_n(F)$ satisfies $\Gamma_{n^2+1} = 0$. The gap between $2n - 1$ and $n^2 + 1$ is closed by an important result that was first proved by Amitsur and Levitzki. They showed that $M_n(F)$ satisfies $\Gamma_{2n} = 0$. We will give a proof of this result that is based on a standard property of matrices.

Lemma b. *Let F be a field of characteristic zero, and suppose that C is a commutative F-algebra. If $\alpha \in M_n(C)$ satisfies $\operatorname{tr}(\alpha^k) = 0$ for $1 \leq k \leq n$, then $\alpha^n = 0$.*

PROOF. We need the following version of Newton's identities:

$$\prod_{i=1}^{n} (\mathbf{x} - \mathbf{x}_i) = \mathbf{x}^n + \Theta_1\left(\sum_1, \ldots, \sum_n\right)\mathbf{x}^{n-1} + \cdots + \Theta_n\left(\sum_1, \ldots, \sum_n\right), \tag{1}$$

where $\Theta_j \in \mathbb{Q}[\mathbf{x}_1, \ldots, \mathbf{x}_n]$ has zero constant term for $1 \leq j \leq n$ and $\sum_k = \mathbf{x}_1^k + \cdots + \mathbf{x}_n^k$. The proof of (1) is outlined in Exercise 5. Since $\operatorname{char} F = 0$, we can assume that $\Theta_j \in F[\mathbf{x}_1, \ldots, \mathbf{x}_n]$. Let X be a set of commuting variables such that there is a surjective F-algebra homomorphism $\phi: F[X] \to C$. Define the surjective F-algebra homomorphism $\psi: M_n(F[X]) \to M_n(C)$ by $\psi([a_{ij}]) = [\phi(a_{ij})]$. Choose $\beta \in M_n(F[X])$ so that $\psi(\beta) = \alpha$. Clearly, $\phi(\operatorname{tr}(\beta^k)) = \operatorname{tr} \alpha^k = 0$, that is, $\operatorname{tr}(\beta^k) \in \operatorname{Ker} \phi$ for $1 \leq k \leq n$. Let $b_1, \ldots, b_n$ be the characteristic roots of β in the algebraic closure of the fraction field of $F[X]$. Note that $b_1^k + \cdots + b_n^k = \operatorname{tr}(\beta^k)$. Therefore, by (1) and the Cayley–Hamilton Theorem,

$$\beta^n + \Theta_1(\operatorname{tr} \beta, \ldots, \operatorname{tr} \beta^n)\beta^{n-1} + \cdots + \Theta_n(\operatorname{tr} \beta, \ldots, \operatorname{tr} \beta^n)\iota = 0.$$

Since the polynomials Θ_i have no constant terms and $\operatorname{tr} \beta^k \in \operatorname{Ker} \phi$, we conclude that all summands except β^n are in $\operatorname{Ker} \psi$. Hence, $\alpha^n = \psi(\beta^n) = 0$. □

Amitsur–Levitzki Theorem. *$M_n(F)$ satisfies the identity $\Gamma_{2n} = 0$.*

PROOF. Since Γ_{2n} is uniform and multilinear, it will suffice by Lemma 20.3a to prove this result when F is a prime field. In fact, we can assume that $F = \mathbb{Q}$ since $M_n(\mathbb{F}_p)$ is a homomorphic image of $M_n(\mathbb{Z})$, which in turn is a subring of $M_n(\mathbb{Q})$. (This argument is justified by a routine extension of Proposition 20.1a to $\mathbb{Z}$-algebras.) Henceforth assume that char $F = 0$. Let A be the exterior (or alternating) F-algebra on a $2n$-dimensional F-space. As is customary, denote the multiplication of A by $(x,y) \mapsto x \wedge y$. It follows from the definition of A (see Exercise 6) that $A = \bigoplus_{j \leq 2n} A_j$, where $A_0 = 1_A F$, $A_1 = z_1 F \oplus \cdots \oplus z_{2n} F$, $z_i \wedge z_i = 0$ for $1 \leq i \leq 2n$, $z_i \wedge z_j = -(z_j \wedge z_i)$ if $j \neq i$, and for $2 \leq k \leq 2n$, the set of elements $z_{i(1)} \wedge \cdots \wedge z_{i(k)}$ with $1 \leq i(1) < i(2) < \cdots < i(k) \leq 2n$ is an F-space basis of A_k. In particular, $A_{2n} = (z_1 \wedge \cdots \wedge z_{2n})F$. The subspace $C = \bigoplus_{j \leq n} A_{2j}$ is obviously a commutative sub-algebra of A. Define $B = M_n(A)$. We can consider B as a free $M_n(F)$-module on the basis $\{z_{i(1)} \wedge \cdots \wedge z_{i(k)}\}$ of A. With this viewpoint, define $\beta = z_1\alpha_1 + z_2\alpha_2 + \cdots + z_{2n}\alpha_{2n}$, where $\alpha_1, \alpha_2, \ldots, \alpha_{2n} \in M_n(F)$. Using part (iv) of Lemma a, it is easy to show by induction on k that

$$\beta^k = \sum z_{i(1)} \wedge \cdots \wedge z_{i(k)} \Gamma_k(\alpha_{i(1)}, \ldots, \alpha_{i(k)}), \tag{2}$$

where the sum is over all sequences $(i(1), \ldots, i(k))$ with $1 \leq i(1) < i(2) < \cdots < i(k) \leq 2n$. In particular, $\beta^2 \in M_n(C)$. If k is even, then Lemma a yields

$$\begin{aligned}\Gamma_k(\alpha_1, \ldots, \alpha_k) &= \sum_{j=1}^{k} (-1)^{j+1}\alpha_j \Gamma_{k-1}(\alpha_1, \ldots, \hat{\alpha}_j, \ldots, \alpha_k) \\ &= \sum_{j=1}^{k} (-1)^j \Gamma_{k-1}(\alpha_1, \ldots, \hat{\alpha}_j, \ldots, \alpha_k)\alpha_j.\end{aligned}$$

Thus, $\Gamma_k(\alpha_1, \ldots, \alpha_k) = (1/2)\sum_{j=1}^{k} (-1)^j(\gamma_j\alpha_j - \alpha_j\gamma_j)$, where γ_j abbreviates $\Gamma_{k-1}(\alpha_1, \ldots, \hat{\alpha}_j, \ldots, \alpha_k)$. In particular, $\operatorname{tr} \Gamma_k(\alpha_1, \ldots, \alpha_k) = 0$, and in general $\operatorname{tr} \Gamma_k(\alpha_{i(1)}, \ldots, \alpha_{i(k)}) = 0$ if k is even and $1 \leq i(1) < \cdots < i(k) \leq 2n$. Therefore, $\operatorname{tr} \beta^{2l} = 0$ for $0 \leq l \leq n$ by (2). It follows from Lemma b that $\beta^{2n} = 0$. That is $\Gamma_{2n}(\alpha_1, \alpha_2, \ldots, \alpha_{2n}) = 0$ by (2). Since $\alpha_1, \alpha_2, \ldots, \alpha_{2n}$ can be any matrices in $M_n(F)$, it follows that $\Gamma_{2n} = 0$ is an identity of $M_n(F)$. □

Corollary. *If F is an infinite field, K/F is a field extension, and $A \in \mathfrak{S}(K)$, then* $\operatorname{Deg} A = n$ *if and only if A_F satisfies $\Gamma_{2n} = 0$ and A_F does not satisfy $\Gamma_{2n-1} = 0$.*

The corollary is an immediate consequence of Proposition 20.3 and the Amitsur–Levitzki Theorem.

EXERCISES

1. Prove Lemma a. Hint. Deduce (iii) from (iv) by induction on n. To prove (iv) write $\Gamma_n = \sum_{j=1}^n \mathbf{y}_j \Phi_j$, where $\deg_j \Phi_j = 0$. Show that $\Phi_1 = \Gamma_{n-1}(\mathbf{y}_2, \ldots, \mathbf{y}_n)$ by computation. Use (ii) to obtain $\Gamma_n = (-1)^{j-1}\Gamma_n(\mathbf{y}_j, \mathbf{y}_1, \ldots, \mathbf{y}_{j-1}, \mathbf{y}_{j+1}, \ldots, \mathbf{y}_n)$, and apply the previous result to find Φ_j.

2. Use Lemma 20.3a and Lemma a to prove that if A is a k-dimensional F-algebra, then A satisfies $\Gamma_m = 0$ for all $m > k$.

3. Prove that if $\Gamma_m = 0$ is a polynomial identity of A, then $\Gamma_n = 0$ is a polynomial identity of A for all $n > m$. Hint. Use Lemma a(iv).

4. Use induction and Lemma a to prove (2).

5. Prove Newton's identities in the form of equation (1). Hint. Write

$$\Psi = \prod_{i=1}^n (\mathbf{x} - \mathbf{x}_i) = \mathbf{x}^n + \Phi_1 \mathbf{x}^{n-1} + \cdots + \Phi_{n-1}\mathbf{x} + \Phi_n,$$

where $\Phi_j \in \mathbb{Q}[\mathbf{x}_1, \ldots, \mathbf{x}_n]$. Note that

$$n\mathbf{x}^{n-1} + (n-1)\Phi_1\mathbf{x}^{n-1} + \cdots + \Phi_{n-1} = d\Psi/d\mathbf{x} = \sum_{i=1}^n (\Psi - \Psi(\mathbf{x}_i))/(\mathbf{x} - \mathbf{x}_i)$$

$$= \sum_{i=1}^n ((\mathbf{x}^n - \mathbf{x}_i^n) + \Phi_1(\mathbf{x}^{n-1} - \mathbf{x}_i^{n-1}) + \cdots + \Phi_{n-1}(\mathbf{x} - \mathbf{x}_i))/(\mathbf{x} - \mathbf{x}_i).$$

Derive the recursion relation $(1-k)\Phi_{k-1} = \sum_1 \Phi_{k-2} + \sum_2 \Phi_{k-3} + \cdots + \sum_{k-1}$ for $2 \le k \le n$. Derive the same relation for $k = n+1$ from the fact that $\Psi(x_i) = 0$ for all i.

6. Let F be a field, and $n \in \mathbb{N}$. Define the exterior algebra on an n dimensional F-space to be

$$A = F\{\mathbf{y}_1, \ldots, \mathbf{y}_n\}/I,$$

where I is the ideal of $F\{\mathbf{y}_1, \ldots, \mathbf{y}_n\}$ that is generated by all elements $(\mathbf{y}_1 a_1 + \cdots + \mathbf{y}_n a_n)^2$, where $a_i \in F$. Denote $z_i = \mathbf{y}_i + I$. Let the multiplication on A be denoted by $(x, y) \mapsto x \wedge y$. Prove the following statements.

(a) $z_i \wedge z_i = 0$ for $1 \le i \le n$; $z_i \wedge z_j = -z_j \wedge z_i$.

(b) A_F is spanned by the set of all elements of the form $1, z_1, \ldots, z_n, \ldots, z_{i(1)} \wedge \cdots \wedge z_{i(k)}, \ldots, z_1 \wedge z_2 \wedge \cdots \wedge z_n$.

Define $A_0 = 1F$, and, for $2 \le j \le n$, $A_j = \sum z_{i(1)} \wedge \cdots \wedge z_{i(j)}F$, where the sum is over all sequences $1 \le i(1) < \cdots < i(j) \le n$. Let B_j be the F-subspace of $F\{\mathbf{y}_1, \ldots, \mathbf{y}_n\}$ that is spanned by the monomials of degree j.

(c) $I \cap (B_0 + B_1) = 0$. Hence $A_0 + A_1 = 1F \oplus z_1F \oplus \cdots \oplus z_nF$.

(d) For $j \ge 2$, $I \cap B_j$ is the F-subspace of $F\{\mathbf{y}_1, \ldots, \mathbf{y}_n\}$ that is spanned by the set of products $x_1x_2 \cdots x_j$, where $x_1, x_2, \ldots, x_j \in B_1$.

(e) Define an F-space homomorphism $\phi\colon B_n \to F$ by $\phi(\mathbf{y}_{\rho(1)} \cdots \mathbf{y}_{\rho(n)}) = 0$ if ρ is not a permutation of $\{1, \ldots, n\}$, or $\phi(\mathbf{y}_{\rho(1)} \cdots \mathbf{y}_{\rho(n)}) = \operatorname{sgn}\rho$ if ρ is a permutation. If $x_1, \ldots, x_n \in B_1$ are given by $x_j = \sum_{i=1}^n \mathbf{y}_i a_{ij}$, then $\phi(x_1 \cdots x_n) = \det[a_{ij}]$.

(f) The mapping ϕ in (e) induces an F-space homomorphism of A_n to F that maps $z_1 \wedge \cdots \wedge z_n$ to 1.

(g) The list of elements in (b) is an F-space basis of A.

20.5. Generic Matrix Algebras

In this section we begin a study of the free $\mathfrak{B}_n$-algebra on the standard alphabet $Y = \{\mathbf{y}_i: i < \omega\}$. These algebras are the center of our attention in the rest of the chapter. To avoid useless repetition of hypotheses, we will assume in this section that the field F is infinite.

To simplify notation we will write C_n (or if necessary $C_n(F)$) instead of $C(\mathfrak{B}_n)$. Thus, $C_n = F\{Y\}/I(\mathfrak{B}_n)$, where $I(\mathfrak{B}_n)$ is the T-ideal of identities that are satisfied by central simple algebras of degree n. The first result of this section gives an alternative description of C_n.

Let $X = \{\mathbf{x}_{ij}^{(k)}: 1 \leq i, j \leq n, k < \omega\}$ be a set of independent commuting variables. As usual, $F[X]$ denotes the integral domain of (commuting) polynomials in all $\mathbf{x}_{ij}^{(k)}$ with coefficients in F, and $F(X)$ is the fraction field of $F[X]$. For each $k < \omega$, define $\xi^{(k)} = [\mathbf{x}_{ij}^{(k)}] \in M_n(F[X]) \subseteq M_n(F(X))$. The matrices $\xi^{(k)}$ are called the standard *generic n by n matrices* over F. We will sometimes use the abbreviations ξ and $[\mathbf{x}_{ij}]$ for $\xi^{(0)}$ and $[\mathbf{x}_{ij}^0]$. Let B_n (or $B_n(F)$) denote the F-subalgebra of $M_n(F(X))$ that is generated by $\{\xi^{(k)}: k < n\}$; B_n is called the *generic matrix algebra* of degree n over F.

Lemma. (i) $B_n \subseteq M_n(F[X])$.
(ii) $B_n F(X) = M_n(F(X))$.
(iii) *$C_n \cong B_n$ by an isomorphism that maps $\mathbf{y}_k$ to $\xi^{(k)}$.*

PROOF. The inclusion (i) is clear from the observation that $M_n(F[X])$ is a subalgebra of $M_n(F(X))$ that contains $\{\xi^{(k)}: k < \omega\}$. The system of equations

$$\xi^{(k)} = \sum_{1 \leq i,j \leq n} \varepsilon_{ij}\mathbf{x}_{ij}^{(k)}, \quad 1 \leq k \leq n^2$$

can be solved for ε_{ij} as a linear combination of $\xi^{(1)}, \ldots, \xi^{(n^2)}$ with coefficients in $F(X)$. Indeed, the coefficient matrix $[\mathbf{x}_{ij}^{(k)}]$ of the system is non-singular by Lemma 19.1. Thus, $M_n(F(X)) = \sum_{k=1}^{n^2} \xi^{(k)}F(X) \subseteq B_nF(X) \subseteq M_n(F(X))$, which proves (ii) and a bit more. Since $B_n \in \mathfrak{B}_n$ (by Lemma 20.3b), it follows from Proposition 20.1c that there is an F-algebra homomorphism $\phi: C_n \to B_n$ such that $\phi(\mathbf{y}_n) = \xi^{(n)}$. The image of ϕ is a subalgebra of $M_n(F(X))$ that includes all $\xi^{(k)}$. Thus, ϕ is surjective. Let $\Phi(\mathbf{y}_0, \ldots, \mathbf{y}_{m-1}) \in F\{Y\}$ be such that $\pi\Phi \in \operatorname{Ker}\phi$, where π is the projection of $F\{Y\}$ to C_n; that is, $\Phi(\xi^{(0)}, \ldots, \xi^{(m-1)}) = 0$. We will show that $\Phi = 0$ is an identity of $M_n(F)$, so that $\Phi \in I(M_n(F)) = \operatorname{Ker}\pi$. Let $\alpha^{(k)} = [a_{ij}^{(k)}] \in M_n(F)$ for $k < m$. There is a homomorphism χ: $F[X] \to F$ such that $\chi(\mathbf{x}_{ij}^{(k)}) = a_{ij}^{(k)}$; and χ induces an F-algebra homomorphism ψ: $M_n(F[X]) \to M_n(F)$ by $\psi([b_{ij}]) = [\chi(b_{ij})]$. Plainly, $\psi(\xi^{(k)}) = \alpha^{(k)}$ for $k < m$. Thus, $\Phi(\alpha^{(0)}, \ldots, \alpha^{(m-1)}) = \psi(\Phi(\xi^{(0)}, \ldots, \xi^{(m-1)})) = 0$. It follows that ϕ is an isomorphism. □

The lemma proves that B_n can be viewed as the free $\mathfrak{B}_n$-algebra with the free generating set $\{\xi^{(k)}: k < \omega\}$. Thus, C_n can be replaced by B_n in all future work. However, the alternative use of B_n or C_n can often clarify proofs: we will write C_n when freeness is to be emphasized, and B_n if matrix properties are of central importance.

Proposition. *If F is an infinite field, then $B_n(F)$ is a non-commutative domain.*

PROOF. We will show that B_n is a prime algebra ($\alpha B_n \beta = 0$ implies $\alpha = 0$ or $\beta = 0$), and C_n has no non-zero nilpotent elements. The proposition will then follow by an elementary computation. Assume that $\alpha, \beta \in B_n$ satisfy $\alpha B_n \beta = 0$ and $\alpha \neq 0$. By the lemma $M_n(F(X))\beta = M_n(F(X))\alpha M_n(F(X))\beta = M_n(F(X))F(X)\alpha B_n \beta = 0$. Thus, $\beta = 0$. The assumption that C_n contains a non-zero nilpotent element is equivalent to the existence of $\Phi \in F\{Y\}$ such that $\Phi = 0$ is not an identity of C_n but $\Phi^k = 0$ is an identity of C_n for some $k \geq 2$. By Corollary 19.7, there is an extension K of F such that $\mathfrak{S}(K)$ contains a division algebra D of degree n. (For example, let $K = F(\mathbf{x}_1, \ldots, \mathbf{x}_n)$, and $D = K((\mathbf{x},\sigma))$, where $\mathbf{x}_i^\sigma = \mathbf{x}_{i+1}$.) By Lemma 20.3b and Proposition 20.3, D satisfies $\Phi^k = 0$, but D does not satisfy $\Phi = 0$. Since D is a division algebra, this situation is impossible. Thus, C_n and B_n have no non-zero nilpotent elements. Finally, if $\alpha, \beta \in B_n$ satisfy $\alpha\beta = 0$, then $(\beta\gamma\alpha)^2 = \beta\gamma\alpha\beta\gamma\alpha = 0$ for all $\gamma \in B_n$. Consequently, $\beta B_n \alpha = 0$ and $\beta = 0$ or $\alpha = 0$. Therefore, B_n is a domain. □

The result of the proposition if false without the hypothesis that F is infinite. See Exercise 2.

EXERCISES

1. Prove that if $\Phi, \Psi \in F\{Y\}$ are such that $\Phi\Psi = 0$ is an identity of $M_n(F)$, then either $\Phi = 0$ or $\Psi = 0$ is an identity of $M_n(F)$.
2. Prove that $\Phi = (\mathbf{y}_0\mathbf{y}_1 + \mathbf{y}_1\mathbf{y}_0)^2 + (\mathbf{y}_0\mathbf{y}_1 + \mathbf{y}_1\mathbf{y}_0)$ is not an identity of $M_2(\mathbb{F}_2)$, but Φ^2 is an identity of $M_2(\mathbb{F}_2)$. Hint. Show that if $\alpha, \beta \in M_2(\mathbb{F}_2)$, then $(\alpha\beta + \beta\alpha)^2$ is a scalar matrix.

20.6. Central Polynomials

The first polynomial identity for $M_2(F)$ was discovered by Wagner in 1936. It is the identity $(\mathbf{y}_0\mathbf{y}_1 - \mathbf{y}_1\mathbf{y}_0)^2\mathbf{y}_2 - \mathbf{y}_2(\mathbf{y}_0\mathbf{y}_1 - \mathbf{y}_1\mathbf{y}_0)^2 = 0$. Viewed differently, Wagner's identity says that in $M_2(F)$, $(\alpha\beta - \beta\alpha)^2$ is always an element of the center.

A polynomial $\Phi \in F\{Y\}$ is *central* for an F-algebra A if $\psi\Phi \in \mathbf{Z}(A)$ for all homomorphisms ψ of $F\{Y\}$ to A. If $\Phi = 0$ is a polynomial identity of A, then Φ is clearly central. In this case, Φ is called trivial; a central polynomial $\Phi(\mathbf{y}_0, \ldots, \mathbf{y}_{m-1})$ is non-trivial if $\Phi(x_0, \ldots, x_{m-1}) \neq 0$ for some $x_0, \ldots,$

$x_{m-1} \in A$. It follows from Wagner's identity that $(\mathbf{y}_0\mathbf{y}_1 - \mathbf{y}_1\mathbf{y}_0)^2$ is central for $M_2(F)$; it is also non-trivial. The constant polynomial that takes a value in F° is also central and non-trivial but not very interesting.

If $\mathfrak{U}$ is a class of F-algebras and $\Phi \in F\{Y\}$, then Φ is central for $\mathfrak{U}$ if Φ is central for all $A \in \mathfrak{U}$. If $\Phi = \Phi(\mathbf{y}_0, \ldots, \mathbf{y}_{m-1})$, then it is clear that Φ is central for $\mathfrak{U}$ if and only if $\Phi\mathbf{y}_m - \mathbf{y}_m\Phi \in I(\mathfrak{U})$. In particular, if $\mathfrak{V}$ is a non-trivial variety of F-algebras and $A \in \mathfrak{V}$ is generic, then Φ is central for $\mathfrak{V}$ if and only if Φ is central for A. Our interest centers on the polynomials that are central for the varieties $\mathfrak{V}_n$.

Lemma a. *For an infinite field F and $n \in \mathbb{N}$, the following properties of $\Phi(\mathbf{y}_0, \ldots, \mathbf{y}_{m-1})$ are equivalent.*

(i) *Φ is central for $\mathfrak{V}_n$.*
(ii) *Φ is central for C_n.*
(iii) *$\pi\Phi \in \mathbf{Z}(C_n)$, where $\pi: F\{Y\} \to C_n$ is the projection.*
(iv) *$\Phi(\xi^{(0)}, \ldots, \xi^{(m-1)}) \in \mathbf{Z}(B_n)$.*
(v) *$\Phi(\alpha_0, \ldots, \alpha_{m-1}) \in \mathbf{Z}(M_n(F))$ for all $\alpha_0, \ldots, \alpha_{m-1} \in M_n(F)$.*
(vi) *Φ is central for $M_n(K)$ for some (all) $K \supseteq F$.*

Proof. The equivalence of (i), (ii), and (v) is a special case of our previous observations. Clearly, (ii) implies (iii). By Lemma 20.5, (iii) and (iv) are equivalent. Finally, (iv) implies (v) because there is a surjective F-algebra homomorphism $\phi: B_n \to M_n(F)$ such that $\phi(\xi^{(k)}) = \alpha_k$ for $k < m$. Since F is infinite and K/F is an extension, $I(M_n(F)) = I(M_n(K)) \cap F\{Y\}$ by Lemma 20.3b. Thus, (v) is equivalent to (vi). □

To simplify matters we will say that Φ is n-central if it satisfies the conditions (i) through (v) of Lemma a. Moreover, it will usually be assumed that F is infinite, so that the property of being n-central is generally independent of the ambient field.

Lemma b. *If $n > 1$, $\Phi(\mathbf{y}_0, \ldots, \mathbf{y}_{m-1}) \in F\{Y\}$ is n-central and $\Phi(0, \ldots, 0) = 0$ then $\Phi = 0$ is an identity of $M_{n-1}(F)$.*

Proof. For $\alpha_0, \ldots, \alpha_{m-1} \in M_{n-1}(F)$, define $\beta_k = \begin{bmatrix} \alpha_k & 0 \\ 0 & 0 \end{bmatrix}$ in $M_n(F)$. Since Φ is n-central, there exists $a \in F$ such that $\Phi(\beta_0, \ldots, \beta_{m-1}) = \iota_n a$. However, it is clear from the form of the β_k and the hypothesis $\Phi(0, \ldots, 0) = 0$ that

$$\Phi(\beta_0, \ldots, \beta_{m-1}) = \begin{bmatrix} \Phi(\alpha_0, \ldots, \alpha_{m-1}) & 0 \\ 0 & 0 \end{bmatrix}.$$

Thus, $a = 0$ and $\Phi(\alpha_0, \ldots, \alpha_{m-1}) = 0$. □

It is natural to ask whether non-trivial, non-constant n-central polynomials exist for all n. It turns out that there is a surfeit of these polynomials, but to exhibit one of them is surprisingly difficult. With a simple non-trivial n-central polynomial in hand, we will be able to show that the center of B_n is very large, which, by Lemma a, implies the existence of numerous n-central polynomials.

The first construction of non-trivial n-central polynomials was given by Formanek in the 1972 paper [33].

Formanek's Theorem. *Let F be an infinite field. For each $n \in \mathbb{N}$ there is a non-trivial n-central polynomial Θ such that Θ is homogeneous of degree n^2.*

PROOF. We can assume that $n \geq 2$; $\mathbf{y}_0$ is plainly a non-trivial 1-central polynomial. Let $\mathbf{x}_1, \ldots, \mathbf{x}_n, \mathbf{x}_{n+1}$ be independent commuting variables. Define

$$\Phi(\mathbf{x}_1, \ldots, \mathbf{x}_n, \mathbf{x}_{n+1}) = \prod_{i=2}^{n} (\mathbf{x}_1 - \mathbf{x}_i)(\mathbf{x}_{n+1} - \mathbf{x}_i) \prod_{2 \leq i < j \leq n} (\mathbf{x}_i - \mathbf{x}_j)^2.$$

We will use the following simple fact.

$$\begin{aligned}&\text{If } d_1, \ldots, d_n \text{ are elements in an extension } K \text{ of } F,\\ &\text{and if } \rho \text{ is a permutation of } \{1, \ldots, n\},\\ &\text{then } \Phi(d_{\rho(1)}, \ldots, d_{\rho(n)}, d_{\rho(1)}) = \textstyle\prod_{i<j} (d_i - d_j)^2.\end{aligned} \tag{1}$$

Write Φ as a sum of distinct monomials,

$$\Phi = \sum \mathbf{x}_1^{l(1)} \cdots \mathbf{x}_n^{l(n)} \mathbf{x}_{n+1}^{l(n+1)} a_l,$$

where the sum is over all sequences $l = (l(1), \ldots, l(n), l(n+1))$ of non-negative integers such that $\sum l(k) = n(n-1)$ and a_l is in the prime field of F. Define $\Theta \in F\{Y\}$ by $\Theta = \Psi_1 + \Psi_2 + \cdots + \Psi_n$, where

$$\Psi(\mathbf{y}_0, \mathbf{y}_1, \mathbf{y}_2, \ldots, \mathbf{y}_n) = \sum \mathbf{y}_0^{l(1)} \mathbf{y}_1 \mathbf{y}_0^{l(2)} \mathbf{y}_2 \cdots \mathbf{y}_n \mathbf{y}_0^{l(n+1)} a_l,$$

and

$$\Psi_k(\mathbf{y}_0, \mathbf{y}_1, \mathbf{y}_2, \ldots, \mathbf{y}_n) = \Psi(\mathbf{y}_0, \mathbf{y}_k, \mathbf{y}_{k+1}, \ldots, \mathbf{y}_n, \mathbf{y}_1, \ldots, \mathbf{y}_{k-1}).$$

Clearly, Θ is homogeneous of degree n^2. We will show that Θ is n-central. Let K be the algebraic closure of the field $F(X)$ that was introduced in Section 20.5, that is, $X = \{\mathbf{x}_{ij}^{(k)} : 1 \leq i,j \leq n, k < \omega\}$. Let

$$\delta = \begin{bmatrix} d_1 & & 0 \\ & \ddots & \\ 0 & & d_n \end{bmatrix}$$

be an arbitrary diagonal matrix in $M_n(K)$, and suppose that ρ and σ are mappings of $\{1, \ldots, n\}$ to itself. Define a homomorphism $\psi : F\{Y\} \to M_n(K)$ such that $\psi(\mathbf{y}_0) = \delta$ and $\psi(\mathbf{y}_k) = \varepsilon_{\rho(k)\sigma(k)}$, the matrix unit, for $1 \leq k \leq n$. The main step of the proof is to show $\psi\Theta = 0$ unless

$$\begin{aligned}&\rho \text{ is a permutation of } \{1, \ldots, n\},\\ &\sigma(1) = \rho(2), \sigma(2) = \rho(3), \ldots, \sigma(n-1) = \rho(n),\\ &\text{and } \sigma(n) = \rho(1);\end{aligned} \tag{2}$$

and $\psi\Theta = \iota_n \prod_{i<j}(d_i - d_j)^2$ if (2) is satisfied. We begin with the observation that $\psi(\mathbf{y}_0^{l(1)}\mathbf{y}_1\mathbf{y}_0^{l(2)}\mathbf{y}_2 \cdots \mathbf{y}_n\mathbf{y}_0^{l(n+1)}) = \varepsilon_{\rho(1)\sigma(1)}\varepsilon_{\rho(2)\sigma(2)} \cdots \varepsilon_{\rho(n)\sigma(n)} d_{\rho(1)}^{l(1)} d_{\rho(2)}^{l(2)} \cdots d_{\rho(n)}^{l(n)} d_{\sigma(n)}^{l(n+1)}$ is zero unless $\sigma(1) = \rho(2)$, $\sigma(2) = \rho(3)$, $\ldots$, and $\sigma(n-1) = \rho(n)$. Thus, $\psi\Psi = 0$ if these conditions aren't met; and if they are, then $\psi\Psi = \varepsilon_{\rho(1)\sigma(n)} \sum d_{\rho(1)}^{l(1)} \cdots d_{\rho(n)}^{l(n)} d_{\sigma(n)}^{l(n+1)} a_l = \varepsilon_{\rho(1)\sigma(n)} \Phi(d_{\rho(1)}, \ldots, d_{\rho(n)}, d_{\sigma(n)})$. Since $\{\rho(1), \ldots, \rho(n), \sigma(n)\} \subseteq \{1, \ldots, n\}$, it follows that

$$\Phi(d_{\rho(1)}, \ldots, d_{\rho(n)}, d_{\sigma(n)}) = \prod_{i=2}^{n} (d_{\rho(1)} - d_{\rho(i)})(d_{\sigma(n)} - d_{\rho(i)}) \prod_{2 \le i < j \le n} (d_{\rho(i)} - d_{\rho(j)})^2$$

is zero unless ρ is a permutation of $\{1, \ldots, n\}$ and $\sigma(n) = \rho(1)$. Thus, (2) is a necessary condition for $\psi\Psi \neq 0$. If (2) holds, then (1) yields $\psi\Psi = \varepsilon_{\rho(1)\rho(1)} \prod_{i<j}(d_i - d_j)^2$. It is clear from the symmetry of (2) that $\psi\Psi_k = 0$ if (2) is not satisfied and $\psi\Psi_k = \varepsilon_{\rho(k)\rho(k)} \prod_{i<j}(d_i - d_j)^2$ when (2) holds. Consequently, $\psi\Theta = 0$ if (2) fails and $\psi\Theta = (\sum_{k=1}^{n} \varepsilon_{\rho(k)\rho(k)}) \prod_{i<j}(d_i - d_j)^2 = \iota_n \prod_{i<j}(d_i - d_j)^2$ when (2) is satisfied. In all cases, $\psi\Theta \in \mathbf{Z}(M_n(K))$. Since Θ is linear in $\mathbf{y}_1, \ldots, \mathbf{y}_n$, it follows that $\Theta(\delta, \beta_1, \ldots, \beta_n) \in \mathbf{Z}(M_n(K))$ for all $\beta_1, \ldots, \beta_n \in M_n(K)$. Let $d_1, d_2, \ldots, d_n$ be the eigenvalues of the generic matrix $\xi^{(0)}$. These elements are distinct (by Exercise 2), so that $\xi^{(0)} = \alpha^{-1}\delta\alpha$ for a suitable $\alpha \in M_n(K)^\circ$. Thus,

$$\begin{aligned}\Theta(\xi^{(0)}, \xi^{(1)}, \ldots, \xi^{(n)}) &= \alpha^{-1}\Theta(\delta, \alpha\xi^{(1)}\alpha^{-1}, \ldots, \alpha\xi^{(n)}\alpha^{-1})\alpha \in \alpha^{-1}\mathbf{Z}(M_n(K))\alpha\\ &= \mathbf{Z}(M_n(K)).\end{aligned}$$

By Lemma a, Θ is n-central. We saw that there is a homomorphism ψ: $F\{Y\} \to M_n(K)$ such that $\psi\Theta = \iota_n \prod_{i<j}(d_i - d_j)^2 \neq 0$. That is, $\Theta = 0$ is not an identity of $M_n(K)$. Since F is infinite $\Theta = 0$ is not an identity of $M_n(F)$. Thus, Θ is non-trivial. □

Razmyslov has given an example of a non-trivial n-central polynomial of degree $2n^2 - 1$ that is uniform and multilinear. (See [65].) These polynomials are also non-trivial on $M_n(F)$ when F is finite by Lemma 20.3a. An exposition of Razmyslov's construction is available in Jacobson's paper [49].

Exercises

1. Use the Cayley–Hamilton Theorem and the observation that $\text{tr}(\alpha\beta - \beta\alpha) = 0$ to prove Wagner's identity $(\mathbf{y}_0\mathbf{y}_1 - \mathbf{y}_1\mathbf{y}_0)^2\mathbf{y}_2 - \mathbf{y}_2(\mathbf{y}_0\mathbf{y}_1 - \mathbf{y}_1\mathbf{y}_0)^2 = 0$ for $M_2(F)$.

2. Prove that if $\xi = [\mathbf{x}_{ij}]$ is a generic matrix, then the characteristic roots of ξ (in the algebraic closure of $F(\mathbf{x}_{ij})$) are distinct. Hint. Let $R = F[\mathbf{x}_{11}, \mathbf{x}_{12}, \ldots, \mathbf{x}_{nn}]$; note that R is a unique factorization domain. The characteristic polynomial of ξ has the form $\Phi = \mathbf{x}^n + a_1\mathbf{x}^{n-1} + \cdots + a_n$ with $a_i \in R$. Use Gauss's Lemma to show that

if ξ has multiple eigenvalues then Φ factors non-trivially in $R[\mathbf{x}]$. Obtain a contradiction by defining a homomorphism ψ of R to an extension of F such that ξ is mapped to a matrix whose characteristic polynomial Ψ is irreducible, and note that $\psi\Phi = \Psi$.

3. Compute the 2-central polynomial that is given by the construction in the proof of Formenak's Theorem.

20.7. Structure Theorems

We have reached a high point of this chapter. It is now possible to show how restrictive polynomial identities can be. The first part of the section presents the Kaplansky–Amitsur Theorem. This result is a cornerstone in the theory of polynomial identities. We will use the Kaplansky–Amitsur Theorem to prove a result that is applicable to the algebras B_n.

Recall that an F-algebra A is primitive if there is a simple right A-module M that is faithful: $Mx = 0$ implies $x = 0$.

Lemma a. *Let A be a primitive F-algebra that satisfies a polynomial identity of degree $n \geq 1$. Suppose that M is a simple, faithful right A-module. If $D = \mathbf{E}_A(M)$, then D is a division algebra,* $\dim_D M = m \leq n/2$, *and* $A \cong M_m(D)$.

PROOF. By Schur's Lemma, D is a division algebra. If $\dim_D M \geq m > n/2$, then there exist elements $u_1, \ldots, u_m \in M$ that are linearly independent over D. Denote $N = Du_1 \oplus \cdots \oplus Du_m$, a submodule of ${}_DM$. Clearly, $B = \{x \in A: Nx \subseteq N\}$ is a subalgebra of A. By the Density Theorem of Section 12.2, the mapping $\phi: B \to M_m(D)$, defined by $\phi(x) = [d_{ij}]$ if $u_i x = \sum_j d_{ij} u_j$, is surjective. A routine calculation shows that ϕ is an algebra homomorphism. By Proposition 20.1a, $M_m(D)$ and its subalgebra $M_m(F)$ satisfy a polynomial identity of degree n. Since $n < 2m$, this conclusion contradicts Corollary 20.3a. Thus, $\dim_D M = m \leq n/2$, and the argument given above (with $u_1, \ldots, u_m$ taken to be a basis of M) shows that $A \cong M_m(D)$—in this case, ϕ is an isomorphism because M is faithful. □

The lemma shows that a primitive algebra can satisfy a non-trivial polynomial identity only if it is Artinian. The Kaplansky–Amitsur Theorem strengthens this conclusion.

Kaplansky–Amitsur Theorem. *Let A be a primitive F-algebra that satisfies a polynomial identity of degree $n \geq 1$. If E is the center of A, then A is a central simple E-algebra and* $\operatorname{Deg} A \leq n/2$.

PROOF. Let M be a simple, faithful right A-module. Denote the division algebra $\mathbf{E}_A(M)$ by D. If $a \in E$, then the right multiplication endomorphism

ρ_a belongs to $\mathbf{Z}(D)$. By Zorn's Lemma there is a maximal subfield K of D that satisfies $\mathbf{Z}(D) \subseteq K$. Define $B = K\rho(A)^* \subseteq \mathbf{E}_K(M)^*$, and consider M as a right B-module. Since M is a simple A-module, it is also a simple B-module. The elements of B are endomorphisms of M, so that M is necessarily a faithful B-module. We will show that $K = \mathbf{E}_B(M)$. First note that $K \subseteq \mathbf{E}_B(M)$ because $K \subseteq \mathbf{C}_B(B) = \mathbf{Z}(B)$. If $\phi \in \mathbf{E}_B(M)$, then $\phi \in \mathbf{E}_A(M) = D$. Moreover, $\alpha\phi = \phi\alpha$ for all $\alpha \in K$. Hence, $K(\phi)$ is a subfield of D that contains K. By maximality, $\phi \in K(\phi) = K$. Next, note that by Lemma 12.4a and Proposition 9.2c, $B \cong A \bigotimes_E K = A^K$. In particular, B satisfies a polynomial identity of degree at most n by Proposition 20.2 and Lemma 20.3a. Therefore, $B \cong M_m(K)$, where $m \leq n/2$. Thus, $\dim_E A = \dim_K B = m^2$. Another application of Lemma a completes the proof of the Kaplansky–Amitsur Theorem. □

Corollary. *Let A be an F-algebra such that $\mathbf{J}(A) = 0$. If A satisfies a polynomial identity of degree n, then there is an injective homomorphism $\phi: A \to \prod_{i \in J} A_i$, where $A_i \in \mathfrak{S}(K_i)$, K_i/F is a field extension,* $\operatorname{Deg} A_i \leq n/2$, *and for all $j \in J$, the projection homomorphism $\pi_j: \prod_{i \in J} A_i \to A_j$ maps $\phi(A)$ onto A_j.*

PROOF. Let $\{N_i: i \in J\}$ be the set of maximal right ideals of A, $J_i = \operatorname{ann}(A/N_i) = \{x \in A: Ax \subseteq N_i\} \lhd A$, and $A_i = A/J_i$. Each A_i is primitive and satisfies a polynomial identity of degree n. By the Kaplansky–Amitsur Theorem $A_i \in \mathfrak{S}(K_i)$ (with $K_i = \mathbf{Z}(A_i)$ of course) and $\operatorname{Deg} A_i \leq n/2$. Map $\phi: A \to \prod_{i \in J} A_i$ by $\phi(x) = (\ldots x + J_i \ldots)$. Plainly ϕ is a homomorphism, $\pi_i\phi(A) = A_i$, and $\operatorname{Ker} \phi = \bigcap_{i \in J} J_i \subseteq \bigcap_{i \in J} N_i = \mathbf{J}(A) = 0$. □

Lemma b. *Assume that F is an infinite field. Let A be an F-algebra such that $\mathbf{J}(A) = 0$. Assume that A satisfies a non-trivial polynomial identity. If I is a non-zero ideal of A then $I \cap \mathbf{Z}(A) \neq 0$.*

PROOF. By the corollary it can be assumed that A is a subalgebra of the product $\prod_{i \in J} A_i$ where each A_i is central simple over an extension K_i of F, $\max\{\operatorname{Deg} A_i : i \in J\} < \infty$, and $\pi_i A = A_i$. The fact that π_i is surjective implies that $\pi_i(I)$ is an ideal of the simple algebra A_i, hence $\pi_i(I) = 0$ or A_i. Since $I \neq 0$, $\pi_i(I) = A_i$ for at least one $i \in J$. Let $m = \max\{\operatorname{Deg} A_i: \pi_i(I) = A_i\}$. Choose a non-trivial m-central polynomial $\Phi(\mathbf{y}_0, \ldots, \mathbf{y}_{k-1}) \in F\{Y\}$ with zero constant term. By the choice of m, there exist $x_1, \ldots, x_k \in I$ and an index $j \in J$ such that $\pi_j\Phi(x_1, \ldots, x_k)$ is a non-zero element of $\mathbf{Z}(A_j)$. Since the constant term of Φ is zero, $\Phi(x_1, \ldots, x_k) \in I$. The proof can be completed by showing that $\pi_i\Phi(x_1, \ldots, x_k) \in \mathbf{Z}(A_i)$ for all $i \in J$ because $\bigcap_{i \in J} \pi_i^{-1}(\mathbf{Z}(A_i)) \subseteq \mathbf{Z}(A)$. If $\pi_i(I) = 0$, then $\pi_i\Phi(x_1, \ldots, x_k) \in \mathbf{Z}(A_i)$ is clear. If $\pi_i(I) = A_i$, then $\operatorname{Deg} A_i \leq m$. If $\operatorname{Deg} A_i < m$, then $\Phi = 0$ is an identity of A_i by Lemma 20.6b. In this case, $\pi_i\Phi(x_1, \ldots, x_k) = \Phi(\pi_i(x_1), \ldots, \pi_i(x_k)) = 0$. If $\operatorname{Deg} A_i = m$, then Φ is central for A_i, so that $\pi_i\Phi(x_1, \ldots, x_k) = \Phi(\pi_i(x_1), \ldots, \pi_i(x_k)) \in \mathbf{Z}(A_i)$. □

Proposition. *Assume that F is an infinite field. Let A be an F-algebra that satisfies a polynomial identity of degree $n \geq 1$. If A is a domain whose center is the field K, then $A \in \mathfrak{S}(K)$ is a division algebra and* $\operatorname{Deg} A \leq n/2$.

PROOF. Let $B = A[\mathbf{x}]$ be the F-algebra of polynomials in the variable $\mathbf{x}$ with coefficients in A. Note that $\mathbf{x} \in \mathbf{Z}(B)$. Since $B \cong A \otimes F[\mathbf{x}] \subseteq A^{F(\mathbf{x})}$ and A satisfies a polynomial identity of degree n, so does B satisfy an identity of degree n. Moreover, $\mathbf{J}(B) = 0$. In fact, if $z \neq 0$ in B, then the degree of $1 - \mathbf{x}z$ is at least one. Hence $1 - \mathbf{x}z$ is not a unit of B. Indeed, the fact that A is a domain implies $\deg w_1 w_2 = \deg w_1 + \deg w_2$, so that $B^\circ = A^\circ$. By Proposition 4.3, $z \notin \mathbf{J}(B)$. It follows from Lemma b that every non-zero ideal of B contains a non-zero element of $\mathbf{Z}(B)$. If I is a non-zero ideal of A, then $I[\mathbf{x}]$ is a non-zero ideal of B. Thus, $(I \cap \mathbf{Z}(A))[\mathbf{x}] = I[\mathbf{x}] \cap \mathbf{Z}(A)[\mathbf{x}] = I[\mathbf{x}] \cap \mathbf{Z}(B) \neq 0$, and $I \cap K = I \cap \mathbf{Z}(A) \neq 0$. Since K is a field, it follows that $I = A$. This argument shows that A is simple and therefore primitive. (The annihilator of any simple A-module is necessarily zero.) The proposition follows from the Kaplansky–Amitsur Theorem. □

EXERCISES

1. Prove the converse of the corollary: if A is a subalgebra of $\prod_{i \in J} A_i$, where each A_i is central simple (over a field K_i) with $\operatorname{Deg} A_i \leq m$ for all $i \in J$, and $\pi_i(A) = A_i$ for all $i \in J$, then $\mathbf{J}(A) = 0$ and A satisfies a polynomial identity of degree $2m$.

2. Let A be an F-algebra that satisfies a non-trivial polynomial identity. Prove that A is Artinian if and only if A is Noetherian and all prime ideals of A are maximal. Hint. See the Exercise of Section 4.5.

3. (a) Let $\Phi \in R[\mathbf{x}]^\circ$, where R is a commutative ring, say $\Phi(\mathbf{x}) = a + \mathbf{x}b_1 + \cdots + \mathbf{x}^n b_n$ with $a, b_1, \ldots, b_n \in R$. Prove that $a \in R^\circ$ and $b_1, \ldots, b_n$ are nilpotent. Hint. Use the results in the exercise of Section 4.5.
   (b) Let A be an F-algebra and suppose that B is a subalgebra of A. Assume that $\Phi \in B[\mathbf{x}]$ satisfies $\Phi(0) = 0$ and $\Psi \in A[\mathbf{x}]$ satisfies $(1 - \Psi)(1 - \Phi) = 1$. Prove that $\Psi \in B[\mathbf{x}]$. Hint. Show $\Psi = (\Psi - 1)\Phi^{n+1} - \Phi - \Phi^2 - \cdots - \Phi^n$ for all $n \in \mathbb{N}$.
   (c) Prove that if A is an F-algebra such that $\mathbf{P}(A) = 0$, then $\mathbf{J}(A[\mathbf{x}]) = 0$. Hint. For $\Phi = a_0 + \mathbf{x}a_1 + \cdots + \mathbf{x}^n a_n$, $a_n \neq 0$, denote the number of non-zero a_i by $w(\Phi)$ and let $l(\Phi) = a_n$. Put $w(0) = l(0) = 0$. Let $r = \min\{w(\Phi) \colon \Phi \in \mathbf{J}(A[\mathbf{x}])\}$. Show that if $\Phi \neq 0$ and $w(\Phi) = r$, then $a_i\Phi = \Phi a_i$ for all coefficients a_i of Φ. Hence, $\Phi \in B[\mathbf{x}]$ for a commutative subalgebra B of A (that depends on Φ). Apply (b) to $\mathbf{x}\Phi$ and use (a) to show that the leading coefficient of Φ is nilpotent whenever $w(\Phi) = r > 0$. Deduce that $\{l(\Phi) \colon w(\Phi) \leq r\} \subseteq \mathbf{P}(A)$.
   (d) Generalize the proposition to F-algebras A such that $\mathbf{P}(A) = 0$.

20.8. Universal Division Algebras

In this section the major theorems of this chapter merge with earlier results on division algebras to settle what was long considered to be the most important problem in the theory of central simple algebras: is every division algebra a crossed product? The expected negative answer turned out to be correct.

It is assumed in this section that the field F is infinite. If F is finite, then every $A \in \mathfrak{S}(F)$ is a crossed product.

Since $C_n(F)$ is a domain by Proposition 20.5, its center $R = \mathbf{Z}(C_n)$ is an integral domain. Let $L_n = L_n(F)$ be the fraction field of R. Define $D_n = D_n(F) = C_n \otimes_R L_n$.

Proposition. (i) *D_n is a central division algebra over L_n with* $\operatorname{Deg} D_n = n$.

(ii) *The mapping $z \mapsto z \otimes 1$ embeds C_n in D_n as an R-subalgebra.*

(iii) *Every element of D_n has the form $w \otimes x^{-1}$ for suitable $w \in C_n$ and $x \in L_n^{\circ}$.*

PROOF. The statement (iii) is a consequence of the observation that $(z_1 \otimes x_1^{-1}) + (z_2 \otimes x_2^{-1}) = (z_1x_2 + z_2x_1) \otimes (x_1x_2)^{-1}$. If $z \neq 0$ in C_n, then $z \otimes 1 \neq 0$ in D_n. (See Exercise 3.) Thus, the R-algebra homomorphism $z \mapsto z \otimes 1$ is injective. If $(z_1 \otimes x_1^{-1})(z_2 \otimes x_2^{-1}) = 0$, then $z_1z_2 \otimes 1 = 0$ and $z_1z_2 = 0$. Since C_n is a domain, it follows that $z_1 = 0$ or $z_2 = 0$. Thus, D_n is also a domain. Moreover, $z \otimes x^{-1} \in \mathbf{Z}(D_n)$ implies that $(yz - zy) \otimes x^{-1} = 0$ for all $y \in C_n$. Hence, $z \in \mathbf{Z}(C_n) = R$, and $z \otimes x^{-1} = 1 \otimes zx^{-1}$. Thus, $\mathbf{Z}(D_n) \cong L_n$ is a field. It is a consequence of Lemma 20.3a and the Amitsur–Levitzki Theorem that C_n and D_n satisfy the standard identity Γ_{2n}, but they do not satisfy Γ_{2n-1}. It follows from Proposition 20.7 that $D_n \in \mathfrak{S}_n(L_n)$ is a division algebra. □

The algebra $D_n(F)$ is called the *universal division algebra* of degree n for the field F. (Of course, F is not the center of $D_n(F)$ if $n > 1$.) It is convenient to identify C_n with its image $C_n \otimes R$ in D_n. By the third part of the proposition, C_n is an essential R-submodule of D_n. That is, if $w \in D_n^{\circ}$, then $wR \cap C_n \neq 0$: there exists $x \in R - \{0\}$ such that $wx \in C_n$.

Lemma a. *Assume that K/F is a field extension and $A \in \mathfrak{S}_n(K)$.*

(i) *If $\phi\colon C_n \to A$ is an F-algebra homomorphism, then $\phi(\mathbf{Z}(C_n)) \subseteq K$.*

(ii) *If $z_1, \ldots, z_m$ are distinct, non-zero elements of C_n, then there is an F-algebra homomorphism $\phi\colon C_n \to A$ such that $\phi(z_1), \ldots, \phi(z_m)$ are distinct elements of A°.*

PROOF. The property (i) is an easy consequence of Lemma 20.6a. The proof of (ii) breaks into three steps. If $x \in \mathbf{Z}(C_n) - \{0\}$, then $\phi(x) \neq 0$ for some

F-algebra homomorphism of C_n to A. In fact, by Lemma 20.6a, $x = \pi(\Phi)$, where $\pi: F\{Y\} \to C_n$ is the projection homomorphism and Φ is a non-trivial, central polynomial for $M_n(F)$. Thus, Φ is a non-trivial, central polynomial for A by Lemma 20.6a. That is, there is a homomorphism $\psi: F\{Y\} \to A$ such that $0 \neq \psi\Phi \in \mathbf{Z}(A) = K$. Since $\operatorname{Ker}\pi = I(\mathfrak{V}_n(F)) \subseteq I(A)$, there is a homomorphism $\phi: C_n \to A$ such that $\psi = \phi\pi$. In particular, $\phi(x) = \phi\pi(\Phi) = \psi\Phi$. Next, assume that $z \in C_n - \{0\}$. By the proposition, $z \in D_n^\circ$, where $D_n \in \mathfrak{S}(L_n)$ is a division algebra. It follows that z is a root of a polynomial $\mathbf{x}^k + \mathbf{x}^{k-1}x_1 + \cdots \mathbf{x}x_{k-1} + x_k \in L_n[\mathbf{x}]$, $x_k \neq 0$. Let $x \in R - \{0\}$ be such that $x_j x \in \mathbf{Z}(C_n)$ for $1 \leq j \leq k$. In particular, $x_k x \neq 0$. Therefore, there is a homomorphism $\phi: C_n \to A$ such that $\phi(x_k x) \neq 0$. Thus, $\phi(z)(\phi(z)^{k-1}\phi(x) + \phi(z)^{k-2}\phi(x_1 x) + \cdots + \phi(x_{k-1}x))\phi(-x_k x)^{-1} = 1$ and $\phi(z) \in A^\circ$. To prove (ii), let $z = z_1 \cdots z_m \prod_{1 \leq i < j \leq m}(z_i - z_j)$. If $\phi: C_n \to A$ is a homomorphism such that $\phi(z) \in A^\circ$, then $\phi(z_1), \ldots, \phi(z_m)$ are obviously distinct, non-zero elements of A°. □

We now prove a technical result that is the heart of Amitsur's Theorem. In order for D_n to be a crossed product, there must be a maximal subfield E of D_n such that E/L_n is Galois. Using Lemma b, this assumption leads to a very strong conclusion about the existence of maximal subfields of central simple division algebras over arbitrary extensions of F. For many choices of n and F, this line of reasoning produces a contradiction. Hence, D_n cannot be a crossed product.

Lemma b. *Assume that E is a subfield of D_n such that E/L_n is Galois. If K/F is a field extension and $D \in \mathfrak{S}_n(K)$ is a division algebra, then there is a subfield L of D such that L/K is Galois and* $\mathbf{G}(L/K) \cong \mathbf{G}(E/L_n)$.

Proof. We will use the following notation: $R = \mathbf{Z}(C_n)$, $G = \mathbf{G}(E/L_n)$. Since E/L_n is Galois, it is simple: $E = L_n[x]$ for a suitable $x \in E$. If the minimum polynomial of x over L_n has degree m, then x is a root of a polynomial $a_0 + \mathbf{x}a_1 + \cdots + \mathbf{x}^m a_m$ in $R[\mathbf{x}]$ with $a_0 \neq 0$. For each $\sigma \in G$, the Noether–Skolem Theorem provides a non-zero element $u_\sigma \in D_n$ such that

$$u_\sigma^{-1} x u_\sigma = x^\sigma. \tag{1}$$

Note that every x^σ is in E because E/L_n is Galois. An essential point of the proof is the fact that x and all u_σ can be found in C_n, and we can assume that $x^\sigma \in R[x]$ for all $\sigma \in G$. Four observations make this claim obvious: $L_n \cap C_n = R$; if $y \in D_n$, then $yb \in C_n$ for some non-zero $b \in R$; only a finite number of elements in D_n require such an adjustment so that a single b will do the job for all of them; the equations (1) remain true when all the entries that appear in them are multiplied by the same b. Henceforth, assume that x and all u_σ are in C_n, and $x^\sigma \in R[x] \subseteq C_n$ for all $\sigma \in G$. Under these conditions, Lemma a provides an F-algebra homomorphism $\phi: C_n \to D$ such that the $\phi(x^\sigma)$ are distinct and not zero, $\phi(a_0) \neq 0$, and $\phi(u_\sigma) \neq 0$ for all $\sigma \in G$.

Define $L = K(\phi(x))$. Since D is a division algebra in $\mathfrak{S}(K)$, L is a subfield of D and $L = K[\phi(x)]$. By Lemma a, $\phi(R[x]) \subseteq L$. In particular, $\phi(x^\sigma) \in L$ for all $\sigma \in G$. The fact that $\phi(x)$ is a root of the non-zero polynomial $\phi(a_0) + \mathbf{x}\phi(a_1) + \cdots + \mathbf{x}^m\phi(a_m) \in K[\mathbf{x}]$ implies that $[L:K] \le m$. The proof will be finished by defining an injective group homomorphism of G to $\mathbf{G}(L/K)$. Once we have such a mapping, it will follow that $m = |G| \le |\mathbf{G}(L/K)| \le [L:K] \le m$, so that L/K is Galois and $\mathbf{G}(L/K) \cong G$. For $\sigma \in G$, define $\chi(\sigma)$ to be the inner automorphism of D that is given by $w \mapsto w^{\chi(\sigma)} = \phi(u_\sigma)^{-1}w\phi(u_\sigma)$. If $w \in K = \mathbf{Z}(D)$, then $w^{\chi(\sigma)} = w$. By (1), $\phi(x)^{\chi(\sigma)} = \phi(u_\sigma^{-1}xu_\sigma) = \phi(x^\sigma) \in L$, so that $\chi(\sigma)|L \in \mathbf{G}(L/K)$. Generalizing this computation, we find that if $y = \sum_{i<k} x^i c_i \in R[x]$, then $\phi(y^\sigma) = \phi(\sum_{i<k}(x^\sigma)^i c_i) = \sum_{i<k}\phi(x^\sigma)^i\phi(c_i) = \sum_{i<k}(\phi(x)^{\chi(\sigma)})^i\phi(c_i) = \phi(y)^{\chi(\sigma)}$. In particular, $\phi(x)^{\chi(\sigma\tau)} = \phi(x^{\sigma\tau}) = \phi(x^\sigma)^{\chi(\tau)} = \phi(x)^{\chi(\sigma)\chi(\tau)}$. Hence, $\chi(\sigma\tau)|L = (\chi(\sigma)|L)(\chi(\tau)|L)$; $\sigma \mapsto \chi(\sigma)|L$ is a homomorphism from G to $\mathbf{G}(L/K)$. This homomorphism is injective because the $\phi(x^\sigma)$ are distinct elements of D. Indeed, if $\phi(x)^{\chi(\sigma)} = \phi(x)$, then $\phi(x^\sigma) = \phi(x)$, so that $\sigma = 1$. □

Amitsur's Theorem. *If $D_n(\mathbb{Q})$ is a crossed product, then $n = 2^s q_1 q_2 \cdots q_r$ where $s \le 2$ and $q_1, q_2, \ldots, q_r$ are distinct odd primes.*

PROOF. Assume that $D_n(\mathbb{Q})$ is a crossed product, say E is a maximal subfield of $D_n(\mathbb{Q})$ with E/L_n Galois. By Theorem 19.9, there is a field extension $K/\mathbb{Q}$ and a division algebra $D \in \mathfrak{S}_n(K)$ such that if L_0 is a maximal subfield of D with L_0/K Galois, then $\mathbf{G}(L_0/K)$ is a product of cyclic groups of prime orders. By Lemma b, $\mathbf{G}(E/L_n)$ must be a product of cyclic groups of prime orders. Let p be a prime such that p does not divide n and $(n,p-1) \le 2$. For example, $p = 2$ will do if n is odd. Otherwise, see Exercise 4. By Theorem 17.10, there is a division algebra $D \in \mathfrak{S}_n(\hat{\mathbb{Q}}_p)$. By Lemma b, D contains a maximal subfield L such that $L/\hat{\mathbb{Q}}_p$ is an abelian extension of degree n and $\mathbf{G}(L/\hat{\mathbb{Q}}_p)$ is a product of cyclic groups of prime orders. By Proposition 17.7 there is a field F between $\hat{\mathbb{Q}}_p$ and L such that $F/\hat{\mathbb{Q}}_p$ is unramified and L/F is totally ramified: $e(F/\hat{\mathbb{Q}}_p) = 1$, $e(L/F) = [L:F] = e(L/\hat{\mathbb{Q}}_p)$. Since $\mathbf{G}(L/\hat{\mathbb{Q}}_p)$ is a product of cyclic groups of prime orders, it is a semisimple $\mathbb{Z}$-module: subgroups are direct summands. In terms of subfields this observation implies that there is a subfield M of L such that $MF = L$ and $M \cap F = \hat{\mathbb{Q}}_p$. It follows that $M/\hat{\mathbb{Q}}_p$ is totally ramified with $[M:\hat{\mathbb{Q}}_p] = e(M/\hat{\mathbb{Q}}_p) = e(L/F) = e(L/\hat{\mathbb{Q}}_p)$. By Corollary 19.8, $e(M/\hat{\mathbb{Q}}_p)$ divides n and $p-1$. Thus, $[M:\hat{\mathbb{Q}}_p] = 1$ or 2. By Proposition 17.8, $\mathbf{G}(F/\hat{\mathbb{Q}}_p)$ is a cyclic homomorphic image of $\mathbf{G}(L/\hat{\mathbb{Q}}_p)$, which implies that $[F:\hat{\mathbb{Q}}_p] = |\mathbf{G}(F/\hat{\mathbb{Q}}_p)|$ is a product of distinct prime factors. Therefore, $n = [M:\hat{\mathbb{Q}}_p][F:\hat{\mathbb{Q}}_p] = 2^s q_1 \cdots q_r$, where $s = 0, 1,$ or 2, and $q_1, \ldots, q_r$ are distinct odd primes. □

EXERCISES

1. Prove that if F is a finite field, then $M_n(F)$ is a crossed product for all $n \in \mathbb{N}$.

2. Prove that tr. deg $L_n(F)/F$ is infinite for all $n \geq 2$.

3. With the notation and hypotheses of the proposition, prove that if $z \otimes 1 = 0$ in D_n, then $z = 0$. Hint. Use the construction in Chapter 9 to show that if $z \otimes 1 = 0$ in D_n, then $z \otimes 1 = 0$ in $C_n \otimes Rx^{-1}$ for some $x \in R - \{0\}$. Verify that $z \mapsto z \otimes x^{-1}$ and $z \otimes yx^{-1} \mapsto zy$ are inverse isomorphisms between C_n and $C_n \otimes Rx^{-1}$, and deduce that $z \otimes 1 = 0$ implies $z = 0$.

4. Let $n \in \mathbb{N}$ be even. Show that there is a prime p such that p does not divide n and the greatest common divisor of n and $p - 1$ is 2. Hint. Use the Dirichlet Density Theorem on $\{rn - 1 : r \in \mathbb{N}\}$.

5. Prove that if F is a local field, M/F is a finite extension, and K and L are intermediate fields such that $M = KL$, $F = K \cap L$, K/F is unramified, and M/K is totally ramified, then M/L is unramified and L/F is totally ramified.

6. Let I be a non-zero proper T-ideal of $F\{Y\}$ such that $\Phi\Psi \in I$ implies $\Phi \in I$ or $\Psi \in I$. Prove that $I = I(\mathfrak{B}_n)$ for some $n \in \mathbb{N}$. Hint. Argue as in the proposition that $F\{Y\}/I$ is generic for some $\mathfrak{B}_n$. The converse of the result in this exercise was given in Exercise 1, Section 20.5.

7. Amitsur's Theorem leaves open two questions: "are algebras of prime degrees crossed products?"; "are algebras of degree 4 crossed products?" The answer to the first of these questions is not known. The purpose of this exercise is to outline the proof of a theorem due to Albert that settles the second question positively.

   Theorem. Every division algebra of degree 4 in $\mathfrak{S}(F)$ contains a maximal subfield E such that $E = K \otimes L$, where K and L are quadratic extensions of F.

   For simplicity we will assume that char $F \neq 2$. The theorem is true without this restriction, but the proof is different.

   (a) Prove that if D contains a quadratic extension of F, then the result of the theorem follows. Hint. Let $K = F(x)$ be a subfield of D with $[K:F] = 2$. Denote the non-identity automorphism of K/F by σ. Choose $u \in D$ according to the Noether–Skolem Theorem so that $u^{-1}xu = u^{\sigma}$. Note that $v = u^2 \in \mathbf{C}_D(K)$. Prove that if $v \notin F$, then $F(v) \cap K = F$ and $F(v)K$ is a subfield of D. Conclude that $L = F(v)$ does the job in this case. Show that if $v \in F$, then (K,σ,v) is isomorphic to a quaternion algebra $A \in \mathfrak{S}(F)$ that is a subalgebra of D. Deduce from the Double Centralizer Theorem that D is a tensor product of two quaternion algebras.

   (b) Prove that if D has exponent 2, then D contains a subfield K such that $[K:F] = 2$. Hint. Show that if Exp $D = 2$, then there is an isomorphism $\phi: D \to D^*$ and $\phi^2 \in \mathrm{Aut}_F D$. Invoke the Noether–Skolem Theorem to find $u \in D^\circ$ such that $\phi^2(x) = u^{-1}xu$ for all $x \in D$. Show that $u^{-1}\phi(x)u = \phi(u)\phi(x)\phi(u)^{-1}$ for all $x \in D$, and conclude that $u\phi(u) \in F$. Prove that if $\phi(u) \neq u$, then $v = \phi(u) - u$ satisfies $\phi(v) = -v$, $F(v^2) \subset F(v)$, and either $F(v)/F$ or $F(v^2)/F$ is quadratic. Show that if $\phi(u) = u$, then either $F(u)/F$ is quadratic or $\phi^2 = id$. In the latter case, choose $y \in D$ so that $\phi(y) \neq y$, and prove that $w = \phi(y) - y$ satisfies $\phi(w) = -w$.

   (c) Prove that if D^K contains a subfield E such that $[E:K] = [K:F] = 2$, then D contains a quadratic extension of F. Hint. Use Corollary 13.4 to reduce the proof to the case in which D^K is a division algebra. Deduce $F(y) = F(y^2)$ for all $y \in D$ from the assumption that D contains no subfield that is a quadratic extension of F. Write $K = F(t)$, where $t^2 = a \in F$, and $E = K(z)$, where $z^2 = b + ct$ with $b, c \in F$. This requires the hypothesis char $F \neq 2$. Use the fact that $z \in D^K$ to write $z = y_1 + y_2 t$ with $y_1, y_2 \in D$. Show that $y_1^2 + y_2^2 a = b, y_1 y_2 + y_2 y_1 = c$. Conclude that $y_1 \in F(y_1^2) \subseteq F(y_2^2) = F(y_2)$, hence $y_1 y_2 = y_2 y_1$ and $2y_1 y_2 = c$. Show that $(y_1^2)^2 + by_1^2 + (1/4)ac = 0$. Deduce a contradiction to the assumption that D has no quadratic subfields.

(d) Use (a), (b), and (c) to complete the proof of Albert's Theorem. Hint. (a) and (b) reduce the proof to the case $\operatorname{Exp} D = 4$. Conclude from Corollary 15.2a that there is a quadratic extension K/F that splits $D \otimes D$. Use (b) and (c) to finish the proof.

Notes on Chapter 20

Anyone who is familiar with the literature of polynomial identities for rings will not have to be told how heavily this chapter leans on the excellent monograph [64] by Procesi. Other sources from which we have drawn ideas and proofs are Jacobson's monograph [48], the book [42] by Herstein, and Amitsur's paper [5].

The material of Section 20.1 admits an enormous generalization to universal algebras. The results in this section are true for varieties of groups, lattices, and so forth. Theorem 20.1 is the specialization to F-algebras of the Birkhoff–Tarski characterization of equational classes. From section two on, the special piquancy of ring theory overwhelms the less robust flavor of universal algebra. Our exposition covers only basic topics in the theory of polynomial identities, and the treatment is more or less standard. The use of Rosset's proof of the Amitsur–Levitzki Theorem in Section 20.4 makes the discussion of this topic considerably shorter than would be possible with earlier proofs. The first seven sections of this chapter seem disconnected from the material that comes before them. The punch line of the chapter doesn't appear until Section 20.8. It is gratifying to find that the hard work of the previous sections has some redeeming value; it provides a solution of an old and difficult problem in classical algebra. It will be surprising if the universal division algebras are not the focus of serious mathematical attention during the next few decades.

References

1. Albert, A. A. 1929. A determination of all normal division algebras in 16 units. *Trans. Amer. Math. Soc.* **31**, 253–260.
2. Albert, A. A. 1932. A construction of non-cyclic normal division algebras. *Bull. Amer. Math. Soc.* **38**, 449–456.
3. Albert, A. A. 1935. *Structure of Algebras*. Amer. Math. Soc. Colloquium Publ. vol. 24. Providence: Amer. Math. Soc.
4. Albert, A. A., Hasse, H. 1932. A determination of all normal division algebras over an algebraic number field. *Trans. Amer. Math. Soc.* **34**, 722–726.
5. Amitsur, S. A. 1972. On central division algebras. *Israel Jour. of Math.* **12**, 408–420.
6. Artin, E. 1957. *Geometric Algebra*. New York: Interscience.
7. Artin, E. 1959. *Theory of Algebraic Numbers*. Göttingen: Mathematisches Institut, Göttingen.
8. Artin, E. 1967. *Algebraic Numbers and Algebraic Functions*. New York: Gordon and Breach.
9. Artin, E., Nesbitt, C., Thrall, R. M. 1944. *Rings with Minimum Condition*. Ann Arbor: Univ. of Michigan.
10. Artin, E., Tate, J. 1967. *Class Field Theory*. New York: Benjamin–Addison Wesley.
11. Auslander, M. 1974. Representation theory of Artin algebras, II. *Comm. Algebra* **1**, 269–310.
12. Auslander, M., Brumer, A. 1968. Brauer groups of discrete valuation rings. *Nederl. Akad. Wetensch. Proc.* Ser. A **71**, 286–296.
13. Auslander, M., Goldman, O. 1960. The Brauer group of a commutative ring. *Trans. Amer. Math. Soc.* **97**, 367–409.
14. Auslander, M., Reiten, I. 1974. Almost split sequences II. In: Proc. ICRA, pp. 9–19. New York–Berlin–Heidelberg: Springer–Verlag.
15. Auslander, M., Reiten, I. 1975–8. Representation theory of Artin algebras, III to VI. *Comm. Algebra* **3**, 239–294; **5**, 443–518; **5**, 519–554; **6**, 257–300.
16. Azumaya, G. 1951. On maximally central algebras. *Nagoya Math. J.* **2**, 119–150.
17. Bass, H. 1968. *Algebraic K–Theory*. Reading: Benjamin–Addison Wesley.
18. Bernstein, I. N., Gel'fand, I. M., Ponomarev, V. A. 1976. Coxeter functors and Gabriel's theorem. *Uspecki Mat. Nauk.* **28**, 1933 [Russian].
19. Brauer, R., Noether, E. 1927. Über minimale Zerfällungskörper irreducibler Darstellungen. *Ak. Berlin S. B.* **27**, 221–226.

20. Brauer, R., Weiss, E. 1950. *Non-commutative Rings*, Part I. Cambridge: Harvard Press.
21. Cartan, H., Eilenberg, S. 1956. *Homological Algebra*. Princeton: Princeton Univ. Press.
22. Cassels, J. W. S., Fröhlich, A. 1967. *Algebraic Number Theory*. Washington, D. C.: Thompson.
23. Colby, R. R. 1966. On indecomposable modules over rings with minimum condition. *Pac. Jour. of Math.* **19**, 23–33.
24. Curtis, C. W., Reiner, I. 1962. *The Representation Theory of Finite Groups and Associative Algebras*. New York–London–Sydney: Wiley–Interscience.
25. De Meyer, F., Ingraham, E. 1971. *Separable Algebras over Commutative Rings*. Lecture Notes in Mathematics, vol. 181. Berlin–Heidelberg–New York: Springer–Verlag.
26. Deuring, M. 1935. *Algebren*. Erg. der Math. Band 4. Berlin: Springer–Verlag.
27. Dickson, S. E. 1969. On algebras of finite representation type. *Trans. Amer. Math. Soc.* **135**, 127–141.
28. Divinski, N. J. 1965. *Rings and Radicals*. Toronto: Univ. of Toronto Press.
29. Dlab, V., Gabriel, P. 1980. *Representation Theory, I, II*. Lecture Notes in Math., vols. 831, 832. Berlin–Heidelberg–New York: Springer–Verlag.
30. Dlab, V., Ringel, C. M. 1976. Indecomposable representations of graphs and algebras. *Mem. Amer. Math. Soc.*, No. 173. Providence: Amer. Math. Soc.
31. Dlab, V., Ringel, C. M. 1978. On algebras of finite representation type. *Jour. of Algebra* **33**, 306–394.
32. Draxl, P., Kneser, M. 1980. SK_1 *von Schiefkörpern*. Lecture Notes in Mathematics, vol. 778. Berlin–Heidelberg–New York: Springer–Verlag.
33. Formanek, E. 1972. Central polynomials for matrix rings. *Jour. of Algebra* **23**, 129–133.
34. Gabriel, P. 1972. Unzerlegbare Darstellungen I. *Manuscripta Math.* **6**, 71–103.
35. Gabriel, P. 1973. Indecomposable representations II. *Symposia Math. Inst. Naz. Alta Mat.* **11**, 81–104.
36. Gerstenhaber, M. 1963. On the cohomology structure of an associative ring. *Ann. of Math.* (2) **78**, 267–288.
37. Gerstenhaber, M. 1964. On the deformation of rings and algebras. *Ann. of Math.* (2) **79**, 59–103.
38. Greenberg, M. 1969. *Lectures on Forms in Many Variables*. New York: Benjamin.
39. Harada, M., Sai, Y. 1971. On categories of indecomposable modules, I. *Osaka J. Math.* **8**, 309–321.
40. Hasse, H., Brauer, R., Noether, E. 1931. Beweis eines Hauptsatzes in der Theorie der Algebren. *Jour. für Math.* **167**, 399–404.
41. Herstein, I. N. 1968. *Noncommutative Rings*. Carus Mathematical Monographs, No. 15. Math. Assoc. of America.
42. Herstein, I. N. 1976. *Rings with Involution*. Chicago and London: Univ. of Chicago Press.
43. Higman, D. G. 1954. Indecomposable representations at characteristic p. *Duke Math. Jour.* **7**, 377–381.
44. Hochschild, G. 1945. Cohomology groups of an associative algebra. *Ann. of Math.* **46**, 58–67.
45. Hungerford, T. W. 1980. *Algebra*. 2nd Edition. Berlin–Heidelberg–New York: Springer–Verlag.
46. Jacobson, N. 1956. *Structure of Rings*. Amer. Math. Soc. Colloquium Publ. vol. 37. Providence: Amer. Math. Soc.
47. Jacobson, N. 1964. *Lectures in Abstract Algebra, vol. 3*. Princeton: Van Nostrand.
48. Jacobson, N. 1975. *P.I.-Algebras, An Introduction*. Lecture Notes in Mathematics, vol. 441. Berlin–Heidelberg–New York: Springer–Verlag.
49. Jacobson, N. 1978. Some recent developments in the theory of P.I. Algebras. In:

Topics in Algebra, pp. 8–46. Lecture Notes in Mathematics vol. 697. Berlin–Heidelberg–New York: Springer–Verlag.
50. Jans, J. P. 1957. On the indecomposable representations of algebras. *Ann. of Math.* (2) **66**, 418–429.
51. Jans, J. P. 1957. The representation type of algebras and subalgebras. *Can. Jour. of Math.* **10**, 39–44.
52. Janusz, G. J. 1973. *Algebraic Number Fields.* New York–San Francisco–London: Academic Press.
53. Kleiner. M. M., Roiter, A. V. 1977. The representations of the differential graded categories. In: *Matric Problems*, pp. 5–70. Kiev [Russian].
54. Lam, T. Y. 1973. *The Algebraic Theory of Quadratic Forms.* Reading: Addison Wesley–Benjamin.
55. Lambek, J. 1966. *Lectures on Rings and Modules.* Waltham: Blaisdell-Ginn.
56. Lang, S. 1970. *Algebraic Number Theory.* Reading: Addison Wesley.
57. Milnor, J. 1971. *Introduction to Algebraic K-Theory.* Annals of Mathematics Studies. Princeton: Princeton Univ. Press.
58. Nakayama, T. 1935. Über die direckte Zerlegung einen Divisions-algebra. *Jap. Jour. of Math.* **12**, 65–70.
59. Nazarova, L. A., Roiter, A. V. 1973. Categorical matrix problems and the Brauer–Thrall conjecture (preprint). Inst. Math. Acad. Sci., Kiev [Russian].
60. Nesbitt, C., Scott, W. M. 1943. Matrix algebras over algebraically closed fields. *Ann. of Math.* (2) **44**, 147–160.
61. Passman, D. S. 1977. *The Algebraic Structure of Group Rings.* New York–London–Sydney: Wiley–Interscience.
62. Peirce, B. O. 1881. Linear associative algebra. *Amer. Jour. of Math.* **4**, 97–229.
63. Platonov, V. P. 1979. Algebraic groups and reduced K-theory. In: *Proceedings of the International Congress of Mathematicians*, Helsinki, 1978, vol. 1, pp. 311–317. Helsinki: Finnish Acad. Sci.
64. Procesi, C. 1973. *Rings with Polynomial Identities.* New York: Marcel Dekker.
65. Razmyslov, J. P. 1973. On a problem of Kaplansky. *Math. USSR Izvestia* **7**, 479–496 [Russian].
66. Reiner, I. 1975. *Maximal Orders.* London–New York–San Francisco: Academic Press.
67. Ringel, C. M. 1978. Representations of *K*-species and bimodules. *J. of Algebra* **41**, 269–302.
68. Roiter, A. V. 1968. Unboundedness of the dimensions of the indecomposable representations of an algebra which has infinitely many indecomposable representations. *Izv. Akad. Nauk SSSR* Ser. Mat. **32**, 1275–1282 [Russian].
69. Roiter, A. V. 1979. Matrix problems. In: *Proceedings of the International Congress of Mathematicians*, Helsinki, 1978, vol. 1, pp. 319–322. Helsinki: Finnish Acad. Sci.
70. Serre, J.-P. 1973. *A Course in Arithmetic.* New York–Heidelberg–Berlin: Springer–Verlag.
71. Serre, J.-P. 1979. *Local Fields.* Graduate Texts in Mathematics, vol. 67. New York–Heidelberg–Berlin: Springer–Verlag.
72. Smalø, S. O. 1978. The inductive step of the second Brauer–Thrall conjecture. *Can. Jour. Math.* **32**, 342–349.
73. Van der Waerden, B. L. 1937. *Moderne Algebra*, vol. 1, 2nd Ed. Die Grundlehren der Math. Wissenschaften Band 33. Berlin: Springer-Verlag.
74. Wallace, D. A. R. 1961. On the radical of a group algebra. *Proc. Amer. Math. Soc.* **12**, 133–137.
75. Wang, S. 1950. On the commutator group of a simple algebra. *Amer. Jour. of Math.* **72**, 323–334.
76. Wedderburn, J. H. M. 1907. On hypercomplex numbers. *Proc. Lond. Math. Soc.* (2) **6**, 77–118.
77. Wedderburn, J. H. M. 1937. Notes on algebras. *Ann. of Math.* **38**, 854–856.

78. Weil, A. 1967. *Basic Number Theory*. Die Grundlehren der Math. Wissenschaften Band 144. New York: Springer-Verlag.
79. Witt, E. 1934. Gegenbeispiel zum Normensatz, *Math. Zeit.* **39**, 462–467.
80. Yamagata, K. 1978. On Artinian rings of finite representation type. *Jour. of Algebra* **50**, 276–283.

Index of Symbols

The list of symbols that we give here is divided into three categories: standard symbols of algebra; fairly standard notation for associative algebras; non-standard symbols that occur in certain chapters of this book. These categories are further subdivided, the first by topics, the second by type fonts, and the third by the chapters in which the notation occurs. Each symbol is accompanied by a verbal translation, or reference to a page of the text, or both.

General Notation

1. Set theory and universal algebra.

$\subseteq$, $\supseteq$	inclusion.
$\subset$, $\supset$	proper inclusion.
$\cap$, $\bigcap_{i \in J}$	intersection.
$\cup$, $\bigcup_{i \in J}$	union.
$\dot{\cup}$	disjoint union.
$\|X\|$	cardinal number of X.
$\aleph_0$	first infinite cardinal number.
ω	first infinite ordinal number.
ϕ^{-1}	inverse (set) mapping.
$\phi\|X$	restriction of ϕ to X.
id_X, id	identity mapping (on X).
$\cong$	isomorphism (for algebras, modules, . . .).

2. Matrix theory.

$M_n(A)$	algebra of n by n matrices, entries in A.
δ_{ij}	Kronecker delta.
ε_{ij}	standard matrix unit; *8*
ι_n, ι	n by n unity matrix; *8*
$\det \alpha$	determinant of α.
$\operatorname{tr} \alpha$	trace of α.

3. Number systems.

$\mathbb{N}$	natural numbers.
$\mathbb{Z}$	integers.
$\mathbb{Q}$	rational numbers.
$\mathbb{F}_p$	field with p elements.
$\mathbb{R}$	real numbers.
$\mathbb{R}^+$	positive real numbers.
$\mathbb{C}$	complex numbers.
$\mathbb{H}$	Hamilton quaternions; *14*
$\hat{\mathbb{Q}}_p$	p-adic numbers; *323*
$\mathbb{Z}(p^\infty)$	rank one, divisible p-group; *58*

4. Algebras.

A°	group of units in A.
A'	commutator $[A^\circ, A^\circ]$ of A°.
A^{ab}	A°/A'.
A^*	opposite algebra to A; *179*
A^e	enveloping algebra $A^* \bigotimes A$ of A; *180*
$I \lhd A$	I is a two sided ideal in A.
A/I	quotient algebra.
$A \dotplus B$	product algebra; *3*
$A \bigotimes B$	tensor product; *163*
$A^{\bigotimes m}$	tensor product of m copies of A.

5. Modules.

M_A, ${}_AM$, ${}_AM_B$	M as a right module, left module, bimodule; *22*
M_θ	module structure induced by homomorphism θ; *22*
$M < N$	M is a submodule of N.
N/M	quotient module.
$M + N$, $\sum_{i \in J} M_i$	lattice sum of submodules; *24*
$M \bigoplus N$, $\bigoplus_{i \in J} M_i$	direct sum of modules.
$\bigoplus n\, M$, $\bigoplus \alpha\, M$	direct sum of n copies (α copies) of M.
uA	$\{ux: x \in A\}$, the cyclic module generated by u.
WY	(where $W \subseteq M$, $Y \subseteq A$) the R-submodule of M that is generated by $\{ux: u \in W, x \in Y\}$.
uY, Wx	special cases $\{u\}Y$, $W\{x\}$.

There is similar notation for left modules and for subsets of an R-algebra.

$M \otimes N$	tensor product over R; *158*
$u \otimes v$	rank one tensor; *158*

6. Fields.

K/F	field extension.
$[K:F]$	field degree, that is, $\dim_F K$.
$N_{K/F}$	field norm.
$T_{K/F}$	field trace.

7. Morphisms.

$\mathrm{Hom}_A(M,N)$	R-module of A-module homomorphisms; *6*
$\mathbf{E}_A(M)$, $\mathbf{E}(M_A)$	R-algebra of A-module homomorphisms; *7*
$\mathrm{Ker}\,\phi$	kernel of ϕ.
$\mathrm{Im}\,\phi$	image of ϕ.
$\mathrm{Coker}\,\phi$	cokernel of ϕ.
$\phi \otimes \psi$	tensor product of module or algebra homomorphisms; *159*

Associative Algebra Notation

1. Lower case Latin.

$e(D/F)$, $e_v(D/F)$	ramification index; *330*
$f(D/F)$, $f_v(D/F)$	relative degree; *330*
$l(M)$	length of the module M; *35*
v_p	valuation corresponding to an irreducible element p (or rational prime p); *318*
v_∞	absolute valuation of $\mathbb{Q}$; *318*
w^σ	*347*

2. Upper case Latin.

$A^m(F)$	affine m space over F.
A^S	$A \otimes S$, that is, scalar extension of A; *169*
$B^n_R(A,M)$, $B^n(G,E^\circ)$	n'th coboundary module; *198*, *254*
$C^n_R(A,M)$, $C^n(G,E^\circ)$	n'th cochain module; *197*, *254*
$D((\mathbf{x},\sigma))$	twisted Laurent series algebra; *383*
$E(D,v)$, $E(D)$, $E(v)$	residue class field; *319*
G_w	decomposition group; *348*
$GL_n(D)$	general linear group, that is, $M_n(D)^\circ$; *302*
$H^n_R(A,M)$, $H^n(G,E^\circ)$	n'th cohomology module; *198*, *254*
J_F	idele group; *361*
$K((\mathbf{x}))$, $K((t))$	Laurent series field; *383*

$K_n(F)$	degree n unramified extension of F; *335*
$M^{(A)}$	$\{u \in M : xu = ux\}$; *185*
$0(D,v)$, $0(D)$, $0(v)$	valuation ring; *319*
$P(D,v)$, $P(D)$, $P(v)$	valuation ideal; *319*
RG	group algebra; *4*
$R\{X\}$	free R-algebra on X; *6*
$R[X]$	free commutative R-algebra (polynomial algebra) on X; *6*
$SK_1(A)$	reduced Whitehead group, that is, Ker $\nu_{A/F}^{ab}$; *300*
$SL_n(D)$	special linear group, that is, Ker (Det); *309*
$Z_R^n(A,M)$, $Z^n(G,E^\circ)$	n'th cocycle module; *198*, *254*

3. Bold face.

$\mathbf{i}, \mathbf{j}, \mathbf{k}$	quaternion units; *14*
$\mathbf{B}(F)$	Brauer group; *228*
$\mathbf{B}(K/F)$	relative Brauer group; *239*
$\mathbf{C}_A(X)$	centralizer of X in A; *164*
$\mathbf{E}_A(M)$	endomorphism algebra.
$\mathbf{G}(K/F)$	Galois group, that is, $\mathbf{E}_F(K)^\circ$; *223*
$\mathbf{I}(A)$	lattice of two sided ideals of A; *26*
$\mathbf{J}(A)$	Jacobson radical of A; *58*
$\mathbf{M}(A)$	multiplication algebra of A; *222*
$\mathbf{P}(A)$	prime radical of A; *64*
$\mathbf{S}(M)$	submodule lattice of M; *24*
$\mathbf{S}(F)$, $\mathbf{S}_p(F)$	normalized valuations of F; *350*
$\mathbf{Z}(A)$	center of A; *218*

4. Lower case Greek.

$\delta^{(n)}$	n'th coboundary homomorphism; *197*
η_k	exponential mapping, that is, $x \mapsto x^k$.
λ_x	left regular representation; *7*, *95*
μ_M, ν_M	*173*
ν_ϕ, $\nu_{A/F}$, ν	reduced norm; *295*, *296*
$\nu_{A/F}^{ab}$, ν^{ab}	abelianized norm; *300*
ρ_x	right regular representation; *7*
σ_F, σ_w	Frobenius automorphism; *335*, *349*
τ_ϕ, $\tau_{A/F}$, τ	reduced trace; *295*, *296*

5. Upper case Greek.

$\Gamma(A)$	quiver of A; *96*
X_ϕ, $X_{A/F}$, X	reduced characteristic polynomial; *295*, *296*
Δ^z, Δ	norm form; *368*

6. Upper case German.

$\mathfrak{S}(F)$	finite dimensional, central simple F-algebras.
$\mathfrak{M}_A$	finitely generated A-modules.
$\mathfrak{N}_A$	finitely generated, indecomposable A-modules.

7. Abbreviations.

ann X	annihilator of X; *23*
char F	characteristic of the field F.
cor	corestriction mapping; *273*
deg Φ	degree of the polynomial Φ.
Deg A	degree of A; *236*
Det	Dieudonné determinant; *308*
dim M	vector space dimension.
Dim A	homological dimension; *206*
Exp A, Exp $[A]$	exponent of A or $[A]$; *260*
Ind A, Ind $[A]$	Schur index of A or $[A]$; *242*
Inf	inflation mapping; *263*
INV, INV_F, INV_v	local and global invariants; *339*, *354*
$\varinjlim$	direct limit; *266*
$\varprojlim$	inverse limit; *268*
rad M	radical of the module M; *37*
res	restriction mapping; *270*
soc M	socle of the module M; *38*
Tr. $\deg_F K$	transcendence degree of K/F.

8. Special symbols.

$N \rtimes A$	split extension of N by A; *213*
$A \sim B$	Morita equivalence (in $\mathfrak{S}(F)$); *228*
$[A]$	Brauer class of A; *228*
(E,G,Φ)	crossed product; *252*
(E,σ,a)	cyclic algebra; *277*
$\left(\frac{a,b}{F}\right)$	quaternion algebra; *14*
$\left(\frac{a,b}{F,\zeta}\right)$	generalized quaternion algebra; *284*
(K,ϕ,ψ)	field compositum; *343*
$\hat{D}_v$, $\hat{D}$, $\hat{F}_v$, $\hat{F}$	completion at v; *323*

Local Notation

1. Representations. (Chapter 5)

$\phi \oplus \psi$;	*83*

Index of Terms

C

D

E

F

Z